# ENZYME HISTOCHEMISTRY

## and Its Application in the Study of Neoplasms

# ENZYME HISTOCHEMISTRY

## and Its Application in the Study of Neoplasms

M. S. Burstone

*National Cancer Institute*
*National Institutes of Health*
*Bethesda, Maryland*

ACADEMIC PRESS • 1962
NEW YORK and LONDON

ACADEMIC PRESS, INC.
111 Fifth Avenue, New York, New York 10003

*United Kingdom Edition published by*
ACADEMIC PRESS, INC. (LONDON) LTD.
24/28 Oval Road, London NW1

LIBRARY OF CONGRESS CATALOG CARD NUMBER: 62-13046

*Second Printing, 1972*

PRINTED IN THE UNITED STATES OF AMERICA

## *FOREWORD*

Histochemical techniques may be considered as extensions of the routine histopathologic diagnostic staining procedures. The latter methods, although providing valuable morphologic information to the pathologist and histologist, lack chemical significance. Aware of the greatly expanded knowledge of biochemistry in recent years, the microscopist inevitably has been confronted with questions concerning the cytologic localization of newly discovered biochemical entities including enzyme systems in tumors. Attention has again focused on morphologic problems with those related to the structural organization of the metabolic machine being of central importance in the new field of molecular biology. The pathologist too has sought to interpret mechanisms based on altered biochemical systems. Of considerable practical importance is the expectation that evidence obtained by histochemical methods would aid in solving problems of diagnostic pathology. Thus modern chemical orientation in the morphologic sciences has demanded precise and highly specific cytochemical methods, particularly for enzyme systems.

The present text covers the subject of enzyme histochemistry in unprecedented depth clearly presenting the basic chemistry involved and evaluating critically the various aspects essential to the successful application of the methods. This work brings together from many sources information of value to both the developmental histochemist and biologist interested in applying new techniques, and furthermore provides insight into the conception and development of the relatively new field of enzyme histochemistry. The thorough compilation of the impressive number of available methods covers the historical development of the field in addition to the advances made, a number of which have come from the author's laboratory. As an interdisciplinary science, histochemistry utilized knowledge in several fields and the author has performed a unique and difficult service in bringing together the essentials of organic chemistry, biochemistry, and morphology upon which the methods are based.

The novel organization of the work with appended compilations of chemical data concerning organic dyes and substrates as well as the detailed outlines of the methods should prove useful for easily accessible reference. The extensive information on the synthesis of substrates and the detailed outlines of the methods including critical appraisal and evaluation will provide a ready aid to those applying the techniques to biological problems. Of particular interest is the chapter dealing with the relationship between histochemical substrates and carcinostatic agents, an area of potential clinical value which as yet has been little explored.

KENNETH M. ENDICOTT

*Director, National Cancer Institute*
*National Institutes of Health*
*Bethesda, Maryland*
*October, 1962*

# *ACKNOWLEDGMENT*

I wish to acknowledge the support of those who encouraged this work, especially Dr. Michael B. Shimkin, Associate Director for Field Studies, National Cancer Institute, and Dr. Eli M. Nadel, Chief, Diagnostic Research Branch, National Cancer Institute. Particular thanks are due Dr. S. S. Spicer, National Institute of Arthritis and Metabolic Diseases, and Dr. Frank B. Johnson, Armed Forces Institute of Pathology, who have read the text in its entirety and who have offered helpful advice and suggestions. Also I should like to express my appreciation to my staff members, including Mrs. N. Joan Archard, Miss Sally Rockwell, and Miss Paula Weaver, who have done the major portion of the typing of the manuscript and related tasks. Much of the photographic data was done by Mr. John W. McGuire and associates. Other photographic data have been generously furnished by my friends and colleagues whose names are appropriately noted in the text. Proofreading of the galleys as well as assistance in the preparation of the subject index have been done by Mrs. Doris Chaney and Miss Linda M. Whiting.

*October, 1962* M. S. BURSTONE

# CONTENTS

# ENZYME HISTOCHEMISTRY
## and Its Application in the Study of Neoplasms

# INTRODUCTION

According to the review edited by Sumner and Myrbäck (1950, p. 2), the term enzyme was introduced by Kühne in 1878. Preliminary studies on enzymes, however, were carried out between 1830 and 1840 by Liebig, Wohler, and Berzelius, who viewed the role of enzymes as similar to that of inorganic catalysts whose function was not necessarily associated with a living organism.

Sumner and Myrbäck (1950, p. 1) state that in one sense life may be defined as an orderly function of enzymes. Disease may manifest itself as a disorder, inhibition, or hyperfunction of enzymes. Greenberg (1960, p. 3) describes enzymes as an invention of nature designed to accelerate and control in a specific manner the numerous chemical reactions that determine the metabolism and vital activity of cells, and thus of the organism. By their unique specificities, by their interlocking assembly line functioning, by changes in the physical environment such as pH and temperature, by their changing content, by the rise and ebb of their substrates, and by the influence of activators and inhibitors, including hormones, enzymes maintain an orderly balance of physiological processes necessary for the preservation of the organism. Thus, enzyme control and regulation is initiated with conception and continues throughout all phases and all functions of the life cycle of the organism in health and disease.

Many aspects dealing with the vast literature on classical enzymology are considered in the reviews by Sumner and Myrbäck (1950, 1951a,b, 1952), Hoffmann-Ostenhof (1953, p. 219), Sumner and Somers (1953), Lindley (1954), Alberty (1956), Mehler 1957, Boyer *et al.* (1959), Best (1960), Boyer (1960), and Clifton (1960, p. 212). Dixon and Webb (1958) emphasize the ramifications of enzymology and its close relation with many sciences including biochemistry, physical chemistry, bacteriology and microbiology, genetics, botany and agriculture, pharmacology and toxicology, pathology, physiology, medicine, and chemical engineering. The application of enzymes in clinical medicines has been the subject of a recent review by Innerfield (1960). Enzyme histochemistry has also become significantly associated with a number of research and clinical areas, including diagnostic and research pathology, oral histology and pathology, and veterinary medicine.

In general, biochemical analysis of enzymatic activity has involved homogenization techniques in which the cells are ruptured. Holter (1952) indicates that this technique is susceptible to various sources of error, the primary ones being the effects of autolysis and elution or resorption during the process of separation. The total enzymatic activity of the homogenate may be assayed, or else the suspension may be separated into various fractions by centrifugation at varying speeds. Of interest is the fact that Bensley and Hoerr (1934), who were the first to isolate mitochondria by differential centrifugation of disrupted guinea pig liver cells, stained and examined microscopically their preparations during various stages of the procedure. Danielli (1953) has termed the technique used to isolate cellular organelles including mitochondria as maceration procedures, and comments that changes in the submicroscopic organization and composition of both nuclear and cytoplasmic bodies may be the result of both the homogenization procedure as well as the solutions which are used in fractionation since the nucleus and cytoplasm contain complex colloidal systems.

By contrast, definitive microscopic histochemical techniques utilize tissue sections the architecture of which is not grossly disturbed, and employ known chemical reactions to produce insoluble chromogenic precipitates precisely at the cytological sites associated with specific enzyme activities. Thus, localization of a variety of enzymes may be accomplished at the cytological level; and even of greater importance, sites of enzyme activity may be delineated to one or two specific cells in a relatively large heterogeneous sample. Although such techniques are not quantitative from a strictly numerical standpoint, the relative activity of the reactive cells may be approximated by the rapidity at which the dye formation or precipitation occurs. The microscopic techniques, however, may be quantitated by the ingenious approach developed by Linderstrøm-Lang and associates (Glick, 1949) in which different but essentially comparable tissue samples are employed for microscopic examination and for quantitative microanalysis, respectively. Another approach to the quantitation of certain types of histochemically produced dyes involves the pooling of tissue sections and extraction of the dye which is then measured colorimetrically.

In applied histochemistry the quality of microscopic localization or in other words the accurate reflection of enzymatic activity as revealed by a dye precipitate is, of necessity, the determining factor as to the value of any technique. Histochemists trained as histologists or histopathologists should expect to see clear-cut microscopic pictures provided that proper histochemical techniques are employed. Available evidence at our present state of knowledge indicates that sharply delineated staining patterns are significant even when there is an overall cytoplasmic activity (Chapter 4). The all too facile description of blurred diffuse dye deposits including small dye crystals (unrelated to any known cellular organelles) as accurate enzymatic localizations must be avoided, since such interpretations merely represent attempts to rationalize poor results obtained with outmoded methods or improper histochemical techniques.

From the standpoint of the experimental and practicing pathologist, however, microscopic enzyme techniques are of potential value in detecting histological and cytological alterations which are not discernible with routine non-specific stains. The development of new and sophisticated procedures makes it reasonable to assume that enzyme histochemical techniques of the future will be employed as diagnostic tools because of their capability of revealing early metabolic alterations which occur concomitantly with the neoplastic process. With reference to present-day procedures, of interest is the suggestion that the aminopeptidase azo-dye technique is of use in detecting lymph node metastases of gastric carcinoma (Chapter 10, p. 415). The well-known application of the alkaline phosphate procedure in the differentiation of "leukemoid" reactions and leukemias (Chapter 5, Section II, D) also warrants further evaluation. The use of the Mylar film or tape technique (Chapter 5, p. 276) may lend itself to the automation of histochemical diagnostic procedures, once appropriate diagnostic histochemical methods have been developed.

An important but relatively unexplored area lies in the application of enzyme techniques in the study of a variety of non-neoplastic pathological conditions. Included are diseases of the musculo-skeletal and nervous systems, to which a number of presently available oxidative enzyme procedures may be applied. In addition to some of the more prosaic applications in experimental histopathology and embryology, enzyme techniques may be of considerable significance in the study of drug metabolism. For example, many hydroxyl-containing aromatic compounds are conjugated *in vivo* as glucuronides which in turn are hydrolyzed by $\beta$-glucuronidase in certain sites (Fishman, 1961, p. 31). It is conceivable that increased activity of the aforementioned sites associated with metabolic transformations may be revealed by histochemical techniques. An analogous situation involves the hydrolysis of bis(2-amino-1-naphthol)phosphate by phosphatase to produce the carcinogenic 1-hydroxy-2-naphthylamine (Troll *et al.*, 1959).

The application of enzyme histochemistry in the study of lesions including tumors which are related to virus or virus-like agents has not received even cursory consideration. Of interest are the findings by Riley (1961) that certain mouse tumors exhibit a virus-tumor synergism which is shown by an accelerated growth of the tumor and by elevation of lactic dehydrogenase in the host plasma. The possible tissue sites of increased dehydrogenase activity could be explored by means of histochemical procedures.

An important application of enzyme histochemistry lies in the study of various enzyme systems which are associated with the calcification process, especially as it relates to bones and teeth of mammalian species. Although at the present time such studies have primarily tended to reveal specific sites of enzyme activity from a morphological standpoint, a considerable amount of evidence has also been gleaned which tends to shed light upon the mechanisms by which the processes of bone apposition and resorption occur. Among the enzyme systems of calcifying tissues which have been delineated by histochemical procedures are phosphatases, esterases, dehydrogenases, and oxidases (Burstone, 1960a,b).

The enzymatic changes which are associated with aging tissue have received considerable interest from a biochemical standpoint (Sinex, 1961). For example, Barrows *et al.* (1960) have reported a decreased succinoxidase activity in rat kidney associated with aging, and correlated it with a decrease in the number of mitochondria per cell. Application of histochemical procedures and their correlation with biochemical methods may augment the evaluation of enzymatic alterations which occur with aging.

The terms constitutive and adaptive enzymes describe those enzymes present in fixed and variable concentrations, respectively (Mehler, 1957, p. 392). A variety of adaptive enzymes has been described in both microorganism and animal tissues. Definitive localization of the distribution and changes in activities of adaptive enzymes may be elucidated by the histochemical approach. Of interest is the observation that liver cytochrome oxidase activity is greatest adjacent to the portal areas where the oxygen tension is highest (Burstone, 1959).

Of particular significance are the newer applications of histochemical techniques in the study of enzyme activity at the electron microscope level, as well as the relationship of histochemical substrates to cancer chemotherapeutic agents, topics which are discussed in Chapters 13 and 15, respectively.

A particularly intriguing aspect of enzyme histochemistry lies in its potential application to *exobiology*, the study of extraterrestrial life. Lederberg (1960) has suggested the application of cytochemical techniques which are adaptable to automation and telemetric recording procedures. Thus, a transparent tape ribbon could be employed to collect samples which would then be incubated in suitable substrate media, following which the cytochemical reaction sites could be detected and evaluated by a microscope-Videcon-transmitter system. A related area involves the application of enzyme histochemistry in the study of the effects of ionizing radiations including those of extraterrestrial origin as well as well controlled laboratory experiments employing known isotopes. In general, enzyme histochemical techniques should be valuable tools in the study of radiation effects and damage in tissue sections.

In large part, the recent advances in the field of enzyme histochemistry have been directly related to the synthesis and development of new substrates and techniques. At the termination of the second World War, only two or three histochemical enzyme procedures were available, while during the relatively short intervening period some forty to fifty new enzyme techniques and modifications have been developed. Although many completely new substrates have been synthesized, cognizance should be taken of the fact that many of those presently employed were first synthesized and reported around the turn of the century. Specific examples are α- and β-naphthyl phosphates which were reported by Kunz in 1894 and α- and β-naphthyl acetates which were synthesized by Miller (1881a,b). In fact, the synthesis of almost all simple esters of α- and β-naphthol including sulfates, glucuronides, and glucosides has been described in the earlier German chemical literature. Indoxyl acetate which was used as the basis of the first so-called indoxyl esterase procedure was first prepared by Vorländer and Drescher in 1901. Even the more complex naphthol AS acetate, whose histochemical application was originally described by Gomori (1952b), was first synthesized in 1909 by Graff. Thus it is apparent that the design and synthesis of really new and specific histochemical reagents has received serious consideration only during recent years. If the present rapid rate of progress of developmental histochemistry continues, a large series of new and specialized techniques for both basic research and diagnosis should be available in the foreseeable future.

The format of the present text includes an appendix at the end of each chapter. In the case of specific enzyme systems, the appendix is divided into two parts, A and B, respectively. Part A gives specific laboratory procedures for the synthesis of various substrates and related compounds, while part B enumerates methods and laboratory procedures for the microscopic demonstration of enzymes. Also included are the various types of fixation and embedding procedures which have proved of value in enzyme histochemistry. A brief review dealing with the application of specific enzyme techniques is incorporated in specific chapters. An integral part of many chapters is a table indicating a large number of potential substrates which either have been applied in enzyme histochemistry or biochemistry, or else may have potential application with reference to new substrate design. These tables also include pertinent data on molecular structure, melting points, and bibliographic references. Although the various appendices may serve as a comprehensive laboratory manual, it is hoped that the organization of the text and general approach to the problem stresses the significant relationship between organic chemistry, and in particular dyestuff chemistry, and microscopic enzyme histochemistry.

Among the previously published texts and reviews on histochemistry including enzyme histochemistry which may be employed as references are those by Glick (1949, Techniques of Histo- and Cytochemistry); Gomori (1952a, Microscopic Histochemistry); Danielli (1953, Cytochemistry, A Critical Approach); Eränkö (1955, Quantitative Methods in Histology and Microscopic Histochemistry); Arvy (1957–1958, Les Techniques Actuelles D'Histoenzymologie); Deane *et al.* (1960, Histochemical Methods for the Demonstration of Enzymatic Activity); Pearse (1960, Histochemistry, Theoretical and Applied); McManus and Mowry (1960, Staining Methods, Histologic and Histochemical); Lison (1960, Histochemie et Cytochemie Animales, Vols. I and II); Lillie (1954, Histopathologic Technic and Practical Histochemistry); and Wachstein (1962, Histochemistry of Enzymes in Tumors).

## References


Alberty, R. A. (1956) *Advances in Enzymol.* **17,** 1–6.

Arvy, L. (1957–1958) "Les Techniques Actuelles d'Histoenzymologie." Collection of reprints from *Biologie méd. (Paris)* 46–47.

Barrows, C. H., Jr., Falzone, J. A., Jr., and Shock, N. W. (1960) *J. Gerontol.* **15,** 130–133.

Bensley, R. R., and Hoerr, N. L. (1934) *Anat. Record* **60,** 449–455.

Best, J. B. (1960) *Intern. Rev. Cytol.* **9,** 129–186.

Boyer, P. D. (1960) *Ann. Rev. Biochem.* **29,** 15–44.

Boyer, P. D., Lardy, H., and Myrbäck, K. (1959) "The Enzymes," 2nd ed., Vol. 1. Academic Press, New York.

Burstone, M. S. (1959) *J. Histochem. and Cytochem.* **7,** 112–122.

Burstone, M. S. (1960a) *In* "Metabolism of Oral Tissues" (R. F. Sognnaes, ed.). *Ann. N. Y. Acad. Sci.* **85,** 431–444.

Burstone, M. S. (1960b) *In* "Calcification in Biological Systems," pp. 217–243. Am. Assoc. Adv. Sci., Washington, D. C.

Clifton. E. E. (1960) *In* "Enzymes in Health and Disease" (D. M. Greenberg and H. A. Harper, eds.), pp. 212–268. Charles C Thomas, Springfield, Illinois.

Danielli, J. F. (1953) "Cytochemistry, a Critical Approach." Wiley, New York.

Deane, H. W., Barrnett, R. J., and Seligman, A. M. (1960) *In* "Handbuch der Histochemie" (W. Graumann and K. Newmann, eds.) Vol. 7, Part 1. Fischer, Stuttgart, Germany.

Dixon, M., and Webb, E. C. (1958) "Enzymes." Academic Press, New York.

Eränkö, O. (1955) "Quantitative Methods in Histology and Microscopic Histochemistry." (In statistical collaboration with Jaakko Kihlberg.) S. Karger, Basel, New York.

Fishman, W. H. (1961) "Chemistry of Drug Metabolism." Charles C Thomas, Springfield, Illinois.

Glick, D. (1949) "Techniques of Histo- and Cytochemistry." Interscience, New York.

Gomori, G. (1952a) "Microscopic Histochemistry." Univ. Chicago Press, Chicago, Illinois.

Gomori, G. (1952b) *Intern. Rev. Cytol.* **1,** 323–335.

Graff, J. (1909) *Ann.* **367,** 253–265.

Greenberg, D. M. (1960) *In* "Enzymes in Health and Disease" (D. M. Greenberg and H. A. Harper, eds.), pp. 3–25. Charles C Thomas, Springfield, Illinois.

Hoffmann-Ostenhof, O. (1953) *Advances in Enzymol.* **14,** 219–260.

Holter, H. (1952) *Advances in Enzymol.* **13,** 1–20.

Innerfield, I. (1960) "Enzymes in Clinical Medicine." McGraw-Hill, New York.

Kunz, P. (1894) *Ber.* **27,** 2559–2565.

Lederberg, J. (1960) *Science* **132,** 393–400.

Lillie, R. D. (1954) "Histopathologic Technique and Practical Histochemistry," McGraw-Hill, New York.

Lindley, H. (1954) *Advances in Enzymol.* **15,** 271–299.

Lison, L. (1960) "Histochimie et Cytochimie Animales," 3rd ed., Vols. I and II. Gauthier-Villars, Paris.

McManus, J. F. A., and Mowry, R. W. (1960) "Staining Methods, Histologic and Histochemical." Hoeber, New York.

Mehler, A. H. (1957) "Introduction to Enzymology." Academic Press, New York.

Miller, O. (1881a) *Ber.* **14,** 1600–1602.

Miller, O. (1881b) *Ann.* **208,** 223–248.

Pearse, A. G. E. (1960) "Histochemistry, Theoretical and Applied," 2nd ed. Little, Brown, Boston, Massachusetts.

Riley, V. (1961) *Science* **134,** 666–668.

Sinex, F. M. (1961) *Science* **134,** 1402–1405.

Sumner, J. B., and Myrbäck, K., eds. (1950) "The Enzymes," Vol. I, Part 1. Academic Press, New York.

Sumner, J. B., and Myrbäck, K., eds., (1951a) "The Enzymes," Vol. I, Part 2. Academic Press, New York.

Sumner, J. B., and Myrbäck, K., eds., (1951b) "The Enzymes," Vol. II, Part 1. Academic Press, New York.

Sumner, J. B., and Myrbäck, K., eds., (1952) "The Enzymes," Vol. II, Part 2. Academic Press, New York.

Sumner, J. B., and Somers, G. F. (1953) "Chemistry and Methods of Enzymes," 3rd ed. Academic Press, New York.

Troll, W., Belman, S., and Nelson, N. (1959) *Proc. Soc. Exptl. Biol. Med.* **100,** 121–122.

Vorländer, D., and Drescher, B. (1901) *Ber.* **34,** 1854–1860.

Wachstein, M. (1962) *In* "Handbuch der Histochemie" (W. Graumann and K. Newmann, eds.), Vol. 7, Part 2, pp. 73–153. Fischer, Stuttgart.

*CHAPTER 1*

# FIXATION AND EMBEDDING TECHNIQUES

## I. INTRODUCTION

Fixation techniques for enzyme histochemistry include the frozen-section procedure, utilization of aqueous fixatives, freeze-drying, and freeze-substitution. Unfixed frozen sections are included in this group since some degree of denaturation may occur during freezing and thawing or subsequent processing.

Although definitive histochemistry usually necessitates proper fixation, most of our knowledge of fixation has been empirically established. Each type of fixation procedure must be carefully selected for the specific problem at hand. Unfortunately, no one type of fixation offers a panacea for the histochemical laboratory.

For enzyme localizations, there exists a close relationship between initial fixation, substrate used, products of hydrolysis, and the final dyestuff pattern. In a general way, an enzyme technique is no better than the state of fixation of a given tissue. For example, a procedure designed to give good histochemical localizations may have limited use with acetone fixed paraffin embedded kidney, since such parenchymatous tissues may be considerably distorted by acetone fixation. The use of unfixed frozen sections in conjunction with techniques for the demonstration of hydrolases may involve considerable hazard, not simply because of loss of enzyme, but also because of solubility and instability of unfixed cytoplasmic constituents. Thus, it may be possible to lose relatively insoluble enzyme simply because of movement of cytoplasm.

Fortunately, many hydrolytic enzymes are well preserved following fixation as well as embedding in paraffin. The most desirable fixative would be one which renders proteins insoluble without resulting in loss of enzyme activity. It is known, for example, that urease treated with alcohol becomes irreversibly insoluble but remains enzymatically active (Sumner, 1948).

Relatively little is known about the mechanisms of histochemical fixation but, in general, chemical fixatives combine with proteins to form intermolecular linkages containing covalent or coordinate bonds.

Sylvén (1951) points out that fresh tissues are extensively altered by common fixation procedures. The changes include:

(1) Dissolution leading to irregular loss of tissue mass.

(2) Chemical reactions leading to changes in chemical reactives as well as new compounds.

(3) Physical alterations of tissue elements leading to abnormal mass distribution.

His experimental data show that specimens fixed for 24 hours in formalin, Zenker's solution, Carnoy's solution, or ethanol reveal variable amounts (10–30%) of nitrogen containing compounds in solution. Among the chemical effects of fixatives such as formaldehyde are loss of reactive groupings (French and Edsall, 1945). Merriam (1958) has also found an appreciable loss of tissue solids following formalin fixation. For example, a 25% reduction in solids was observed in formalin fixed liver. Frozen-dried specimens also exhibited a weight loss which was of a lesser magnitude. In the case of the frozen-dried specimens the processing procedure was not entirely the same as employed under histochemical conditions in that blocks rather than 6–10 $\mu$ sections were utilized. Berenbom *et al.* (1952) found that frozen-dried as well as acetone fixed specimens lost nitrogen and phosphorus during subsequent stages of processing. The various findings tend to point out the potential and serious losses which may occur when completely unfixed frozen sections are employed.

Danielli (1953, p. 17) has also reported on the actions of various chemical fixatives. Some of his data are indicated in Table 1. Of course, the deleterious effect of these reagents may be minimized by fixing at low temperatures. From time to time various well known routine fixatives have been recommended for histochemical purposes with the proviso that they be used in the cold. Although useful results may be obtained with them in conjunction with a number of enzyme techniques, the success may be due in large part to the fact that the enzyme is considerably inhibited. Thus the subsequent reaction rate is greatly decreased and the reaction products show less tendency to diffuse under such conditions.

How hopeful are the prospects of retaining most of an enzyme in a section for microscopic study? First, undue emphasis should not be placed upon the soluble state of proteins including enzymes. According to Haurowitz (1950, p. 234), the insoluble state of enzymes seems to be essential for their action in cells. He points out that if catalytically active small groups were present as free molecules, they could diffuse and would be readily lost by the cell. Haurowitz continues, "A further consequence of

TABLE 1

EFFECT OF VARIOUS CHEMICAL FIXATIVES UPON TISSUE GROUPINGS

| Fixative | CHOH | $NH_2$ | SH | COOH | Histidine | Tryptophan | Tyrosine |
|---|---|---|---|---|---|---|---|
| Formaldehyde | + | − | ? | + | + | + | + |
| Ethanol | + | + | + | ? | + | + | + |
| Acetic acid | − | − | − | − | ? | ? | − |
| Osmic acid | ? | ? | − | ? | ? | − | ? |
| Acetone | + | + | + | + | + | + | + |

(+) indicates that this grouping is relatively unaffected by the reagent while (−) indicates that the grouping may be seriously altered.

the insoluble state of the enzyme is the presence of an interface between the aqueous solution and the insoluble protein phase. This heterogeneous character of the enzyme system in some instances favors the action of the enzyme." He points out that an enzyme may be a very active lipoxidase in heterogeneous water-oil emulsions, but is quite inefficient when the same system is made homogeneous by the addition of ethanol or bile salts. According to Morton (1955, p. 25), "Even those [enzymes] which occur in true solution in the cytoplasm are probably associated in molecular aggregates, whereas those of cytoplasmic particles belong to complexes composed of several groups of compounds and contain relatively large amount of lipid. These multimolecular aggregates are thus generally insoluble in distilled water or dilute salt solutions." The large molecular size of some enzymes is exemplified by swine kidney phosphatase which has a molecular weight of over 500,000 (Mathies and Goodman, 1953). Sumner and Somers (1953, p. 8) state, "As they exist in nature, enzymes may be surrounded by other proteins and may be liberated completely only when the protein is dissolved, since they cannot diffuse through it." Thus, with adequate fixation, or in some cases with no fixation whatsoever, it should be possible to visualize the sites of some enzyme activity as discrete aggregates.

Available evidence at our present state of knowledge would indicate that sharp histochemical localizations are significant even when there is an overall cytoplasmic activity (Fig. 3, Chapter 5). Indistinct or fuzzy localizations invariably indicate poor fixations (or none), or diffusion of the reaction products. Chessick (1953) has reported that diffusion of enzyme (esterase) may be a cause of blurring of staining if unfixed tissues are used. However, in paraffin sections the enzyme is non-diffusible, as can be shown by the fact that prolonged soaking of slides in water or buffer prior to incubation does not affect the final staining. Similar data is also presented in Gomori's (1950) classic paper "Sources of Error in Enzyme Histochemistry." Moog (1951) has similarly noted that alkaline phosphatase does not diffuse in fixed sections. Mounted sections of frozen-dried paraffin embedded material are also relatively stable when carefully dewaxed and hydrated.

A significant factor directly related to fixation is the possible effect of postmortem changes upon enzyme reactions. Empirical observations appear to indicate that immediate fixation is not a prerequisite to the obtaining of useful histochemical locations at routine magnifications. In fact, autopsy tissues appear to be entirely satisfactory for the demonstration of many hydrolytic enzymes. On the other hand, Gallagher *et al.* (1956) investigated *in vivo* and *in vitro* autolytic changes in rat liver and found that the oxidative phosphorylation systems were very labile and that loss of soluble respiratory cofactors occurred during early stages of autolysis. When rat liver was autolyzed at 38° under anaerobic conditions there occurred a rapid fall in the ability of tissue homogenates to oxidize tricarboxylic acid cycle intermediates, with the exception of succinate.

Smith *et al.* (1956, 1957) studied, by means of quantitative methods, seven enzymes in the brains of rabbits 0, 2, and 6 hours after death. Included among the enzymes were acid phosphatase, alkaline phosphatase, hexokinase, phosphofructokinase, and several dehydrogenases. The results for all enzymes with the exception of phosphofructokinase showed changes in the order of ±15% in each of the three cerebellar layers. Activity of the latter enzyme, however, showed an approximate decrease of 70% after 6 hours. They feel that with reference to the tissue of study (brain), six of the seven enzymes exhibit a stability which suggests that significant and reproducible results may be obtained with routine autopsy tissues. However, they offer a note of caution and point out that autolytic and proteolytic activities in sites other than brain may be considerably more active. It is conceivable that proteolytic and subsequent autolytic activities may be very high in invasive tumors and adjacent tissues. Robins *et al.* (1958) extended the aforementioned studies by noting effects of experimentally induced fever or uremia. The findings of both studies indicate that 14 stable enzymes are present in the brain. Among these there is at least one enzyme involved in each of the following metabolic transformations: the glycolytic cycle, the tricarboxylic acid (Krebs) cycle, glutamate metabolism, purine metabolism, the hexose "shunt," overall energy metabolism (adenosinetriphosphatase), hydrolysis of phosphate bonds, primary phosphorylation of glucose (hexokinase), and hydrolysis of glucuronides.

Van Lancker and Holtzer (1959) observed the release of acid phosphatase and β-glucuronidase (from mouse liver) after 30, 60, and 120 minutes. In addition, it was found that after 2 hours a 20% drop in cytochrome oxidase and glucose-6-phosphate occurred. Taft (1960) followed the activities of a number of enzymes (alkaline and acid phosphatases, adenosinetriphosphatase, 5-nucleotidase, and glucose-6-phosphate) in rat liver 30, 60, 90, 180, 300, and 360 minutes after death. No significant loss in activity was found. An increased activity of glucose-6-phosphatase and 5-nucleotidase was observed with reference to the longer autolysis periods. Gössner (1960) by means of microscopic histochemical techniques found changes in the distribution of enzyme activities of liver implants in which circumscribed zones of heat and cold necrosis had been induced. Although not directly related to fixation, this type of experiment serves to indicate the degree to which enzyme systems may survive adjacent to zones of frank necrosis. From the incomplete experimental evidence as well as general experience, it would appear that tissues fixed within 15 to 30 minutes following surgical removal or autopsy should be adequate for use with most histochemical techniques. On the other hand, because of the semi-quantitative nature of histochemical methods, human autopsy tissues or other specimens subjected to longer postmortem periods may also be of use.

Holt (1956) was one of the first to show discrete histochemical localization of certain sites of enzymatic activity. Similar localizations have also been demonstrated in highly active frozen-dried tissues (Burstone, 1956c, 1958c; Stowell *et al.*, 1961). However, discrete localizations in themselves do not necessarily indicate accurate localizations. For example, if unfixed frozen sections are used in conjunction with certain azo-dye methods, one may at times obtain a discrete pattern of staining. Upon closer examination, it can be seen that this pattern of staining consists for the most part of disorganized clumps of dye and dye crystals (Nachlas *et al.*, 1957). One certainly would not wish to equate random dye precipitates in the azo-dye and tetrazole methods with cellular organelles.

In general, azo-dye techniques give better results with paraffin embedded as contrasted with frozen sections. This is due to the fact that azo-dyes are lipid soluble to a greater or lesser extent; and also lipid may interfere with the penetration of some of the more complex naphtholic substrates. On the other hand, the metal salt techniques such as the glycerophosphate procedure for acid phosphatase appear to give better results with frozen sections.

It is obviously desirable to correlate histochemical and biochemical data to elucidate not only inhibitory effects of fixation, but also effects upon localization. Unfortunately, the marked difference in sample characteristics (homogenate or extract, versus tissue section) may render correlation difficult. In addition, the damage of cells incidental to biochemical or histochemical processing poses new problems. For example, Tarver (1954) states that there is little evidence bearing on the activity of cathepsins in intact cells, but during the last few years it has become increasingly clear that the enzymatic activities exhibited by cells depend upon their state of organization. When the cell structure is in any way disrupted, activities which were formerly negligible (may) come to the fore. For example, in leaf cells the activity of invertase may be masked, although macerated leaf tissue exhibits high activity. At present, it would appear that some differential centrifugation techniques offer considerable promise for correlation of histochemical and biochemical aspects of intracellular localization of enzymes.

## II. FROZEN-SECTION PROCEDURES

Unfixed frozen sections were among the first to be employed for enzyme histochemistry because they were highly active and readily prepared. Their use, however, in enzyme histochemistry must be carefully evaluated in terms of the specific problem at hand because of the known disruptive effects of freezing and thawing as well as loss of soluble enzyme and cofactors. According to Gomori (1950, 1952) diffusion of enzyme in unfixed frozen sections is a significant factor in the production of false localizations. Porter *et al.* (1953) have found that freezing results in a marked reduction in the capacity of rat liver homogenates to oxidize many substances. This effect is believed to be a purely mechanical one, in which the enzyme-bearing particles are altered by compression and shearing between ice crystals. Under optimal conditions only 7 out of 15 oxidative activities of liver showed recoveries of better than 50% after freezing. Freezing in the presence of sucrose prevented most of the inhibitory effects. The aforementioned experiment actually includes the effects of freezing and thawing, and it is possible that the deleterious effects were in no small part due to the subsequent thawing.

The freezing of tissues so as to make it possible to cut thin sections is one of the oldest histological techniques and was introduced by Raspail and by Stilling (Baker, 1950, p. 104; 1958). Subsequently freezing microtomes in which the tissue holder was cooled by $CO_2$ gas were designed. In 1931, Schultz-Brauns reported the use of carbon dioxide in order to cool the knife as well as the tissue block. Adamstone and Taylor (1947, 1948) devised a "cold knife" technique which has subsequently influenced most frozen-section procedures. They were stimulated by the revival of the Altmann freeze-dry technique by Gersh (1932) and wished to modify the ordinary freezing methods so that a cut section, still frozen, could be manipulated onto a glass slide and immersed in a Coplin jar of the desired reagent just prior to melting. They suggested that with this technique sections showing a minimum of cellular distortion could be obtained and that they facilitated the application of enzymatic studies such as the Gomori technique. Carbon dioxide or liquid air was used for freezing the tissue while the microtome knife itself was cooled by taping blocks of dry ice to its surface. The frozen cut sections were manipulated by a brush so that they lay flat on the knife and were then transferred to a glass slide by a special section lifter which also contained dry ice, so that the section remained frozen while it was transferred to a glass slide. This procedure constituted a vast improvement over the usual frozen-section technique in which the section is floated on a water bath. Adamstone and Taylor (1948) emphasize that the melting of the ice in the section must be prevented until chemical tests are started or fixative applied to preserve the section for subsequent processing. Immersion of the slide containing the tissue section must be done at *exactly the right moment,* which is the instant at which the section just begins to melt and flatten on the slide. If the section is allowed to melt completely the cells are ruptured or distorted. They also point out it is not feasible to attempt a procedure in a warm or humid room, and that certain tissues are inherently very difficult to section. They point out that the technique may be modified so that the tissue is first fixed before frozen sections are cut. This method negates the advantage of working with unfixed tissue but does obviate the necessity for handling the sections with extreme care. In order to facilitate the immediate immersion in a fixative or other solution just as melting occurs, a mechanical system was devised (Adamstone, 1951). With this system the slide was held in a chute above the jar of fixative and as soon as the first frozen section was placed upon it the slide was released by a magnetic system and rapidly immersed into the solution by a spring-actuated mechanism.

Taft (1949) evaluated the Adamstone-Taylor frozen-section method for use in routine surgical pathology. He found that sections prepared according to their method could be fixed, stained, and mounted within 15 minutes from the time that the block of tissue was frozen. He also confirmed their observation that sections cannot be made if the temperature of the room is too warm. Taft found that tissues containing a large amount of fat, and also those containing large amounts of dense collagen, cut with difficulty. The most satisfactory sections were obtained from liver, brain, and kidney. White and Allen (1951) modified the Adamstone-Taylor method by using a Bausch and Lomb rotary freezing microtome. A "fence" made from sheet copper or tin was packed with dry ice and placed upon the knife to keep it cold. They were able to obtain serial sections of 6–8 $\mu$ in thickness. Wachstein and Meisel (1953) sectioned unfixed frozen tissues with a Sartorius freezing microtome provided with a system for the simultaneous cooling of the knife and freezing stage by carbon dioxide. They reported that with some experience the most favorable degree of cooling of the knife and of freezing of the specimen can be obtained. The knife could be cooled below zero degrees within a few seconds and a temperature as low as $-30°$ could be maintained. In contrast to the Adamstone-

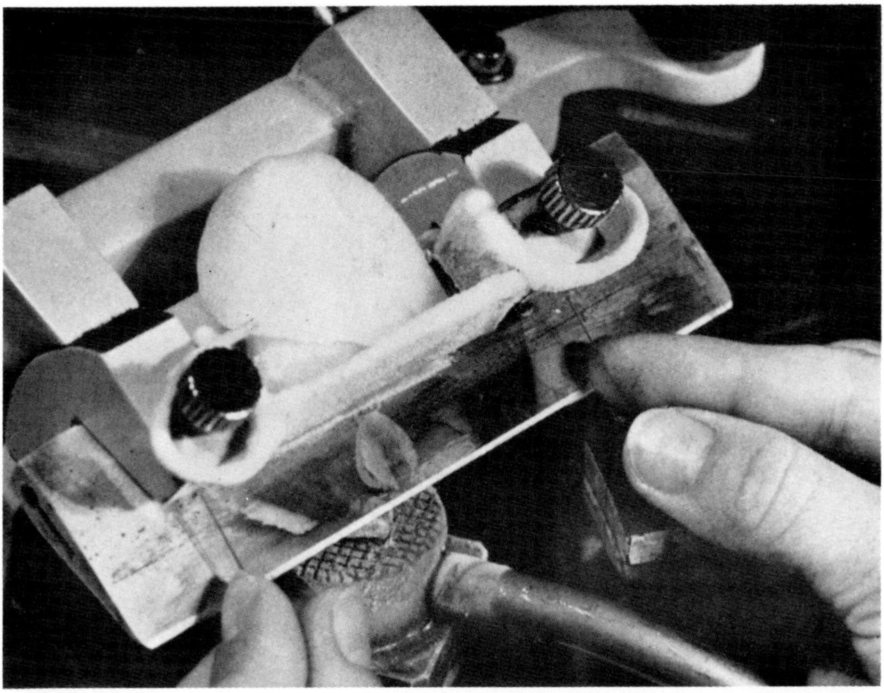

FIG. 1. Freshly cut section of heart. Note the simple wire bracket dry ice holder for cooling the knife.

Taylor procedure, sections were directly immersed in the incubating solution or fixing fluid. They were subsequently placed on clean slides and dried for approximately 7 minutes at 56°. From the standpoint of enzyme histochemistry the work of Wachstein and Meisel is of interest because they used the frozen-section technique in the demonstration of alkaline phosphatase, acid phosphatase, and esterase in a number of tissues. A modified Adamstone-Taylor system with the knife cooled by dry ice held in place by a metal "fence" is shown in Fig. 1.

Enclosing the microtome in a refrigerated chamber whose temperature is near the optimal for cutting tissues has extended the usefulness of the cold knife method (Fig. 2). These refrigerated chambers or cryostats maintain their low temperature by means of dry ice or mechanical refrigeration. The first of these devices was designed by Linderstrøm-Lang and Morgensen (1938). Coons *et al.* (1951) modified their technique and cut sections in a cryostat chamber maintained at a temperature of around −16° after which the sections were placed on glass slides slowly thawed and dried. Shimizu *et al.* (1956) used a sliding microtome in a refrigerated chamber and employed a device to prevent sections from curling. This

FIG. 2. Enclosed electrically refrigerated chamber which contains microtome. Note external handle for turning microtome drive (courtesy Dr. B. Klionsky).

attachment was a modification of the type originally introduced by Linderstrøm-Lang and Morgensen (1938). They found that by preventing curling of the sections adjacent to the cooled knife, the refrigerated chamber could be maintained at higher temperature (0°). Cut sections were transferred to glass slides, allowed to dry at room temperature, and subjected to various histochemical methods. These authors reported satisfactory demonstration of succinic dehydrogenase, cytochrome oxidase, and adenosinetriphosphatase. The anti-roll plate has also been employed in the cryostat of Klionsky and Smith (1960), Pearse (1960, p. 19), and Chang *et al.* (1961a, b).

An analysis of frozen-section techniques was made by Thornburg and Mengers (1957) who ascertained optimal conditions for cutting thin sections from unfixed frozen tissues. They reported a set of optimal temperatures for the tissue block, knife, and refrigerated chamber and suggest that there is a zone of "micro-melting" at the edge of the knife.

A variation of the frozen-section technique has been reported by Chang and Hori (1961) who immersed frozen sections directly into "substituting" fluids in a modified freeze-substitution procedure.

An excellent discussion dealing with the technical aspects of the preparation of frozen sections is described in the recent book by Steedman (1961).

The inherent difficulty with the frozen-section or cryostat technique is that tissues must usually be thawed before being treated with appropriate histochemical reagents. This thawing may exert a serious disruptive effect upon cellular organelles including those containing sites of enzyme activity. Some structures, however, such as the mitochondria of the human heart, are fairly resistant to freezing and thawing so that useful mitochondrial localizations may be obtained with fresh frozen sections mounted on glass slides. As previously mentioned, thawing is also believed to occur during sectioning due to the pressure of the knife passing across the block (Thornburg and Mengers, 1957), so that thawing damage to cells may occur even before the sections are immersed in the substrate solution.

From both histological and cytological standpoints it is obvious that unfixed frozen sections do not usually exhibit a quality of detail as are seen in frozen-dried and freeze-substituted specimens. Thus enzyme localizations in frozen sections may be variable and not always reproducible. Some improvement in the quality of preparations may be obtained by briefly fixing unfixed frozen sections mounted on glass slides with acetone or other solvents. This procedure has been used in the demonstration of aminopeptidase (Willighagen and Planteydt, 1959) and alkaline and acid phosphatases (Burstone, 1961).

## III. FIXATION IN AQUEOUS AND NON-AQUEOUS MEDIA

Gomori (1939) employed alcohol fixed paraffin embedded tissues for his original histochemical demonstration of alkaline phosphatase. He reported on the basis of studies with tissue extracts that phosphatase is not destroyed either by alcohol or exposure to 56° heat in the completely dehydrated state. The maximum decrease in activity was less than 20%. Since that time a large number of empirical fixation procedures have been devised and applied to the study of enzyme histochemistry. Analyses of these have attempted to determine the degree to which a given enzyme will be inactivated by means of bioassay of the tissue before and after fixation. Also of significance are the solvents to which the specimen is subjected during dehydration, clearing, paraffin embedding, and dewaxing. As previously indicated, the purpose of fixation in enzyme histochemistry is to render enzyme molecules insoluble and non-diffusible by establishing appropriate cross linkages and other bonds without markedly affecting the active groups. The effect of fixation upon enzyme cofactors unfortunately had received negligible attention. Baker (1950, 1958, 1960) has reviewed techniques and known physical-chemical mechanisms of fixation. Numerous data illustrate the relatively slow rate of penetration of most fixative solutions. Dempster (1960) has compared the rates of penetration of a number of routine fixatives and has found that formalin exhibited one of the lowest rates of penetration into liver (Chart 1). This rate would be expected to be even slower at the low temperature employed for histochemical processing.

Although the penetration of solutes (including substrates) into thin tis-

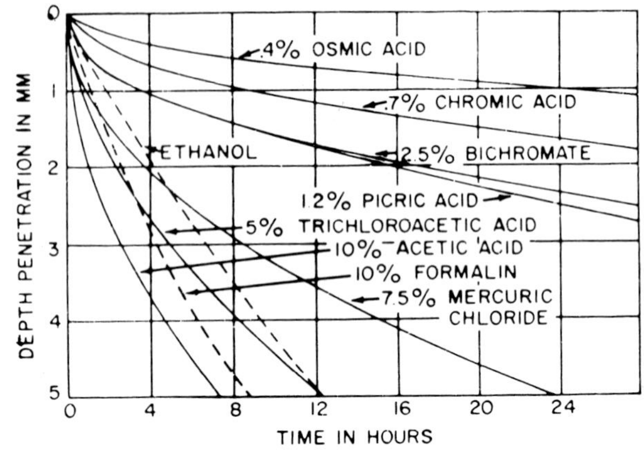


CHART 1. Plot of depth of penetration of reagents into rabbit liver relative to time of fixing (Dempster, 1960).

sue sections (6–10 mμ) does not usually present any problems, the employment of thick tissue slices gives rise to certain complexities which have been studied by Cascarano and Zweifach (1961). These authors have utilized tetrazolium salt experiments to demonstrate that the visual by-products of histochemical procedures provide a means of study of the variables encountered in the metabolic slice technique. Experimental determinations of the penetration and tissue activity agree closely with theoretically expected values which were calculated from previously established mathematical relationships.

A relatively recent innovation employs the use of hypertonic fixing solutions containing colloids or other substances. This is based upon the widely accepted biochemical procedure of utilizing sucrose solution as suspending media in the preparation of tissue homogenates. Polyvinylpyrrolidone, sucrose, gum acacia, dextran, methyl cellulose, or combinations of these have been employed (Birbeck and Reid, 1956; Greenfield and Price, 1956; Novikoff, 1957; Holt, 1958; Holt and Withers, 1958; Pearse, 1958; Scarpelli and Pearse, 1958; Hannibal and Nachlas, 1959; Pearse, 1960, p. 336). It should be noted, however, that the addition of these agents has been primarily used in an attempt to improve preservation of certain organelles (e.g. mitochondria) at the cytological level. In specific instances, certain of these agents appear to have improved the quality of fixation, especially if they are used during fixation and not after frozen sections have been cut. However, in other cases, particularly those involving the use of unfixed frozen sections, there appears to be no justification of their use according to Walker and Seligman (1961). They observed that the use of sucrose at isotonic or hypertonic concentrations, 7.5% PVP, divalent cations, and physiological salt solutions failed to improve the quality of histochemical preparations once fresh-frozen sections were prepared. These authors aptly point out that however much is known about the behavior of mitochondria in cell fractions and how well defined may be the condition for maintaining the ultrastructural integrity of mitochondria in such suspensions, this information cannot be directly applied to prevent damage sustained by fresh tissue first subjected to freezing and thawing. The only reasonable method for minimizing mitochondrial damage is to incubate very small blocks of tissue in the presence of the so-called "protecting agents" and substrate at the time. This is followed by formalin fixation and sectioning. Another technique involves the incubation of fragments of tissue in substrate solutions followed by freeze-drying (Burstone, 1959c; Burstone *et al.*, 1959) and paraffin embedding. With this technique one may obtain thin sections which exhibit excellent morphological preservation of cellular components (Fig. 3). Stained blocks of small size may also be fixed in osmium tetroxide followed by preparation for electron microscopy (Karmarkar *et al.*, 1959).

Various types of 10% formalin solutions, either buffered or containing calcium or sodium salts, have been employed for enzyme histochemistry (Lillie, 1954, p. 34). With reference to the mode of action of formaldehyde upon tissues, it is known that this fixative forms primary additive compounds with groupings containing active hydrogen atoms such as amines. Subsequently a cross-linking reaction involving the formation of methylene bridges occurs. Among the chemical groups of tissue protein which are known to react with formaldehyde are: amine, amide, peptide, carboxyl, hydroxyl, imine, thiol, disulfide, and indole (French and Edsall, 1945; Fraenkel-Conrat and Olcott, 1945). Thus it would appear that formaldehyde combines with many groupings associated with known enzyme systems. This is apparent from the fact that tissues fixed in formalin at room temperature according to standard procedures are essentially worthless for enzyme histochemistry. On the other hand, when fixation is carried out at a low temperature, certain enzymes may be preserved to a greater or lesser degree. However, it would appear under these conditions the degree of fixation might not be optimal.

Stafford and Atkinson (1948) were among the first to report that alkaline and acid phosphatases are preserved following fixation in formalin at 4° for a period of 24 hours. Eränkö (1951, 1952) was the first to employ the calcium chloride-formalin of Baker (1944) as a fixative for the demonstration of phosphatases. He specifically reported that frozen sections of fixed tissues gave reproducible results with the Gomori acid phosphatase technique. Calcium-formol has subsequently been employed by a number of investigators as a fixative in enzyme histochemistry. Baker (1944) originally observed that the calcium salt $CaCl_2$ acted osmotically and

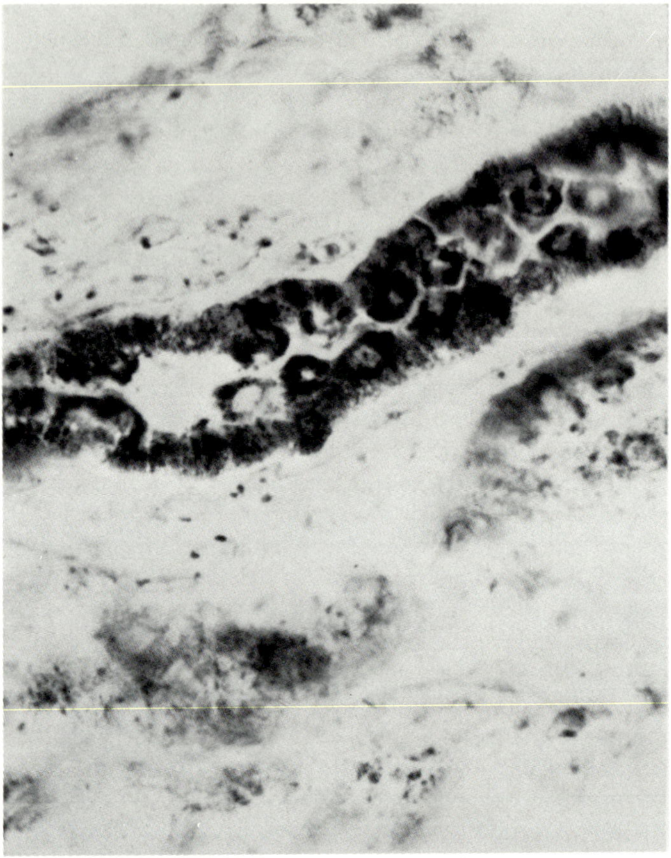

FIG. 3. Thin tissue block incubated in aminodiphenylamine cytochrome oxidase reagent, frozen and dried and embedded in paraffin and sectioned. Note localization of oxidase to mitochondria of kidney tubule (× 835).

prevented loss of phospholipids. Seligman *et al.* (1951) studied the effects of formalin fixation upon five different enzyme systems of rat liver homogenates. Varying degrees of inactivation resulted, with shorter fixation periods at lower temperatures giving the most satisfactory results. Gomori (1952) noted that cold neutral formalin may be used as a fixative in the demonstration of a number of hydrolytic enzymes. He subsequently reported the use of 10% formalin containing 2% sodium acetate in the study of esterases of the central nervous system (Gomori and Chessick, 1953). More recently cold formol-calcium has been applied for the demonstration of hydrolytic enzymes by Holt (1958), Holt and Withers (1958), and Holt *et al.* (1960). They employed formol-calcium containing 0.88 *M* sucrose with 1% gum acacia (gum-sucrose) in a fixation procedure in conjunction

TABLE 2

RELATIVE PRESERVATION OF ACID PHOSPHATASE AND ESTERASE FOLLOWING FIXATION FOR 24 HOURS IN COLD (2°) FORMOL-CALCIUM AND ACETONE[a, b]

| Tissue treatment (Fixation and after-treatment) | Kidney | | Liver | |
|---|---|---|---|---|
| | Acid phos-phatase | Esterase | Acid phos-phatase | Esterase |
| No fixation | | | | |
| No after-treatment | 100 | 100 | 100 | 100 |
| Formol-calcium | | | | |
| Blotted | 8.0 | 20 | 8.5 | 24 |
| Washed 12 hours | 41 | 42 | 38 | 39 |
| Gum-sucrose 24 hours | 43 | 48 | 40 | 52 |
| Gum-sucrose 7 days | 57 | 62 | 55 | 55 |
| Paraffin embedding followed by dewaxing | 1.8 | 4.6 | 1.7 | 7.0 |
| Acetone | | | | |
| Paraffin embedding followed by dewaxing | 20 | 30 | 28 | 26 |
| Frozen-dried, paraffin embedding followed by dewaxing | 98 | 99 | 97 | 99 |

[a] Holt *et al.* (1960).
[b] Comparison is made with frozen-dried specimens.

with the indoxyl substrates for esterase. Holt (1959) also used this technique in the study of acid phosphatase in rat kidney and liver and obtained good localizations at the cytological level. It was reported that formol-calcium fixed tissues impregnated with gum-sucrose for 7 to 30 days at low temperature still retained a high degree of phosphatase and esterase activity (Holt *et al.*, 1960), which was correlated with phospholipid preservation. Of interest is their observation that frozen-dried tissues exhibited the highest activity (Table 2).

Novikoff and Masek (1958) and Novikoff and Arase (1958) reported the survival of lactic dehydrogenase and diphosphopyridine nucleotide diaphorase (DPND) following formol-calcium fixation. Succinic dehydrogenase activity was completely inhibited. Subsequently, Novikoff *et al.* (1961) compared staining results in cold formol-calcium fixed tissues with two different tetrazoles which were employed in conjunction with DPNH (diphosphopyridine nucleotide) as the substrate. These authors clearly show that the morphological structure of the kidney or liver cell mitochondrion is revealed by adequate fixation in conjunction with a suitably designed substrate. Walker and Seligman (1961) have found that succinic dehydrogenase could be demonstrated in mitochondria of a selected group of tissues following very brief fixation (8–16 minutes) in cold, unneutralized formalin. However, such techniques probably are not desirable for the routine demonstration of more sensitive enzyme systems including cytochrome oxidase. Fishman and Baker (1956) and Baker *et al.* (1958) found that incorporation of 0.1% chloral hydrate in 10% formalin resulted in improved fixation of $\beta$-glucuronidase, acid phosphatase, and lipase. The mechanism of action of chloral hydrate is unknown.

In general, formalin fixed frozen sections should be incubated free floating in the substrate solution. In some cases sections may be mounted on glass slides and allowed to dry, but good adherence between the section and slide is not always obtained. According to Gomori (1952, p. 139), if paraffin embedding is to be employed the formalin must be washed out carefully. Formalin fixed tissues may be impregnated with 10–15% gelatin at 37° prior to the cutting of frozen sections of muscle in the study of cholinesterase (Chessick, 1954). Fishman and Baker (1956) found that a modified gelatin embedding technique was of considerable value in the demonstration of $\beta$-glucuronidase in friable tissues (Appendix 1, Section 9). Alginate gels (Lewis and Shute, 1960) and polyvinyl alcohol (Feder, 1959) (Appendix 1, Section 8) have also been similarly employed. Becker *et al.* (1960) studied a number of hydrolytic and oxidative enzymes in frozen sections cut from formol-calcium fixed tissues embedded in cold polyvinyl alcohol.

The use of water-soluble methacrylates as embedding media following formalin, as well as other types of fixation, has been reported (Wichterle *et al.*, 1960; Bernhard *et al.*, 1961). Two water-soluble monomers were employed, glycol monomethacrylate and triethyleneglycol dimethacrylate. In the presence of suitable catalysts, the polymerization reaction was found to proceed at a relatively low temperature. Although the techniques have been primarily applied to the study of ultrastructure at the electron microscope level, modifications of the procedure should make it adaptable for routine tissue sectioning. Recently, Feder (1960) employed acrolein and other aldehydes which gave excellent morphological preservation. These when used at low temperature may be useful histochemical fixatives. Sabitini *et al.* (1961) have found that a number of aldehyde compounds may be successfully employed as fixatives in conjunction with histochemical methods for both the light and electron microscope. Among the new fixatives are glutaraldehyde (6½%), hydroxyadipaldehyde (12½%), crotonaldehyde (10%), pyruvic aldehyde (5%), acetaldehyde (10%), and methacrolein (5%). These aldehyde compounds are buffered with phosphate or cacodylate buffers. Of particular interest is hydroxyadipaldehyde which permits preservation of a number of oxidative enzymes including cytochrome oxidase, succinic dehydrogenase, and DPN diaphorase.

Acetone and alcohol (ethanol) have been most frequently employed as fixatives in conjunction with the older enzyme techniques. Novikoff (1959) has reviewed the data of various workers on the percentage of enzyme activities surviving acetone and ethanol fixation. He pointed out that the results of different investigators are not entirely comparable since there were variations in tissue size and details of histological technique. In addition, these studies were principally carried out with one tissue (liver). The findings of Danielli (1953, p. 48) with reference to the effect of alcohol fixation upon rat kidney alkaline phosphatases are not entirely in agreement with some of the aforementioned observations. He was unable to find any significant difference between phosphatase activity of unfixed specimens and those fixed in alcohol for several hours or longer. Danielli also studied the preservation and distribution of alkaline phosphatase following fixation in some twenty different solutions, and concluded that although the broad outline of results obtained with certain chemical fixatives may be satisfactory, freeze-drying should be employed whenever possible.

Gomori (1952, p. 179) tested the effect of formalin, acetone, and ethanol upon slides with markings (Coujard) containing purified intestinal phosphatase. He found that cold formalin, acetone, or ethanol causes relatively little loss of activity. On the other hand, neutral formalin at room temperature destroys over 75% of the enzyme in less than 24 hours. Gomori and Coutinho (1953) found that cold ethanol selectively inactivates aliesterase while it preserves cholinesterase reasonably well. Gomori (1952, p. 138) recommended cold acetone as a fixative for a large number of hydrolytic enzymes including alkaline and acid phosphatases, cholinesterase, and non-specific esterase. Since some acetone fixed tissues exhibit a tendency to fragment during sectioning, he recommended infiltration of the tissues with dilute celloidin prior to paraffin embedding (Appendix 1, Section 4). Novikoff *et al.* (1960) reported that frozen sections cut from cold acetone fixed blocks were useful for the demonstration of aminopeptidase and some dehydrogenases. In the opinion of the author, acetone distinctly inhibits aminopeptidase so that lower and possibly more sensitive sites of activity may not be discernible. Wolman and Behar (1952) preferred chilled ethanol for the preservation of various enzymes and pointed out the limitation of acetone as a good cytological fixative. They reported that the use of absolute ethanol containing 5% acetic acid at a temperature of −8° for 4 hours followed by dehydration in absolute ethanol at room temperature resulted in good preservation of alkaline and acid phosphatases, lipase, phosphamidase, and cholinesterase in rat and human material.

Much of the distortion which occurs with cold acetone fixation is caused by the great temperature differential between warm tissue and cold acetone into which it is immersed. Remarkably good fixation at the histological and in some cases at the cytological level may be obtained by first prechilling (not freezing) thin slices of tissue approximately 1 to 2 mm. thick before placing in cold acetone. Following overnight fixation the tissues are dehydrated, rapidly cleared in petroleum ether, and embedded in paraffin (Appendix 1, Section 5). This simple and convenient process requires a minimum of time and may be used for the demonstration of a number of hydrolytic enzymes. Of course, some of the more sensitive hydrolytic enzyme systems such as aminopeptidase will not be preserved unless the tissue exhibits rather high initial activity.

The paraffin embedding procedure subsequent to fixation, dehydration, and clearing has been stated by some authors to result in considerable enzyme inactivation (Stafford and Atkinson, 1948; Doyle, 1948; Novikoff, 1952; Berenbom *et al.*, 1952). However, it is not clear whether the inactivation was due to the effect of hot paraffin or to the presence of residual solvents and fixatives acting upon the tissue at relatively high temperatures. Some authors, however, report that paraffin embedding causes little or no inactivation with reference to alkaline phosphatase and several other hydrolases (Cappelin, 1947; Cleland, 1950; Danielli, 1953, p. 33; Doyle, 1948, 1955; Doyle and Liebelt, 1954; Pratt, 1953, 1954). Gomori (1952, p. 180) found that in Coujard type preparations, embedding did not inactivate the slides except after formalin fixation. This was probably due to the presence of residual formalin. Dalgaard (1956) conducted model experiments with slices of gelatin containing dog sera, and found that phosphatase activity was unaffected by fixation in 80% ethanol and by subsequent dehydration and embedding techniques. He did, however, find that embedding temperatures over 53° resulted in a distinct loss of activity. These observations tend to confirm the original findings of Mowry (1949) who recommended that fixed blocks be cleared in petroleum ether and embedded in low-melting (42–44°) paraffin.

Whenever feasible, undecalcified tissues should be used for enzyme histochemistry. Histochemical demonstration of a number of hydrolytic enzymes in frozen-dried undecalcified teeth and bones have been reported by Burstone (1957a, 1960a,b) and Burstone and Keyes (1957). Cobb (1953) reported the demonstration of phosphorylase in undecalcified frozen-dried bones. Albert and Lindner (1960) sectioned undecalcified molar teeth and knee joints of rats 7 to 21 days in age and employed them in a study of phosphatase using an azo-dye method. Their technique involved the use of celloidin and solvents which may inactivate enzyme systems. Other techniques for sectioning of undecalcified bone and tooth structure have been reported by Sognnaes (1947), Palmgren (1954), Duthie and Barker (1955), Roberts and Hoyte (1958), and Frost (1958, 1959).

Demonstration of alkaline phosphatase in decalcified bone was first reported by Lorch (1946) who used a buffered citric acid solution. Morse (1945) and Greep *et al.* (1948) employed a buffered formic acid (Appendix 1, Section 6). Study of phosphatases in hard tissues with similar techniques have also been made by Zorzoli (1948), Lillie *et al.* (1951), Majno and Rouiller (1951), and Schajowicz and Cabrini (1959). The previously cited work of Lillie *et al.* (1951) is of interest in that they found that some strong mineral acids could be used in place of formic acid without impairing the subsequent demonstration of alkaline phosphatase. Dotti *et al.* (1951) claimed that superior and more rapid results could be obtained by use of a cation exchange resin in conjunction with formic acid.

More recently, chelating agents, such as ethylenediaminetetraacetic acid (EDTA), have been employed (Sreebny and Nikiforuk, 1951; Birge and Imhoff, 1952; Hilleman and Lee, 1953; Freiman, 1954; Schajowicz and Cabrini, 1958). With all of these techniques the decalcification times are often rather long and the degree of enzyme inhibition has not been ascertained (Appendix 1, Section 7). Reactivation of alkaline phosphatase and nucleotidase following treatment of tissue with EDTA has also been studied by Freiman (1953, 1956). In the case of alkaline phosphatase, paraffin sections are dewaxed and immersed in a solution of 0.1% magnesium acetate prior to incubation in the substrate. Calcium, cobalt, and zinc ions are also suitable as reactivators. 5-Nucleotidase, on the other hand, was activated by nickel and cadmium, which had no appreciable effect upon nonspecific alkaline phosphatase.

## IV. FREEZE-DRYING

Freezing and drying as a histological procedure was first described by Altmann (1890). He claimed that this technique eliminated shrinkage of histological specimens, and furthermore offered a means of studying intravital dyes. Of historical interest is the work of Mann (1902), Kossel (1913), Bayliss (1924), and Behrens (1932). Modern freeze-drying for histological and histochemical purposes was developed by Gersh (1932), who pointed out that the freeze-dry procedure facilitated the retention of tissue constituents in a reactive state. It is apparent that the use of routine fixatives for histochemical studies represents a compromise if not somewhat equivocal procedure. Although the exact details of tissue changes which occur during freezing and drying are not completely known, the fact that various frozen and dried biological and pharmaceutical products retain their native protein and other characteristics would recommend this process as an optimal type histochemical fixation for enzyme preservation since initial fixation and dehydration are carried out at low temperature. Hoerr, in 1936, noted that the "glycolytic enzyme" of the liver was not destroyed by freeze-drying. Gomori (1952) suggested that freeze-drying is useful for enzyme histochemistry but felt at the time that the equipment was too bulky and expensive for routine use.

Once dried, tissues exhibit considerable resistance to denaturation by heat and to some extent by chemical agents. Although frozen-dried tissues are usually embedded in paraffin the heat encountered does not appear to be detrimental provided the specimens are sufficiently dehydrated. Greaves (1954) points out that when the water content of frozen-dried products is very low they exhibit remarkable resistance to high temperature. He states that water seems to be the "catalyst" which is necesesary for marked denaturation to occur. Of course, from a traditional as well as modern industrial standpoint, drying has been a method for preserving natural substances as well as pharmaceutical products (Flosdorf, 1954; Rowe, 1960).

It is erroneous to equate frozen-dried tissues with unfixed frozen sections. Freeze-drying is the removal of water from a tissue specimen after it has been frozen. Thus the disruptive effects of thawing which are encountered in the use of unfixed frozen sections are avoided. In addition, the techniques employed in the freeze-dry procedure usually result in a smaller ice-crystal size than those employed in the preparation of frozen sections. Some components of frozen-dried tissues also appear to exhibit a mild state of denaturation. Sylvén (1951) points out that freezing and drying results in some decrease in the solubility of certain cytoplasmic proteins. A similar decrease in aqueous solubility has been observed in the case of frozen-dried serum lipoproteins (Potthoff, 1949; Oncley *et al.*, 1950). Of course, any decrease in the solubility of cytoplasmic constituents would be of advantage in conjunction with histochemical procedures. It is of interest that zymograms of fresh and frozen-dried liver are essentially identical while those of cold formalin fixed liver show suppression of certain bands (Fig. 4).

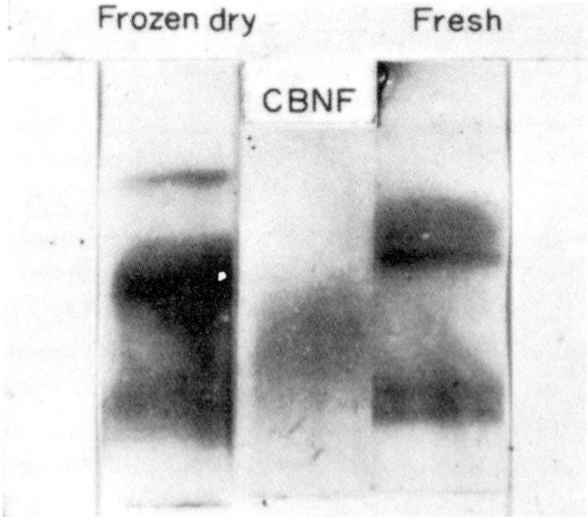


FIG. 4. Zymogram of esterase activity comparing fresh, frozen-dried, and cold buffered formalin fixed rat liver with naphthol AS acetate. The fresh and frozen-dried are comparable while the formalin fixed liver shows suppression of some bands.

The freeze-dry procedure may be divided into the following phases: (a) freezing (quenching); (b) dehydration or drying (usually *in vacuo*); and (c) embedding and subsequent processing for histochemical study.

## A. Freezing

When a tissue specimen is frozen, the free water is removed from the organic matrix and deposited in the form of ice crystals. The organic matrix which is compressed by the forming crystals appears as a honeycomb and in tissue sections is referred to as the ice-crystal artefact. In general, freezing should occur as rapidly as possible in order to minimize the size of ice-crystal artefacts. However, in many instances, particularly involving tissues which have a high water content, rapid freezing does not appear to appreciably reduce the size of the crystals. Goodspeed and Uber (1935) called attention to the fact that various conditions including unknown factors might determine the rate of freezing of tissues.They noted that certain centers of crystallization occurred, and that slow freezing gave rise to intercellular nucleocytoplasmic surfaces. On the other hand, rapid freezing increased the number of centers of crystallization and determined the degree of reticulation (ice-crystal artefact) produced.

The ideal situation would be the direct transformation of water in tissues to a vitreous or completely amorphous state with no evidence of crystalline pattern. In recent years the transfer of water out of solution into ice crystals has been studied rather intensively. Pioneering work in this area has been done by Luyet (1960) who has studied transitions from the liquid or amorphous state into a crystalline state, and also other transitions such as solidification into the amorphous state, vitrification, and transformations within the amorphous state in the form of incomplete crystallization. The possibility of obtaining complete vitrification has been considered (Luyet, 1951). With regard to obtaining an amorphous state of protoplasm after freezing, he feels that rapid cooling can hinder the crystallization process somewhat but does not prevent it entirely. Electron micrographs of frozen-dried tissues often reveal obvious ice-crystal artefacts or else areas showing no apparent crystallization. Stephenson (1956) feels that samples which show no obvious ice-crystal artefacts are vitrified. It is reasonable to assume that some areas of small tissue specimens may be frozen in the vitreous state. At approximately $-150°$ crystal formation ceases and any remaining water remains uncrystallized. In order to completely avoid crystal formation it has been estimated that the specimen must be cooled below $-100°$ in less than .02 seconds.

The earliest form of an ice crystal is called a crystal nucleus. This nucleus is made up of an aggregation of molecules which may grow to form a larger crystal or series of crystals. Thus in order to obtain a small crystal size, a rapid rate of nucleation and a low rate of growth are to be desired. The addition of glycerin or similar substances may reduce the amount of water available for crystallization. However, this approach has not as yet found application in the histochemical use of freeze-drying. Meryman (1956, 1957) has carefully reviewed physical principles of ice-crystal growth *in vitro* and *in vivo*. Of interest is the fact that in tissues which are subjected to slow freezing, crystal nucleation is generally confined to the extracellular spaces, while in the case of rapid freezing the tendency for preferential extracellular nucleation is subordinated and nuclei appear uniformly throughout the specimen. Although the crystal size which occurs during rapid freezing is generally many times smaller than that occurring with slow freezing, experience has shown that viability of tissue is enhanced by slow freezing.

Mathematical considerations of factors which influence ice-crystal formation during rapid freezing of biological material have been made by Stephenson (1960). He points out that two closely related problems arise during the rapid freezing of tissue specimens. The first is the dependence of the cooling rate on the sample size, shape, and thermal properties, on the velocity of immersion, and on the temperature and physical properties of the coolant. The second is the relation of the size and number of ice crystals to the thermal history of the sample. Analysis of these problems is not simple because of the fact that as ice crystals grow they release latent heat of crystallization. This heat modifies the course of cooling and consequently the temperature of the sample. Thus as soon as ice crystals begin to form throughout a specimen they may act as numerous small heat sources evolving heat as they grow in size.

Recently Fernández-Morán (1960) has studied the formation of submicroscopic ice crystals by means of electron microscopy. It was possible to stabilize thin ice crystals at low temperatures and to obtain high resolution electron micrographs and electron diffraction patterns of these. By oriented deposition of ice crystals on films of mica, characteristic moiré patterns were observed. These ingenious techniques may elucidate the changes in submicroscopic structure which occur in tissues during the freezing procedure.

Danielli (1953) has determined some of the time relationships in freeze-drying as compared with routine aqueous fixation by plunging a specimen containing a thermocouple into a freezing mixture. It was found that the temperature in the center of a 2 mm. section of tissue fell to below $-50°$ in less than 10 seconds and below $-170°$ in less than 30 seconds. A 2 mm. specimen is thoroughly frozen in 2 to 3 seconds as compared with a minimum of over 300 seconds which are required when a rapidly penetrating fixative solution is used. Thus freezing in isopentane cooled by liquid nitrogen results in a fixation time which is approximately

1% of that required when a chemical fixative is used. With the application of materials exhibiting thermal superconductivity such as liquid helium II, even shorter fixation times may be achieved (Fernández-Morán, 1960). Even though freeze-drying presents great advantages over chemical fixation, artefacts at the submicroscopic and microscopic levels may still arise. It has been observed, for example, that gels frozen at low temperature exhibit a laminated structure; and at times similar laminations may be seen in frozen-dried tissues at high optical magnification (Bell, 1952a). When specimens over several millimeters in size are frozen, gross cracks may be seen. These, however, do not as a rule interfere with subsequent processing or study. Although small pieces of tissue are to be preferred, they need not be excessively tiny when used in conjunction with routine histochemical procedures.

Since the probability of formation of ice-crystal nuclei is low below −33° and increases considerably between −33° and −43° (Fisher *et al.*, 1949), various rapid cooling procedures have been employed since the inception of freeze-drying. These include direct cooling in liquid oxygen, nitrogen, hydrogen, and helium or else fluids cooled by the aforementioned liquefied gases (Luyet, 1951). Gersh (1932) used liquid air which despite its low temperature offered relatively poor surface fixation (Gersh and Stephenson, 1954). Although the low temperature of liquid nitrogen was satisfactory in establishing a large temperature gradient, poor conductivity resulted because of the formation of a vaporized air film adjacent to the specimen. Other quenching liquids and mixtures have also been employed. These include: isopentane cooled by liquid nitrogen (Hoerr, 1936; Simpson, 1941; Stowell, 1951; Glick and Malmström, 1952), isopentane containing dry butane cooled by nitrogen (Emmel, 1946), propane and isopentane cooled by nitrogen (Bell, 1952a), propane cooled by nitrogen (Bell, 1952a; Moberger *et al.*, 1954), Freon 12 cooled by nitrogen (Bell, 1952b), dry ice-pentane mixture (Zlotnik, 1960), and liquid helium II (Fernández-Morán, 1960). The temperatures of the aforementioned quenching media are given below:

Liquid nitrogen, −195°
Isopentane, −165° (approx.)
Isopentane-butane, −190°
Propane-isopentane mixture, −190°
Propane, −185°
Freon 12, −158°
Dry ice-pentane, −125°
Liquid helium II, −272°

Freezing in isopentane cooled by liquid nitrogen has found widespread use for routine histological and histochemical studies. For more specialized cytochemical work, propane or propane-isopentane mixtures are recommended. Stephenson (1956) found that propane gave an appreciable advantage with respect to heat transfer as compared with isopentane. Over a wide range, the velocity of immersion did not significantly affect the cooling rate. The employment of liquid helium II represents an innovation because of its thermal superconductivity and its superfluidity.

In general, it is advisable to place tissue specimens on a small piece of aluminum foil and drop this directly into the agitated quenching bath. Specific metal holders have also been devised for this purpose but their use may entail considerable heat loss. Eränkö (1954a) recommended the use of the very high heat conductivity of copper by first cooling copper discs and freezing thin slices of tissue between the cold metal surfaces. He points out that under these conditions the same amount of heat is transferred about several thousand times more quickly through copper than through an organic liquid. The specimens which he used were less than a millimeter thick and the metal temperature was −192°. He reported that microscopic examination of the subsequently freeze-dried and stained specimens did not reveal ice-crystal artefacts at very high optical magnification. It would seem, however, that fluids are more suitable for the freezing of specimens of irregular size where close apposition of the freezing solution is necessary.

According to Gersh (1932), −20° is below the freezing point of tissue fluids. However, due to concentration of inorganic substances during the drying process, a lower temperature must be used. Goodspeed and Uber (1934) found that dehydration at −32° was more satisfactory than at −23°, and also commented that dehydration time varies and must be arbitrarily determined. Hoerr (1936), Simpson (1941), Scott and Hoerr (1950), and Gersh and Stephenson (1954) are in agreement that drying temperatures above −30° do not give as satisfactory results as the lower temperatures. Simpson (1941) found that tissues dehydrated at −40° exhibited a more uniform quality of preservation than those dehydrated at −30° or above. He reported, however, that there did not appear to be any advantage in employing temperatures below −40°. Scott (1943), Copeland (1951), and Kulenkampff (1954) favor lower drying temperatures. With reference to preparation for electron microscopy, Gersh (1956) found that morphology was the same when specimens were dried at −30° or −78°. On the other hand, Seno and Yoshizawa (1960) feel that a drying temperature of −55° to −60° is necessary. Of interest, however, is the fact that drying of pharmaceutical or other industrial products is usually carried out at −30° or even higher (Flosdorf, 1954).

### B. Drying

The sublimation of water vapor from the frozen eutectic mixtures in tissues occurs when the partial pressure of water vapor at the frozen surface exceeds that of the atmosphere adjacent to it. The eutectic or lowest temperature at which a liquid phase is present in tissues is not known. Thus many workers have felt that it would be desirable to dehydrate at the lowest temperature possible. At very low temperatures (e.g. below −40°) the vapor pressure of ice falls rapidly so that excessively long drying times are required. In general −30° to −40° is employed for the drying of tissues. Packer and Scott (1942), in a study of the location of inorganic salts, used a temperature below −55° which is the eutectic point of calcium chloride and is the lowest of any salt which may be present. It is known, however, that in tissues complex systems exist which have very low eutectic points and it is not possible at present to determine what the eutectic point of so complex a system as living tissue may be (Scott and Hoerr, 1950). Figure 6 shows the temperature and eutectic ranges which are believed to be encountered in the freeze-dry process (Gustafsson, 1954).

At atmospheric pressure and those down to a few millimeters of mercury, collisions between evaporating water molecules and those of the surrounding air retard evaporation. Under higher vacuum the partial pressure of the air not only immediately above the surface, but throughout the entire system, is kept to a value at least ten times below the local pressure of the water vapor. As previously mentioned, when tissue is frozen, ice crystals form and separate from the organic matrix compressing it, so that a network or wall is formed around each ice crystal. Tissue dries from the exterior inward. As water molecules arising from sublimation traverse the specimen, they must pass through dried areas previously occupied by ice crystals in order to be removed by the vacuum system. Thus the dry organic shell, which increases in thickness as drying proceeds, offers resistance to water molecules attempting to pass through it to the outside chamber. Figure 5

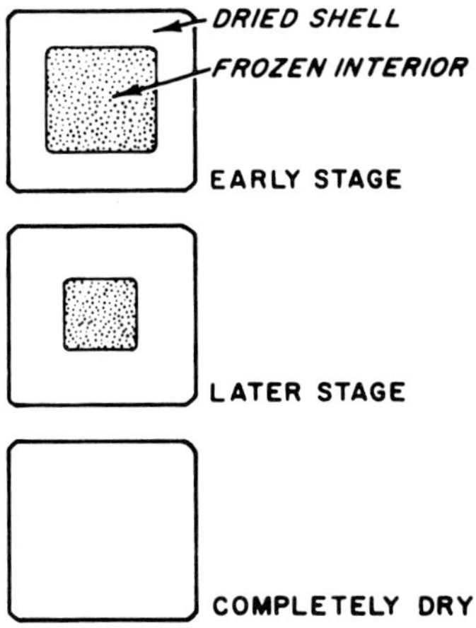


FIG. 5. Schematic representation of increase in thickness of dried shell during the vacuum drying process (Stephenson, 1953).

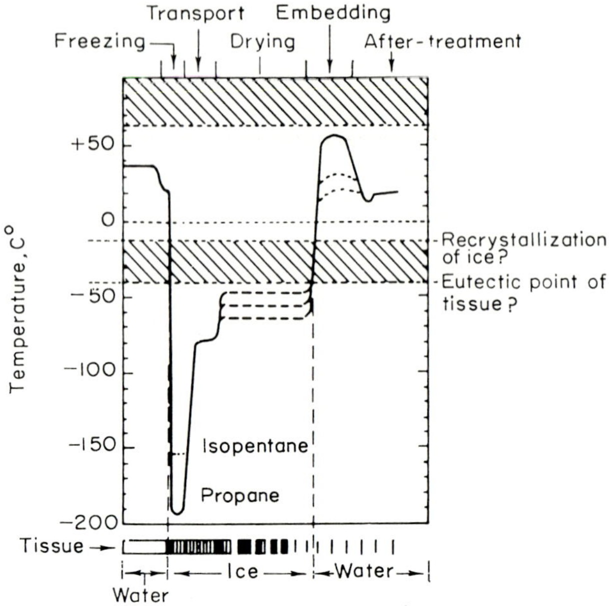


FIG. 6. Temperature ranges encountered during the various stages of freeze-drying and subsequent processing of tissues for histological application (Gustafsson, 1954).

shows the progressive increase in the thickness of the dry shell as drying proceeds. By interrupting the drying of a large tissue block, the sharp interface between the dried shell and the frozen interior is easily discernible (Gustafsson, 1954). The wide temperature range encountered in tissue during freezing and drying is shown in Fig. 6. Drying occurs because heat is supplied to ice crystals in the tissue, thus causing water molecules to vaporize from their free surfaces. According to Stephenson (1953), after a water molecule sublimes it either passes through the dried shell into the drying chamber and eventually reaches the vapor trap, or else it may be returned to the interface between the dried shell and frozen interior and be recaptured by an ice crystal (Fig. 7). He suggests that both paths may have variations. For example, a molecule which escapes from an ice crystal may enter a tissue space partially occupied by another ice crystal. After making an indeterminate number of collisions it is recaptured by the ice crystal or passes through one of the holes in the walls to a neighboring space. In general, a molecule which reaches the surface does so only after having made a large number of collisions within the dried shell. Once a water molecule is free of the dry shell it may be readily removed by means of the trap or vacuum system. It has been shown that for a given temperature of subliming ice crystals, the minimum drying time of guinea pig liver is greater than the minimum sublimation time of a piece of ice of the same size by a factor of the order of one thousand (Stephenson, 1953).

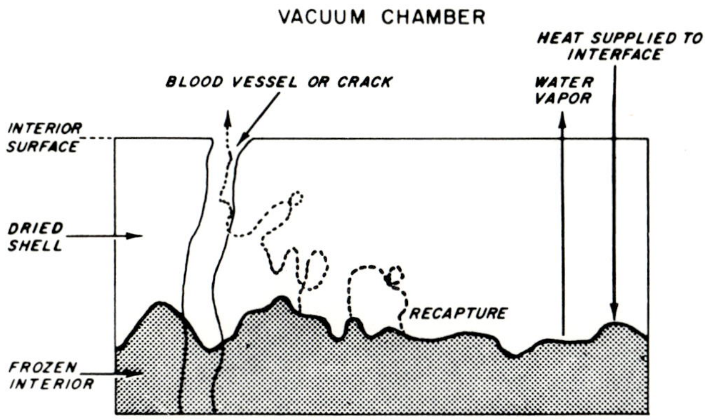


FIG. 7. Schematic representation of paths of water molecules leaving the frozen interface of a specimen in a vacuum chamber (Stephenson, 1953).

This was also demonstrated by the work of Jansen (1954) who constructed an ingenious vacuum tight torsion balance in which variations in the weights of various drying materials could be accurately determined. He compared the drying rates of pure water (ice), pure water absorbed by filter paper, and rat liver. It was found that the drying rates of rat liver and filter paper were longer than that of the pure water. Of interest is the fact that water frozen after being absorbed by filter paper dries at a different rate than that of ice itself. Yamaida (1953) has also studied dehydration rates in a vacuum system at low temperatures.

The change from the solid to vapor state requires energy in the form of heat which must be constantly supplied to the frozen surface of the drying specimen. As each molecule changes from ice to vapor a discrete amount of heat is removed from the environment and unless it is replaced the temperature of the environment will fall. A temperature reduction in turn reduces the degree of sublimation. Thus, heat transfer to the drying specimen must be constant and reasonably efficient if the drying process is to proceed at an appreciable rate. Heat transfer is brought about by radiation and to some extent by conduction. A drying object will receive the maximum amount of heat through radiation if it is surrounded as completely as possible by a radiating shield of the temperature selected. Freeze-dry devices in which the tissue is exposed at close proximity to the very low temperature of a cold trap may be less efficient with regard to drying rate than those in which the cold trap is placed at some distance from the specimen.

Once water molecules have left the dried shell they must be removed by the adjacent vapor trap. The vapor trap may consist of a chamber cooled by alcohol-dry ice, or liquid nitrogen. A trap containing phosphorus pentoxide at room temperature has proved to be very satisfactory in many systems. At atmospheric pressure, the mean free path of a water molecule is about 0.05 $\mu$. However, when the partial pressure of air is $10^{-4}$ mm. Hg this corresponds to a mean free path of approximately 50 cm. Thus the collision of water molecules with air molecules should be relatively infrequent. Mathematical analysis of this and other factors, as well as practical experience, indicates that the mean free path distance is not critical (Goodspeed and Uber, 1934; Stephenson, 1954; Gersh and Stephenson, 1954). Attempts to radically decrease the drying time have met with limited success for the following reasons:

(1) The major impediment of fast drying lies in the object itself, namely in the resistance of the drying shell.

(2) It is difficult to prevent drying tissues from cooling too much below the optimal temperature because of withdrawal of latent heat of evaporation, although heat can be supplied to the drying tissue through conduction or radiation.

It is apparent that tissues are rather poor thermal conductors, and in placing a slice of tissue upon a heated flat surface, the surface available for evaporation is immediately decreased by half. In recent years several innovations have been suggested which may facilitate more rapid drying. It was suggested that heat could be supplied to the drying object by filling the drying tube with a heat conducting gas at an appropriate pressure. Jansen (1954) found that this type of system was workable with reference to sublimation of ice. Treffenberg (1953), Jensen (1954), and Meryman (1959) employed systems which removed water from the specimen surface by a stream of dry air. Such systems may operate at atmospheric pressure or under a partial vacuum. These studies tend to show that the passage of water vapor from the interface between the frozen center and dried shell is facilitated by the vapor pressure gradient rather than by the absolute pressure of the system.

Another method of providing heat is the application of high frequency electromagnetic radiation in the microwave range (Copson, 1958). He was able to dry large samples of dense structure, including bone, in a short period of time because of the heat energy applied directly to the inner frozen core of the specimen. As yet, this approach has not been applied to histological specimens.

On the basis of experience it would appear that a vacuum of $10^{-3}$ to $10^{-4}$ mm. Hg is adequate for freeze-drying. Although such pressures can be obtained with a good mechanical pump under optimal conditions, it is invariably best to have a diffusion pump included in the system in order to compensate for minor leaks or slight contamination of pump oil. The determination of the end point of drying can be made by the use of various types of vacuum gauges and also on the basis of experience with the particular freeze-dry system in use. A McLeod gauge may be used to measure the partial pressure in the system, but this gauge, of course, does not measure water vapor. Gauges employing hot filaments may be employed in the detection and measurement of water vapor, but contamination of the filament may result in inaccurate readings. The use of a Tesla coil may be of value to ascertain the final stages of drying. At pressures below $10^{-3}$ mm. Hg the glow discharge will disappear from the interior of the system. The maintenance of a pressure this low or lower coupled with what appears to be a reasonable drying time, as determined by trial and error for a particular apparatus, should serve as an adequate indication for the end point of drying.

## C. Processing of Frozen-Dried Specimens

Dried specimens are usually warmed up to room temperature and directly embedded in paraffin. Many freeze-dry systems have provision for paraffin infiltration within the vacuum chamber (Scott and Williams, 1936; Stowell, 1951; Moberger *et al.*, 1954). Infiltration within the vacuum chamber is apt to be a cumbersome procedure in that the specimens usually must be reoriented and re-embedded. In most cases it is entirely satisfactory to remove the dried specimens and embed them in a vacuum oven or suitable chamber. Figure 8 illustrates a modified vacuum oven which is used to infiltrate large numbers of frozen-dried specimens. Glass high vacuum stopcocks have been used to replace the original metal ones and the inlet tube has been rebored. With this chamber an internal pressure of less than $10 \times 10^{-3}$ mm. Hg can be obtained. Dried tissues are known to be difficult to infiltrate unless a suitable vacuum is present. At atmospheric pressure specimens tend to float on the surface of the melted paraffin for a long period of time. Although dried specimens are hygroscopic to some extent, they are sufficiently stable to allow exposure to the atmosphere and further treatment outside of the drying tube. If specimens removed from the drying tube and exposed to the atmosphere become wet in appearance, it indicates that the tissue was not dried and not that atmospheric moisture was absorbed. Fat, which is very difficult to dry, often shows this characteristic. Frozen-dried specimens briefly infiltrated and embedded in paraffin are highly suitable for use with a number of enzyme techniques including those for aminopeptidase. The latter enzyme is markedly sensitive to aqueous fixation followed by routine embedding procedures. Some of the oxidative enzyme systems (e.g. succinic dehydrogenase) withstand freezing, drying, and paraffin embedding. Subsequent dewaxing with paraffin solvents results in very low activity, probably due to the loss of cofactors. When adequately infiltrated and embedded, frozen-dried specimens generally section fairly easily. Because of their hygroscopic nature the sections

FIG. 8. Modified vacuum oven for paraffin embedding of a large number of frozen-dried specimens (Burstone, 1956c).

cannot be floated on a water bath to remove wrinkles but must be dry mounted on a clean glass slide. Following dry mounting, the paraffin is melted for a short period of time on a thermostatically controlled hot plate in order to assure close approximation between the tissue section and the glass slide. On the other hand, the double embedding technique is a useful adjunct in that it enables frozen-dried sections to be floated out on water without the loss of enzyme activity (Burstone, 1958c, 1960c). Double embedding in celloidin at low temperature followed by paraffin also facilitates the sectioning of larger and more brittle dried specimens.

## D. Preservation of Enzymes in Frozen-Dried Tissues

The vast experience of the pharmaceutical and food industries has demonstrated the good state of preservation of proteins including enzymes in frozen-dried tissues (Harris, 1954; Sizer and Josephson, 1942; Joslyn, 1949). The elegant preservation of histological and cytological structures in properly processed frozen-dried material has been demonstrated over the years by the work of Gersh and others (Gersh and Stephenson, 1954). Mucin and zymogen granules, thyroid colloid (Fig. 9), and mitochondria

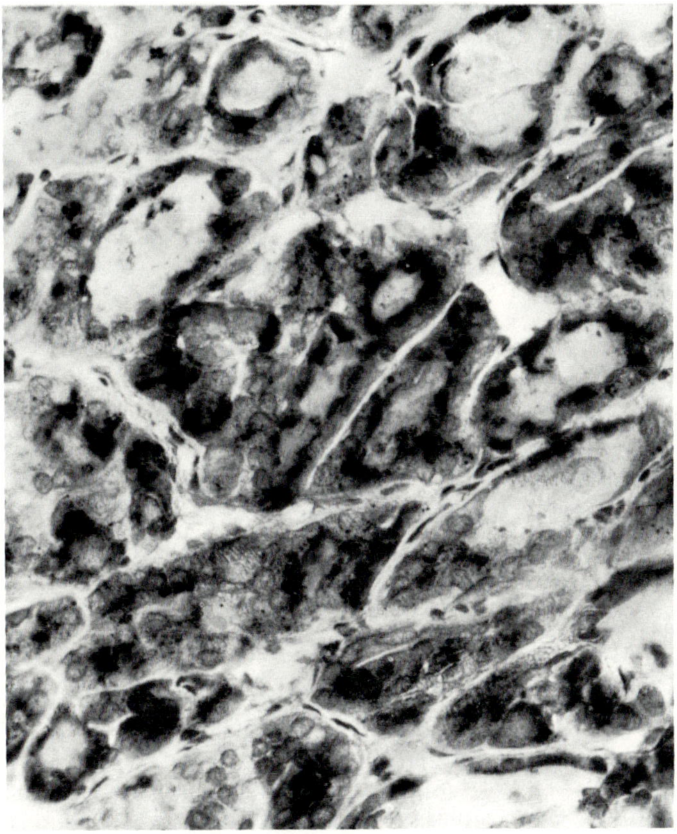

FIG. 9. Frozen-dried human thyroid showing preservation of colloid material. Note the absence of artefactual "scalloping" effect (× 143).

(Figs. 10 and 11) exhibit a good state of preservation. Phase contrast and ultraviolet microscopy have shown close similarities between frozen-dried and living cells (Danielli, 1953; Caspersson, 1950).

Until very recent years frozen-dried tissues have been applied to either quantitative studies or, in the case of microscopic histochemistry, to the demonstration of enzymes by the older metal-salt techniques. Dounce (1950) and Mirsky (1951) used frozen-dried tissues in quantitative studies of several enzymes including catalase, arginase, esterase, and lipase. Lowry *et al.* (1954) have employed frozen-dried specimens in their microquantitative techniques for the study of a number of enzymes in the central nervous system. Doyle (1950, 1954, 1955) and Doyle and Liebelt (1955) have studied the preservation of phosphatase, esterase, and peptidase in frozen-dried paraffin embedded blocks by quantitative techniques and observed that these enzymes are well preserved in paraffin for an indefinite period of

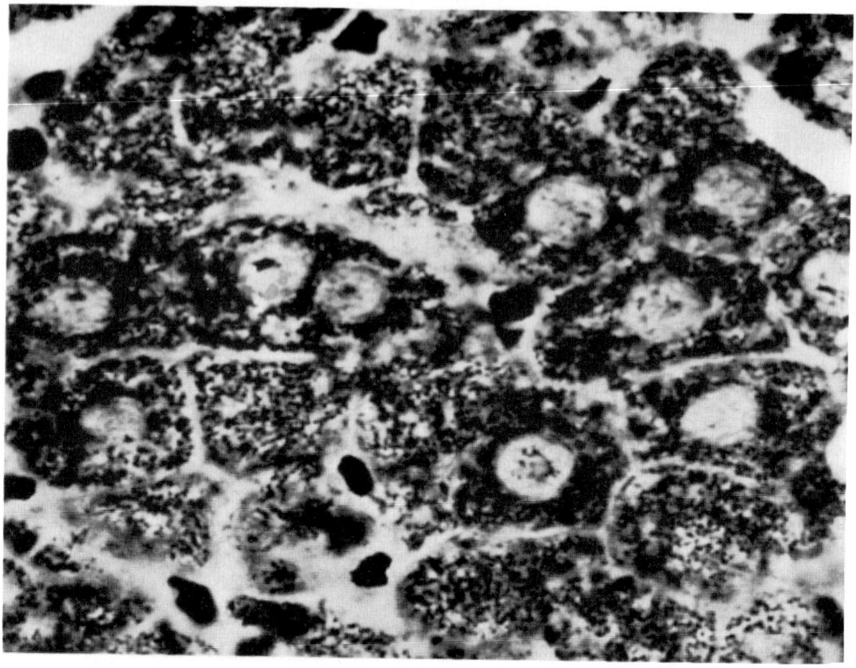

FIG. 10. Regaud mitochondrial stain. Frozen-dried rat liver (× 1100).

time. It was found that freeze-drying in liquid nitrogen, followed by vacuum dehydration at −28° and paraffin embedding, resulted in preservation of essentially the full activity of labile as well as more stable peptidases (Doyle, 1950). With reference to esterase, Doyle (1954) found that gastric tissue could be frozen and dried and embedded in paraffin with negligible loss of activity. It was also observed that lymphatic tissue fixed for one hour in cold formalin exhibited only 5 to 8% of the activity of frozen-dried specimens (Doyle and Liebelt, 1954).

The earlier attempts to apply frozen-dried material to microscopic enzyme histochemistry dealt almost exclusively with the demonstration of phosphatases by the glycerophosphate technique (Bevelander and Johnson, 1946, 1950; Yokoyama and Stowell, 1951; Wang and Grossman, 1949b; Emmel, 1946; Reale, 1956; Neumann, 1952; Firket and Michel, 1954). It should be pointed out that the metal-salt methods are in many instances not entirely ideal for application with paraffin embedded tissues, so that frozen-dried specimens may not exhibit their optimal qualities with such techniques. Naidoo and Pratt (1951, 1952) studied, in addition to acid and alkaline phosphatases, ATPase, 5-nucleotidase, and thiamine pyrophosphatase of the central nervous system. Comparison of frozen-dried

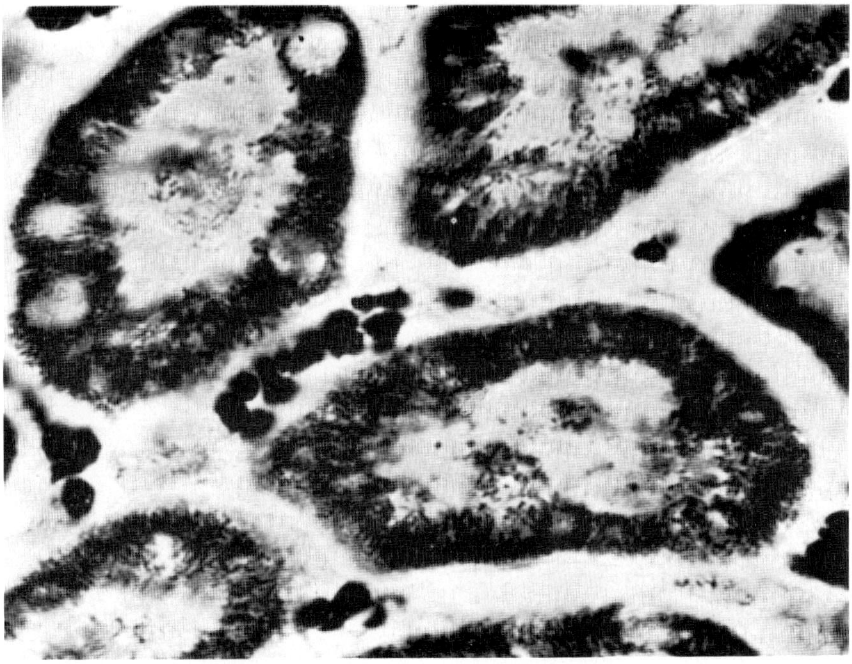

FIG. 11. Regaud mitochondrial stain. Frozen-dried rat kidney (× 1100).

material with homogenates revealed no loss of enzyme activity. Demonstration of other enzymes in frozen-dried tissue include cholinesterase (Shen *et al.*, 1952), polyphenol oxidase (Smyth, 1954), phosphamidase (Neumann *et al.*, 1954), phosphorylase (Cobb, 1953), and DOPA oxidase (Rappaport, 1955).

More recently, newer azo-dye procedures have been used to study extensively the distribution of a number of hydrolytic enzyme systems in frozen-dried tissues (Burstone, 1956a, b, 1957a, b, c, 1958a, b, c, 1959a, b, 1960a, b, c; Burstone and Folk, 1956; Burstone and Keyes, 1957; Glenner and Burstone, 1958; Glenner *et al.*, 1959).

## E. Application of Frozen-Dried Tissue to Electron Microscopy

There have been several studies on the application of frozen-dried material to electron microscopy but few of these related to the histochemical demonstration of enzymes at the time of this writing. The possibility of preserving tissues by purely physical techniques involving almost instantaneous fixation should facilitate the study of the localization of enzymes at the submicroscopic level. Sjöstrand and Baker (1958) were among the first to use freeze-drying for electron microscopy of cells. They observed vacuolization due to ice crystal formation but found that it did not interfere in a critical way with the observation of certain submicroscopic structures, such as mitochondria and α-cytomembranes. It was found that serious artefacts were produced when the dried tissue was infiltrated with plastic monomer according to routine procedures. Artefacts were minimized by infiltration and polymerization with plastic at low temperatures according to the technique of Müller (1957). Müller also pointed out the value of propane as a quenching medium, as well as the significance of low temperature dehydration. Hanzon and Hermodsson (1960) extended and confirmed the aforementioned work of Sjöstrand and Baker and also stressed the importance of very low temperature dehydration in order to avoid artefacts (vacuoles) due to ice crystal formation. Exocrine pancreas was used as a test object, and morphological structures observed with conventional osmium-fixed material could be confirmed to a great extent. Gersh *et al.* (1957, 1960) employed sophisticated freeze-dry techniques in conjunction with electron microscope stains, some of which were capable of demonstrating protein groups. Seno and Yoshizawa (1960) employing the basic techniques of Müller (1957) studied the submicroscopic structure of frozen-dried liver and pancreas. These authors concluded that routine osmium fixation gives satisfactory preservation of a number of cell structures which are clearly delineated in carefully processed frozen-dried specimens. Grunbaum and Wellings (1960a, b) have also investigated frozen-dried material for the study of cellular ultrastructure. With reference to enzyme systems, Nelson (1958, 1959) has studied adenosinetriphosphatase and succinic dehydrogenase in frozen-dried rat sperm. Yaeger (1961) has applied a tetrazole method to localize succinic dehydrogenase in muscle cells of mouse diaphragm. The widespread application of frozen-dried tissue to definitive enzyme histochemistry awaits the development and synthesis of new substrates and new embedding techniques and modifications.

### F. Design of Freeze-Dry Equipment

The first and pioneering modern approach to the freeze-drying of tissues for histological purposes was reported by Gersh in 1932. Since that time a variety of freeze-dried devices have been described which in large part derive from the original design. Gersh employed liquid ammonia in a Dewar flask which was hermetically sealed to the dehydrating chamber. The temperature inside the chamber was found to be close to −22°, the temperature of ammonia. Phosphorus pentoxide was employed as a drying agent and the vacuum pumps consisted of a mechanical and mercury vapor diffusion pump. Incorporated in the system was a special chamber for vacuum embedding. In 1934, Goodspeed and Uber reported the design

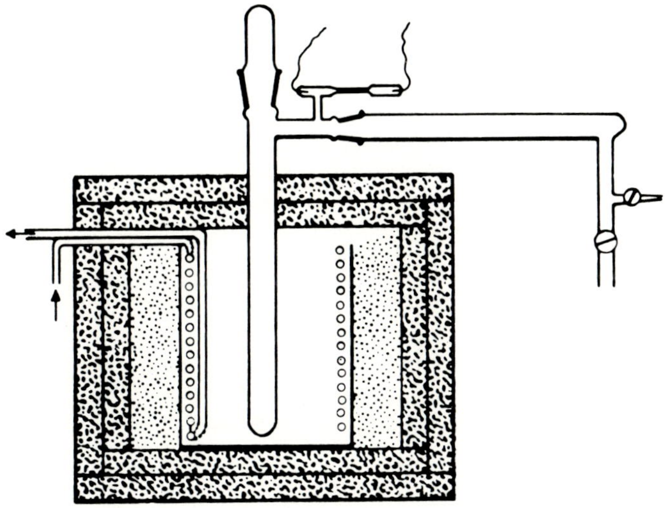

FIG. 12. Freeze-dry apparatus of Goodspeed and Uber (1934). Note the simple drying tube and the electrically refrigerated chamber.

of a freeze-dry system which is very similar to many modern counterparts (Fig. 12). It consisted of a drying tube inserted vertically in a mechanically refrigerated chamber (−22° to −32°). A single mechanical vacuum pump was used in conjunction with a phosphorus pentoxide drying chamber. A glow discharge tube was built into the system in order to check the approximate vapor pressure. Goodspeed and Uber stated that there is no advantage to using high vacuum systems and that a good rotary pump is adequate. They were among the very first to offer a mathematical analysis of drying mechanisms, and pointed out that the mean free path of water molecules in their vacuum system was over 5 cm. As contrasted with other systems, they removed dried tissues from the vacuum chamber and embedded them in a small vacuum bell jar in a paraffin oven. In 1936, Scott and Williams reported on a system with a very short mean free path. It contained a variable temperature control and permitted embedding in the dehydrating chambers. Packer and Scott (1942) also employed a variable temperature control. In addition, they utilized two vacuum gauges in order to help ascertain drying time. Two drying traps were also employed: (1) phosphorus pentoxide; (2) liquid nitrogen. Sjöstrand (1944) used a number of dehydrating chambers with a permanent refrigeration system operating at −47° (Fig. 13). In 1948, Gersh published a design in which fixed refrigeration and multiple drying chambers were used. A drying temperature of −32° to −40° was recommended. Other developments were made by Pease and Baker (1949) who used a drying chamber of wide diameter connected to a vacuum pump with no vapor trap intervening.

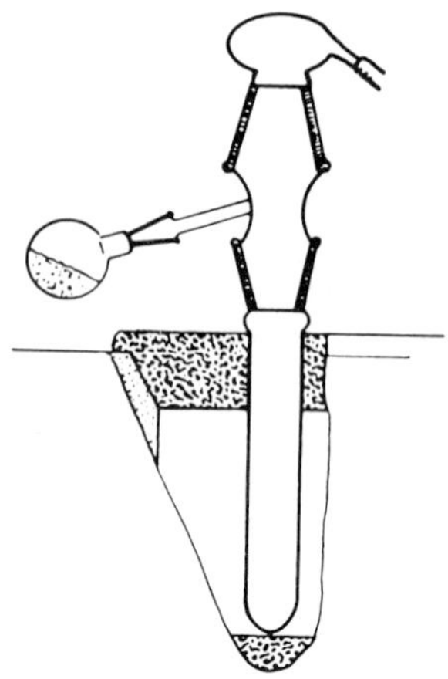

FIG. 13. Freeze-dry apparatus of Sjöstrand (1944). The drying tube is inserted into a refrigerated chamber, while the dehydrating agent is contained in a separate flask at room temperature.

Modified freeze-dry systems have also been reported by Taylor (1944), Wang and Grossman (1949a), and Harris *et al.* (1950). Mendelow and Hamilton (1950) used a metal system in which the liquid nitrogen vapor trap was close to the tissue. They employed ethyl oxalate kept frozen at −40° by dry ice. The vapor trap or cold finger contained liquid nitrogen and was the forerunner of systems based on a short mean free path between drying tissue and vapor trap. Dried tissues were removed and embedded in a vacuum oven. The design of Stowell (1951) employed a drying chamber which was entirely surrounded by liquid nitrogen which acted as both a refrigerant and a vapor trap. The temperature of the drying tissue was maintained by a thermostatically controlled heating element; this system likewise was characterized by a very short mean free path. Dry ice or other mixtures may also be used as the refrigerant. Glick and Malmström (1952) described a freeze-dry apparatus which embodied a similar principle. A liquid nitrogen vapor trap was employed while dry ice-acetone mixtures were used to maintain the temperature of the frozen tissue. It was pointed out that the very low temperature of the frozen specimen would result in prolonged dehydration time, and therefore a mixture of a number of ethers and dry ice was suggested. The aforementioned system was modified by Moberger *et al.* (1954). A vacuum of approximately $10^{-5}$ mm. Hg was obtained in 15 minutes. They also reported that dehydration could be accomplished at a temperature below −60° in 6 to 12 hours depending upon the size and nature of the tissue specimens. As with most systems of this type, the tissues were embedded in the drying tube. Glick and Bloom (1956) subsequently reported another modification in which they pointed out that elaborate high vacuum systems were not necessary. Butler and Bell (1953), Gustafsson (1953), Danielli (1953), and Lacy and Blundell (1955) have reported devices characterized by a short mean free path and a drying chamber partly or completely surrounded by a cold trap. Eränkö (1954b) described a relatively simple inexpensive system which consisted of an inner glass tube connected by a ground joint to an outer tube, and a two stage mechanical pump for evacuating the space between the tubes. This design, too, was based on the principle of short path condensation. Diethyl oxalate and solid carbon dioxide were used to produce a temperature of $-44°$ in the drying chamber. The trap was cooled either with liquid air or a mixture of dry ice and ethanol or acetone. A somewhat similar system with an exceedingly short mean free path was described by Freed in 1955. A simplified apparatus based on the design of Gersh was introduced by Arcadi and Tesar (1954). The refrigerated chamber consisted of an insulated box made of plywood packed with rock wool. Dry ice was employed to maintain a temperature of $-30°$ in the refrigerated chamber. Phosphorus pentoxide was employed as the drying agent and the vacuum system consisted of a single rotary oil pump which maintained a pressure of $10^{-3}$ mm. Hg. An interesting innovation in simplified designs was reported by Jansen (1954) (Fig. 14). The apparatus consisted of a U-shaped glass tube, one leg of which is connected to a vacuum pump; part of the other leg which is the drying tube is surrounded by a copper mantle which in turn is enclosed by a sealed vacuum jacket. A piece of movable copper tubing surrounds this jacket from about the middle to

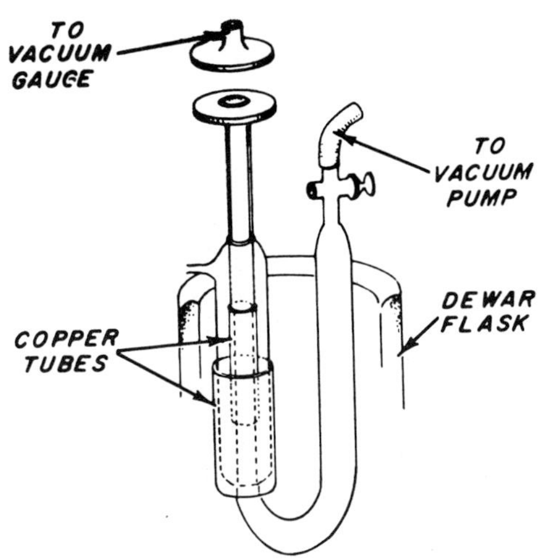


FIG. 14. Schematic representation of freeze-dry apparatus of Jansen (1954). The temperature range is varied by adjusting the height of the external copper tube.

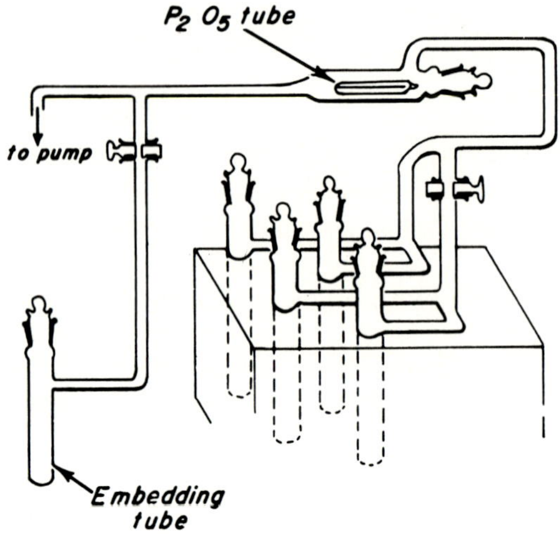


FIG. 15. All-glass freeze-dry apparatus of Gersh (1948).

well below the lower end. The lower part of the tube is cooled by a dry ice-acetone mixture in the Dewar flask. The portion of the tube immersed in the dry ice-acetone mixture will, of course, be at a temperature of approximately −70°; but the temperature of the drying tube which is isolated by the vacuum jacket can be varied between −70° and room temperature by adjusting the relative heights of both copper tubes. A temperature of −40° to −45° is obtained within the drying tube. Jansen reports that once set, the temperature has been found to remain within this range in spite of changes in room temperature and pressure in the drying tube. Naidoo and Pratt (1956) described multiple drying tubes with a phosphorus pentoxide drying trap. Temperature was maintained by a mixture of dry ice and ethyl phenyl ether (−34°).

A system based on the design of Goodspeed and Uber (1934) and Gersh (1948) (Fig. 15) has been in operation in the author's laboratory for over 5 years (Burstone, 1956c). The unit utilizes mechanical refrigeration and the drying tubes are of large capacity. Ball joints are used throughout to relieve strain on the glass. This type of system has proved to be of use in processing large numbers of surgical pathology specimens (Fig. 16). Each of the drying tubes is surrounded by a loosely fitting copper well one-eighth inch thick in the refrigerated chamber. Figure 17 shows the details of one of the combined drying tubes and phosphorus pentoxide trap. One of these tubes inserted into a suitable refrigerating unit constitutes a rugged and inexpensive drying unit of fairly large capacity. A rebuilt vacuum oven is employed for the simultaneous embedding of a large number of thick

FIG. 16. Freeze-dry apparatus employed in the author's laboratory for the past five years. Design is characterized by an all-glass system employing ball joints and tubes of wide diameter.

specimens (Fig. 8). Benditt *et al.* (1961) have also described a somewhat similar system consisting of a three barreled glass vacuum unit which fits into a refrigerated chamber (Fig. 18). A ⅓ h.p. refrigerating unit using Freon 22 is capable of maintaining a temperature of −40°. Provision is made for infiltration of the tissue in the apparatus if desired. Systems such as this are advantageous to the pathologist in that large numbers of specimens may be dried simultaneously.

A few designs of drying devices based on the use of a dry gas to remove water evaporating from the tissue specimen have been reported (Treffenberg, 1953; Jansen, 1954; Kramer and Hill, 1956). Although high vacuum equipment may not be needed for these systems, some provision is required for drying the air and maintaining the necessary low temperature of the air in specimens during the dehydration. The system of Kramer and

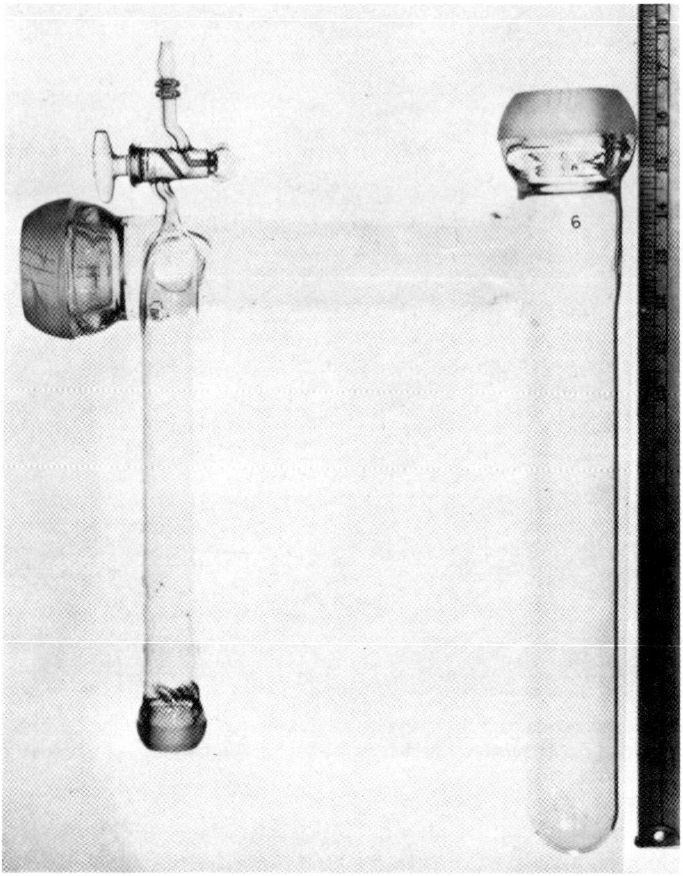

FIG. 17. Photograph of single tube as employed in apparatus shown in Fig. 16.

Hill appears to circumvent some of the aforementioned difficulties. A mixture of carbon dioxide snow and Cellosolve contained in a Dewar flask serves to cool the chamber and provide a stream of dry gas. Drying time of 10 to 24 hours is claimed for tissue blocks of small size.

The specialized freeze-dry systems developed for electron microscopy often include a means for embedding the specimens in a vacuum chamber, usually the drying chamber (Müller, 1957; Sjöstrand and Baker, 1958; Gersh *et al.*, 1960; Grunbaum and Wellings, 1960a; Hanzon and Hermod-

FIG. 18. Freeze-dry apparatus of Benditt *et al.* (1961) (courtesy Dr. E. P. Benditt).

sson, 1960; Seno and Yoshizawa, 1960). Several of these systems have a provision for reacting various anhydrous vapors *in vacuo* with the dried specimen.

In summary, the various freeze-dry systems may be grouped into three types:

(1) Vacuum chambers usually cooled by mechanical refrigeration and characterized by a long mean free path to the vapor trap.

(2) Vacuum chambers cooled by liquid and/or solid refrigerants with a short mean free path to the vapor trap.

(3) Refrigerated drying tube through which passes a stream of dry air or gas.

Further details of design and general application of freeze-drying in histochemistry may be referred to in the original papers and reviews by Kanehira (1953), Gersh and Stephenson (1954), and Neumann (1958).

## V. FREEZE-SUBSTITUTION

The freeze-substitution technique was originated by Simpson (1941) in the course of an analysis of the Altmann-Gersh freeze-dry procedure. In an attempt to isolate the effects of freezing from the effect of subsequent vacuum dehydration he froze tissues and placed them in organic solvents at low temperatures of from −40° to −78°. Included among the solvents were methyl Cellosolve, ethanol, diethyl ether, and chloroform. Lison (1949) employed the technique in the study of glycogen while Blank *et al.* (1951) used freeze-substitution as a means of studying radioactive $P^{32}$ and referred to their method as non-vacuum freeze-drying. They found that by using propylene-glycol followed by polyethylene-glycol embedding (Blank, 1949), losses of radioactive material were minimized as compared with the ordinary paraffin technique. Various modifications as well as applications of the procedure have been employed by an number of workers (Russell *et al.*, 1949; Lillie and Laskey, 1951; Taft, 1951; Baud, 1952; Persson, 1952; Gourévitch, 1953; Davies, 1954; Pollister and Ornstein, 1959; Deitch and Godman, 1955; Freed, 1955; Freed and Green, 1955; Woods and Pollister, 1955; Chang, 1956; Taylor, 1956; Davies *et al.*, 1957; Hancox, 1957; Fernández-Morán, 1957, 1959; McMaster-Kaye and Taylor, 1958; Nair, 1958; Patten and Brown, 1958; Davis *et al.*, 1959).

Study of the physical mechanisms involved in freeze-substitution has received little attention. Persson (1952) studied the dehydration time of cylinders of gelatin which were frozen at different temperatures and found that the shortest drying time was correlated with the lowest initial freezing temperature. Hancox (1957) found that protein denaturation was minimal in rat tissues dehydrated in butanol at −38°. Johnson (1959) compared substitution rates using a number of anhydrous solvents at −45°. Acetone and ethanol-acetone showed the most rapid dehydration action, while iso-amyl alcohol, butanol, and tetrahydrofuran were very slow. Less than half of a 5 μl. sample of frozen synthetic tissue fluid could be dehydrated in 10 ml. of the aforementioned solvents in 12 days. None of the solvents causes as rapid dehydration as obtained at −30° with a simple freeze-dry apparatus. The advantage claimed for freeze-substitution is that following quenching, the solvents employed for dehydration dissolve the ice out of the specimen and the tissue thus retains its relatively undenatured state. Bell (1952a) points out that it is not certain at what stage the substitution of solvent for water occurs. It may well take place to some degree during the warming-up stage. This suggests that diffusion artefacts may occur and that the fixation may in part be chemical and similar to the cold acetone technique of Gomori (1952).

The most recent extensive consideration of methods and principles of fixation by freeze-substitution has been reported by Feder and Sidman (1958). They rapidly froze tissues in propane-isopentane cooled by liquid nitrogen. The frozen tissue was placed in a substituting fluid at −70° (for approximately one week) consisting of 1% mercuric chloride or picric acid in ethanol or 1% osmium tetroxide or chromic acid in acetone. After substitution, the specimens were washed for 12 hours or more followed by three changes of ethanol or other solvents and placed in chloroform for 6–12 hours. They were then brought up to room temperature, transferred to xylene, and embedded in paraffin. Their findings confirmed the original observations of Lison (1949) who found that the quality of preservation was vastly improved by incorporation of a fixative with the dehydrating solvent. Feder and Sidman (1958) point out that most investigators have used non-aqueous solvents as substituting fluids and that the relatively unfixed specimens were subject to distortion during subsequent processing and embedding. They believe that the fixing agent in the substituting fluid actually exerts its effect at −70° and not when the specimen is warmed prior to embedding. This contention is supported by several model experiments. These authors did not report any studies on the use of their technique for enzyme histochemistry.

Davis and Ornstein (1959) have reported the histochemical demonstration of esterase and alkaline phosphatase at both routine optical and electron microscope levels in freeze-substituted tissues using azo-dye methods. Although their results were satisfactory from cytological standpoint it is not known how much inactivation of the enzymes studied had occurred. Bullivant (1960) froze mouse pancreas in liquid helium II, and substituted the tissue ice with methanol at −75°. Specimens were substituted for 2 weeks at low temperature and then embedded and polymerized in methacrylate plastic at low temperature. Sections were treated to reveal acid phosphatase according to the Gomori glycerophosphate-lead nitrate method. It was pointed out that although the resulting precipitate may not reveal the true site of the enzyme, this type of histochemical approach is feasible for the resolution of sites of enzyme activity at the electron microscope level.

Protagonists of the freeze-substitution technique stress that the procedure should be as effective as freeze-drying and that freeze-substitution requires no vacuum system. However, it should be pointed out that freeze-substitution (just as with most freeze-dry systems) requires a low temperature (−50 to −70°) ice chest or mechanical refrigerator. These cold boxes are relatively expensive to maintain over a period of time. In addition, substitution of rather small pieces of tissue requires considerable solvent and is very time consuming. In view of the prolonged periods during which freeze-substituted tissues are subjected to organic solvents at low and at room temperatures, as well as of the potential hazard of inactivation of enzymes due to the presence of residual fixatives, it would appear that at the present time freeze-drying is a superior technique for enzyme histochemistry.

## APPENDIX 1

# FIXATION AND EMBEDDING METHODS

1. *"Neutral" Formalin Solutions*

*a. Formol-calcium (Baker, 1944, 1949).*

| | |
|---|---|
| Formalin (37–40% formaldehyde) | 10.0 ml |
| $ClCl_2$ (anhydrous), 10% aqueous | 10.0 ml. |
| Distilled water | 80.0 ml. |

Baker points out that calcium chloride acts osmotically and appears to prevent disruption of phospholipid containing structures. This fixative at 0–5° was first used by Eränkö (1951, 1952) in enzyme histochemistry for the preservation of acid phosphatase.

*b. Neutral formalin (Lillie, 1947, 1954).*

| | |
|---|---|
| Formalin | 100.0 ml. |
| Distilled water | 900.0 ml. |

Calcium or magnesium carbonate to excess

*c. Neutral buffered formaldehyde (Lillie, 1947, 1954).*

| | |
|---|---|
| Formalin | 100.0 ml. |
| Distilled water | 900.0 ml. |
| Sodium dihydrogen phosphate monohydrate | 4.0 g. |
| Disodium phosphate, anhydrous | 6.5 g. |

*d. Sodium acetate-formalin (Gomori and Chessick, 1953).*

| | |
|---|---|
| Formalin | 100.0 ml. |
| Distilled water | 900.0 ml. |
| Sodium acetate | 20.0 g. |

*e. Formalin-chloral hydrate (Fishman and Baker, 1956; Baker et al., 1958).*

| | |
|---|---|
| Formalin | 20.0 ml. |
| Distilled water | 80.0 ml. |
| Chloral hydrate | 0.1 g. |

The solution is stored at 5–8° in bottles containing marble chips (calcium carbonate). The use of calcium chloride would also appear to be satisfactory. Tissues are fixed overnight to 24 hours and washed in running water for approximately 1 hour.

*f. Formol-calcium gum-sucrose (Holt and Withers, 1958).* The fixative is freshly prepared by dissolving 1% of dried calcium chloride in a 4% solution of formaldehyde made by accurate dilution of 36 to 40% reagent grade formalin. Granular calcium carbonate is added at the rate of 5.0 g. per 500.0 ml. of fixative. The mixture is shaken and cooled to approximately 2°. The supernatant is decanted into a pre-cooled beaker having a stainless steel wire mesh about 1.0 cm. from the bottom. Thin slices of tissue are fixed for 24 hours. Not more than 1.0 g. of tissue is used per 100.0 ml. of fixative. After fixation the tissues are blotted and transferred without washing to cold 0.88 *M* sucrose solution containing 10.0 g. per liter of gum acacia for a period of 24 hours. Frozen sections are subsequently prepared and incubated "free-floating." The formalin solution should be used the same day as prepared, while the sucrose-gum acacia may be stored in a refrigerator if 0.1 g. of thymol per liter is present.

2. *Preparation of Smears, Mylar Film Method (Burstone and Fleming, 1959)*

This technique utilizes a polyester (polyethylene terephthalate) plastic film (Mylar®)* to which cells readily adhere. Mylar may be obtained in sheets or rolls. In general, rolls of 3/4-inch type C Mylar of 25 mil thickness (6.25 $\mu$) are used. Four-inch lengths of Mylar are cut and placed on a clean piece of white paper on a masonite board. Mylar tends to accumulate a static charge and should be protected from dust and dirt. After routine smearing techniques have been used on the short length of Mylar, 1-inch lengths may be cut for incubation or suitable staining. The 1-inch pieces are placed face down into substrate solutions. They generally float on top of the solution. Following incubation they are transferred face down into tap or distilled water, followed by counterstaining if desired. For mounting in a synthetic medium, the films may be dehydrated in alcohols followed by xylene, since Mylar is unaffected by most solvents. In the case of dyes sensitive to alcohols, the film is placed smear side up on a glass slide, and then mounted in glycerol-gelatin or water soluble plastic.

3. *Frozen-Section Fixation Method for the Demonstration of Enzymes (Burstone, 1961)*

Unfixed frozen sections mounted on glass slides are immersed in acetone (or diacetone alcohol) and acetone-water mixtures at room temperature according to the following schedule:

1. Absolute acetone, 1–5 minutes.
2. 90–95% Acetone-water, 1–3 minutes.
3. 80–85% Acetone-water, ½–1 minute.
4. 70% Acetone-water, ½ minute.

The slides are then transferred to distilled water for 10–20 seconds and then immersed in the appropriate substrate solution. Although the quality of fixation does not compare with that obtained with freeze-dry and similar

* E. I. du Pont de Nemours and Co., Wilmington, Delaware.

methods, the simplicity of the method lends it to general application where cytochemical detail is not of primary importance.

4. *Cold Acetone Fixation, Double Embedding Method* (*Gomori, 1952*)

1. Chill (do not freeze) thin slices (1–2 mm.) of tissue for 10–15 minutes and then fix in cold acetone (0–5°) overnight. Transfer to another change of acetone for a total fixation time of 24 hours.
2. Dehydrate in 2 changes of cold acetone, approximately 12 hours each.
3. One or 2 changes of absolute ethanol (overnight) may be employed.
4. Transfer specimens to cold ethanol-ether (1:1) for several hours.
5. Transfer specimens to a cold 2–3% solution of celloidin (USP collodion diluted with 1 volume of ethanol-ether for 12–24 hours.
6. Drain specimens rapidly and carry them through 2 changes of chloroform or (benzene) at room temperature.
7. Embed in paraffin under vacuum.

5. *Cold Acetone or Ethanol Fixation, Rapid Method*

1. Chill thin slices of tissues (1–2 mm. in thickness) for 10–15 minutes and fix in cold acetone (0–5°) overnight.
2. Transfer through 2 changes of cold petroleum ether 1 hour each.
3. Transfer to another change of petroleum ether at room temperature.
4. Embed in paraffin under vacuum.

6. *Decalcification with Formic Acid-Sodium Citrate* (*Morse, 1945*)

1. Fix specimens in 10% neutral formalin or calcium-formol.
2. Wash specimens in running tap water for 2–5 hours.
3. Decalcify in the following formic acid-sodium citrate reagent until a chemical test is negative for calcium or until specimens can be cut with a razor blade.

*Formic acid-sodium citrate reagent*

| | | |
|---|---|---|
| Solution A: | 90% Formic acid C.P. | 1 part |
| | Distilled water | 1 part |
| Solution B: | Sodium citrate C.P. | 20 g. |
| | Distilled water | 100 ml. |

Just before use combine equal portions of solution A and solution B.

4. Morse (1945) recommends dehydration with ethyl and butyl alcohols. However, for histochemical purposes rapid dehydration in cold acetone may be advisable (Appendix 1, Section 5).

5. Since decalcification may take days or weeks considerable enzyme inactivation must occur. However, Schajowicz and Cabrini (1959) feel that acid phosphatase can be satisfactorily demonstrated in decalcified specimens.

7. *Demineralization of Bone and Tooth Structure with EDTA (Freiman, 1954)*

1. Fix thin pieces of bone overnight in cold (0–5°) 95% ethanol, neutral formalin, formol-calcium, or acetone.

2. Transfer to cold 5% solution of ethylenediaminetetraacetic acid (EDTA) neutralized with NaOH to give a pH of 6.0–6.5. The solution is prepared by first dissolving 20.0 g. NaOH in 600.0 ml. $H_2O$. Fifty grams of EDTA are added with constant stirring. When solution is effected, the distilled water is added to make a total volume of 1 liter. The disodium salt of EDTA is preferred because it is more readily soluble and less NaOH is required.

3. After 24 hours the tissues are transferred to a fresh chelating solution at room temperature where they are allowed to remain until demineralized. Cancellous bone is usually decalcified in approximately three days while cortical bone may require several weeks. The decalcified tissue is thoroughly washed in running water.

4. A rapid dehydration and embedding procedure as described under the cold acetone fixation method should be employed.

5. In the case of alkaline phosphatase, dewaxed sections are immersed in 0–1% magnesium acetate at neutral pH for several hours in order to reactivate the enzyme.

8. *Gelatin and PVA (Polyvinyl Alcohol) Embedding*

Following formol-calcium or other types of formalin fixation in the cold, tissues may also be embedded in gelatin (Chessick, 1954) or polyvinyl alcohol (PVA) (Feder, 1959). Prior to the cutting of frozen sections the fixed tissues are first washed in running tap water for 1 hour. Alternatively the specimens may be briefly rinsed in cold distilled water and transferred to a fairly large volume of cold distilled water for an overnight period.

Gelatin embedding of tissues for the subsequent study of cholinesterases has been described by Chessick (1954). Following overnight fixation in cold neutral formalin, the tissues are washed for 2 hours in running tap water and then impregnated for 4–8 hours at 37° with 10–15% aqueous gelatin. Following this step the blocks are refixed in formalin preparatory to cutting frozen sections. This technique is of use with tissues such as muscle in which the fibers show a loss of coherence following cutting of frozen sections.

With the PVA technique tissues are fixed in cold neutral formalin (e.g. formol-calcium) 18–24 hours, rinsed in running water, and then embedded in polyvinyl alcohol (Feder, 1959). After approximately 10 days or more in a desiccator at 2–4° the plastic is hardened sufficiently to allow sectioning at 5–10 $\mu$.

9. *Chloral Hydrate-Glycerol Gelatin Embedding* (*Fishman and Baker, 1956*)

This technique was originally recommended for friable tissues to be used in conjunction with the β-glucuronidase technique.

1. Fix tissues in formol-chloral hydrate for 4–24 hours.
2. Wash out fixative in running water for 1–2 hours.
3. Embed in chloral hydrate-glycerol gelatin for 2–4 hours at 37°. This embedding medium is made by dissolving 15.0 g. of gelatin in 70.0 ml. water by heating in a water bath. When completely dissolved, 15.0 ml. glycerol and 100.0 ml. chloral hydrate are added, and the entire mixture poured into jars, allowed to solidify, and stored in the cold. Dissolve for use by rotating the container under a hot water tap.
4. Cast in watch glasses or other suitable containers. Cool entire mold in a refrigerator.
5. Transfer to fixative (same as in step 1) for 1–3 days at 8°.
6. Remove cast from mold by running a needle or scalpel around periphery of gelatin. Trim excess gelatin, especially on surface to be attached to freezing microtome.
7. The block is placed upon a damp piece of filter paper which is frozen to the stage. The block should appear opaque-white. Sections may be cut at 5–15 μ.

10. *Freezing of Tissues at Very Low Temperatures*

1. Tissues are rapidly frozen at low temperatures preparatory to vacuum dehydration, "substitution" (of ice) in the presence of an organic solvent, or in an attempt to assure better morphological preservation of tissue components in the preparation of cryostat or frozen sections.
2. Thin slices of freshly removed tissue are made with razor blade. Very small specimens are obtained by using two razor blades separated by a thin (0.5 mm.) steel plate. The specimens are placed upon a small piece of aluminum foil which is subsequently dropped into the quenching medium in a 200.0 ml. beaker. The beaker is cooled by partly suspending it in a Dewar flask of liquid nitrogen. Quenching liquids and the temperatures employed are as follows:

a. Isopentane −160° (Hoerr, 1936). He points out that some crude grades of isopentane do not freeze at −195°. Reasonably pure isopentane starts to become viscous at approximately −160°. Although the exact temperature may be verified with a thermocouple, completely consistent results are obtained by quenching in a slightly viscous isopentane medium.

b. Freon 12 (dichlorodifluoromethane) −160° (Bell, 1952b).

c. Propane −175° (Gersh, 1956). Gaseous propane is condensed into a liquid state in a long necked flask immersed in liquid nitrogen. The liquid propane is then transferred to a beaker kept at the desired temperature by liquid nitrogen. If the temperature of propane falls below −180° liquid oxygen may be condensed into the quenching medium (see Stephenson, 1956, for precautions).

d. Propane: isopentane (3:1) −175°. The detailed instructions given by Feder and Sidman (1958) should be consulted for employing the aforementioned quenching medium for use with freeze-dry or freeze-substitution techniques. Propane: isopentane may form explosive mixtures with liquid oxygen which may be present as a condensate.

3. Prior to vacuum dehydration the specimens may be placed in gelatin capsules or plastic centrifuge tubes and stored in liquid nitrogen for an indefinite period of time.

11. *Double-Embedding of Frozen-Dried Tissues* (*Burstone, 1960a*)

1. Dried tissues are placed in cold acetone (0–4°) for approximately 4 hours followed by cold ethanol for approximately 16 hours (overnight).

2. Tissues are placed in cold (0–5°) alcohol-ether (equal volumes of each) for 3 hours.

3. Tissues are transferred to U.S.P. flexible collodion (5%) (0–5°), which may be diluted with an equal volume of alcohol-ether, for approximately 16 hours (overnight).

4. Excess collodion is removed by gently blotting with Kleenex, and tissues are transferred to two changes of chloroform (1 hour each at room temperature).

5. Tissues are then transferred to melted paraffin (U.S.P. m.p. 56–58°) with 0.5% beeswax for 15 minutes. Tissues are transferred to another change of paraffin and infiltrated under vacuum for 30–45 minutes.

6. Sections may be cut on a rotary microtome and flattened on a water bath.

This technique is useful in preparing frozen-dried undecalcified specimens and also frozen-dried soft tissues for multiple blocks.

12. *Carbowax Embedding* (*Blank, 1949; Blank et al., 1951*)

1. Tissue slices 2–3 mm. thick are dropped directly into cold propylene glycol (−20 or lower) stored in a deep freeze. It is best to pre-chill (not freeze) the tissue specimens for approximately 3–5 minutes before immersing them in the cold solvent.

2. After a minimum of 1–2 hours at the low temperature, the specimens are transferred to a mixture of polyethylene glycols (Carbowax) for infiltration (9 parts Carbowax 4000 and 1 part Carbowax 1500).

3. Tissues are infiltrated for 2–3 hours and blocks prepared with the water-soluble wax.

4. Sections are cut and floated onto slides, and treated with appropriate substrates or staining reagents.

13. *Freeze-Substitution for Enzyme Histochemistry* (*Feder and Sidman, 1958; Davis* et al., *1959*)

1. Very small specimens are frozen on aluminum foil according to any of the techniques described under freezing of tissue at very low temperatures (Appendix 1, Section 10).

2. The frozen specimens are transferred to a substituting fluid at $-70°$ and stored in a dry ice chest or suitable deep freeze for approximately one week. Substituting solutions consist of osmium tetroxide in acetone, and mercuric chloride or picric acid in ethanol.

3. Small specimens require about a week for substitution while larger specimens require a considerably longer period of time.

4. Following substitution the container with the substituting fluid and specimens is transferred to refrigerator and the specimens washed for 12 hours or more in 3 changes of ethanol (or other solvent). Specimens treated with osmium or chromium containing solutions are washed in 2–3 changes of acetone at $-70°$ to avoid a reaction between the fixative and solvent at room temperature. The tissues are then placed in chloroform for 6–12 hours, brought to room temperature, transferred to xylene or benzene for 1 hour, and embedded in paraffin.

The degree to which enzymes are preserved by the aforementioned procedures is not clear. It is apparent that exposure of tissues to organic solvents and residual traces of fixatives at room temperature must have a deleterious effect.

14. *Freezing in Helium II Followed by Freeze-Substitution for Electron Microscopy and Enzyme Histochemistry* (*Fernández-Morán, 1959; Bullivant, 1960*)

1. Liquid helium I is contained in a double Dewar flask with liquid nitrogen in the outside vessel. Under vacuum (less than 38 mm. Hg) helium gas is removed and the temperature of the remaining liquid is lowered. The transition to helium II is noted at the point at which the liquid ceases to boil.

2. Very small pieces of tissue (less than 1 mm. thick) are placed in small cups of aluminum foil and these introduced into the helium II through a large bore stopcock in the top of the Dewar flask. Following freezing, the specimens in liquid helium are poured directly into liquid nitrogen. The specimens are transferred from the foil to gelatin capsules of the type normally used for methacrylate embedding. This transfer process is carried out under liquid nitrogen.

3. The capsules are quickly lifted out and placed in clean cold (−75°) methanol dried over a Linde molecular sieve. Substitution of the small specimens at low temperature is continued for 2 weeks, and three changes of methanol are employed.

4. The specimens are impregnated with a mixture of 90% butyl methacrylate and 10% methyl methacrylate at −75°. Polymerization is effected with ultraviolet light at −5°.

Acid phosphatase has been demonstrated in mouse pancreas with tissues processed in the aforementioned manner. Modifications should facilitate the demonstration of other hydrolytic enzymes.

## References

Adamstone, F. B. (1951) *Stain Technol.* **26,** 157–161.
Adamstone, F. B., and Taylor, A. B. (1947) *Anat. Record* **99,** 639.
Adamstone, F. B., and Taylor, A. B. (1948) *Stain Technol.* **23,** 109–116.
Albert, D. L., and Lindner, J. E. (1960) *Stain Technol.* **35,** 277–281.
Altmann, R. (1890) "Die Elementarorganismen und ihre Beziehungen zuden Zellen." Veit, Leipzig, Germany.
Arcadi, J. A., and Tesar, C. (1954) *J. Lab. Clin. Med.* **43,** 479–481.
Baker, J. R. (1944) *Quart. J. Microscop. Sci.* **85,** 1–71.
Baker, J. R. (1949) *Quart. J. Microscop. Sci.* **90,** 293–307.
Baker, J. R. (1950) "Cytological Technique," 3rd ed. Methuen, London.
Baker, J. R. (1958) "Principles of Biological Microtechnique." Methuen, London.
Baker, J. R. (1960) "Cytological Technique," 4th ed. Methuen, London.
Baker, J. R., Hew, H., and Fishman, W. H. (1958) *J. Histochem. and Cytochem.* **6,** 244–250.
Baud, C. A. (1952) *Bull. microscop. appl.* [2] **2,** 158–160.
Bayliss, W. M. (1924) "Principles of General Physiology." Longmans, Green, London.
Becker, N. H., Goldfischer, S., Shin, W.-Y., and Novikoff, A. B. (1960) *J. Biophys. Biochem. Cytol.* **8,** 649–663.
Behrens, M. (1932) *Z. physiol. Chem. Hoppe-Seyler's* **209,** 59–74.
Bell, L. G. E. (1952a) *Intern. Rev. Cytol.* **1,** 35–63.
Bell, L. G. E. (1952b) *Nature* **170,** 719.
Benditt, E., Lagunoff, D., and Johnson, F. B. (1961) *Arch. Pathol.* **72,** 546–549.
Berenbom, M., Yokoyama, H. O., and Stowell, R. E. (1952) *Proc. Soc. Exptl. Biol. Med.* **81,** 125–128.
Bernhard, W., Granboulan, N., Barski, G., and Tournier, P. (1961) *Compt. rend. acad. sci.* **252,** 202–204.
Bevelander, G., and Johnson, P. L. (1946) *J. Dental Research* **25,** 381–385.
Bevelander, G., and Johnson, P. L. (1950) *Anat. Record* **108,** 1–21.
Birbeck, M. S. C., and Reid, E. (1956) *J. Biophys. Biochem. Cytol.* **2,** 609–624.
Birge, E. A., and Imhoff, C. E. (1952) *Am. J. Clin. Pathol.* **22,** 192–193.
Blank, H. (1949) *J. Invest. Dermatol.* **12,** 95–99.
Blank, H., McCarthy, P. L., and deLamater, E. D. (1951) *Stain Technol.* **26,** 193–197.
Bullivant, S. (1960) *J. Biophys. Biochem. Cytol.* **8,** 639–647.
Burstone, M. S. (1956a) *J. Histochem. and Cytochem.* **4,** 130–139.

Burstone, M. S. (1956b) *J. Natl. Cancer Inst.* **16,** 1149–1161.
Burstone, M. S. (1956c) *J. Natl. Cancer Inst.* **17,** 49–63.
Burstone, M. S. (1957a) *Arch. Pathol.* **63,** 164–167.
Burstone, M. S. (1957b) *J. Natl. Cancer Inst.* **18,** 167–173.
Burstone, M. S. (1957c) *Oral Surg. Oral Med. Oral Pathol.* **10,** 296–303.
Burstone, M. S. (1958a) *J. Natl. Cancer Inst.* **20,** 601–614.
Burstone, M. S. (1958b) *J. Natl. Cancer Inst.* **21,** 523–539.
Burstone, M. S. (1958c) *J. Histochem. and Cytochem.* **6,** 322–339.
Burstone, M. S. (1959a) *J. Histochem. and Cytochem.* **7,** 39–41.
Burstone, M. S. (1959b) *J. Histochem. and Cytochem.* **7,** 147–148.
Burstone, M. S. (1959c) *J. Histochem. and Cytochem.* **7,** 112–122.
Burstone, M. S. (1960a) *N. Y. Acad. Sci.* **85,** 431–444.
Burstone, M. S. (1960b) *J. Natl. Cancer Inst.* **24,** 1199–1217.
Burstone, M. S. (1960c) *In* "Calcification in Biological Systems" (R. F. Sognnaes, ed.), pp. 217–243. Am. Assoc. Advance. Sci., Washington, D. C.
Burstone, M. S. (1961) *J. Histochem. and Cytochem.* **9,** 146–153.
Burstone, M. S., and Fleming, T. J. (1959) *J. Histochem. and Cytochem.* **7,** 203.
Burstone, M. S., and Folk, J. E. (1956). *J. Histochem. and Cytochem.* **4,** 217–226.
Burstone, M. S., and Keyes, P. H. (1957) *Am. J. Pathol.* **33,** 1229–1235.
Burstone, M. S., Glenner, G. G., and Lillie, R. D. (1959) *J. Histochem. and Cytochem.* **7,** 299–300.
Butler, L. O., and Bell, L. G. (1953) *Nature* **171,** 971–972.
Cappelin, M. (1947) *Bull. histol. appl. et tech. microscop.* **24,** 155–162.
Cascarano, J., and Zweifach, B. W. (1961) *Am. J. Physiol.* **200,** 1285–1292.
Caspersson, T. O. (1950) "Cell Growth and Cell Function, a Cytochemical Study." Norton, New York.
Chang, J. P. (1956) *Exptl. Cell Research* **11,** 643–646.
Chang, J. P., and Hori, S. H. (1961) *J. Histochem. and Cytochem.* **9,** 292–300.
Chang, J. P., Russell, W. O., Moore, E. B., and Sinclair, W. K. (1961a) *Am. J. Clin. Pathol.* **35,** 14–19.
Chang, J. P., Russell, W. O., and Moore, E. B. (1961b) *J. Histochem. and Cytochem.* **9,** 208.
Chessick, R. D. (1953) *J. Histochem. and Cytochem.* **1,** 471–485.
Chessick, R. D. (1954) *J. Histochem. and Cytochem.* **2,** 258–273.
Cleland. K. W. (1950) *Proc. Linnean Soc. N.S. Wales* **75,** 35–53.
Cobb, J. D. (1953) *Arch. Pathol.* **55,** 496–502.
Coons, A. H., Leduc, E. H., and Kaplan, M. H. (1951) *J. Exptl. Med.* **93,** 173–188.
Copeland, D. E. (1951) *J. Natl. Cancer Inst.* **12,** 224–225.
Copson, D. A. (1958) *Food Technol.* **12,** 270–272.
Dalgaard, J. B. (1956) *J. Histochem. and Cytochem.* **4,** 14–16.
Danielli, J. F. (1953) "Cytochemistry, a Critical Approach." Wiley, New York.
Davies, H. G. (1954) *Quart. J. Microscop. Sci.* **95,** 433–457.
Davies, H. G., Deeley, E. M., and Denby E. F. (1957) *Exptl. Cell Research Suppl.* **4,** 136–149.
Davis B. J. and Ornstein, L. (1959) *J. Histochem. and Cytochem.* **7,** 297–298.
Davis, B. J., Ornstein, L., Taleporos, P., and Koulish, S. (1959) *J. Histochem. and Cytochem.* **7,** 291–292.
Deitch, A. D., and Godman, G. C. (1955) *Anat. Record* **123,** 1–18.
Dempster, W. T. (1960) *Am. J. Anat.* **107,** 59–72.
Dotti, L. B., Paparo, G. P., and Clarke, B. E. (1951) *Am. J. Clin. Pathol.* **21,** 475–479.

DOUNCE, A. L. (1950) Cytochemical foundations of enzyme chemistry. *In* "The Enzymes" (J. B. Sumner and K. Myrbäck, eds.), Vol. I, Part 1, pp. 187–266. Academic Press, New York.
DOYLE, W. L. (1948) *Proc. Soc. Exptl. Biol. Med.* **69,** 43–44.
DOYLE, W. L. (1950) *Federation Proc.* **9,** 34.
DOYLE, W. L. (1954) *J. Gen. Physiol.* **38,** 141–144.
DOYLE, W. L. (1955) *J. Biophys. Biochem. Cytol.* **1,** 221–236.
DOYLE, W. L., AND LIEBELT, R. (1954) *Anat. Record* **118,** 384.
DOYLE, W. L., AND LIEBELT, R. (1955) *J. Histochem. and Cytochem.* **3,** 50–60.
DUTHIE, R. B., AND BARKER. A. N. (1955) *J. Bone and Joint Surg.* **37B,** 304–323.
EMMEL V. M. (1946) *Anat. Record* **95,** 159–175.
ERÄNKÖ, O. (1951) *Ann. Med. Exptl. et Biol. Fenniae Helsinki* **29,** 287–289.
ERÄNKÖ, O. (1952) *Acta Anat.* **16** *Suppl. 17,* 3–60.
ERÄNKÖ, O. (1954a) *Acta Pathol. Scand.* **35,** 426–432.
ERÄNKÖ, O. (1954b) *Acta Anat.* **22,** 331–336.
FEDER, N. (1959) *J. Histochem. and Cytochem.* **7,** 292–293.
FEDER, N. (1960) *J. Histochem. and Cytochem.* **8,** 309–310.
FEDER, N., AND SIDMAN, R. L. (1958) *J. Biophys. Biochem. Cytol.* **4,** 593–600.
FERNÁNDEZ-MORÁN, H. (1957) *In* "Metabolism of the Nervous System" (D. Richter, ed.), pp. 1–34. Pergamon Press, New York.
FERNÁNDEZ-MORÁN, H. (1959) *Science* **129,** 1284–1285.
FERNÁNDEZ-MORÁN, H. (1960) *Ann. N. Y. Acad. Sci.* **85,** 689–713.
FIRKET, H., AND MICHEL, J. P. (1954) *Bull. microscop. appl.* [2]**4,** 1–5.
FISHER, J. C., HOLLOMON, J. H., AND TURNBULL, D. (1949) *Science* **109,** 168–169.
FISHMAN, W. H., AND BAKER, J. R. (1956) *J. Histochem. and Cytochem.* **4,** 570–587.
FLOSDORF, E. W. (1954) *In* "Biological Application of Freezing and Drying" (R. J. C. Harris, ed.), pp. 63–86. Academic Press, New York.
FRAENKEL-CONRAT, H., AND OLCOTT, H. (1945) *J. Biol. Chem.* **161,** 259–268.
FREED, J. J. (1955) *Lab. Invest.* **4,** 106–121.
FREED, J. J., AND GREEN, E. U. (1955) *Nature* **176,** 116–117.
FREIMAN, D. G. (1953) *Proc. Soc. Exptl. Biol. Med.* **84,** 338–341.
FREIMAN, D. G. (1954) *Am. J. Clin. Pathol.* **24,** 227–231.
FREIMAN, D. G. (1956) *Lab. Invest.* **5,** 338–347.
FRENCH, D., AND EDSALL, J. T. (1945) *Advances in Protein Chem.* **2,** 277–335.
FROST, H. M. (1958) *Stain Technol.* **33,** 273–277.
FROST, H. M. (1959) *Stain Technol.* **34,** 135–146.
GALLAGHER, C. H., JUDAH, J. D., AND REES, K. R., (1956) *J. Pathol. and Bacteriol.* **52,** 247–256.
GERSH, I. (1932) *Anat. Record* **53,** 309–337.
GERSH, I. (1948) *Bull. Intern. Assoc. Med. Museums* **28,** 179–185.
GERSH, I. (1956) *J. Biophys. Biochem. Cytol. Suppl.* **2,** 37–43.
GERSH, I., AND STEPHENSON, J. L. (1954) *In* "Biological Applications of Freezing and Drying" (R. J. C. Harris, ed.), pp. 329–384. Academic Press, New York.
GERSH, I., ISENBERG, I., STEPHENSON, J. L., AND BONDAREFF, W. (1957) *Anat. Record* **128,** 91–112.
GERSH, I., VERGARA, J., AND ROSSI, G. I. (1960) *Anat. Record* **138,** 445–459.
GLENNER, G. G., AND BURSTONE, M. S. (1958) *Anat. Record* **130,** 243–252.
GLENNER, G. G., BURSTONE, M. S., AND MEYER, D. B. (1959) *J. Natl. Cancer Inst.* **23,** 857–873.
GLICK, D., AND BLOOM, D. (1956) *Exptl. Cell Research* **10,** 687–696.
GLICK, D., AND MALMSTRÖM, B. G. (1952) *Exptl. Cell Research* **3,** 125–135.

Gössner, W. (1960) *Verhandl. deut. Ges. Pathol.* **44,** 204–208.
Gomori, G. (1939) *Proc. Soc. Exptl. Biol. Med.* **42,** 23–26.
Gomori, G. (1950) *J. Lab. Clin. Med.* **35,** 802–809.
Gomori, G. (1952) "Microscopic Histochemistry; Principles and Practice." Univ. Chicago Press, Chicago, Illinois.
Gomori, G., and Chessick, R. D. (1953) *J. Neuropathol. Exptl. Neurol.* **12,** 387–396.
Gomori, G., and Coutinho, H. B. (1953) *Proc. Soc. Exptl. Biol. Med.* **83,** 809–811.
Goodspeed, T. H., and Uber, F. M. (1934) *Proc. Natl. Acad. Sci. U. S.* **20,** 495–501.
Goodspeed, T. H., and Uber, F. M. (1935) *Univ. Calif. (Berkeley) Publ. Botany* **18,** 23–44.
Gourévitch, A. (1953) *Bull. microscop. appl.* [2] **3,** 130–137.
Greaves, R. I. N. (1954) *In* "Biological Applications of Freezing and Drying" (R. J. C. Harris, ed.), pp. 87–127. Academic Press, New York.
Greenfield, R. E., and Price, V. E. (1956) *J. Biol. Chem.* **220,** 607–618.
Greep, R. O., Fischer, C. J., and Morse, A. (1948) *J. Am. Dental Assoc.* **36,** 427–442.
Grunbaum, B. W., and Wellings, S. R. (1960a). *J. Ultrastruct. Research* **4,** 73–80.
Grunbaum. B. W., and Wellings, S. R. (1960b) *J. Ultrastruct. Research* **4,** 117–126.
Gustafsson, B. E. (1953) "Apparatuses and Techniques for Freezing-drying of Tissues" *Lunds Univ. Ärsskr.* **49,** 1–9.
Gustafsson, B. E. (1954) *Kgl. Fysiograf. Sällskap. i Lund Förh.* **24,** 12–20.
Hancox, N. M. (1957) *Expt. Cell Research* **13,** 263–275.
Hannibal, M. J., and Nachlas, M. M. (1959) *J. Biophys. Biochem. Cytol.* **5,** 279–288.
Hanzon, V., and Hermodsson, L. H. (1960) *J. Ultrastruct. Research* **4,** 332–348.
Harris, J. E., Sloane, J. F., and King, D. T. (1950) *Nature* **166,** 25–26.
Harris, R. J., ed. (1954) "Biological Application of Freezing and Drying." Academic Press, New York.
Haurowitz, F. (1950) "Chemistry and Biology of Proteins." Academic Press, New York.
Hilleman, H. H., and Lee, C. H. (1953) *Stain Technol.* **28,** 285–287.
Hoerr, N. L. (1936) *Anat. Record* **65,** 293–317.
Holt, S. J. (1956) *J. Histochem. and Cytochem.* **4,** 541–554.
Holt, S. J. (1958) Indigogenic staining methods for esterases. *In* "General Cytochemical Methods" (J. F. Danielli, ed.), Vol. 1, pp. 375–398. Academic Press, New York.
Holt, S. J. (1959) *Exptl. Cell Research Suppl.* **7,** 1–27.
Holt, S. J., and Withers, R. F. J. (1958) *Proc. Roy. Soc.* **B148,** 520–532.
Holt, S. J., Hobbinger, E. E., and Pawan, G. L. S. (1960) *J. Biophys. Biochem. Cytol.* **7,** 383–386.
Jansen, M. T. (1954) *Exptl. Cell Research* **7,** 318–326.
Jensen, W. A. (1954) *Exptl. Cell Research* **7,** 572–574.
Johnson, F. B. (1959) *J. Histochem. and Cytochem.* **7,** 293.
Joslyn, M. A. (1949) *Advances in Enzymol.* **9,** 613–652.
Kanehira, S. (1953) *Low Temp. Sci. Ser. A.* **10,** 137–156.
Karmarkar, S. S., Barrnett, R. J., Nachlas, M. M., and Seligman, A. M. (1959) *J. Am. Chem. Soc.* **81,** 3771–3775.
Klionsky, B., and Smith, O. D. (1960) *Am. J. Clin. Pathol.* **33,** 144–151.
Kossel, A. (1913) *Z. physiol. Chem. Hoppe-Seyler's* **84,** 354–358.
Kramer, H., and Hill, R. G. (1956) *Quart J. Microscop. Sci.* **97,** 313–317.
Kulenkampff, H. (1954) *Verhandl. anat. Ges.* **101,** 166–169.
Lacy, D., and Blundell, M. (1955) *J. Roy. Microscop. Soc.* [3] **75,** 48–57.
Lewis, P. R., and Shute, C. C. D. (1960) *J. Histochem. and Cytochem.* **8,** 334–335.

LILLIE, R. D. (1947) "Histopathologic Technic." McGraw-Hill, New York.

LILLIE, R. D. (1954) "Histopathologic Technique and Practical Histochemistry." McGraw-Hill, New York.

LILLIE, R. D., AND LASKEY, A. (1951) *Bull. Intern. Assoc. Med. Museums*, **32,** 80–82.

LILLIE, R. D., LASKEY, A., GRECO, J., AND BURTNER, H. J., AND JONES, P. (1951) *Am. J. Clin. Pathol.* **21,** 711–722.

LINDERSTRØM-LANG, K., AND MORGENSEN, K. R. (1938) *Compt. rend. trav. lab. Carlsberg Sér. chim.* **23,** 27–34.

LISON, L. (1949) *Compt. rend. soc. biol.* **143,** 115–116.

LORCH, I. J. (1946) *Nature* **158,** 269.

LOWRY, O. H., ROBERTS, N. R., WU, M. L., HIXON, W. S., AND CRAWFORD, E. J. (1954) *J. Biol. Chem.* **207,** 19–37.

LUYET, B. J. (1951) *In* "International Symposium on Freezing and Drying" (R. J. C. Harris, ed.), pp. 77–96. Hafner, New York.

LUYET, B. J. (1960) *Ann. N. Y. Acad. Sci.* **85,** 549–569.

McMASTER-KAYE, R., AND TAYLOR, J. H. (1958) *J. Biophys. Biochem. Cytol.* **4,** 5–11.

MAJNO, G., AND ROUILLER, C. (1951) *Virchow's Arch. pathol. Anat. u. Physiol.* **321,** 1–61.

MANN, G. (1902) "Physiological Histology Methods and Theory." Oxford Univ. Press (Clarendon), London and New York.

MATHIES, J. C., AND GOODMAN, E. D. (1953) *J. Am. Chem. Soc.* **75,** 6061–6062.

MENDELOW. H., AND HAMILTON, J. B. (1950) *Anat. Record* **107,** 443–451.

MERRIAM, R. W. (1958) *J. Histochem. and Cytochem.* **6,** 43–51.

MERYMAN, H. T. (1956) *Science* **124,** 515–521.

MERYMAN, H. T. (1957) *Proc. Roy. Soc.* **B147,** 452–459.

MERYMAN, H. T. (1959) *Science* **130,** 628–629.

MIRSKY, A. E. (1951) *Cold Spring Harbor Symposia Quant. Biol.* **16,** 481–482.

MOBERGER, G., LINDSTRÖM, B., AND ANDRESSON, L. (1954) *Exptl. Cell Research* **6,** 228–237.

MOOG, F. (1951) *Science* **114,** 109.

MORSE, A. (1945) *J. Dental Research* **24,** 143–153.

MORTON, R. K. (1955) *In* "Methods in Enzymology" (S. P. Colowick and N. O. Kaplan, eds.), Vol. I, pp. 25–51. Academic Press, New York.

MOWRY, R. W. (1949) *Bull. Intern. Assoc. Med. Museums* **30,** 95–98.

MÜLLER, H. R. (1957) *J. Ultrastruct. Research* **1,** 109–137.

NACHLAS, M., YOUNG, A., AND SELIGMAN, A. M. (1957) *J. Histochem. and Cytochem.* **5,** 565–583.

NAIDOO, D., AND PRATT, O. E. (1951) *J. Neurol. Psychiat.* **14,** 287–294.

NAIDOO, D., AND PRATT, O. E. (1952) *J. Neurol. Psychiat.* **15,** 164–168.

NAIDOO, D., AND PRATT, O. E. (1956) *Acta Histochem.* **3,** 85–103.

NAIR, K. K. (1958) *Experientia* **14,** 172–173.

NELSON, L. (1958) *Biochim. et Biophys. Acta* **27,** 634–641.

NELSON, L. (1959) *Exptl. Cell Research* **16,** 403–410.

NEUMANN, K. (1952) *Z. wiss. Mikroskop.* **60,** 449–462.

NEUMANN, K. (1958) *In* "Handbuch der Histochemie" (W. Graumann and K. Neumann, eds.), Vol. 1, pp. 1–77. Fischer, Stuttgart, Germany.

NEUMANN, K., OEHLERT, G., AND HANSMANN, H. (1954) *Z. Geburtshilfe u. Gynäkol.* **141,** 109–130.

NOVIKOFF A. B. (1952) *Exptl. Cell Research Suppl.* **2,** 123–143.

NOVIKOFF. A. B. (1957) *Symposia Soc. Exptl. Biol.* **10,** 92–109.

Novikoff, A. B. (1959) *In* "Analytical Cytology" (R. C. Mellors, ed.), 1st ed., pp. 69–145. McGraw-Hill, New York.
Novikoff, A. B., and Arase, M. M. (1958) *J. Histochem. and Cytochem.* **6,** 397.
Novikoff, A. B., and Masek, B. (1958) *J. Histochem. and Cytochem.* **6,** 217.
Novikoff, A. B., Shin, W-Y., and Drucker, T. (1960) *J. Histochem and Cytochem.* **8,** 37–40.
Novikoff, A. B., Shin, W-Y., and Drucker, T. (1961) *J. Biophys. Biochem. Cytol.,* **9,** 47–61.
Oncley, J. L., Gurd, F., and Melin, M. (1950) *J. Am. Chem. Soc.* **72,** 458–465.
Packer, D. M., and Scott G. H. (1942) *J. Tech. Meth.* **22,** 85–96.
Palmgren, A. (1954) *Nature* **174,** 46.
Patten, S. F., and Brown, K. A. (1958) *Lab. Invest.* **7,** 209–223.
Pearse, A. G. E. (1958) *J. Clin. Pathol.* **11,** 520–534.
Pearse, A. G. E. (1960) "Histochemistry, Theoretical and Applied," 2nd ed., p. 19. Little, Brown, Boston, Massachusetts.
Pease, D. C., and Baker, R. F. (1949) *Am. J. Anat.* **84,** 175–200.
Persson, B. H. (1952) *Acta Soc. Med. Upsaliensis* **57,** 155–160.
Pollister, A. W., and Ornstein, L. (1959) *In* "Analytical Cytology" (R. C. Mellors, ed.), pp. 431–510. McGraw-Hill, New York.
Porter, V. S., Deming, N. P., Wright, R. C., and Scott, E. M. (1953) *J. Biol. Chem.* **205,** 883–891.
Potthoff, C. J., ed. (1949) "Preservation of the Formed Elements and of the Proteins of the Blood." Harvard Med. School, Boston, Massachusetts.
Pratt, O. E. (1953) *Biochem. J.* **55,** 140–145.
Pratt, O. E. (1954) *Biochim. et Biophys. Acta* **14,** 380–389.
Rappaport, B. Z. (1955) *Arch. Pathol.* **60,** 1–9.
Reale, E. (1956) *Monit. zool. ital.* **63,** 188–196.
Roberts, P. A., and Hoyte, D. (1958) *J. Anat.* **92,** 621–622.
Robins, E.. Smith, D. E., Daesch, G. E., and Payne, K. E. (1958) *J. Neurochem.* **3,** 19–27.
Rowe, T. W. G. (1960) *Ann. N. Y. Acad. Sci.* **85,** 641–681.
Russell, R. S., Sanders, F. K., and Bishop, O. N. (1949) *Nature* **163,** 639–640.
Sabitini, D., Bensch, C., and Barrnett, R. (1961) *Anat. Record* **142,** 274.
Scarpelli, D. G., and Pearse, A. G. E. (1958) *J. Histochem. and Cytochem.* **6,** 369–376.
Schajowicz, F., and Cabrini, R. L. (1958) *Science* **127,** 1447–1448.
Schajowicz, F., and Cabrini, R. L. (1959) *Stain Technol.* **34,** 59–64.
Schultz-Brauns, O. (1931) *Klin. Wochschr.* **10,** 113–116.
Scott, G. H. (1943) *Biol. Symposia* **10,** 277–289.
Scott, G. H., and Hoerr, N. L. (1950) *In* "Medical Physics (O. Glasser, ed.), Vol. II, pp. 292–296. Year Book Publishers, Chicago, Illinois.
Scott, G. H., and Williams, P. S. (1936) *Anat. Record* **66,** 475–481.
Seligman, A. M., Chauncey, H. H., and Nachlas, M. M. (1951) *Stain Technol.* **26,** 19–23.
Seno, S., and Yoshizawa, K. (1960) *J. Biophys. Biochem. Cytol.* **8,** 617–638.
Shen, S. C., Greenfield, P., and Sippel, T. (1952) *Proc. Soc. Exptl. Biol. Med.* **81,** 452–455.
Shimizu, N., Kubo, Z., and Morikawa, N. (1956) *Stain Technol.* **31,** 105–109.
Simpson, W. L. (1941) *Anat. Record* **80,** 173–189.
Sizer, I. W., and Josephson, E. S. (1942) *Food Research* **7,** 201–209.
Sjöstrand, F. (1944) *Acta Anat.* **1,** *Suppl. 1,* 1–63.
Sjöstrand, F. S., and Baker, R. F. (1958) *J. Ultrastruct. Research* **1,** 239–246.

Smith, D. E., Robins, E., Eydt, K. M., and Daesch, G. E. (1956) *J. Histochem. and Cytochem.* **4,** 420–421.
Smith, D. E., Robins, E., Eydt, K. M., and Daesch, G. (1957) *Lab. Invest.* **6,** 447–457.
Smyth, J. D. (1954) *Quart. J. Microscop. Sci.* **95,** 139–152.
Sognnaes, R. F. (1947) *Anat. Record* **99,** 133–144.
Sreebny, L. M., and Nikiforuk, G. (1951) *Science* **113,** 560.
Stafford, R. O., and Atkinson, W. B. (1948) *Science* **107,** 279–281.
Steedman, H. F. (1961) "Section Cutting in Microscopy." Charles C Thomas, Springfield, Illinois.
Stephenson, J. L. (1953) *Bull. Math. Biophys.* **15,** 411–429.
Stephenson, J. L. (1954) *Bull. Math. Biophys.* **16,** 23–43.
Stephenson, J. L. (1956) *J. Biophys. Biochem. Cytol.* **2,** 45–52.
Stephenson. J. L. (1960) *Ann. N. Y. Acad. Sci.* **85,** 535–540.
Stowell, R. E. (1951) *Stain Technol.* **26,** 105–108.
Stowell, R. E., Chang, J. P., and Berenbom, M. (1961) *Lab. Invest.* **10,** 111–128.
Sumner, J. B. (1948) *Science* **108,** 410.
Sumner, J. B., and Sommers, G. F. (1953) "Chemistry and Methods of Enzymes." 3rd ed. Academic Press, New York.
Sylvén, B. (1951) *Acta Unio Intern. contra Cancum* **7,** 708–712.
Taft, E. B. (1949) *Am. J. Pathol.* **25,** 824–825.
Taft, E. B. (1951) *Stain Technol.* **26,** 205–212.
Taft, E. B. (1960) *Lab. Invest.* **9,** 169–173.
Tarver, H. (1954) *In* "The Proteins" (H. Neurath and K. Bailey, eds.), Vol. II, Part B, pp. 1199–1296. Academic Press, New York.
Taylor, A. (1944) *J. Lab. Clin. Med.* **29,** 657–663.
Taylor, J. H. (1956) *In* "Physical Techniques in Biological Research" (G. Oster and A. W. Pollister, eds.), Vol. III, pp. 545–576. Academic Press, New York.
Thornburg, W., and Mengers, P. E. (1957) *J. Histochem. and Cytochem.* **5,** 47–52.
Treffenberg, L. (1953) *Arkiv Zool.* **4,** 295–296.
Van Lancker, J. L., and Holtzer, R. L. (1959) *Am. J. Pathol.* **35,** 563–573.
Wachstein, M., and Meisel, E. (1953) *Stain Technol.* **28,** 135–139.
Walker, D. G., and Seligman, A. M. (1961) *J. Biophys. Biochem. Cytol.* **9,** 415–427.
Wang, K. J., and Grossman, M. I. (1949a) *J. Lab. Clin. Med.* **34,** 292–296.
Wang, K. J., and Grossman, M. I. (1949b). *Anat. Record* **104,** 79–87.
White, R. T., and Allen, R. A. (1951) *Stain Technol.* **26,** 137–138.
Wichterle, O., Bartl, P., and Rosenberg, M. (1960) *Nature* **186,** 494–495.
Willighagen, R. G., and Planteydt, H. T. (1959) *Nature* **183,** 263–264.
Wolman, M., and Behar, A. (1952) *Exptl. Cell Research* **3,** 619–621.
Woods, P. S., and Pollister, A. (1955) *Stain Technol.* **30,** 123–131.
Yaeger, J. A. (1961) *Exptl. Cell Research* **22,** 493–502.
Yamaida, M. (1953) *Chem. Abstr.* **47,** 8814.
Yokoyama, H. O., and Stowell, R. E. (1951) *J. Natl. Cancer Inst.* **12,** 211–213.
Zlotnik, I. (1960) *Quart. J. Microscop. Sci.* **101,** 251–254.
Zorzoli, A. (1948) *Anat. Record* **102,** 445–467.

## CHAPTER 2

# NAPHTHALENE AND OTHER AROMATIC DERIVATIVES AS THE BASIS OF HISTOCHEMICAL REAGENTS

The so-called aromatic ring compounds possess a cyclic structure, and exhibit a type of unsaturation which enables them to undergo substitution reactions. Among the simpler aromatic compounds are benzene ($C_6H_6$), toluene, and analogues. Also included are heterocycles such as pyridine and thiophene, as well as more complex derivatives. Aromatic compounds such as benzene which contain one ring or nucleus are referred to as mononuclear while compounds containing two or more rings are polynuclear in structure (Chart 1). Of the latter group the naphthols and analogues furnish the basis of a large number of ester substrates, as well as being directly applicable in the demonstration of certain enzymes (e.g. oxidases). The structure of naphthalene, which at one time was an undesirable by-product of coal gas, is shown in Chart 1 (8). The value of this compound as an intermediate in industrial chemistry led to the synthesis of a large number of substituted naphthols and naphthylamines during the 19th and present century (Fierz-David and Blangey, 1949; Venkataraman, 1952; Lubs, 1955; Colour Index, 1956). Also of interest are other complex constituents obtained from coal tar, or petroleum by-products, which form the basis of several histochemical reagents (Chart 1). Of this group, derivatives of quinoline [Chart 1 (10)], carbazole [Chart 1 (14)], and diphenylene oxide [Chart 1 (15)] have found application in enzyme histochemistry. Some of the other nuclei represented have had less application in substrate design in the past, but may offer suggestions for newer compounds and reactions. The aforementioned aromatic hydrocarbons are converted into specific histochemical intermediates or reagents by a series of reactions including sulfonation, alkali fusion, nitration and subsequent reduction, halogenation, and condensation (de La Mare and Ridd, 1959). By means of these techniques, hydroxyl, amino, and other groupings are introduced into the aro-

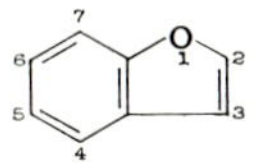

1

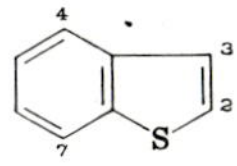

2

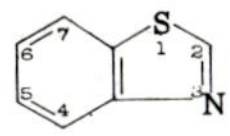

3

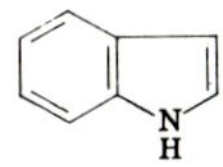

4

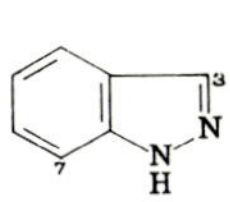

5

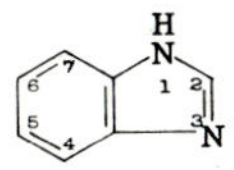

6

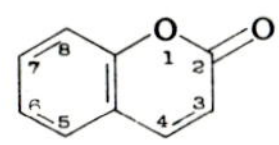

7

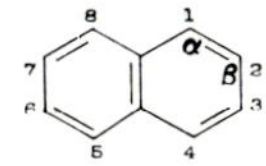

8

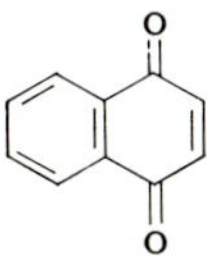

9

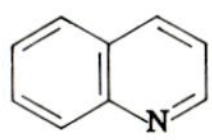

10

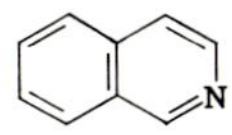

11

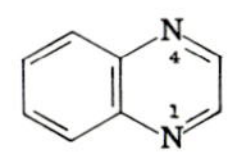

12

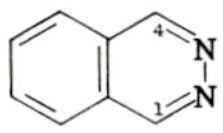

13

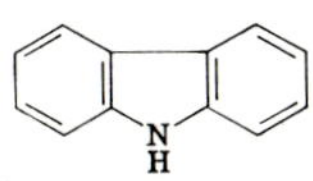

14

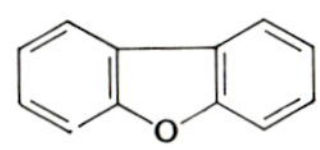

15

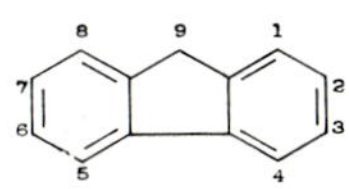

16

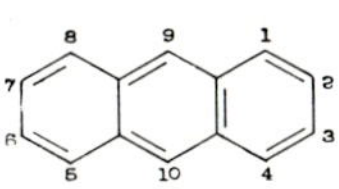

17

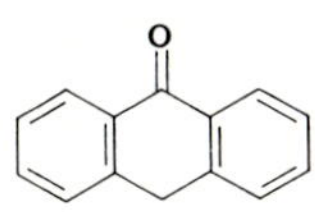

18

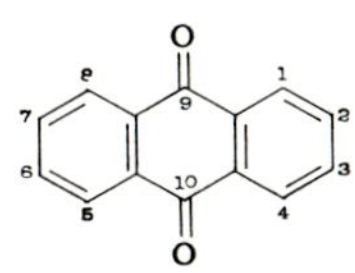

19

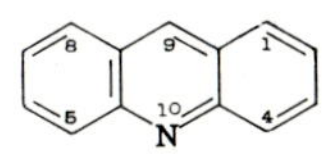

20

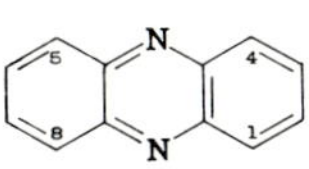

21

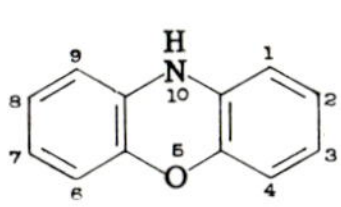

22

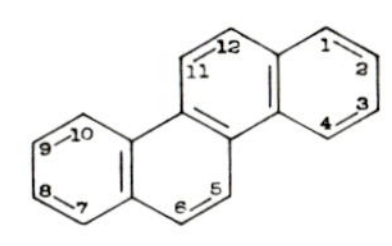

23

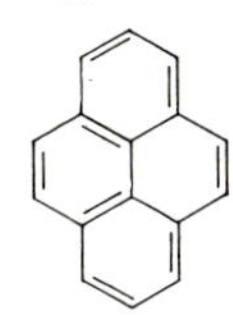

24

CHART 1. Basic ring structure of aromatic compounds including heterocyclic nuclei which may have application as the basis of histochemical substrates. (1) benzofuran (coumarone), (2) thionaphthene, (3) benzothiazole, (4) indole, (5) indazole (benzopyrazole), (6) benzimidazole, (7) coumarin (1,2-benzopyrrole), (8) naphthalene, (9) 1,4-naphthoquinone, (10) quinoline, (11) isoquinoline, (12) quinoxaline, (13) phthalazine, (14) carbazole, (15) diphenylene oxide (dibenzofuran), (16) fluorene, (17) anthracene, (18) anthrone, (19) anthraquinone, (20) acridine, (21) phenazine, (22) phenoxazine, (23) chrysene, (24) pyrene.

matic nucleus. In addition, more complex derivatives may also be prepared. Subsequently, these compounds may be esterified or treated in other ways to form special substrates which are described in subsequent chapters.

Naphthols are initially synthesized in industry by alkali fusion (NaOH or KOH) with a naphthalenesulfonic acid: The naphtholate produced is subsequently treated with acid to form the naphthol. Naphthylamines are hydrolyzed to naphthols in the presence of water and sodium bisulfite under high temperature and pressure:

$$C_{10}H_7NH_2 + HOH + (NH_4)NaSO_3 \rightleftarrows C_{10}H_7OH + 2NH_3$$

Naphthols may also be prepared from halogenated naphthalenes by alkali hydrolysis in the presence of catalysts. More complex hydroxyl derivatives such as anthrols (hydroxyanthracenes) are prepared by reduction of anthraquinone followed by alkali fusion. Hydroxydibenzofuran compounds are derived from halogenated dibenzofurans which have been subjected to the aforementioned alkali fusion process. In the case of hydroxycarbazole, the starting material is carbazole which is sulfonated and then treated with alkali under specified conditions.

$\beta$-Naphthol (2-naphthol) has been employed by the dyestuff industry for over 50 years in the production of so-called developed azo-dyes. Simple bromo-naphthols or benzoyl naphthols were also developed and applied for the same industrial application. These simpler naphthalene derivatives were among the first to be employed as the basis of histochemical substrates.

One of the very important naphthols is 3-hydroxy-2-naphthoic acid (BONA: $\beta$-oxynaphthoic acid). It is made by the Kolbe-Schmitt reaction in which sodium naphtholate is treated with carbon dioxide under elevated temperature and pressure. BONA in itself has not been of value as the basis of histochemical reagents but is used to produce the "so-called" naphthol AS (naphthol anilid-säure) compounds. Substrates derived from naphthol AS and analogues have not only helped elucidate our state of knowledge on the precise sites of enzyme at the cytological level, but constitute a valuable group of reagents for the applied histochemist and pathologist.

Naphthol AS compounds are defined as the arylamides (arylides) of 3-hydroxy-2-naphthoic acid or acylacetic acids. The latter, of course, are not naphthols but compounds containing reactive methylene groups capable of coupling with diazonium salts. The first series of the 3-hydroxy-2-naphthoic acid arylamide group was naphthol AS which was synthesized in 1892 by Schöpff. It may be prepared by refluxing BONA and aniline with $PCl_3$ (I) (see Appendix 2). Naphthol AS compounds may also be synthesized by preparing the acid chloride of BONA and subsequently reacting it with a suitable amine. In addition to the BONA, the anilides derived from

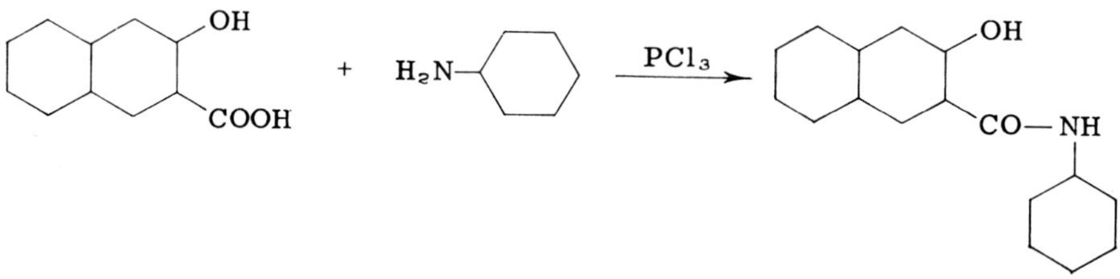


(I) Synthetic routes indicating preparation of naphthol AS

TABLE 1

TRADE DESIGNATION OF COMMERCIALLY AVAILABLE AS NAPHTHOLS[a]

| Naphthol | Brenthol |
|---|---|
| AS | AS |
| AS-D | OT |
| AS-OL | FR |
| AS-RL | PA |
| AS-LT | MA |
| AS-BG | FO |
| AS-BS | MN |
| AS-AN | — |
| AS-E | BB |
| AS-TR | CT |
| AS-Cl (RC) | — |
| AS-KB | — |
| AS-P | — |
| AS-RP | — |
| — | BA |
| AS-ITR | — |
| AS-LC | — |
| AS-BO | AN |
| AS-SW | BN |
| AS-S | — |
| AS-BR | DA |
| AS-BM | — |
| AS-GR | NG |
| AS-BT (DB) | — |
| AS-LB | BT |
| AS-SG | GB |
| AS-SR | RB |
| — | BC |
| AS-BI | — |
| AS-BB | — |
| AS-BN | — |

[a] American and German AS naphthols and their Brenthol (I.C.I.) equivalent.

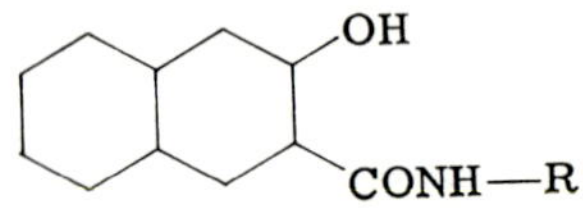


(II) General configuration of typical naphthol AS compounds. R = aryl radical

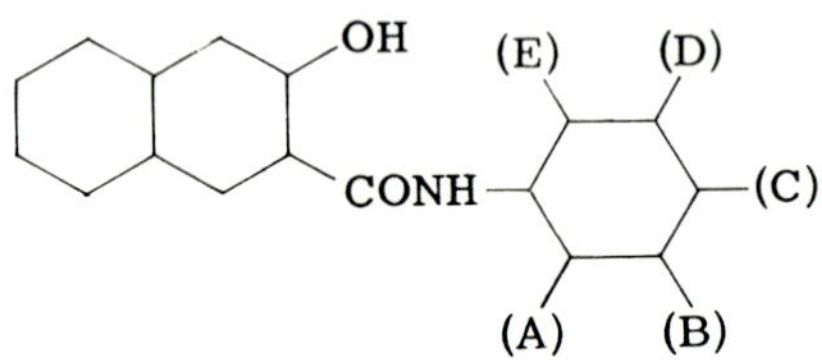


(III) General configuration of naphthol AS compounds. A through E indicate the positions of electropositive and/or electronegative groupings

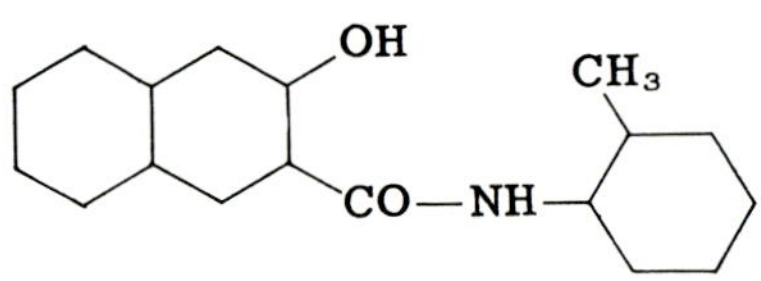


(IV) Naphthol AS-D

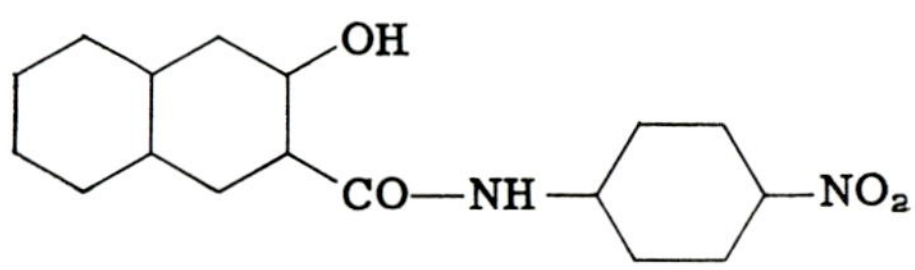


(V) Naphthol AS-AN

*o*-hydroxycarboxylic acids of anthracene, carbazole, benzocarbazole, and dibenzofuran have been classified under the naphthol AS category although they are not naphthols (Dorman, 1939). The trade designations of the various AS naphthols are indicated in Table 1. Brenthol is the I.C.I. (Imperial Chemical Industries) equivalent of certain AS naphthols.

AS naphthols containing the naphthalene nucleus may be designated as follows, with R representing the substituted aryl radical (II). This radical may consist of benzene or more complex nuclei such as naphthalene, pyrene, or chrysene. In most cases R consists of a substituted benzene ring containing electropositive or electronegative substituents as represented by A

TABLE 2

α-NAPHTHOL AND DERIVATIVES

| Chemical name and formula | Mol. wt. | Melting point (°C.) |
|---|---|---|
| 1. α-Naphthol, 1-naphthol, 1-hydroxynaphthalene, $C_{10}H_8O$ | 144.17 | 95.8–96.0 |
| 2. Sodium 1-naphtholate, $C_{10}H_7ONa$ (Meldrum and Patel, 1928) | 166.16 | 44–45 |
| 3. Potassium 1-naphtholate, $C_{10}H_7OK$ (Meldrum and Patel, 1928) | 182.27 | Liquid |
| 4. 4-Benzoyl-1-naphthol, 1-hydroxy-4-naphthyl phenyl ketone, $C_{17}H_{12}O_2$ (Scholl and Seer, 1912; Lederer, 1931) | 248.28 | 164–165 |
| 5. 2-Chloro-1-naphthol, $C_{10}H_7OCl$ | 178.62 | 65 |
| 6. 4-Chloro-1-naphthol, $C_{10}H_7OCl$ | 178.62 | 120–121 |
| 7. 6-Chloro-1-naphthol, $C_{10}H_7OCl$ | 178.62 | 94 |
| 8. 2,4-Dichloro-1-naphthol, $C_{10}H_6OCl_2$ | 213.07 | 107–108 |
| 9. 5,7-Dichloro-1-naphthol, $C_{10}H_6OCl_2$ | 213.07 | 132 |
| 10. 2-Bromo-1-naphthol, $C_{10}H_7OBr$ | 223.08 | 45 |
| 11. 4-Bromo-1-naphthol, $C_{10}H_7OBr$ | 223.08 | 129 |
| 12. 6-Bromo-1-naphthol, $C_{10}H_7OBr$ | 223.08 | 129 |
| 13. 2,4-Dibromo-1-naphthol, $C_{10}H_6OBr_2$ | 301.99 | 111 |
| 14. 5-Nitro-1-naphthol, $C_{10}H_7ONO_2$ | 189.17 | 172 |
| 15. 1-Hydroxy-2-naphthoic acid, $C_{11}H_8O_3$ | 188.19 | 194 |
| 16. 1-Hydroxy-2-acetonaphthone, $C_{12}H_{10}O_2$ | 186.21 | 99–101 |
| 17-Phenyl-1-hydroxy-2-naphthoate, $C_{12}H_{12}O_3$ | 204.23 | 90–93 |
| 18. 1-Hydroxy-2-naphthanilide, $C_{17}H_{13}O_2N$ (Anschütz *et al.*, 1906) | 263.30 | 154 |
| 19. 4-Acetamino-1-naphthol, $C_{12}H_{11}O_2N$ (Witt and Dedichen, 1896) | 201.23 | 187 |
| 20. 8-Hydroxyquinoline, $C_9H_7NO$ | 145.16 | 76 |

through E (III) (e.g. IV and V). Because of their importance in histochemistry as reagents and intermediates in the synthesis of substrates a list of hydroxyl containing aromatic compounds including the AS derivatives is given in Tables 2 through 5. Although many have not been applied in histochemical studies, their structure may offer suggestions for the design of new reagents. In specific cases bibliographical notation is given with particular reference to the original synthesis of the compound.

Table 5 lists the commercially available members of the naphthol AS class. As previously mentioned, it should be noted that many are not naphthols *per se* but complex arylamides or other compounds. The melting points are in most instances those given in the Colour Index (1956).

A number of additional arylamides of 3-hydroxy-2-naphthoic acids have been reported in the literature. These are not commercially available and must be synthesized (Appendix 2). Several have already found appli-

TABLE 3

β-NAPHTHOL AND DERIVATIVES

| Chemical name and formula | Mol. wt. | Reference | Melting point (° C). |
|---|---|---|---|
| 1. β-Naphthol, 2-naphthol, 2-hydroxynaphthalene, $C_{10}H_8O$ | 144.17 | — | 122 |
| 2. Sodium 2-naphtholate, $C_{10}H_7ONa$ | 166.16 | Meldrum and Patel, 1928 | 120 |
| 3. Potassium 2-naphtholate, $C_{10}H_7OK$ | 182.27 | Meldrum and Patel, 1928 | 38–40 |
| 4. 6-Benzoyl-2-naphthol, 2-hydroxy-6-naphthyl phenyl ketone, 6-hydroxy-2-benzoylnaphthalene, $C_{17}H_{12}O_2$ | 248.28 | Sen and Ray, 1932; Anderson and Thomas, 1943; Schönberg and Mustafa, 1946 | 158–159 |
| 5. 1-Chloro-2-naphthol, $C_{10}H_7OCl$ | 178.62 | Zincke, 1888 | 70 |
| 6. 1-Bromo-2-naphthol, $C_{10}H_7OBr$ | 223.08 | Smith, 1879 | 85 |
| 7. 6-Chloro-2-naphthol, $C_{10}H_7OCl$ | 178.62 | Claus and Zimmermann, 1881; Ruggli *et al.*, 1929 | 115 |
| 8. 6-Bromo-2-naphthol, $C_{10}H_7OBr$ | 223.08 | Tilden and Armstrong, 1901; Franzen and Stäuble; 1921–1922; Koelsch, 1940 | 129 130 |
| 9. 1,6-Dibromo-2-naphthol, $C_{10}H_6OBr$ | 222.07 | Tilden and Armstrong, 1901; Leman, 1938 | 106 |
| 10. 1,6-Dichloro-2-naphthol, $C_{10}H_6OCl_2$ | 213.07 | Ruggli *et al.*, 1929 | 119.5 |
| 11. 5,8-Dichloro-2-naphthol, $C_{10}H_6OCl_2$ | 213.07 | — | 143 |
| 12. 5,8-Dibromo-2-naphthol, $C_{10}H_6OBr_2$ | 301.99 | — | 147 |
| 13. 3-Hydroxy-2-naphthoic acid (2-hydroxy-3-naphthoic acid), $C_{11}H_8O_3$ | 188.19 | — | 222–223 |
| 14. 1,6-Dibromo-2-hydroxy-3-naphthoic acid, $C_{11}H_6O_3Br_2$ | 346.00 | FIAT, 1948, p. 157 | 245–250 |
| 15. 6-Bromo-3-hydroxy-2-naphthoic acid (7-bromo-3-hydroxy-2-naphthoic acid), $C_{11}H_7O_3Br$ | 267.09 | FIAT, 1948, p. 157 | 265–266 |
| 16. 5-Acetamino-2-naphthol, $C_{12}H_{11}O_2N$ | 201.23 | Kehrmann and Denk, 1901 | 213–214 |
| 17. 8-Acetamino-2-naphthol, $C_{12}H_{11}O_2N$ | 201.23 | Friedländer and Zinberg, 1896 | 165 |
| 18. Phenyl-3-hydroxy-2-naphthoate, $C_{12}H_{12}O_3$ | 204.23 | — | 128–129 |

TABLE 4

MORE COMPLEX HYDROXYL-CONTAINING ARYL COMPOUNDS

| Chemical name and formula | Mol. wt. | Reference | Melting point (° C.) |
|---|---|---|---|
| 1. 1-Anthranol, 1-hydroxyanthracene, $C_{14}H_{10}O$ | 194.23 | — | 158 |
| 2. 2-Anthranol, 2-hydroxyanthracene, $C_{14}H_{10}O$ | 194.23 | — | 255 |
| 3. 3-Hydroxy-2-anthracene carboxylic acid, $C_{15}H_{10}O_3$ | 238.25 | Hopper *et al.*, 1939 | 298 |
| 4. 2-Hydroxydibenzofuran, $C_{12}H_{18}O_2$ | 194.28 | Gilman *et al.*, 1935 | 139 |
| 5. 3-Hydroxydibenzofuran, $C_{12}H_{18}O_2$ | 194.28 | Gilman and Van Ess, 1939 | 134 |
| 6. 2-Hydroxycarbazole, $C_{12}H_9ON$ | 183.21 | Ballauf and Schmelzer, 1932 | 276 |
| 7. 3-Hydroxycarbazole, $C_{12}H_9ON$ | 183.21 | Ruff and Stein, 1901 | 255–256 |

cation in the synthesis of substrates for the so-called post-coupling enzyme techniques (Burstone, 1960). Partial review of many complex naphthols have been made by Bhat *et al.* (1940) Rath and Burkhardt (1940), Spiegler (1953), and Bhat and Venkataraman (1942). Table 6 is a list of some of the aforementioned special complex arylamides of 3-hydroxy-2-naphthoic acid. Anilides from *p*-aminoazobenzene (m.p. 251°), and dehydrothio-*p*-toluidine have also been prepared but not fully characterized.

There are few reports on the physical characteristics of naphthol AS derivatives including infrared, ultraviolet, and x-ray spectra. Infrared and ultraviolet spectra of two naphthols and histochemical substrates derived from them are shown in Figs. 1–4. Several naphthol AS compounds are believed to exist in isomeric forms on the basis of x-ray diffraction studies (FIAT, 1948, p. 398). The x-ray pattern was transformed into a "curve" by means of a recording photometer which represented each diffraction line as a peak (Fig. 5). Thus, it was possible to distinguish between $\alpha$ and $\beta$ forms of naphthols AS-SW and AS-SG. The relationship of these findings to the specificity of histochemical substrates is not known. X-ray diagrams of naphthols AS-D and AS-BI, which have not been previously reported, are shown in Figs. 6 and 7.

As previously mentioned, naphthylamines are hydrolyzed to naphthols in the presence of bisulfite under specified conditions. Conversely, naphthylamines can be produced from naphthols by treating them with am-

TABLE 5

COMMERCIALLY AVAILABLE NAPHTHOL AS DERIVATIVES

| Designation | Colour Index (C.I.) No. | Naphthol | Mol. wt. | Melting point (° C.) |
|---|---|---|---|---|
| AS | 37505 | 3-Hydroxy-2-naphthanilide, $C_{17}H_{13}O_2N$ (Schöpff, 1892; Rowe and Stafford, 1924; Lubs, 1955) | 263.30 | 240, 244, 249 |
| AS-AN | 37516 | 3-Hydroxy-4′-nitro-2-naphthanilide, $C_{17}H_{12}O_4N_2$ | 308.30 | 277 |
| Brenthol BA | 37532 | 5′-Bromo-3-hydroxy-2-naphth-*o*-anisidide, $C_{18}H_{14}O_3NBr$ (Hopper *et al.*, 1939) | 372.23 | 216–217 |
| AS-BB | — | This is an acid coupling of 4-chloro-2-anisidine, with 1-amino-7-naphthol in a position *para* to the amino group. | — | — |
| AS-BG | 37545 | 3-Hydroxy-2′,5′-dimethoxy-2-naphthanilide, $C_{19}H_{17}O_4N$ (Hopper *et al.*, 1939) | 323.35 | 185 |
| AS-BI | 37566 | 7-Bromo-3-hydroxy-2-naphth-*o*-anisidide, $C_{18}H_{14}O_3NBr$ (6-bromo-2-hydroxy-3-naphthoyl-*o*-anisidine) (FIAT, 1948, p. 154) | 372.23 | 189–190 |
| AS-BM | 37570 | 3,3″-Dihydroxy-4′,4‴-bi-2-naphtho-*o*-toludide, $C_{36}H_{28}O_4N_2$ | 552.64 | 327 |
| AS-BN | — | This is 4-chloro-2-aminophenol coupled to resorcinol and then metallized with copper. | — | — |
| AS-BO | 37560 | 3-Hydroxy-*N*-1-naphthyl-2-naphthamide, $C_{21}H_{15}O_2N$ | 313.36 | 222–223 |
| AS-BR | 37575 | 3,3″-Dihydroxy-4′,4‴-bi-2-naphth-*o*-anisidide, $C_{36}H_{28}O_6N_2$ | 584.64 | 348 |
| AS-BS | 37515 | 3-Hydroxy-3′-nitro-2-naphthanilide, $C_{17}H_{12}O_4N_2$ | 308.30 | 246–247 |
| AS-CL (RC) | 37531 | 5′-Chloro-3-hydroxy-2-naphth-*o*-anisidide, $C_{18}H_{14}O_3NCl$ | 327.78 | 218 |
| AS-D | 37520 | 3-Hydroxy-2-naphtho-*o*-toluidide, $C_{18}H_{15}O_2N$ | 277.33 | 195–196 |
| AS-DB (BT) | 37605 | 3-Hydroxy-2′,5′-dimethoxy-2-dibenzofurancarboxanilide, $C_{21}H_{17}O_5N$ | 363.38 | 217 |
| AS-E | 37510 | 4′-Chloro-3-hydroxy-2-naphthanilide, $C_{17}H_{12}O_2NCl$ | 297.75 | 258–259 |
| AS-G | 37610 | 4,4′-Bi-*o*-acetoacetotoluidide, $C_{22}H_{24}O_4N_2$ | 380.45 | 204–205 |
| AS-GR | 37585 | 3-Hydroxy-2-*o*-anthrotoluidide, $C_{22}H_{17}O_2N$ | 327.39 | 270 |
| AS-ITR | 37550 | 5′-Chloro-3-hydroxy-2′,4′-dimethoxy-2-naphthanilide, $C_{19}H_{16}O_4NCl$ (Hopper *et al.*, 1939) | 357.80 | 198 |
| AS-KB | 37526 | 5′-Chloro-3-hydroxy-2-naphtho-*o*-toluidide, $C_{18}H_{14}O_2NCl$ | 311.78 | 237 |

TABLE 5—*Continued*

| Designation | Colour Index (C.I.) No. | Naphthol | Mol. wt. | Melting point (° C.) |
|---|---|---|---|---|
| AS-LB | 37600 | 4′-Chloro-2-hydroxy-3-carbazole-carboxanilide, $C_{19}H_{13}O_2N_2Cl$ | 336.79 | 258–259 |
| AS-LC | 37555 | 4′-Chloro-3-hydroxy-2′,5′-dimethoxy-2-naphthanilide, $C_{19}H_{16}O_4NCl$ | 357.80 | 188–189 |
| AS-LG | 37615 | α,α′-Terephthaloylbis(5-chloro-2,4-dimethoxyacetanilide) $C_{28}H_{26}O_8N_2Cl_2$ | 589.45 | 253–254 |
| AS-L3G | 37620 | α,α′-Terephthaloylbis(4-chloro-5-methyl-*o*-acetanisidide) $C_{28}H_{26}O_6N_2Cl_2$ | 557.45 | 245–246 |
| AS-L4G | 37625 | 2-α-Acetylacetamido-6-ethoxybenzothiazole, $C_{13}H_{14}O_3N_2S$ (Hopper *et al.*, 1939) | 278.34 | 208 |
| AS-LT | 37540 | 3-Hydroxy-2′-methyl-2-naphth-*p*-anisidide, $C_{19}H_{17}O_3N$ | 307.35 | 196–197 |
| AS-MCA | 37511 | 3′-Chloro-3-hydroxy-2-naphthanilide, $C_{17}H_{12}O_2NCl$ (Hopper *et al.*, 1939) | 297.75 | 194 |
| AS-MX | 37527 | 3-Hydroxy-2-naphtho-2,4-xylidide, $C_{19}H_{17}O_2N$ | 291.35 | 223–225 |
| AS-OL | 37530 | 3-Hydroxy-2-naphth-*o*-anisidide, $C_{18}H_{15}O_3N$ | 293.33 | 167–168 |
| AS-P(H) | 37558 | 3-Hydroxy-2-naphtho-*o*-phenetidide, $C_{19}H_{17}O_3N$ | 307.35 | 157–158 |
| AS-RL | 37535 | 3-Hydroxy-2-naphth-*p*-anisidide, $C_{18}H_{15}O_3N$ | 293.33 | 229–230 |
| AS-RP(H) | 37559 | 3-Hydroxy-2-naphtho-*p*-phenetidide, $C_{19}H_{17}O_3N$ | 307.35 | — |
| AS-RT | 37521 | 3-Hydroxy-2-naphtho-*p*-toluidide, $C_{18}H_{15}O_2N$ | 277.33 | 222 |
| AS-S | 37580 | 3-Hydroxy-*N*-(2-methoxy-3-dibenzofuryl)-2-naphthamide, $C_{24}H_{17}O_4N$ | 383.41 | 207–211 |
| AS-SG | 37595 | 2-Hydroxy-11*H*-benzo[*a*]carbazole-3-carbox-*p*-anisidide, $C_{24}H_{18}O_3N_2$ | 382.42 | 324–326 |
| AS-SR | 37590 | 2-Hydroxy-2′-methyl-11*H*-benzo[*a*]carbazole-3-carbox-*p*-anisidide, $C_{25}H_{20}O_3N_2$ | 396.45 | 306–308 |
| AS-SW | 37565 | 3-Hydroxy-*N*-2-naphthamide, $C_{21}H_{15}O_2N$ | 313.36 | 243–244 |
| AS-TR | 37525 | 4′-Chloro-3-hydroxy-2-naphtho-*o*-toluidide, $C_{18}H_{14}O_2NCl$ | 311.78 | 244–245 |
| Undesignated | — | 3-Hydroxy-*N*-β-5,6,7,8-tetrahydro-β-naphthylamide, $C_{21}H_{19}O_2N$ (Morschel, 1936; Hopper *et al.*, 1939) | 317.39 | 200 |
| Undesignated | — | 3-Hydroxy-*o*-ethoxy-2-dibenzothiophenecarboxylanilide, $C_{21}H_{17}O_3NS$ (Venkataraman, 1952) | 363.44 | — |

TABLE t

SPECIAL COMPLEX ARYLAMIDES OF 3-HYDROXY-2-NAPHTHOIC ACID

| Arylamide of 3-hydroxy-2-naphthoic acid | Mol. wt. | Melting point (° C.) | Reference |
|---|---|---|---|
| 2-Methyl-4-bromoanilide, $C_{18}H_{14}O_2NBr$ | 356.23 | 282–285 | Ryberg, 1930 |
| 2,5-Dimethylanilide, $C_{19}H_{17}O_2N$ | 291.35 | 200 | Wahl, 1936 |
| 3,6-Dimethyl-2-chloroanilide, $C_{19}H_{16}O_2NCl$ | 325.80 | 230 | Wahl, 1936 |
| 2,5-Dimethyl-3-chloroanilide, $C_{19}H_{16}O_2NCl$ | 325.80 | 180 | Wahl, 1936 |
| 2,5-Dimethyl-4-chloroanilide, $C_{19}H_{16}O_2NCl$ | 325.80 | 258 | Wahl, 1936 |
| 3,6-Dimethyl-2,4-dichloroanilide, $C_{19}H_{15}O_2NCl_2$ | 360.25 | 223 | Wahl, 1936 |
| 3,6-Dimethyl-2,5-dichloroanilide, $C_{19}H_{15}O_2NCl_2$ | 360.25 | 250 | Wahl, 1936 |
| 2,5-Dimethyl-3,4-dichloroanilide, $C_{19}H_{15}O_2NCl_2$ | 360.25 | 233 | Wahl, 1936 |
| 3,6-Dimethyl-2-nitroanilide, $C_{19}H_{16}O_4N_2$ | 336.35 | 150 | Wahl, 1936 |
| 2,5-Dimethyl-3-nitroanilide, $C_{19}H_{16}O_4N_2$ | 336.35 | 145 | Wahl, 1936 |
| 2,5-Dimethyl-4-nitroanilide, $C_{19}H_{16}O_4N_2$ | 336.35 | 255 | Wahl, 1936 |
| 2,3,5,6-Tetrachloroanilide, $C_{17}H_9O_2NCl_4$ | 401.10 | 232 | Peters *et al.*, 1943 |
| 2,3,5,6-Tetrachloro-4-nitroanilide, $C_{17}H_8O_4N_2Cl_4$ | 446.10 | 269–270 | Peters *et al.*, 1943 |
| 2,3,5,6-Tetrachloro-4-methoxyanilide. $C_{18}H_{11}O_3NCl_4$ | 431.12 | 208 | Peters *et al.*, 1943 |
| *o*-Nitroanilide, $C_{17}H_{12}O_4N_2$ | 308.30 | 195–196 | Rowe and Giles, 1935 |
| 3-Aminobenzoylthiophene arylamide, $C_{22}H_{15}O_3NS$ | 373.44 | — | Spiegler, 1953 |
| 4-Aminobenzoylthiophene arylamide, $C_{22}H_{15}O_3NS$ | 373.44 | — | Spiegler, 1953 |
| 2-Aminoanthraquinone arylamide, $C_{25}H_{15}O_4N$ | 393.40 | 275–280 | Ferber, 1942 |
| *p*-Aminobenzoic acid anilide, $C_{18}H_{13}O_4N$ | 307.31 | 315–316 | Rath and Burkhardt, 1940 |
| *o*-Sulfonic acid anilide, $C_{17}H_{13}O_5NS$ | 343.37 | — | Vorozhtsov and Genkin, 1938 |
| *p*-Sulfonic acid anilide, $C_{17}H_{13}O_5NS$ | 343.37 | — | Vorozhtsov and Genkin, 1938 |
| *p*-Acetylanilide, $C_{19}H_{15}O_3N$ | 305.34 | 263–264 | Bhat and Venkataraman, 1942 |
| *p*-Benzoylanilide, $C_{24}H_{17}O_3N$ | 367.41 | 255 | Bhat and Venkataraman, 1942 |
| *p*-Aminobenzanilide, $C_{24}H_{18}O_3N_2$ | 382.42 | 290–291 | Bhat and Venkataraman, 1942 |
| 4,6-Dimethyl-3-benzoyl-aminoanilide, $C_{26}H_{22}O_3N_2$ | 410.48 | 281–282 | Bhat *et al.*, 1940 |
| *p*-Anilinoanilide, $C_{23}H_{18}O_2N_2$ | 354.41 | 221–222 | Bhat *et al.*, 1940 |
| *m*-Octanoylaminoanilide, $C_{25}H_{28}O_3N_2$ | 404.52 | 350 | Bhat and Venkataraman, 1942 |
| *p*-Lauroylaminoanilide, $C_{29}H_{36}O_3N_2$ | 460.62 | 244–245 | Rath and Burkhardt, 1940 |
| *p*-Stearoylaminoanilide, $C_{35}H_{48}O_3N_2$ | 544.79 | 221 | Rath and Burkhardt, 1940 |
| *p*-Cinnamoylaminoanilide, $C_{26}H_{20}O_3N_2$ | 408.46 | 290–291 | Bhat *et al.*, 1940 |
| [illegible] | 432.51 | 250–251 | Bhat *et al.*, 1940 |

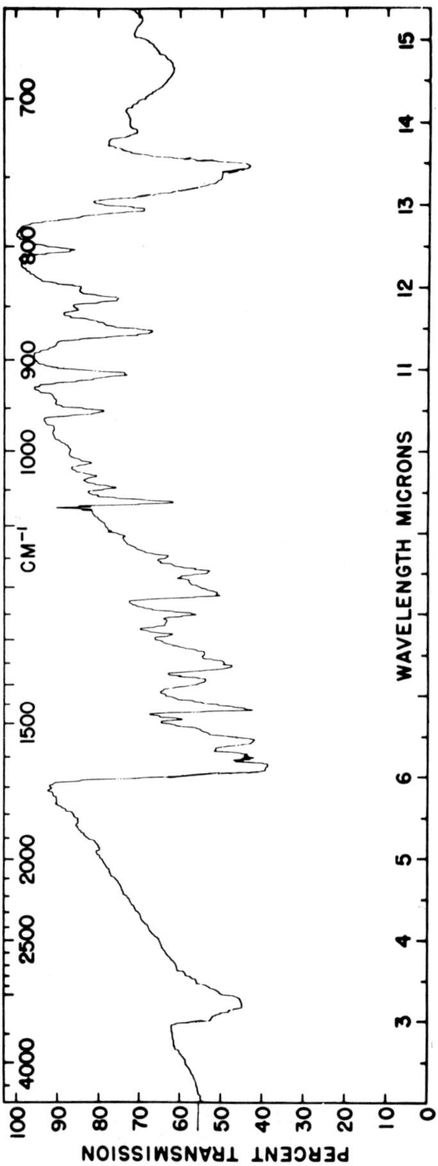


FIG. 1. Infrared spectrum of naphthol AS-D.

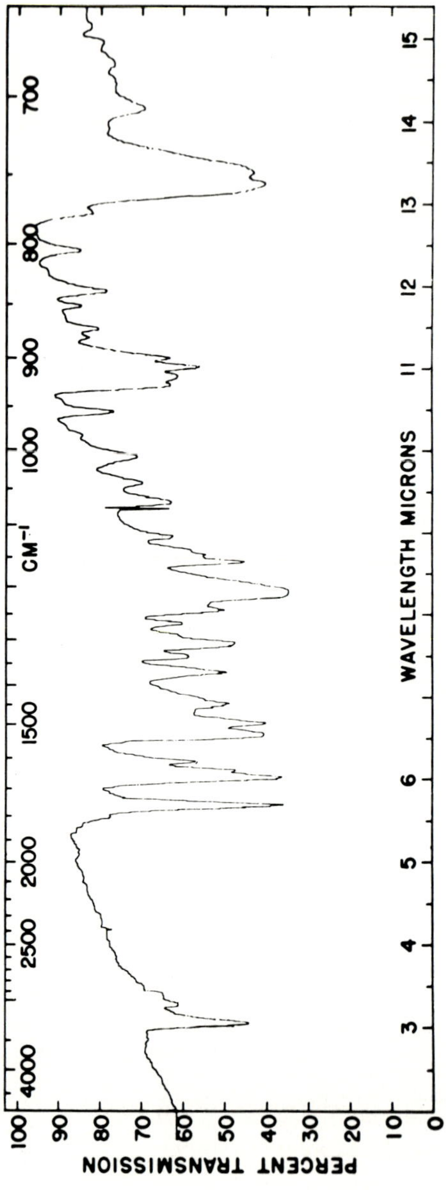


FIG. 2. Infrared spectrum of naphthol AS-D acetate.

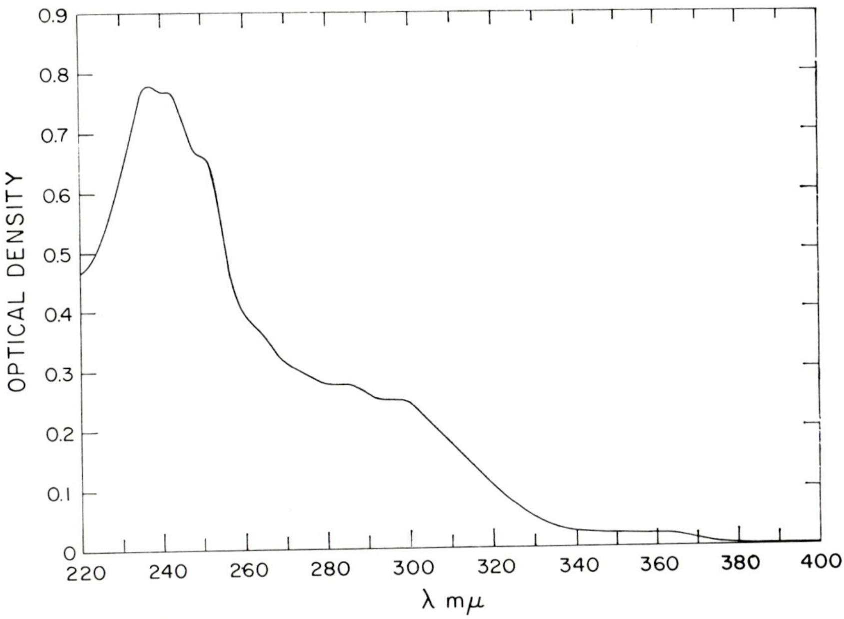


FIG. 3. Ultraviolet spectrum of naphthol AS-D.

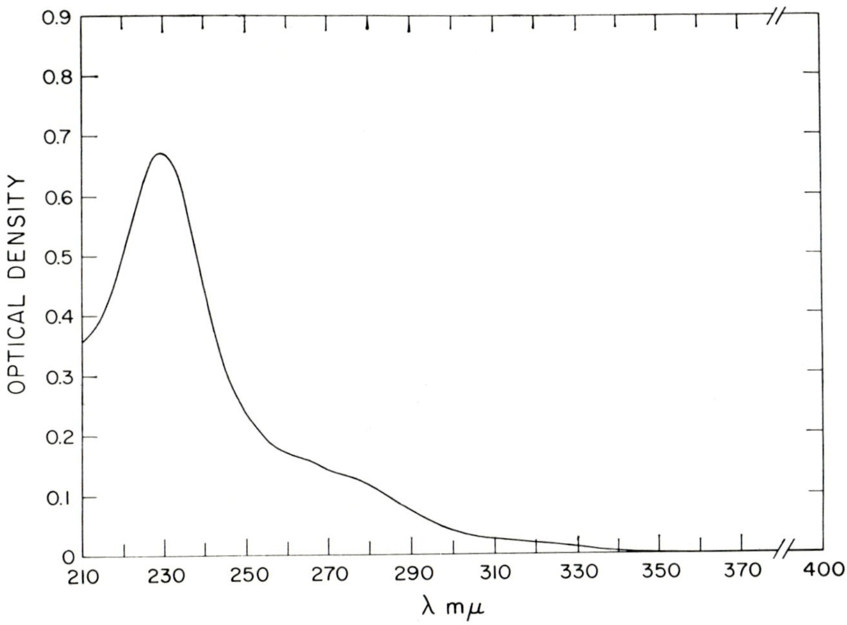


FIG. 4. Ultraviolet spectrum of naphthol AS-D acetate.

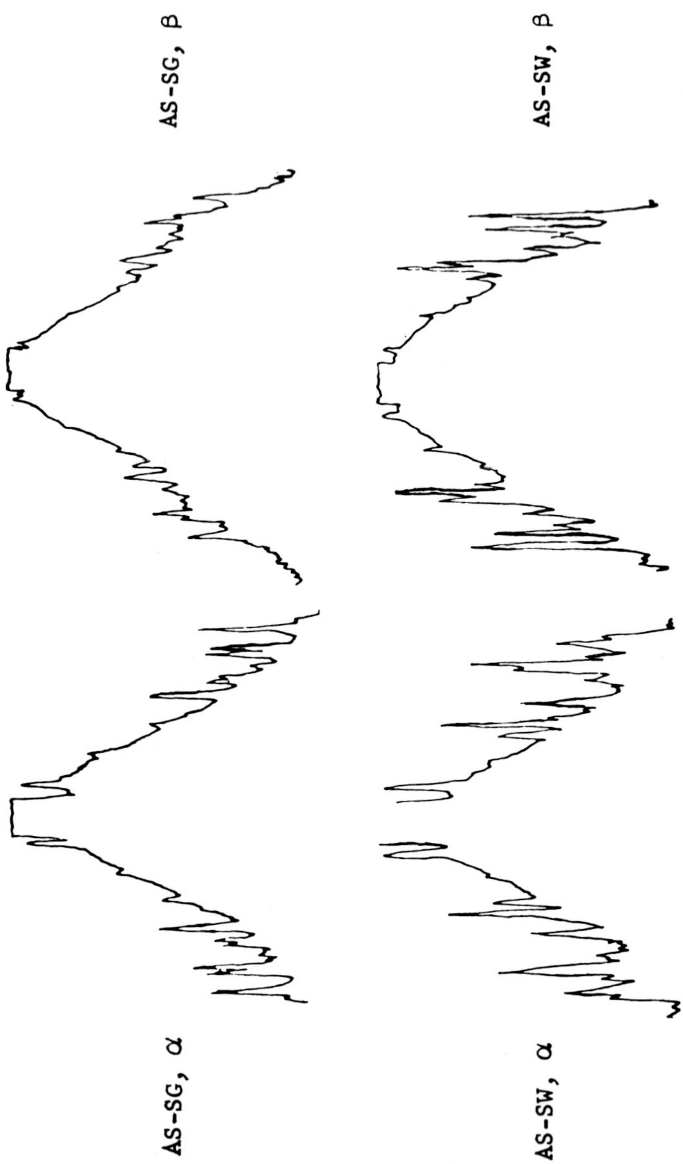


FIG. 5. X-ray diffraction pattern of naphthol AS-SG and AS-SW.

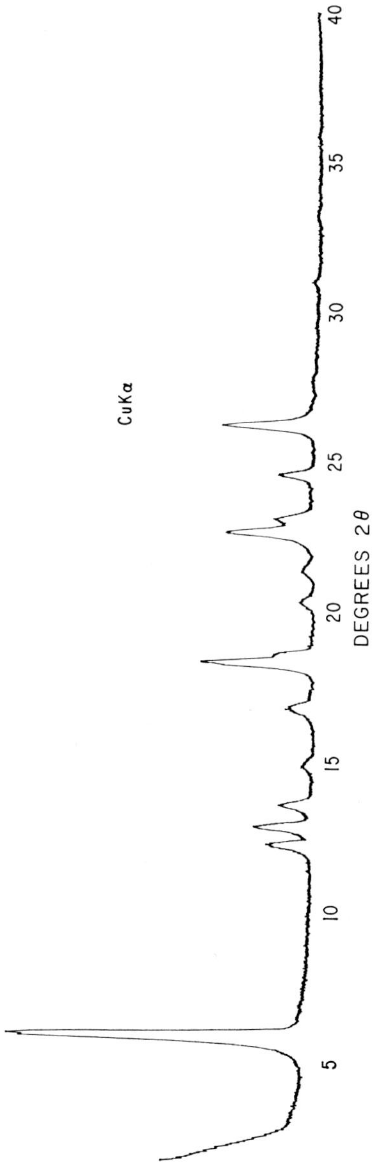


FIG. 6. X-ray diagram of naphthol AS-D (courtesy H. F. McMurdie).

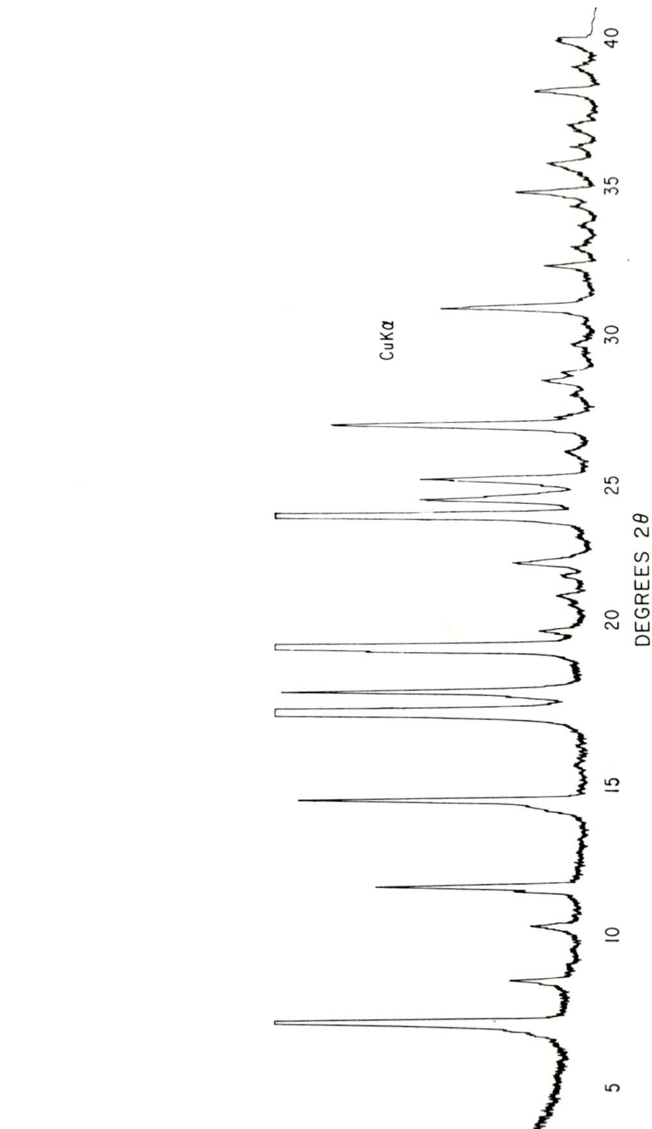


FIG. 7. X-ray diagrams of naphthol AS-BI (courtesy H. F. McMurdie).

TABLE 7

ARYLAMINES

| Chemical name and formula | Mol. wt. | Reference | Melting point (° C.) |
|---|---|---|---|
| 1. *N*,*N*-Dimethyl-*p*-phenylenediamine, $C_8H_{12}N_2$ | 136.2 | Winans and Adkins, 1932 | 53 |
| 2. *N*,*N*-Dimethyl-*p*-naphthalenediamine (4-amino-1-*N*,*N*-dimethylnaphthylamine) $C_{12}H_{14}N_2$ | 186.26 | Friedländer and Welmans, 1888; Cohn, 1895; Cobenzl, 1915; Gokhle and Mason, 1930; Hodgson and Crook, 1936 | 22 |
| 3. Benzidine (*p*-diaminodiphenyl), $C_{12}H_{12}N_2$ (carcinogen) | 184.24 | — | 115–120 |
| 4. *o*-Tolidine, $C_{14}H_{16}N_2$ | 212.30 | — | 121–131 |
| 5. *p*-Aminodiphenylamine, $C_{12}H_{12}N_2$ | 184.24 | Meyer, 1920 | 75 |
| 6. *o*-Aminodiphenylamine, $C_{12}H_{12}N_2$ | 184.24 | — | 80 |
| 7. *p*-Methoxy-*p*-aminodiphenylamine, $C_{13}H_{14}N_2O$ | 214.27 | — | 102 |
| 8. 2-Aminocarbazole, $C_{12}H_{10}N_2$ | 182.23 | Blank, 1891; Chem. Abstr., 1932 | 238 |
| 9. 3-Aminocarbazole, $C_{12}H_{10}N_2$ | 182.23 | Whitner, 1924 | 254 |
| 10. 3-Amino-9-ethylcarbazole, $C_{14}H_{14}N_2$ | 210.28 | Lindemann, 1924 | 113–114 |
| 11. 2-Aminodibenzofuran, 2-aminodiphenylene oxide, $C_{12}H_9ON$ | 183.21 | (Kirkpatrick and Parker, 1935) | 220 (dec.) |
| 12. 3-Aminodibenzofuran, $C_{12}H_9ON$ | 183.21 | Cullinane, 1932; Gilman *et al.*, 1934 | 98 |
| 13. 2-Aminofluorenone, $C_{13}H_9ON$ | 195.22 | — | 163 |
| 14. α-Naphthylamine, 1-naphthylamine, $C_{10}H_9N$ | 143.19 | — | 49 |
| 15. 5,6,7,8-Tetrahydro-1-naphthylamine, $C_{10}H_{13}N$ | 147.22 | — | Oil |
| 16. β-Naphthylamine, 2-naphthylamine (carcinogen) $C_{10}H_9N$ | 143.19 | — | 111–113 |
| 17. 5,6,7,8-Tetrahydro-2-naphthylamine, $C_{10}H_{13}N$ | 147.22 | — | 38.5–39.5 |
| 18. 5-Nitro-1-naphthylamine, $C_{10}H_8O_2N_2$ | 188.19 | — | 119–120 |
| 19. 4-Methoxy-1-naphthylamine, $C_{11}H_{11}ON$ | 173.22 | Slotta *et al.*, 1933 | Oil |
| 20. 3-Methoxy-1-naphthylamine, $C_{11}H_{11}ON$ | 173.22 | Baddar and Gindy, 1944 | Oil |
| 21. 4-Methoxy-2-naphthylamine, $C_{11}H_{11}ON$ | 173.22 | Sannie and Vertalier, 1954; Rosenblatt *et al.*, 1958 | 57.5–58.5 |
| 22. 4-Methoxy-5,6,7,8-tetrahydro-2-naphthylamine, $C_{11}H_{15}ON$ | 177.25 | Schroeter, 1938 | 102–103 |

TABLE 7—*Concluded*

| Chemical name and formula | Mol. wt. | Reference | Melting point (° C.) |
|---|---|---|---|
| 23. 5-Methoxy-2-naphthylamine, $C_{11}H_{11}ON$ | 173.21 | Mueller and Hamilton, 1944 | 71–72 |
| 24. 8-Amino-1,2,3,4-tetrahydroquinoline, $C_9H_{12}N_2$ | 148.21 | Hazelwood *et al.*, 1938; Elderfield and Claflin, 1952 | Oil |
| 25. 3-Amino-2-naphthoic acid, $C_{11}H_9O_2N$ | 187.20 | Allen and Bell, 1942 | 216–217 |
| 26. 3-Amino-2-naphthanilide, $C_{17}H_{14}ON_2$ | 262.32 | Heller, 1929 | 192 |
| 27. 1-Anthramine, 1-aminoanthracene, $C_{14}H_{11}N$ | 193.25 | — | 127 |
| 28. 2-Anthramine, $C_{14}H_{11}N$ | 193.25 | — | 238 |
| 29. 2-Aminochrysene, $C_{18}H_{13}N$ | 243.31 | — | 201–203 |

monia and ammonium sulfite (Bucherer reaction). Naphthylamines are also produced by reduction of nitronaphthalenes. Aminocarbazoles are similarly prepared by the reduction of nitrocarbazoles while aminodibenzofuranes may be synthesized by reduction of the corresponding halogen compound. With regard to substrates, arylamines have found somewhat less application in histochemistry than corresponding hydroxyl compounds. Of note, however, are the use of arylamines as intermediates in the preparation of peptidase substrates as well as their direct application as cytochrome oxidase substrates.

Arylamines also constitute the basis of another group of valuable histochemical reagents, the diazonium salts. Anilines, toluidines, xylidines, as well as more complex phenyl ethers and anilides, are used as the diazonium component in a large number of histochemical reactions. The arylamines listed in Table 7 have found either direct application as oxidase substrates or have been employed as intermediates in the synthesis of substrates for the demonstration of proteolytic enzymes. Those amines used primarily for diazonium salts will be considered in Chapter 3.

It must be emphasized that of the large group of compounds mentioned in this chapter only a very few were employed in histochemistry just ten years ago. Continuation of the present rate of progress in the development and application of organic reagents should vastly augment the histochemical armamentarium in the foreseeable future.

## APPENDIX 2

# SYNTHESIS AND PURIFICATION OF NAPHTHOLIC COMPOUNDS

### 1. *Preparation of Phenolates and Naphtholates*

Sodium and potassium phenolates and naphtholates are water soluble, and are employed without isolation in many instances in some esterification reactions. Naphtholates of the naphthol AS series are highly fluorescent in solution. Phenolates and naphtholates may be prepared by treating the appropriate compound with sodium, potassium, or sodium methylate. Meldrum and Patel (1928) have reported a simple procedure for the preparation of phenolates. The aromatic hydroxyl compound is dissolved in an aqueous solution of potassium hydroxide to which a concentrated solution of the same alkali is added. The crystalline phenolate is precipitated and can be isolated in a relatively pure state. The phenol is initially dissolved in a minimum of 20% aqueous NaOH or 28% aqueous KOH. This type of solution may be directly used in certain esterification reactions. If the phenolate is isolated in crystalline form, it should be stored in a desiccator as there is a gradual transition back to the original hydroxyl derivative.

### 2. *Laboratory Synthesis of Naphthol AS Derivatives*

*a. Procedure 1 (after Rowe, 1930).* Equipment: 1 liter 3-neck flask, heating mantle, thermometer (−5 to 250°), stirrer, reflux condenser. 2-Hydroxy-3-naphthoic acid (0.23 moles), suitable arylamine (0.2 moles), and 400.0 ml. xylene are added to the reaction flask. (Toluene is used instead of xylene when nitroanilines or other heat sensitive amines are used. Monochlorobenzene is used when the amines are of very low solubility.) The mixture is heated to 80° (with agitation) and 8.0 ml. (12.0 g.) phosphorus trichloride ($PCl_3$) added. The mixture is refluxed for 2 hours, and the reaction mixture decanted while hot. The decanted mixture is cooled and filtered, and the precipitate washed with xylene. The precipitate is added to 1 liter of water, neutralized with a 2% sodium carbonate solution, and boiled until the xylene has been driven off. The mixture is adjusted to approximately pH 9 with a 2% sodium carbonate solution and again filtered and cooled. The precipitate is washed with water and then added to 1 liter of 3% HCl and boiled for 10 minutes. The solution is filtered and then cooled. The resulting precipitate is washed with hot water and dried. Recrystallization techniques are described under purification procedures for naphthol AS derivatives.

*b. Procedure 2 (Hopper et al., 1939).* This simplified procedure was described for the preparation of the 5-chloro-2,4-dimethoxyanilide and other derivatives. In the case of the aforementioned chloroanilide, an equimolar quantity (18.0 g.) of it and 3-hydroxy-2-naphthoic acid (18.0 g.) are added to 10.0 ml. pyridine and heated to 70°. Then 6.0 ml. $PCl_3$ are added over a period of 30 minutes with constant stirring. The mixture is subsequently heated with frequent stirring in an oil bath at 115° for 4 hours. The viscous mass is cooled, mixed with ethanol, and reheated until dissolved. Upon cooling and standing overnight a brown solid separates. Further yield was obtained by diluting the mother liquid with water. Benzene-nitrobenzene (1:1) was employed for recrystallization.

3. *Synthesis of β-Hydroxynaphthoic Acid-β-Aminoanthraquinonylamide (Ferber, 1942; Burstone, 1960)*

The synthesis of this special naphthol is given because it is the basis for an interesting post-coupling technique for the demonstration of alkaline phosphatase which gives exceedingly sharp localizations. An analogous group of quinone derivatives is also being prepared and evaluated as the basis of esterase substrates.

Twenty-two grams of β-aminoanthraquinone and 18.0 g. of 3-hydroxy-2-naphthoic acid are triturated with a small amount of xylene, and 300.0 ml. of xylene are then added and the mixture slowly heated with constant stirring. When a temperature of 60° is reached, 4.5 g. of $PCl_3$ are slowly added through separatory funnel (excess $PCl_3$ is undesirable). There is a gradual color change from orange to greenish-yellow to light yellow. In the course of an hour the temperature is raised to the boiling point of xylene. After 5–6 hours at reflux, HCl production stops and the color of the reaction mass becomes considerably darker. The final reaction product is insoluble in xylene. The remaining xylene is removed by drying at 140° and the reaction product is purified with 2% sodium carbonate and 5% HCl. As previously described in the general procedure for preparation of naphthol AS derivatives the product is further purified by recrystallization from nitrobenzene or from a mixture of *N*,*N*-dimethylformamide (DMF)-methanol. The yield was approximately 80%.

4. *Purification of Naphthol AS Derivatives*

Commercially available samples of the naphthol AS series vary in purity from 30 to 95%. The purity of some commercial samples (Pfister Chemical Works) is indicated as follows: AS, 95%; AS-AN, 95%; AS-BS, 95%; AS-D, 95%; AS-DB, 50%; AS-GR, 30%; AS-LC, 85%; AS-LG, 40%; AS-MCA, 94%; AS-MX, 95%; AS-TR, 95%; and AS-SG, 50%. Even in the case of samples of ostensibly high initial purity it is usually advisable to recrystallize them prior to synthesis. This is necessitated by variations in industrial chemicals including the presence of some of the original intermediates as well as tarry water soluble diluents.

Naphthols may be recrystallized from nitrobenzene or nitrobenzene-benzene mixtures. The author has usually employed a mixture of *N,N*-dimethylformamide (DMF)-ethanol or methanol (1:3) (Burstone, 1958, 1960). The amount of DMF may be increased in the case of the more insoluble naphthols.

### 5. *Synthesis of Brominated Naphthoic Acid Derivatives (FIAT, 1948, p. 157)*

These compounds are intermediates in the preparation of naphthol AS-BI (Table 5), and their synthesis does not appear to be described in the more readily available literature. The quantity of starting materials is that employed by the Germans for an industrial synthesis, and may be reduced accordingly for laboratory preparations.

*a. 1,6-Dibromo-2-hydroxy-3-naphthoic acid.* Twenty-three hundred grams of 2-hydroxy-3-naphthoic acid are added at 0.5° to 11,500 g. of $H_2SO_4$ 66° Bé (Baumé). The mixture is stirred constantly and bromine (2000 g.) is added over a period of 2 to 3 hours during which the temperature is kept at −10 to −2°. The evolved HBr is collected in 7000 g. of 20% $H_2SO_4$. After all of the bromine has been added, the $H_2SO_4$-HBr solution is added at low temperature to the reaction mixture. The reaction mixture temperature is slowly raised over a 24-hour period to 20°. It is necessary to keep the reaction mixture well agitated during the entire operation. [For smaller laboratory syntheses, the bromination probably could be carried out using glacial acetic acid as the solvent medium and HBr gas or HBr in acetic acid (author)]. The reaction mass is then added to 30.0 kg. of ice and 20.0 liters of water, filtered and washed until free of acid. The moist cake (containing 50% water) is then converted to the 6-bromo derivative.

*b. 6-Bromo-2-hydroxy-3-naphthoic acid.* Eight thousand grams of 1,6-dibromo-2-hydroxy-3-naphthoic acid (50%) is pasted with 15,000 ml. hot water and converted into the monosodium salt by the addition of 1,200 ml. of NaOH (36° Bé). Twenty-one hundred grams of anhydrous sodium sulfite and 660 g. of anhydrous sodium carbonate are added. The mixture is placed in an autoclave and heated for 8 to 9 hours at 150–160° at a pressure of 7–8 atmospheres. The reaction mixture is allowed to cool while being agitated. The sodium salt of 6-bromo-2-hydroxy-3-naphthoic acid is filtered off and washed with 5% NaCl solution until the filtrate is colorless. The compound is dissolved in about 50.0 liters of boiling water, filtered, and the clear solution converted into the free acid with HCl at 80–90°. The product is then filtered hot, washed free of acid and dried. The m.p. is 263°, and the yield (98.5% pure) is 87% of theory. The 6-bromo-2-hydroxy-3-naphthoic acid may be converted into suitable anilides by the techniques described under "Laboratory Synthesis of Naphthol AS Derivatives" (Appendix 2, Section 2).

## Commercial Sources of Reagents

1. Aldrich Chemical Company, Milwaukee, Wisconsin
2. British Drug House, Poole, England (U.S.A. Distributor, The Ealing Corp., Natick, Massachusetts)
3. Dajac Laboratories, The Borden Chemical Company, Philadelphia, Pennsylvania
4. Eastman Kodak, Rochester, New York
5. General Analine and Film Corp., New York, New York
6. Imperial Chemical Industries Ltd., Manchester, England
7. Pfister Chemical Works, Ridgefield, New Jersey
8. Sigma Chemical Company, St. Louis, Missouri

## References

Allen, C. F. H., and Bell, A. (1942) *Org. Syntheses* **22,** 19–23.

Anderson, L. C., and Thomas, D. G. (1943) *J. Am. Chem. Soc.* **65,** 234–238.

Anschütz, L., Weber, E., and Runkel, K. (1906) *Ann.* **346,** 360–381.

Baddar, F. G., and Gindy, M. (1944) *J. Chem. Soc.* pp. 450–452.

Ballauf, F., and Schmelzer, A. (1932). *Chem. Abstr.* **26,** 1000.

Bhat, R. V., and Venkataraman, K. (1942) *J. Soc. Dyers Colourists* **58,** 155–161.

Bhat, R. V., Forster, R. B., and Venkataraman, K. (1940) *J. Soc. Dyers Colourists* **56,** 166–178.

Blank, A. (1891) *Ber.* **24,** 306.

Burstone, M. S. (1958) *J. Natl. Cancer Inst.* **21,** 523–539.

Burstone, M. S. (1960) *J. Natl. Cancer Inst.* **24,** 1199–1217.

Chem. Abstr. (1932) **26,** 2469. DRP 542, 422.

Claus, A., and Zimmermann, O. (1881) *Ber.* **14,** 1477–1485.

Cobenzl, A. (1915) *Chem. Zentr.* **39,** 859–860.

Cohn, P. (1895) *Monatsh. Chem.* **16,** 798–806.

"Colour Index" (1956) 2nd ed., Vol. 3, 3326–3333. Chorley & Pickersgill, Leeds, England.

Cullinane, N. M. (1932). *J. Chem. Soc.* pp. 2365–2367.

de La Mare, P. B. D., and Ridd, J. H. (1959) "Aromatic Substitution." Butterworths, London.

Dorman, K. L. (1939) *Am. Dyestuff Reptr.* **28,** 79–84, 101–106.

Elderfield, R. C., and Claflin, E. F. (1952) *J. Am. Chem. Soc.* **74,** 2953–2959.

Ferber, E. (1942) *Melliand Textilber.* **23,** 598–600.

FIAT (Field Information Agency Technical) (1948) Armed Services Technical Information Agency. Vol. 3, ATI-64331. Arlington, Virginia.

Fierz-David, H. E., and Blangey, L. (1949) "Fundamental Processes of Dye Chemistry." Interscience, New York.

Franzen, H., and Stäuble, G. (1922) *J. prakt. Chem.* **103**(2), 352–390.

Friedländer, P., and Welmans, P. (1888) *Ber.* **21,** 3123–3130.

Friedländer, P., and Zinberg, S. (1896) *Ber.* **29,** 37–42.

Gilman, H., and Van Ess, P. R. (1939) *J. Am. Chem. Soc.* **61,** 1365–1371.

Gilman, H., Brown, G. E., Bywater, W. G., and Kirkpatrick, W. H. (1934) *J. Am. Chem. Soc.* **56,** 2473–2477.

Gilman, H., Bywater, W. G., and Parker, P. T. (1935) *J. Am. Chem. Soc.* **57,** 885–887.

GOKHLE, B., AND MASON, F. A. (1930) *J. Chem. Soc.* pp. 1757–1758.
HAZLEWOOD, S. J., HUGHES, G. K., AND LIONS, F. (1938) *J. Proc. Roy. Soc. N.S. Wales* **71,** 462–474.
HELLER, K. (1929) *J. prakt. Chem.* **121**(2), 193–203.
HODGSON, H. H., AND CROOK, J. H. (1936) *J. Chem. Soc.* pp. 1500–1503.
HOPPER, I. V., MCGREGOR, J. H., AND WILSON, F. J. (1939) *J. Soc. Dyers Colourists* **55,** 449–453.
KEHRMANN, F., AND DENK, A. (1901) *Ber.* **33,** 3295–3300.
KIRKPATRICK, W. H., AND PARKER, P. T. (1935) *J. Am. Chem. Soc.* **57,** 1123–1126.
KOELSCH, C. F. (1940) *Org. Syntheses* **20,** 18–20.
LEDERER, H. (1931) *Melliand Textilber.* **12,** 461–462.
LEMAN, A. (1938) *Ann. chim.* (*Paris*) **9**(11), 430–447.
LINDEMANN, H. (1924) *Ber.* **57,** 555–559.
LUBS, H. A., ed. (1955) "The Chemistry of Synthetic Dyes and Pigments." Reinhold, New York.
MELDRUM, A. N., AND PATEL, M. M. (1928) *J. Indian Chem. Soc.* **5,** 91–94.
MEYER, R. (1920) *Ber.* **53,** 1265–1276.
MORSCHEL, H. (1936) U. S. Pat. 2,040,397.
MUELLER, A. C., AND HAMILTON, C. S. (1944) *J. Am. Chem. Soc.* **66,** 860–862.
PETERS, A. T., ROWE, F. M., AND STEAD, D. M. (1943) *J. Chem. Soc.* pp. 233–235.
RATH, H., AND BURKHARDT, R. (1940) *Ber.* **73,** 701–708.
ROSENBLATT, D. H., NACHLAS, M. M., AND SELIGMAN, A. M. (1958) *J. Am. Chem. Soc.* **80,** 2463–2465.
ROWE, F. M. (1930) *J. Soc. Dyers Colourists* **46,** 227–230.
ROWE, F. M., AND GILES, C. H. (1935) *J. Soc. Dyers Colourists* **51,** 278–283.
ROWE, F. M., AND STAFFORD, F. R. (1924) *J. Soc. Dyers Colourists* **40,** 228–230.
RUFF, O., AND STEIN, V. (1901) *Ber.* **34,** 1668–1684.
RUGGLI, P., KNAPP, F., MERZ, E., AND ZIMMERMANN, A. (1929) *Helv. Chim. Acta* **12,** 1034–1051.
RYBERG, B. A. (1930) *Chem. Abstr.* **24,** 1984.
SANNIE, C., AND VERTALIER, S. (1954) *Bull. soc. chim.* (*France*) 238–240.
SCHOLL, R., AND SEER, C. (1912) *Ann.* **394,** 111–177.
SCHÖNBERG, A., AND MUSTAFA, A. (1946) *J. Chem. Soc.* p. 642.
SCHÖPFF, M. (1892) *Ber.* **25,** 2740–2745.
SCHROETER, G. (1938) *Ber.* **71,** 1040–1056.
SEN, R. N., AND RAY, S. K. (1932) *J. Indian Chem. Soc.* **9,** 181–184.
SLOTTA, K. H., FRANKE, W., AND HABERLAND, G. (1933) *Ber.* **66,** 108–112.
SMITH, A. J. (1879) *J. Chem. Soc.* **35,** 789–793.
SPIEGLER, L. (1953) *J. Org. Chem.* **18,** 1292–1316.
TILDEN, W. A., AND ARMSTRONG, H. E. (1901) *Brit. Assoc. Advance. Sci. Rept.* pp. 152–154.
VENKATARAMAN, K. (1952) "The Chemistry of Synthetic Dyes," Vol. 2. Academic Press, New York.
VOROZHTSOV, N. N., AND GENKIN, N. D. (1938) *Chem. Abstr.* **32,** 5393.
WAHL, M. H. (1936) *Ann. chim.* (*Paris*) [11] **5,** 64–82.
WHITNER, T. C., JR. (1924) *J. Am. Chem. Soc.* **46,** 2326–2329.
WINANS, C. F., AND ADKINS, H. (1932) *J. Am. Chem. Soc.* **54,** 306–312.
WITT, O. N., AND DEDICHEN, J. (1896) *Ber.* **29,** 2945–2954.
ZINCKE, T. (1888) *Ber.* **21,** 3378–3390.

*CHAPTER 3*

# DIAZONIUM SALTS AND AZO-COUPLING

## I. CHARACTERISTICS AND FORMATION OF DIAZO COMPOUNDS

In 1890 Peter Greiss discovered a new class of chemical compounds which constitute the diazo group. He subsequently synthesized the first azo-dye by coupling a phenolic compound with a diazonium component. A review of the history, properties, and constitution of diazo compounds may be found in several sources (Holzach, 1947; Fierz-David and Blangey, 1949; Saunders, 1949; Lubs, 1955; Zollinger, 1961). This chapter will primarily deal with the theoretical and practical application of the aforementioned compounds in enzyme histochemistry. Diazo compounds contain the $N_2$ grouping combined to only one hydrocarbon radical instead of two as in azo compounds. The formation of diazo compounds depends upon the replacement of three hydrogen atoms of the salt of an aryl primary amine by nitrogen from nitrous acid to form the so-called diazonium salts.

$$ArNH_2Cl + HONO \rightarrow ArN_2Cl + 2H_2O + NaCl$$

Thus, when aniline hydrochloride in an excess of cold HCl is treated with sodium nitrite, benzenediazonium chloride is formed. Earl and Hall (1933) have suggested that aniline nitrite is a transition stage between aniline hydrochloride and the diazonium salt. Fieser and Fieser (1956) suggest that an *N*-nitroso derivative may be an intermediate. More recently Hughes *et al.* (1958) have indicated that diazotization depends kinetically upon a process ending in *N*-nitrosation to form a nitrosammonium ion, and that the processes leading to nitrosation may consist of several steps.

In the presence of a base such as NaOH, diazonium chlorides are transformed momentarily into diazonium hydroxides and then into isodiazotates. The latter exhibit geometric isomerism (Hodgson and Marsden, 1945). The *syn* form (labile) couples actively, while the *anti* form (stable) or antidiazotate (nitrosamine) does not couple. The isomerism is reversible, however, and the diazonium salt may be regenerated by treatment of the anti-

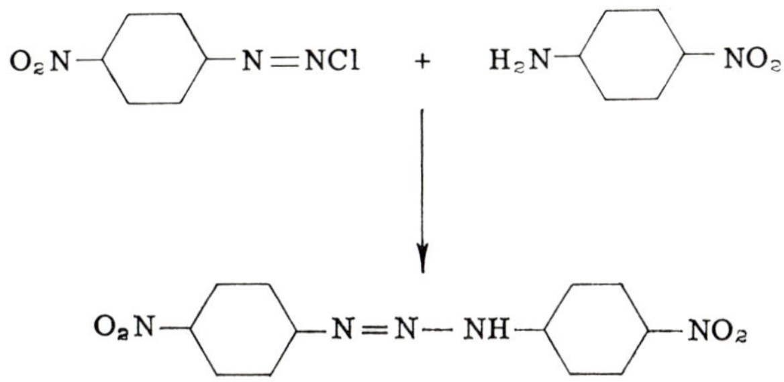


Formation of diazoamino compound from diazotized *p*-nitroaniline

(I)

diazotate with acid. The dyestuff industry has utilized the stable diazotates in conjunction with various naphtholates as printing and dying mixtures known as "Rapid Fast Colors." The dye is formed by activation of the wet mixture by heat and dilute acid. Diazo compounds react with certain amines to form diazoamino or trizene compounds. If insufficient nitrite is present during the diazotization process certain amines such as *p*-nitroaniline may form a diazoamino compound (I). The diazoamino compounds may undergo isomeric rearrangement to form aminoazo derivatives as indicated by diazoaminobenzene ($C_6H_5$—N═N—H—$C_6H_5$) and aminoazobenzene ($C_6H_5$—N═N—$C_6H_4$—$NH_2$). The so-called "Rapidogens" are commercial products containing naphthol and diazoamino compound. Upon treatment with acid, the diazoamino compound is split and the subsequently formed diazo compound couples with the naphthol. When diazo compounds are treated with sodium sulfite in neutral or alkaline solution diazosulfonates are formed. The stable antidiazosulfonates do not couple with naphthols but are readily activated by mild oxidizing agents or in some cases by the action of light. With reference to the latter, diazosulfonates are employed in some photographic processes to produce various azo-dyes in areas of photographic paper exposed under a negative. Aminodiphenylamine and analogues as well as monoacyl-*p*-phenylenediamine derivatives are known to form stable diazosulfonates. Preliminary attempts by the author to employ diazosulfonates as histochemical reagents for the demonstration of tissue oxidase have not been successful. This approach involves incubation of a tissue section with a diazosulfonate and coupling agent (naphthol). If the tissue oxidase is able to act upon the sulfonate, a diazo compound is produced which subsequently couples with the naphthol to form an insoluble dye.

An interesting approach to the stabilization of diazonium cations has been reported by Lambert and Cates (1957). A freshly prepared diazonium salt solution is treated with a sulfonic acid ion exchange resin, dried, and stored in the dark at room temperature. This type of complex is reported to be stable for months. The diazonium chloride may be released by treatment of the resin with an inert salt such as potassium chloride. This type of procedure offers possibilities by which inhibitor-free diazonium chlorides may be made available for immediate use.

A number of techniques have been employed in the preparation of diazo compounds. Details of some specific synthetic procedures are given in Appendix 3. Among the general methods of preparation are:

a. The "direct method" in which a solution of metallic nitrite (e.g. $NaNO_2$) is added to a cold solution of arylamine in aqueous mineral acid. Poorly soluble amines may be dissolved in an organic solvent such as *N*,*N*-dimethylformamide (Burstone and Weisburger, 1961a).

b. The "inverted method" in which mixed alkaline solutions of metallic nitrite and a salt of a sulfonated (e.g. sulfanilic acid) or carboxylated arylamine are run into an excess of cold mineral acid.

c. The "acid-solution method" in which an amine is dissolved in concentrated acid (glacial acetic, sulfuric, or phosphoric) and diazotized with nitrosylsulfuric acid. This technique is useful for diazotization of weakly basic amines such as α-aminoanthraquinone.

d. The "method of Witt" in which the amine is dissolved in concentrated nitric acid in the presence of a metabisulfite which forms nitrous acid by reduction. This method is not suitable for amines which are easily oxidized.

e. The "method of Knoevenagel" in which the arylamine salt is dissolved in water, or suspended in an inert solvent and treated with an alkyl nitrite (e.g. amyl or octyl nitrite). Modification of this technique has been applied in the preparation of histochemical reagents (Burstone and Weisburger, 1961a).

In addition to the traditional method of treating an arylamine with nitrous acid, diazonium salts may also be prepared by oxidation, reduction, condensation, or scission reactions involving somewhat diverse compounds (Saunders, 1949, p. 50).

a. Treatment of arylhydrazines with an oxidizing agent (mercuric oxide, mercuric acetate, or bromine).

b. Treatment of anthraquinone with an oxidizing agent (mercury catalyst).

c. Treatment of a nitrosophenol; *p*-nitrosodiphenylamine, *p*-nitrosodimethylaniline, or nitrosobenzene with a reducing agent (nitrous acid).

d. Condensation of nitrosobenzene with hydroxylamine, or condensation of thionylphenylhydrazine with acetyl chloride.

When attempts are made to diazotize *o*-aminophenols or *o*-aminonaphthols according to routine procedures, these compounds may undergo oxidation. Successful diazotization may be effected in the presence of copper sulfate (Appendix 3, Section 6). The properties of such compounds differ from the usual diazo salts in that they are *o*-azo compounds or diazo oxides as illustrated by the diazotization of 1-amino-2-naphthol-4-sulfonic acid (II). Although they are slow couplers, the diazo oxides are of interest because of their hydroxyl group which offers a coordinating site for the formation of metal complexes of *o*,*o'*-dihydroxyazo dyes.

Arylamine diazonium chlorides and analogous simple salts are generally unstable, and require refrigeration for storage. Even at low temperature, however, slow decomposition occurs over a period of time. On the other hand, some diazo compounds are remarkably heat stable. It has been reported that diazotized *p*-aminodiphenylamine can be recrystallized from hot water (Saunders, 1949, p. 24), while α-aminoanthraquinone can be diazotized at room temperature (Saunders, 1949, p. 13). Commercially available diazo compounds are usually stabilized as acids, simple salts, and double salts (Table 1).

Examples of commercially stabilized diazonium salts are represented in Chart 1.

Commercial techniques of stabilization of industrial products may be subject to modification, particularly with reference to non-active diluents. The latter, which include aluminum, sodium, magnesium, and zinc sulfates, magnesium oxide, magnesium bicarbonate, and disodium naphthalene-1,6-disulfonate, may be present in order to enhance stability and storage life. Even though small quantities of diazonium salts are used in enzyme reactions (10–30 mg. per Coplin jar), the metallic and other stabilizers may exert a marked inhibitory effect upon certain enzyme systems. Among these are aminopeptidase and certain oxidative enzymes. Diazonium fluoborates and naphthalene disulfonates are markedly inhibitory. Zinc chloride double salts are less inhibitory, and diazonium chlorides and sulfates exhibit the least effect. Diazo compounds are also known to directly couple with certain protein constituents, but the effect upon enzymatic reactions is not clear. According to Pearse (1960, p. 872), it is possible to remove aluminum sulfate by the addition of a few milliliters

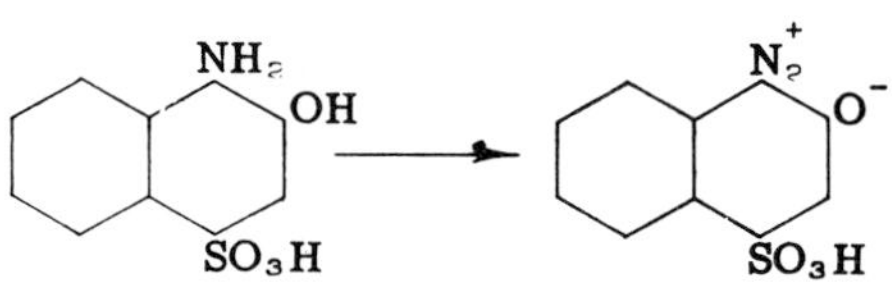


Diazotization of 1-amino-2-naphthol-4-sulfonic acid to form a diazo oxide

(II)

TABLE 1

TYPES OF STABILIZED DIAZONIUM SALTS

| Diazonium salt | Formula |
|---|---|
| Diazonium chloride—zinc chloride double salts (Also double salts with $MnCl_2$, $CoCl_2$, and $CdCl_2$) | $R\cdot N\vdots N\}Cl, \frac{1}{2}ZnCl_2$ |
| Diazonium acid sulfates | $R\cdot N\vdots N\}SO_4H$ |
| Diazonium alkyl sulfates | $R\cdot N\vdots N\}SO_3\cdot O\cdot Alk$ |
| Diazonium chlorides | $R\cdot N\vdots N\}Cl$ |
| Diazonium fluoborates | $R\cdot N\vdots N\}BF_4$ |
| Diazonium benzenesulfonates (Also with $NO_2$, Cl, and $NH\cdot CO\cdot CH_3$ as substituents) | $R\cdot N\vdots N\}SO_3\cdot C_6H_5$ |
| Diazonium acid 1,5-naphthalenedisulfonates | 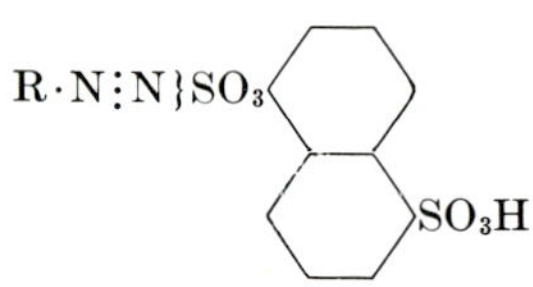 |



of a saturated aqueous solution of purpurin. The resultant aluminum lake can be removed by filtration. Pearse claims that this technique will appreciably increase the speed of the azo-dye alkaline phosphatase reaction. Another approach to the removal of diluents employs ion exchange resins (Burstone, 1958) (Appendix 3, Section 1).

With the advent of successful post-coupling methods, practically all commercially available diazonium salts may be employed, since any possible inhibitory effect is of no consequence. In addition, diazonium salts possessing inhibitory qualities may also be employed in colorimetric studies in which the diazo component is added following the enzymatic reaction. Commercially available diazotizable amines (bases) and diazonium salts usually consist of anilines, toluidines, and anisidines substituted by chloro, nitro, cyano, trifluoromethyl, arylamino, or acylamino groups. Water-solubilizing groups such as OH, COOH, and $SO_3H$ are usually absent. A large percentage of diazotizable amines have electropositive substituents in the 2-position and electronegative substituents in the 5-position (III) (Maynard, 1955). This type of configuration generally enhances brilliance of shade as well as fastness of the final azo-dye. When a single substituent is present in an aniline molecule, the effect ranges from hypsochrome to bathochrome as the substituent is shifted from *ortho* to the *meta* and *para* positions. Alkyl and alkoxy groupings produce more of a bathochrome effect than do nitro and chloro substituents.

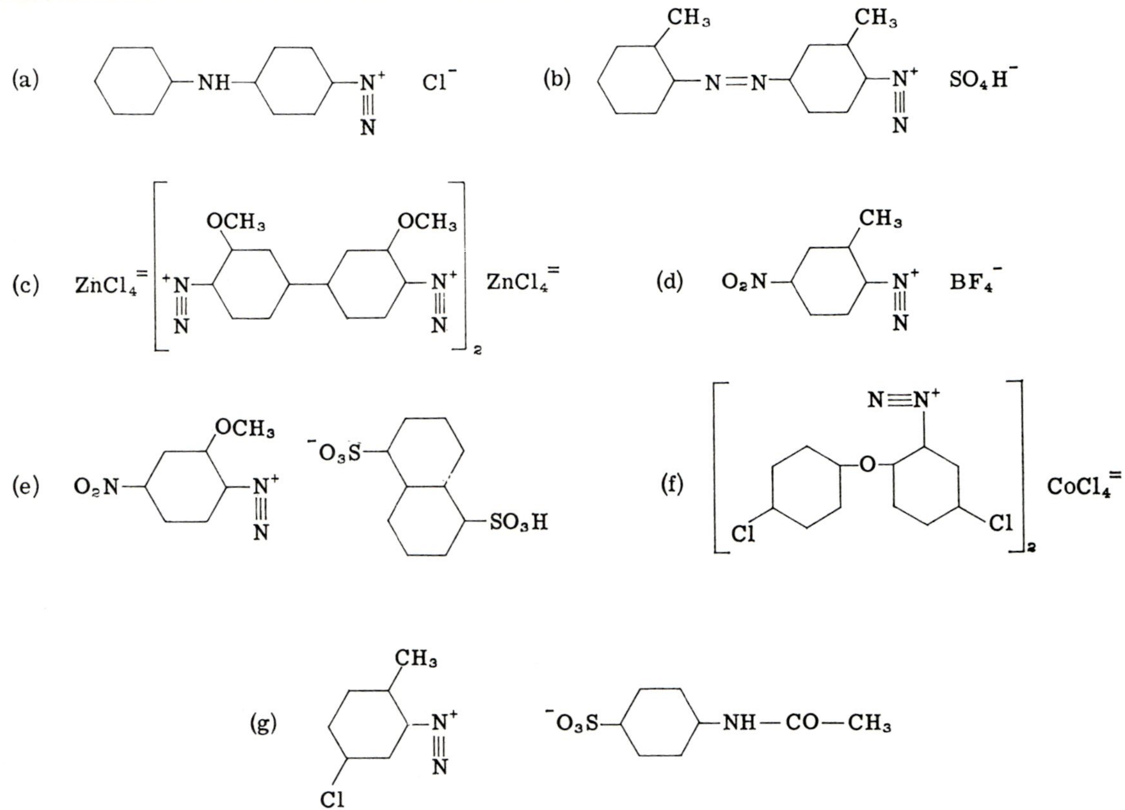


CHART 1. Specific examples of commercially stabilized diazonium salts. (a) Variamine Blue RT, diazonium chloride; (b) Garnet GBC, diazonium acid sulfate; (c) Blue B, double zinc salt; (d) Red RL, diazonium fluoborate; (e) Red B, diazonium acid 1,5-naphthalenedisulfonate; (f) Red FR, diazonium colbalt chloride complex; (g) Red KB, diazonium *N*-acetylsulfanilate.

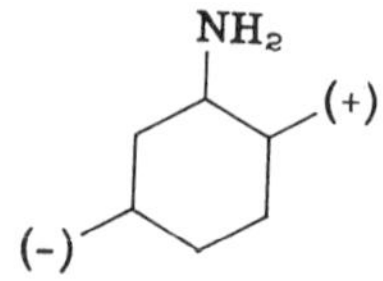

Characteristic location of electropositive and electronegative substituents in many commercially available diazonium salts.

(III)

Nitro groups tend to dull a shade while chloro and cyano groups tend to brighten it.

Chart 2 shows the types of diazotizable amines (bases) employed by the dyestuff industry. Diazo salts and bases of American manufacture are usually preceded by the word "Fast," or "Diazo," while those of British manufacture are preceded by the word "Brentamine." German diazo compounds may contain the prefix "Echt." With reference to the diazonium salts, the amount of active diazotate is usually about 20%, but more concentrated preparations containing over 40% active ingredients are also available. As previously indicated, there may be variations in the stabilization techniques.

The following diazo oxides and diazonium salts are also commercially available: 1-diazo-2-naphthol-4-sulfonic acid, 2-diazo-1-naphthol-4-sulfonic acid, 4-diazo-*N*-ethyl-*N*($\beta$-hydroxyethyl)aniline, 4-diazo-*N*-monoethyl-*o*-toluidine, 4-diazo-*N*,*N*-diethyl-*m*-toluidine, 4-diazo-*N*,*N*-dimethylaniline, and 4-diazo-*N*,*N*-diethylaniline.

Several additional diazonium salts and bases are also available but their structure has not been disclosed. They include Fast Scarlet Salt AFA, Brentamine Fast Scarlet Salt AT, Variogen Bases I and II, and Variogen Salt II.

In recent years newer diazotizable amines have been suggested by industry but do not apear to be readily available (Venkataraman, 1952, p. 700). Formulae of some are shown in Chart 3. These compounds result in the formation of highly insoluble, highly chromogenic dyes which may facilitate histochemical localizations and enhance permanency of the stained preparations.

In 1953, Lillie *et al.* reported the diazotization of safranine, a diamino phenylphenazonium dye, which was employed to detect certain phenolic substances in tissues. This work was based upon the fact that the dye Janus Black is commercially synthesized by coupling diazotized safranine with phenol. Subsequently, Davis and Ornstein (1959) and Davis (1959) found that pararosaniline (IV) could be diazotized (Appendix 3, Section 3,*b*) and in conjunction with azo-dye methods for the demonstration of hydrolytic enzymes. Burstone and Weisburger (1961a, b) have prepared diazotates of a large number of arylamines for use with post-coupling and in some cases with simultaneous coupling methods (Appendix 3, Section 3, *c*). Among the amines are the following:

1. 3-aminofluoranthene
2. 2-aminofluorenone
3. *p*-(acetoxymercuri)aniline
4. 2-amino-3-nitro-9-fluorenone
5. 5*H*-aminotetrazole
6. sulfaquinoxaline
7. *N*-phenylsulfanilamide
8. 4-aminoantipyrine HCl
9. 8-aminoquinoline
10. 6-aminoquinoline
11. 6-amino-2-mercaptobenzothiazole
12. 8-amino-1,2,3,4-tetrahydroquinoline
13. 5,6,7,8-tetrahydro-1-naphthylamine
14. 3-aminodibenzofurane
15. 2-aminobenzophenone
16. 4-phenylazo-1-naphthylamine
17. 4-nitro-1-naphthylamine
18. 2-amino-3-naphthoic acid
19. 2-amino-3-naphthanilide
20. *m*-aminocarboxypyrazolone
21. 1-(*m*-aminophenyl)-3-methyl-5-pyrazolone
22. 1-(*p*-aminophenyl)-3-methyl-5-pyrazolone
23. 3-amino-9-ethylcarbazole
24. methylenedianiline
25. diaminodiphenylsulfone
26. 4-phenylazo-1-naphthylamine
27. 4-(3,5-dimethylphenylazo)-3,5-dimethylaniline
28. 1-amino-5-chloroanthraquinone
29. 1-amino-4-bromo-3-methylanthraquinone
30. 1-amino-2-bromo-4-toluidinoanthraquinone
31. 1-amino-4-hydroxyanthraquinone

The diverse aromatic ring structures of the aforementioned compounds include the following: fluoranthene, fluorenone, anthraquinone, tetrazole, sulfanilamide and analogues, antipyrine, quinoline, benzothiazole, dibenzofurane, benzophenone, naphthoic acid, pyrazolone, and carbazole. Many of the resulting azo-dyes form stable chelates of cobalt and other metals. Diazotized 3-aminofluoranthene was found to be highly stable in solution, and quite suitable for use with some simultaneous coupling procedures. The anthraquinones produce very insoluble dyes but are more inhibitory than most of the other diazonium components. During the process of diazotization, nitrosation may also take place. A discussion of the kinetics and relationship between nitrosation and diazotization has been reviewed in a recent article by Ridd (1961).

## II. STABILITY AND COUPLING RATE OF DIAZO COMPOUNDS

With an azo-dye technique the potential qualities of a naphthol may be nullified if coupling with the diazonium salt does not take place promptly. Although there is no actual knowledge about the submicroscopic nature of a "capture reaction" involving a stain precursor (e.g. liberated naphthol), we can assume that this reaction should take place as rapidly as possible. This requirement demands the use of a diazonium salt which is a fast coupler, since these compounds are known to vary greatly in their coupling energies. In the past, the selection of diazonium salts for enzyme histochemistry has, in most instances, simply been made at ran-

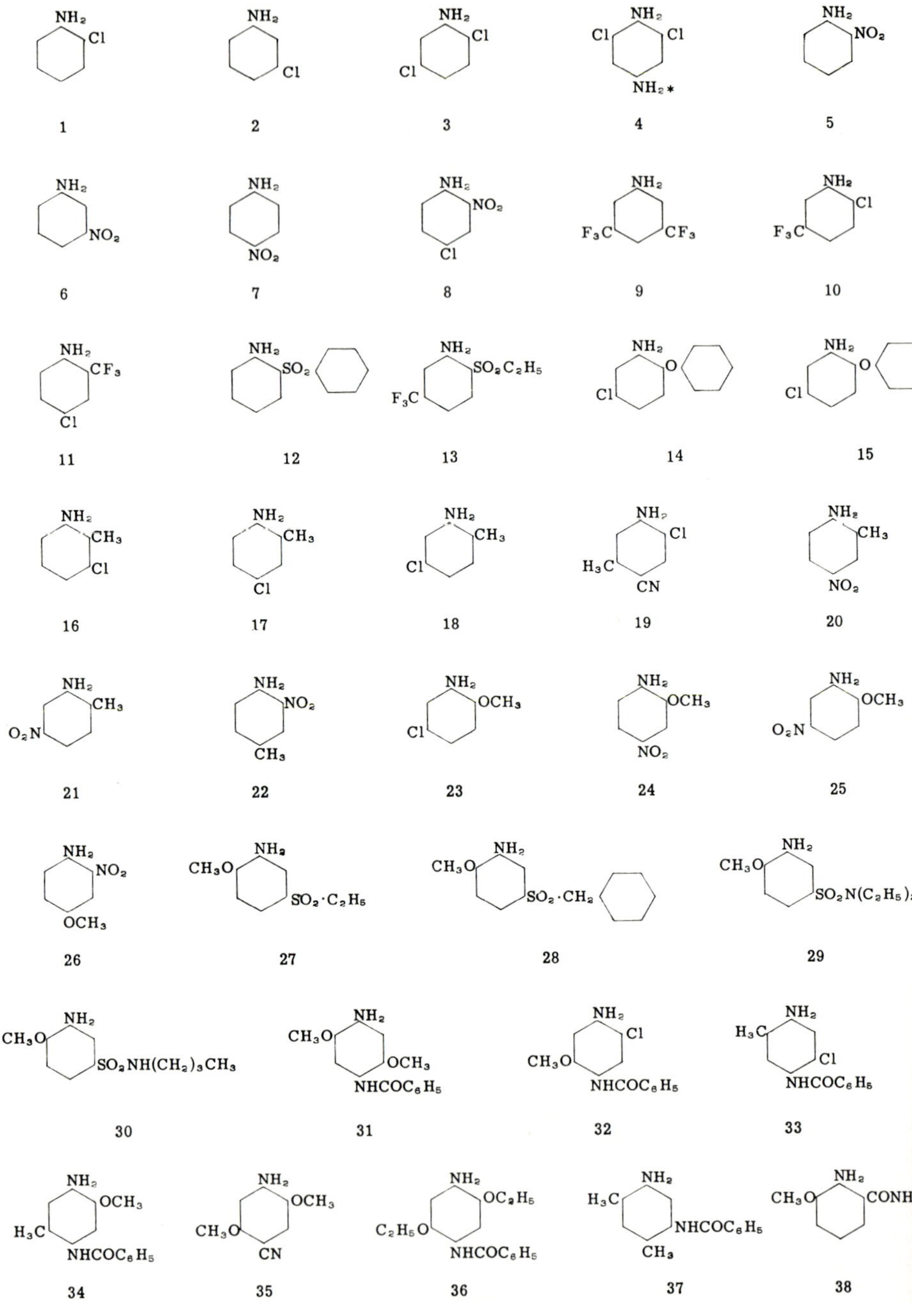


CHART 2. Commercial diazotates. The structures are identified according to number in the tabulation. Key to chart on pages 98 and 99.

CHART 2.—*Continued**

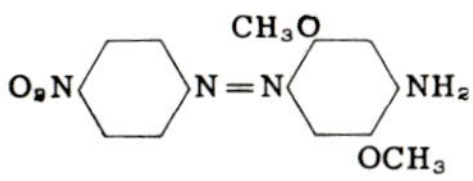


39

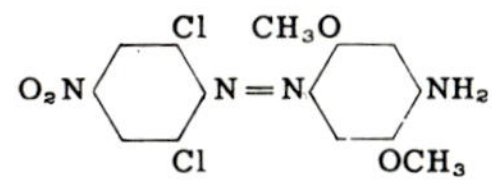


40

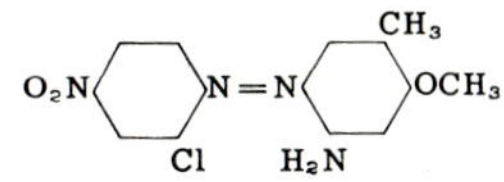


41

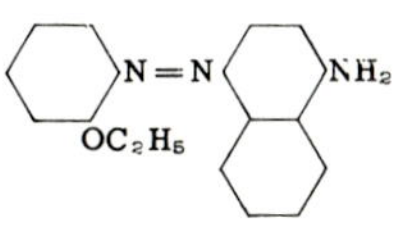


42

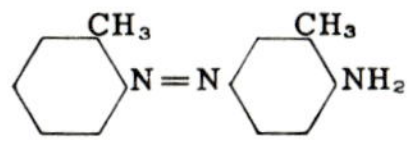


43

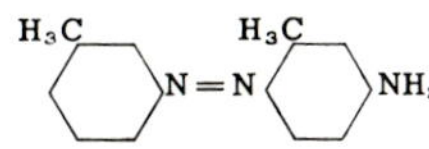


44

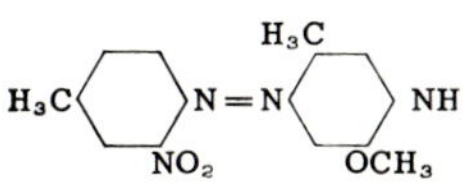


45

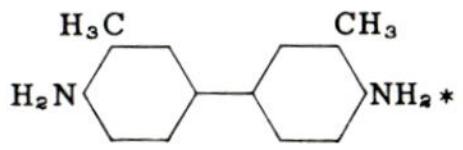


46

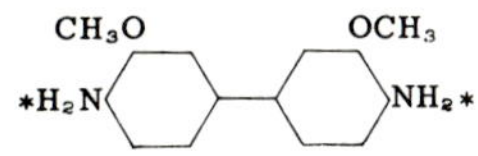


47

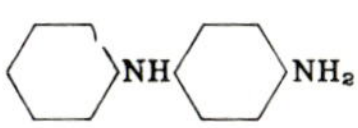


48

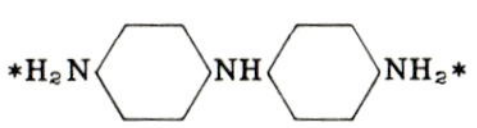


49

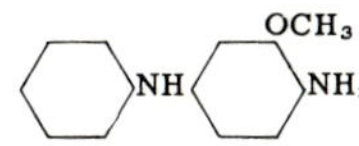


50

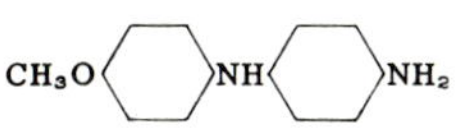


51

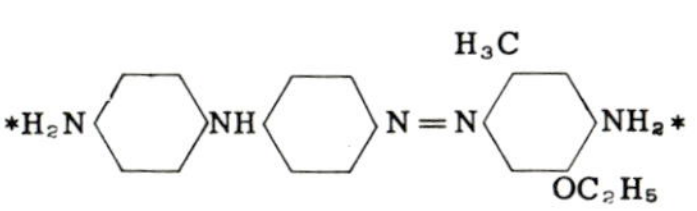


52

* See following page for key to chart 2.

### KEY TO CHART 2[a,b]

| Chart No. | C.I. Azoic Diazo No. | C.I. No. | Designation and chemical composition | |
|---|---|---|---|---|
| 1 | 44 | 37000 | Yellow GC | *o*-Chloroaniline (c) |
| 2 | 2 | 37005 | Orange GC | *m*-Chloroaniline (d) |
| 3 | 3 | 37010 | Scarlet GG | 2,5-Dichloroaniline (c) |
| 4 | — | 37020 | Brown RR | 2,6-Dichloro-*p*-phenylenediamine (c) |
| 5 | 6 | 37025 | Orange GR | *o*-Nitroaniline (c) |
| 6 | 7 | 37030 | Orange R | *m*-Nitroaniline (e) |
| 7 | 37 | 37035 | Red GG | *p*-Nitroaniline (d) |
| 8 | 9 | 37040 | Red 3GL | 4-Chloro-2-nitroaniline (c) |
| 9 | 16 | 37045 | Orange GGD | 3,5-bis(Trifluoromethyl)aniline (c) |
| 10 | 49 | 37050 | Orange RD | 2-Chloro-5-trifluoromethylaniline (e) |
| 11 | 17 | 37055 | Scarlet VD | 4-Chloro-2-trifluoromethylaniline (e) |
| 12 | 18 | 37060 | Orange LG | *o*-(Phenylsulfonyl)aniline (c) |
| 13 | 19 | 37065 | Golden Orange GR | 2-Ethylsulfonyltrifluoromethylaniline (c) |
| 14 | 26 | 37070 | Red FG | 5-Chloro-2-phenoxyaniline (c) |
| 15 | 33 | 37075 | Red FR | 5-Chloro-2-(*p*-chlorophenoxy)aniline (c) |
| 16 | 46 | 37080 | Scarlet TR | 3-Chloro-*o*-toluidine |
| 17 | 11 | 37085 | Red TR | 4-Chloro-*o*-toluidine (e) |
| 18 | 32 | 37090 | Red KB | 5-Chloro-*o*-toluidine (g) |
| 19 | — | 37095 | Red F3G | 4-Amino-5-chloro-*o*-tolunitrile |
| 20 | 34 | 37100 | Red RL | 4-Nitro-*o*-toluidine (d) |
| 21 | 12 | 37105 | Scarlet G | 5-Nitro-*o*-toluidine (e) |
| 22 | 8 | 37110 | Red GL | 2-Nitro-*p*-toluidine (e) |
| 23 | 10 | 37120 | Red RC | 5-Chloro-*o*-anisidine (c) |
| 24 | 5 | 37125 | Red B | 4-Nitro-*o*-anisidine (e) |
| 25 | 13 | 37130 | Scarlet R | 5-Nitro-*o*-anisidine (c) |
| 26 | 1 | 37135 | Bordeaux GP | 2-Nitro-*p*-anisidine (c) |

KEY TO CHART 2—*Continued*

| | | | | |
|---|---|---|---|---|
| 27 | 29 | 37140 | Red GTR | 4-Ethylsulfonyl-2-aminoanisole (c) |
| 28 | 31 | 37145 | Scarlet LG | 5-Benzenesulfonyl-*o*-anisidine (c) |
| 29 | 42 | 37150 | Red ITR | 4-(*N*,*N*-Diethylsulfamyl)-2-aminoanisole (c) |
| 30 | 14, 25, and 28 | 37151 | Red PDC | 4-(*N*-Butylsulfamyl)-2-aminoanisole |
| 31 | 24 | 37155 | Blue RR | 4′-Amino-2′,5′-dimethoxybenzanilide (c) |
| 32 | 43 | 37160 | Corinth LB | 4′-Amino-5′-chloro-*o*-benzanisidide (c) |
| 33 | 30 | | Red Violet LB (Bordeaux OL) | 4′-Amino-2′-chloro-*o*-toluidide (a) |
| 34 | 41 | 37165 | Violet B | 4′-Amino-6′-methyl-*m*-benzanisidide (c) |
| 35 | 40 | 37170 | Bordeaux BD | 4-Amino-2,5-dimethoxybenzonitrile (c) |
| 36 | 20 | 37175 | Blue BB (BBN) | 4′-Amino-2′,5′-diethoxybenzanilide (c) |
| 37 | 15 | 37180 | Red RBE | 5′-Amino-2,4-benzoxyxylidide |
| 38 | — | — | Red KL | 5-Methoxy-2-aminobenzamide |
| 39 | 38 | 37190 | Black K | 2,5-Dimethoxy-4(*p*-nitrophenylazo)aniline (c) |
| 40 | 51 | 37195 | Dark Blue R | 2,5-Dimethoxy-4-(2,6-dichloro-*p*-phenylazo)aniline |
| 41 | 21 | 37200 | Brown V | 6-(2-Chloro-4-nitrophenylazo)-4-methyl-*m*-anisidine (a) |
| 42 | 23 | 37205 | Black LB | 2-Ethoxyphenylazo-1-naphthylamine |
| 43 | 4 | 37210 | Garnet GBC | *o*-Aminoazotoluene (b) |
| 44 | 27 | 37215 | Garnet GC | *m*-Aminoazotoluene (b) |
| 45 | 39 | 37220 | Corinth V | 2-Nitro-4-methylphenylazo-5-methyl-*o*-anisidine (c) |
| 46 | — | — | Navy Blue RA | *o*-Tolidine |
| 47 | 48 | 37235 | Blue B | *o*-Dianisidine (c) |
| 48 | 22 | 37240 | Variamine Blue RT (Blue VRT) | *p*-Aminodiphenylamine (a) |
| 49 | — | 37245 | Black B | 4,4′-Diaminodiphenylamine (c) |
| 50 | 47 | 37250 | Variamine Blue FG (Blue GG) | 4-Amino-3-methoxy-diphenylamine (a) |
| 51 | 35 | 37255 | Variamine Blue B (Blue VB) | 4-Amino-4′-methoxy-diphenylamine (a) |
| 52 | 45 | 37260 | Black G | 2-Methyl-4′-*p*-aminoanilinoazobenzene (c) |

[a] These compounds are obtained as bases and/or salts. In the case of diamines an asterisk in the structural formula in Chart 2 indicates which of the amino groups are diazotized.

[b] Letters (a) through (f) indicate form of stabilization which may be employed with stable salts: (a) diazonium chloride, (b) diazonium sulfate, (c) double zinc salt, (d) fluoborate, (e) acid 1,5-naphthalene disulfonate, (f) cobaltous chloride complex, and (g) *N*-acetylsulfanilate.

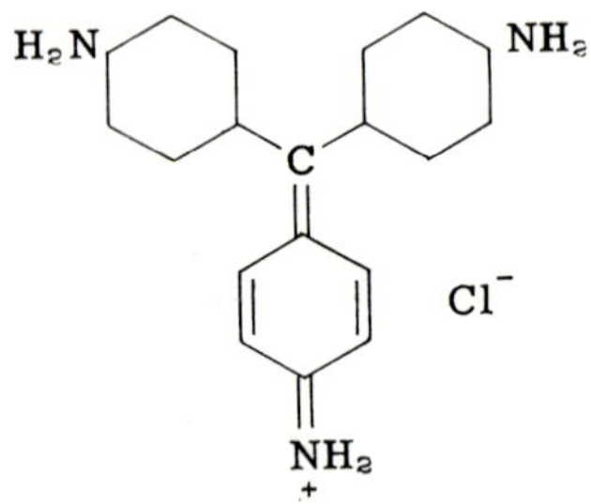


Structure of pararosaniline

(IV)

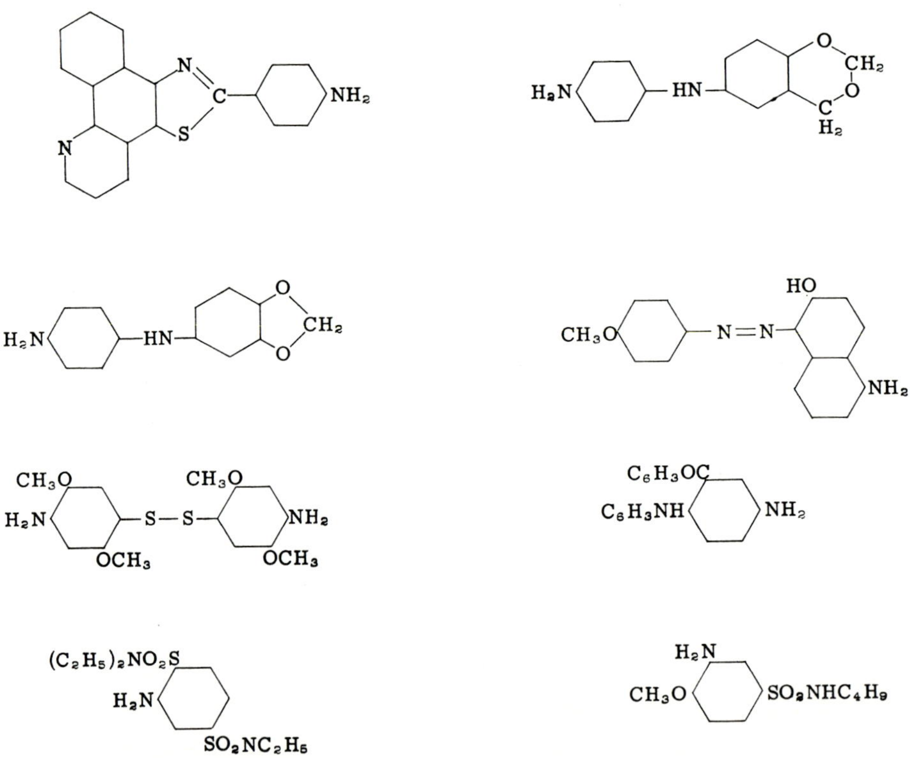


CHART 3. Structure of more complex diazotizable bases.

dom. Pearse (1960, p. 407) and Gössner (1958) have empirically tested a number of diazonium salts with several hydrolase techniques and noted whether or not the dye was diffuse or particulate, and if non-specific staining due to adsorption of diazo compound occurred. Although such observations are of some practical use, they are limited by major differences which occur when different types of fixation are employed. For example, a particulate or granular distribution of dye frequently occurs when frozen sections are used but not with their paraffin embedded counterpart. Although there have been certain generalizations relating to the coupling of phenols and amines at different pH levels, optimal coupling under histochemical conditions must take into account a number of factors. In general, phenols couple more readily than amines, and naphthalene more readily than benzene derivatives. Diazonium salts exhibit varying degrees of instability and should be stored under refrigeration. Solutions of diazonium salts should be used immediately, and their storage is not recommended. The stability of diazonium solutions varies. In general, it is not desirable to subject them to extremes of alkalinity or acidity. For example, it may be advisable to run an azo-dye method for phosphatase at pH 8.3–8.5 instead of pH 9.0, since the stability of diazonium salts is poor under the latter condition. Thus, the coupling reaction may be considered as a race between azo-dye formation and decomposition of the diazo compound (to the stable isodiazotate which does not couple). *According to Hees (1949b), a pH value of aproximately 8.2 should be regarded as the limit of stability of diazonium salts in the alkaline range.*

When longer incubations are necessary, cooling of the solution or changing the solution at hourly intervals should be considered. Diazo compounds can be divided into three groups: those exhibiting imperceptible decomposition at room temperature, those exhibiting an intermediate rate of decomposition, and those which readily decompose. Diazonium chlorides containing electropositive groups are more sensitive to elevated temperatures than those containing electronegative groups (Snow, 1932). The presence of electronegative groups is also known to enhance the stability of diazo solutions. The decomposition rate of various diazonium salts is indicated in the following order of decreasing stability:

*p*-chloroaniline, *m*-nitroaniline, *m*-4-nitroxylidine, *o*-nitroaniline, *o*-anisidine, *p*-nitroaniline, *p*-bromoaniline, *o*-phenetidine, *p*-iodoaniline, 5-aminosalicylic acid, *p*-toluidine, aminoazobenzene, *m*-aminophenol, *p*-phenylenediamine, picramic acid, sulfanilic acid, *p*-aminobenzoic acid, α-naphthylamine, *p*-anisidine, β-naphthylamine, *m*-4-xylidine, anthranilic acid, aminoazotoluene, *m*-aminobenzoic acid, *o*-4-xylidine, *o*-3-

xylidine, aniline, *o*-toluidine, *m*-toluidine, *m*-anisidine, *p*-xylidine, and *p*-phenylenediamine (tetrazotized).

Of these almost all *para* substituted compounds constitute the most stable group, while singly substituted *ortho* and *meta* compounds fall into either the most stable or least stable group depending upon the nature of the substituents. In the case of the nitro substituent, its presence in either *ortho*, *meta*, or *para* positions does not produce appreciable differences in stability. On the other hand, the location of methyl or methoxy groups distinctly alters stability. For example, an *o*-methoxy diazo compound may be much more stable than its *meta* or *para* counterpart. When several substituents are present, the effect is either additive or the influence of one group overshadows the others. Thus xylidine diazos are more stable than toluidines.

The decomposition of solutions of diazonium salts such as benzenediazonium chloride is a first-order reaction which follows the ordinary law for a unimolecular reaction,

$$K = \frac{1}{t} \log \frac{A}{A - X}$$

Crossley *et al.* (1940) confirmed some of the aforementioned findings of Snow (1932) with reference to the stabilities of substituted benzenediazonium chlorides. Nitro, halogen, and carboxyl substituents as well as *p*-methoxy or *p*-methyl groups enhanced stability. In contradistinction to the findings of Snow, stability was decreased by an *o*-methoxy substituent.

Practically no information is available on coupling rates as they occur under histochemical conditions. Gomori (1952, p. 170) reported that the azo-coupling reaction (with older naphtholic substrates) is sluggish below pH 6, and that this condition can give rise to stained sites lacking in detail and surrounded by a dye halo. He also pointed out (1952, p. 169) that "any naphthol will not couple indiscriminately with any diazonium salt. Each new combination must be tried out in the test tube before its application to histochemistry can be considered." Thus, sweeping generalizations on rapidity of coupling rates should not be based upon isolated experiments related to azo-coupling of enterochromaffin or of special naphtholic components not directly applicable as substrates.

Defendi and Pearse (1955) employed a flowmeter device of the type originally designed by Hartridge and Roughton (1923) in order to measure coupling rates *in vitro*. The system consisted of two bottles connected, with a minimum of dead space, to a long capillary tube and then to a collecting bottle. After mixing naphthol solution and diazonium salts solution, their rate of flow through the capillary tube was controlled by negative pressure. The amount of final reaction product (FRP), the azo-dye, was measured photometrically at various points along the capillary tube and curves were made in which the percentage of dye produced formed the ordinate and the reaction time the abscissa. Although several different naphthols were employed, the diazo component was *o*-dianisidine (Diazo Blue B). The more rapid coupling halogen or nitro containing diazonium salts were not studied. Defendi and Pearse found that several Brenthols (I.C.I. equivalent of AS naphthols) coupled very slowly with Brentamine Fast Blue B. This simply confirms the I.C.I. findings listed in Table 2, and it should not be inferred that all naphthol AS derivatives are slow couplers with all diazonium salts. These authors feel that the insoluble nature of the primary reaction product (PRP) is of greater significance to the accuracy of localization than is the rapidity of coupling. The kinetic evaluation of the coupling reactions was based upon the earlier work of Conant and Peterson (1930) and Wistar and Bartlett (1941) who indicated that coupling is a bimolecular reaction and that the velocity constant is a simple function of the hydrogen ion concentration:

$$\log k = \log k_0 + \text{pH}$$

Defendi and Pearse (1955) found that the usual straight-line relationship between the bimolecular velocity constant of coupling and pH became invalid at certain "critical points." The "critical point" for $\beta$-naphthol was found to be between pH 7.4 and 7.0, while that for $\alpha$-naphthol lay between pH 5.3 and 5.0. These observations tend to support the view that $\alpha$-naphthol substrates are more satisfactory for use in an acid pH range than corresponding $\beta$-naphthol derivatives. More recently Nachlas *et al.* (1959) elaborated on the Pearse technique to study coupling rates of a large number of diazonium salts with naphthol- and naphthylaminesulfonic acids. Unfortunately, these sulfonic acid derivatives are never employed as the basis of histochemical reagents, and were used in the *in vitro* study only because of their aqueous solubility. Although the aforementioned experiments offer interesting data, they do not elucidate the submicroscopic nature of the "capture reaction" and subsequent dye formation.

From a purely *in vitro* standpoint, the kinetics of coupling reactions involving most diazonium salts have been carefully investigated by dyestuff chemists (Saunders, 1949; Hees, 1949a, b, 1950; Pütter, 1951; Zollinger, 1952). The experimental techniques of kinetic measurement involve the determination of unreacted diazo component following a coupling reaction, the colorimetric determination of azo-dye following a coupling reaction, and changes in concentration of diazo component as determined by polarography. The latter technique is accurate, and rapid coupling reactions are relatively easy to follow (Elofson *et al.*, 1950). Polarographic techniques have not as yet been applied in the study of histochemical reagents. The work of Zollinger (1952) offers an excellent review of the kinetics of coupling, including quantitative data on the reactivity of various diazo and coupling components.

The formation of insoluble azo-dyes from coupling components and diazonium salts in solution is a reaction which runs according to stoichiometric laws. Coupling speed is inhibited by both acids and alkalies. The speed of dye formation is inversely proportional to the hydrogen ion or hydroxyl ion concentration of the reaction medium. Thus, the theoretical pH optimum is at neutrality. In practice, however, all reactions obviously cannot be run at pH 7 since the pH optimum of enzymes covers a fairly wide range. Table 2 adapted from data of the Imperial Chemical Industries (Anonymous, 1951) indicates the relative stabilities and coupling rates of many of the diazonium salts employed in histochemical reactions. It is interesting to note that the commonly used Fast Blue B salt (diazotized *o*-dianisidine) is classified as a "very slow" coupler, while Fast Red RC (diazotized 4-chloro-2-aminoanisole) and Fast Garnet GBC salts are classified as "rapid" and "very rapid" couplers, respectively.

Coupling energy may be defined as the optimal reactive ability of a diazo compound which is combined with a naphthol at a certain pH. Coupling speed or rate is defined as time in which the combination or coupling of a diazo salt with a naphthol occurs. The presence of negative substituents is known to increase coupling energy (and rate of dye formation), while positive substituents reduce coupling energy. Hees (1949a) has classified "negative" substituents (Table 3), and "positive" substituents on diazonium salts (Table 4) according to their influence upon coupling energy. For example, the electronegative groupings in row A of Table 3 increase the speed of dye formation more than those in row C. The groupings in row D and E of Table 4 exert more of a retarding effect upon coupling than do those in rows A and B. The higher the coupling energy, the more acid the pH of the medium may be and yet assure rapid azo-dye formation. As indicated by Tables 3 and 4, the fast couplers are those which possess strongly electronegative substituents, while the very slow couplers are those which are distinctly electropositive in nature. Aromatic derivatives characterized by high coupling energy (Table 3, row A), and which will couple efficiently with naphthols in a more strongly acid medium, are of the following types:

A. Dinitroaniline, nitroaniline, nitrochloroaniline, chloroaniline, dichloroaniline, nitrotoluidine, nitroanisidine, dichlorotoluidine, dichloroanisidine, trifluoromethylaniline, and aminobenzonitrile.

TABLE 2

COUPLING RATE AND STABILITY OF VARIOUS COMMERCIAL DIAZOTATES

| Azoic Diazo No. | Colour Index No. | Brentamine | Per-cent of amine | Coupling rate | Solution stability |
|---|---|---|---|---|---|
| 38 | 37190 | Fast Black K (salt) | — | Slow | Poor |
| 48 | 37235 | Fast Blue B (salt) | 20 | Very slow | Good |
| 20 | 37175 | Fast Blue BB (salt) | 40 | Slow (Ac)[a] | Good |
| 35 | 37255 | Fast Blue VB (salt) | — | Very slow (N)[b] | Good |
| 1 | 37135 | Fast Bordeaux GP (salt) | 20 | Rapid | Fair |
| 39 | 37220 | Fast Corinth V (salt) | — | Fairly rapid (Ac) | Very good |
| 4 | 37210 | Fast Garnet GBC (salt) | 20 | Very rapid | Fair |
| 2 | 37005 | Fast Orange GC (salt) | 20 | Rapid | Fair |
| 6 | 37025 | Fast Orange GR (salt) | 20 | Very rapid | Very good |
| 36 | 37275 | Fast Red AL (salt) | — | Rapid | Good |
| 5 | 37125 | Fast Red B (salt) | 20 | Very rapid | Very good |
| 37 | 37035 | Fast Red GG (salt) | — | Very rapid | Good |
| 8 | 37110 | Fast Red GL (salt) | 20 | Rapid | Fair |
| 9 | 37040 | Fast Red 3 GL (salt) | 40 | Rapid | Fair |
| 32 | 37090 | Fast Red KB (base only) | — | Rapid (Ac) | Fair |
| 42 | 37150 | Fast Red LTR (salt) | 40 | Fairly rapid | Good |
| 10 | 37120 | Fast Red RC (salt) | 20 | Rapid (Ac) | Very good |
| 34 | 37100 | Fast Red RL (salt) | 20 | Very rapid | Good |
| 46 | 37080 | Fast Red TR (salt) | 20 | Rapid (Ac) | Very good |
| 12 | 37105 | Fast Scarlet G (salt) | 20 | Very rapid | Fair |
| 3 | 37010 | Fast Scarlet GG (salt) | 25 | Very rapid | Very good |
| 13 | 37130 | Fast Scarlet R (salt) | 25 | Very rapid | Fair |
| 11 | 37085 | Fast Scarlet TR (base only) | — | Rapid (Ac) | Poor |
| 41 | 37165 | Fast Violet B (salt) | 20 | Slow (Ac) | Good |
| 44 | 37000 | Fast Yellow GC (salt) | 20 | Rapid | Good |

[a] (Ac), sensitive to excess acetic acid.
[b] (N), neutrality required.

Those of somewhat less coupling energy (Table 3, row B) are:

B. Chlorotoluidine, chloroanisidine, chloraminodiphenyl ether, alkylsulfone aniline and analogues, arylsulfone aniline and analogues, aralkylsulfone aniline and analogues, alkylsulfonamido aniline and analogues.

The following types (Table 3, row C) exemplify the least energetic of the negatively substituted diazos:

C. Carboxylic acid alkyl ester arylamines, carboxylic acid aryl ester arylamines, acetylamino arylamines, and benzoylamino arylamines.

TABLE 3

EFFECT OF NEGATIVE SUBSTITUENTS UPON COUPLING RATE (HEES, 1949a)

| Negative substituents (effect decreasing from left to right and from A to C) |
|---|
| A. —$NO_2$, —Cl, —Br, —$CF_3$, —C≡N |
| B. —$SO_2$—Alkyl, —$SO_2$—Aryl, —$SO_2$—NH—Alkyl, —$SO_2$—NH—Aryl |
| C. —CO—Alkyl, —CO—Aryl, —COO—Alkyl, —COO—Aryl |

TABLE 4

EFFECT OF POSITIVE SUBSTITUENTS ON COUPLING ENERGY (HEES, 1949a)

| Positive substituents[a] | Formula |
|---|---|
| A. Alkyls and aralkyls | $CH_3$, $C_2H_5$, $CH_2$—$C_6H_5$ |
| B. Alkoxy groups | $OCH_3$, $OC_2H_5$ |
| C. Aryloxy groups | $OC_6H_5$ |
| D. Alkyl amino groups | NH—Alkyl |
| E. Aryl amino groups | NH—$C_6H_5$ |

[a] Effect increasing from A to E.

With reference to Table 4, compounds exhibiting pronounced coupling sluggishness are exemplified by:

D. Aminodiphenyl, aminodiphenylmethane, and aminodiphenylamine derivatives.

These diazo compounds couple satisfactorily under histochemical conditions in an alkaline medium.

Although electronegative groupings increase coupling energy when incorporated into the diazonium component, they have a converse effect when present in the naphthol. On the other hand, the presence of electropositive substituents (e.g. methyl, methoxy, ethoxy) in the naphthol tend to accelerate coupling. According to Hees (1949a), the influence of naphtholic substituents upon coupling energy is less marked and is almost always superseded by the effect of the structure of the diazotized amine. This is confirmed by the fact that attempts to increase coupling rates with reference to most industrial dyeing methods have been with reference to altering the structure of the diazo component and not the naphthol.

The introduction of an electropositive substituent into a negatively substituted amine always results in a decline in coupling speed. The magnitude of the decline depends upon the positivizing strength of the new substituent and its positional relationship to the diazotizable amino group. For example, the addition of an *o*-methyl group to *m*-nitroaniline results in an appreciable decrease in coupling energy and speed. Benzoylamino diazo compounds are of particular interest because of their frequent application in histochemical techniques (e.g. Blue BB, Blue RR, and Red Violet LB salts). These compounds contain a negativizing influence, the benzoyl grouping. When the benzoylamino group is in a position *para* to the diazotized amino group, the compound exhibits good coupling speed

TABLE 5A

Diazonium Salts of High Coupling Energy at pH 4.5[a] (Hees, 1949b)

| Coupling speed | | |
|---|---|---|
| High | Medium | Low |
| Orange R | Yellow GC | Golden Orange GR |
| Orange GR | Orange GC | Orange RD |
| Scarlet GGS | Orange LG | Scarlet VD |
| Red GG | Orange GGD | Red B |
| Red 3GL | Scarlet G | Bordeaux GP |
| | Red RL | Garnet GC |
| | Red GL | Garnet GBC |
| | Red AL | Brown V |
| | Brown RR | Dark Blue R |
| | | Black K |

[a] All of these would be expected to be rapid couplers at pH 5.5–6.0.

TABLE 5B

Diazonium Salts of Medium Coupling Energy at pH 5.5–6.5 (Hees, 1949b)

| Coupling speed | | |
|---|---|---|
| High | Medium | Low |
| Scarlet TR | Red FG | Scarlet LG |
| | Red KB | Red ITR |
| | RED FR | |
| | Red TR | |
| | Red RC | |
| | Corinth V | |

TABLE 5C

Diazonium Salts of Low Coupling Energy at pH 6.0–7.0 (Hees, 1949b)

| Coupling speed | | |
|---|---|---|
| High | Medium | Low |
| Corinth LB | Violet B | Red RBE |
| | Blue RR | |
| | Blue BB | |
| | Black G | |

TABLE 5D

DIAZONIUM SALTS OF VERY LOW COUPLING ENERGY EVEN AT pH 7.0–8.2 (HEES, 1949b)

| Coupling speed | |
|---|---|
| Medium | Low |
| Variamine Blue B | Variamine Blue FG |
| Variamine Blue RT | Diazo Blue B |

(e.g. Red Violet LB salt). This is enhanced by the presence of halogen or similar substituents in the same nucleus as the diazotized amino group. On the other hand, if the benzoylamino group is *meta* to the amino group the coupling speed is low.

Arylamines containing the strongly electropositive phenylamine group are slow couplers. These are exemplified by 4-aminodiphenylamine and 4-amino-4′-methoxydiphenylamine. According to Hees (1949a), if the methoxy group is *ortho* to the amine, there is a perceptible decline in coupling speed. Hees states that of all commercial diazonium salts studied Blue B (diazotized *o*-dianisidine) exhibits the lowest coupling speed. On the other hand, if aminodiphenylamine derivatives are substituted at the secondary amino group by a nitroso or sulfonic acid group, the coupling energy and speed is increased. With reference to diazotized aminoazo arylamines, the arylazo group has a negativizing influence and the coupling rate is reasonably rapid.

Tables 5A through 5D are classifications of commercial diazonium salts according to coupling energy and coupling speed with naphthol AS derivatives at an optimal pH (Hees, 1949b). These tables will assist in the determination of appropriate diazonium salts to be used in various pH ranges which correspond to both the optima of the enzyme system and the azo-coupling rate.

the same dyes in a highly aggregated state. The shape and configuration of the crystal is also known to affect the color values of pigments (Adam, 1938). A dyestuff may sometimes exhibit different colors on different crystal faces, depending upon the rate of formation of the crystals and other factors. Chelation is also known to produce color changes in azo-dyes. These changes include darkening as well as dulling of the shade.

The spectrophotometric and other physical characteristics of routine histological dyes have received fairly careful consideration in the literature (Conn, 1953; Baker, 1958, 1960). This is in no small part due to the fact that most histological stains were developed as textile dyestuffs, and were thus rather carefully studied by the dyestuff industry. On the other hand, dyes resulting from histochemical reactions have not, for the most part, been so characterized. Comparison of infrared and ultraviolet characteristics of substrates and resulting dyes may be of value from both fundamental and applied standpoints. Application of spectroscopy may also be of value in elucidating the relationship between biological activity and structure (Sadler, 1960).

Luminescence occurs when energy absorbed by a compound is emitted as light. Fluorescence is a type of luminescence which results when absorbed light energy (visible or invisible) is re-emitted as visible light. The fluorescence is in the form of a longer wavelength than that of the absorbed radiation, and persists only as long as the source of excitation is present. The exciting rays have wavelengths of less than 420 $m\mu$ and extend down to the x-ray spectrum. Phosphorescence, by contrast, occurs when luminescence is still produced after cessation of irradiation. The fluorescent state is believed to be produced by the transition of a molecule from a state of excitation to a lower energy level. The duration of this is less than one second, after which the molecule returns to its original physical state. Certain groupings, such as OH, $OCH_3$, $CH_2$, and $NH_2$, tend to enhance the fluorescence of a compound, while other groupings (e.g. $CH_3$, CO, and COOH) weaken or completely inhibit fluorescence. In certain instances, the fluorescence phenomenon is not characteristic of the substance itself but is related to small amounts of impurities. Of interest is the observation that there is an approximate mirror image relationship between the fluorescence spectra of some polycyclic hydrocarbons and certain wavelengths of their ultraviolet absorption spectra. The fluorescent labeled antibody procedure has been used in several instances to demonstrate enzymes in tissue sections. In addition, techniques for the demonstration of hydrolytic enzymes may be modified so that one of the products of hydrolysis is a fluorochrome (Chapter 5). One of the fluorescence techniques for the demonstration of hydrolytic enzymes depends upon the fact that the enzymatically released naphthol is fluorescent. Since a large number of naphthols are highly fluorescent under the influence of ultraviolet light, this approach warrants further exploration.

A recent advance as well as innovation in analytical techniques has been the development of the electron probe x-ray microanalyzer (Tousimis and Adler, 1963; Tousimis, 1962). This instrument employs a special electron optical system to focus an extremely fine beam of electrons (1 $\mu$ or less in diameter) on a tissue section. The elements present within the focus of the beam emit x-rays when excited by the high energy electrons (1–50 kv.) of the electron probe. These x-rays are analyzed either by an x-ray spectrograph or electronically. Exceedingly small amounts ($10^{-15}$ g.) of a number of elements may be detected and quantitatively measured. At present, the method has been primarily used to delineate the distribution of Cu, Fe, Ag, and Ca in tissue sections of normal and pathological tissues. The technique should also be applicable to the detection of chelates including azo-dye metal complexes produced under histochemical conditions. Hall (1961) has reviewed the biological potential of x-ray fluorescence and related methods as adapted to the study of tissue sections. These as well as analogous micro-methods employing x-ray diffraction, ultraviolet, and infrared techniques may also be of value in exploring the nature of histochemical dyes in sections. Another potential tool is the optical *maser* (microwave amplification by stimulated emission of radiation) (Javan, 1960, p. 564; Dacey, 1962). *Maser* emission is coherent and can be focused into an area of dimensions comparable to wavelengths of light. Thus dyestuff localized to portions of a single cell could be selectively studied.

## II. AZO-DYES

Azo-dyes are classified as a monoazo, disazo, trisazo, and tetrakisazo depending upon the number of azo groupings (—N=N—) present. They are generally prepared by reacting a diazonium salt with another compound referred to as a coupling component. The latter group consists of the following:

1. Compounds with phenolic hydroxyl groups such as naphthols and analogues.
2. Primary and some secondary arylamines.
3. Aromatic compounds which possess a reactive methylene group capable of enolization. These include derivatives of acetoacetic ester and arylamines.
4. Certain heterocycles such as pyrrole, indole, and imidazole.
5. Compounds such as *o*-alkylphenols and naphthols, and *N*-acylamines. These include phenol and naphthol ethyl ethers and acetylnaphthylamines.
6. Aliphatic enols such as ethyl acetoacetate, and also compounds containing an alicyclic carbon such as ascorbic acid, or dimethyldihydroresorcinol (dimedone).

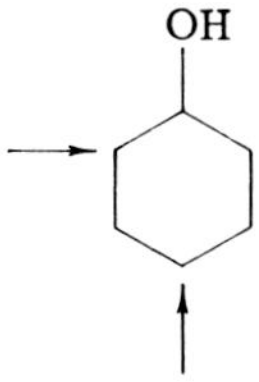

Phenol

1

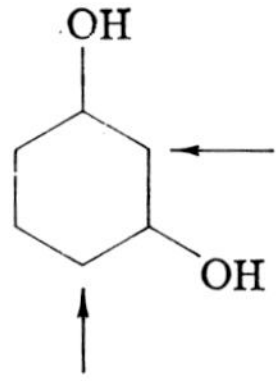

Resorcinol

2

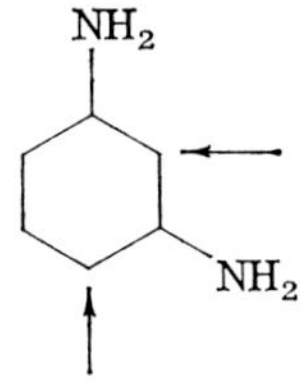

*m*-Phenylenediamine

3

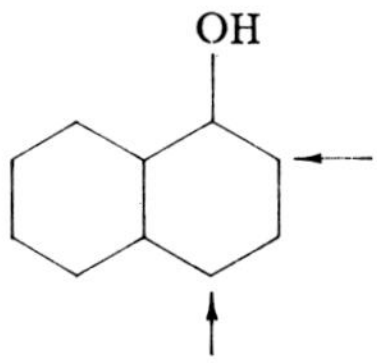

1-Naphthol

4

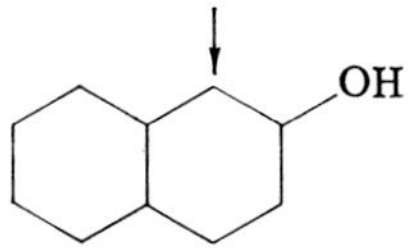

2-Naphthol

5

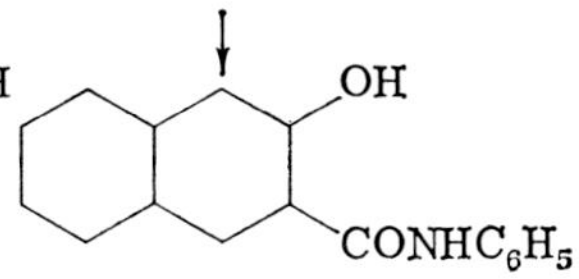

3-Hydroxy-2-naphthoic acid anilide

6

Acetoacetanilide

7

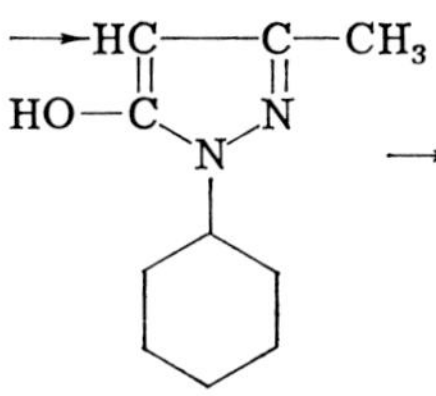

3-Methyl-1-phenyl-5-pyrazolone (enol)

8

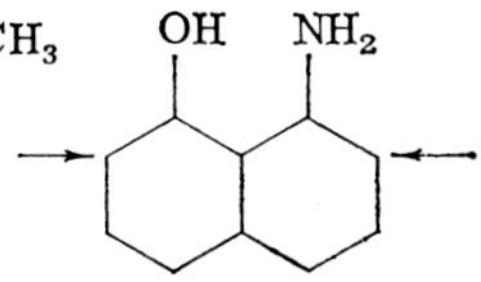

8-Amino-1-naphthol

9

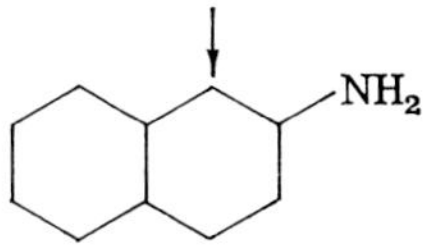

2-Naphthylamine

10

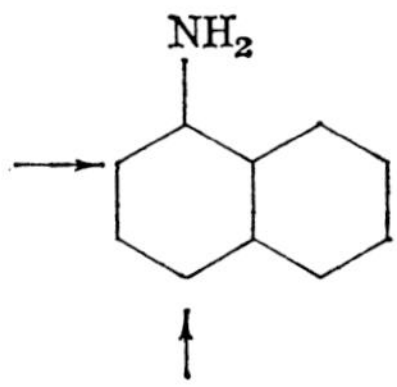

1-Naphthylamine

11

CHART 1. Coupling positions of arylamines and phenolic compounds.

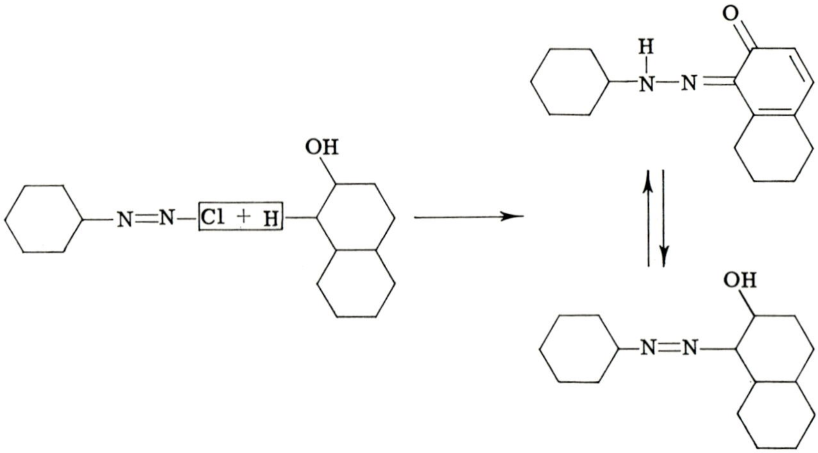

(I) Reaction between benzenediazonium chloride and β-naphthol

Azo-dyes are also formed by the alkaline reduction of aromatic nitro compounds, by the condensation of aromatic nitroso compounds with arylamines, and by the oxidation of primary arylamines. With reference to the coupling activity of the aforementioned components, phenols couple more readily than arylamines, and members of the naphthalene group couple more readily than those of the benzene series. Nitro, halogen, carboxyl, sulfo, carbonyl, and other negative substituents in the coupling component tend to retard the reaction. On the other hand, the presence of electropositive substituents tends to accelerate coupling. However, it should be pointed out that the effect of positive substituents is minimal and that when an increase in coupling rate is desired, it is more feasible to use an appropriate diazonium salt which contains electronegative substituents. The coupling positions in a number of agents are shown in Chart 1. In the benzene series, coupling usually occurs *para* to the hydroxyl or amino group. If the *para* position already is occupied the *ortho* position is used.

---

CHART 2. Structures of a number of azo-dyes derived from hydroxyl, amino, and methylene compounds.

1. Diazotized aniline + phenol (C.I. 11800)
2. Diazotized aniline + resorcinol (C.I. 11920)
3. Diazotized aniline + aniline (C.I. 11000)
4. Diazotized aniline + 1-naphthylamine (C.I. 11350)
5. Diazotized aniline + 2-naphthylamine (C.I. 11380)
6. Diazotized *p*-nitroaniline + acetoacetanilide (C.I. 11665)
7. Diazotized aniline + 1-naphthol (C.I. 12000)
8. Diazotized aniline + 2-naphthol (C.I. 12055)
9. Diazotized toluidine + 3-hydroxy-2-naphthanilide
10. Diazotized aniline + 3-methyl-1-phenyl-5-pyrazolone (C.I. 12700)

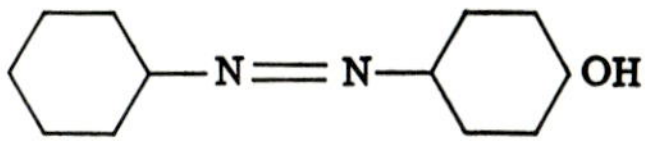

1

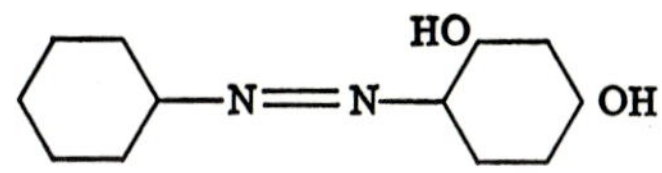

2

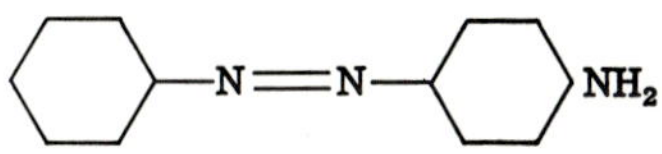

3

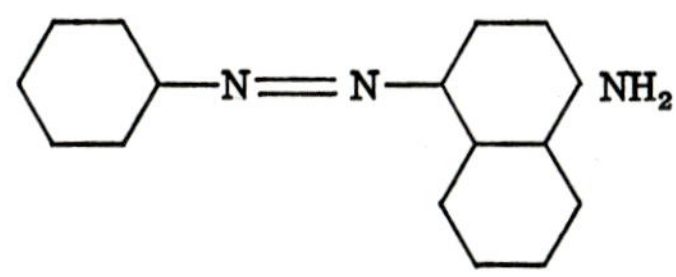

4

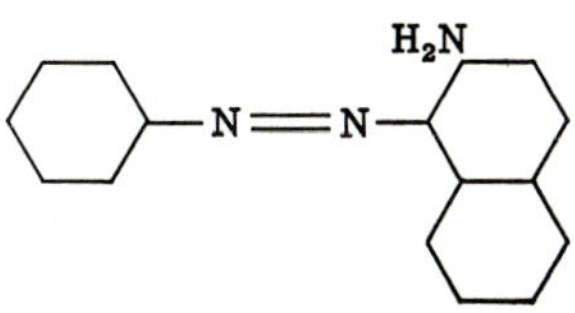

5

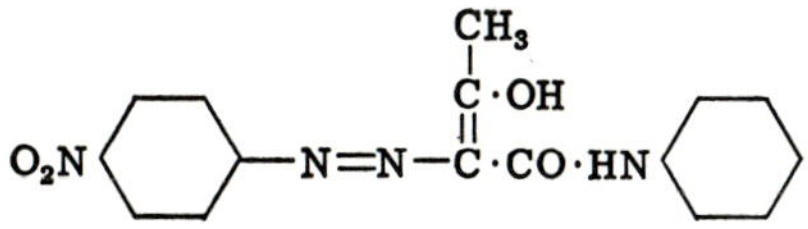

6

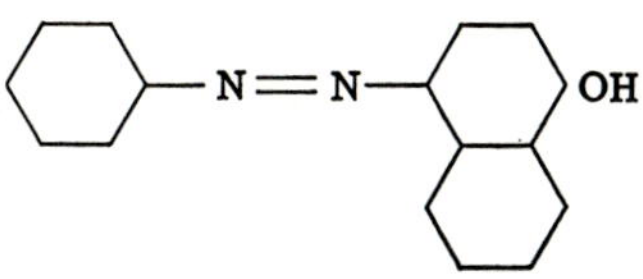

7

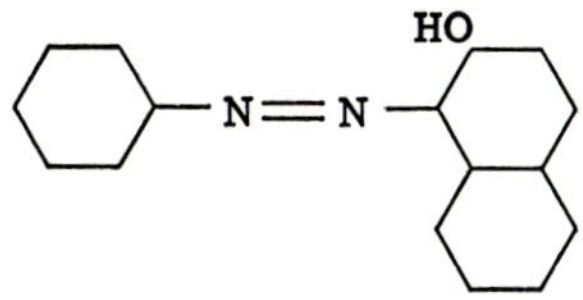

8

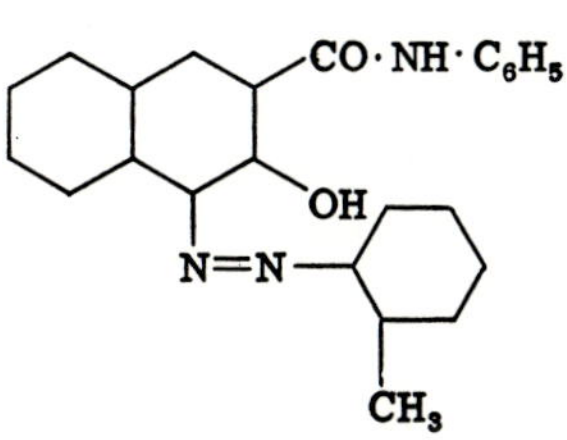

9

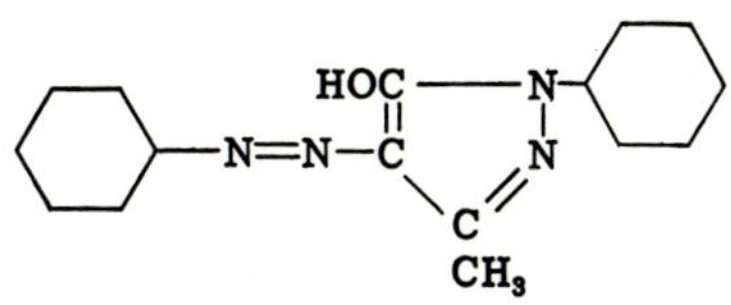

10

CHART 2

In the case of amino-naphthols where two coupling positions are available, the point of coupling is determined by the pH of the medium. Thus, coupling occurs adjacent to the amino group in an acid medium and adjacent to the hydroxyl group in an alkaline medium. The reaction between benzenediazonium chloride and β-naphthol may be indicated in (I). The resulting 4-benzeneazo-1-naphthol exists in the tautomeric forms shown, with the equilibrium dependent upon the solvents. In an aqueous alkaline medium the predominant tautomer is the azo form. Chart 2 illustrates the structural formulae of azo-dyes resulting from coupling of diazonium components with several naphthols and naphthylamines, while Table 3 characterizes some histochemically derived azo-dyes.

The relationship between the chemical constitution and color of an azo-dye is important from the standpoint of applicability of commercially available substrates as well as of substrate design. Although azo-dyes derived from phenolic derivatives are lighter in shade than their naphtholic counterpart, it does not necessarily follow that increasing the complexity of the molecule always deepens the color of the azo-dye. In general, certain azo-naphthols are known to give red dyes, while the analogous azo-phenols give yellow or orange dyes (Sircar, 1916). According to Sircar (1916) the longer the chain of alternating double and single bonds between the part of the molecule containing the auxochrome and the azo linkage, the deeper the color of the dye. A study of the absorption spectra of a number of dyes revealed that as the length of the aforementioned chain is increased, so is the peak of absorption band gradually shifted toward the red end of the spectrum. Other studies dealing with the relationship between the absorption spectra and chemical constitution of azo-dyes have been described by Brode (1929) and Funkhouser and Brode (1934).

There have been but few applications of azo-dyes per se as reagents in enzyme histochemistry. Loveless and Danielli (1949) obtained a phosphatase substrate by phosphorylating 4-(*p*-nitrobenzeneazo)-1-naphthol (Chapter 5). An analogous procedure was employed by Friedenwald and Becker (1948) who synthesized a hydroxyphenylazo naphthyl glucuronide. When azo-dyes are oxidized under certain conditions azoxy compounds are formed, and these may in turn be reduced to the original azo-dye. There has not as yet been any report on the application of azoxy compounds to demonstrate enzymatic reducing sites in tissues. Many azo-dyes may be reduced to form hydrazo compounds, which can, in turn, be oxidized to re-form the azo-dye.

$$RNHNHR \xrightarrow{(O)} RN{=}NR + H_2O$$

Preliminary experiments in this laboratory have shown that azo-dyes derived from naphthol AS and derivatives are readily reduced in the

TABLE 3

MOLECULAR CONSTITUTION AND CHARACTERISTICS OF SOME HISTOCHEMICALLY DERIVED AZO-DYES

| Compound | Reference | Appearance | Melting point (° C.) |
|---|---|---|---|
| 1. 4-Benzeneazo-1-naphthol, $C_{16}H_{12}ON_2$ (tautomeric with monophenylhydrazone of 1,4-naphthoquinone) | Conant *et al.*, 1941 | dark needles with a metallic luster | 208 (dec.) |
| 2. 4-(*m*-Nitrobenzeneazo)-1-naphthol, $C_{16}H_{11}O_3N_3$ | Van Alphen and Drost, 1949 | dark red needles | 242–243 |
| 3. 4-(*o*-Aminoazotoluene)-1-naphthol, $C_{24}H_{20}ON_4$ | May and Hunt, 1928 | dark red fine crystals | 153–156 |
| 4. 1-Benzeneazo-2-naphthol, $C_{16}H_{12}ON_2$ (tautomeric with 1-phenylhydrazone of 1,2-naphthoquinone) | Hauser and Breslow, 1941 | reddish leaflets with golden luster | 134 |
| 5. 1-(*o*-Aminoazotoluene)-2-naphthol (Sudan IV), $C_{24}H_{20}ON_4$ | Ruggli and Courtin, 1932 | dark red needles | 186 |
| 6. 4-Benzeneazo-3-hydroxy-2-naphthanilide, $C_{23}H_{17}O_2N_3$ | Brass and Sommer, 1928 | red needles | 237 |
| 7. 4-(4-Benzoylamino-2,5-diethoxybenzeneazo)-3-hydroxy-2-naphthanilide, $C_{34}H_{30}O_5N_4$ | Gränacher *et al.*, 1941 | blue-black crystals | 294–295 |
| 8. 4-Benzeneazo-1-naphthylamine, $C_{16}H_{13}N_3$ | Griess, 1866 | red needles | 123 |
| 9. 1-Benzeneazo-2-naphthylamine, $C_{16}H_{13}N$ | Hodgson and Foster, 1942 | red leaflets | 103 |

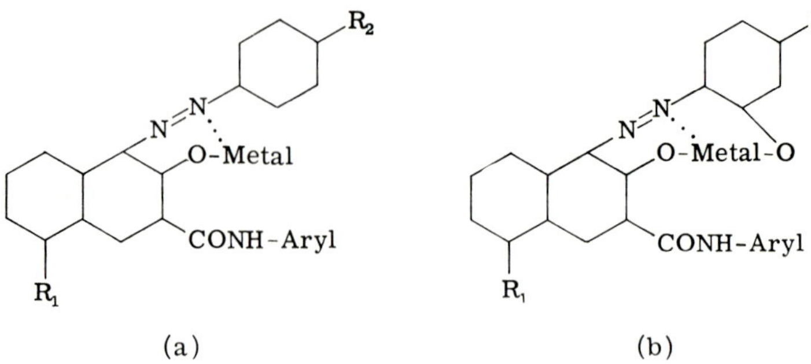


$R_1$ = AS-naphthol

$R_2$ = diazonium component

(II) Potential metal complexes of *o*-hydroxy (a) and *o,o'*-dihydroxy (b)

presence of sodium amalgam but, unfortunately, they are also readily oxidized back to the original azo compound by the presence of atmospheric oxygen. It would be interesting to prepare relatively stable hydrazo compounds which would form dyes only under the influence of certain enzyme systems.

Azo-dyes may contain groups which are capable of coordinating with metals to form a stable complex. The presence of amino, hydroxyl, methoxy, or other groups in an *o*-position to the azo grouping permits intramolecular chelation to occur. Potential metal complexes of *o*-hydroxy and *o,o'*-dihydroxy naphthol AS azo-dyes are shown in (II). With reference to the *o,o'*-dihydroxy azo-dyes, the diazonium component is generally a diazo oxide. Because of the inherently slow coupling rate of these compounds, their use is limited to the post-coupling procedures. Intermolecular complexes may also be produced such as chelates of azo-dyes derived from β-naphthylamine and β-naphthol (III and IV). In general, copper, nickel, and cobalt intramolecular chelates may be prepared from *o*-hydroxy azo compounds. For example, an inner complex of this type may be prepared

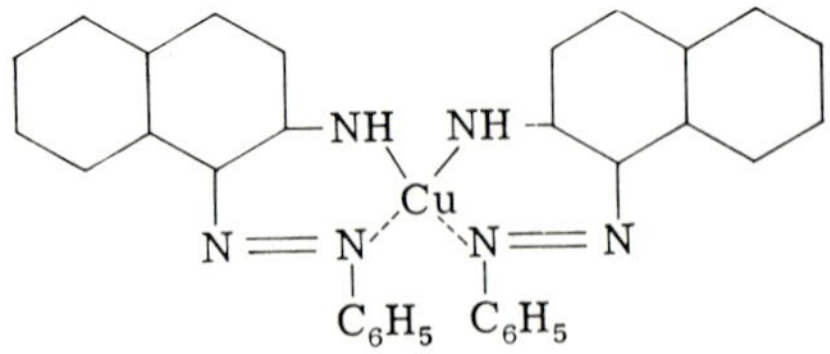


(III) Copper chelate of azo-dye from β-naphthylamine

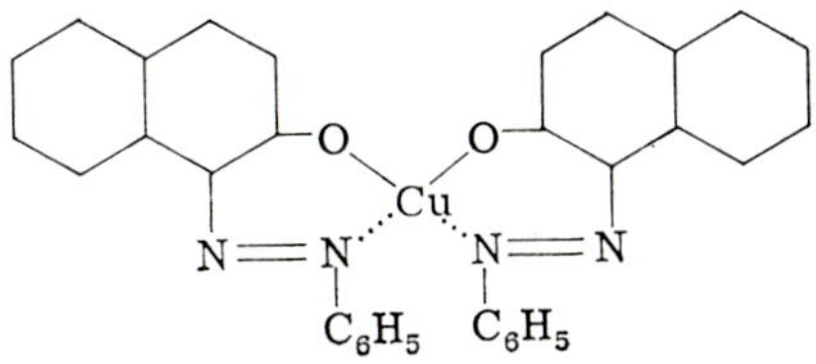

(IV) Copper chelate of azo-dye from β-naphthol

by treating *p*-nitrobenzeneazo-2-naphthol with a suitable copper salt to give a dye known as Paranitraniline Brown (Pratt, 1947, p. 94). Another chelate complex which dates back to the earlier history of the dyestuff industry is dianisidine blue, a copper complex containing one molecule of *o*-dianisidine and two of β-naphthol. Details on the preparation of physical characteristics of similar compounds are given by Elkins and Hunter (1935) and Haendler and Smith (1940). On the other hand, in order to form a stable combination with chromium, iron, manganese, or zinc, an additional hydroxyl or other coordinating group is required. This additional grouping may be present in the same molecule as part of the diazo component or may be furnished by an adjacent molecule (Drew and Landquist, 1938; Zollinger, 1958; 1961, p. 338). This type of reaction has been employed in an attempt to stabilize azo-dyes derived from β-naphthylamine (e.g. in the aminopeptidase reaction) by Nachlas *et al.* (1957a, b) and Glenner *et al.* (1959). From a theoretical standpoint, chelate complexes involving four naphthol or naphthylamine molecules may also result, but the existence of such structures has not been established with reference to dyes produced under histochemical conditions.

Unfortunately, some of the chelate complexes of histochemically derived dyes have not proved to be particularly stable; this phenomenon is related to the fact that the complexing procedure is usually performed at room temperature and at a neutral pH. The production of stable cyclic chelates according to the techniques of industrial technology is generally carried out at an elevated temperature and in an alkaline medium (e.g. in the presence of ammonia or ethanolamine). Azo-dye metal complexes are of value not only from the standpoint of dye stabilization but also in their potential application to the study of enzyme localizations at the electron microscope level.

## III. OXIDATION DYES

### A. Azomethine Dyes

These compounds which are characterized by the grouping —CH═N— are generally formed by the condensation of an aromatic aldehyde com-

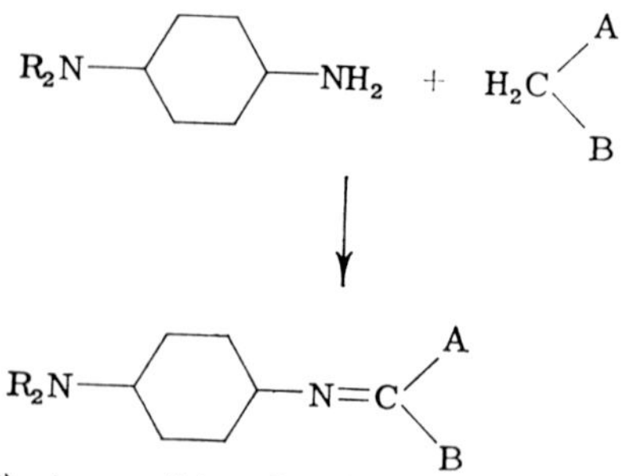

(V) Azomethine dye derived from methylene containing compound

pound with an arylamine. Compounds containing reactive methylene groups may form azomethine dyes when oxidized in the presence of an arylamine (V). The methylene compounds may be classified as:

a. Open chain methylene compounds. These include acylacetonitriles, acylacetamides, and certain ketones.

b. Cyclic methylene compounds. These include pyrazolones and coumarone derivatives.

The mechanism of oxidative formation of azomethine dyes has not been completely elucidated. It is assumed that the sequence of events involves the oxidation of the amine to a semiquinone which is in equilibrium with the quinonediimine. Subsequent condensation of the quinonediimine with the methylene coupler carbanion results in formation of the leuco dye (Thomas and Weisburger, 1932). Azomethine dyes may also be formed from arylazopyrazolones which readily undergo oxidative coupling with certain phenylenediamines. In this reaction, the azo groups are replaced by the amine (Vittum *et al.*, 1950). The only application of azomethine dye formation to enzyme histochemistry has been in the demonstration of cytochrome oxidase (Chapter 11).

## B. Quinoneimine Dyes

These dyes are derived from *p*-quinone and analogues. They include the indamines (VI), indoanilines (VII), and indophenols (VIII). Indamine dyes are derived from the oxidative condensation of an *N*-substituted primary amine and another primary amine, indophenols are derived from an aminophenol and a phenol, while indoanilines are derived from a primary arylamine or *N*-substituted amine and a phenol. The term "indophenol" has often been employed to include the indoanilines. This, however, is not correct terminology. Thus "nadi" type reactions employing α-naphthol and phenylenediamine derivatives result in the formation of indoaniline and not indophenol dyes. Of interest is the reaction of naphthols with a sub-

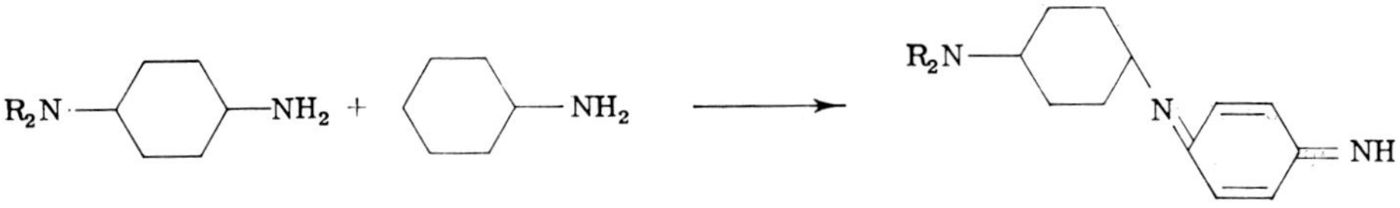

(VI) Indamine

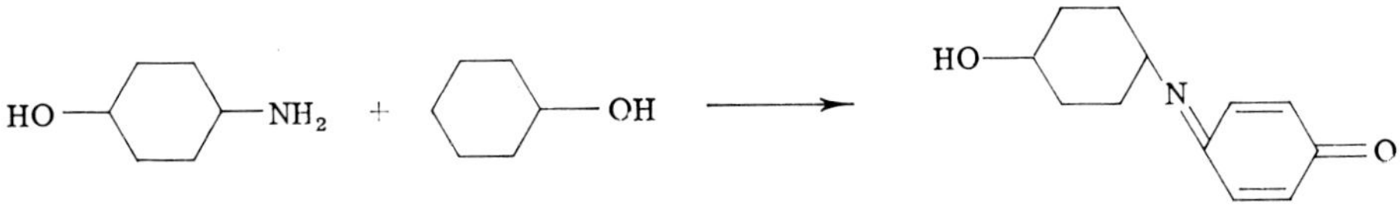

(VII) Indophenol

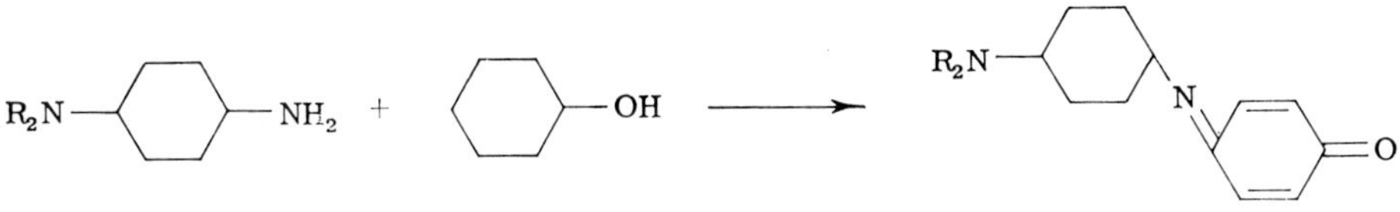

(VIII) Indoaniline

stituent in the 4-position (e.g. 2,4-dibromo-1-naphthol), in which this substituent is ejected with the formation of a *p*-indoaniline dye (Vittum and Brown, 1949). With reference to the phenolic and naphtholic components, the mechanism of dye formation is believed to involve oxidation of the amine to a semiquinone ion which reacts with the naphtholate ion to produce leuco dye. The leuco dye is presumably oxidized to the final dyestuff by another semiquinone ion (Vittum and Brown, 1946). Of the above-mentioned systems, the indamine and indoanilines are of great value for the improved demonstration of cytochrome oxidase (Chapter 11).

It should also be pointed out that many primary arylamines can be oxidized to form insoluble dyes. In fact, one of the earliest dyestuffs was aniline black which was produced by the oxidation of the aniline in solution. The chemical mechanisms, which have been intensively studied in the past, indicate that the oxidation of aniline results in the formation of intermediate *p*-quinoneimines which subsequently further react with aniline to give an azine type dye (IX) (Green and Wolff, 1913). The early oxidation products of *p*-diamines, the Wurster's salts, are free radicals which may polymerize in solution. Their stability and other physical characteristics have been investigated by Michaelis and collaborators (1933, 1939). The so-called Wurster's red is an addition compound derived from the oxidation

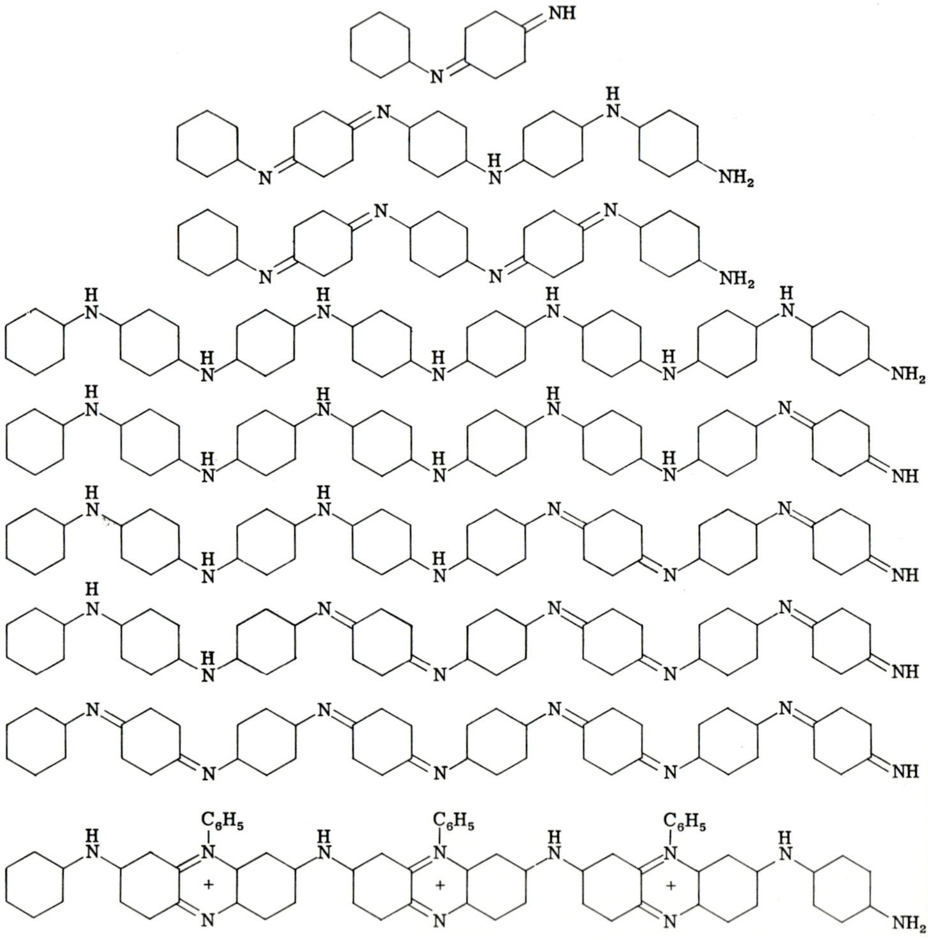

Azine type dyes produced by oxidation of aniline

(IX)

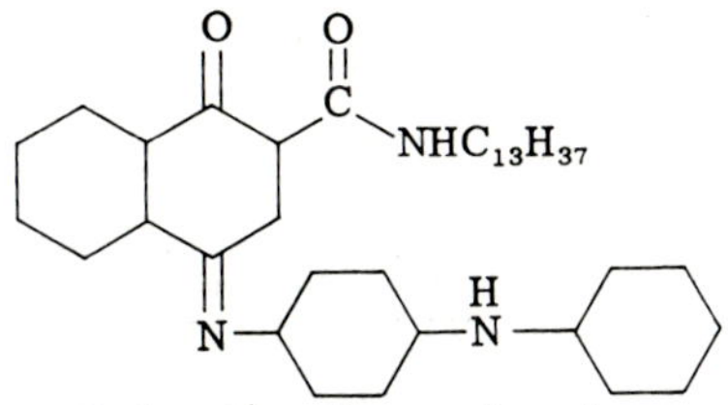

Indoaniline dye produced
oxidation of 1-hydroxy-2-naphth-
anilide and *p*-aminodiphenylamine
(X)

product of *N,N*-dimethyl-*p*-phenylenediamine and its base (Baker and Johnson, 1952). More recently, the role of free radicals of the Wurster's type in the histochemical reaction for the demonstration of cytochrome oxidase ("nadi" reaction) has been considered by Person and Fine (1961). Of interest is a somewhat more complex oxidation product of *p*-phenylene-diamine which is known as Bandrowski's base. This base is believed to be formed by quinonoid addition of quinonediimine and *p*-phenylenediamine (Heiduschuka and Goldstein, 1916; Forster and Soyka, 1931; Feigl and Casto-Neto, 1956). Little is known regarding the constitution of the products subsequently formed but it is probable that they are of the complex azine type. Homologues of aniline and related compounds are also capable of producing oxidation dyes. These include *o*-tolidine, *p*-phenylenediamine, 2-amino-5-diethylaminotoluene, *p*-aminodiphenylamine, 3-aminocarbazole, 3-amino-9-ethylcarbazole, and a series of *N*-alkyl substituted amines. Of these, the *p*-aminodiphenylamine and carbazole derivatives have proved most useful as histochemical reagents for the demonstration of cytochrome oxidase (Chapter 11). In general, these amines are not used singly but are employed in combination with other amines or hydroxyl or methylene compounds: a procedure which tends to minimize the crystalline structure of the dye. The following is a complex indoaniline dye produced by the action of cytochrome oxidase in tissue sections upon a 1-hydroxy-2-naphthanilide derivative and *p*-aminodiphenylamine (X). Aminopeptidase may also be demonstrated by oxidation techniques involving complex amines (Burstone and Weisburger, 1961a, b) (Chapter 10). A potential method for the histochemical demonstration of esterase activity has been suggested by Kramer and Gamson (1958) who employed indophenol acetate (Heller, 1912) as a substrate. Upon enzymatic hydrolysis, the highly colored blue indophenol is formed. More complex indophenol esters have been described by Master (1960).

## C. Azine Dyes

The first azine dye was synthesized in 1856 by Perkin who oxidized an impure mixture of aniline and toluidine with potassium dichromate.

Azine dyes are produced by the oxidation of an indamine in the presence of an aminophenol, or by condensation of an *o*-quinone with an aromatic diamine. Oxidation of indamines in the presence of aniline yields diamino-*N*-phenyl phenazines known as safranines. Oxidation of *N*-substituted amines also results in the formation of azine dyes. *o*-Aminophenols are capable of autocoupling in the presence of an oxidizing agent to form azines, and a similar reaction also occurs with 2-amino-6-phenylphenol (Henn, 1953). Although azine dyes are commercially synthesized in the presence of strong oxidizing agents and at high temperature, they may be formed under relatively mild conditions of oxidation (Schmidt *et al.*, 1953). For example, oxidized 3′-(diethylamino)methanesulfonanilide and *N,N*-diethyl-*p*-phenylenediamine initially form a blue indamine dye which upon ring closure changes to an orange-brown dye, and finally to a bluish red (magenta) shade (XI). Further exploration of such techniques is being undertaken in this laboratory. It should also be pointed out that it may be possible to produce azine dyes by post-incubation treatment of stained sections with an arylamine in the presence of an oxidizing agent. The value of azine dyes over the simpler indamines and indoanilines lies in their greater stability and higher chromogenicity. Because of these inherent advantages, applications of azine dye technique may facilitate cytochemical localization of oxidases at both the cytological and electron microscope levels.

## D. Indigoid Dyes

Indigo is one of the oldest dyes, having been employed in India for at least 3,500 years. The organic synthesis of indigo and analogues is considered to be one of the outstanding achievements of modern organic chemistry and has been reviewed by Thorpe and Ingold (1923), Mayer (1943), and Venkataraman (1952). Indigoid dyes may be classified as:

a. Symmetrical derivatives of indigo and thioindigo.

b. Asymmetrical derivatives of indigo and thioindigo.

c. Indirubins which are 2,3′-bisindole indigos.

d. Indigos and indirubins where one imino grouping is substituted by a sulfur atom.

e. Dyes containing one ketone-containing heterocyclic group joined by a double bond to a carbon atom in an α-position to a keto or oxo grouping of an aromatic ring.

Indigo may be formed by oxidation of indoxyl (XII) which exists in two tautomeric forms. The indigo (XIII) may subsequently be reduced to leuco indigo or Indigo White. Among the naturally occurring derivatives of indoxyl are its glucoside (XIV), indican, and indoxyl sulfate (XV). The glucoside is found in the leaves and stems of a number of plants including *Indigofera tinctoria*, while indoxyl sulfate has been isolated from

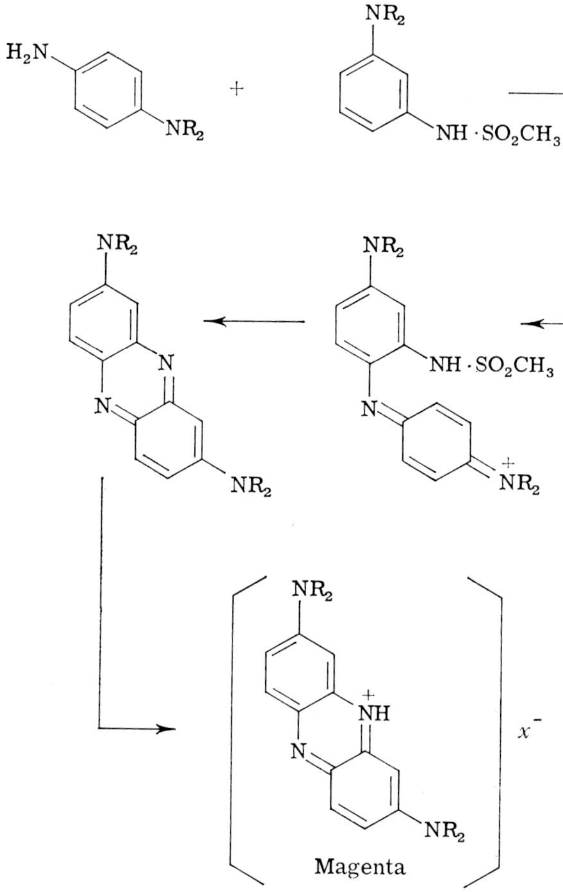

Formation of a magenta azine oxidation dye

(XI)

the urine of mammals and is formed by oxidation of indole to indoxyl by the liver and subsequent conjugation with sulfuric acid. When the indigo plant is extracted with water and the glucoside is hydrolyzed the released indoxyl is then oxidized by means of air to indigo. This technique has been one of the most valuable dyestuff procedures of ancient, medieval, and modern times. Indigo is known to form metal complexes with copper and iron (Kunz, 1922; Kuhn and Machemer, 1928). When indigo is halogenated with chlorine or bromine; the 5,5′-positions are first occupied, then the 7,7′-positions followed by the 4,4′-positions, and finally the 6,6′-positions (Johnson, 1955).

Of interest is Tyrian Purple, a dye of remarkable stability. Accord-

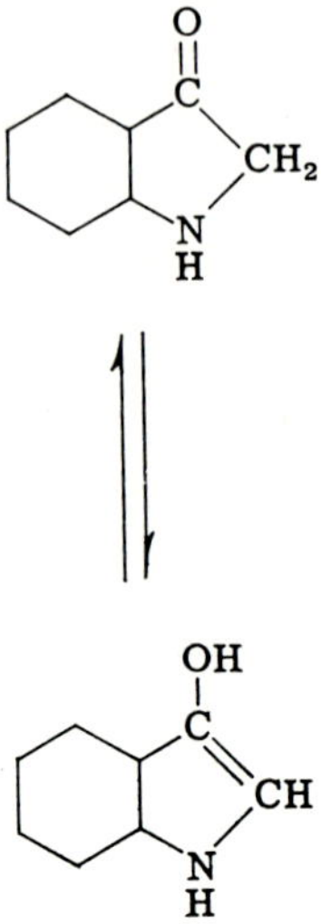

Tautomeric forms of indoxyl

(XII)

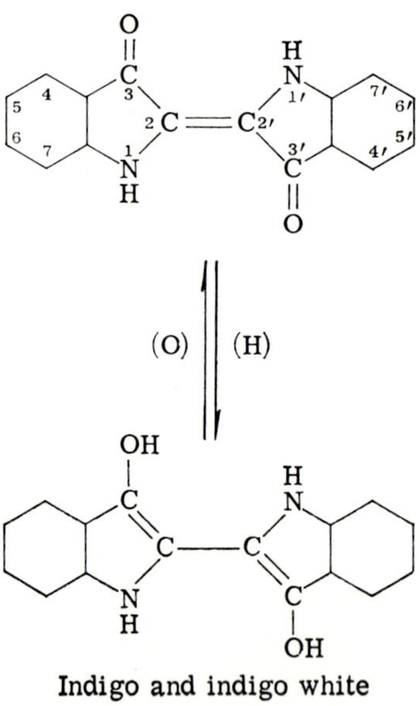

Indigo and indigo white

(XIII)

ing to Pliny the best quality was made at Tyre, and hence its name. The structure of this compound was elucidated by Friedländer (1909, 1922) who obtained only 1.4 g. of pure dye from 12,000 mollusks collected at Trieste. The dye which was found to be 6,6′-dibromoindigo

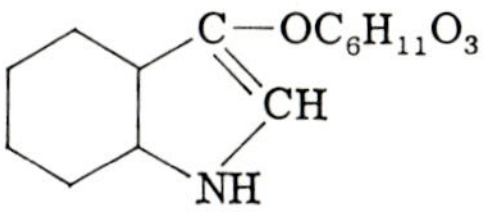


Indoxyl glucoside

(XIV)

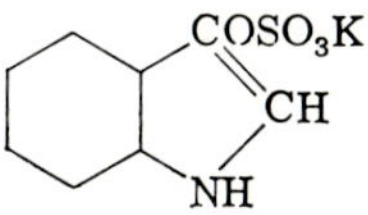


Indoxyl sulfate

(XV)

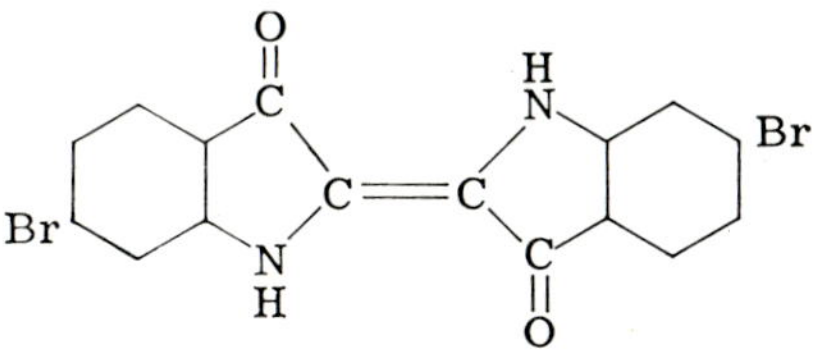


Tyrian purple

(XVI)

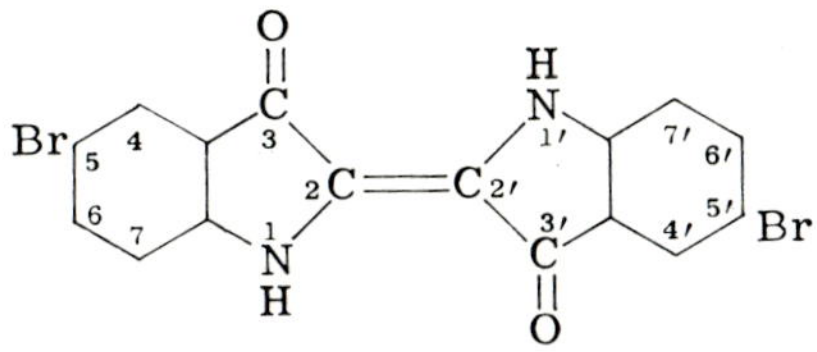


5,5′-Dibromoindigo

(XVII)

originally was synthesized by Baeyer and Drewsen (1883). It was highly valued by the ancients and used as a symbol of royalty, and was obtained by air oxidation of a colorless fluid isolated from a certain shellfish (*Murex brandaris*) which was found throughout the Near East. It is stated that heaps of shells still are to be found in the vicinity of the ancient dye works. It is interesting to note the similarity between Tyrian Purple (XVI) and some of the dyes produced by modern histochemical reactions, such as (XVII) formed in Holt's reaction (1958) for the demonstration of esterase.

In 1883 Baeyer published the structural formula of indigo as the *cis* form of the presently accepted *trans* structure. Although indigo and thioindigo derivatives occur in both *cis* and *trans* forms, they are predominantly in the *trans* form when in the solid state. The earlier syntheses of indigo were accomplished by the oxidation of indole or by treatment of isatin with phosphorus, phosphorus trichloride, and acetyl chloride followed by air oxidation. Subsequently, *o*-nitrocinnamic acid was converted into *o*-nitrophenylpropiolic acid, which gave indigo by reduction in the presence of alkali. A later technique involved the aldol condensation of *o*-nitrobenzaldehyde and *o*-nitrocinnamic acid. Another approach involved the condensation of aniline with chloroacetic acid to give *N*-phenylglycine which, in turn, was treated with sodium hydroxide and sodium amide to give indoxyl. In the work of Holt (1958) and Holt and Sadler (1958a) a large number of substituted indoxyl derivatives were prepared by ring closure of phenylglycine-*o*-carboxylic acid. Substitutions on the aforementioned acid were obtained by a direct substitution reaction or else by direct condensation of substituted anthranilic acids with chloroacetic acid. Chart 3 indicates the schematic representation by which indoxyl and indigoid dyes are prepared. These routes may be employed in the preparation of indoxyl intermediates for histochemical substrates. The following represents the steps in the synthesis of 5-bromo-4-chlorodiacetylindoxyl which is reported by Holt (1958) to be the best substrate with reference to facilitation of accurate cytochemical localizations:

*m*-Chloroaniline → chloroisonitrosoanilide → 4-chloroisatin → 6-chloroanthranilic acid → 3-chlorophenylglycine

Carboxylic acid (Na salt) → 5-bromo-4-chlorophenylglycine carboxylic acid → 5-bromo-4-chlorodiacetylindoxyl

Details of the preparation of intermediates and indoxyl substrates will be described in Chapter 6.

Application of indoxyl esters as substrates was first reported by Derrien (1911) who noted the enzymatic hydrolysis of an indoxyl sulfate. During the last five years the detailed studies of the aforementioned Holt group have resulted in the synthesis of a large series of original and sophisticated indoxyl esters.

## E. Miscellaneous Oxidation Dyes

Several α-naphthol derivatives may be oxidized to form insoluble pigments. Among these is 4-methoxy-1-naphthol which can be oxidized by ferric chloride to form a blue dye (Russig, 1900). α-Naphthol may similarly be oxidized to form bluish dyes. The author has employed this principle for the demonstration of phosphatases or esterases by employing α-naphthyl phosphate or α-naphthyl acetate in the presence of ferricyanide, periodate, or other oxidizing agents. The quality of the histochemical localizations was inferior, however, to those obtained with the usual azo-dye methods. Certain cyclic methylene compounds, such as 3-methyl-1-phenyl-5-pyrazolone, may also be oxidized to form insoluble dyes (Knorr, 1887). Aromatic aldehydes (e.g. benzaldehyde) may condense with two moles of antipyrine to form leuco compounds of dyes which are similar to the basic triphenylmethane dyes (Poray-Koshits *et al.*, 1947). Oxidation of

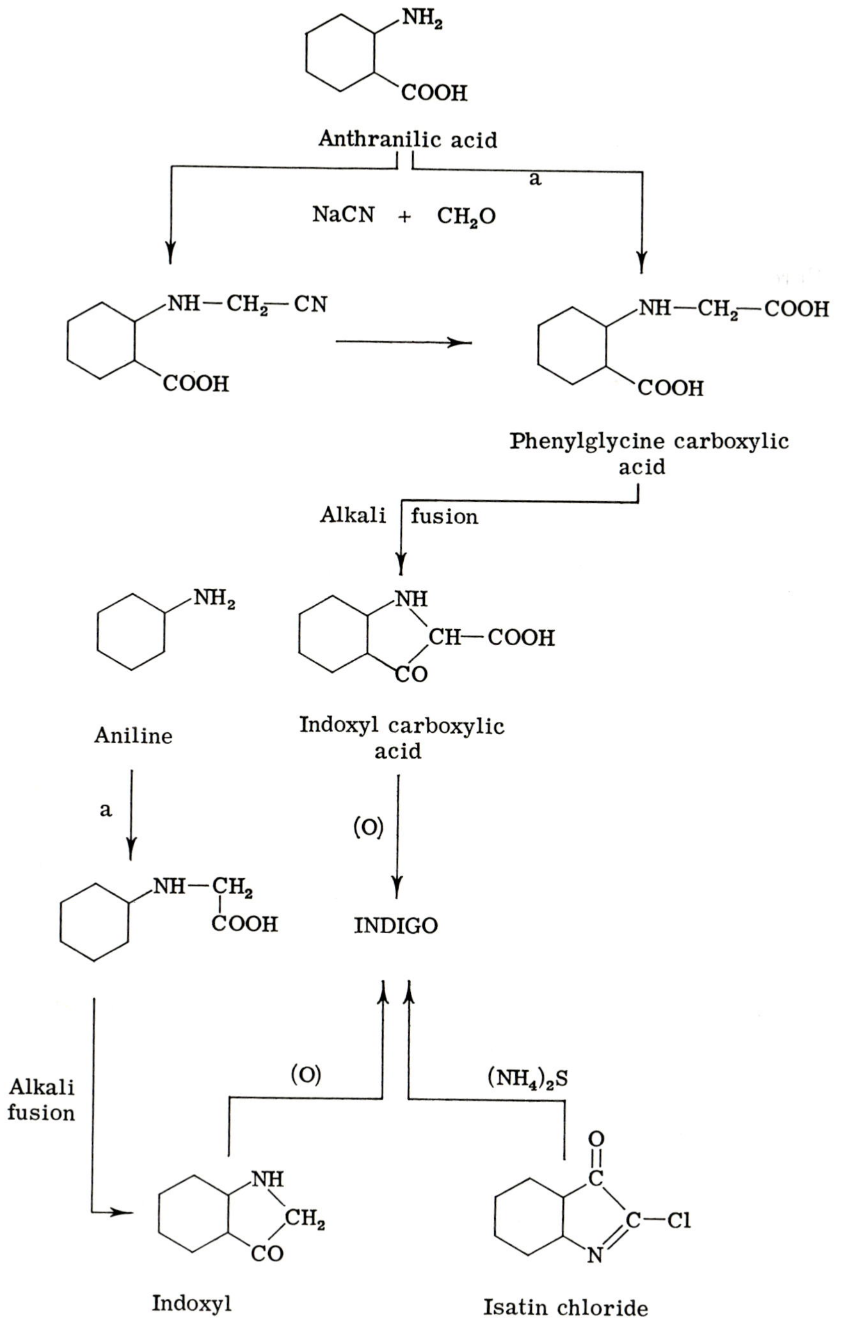


a = chloroacetic acid

CHART. 3. Schematic representation of methods by which indoxyl and indigoid dyes are prepared.

with diazonium salts to produce green dyes. Exploration of the phthalocyanine group may yield new histochemical reagents for optical as well as electron microscopy.

## V. SUBSTANTIVITY AND HISTOCHEMICAL LOCALIZATIONS

The term "substantivity" as originally coined by the dyestuff industry indicates the affinity of dyes for textiles or, in the case of naphthols, the amount of naphthol adsorbed from solution onto a given sample of cellulose or other material. The histochemical connotation of substantivity relates to the affinity or insoluble nature of the reaction products formed by substrates *in situ*. The physical mechanisms by which azo, azomethine, indoaniline, indigoid, and other dyes produced under histochemical conditions stain cellular components have not been elucidated. Three types of dyestuffs may result from histochemical reactions:

1. Dyes which are amorphous in appearance even at the highest optical level (e.g. dyes derived from naphthol AS derivatives and 4,4′-dichloroindigo and nitro-blue tetrazolium).

2. Dyes which are initially amorphous or of fine crystalline pattern but which subsequently crystallize (e.g. derivatives of 6-benzoyl-2-naphthol and 6-bromo-2-naphthol, and indoaniline dyes from $\alpha$-naphthol and phenylenediamines).

3. Dyes which initially exhibit a grossly crystalline pattern (e.g. unsubstituted indigo and simple formazan dyes).

Dyes may be held in tissue sections in several ways: by chemical combination with tissue proteins, by physical forces (e.g. van der Waals) and hydrogen bonding, or simply by the formation of insoluble pigment particles which are mechanically held in the tissue section. It is also conceivable that the reaction product may be a dye solution which is held in close proximity to tissue proteins as a solid solution. The formation of dyes from histochemical enzyme reactions contrasts with ordinary staining procedures in that in the former, dye formation from colorless precursors proceeds at a known rate, while in the latter the tissue section is initially subjected to a concentrated dye solution. The formation of a precipitate may be divided into two phases: the first consists in the inception or nucleation of submicroscopic particles of the reaction product. The second phase consists in growth of grains by aggregation of these particles. With the occurrence of intermolecular hydrogen bonding as well as other affinities, the particle size may grow and distinct crystals may become visible at the microscopic level. However, by altering the configuration of the molecule with certain substituents, the tendency toward crystallization is minimized or eliminated and optically homogeneous dyes are formed.

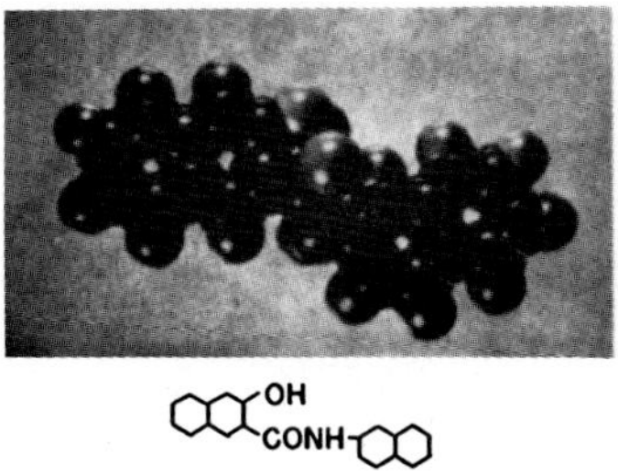


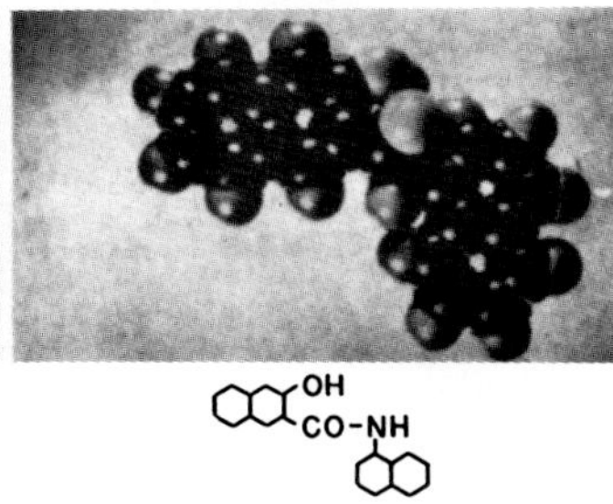


FIG. 2. Model of naphthol AS-SW (*left*).
FIG. 3. Model of naphthol AS-BO (*right*).

Certain geometric molecular configurations are also considered to be related to enhanced substantivity, but most of the data pertaining thereto relates to the dyeing of cellulose fibers (Vickerstaff, 1950). Dye molecules which exhibit a linear or long chainlike structure have been found to be more substantive than those which exhibit an angular or less linear configuration. On the other hand, highly substantive anthraquinone dyes often exhibit an angular configuration. The following isomeric naphthols, AS-SW (Fig. 2) and AS-BO (Fig. 3), exhibit different degrees of adsorption with regard to cellulose fibers, with the linear compound exhibiting almost twice the substantivity of the more angular one. Coplanar position of aromatic nuclei in relation to the object being stained and possibly with reference to adjacent dye molecules is also a consideration. Large flat molecules may tend to polymerize due to intermolecular resonance, but this feature may also lead to undesirable crystallization. On the other hand, lack of planar configuration may lessen the chemical reactivity of dyes and render them more stable. Van der Waals attraction between the aromatic nuclei of dyes and hydrophobic protein chain molecules of tissues may be related to adsorption of dyes, and this factor may explain the observation that dry wool and nylon can adsorb appreciable amounts of benzene or stilbene from dry solvents (Chipalkatti *et al.*, 1954). With regard to linearity and coplanarity of dye molecules as related to the staining of proteins, it is known that a polypeptide chain may be folded in one or more ways and the side chains may assume different configurations, thus rendering steric problems more complicated. Robinson (1954) and Bamford *et al.* (1954) studied the mechanisms of the dyeing of synthetic polypeptides such as polyglycine. Factors such as attachments of dye molecules to side chains and the backbone of the Pauling helix were considered. Two possible sites of attachment of dye molecules were polypeptide backbones and side chain groupings. The significant reviews of Frey-Wyssling (1948, 1953) and Howitt (1949) should be consulted for detailed descriptions of the configuration and specific groupings of protein polypeptide chains and

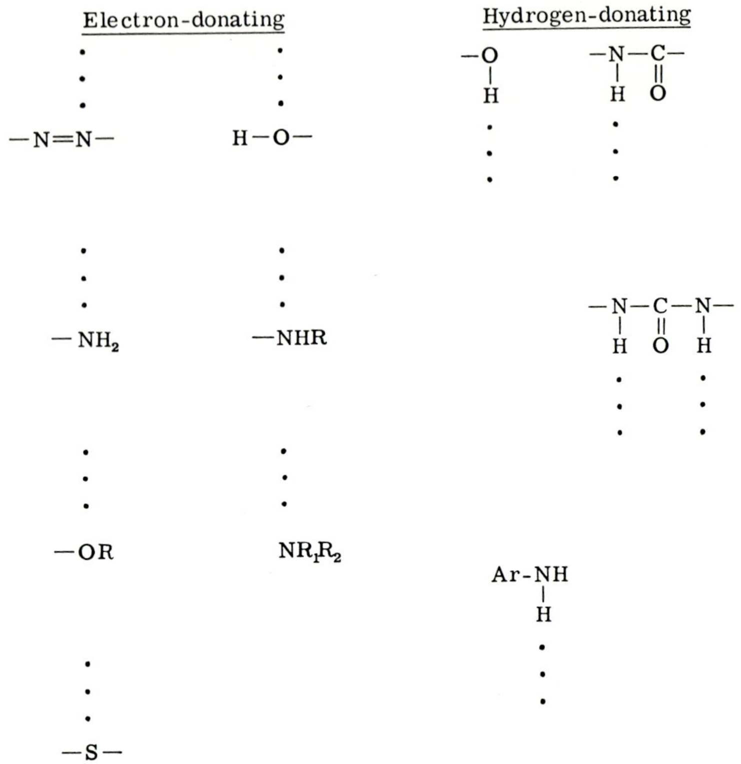


CHART 4. Electron-donating and hydrogen-donating groups of dyes or tissues capable of forming coordinate links (Evans, 1949, p. 257; Spiegler, 1953).

side chains which may enter into the dye binding process. The possibility of participation of insoluble lipoprotein complexes should also be considered. Electron-donating and hydrogen-donating groupings of both dye and tissue which are capable of coordinating to form hydrogen bonds are shown in Chart 4.

The affinity of azo-dyes for proteins and synthetic polypeptides is believed to be related in part to hydrogen bonding between the peptide groupings of the protein and hydroxyl, amino, or azo groups of the dye (Eistert, 1938; Valko, 1939; Evans, 1949, p. 261; Venkataraman, 1952, p. 1303). Valko (1939) noted that certain synthetic fibers which contain only peptide bonds, but no free amino groups, are readily stained by acid wool dyes. He concluded that the dyeing process in this case obviously depended upon formation of hydrogen bonds between the dye and the fiber. More recently, dyeing experiments with models of urea resin and polyurethans further indicate a reaction between the dye and the amide grouping (Elöd and Einsele, 1960). The peptide grouping of naphthol AS compounds, in addition to exerting a bathochrome effect, is also believed to be involved in hydrogen bonding. The possible sites of hydrogen bonding between a naphthol AS dye and tissue protein are shown in (XXII). It should be noted that the diazonium component also contains a peptide bond which may enter into the coordinating effect. Indigoid dyes are also capable of entering into intermolecular hydrogen bonding involving two or more molecules (XXIII) (Weinstein and Wyman, 1956). The presence of halogen or other substituents in the 4- and 7-positions would tend to interfere with the close approach of —CO— and —NH— groups of neighboring molecules and thus prevent association. Thus, the larger

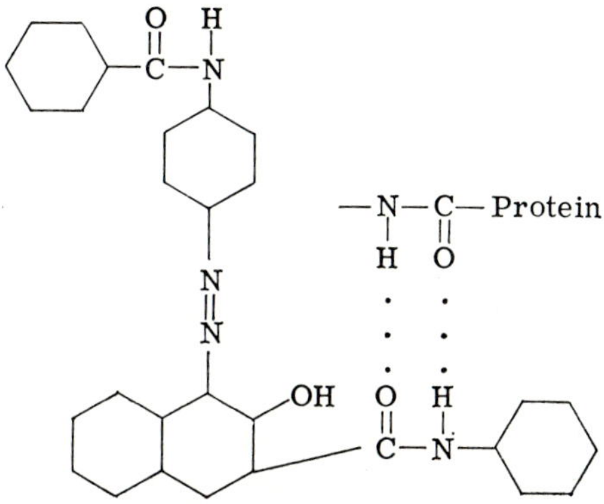


Possible sites of hydrogen bonding between a naphthol AS dye and tissue protein

(XXII)

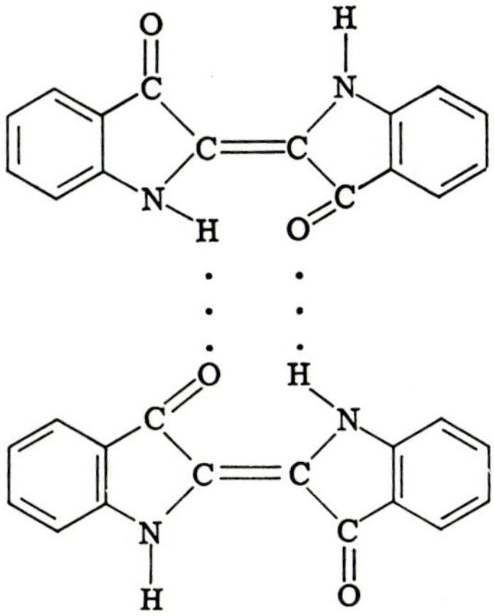


Intermolecular hydrogen bonding involving two molecules of indigo

(XXIII)

the substituent group, the more effective it will be in preventing association, but this does not imply that association of the dye molecule with tissue protein will be diminished. Although the precise meaning of substantivity and the relationship between substantivity and substrate design is obscure, a certain amount of information has been gleaned which should be of advantage in substrate design. Specifically, the presence of peptide bonds as exist in naphthol AS derivatives appears to be of distinct value.

Covalent bonds are formed between mordanting metals and azo-dyes as has been previously described in this chapter. In addition to making the azo-dye more insoluble and minimizing crystallization, the presence of a metal may also confer a mordanting action. The latter, of course, has been employed with reference to histological stains since their inception. Little consideration, however, has been given to the development of histochemical reagents which facilitate localization, not only because of their insoluble nature but also because of the mordant effect of certain metals. Utilization of principles of mordant dye chemistry, which have proved to be so valuable in both textile dyeing and histological staining, may offer insight into improved techniques for visualization of sites of enzyme activity at the electron microscope level. Giles (1944) has written an excellent account on the mechanism of mordanting of animal fibers from historical, theoretical, and applied standpoints. Included is a detailed bibliography relating to the various types of metal salts employed from antiquity to modern times. With reference to the dyeing of wool, it is believed that chromium forms electrovalent bonds with wool and coordinate bonds with groupings of the azo-dye. Giles and MacEwan (1959) have conducted model experiments on the mechanism of dyeing of nylon and edestin with unsulfonated azo compounds, and believe that both van der Waals forces and hydrogen bonding are involved. It is also known that organic molecules not ordinarily bound to proteins may form strong complexes if suitable metals are present (Klotz and Loh-Ming, 1954). Experimental data indicate that the metal acts as a bridge and must be capable of coordinating with both the protein and the molecules. Of the metals used, $Hg^{++}$, $Cu^{++}$, $Ag^{+}$, $Ni^{++}$, $Zn^{++}$, $Co^{++}$, and $Mn^{++}$ are effective mediators, while $Ca^{++}$ and $Mg^{++}$ are not. More recently, Koltun and associates (1958, 1959) have studied coordination complexes of proteins and related substances with reference to $Cu^{++}$ and $Zn^{++}$ ions. With reference to histochemical procedures, the substituted monotetrazolium salt technique of Pearse (1957) employs $Co^{++}$ ions in the substrate solution for simultaneous chelation and for enhanced stability of the final formazan dye. Whether or not the cobalt is coordinated with tissue protein in this instance is not known.

The presence of conjugated double bonds, heterocyclic and saturated ring structures, sulfur atoms, and quinone groupings has been utilized in substrate designs with particular emphasis on the so-called post-coupling procedures (Chapter 5, p. 197). In connection with conjugated double bonds, it is interesting to note that the substantivity of curcumin, the coloring matter of turmeric, is believed to be related to its double bond structure (Bhat *et al.*, 1940). The conjugated structure of certain carotenoids such as bixin and crocetin should also be noted (Zollinger, 1959). With these and analogous compounds, it was found that the substantivity was directly related to the length of the polyene chain. Application of heterocyclic compounds to substrate design has found little application in histochemistry except for cytochrome oxidase substrates such as cyanoacetylcoumarone and pyrazolones (Chapter 11). The quinone grouping in the form of anthraquinone is an important feature of one of the most useful post-coupling substrates, a derivative of hydroxynaphthoic acid and aminoanthraquinone (see Chart 3 in Chapter 5). It should be noted that many anthraquinone dyes are characterized by high substantivity. With reference to saturation of aromatic rings, a tetrahydronaphthalene derivative is also advantageous for a non-coupling procedure (see Chart 4 in Chapter 5), but the reason for this is obscure.

As regards the formation of an insoluble precipitate from a stain precursor such as occurs with the metal-salt techniques for phosphatases, Johansen and Linderstrøm-Lang (1952) have offered a mathematical analysis relating to the diffusion of phosphate ions and the precipitation of calcium phosphate. They conclude that calcium phosphate exhibits a marked tendency to form supersaturated solutions and the time required for the phosphate concentration to reach a critical level may vary from approximately 0.1 to 2.5 seconds, which is sufficient time to permit gross diffusion artefacts to occur. These diffusion artefacts include preferential precipitation in the sites of performed crystal nuclei and adsorption related to special chemical or physical configuration of cellular organelles. They thus feel that calcium phosphate tends to precipitate at such preferential sites rather than at centers of high enzymatic activity. Gomori and Benditt (1953) point out that the conclusions of Johansen and Linderstrøm-Lang are primarily based on theoretical assumptions, some of which are not borne out by actual experiment. The major diffusion artefacts are primarily associated with non-optimal conditions of incubation such as non-optimal pH or low calcium ion concentration. Gomori (1952, p. 145) states that "further experiments will have to decide whether the diffusion artefacts persisting [with the glycerophosphate method] in spite of optimal conditions are within or beyond the resolving power of high powered dry objectives." In view of the fact that further experiments have not been reported, it is interesting to note the widespread and, in some cases, uncritical application of the glycerophosphate method to electron microscopy, when many of its shortcomings at the optical level have not been elucidated.

In the case of precipitates which are relatively soluble, diffusion currents may simply wash away part of the reaction product and prevent its deposition in the tissue section. On the other hand, the relative sluggishness of these currents in staining solutions may cause a somewhat different type of artefact if the precipitate is relatively insoluble (Gomori, 1952, p. 146). Gomori mentioned that this type of artefact is observed mainly with some azo-dyes but also may occur with phosphate precipitates. Under such conditions, the reagent may become exhausted in the vicinity of centers of very high activity more quickly than it can be supplied by diffusion currents from the rest of the solution. With this lack of the precipitating agent, the product of hydrolysis will remain unbound and free to diffuse until it encounters a fresh supply of reagent some distance away. This situation will result in too little staining at the sites of genuine activity and an irregular precipitate of azo-dye (or in the case of calcium phosphate, staining of nuclei) around them. This type of artefact can be minimized by increasing the interchange of fluid around the tissue section by agitating the slide or stirring the solution during incubation. Gomori (1952, p. 172) specifically notes that esters of $\alpha$- and $\beta$-naphthol are usually so rapidly hydrolyzed by highly active tissues that diffusion artefacts result unless the incubating mixture is agitated all the time (Fig. 4). This type of diffusion phenomenon was observed even when the most rapid coupling diazonium salts were employed.

The kinetics of enzyme staining reactions have been primarily studied with reference to indigogenic methods. O'Sullivan (1955) has described a method for obtaining the solutions to equations characterizing systems in which a reacting substance diffuses between two adjacent regions. This substance may be produced or removed in the two regions at rates which may be expressed as polynomial functions of time, and may be removed throughout both regions by a first-order reaction. With the rapidity at which penetration of a substrate occurs, it is likely that the product of the enzymatic reaction will be produced uniformly throughout the site by a zero-order reaction. This product then diffuses from the site and reacts with the developing or capturing agent. Further analyses are considered in which the intermediate produced by a zero-order reaction enzymatic reaction is converted by an irreversible first-order reaction into a non-diffusible colored product (O'Sullivan, 1956). The localization factor, or degree of localization, is independent of the activity per unit volume of the site and also of the staining time, but increases with the size of the site. Thus the precision of localization depends upon complex relationships between three factors:

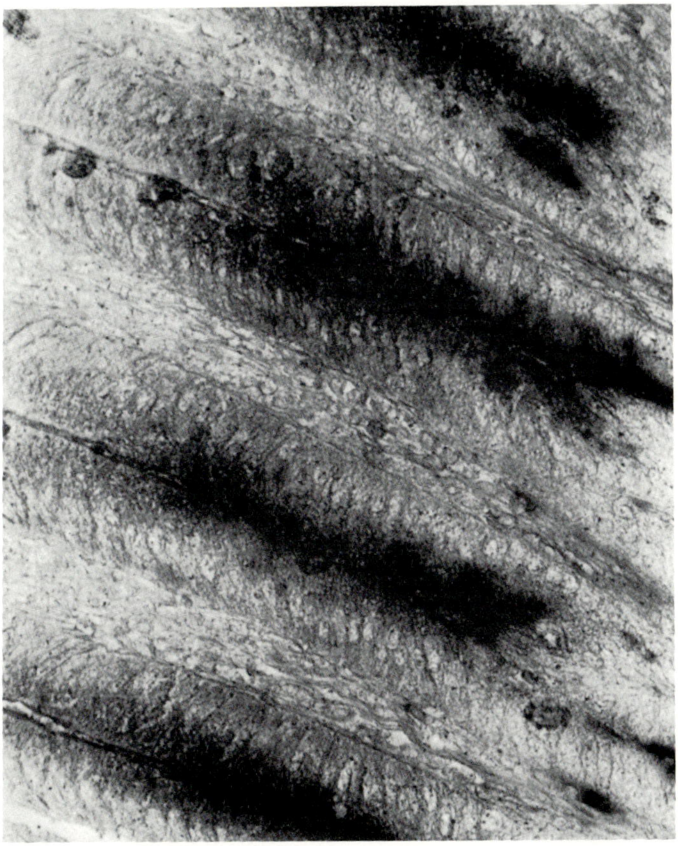

FIG. 4. Diffusion artefact in rat intestine produced by slow coupling of diazonium salt with enzymatically released naphthol. Alkaline phosphatase technique employing α-naphthol phosphate.

the size of the enzyme site, the diffusion coefficient of the stain precursor, and the velocity constant of the capture reaction. Further theoretical studies indicate that the indoxyl staining methods may give a direct quantitative representation of levels of enzyme activity in terms of stain densities only when the capture reaction is kinetically of the first order (Holt and O'Sullivan, 1958).

We must conclude that relatively little is known about the mechanism of stain immobilization in tissue sections. From a practical standpoint, there is little question that increased molecular size (and related insolubility) of enzymatically released products enhances substantivity. This may simply be related to the more insoluble nature of the dye formed, hydrogen bonding or other physical forces, or a combination of several of these factors.

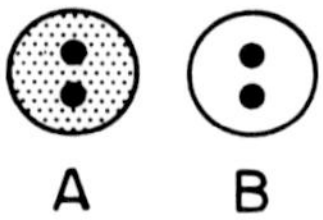


FIG. 5. Diagrammatic representation of cytoplasmic enzyme activity of two cells, A and B. Cell A shows an overall cytoplasmic activity as indicated by gray stippling, as well as activity represented by two discrete black bodies. The latter are just as sharply defined in both A and B despite differences in contrast due to background.

The incorporation of halogen, nitro, or other groupings into substrate design tends to minimize intermolecular association and enhances the production of more homogeneous and less crystalline dyes.

The sites of enzyme activity in a tissue section when visualized by means of good histochemical techniques should be sharp and clear cut. All available evidence at our present state of knowledge indicates that sharp histochemical localizations are significant even when there is an overall cytoplasmic activity (Figs. 5 and 6). *Histochemists trained as histologists or pathologists should always expect to see clear cut microscopic localizations provided proper techniques have been employed.* Fuzzy localizations, including halos of dye and disorganized clumps of dye crystals, invariably indicate poor fixation (or none), diffusion of enzyme (or cytoplasm containing enzyme), and/or diffusion of the products of hydrolysis. Pertinent to this have been some of the differences in opinion between the Gomori and the Seligman group as to the significance of "sharpness' in histochemical localizations. Ravin *et al.* (1953) stated that the azo-dye method for esterase was not improved by using α-naphthyl acetate as a substrate in the place of β-naphthyl acetate. They claimed that the azo-dye from α-naphthol "sticks tenaciously to protein" and on this basis they state that "the sharpness, therefore, has no special significance." Chessick (1953), working under the supervision of the late George Gomori, pointed out that even if it were true that azo-dyes from α-naphthol stick tenaciously to protein (which has not been demonstrated under conditions of histochemical technique), this would be an advantage rather than a disadvantage, since it would tend to prevent the diffusion of the dye from its site of original precipitation. Thus, the sharpness would be of great significance. In any event, the test of time has confirmed the original contentions of the Gomori group that α-naphthyl acetate is by far a superior substrate to the β-isomer (Pearse, 1960, p. 464).

O'Sullivan (1955) has proposed the following conditions for the obtaining of successful cytochemical localizations:

1. The preparatory processing and subsequent treatment of tissues should not affect the distribution of enzyme activity.

2. Rapid penetration of the substrate and developing agent should occur with reference to all cellular components.

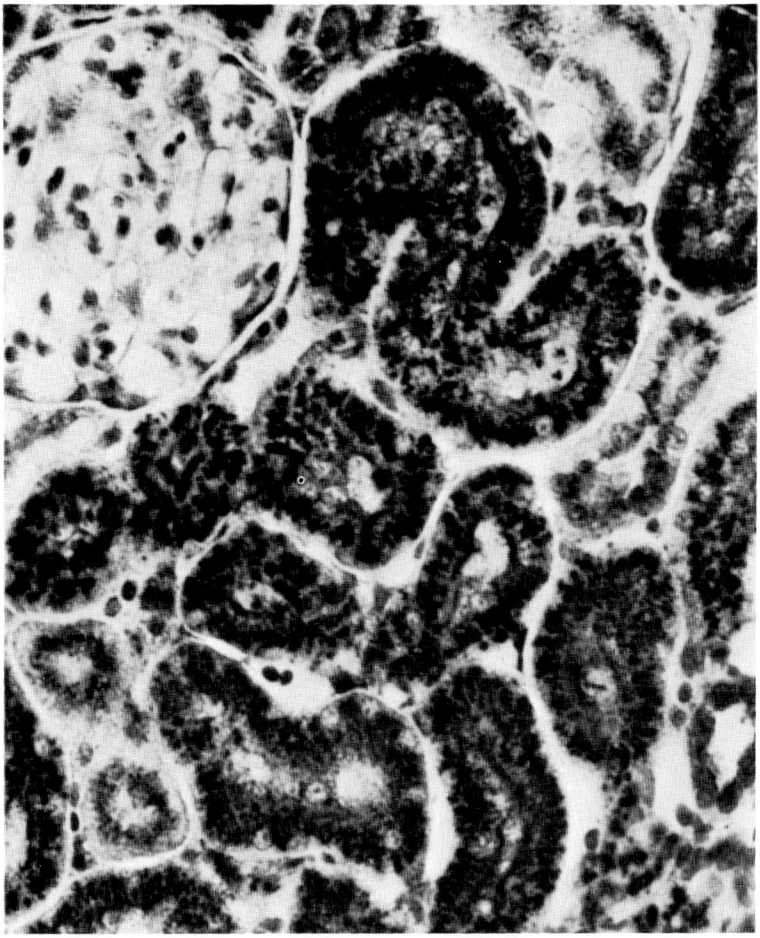

FIG. 6. Localization of acid phosphatase in rat kidney with naphthol AS-BI phosphate. Note both "droplets" and background staining.

3. The substrate should specifically interact with one enzyme or group of enzymes.

4. The developing agent should not affect the enzymatic process or the penetration of the substrate.

5. The product of the enzyme-catalyzed reaction should combine very rapidly with the developing agent and the velocity constant should not be affected by different cellular environment.

6. The colored product or dye should be immediately precipitated, i.e., it should have a very low solubility, should not supersaturate, should deposit in an amorphous or sub-microcrystalline state, and should be stable.

7. No preferential adsorption of any of the substances involved should occur in special sites which are unrelated to enzyme distribution.

Needless to say, such criteria are difficult to meet in their entirety, but the post-coupling, non-coupling, and oxidation techniques offer the best possibilities for achieving the aforementioned standards. More recently, Nachlas *et al.* (1957b) have also considered problems of enzymatic localization by reactions applied to tissue sections and confirm many of the conclusions previously outlined by the aforementioned work of Holt and O'Sullivan. They point out that discrete deposition of dye within a cell does not necessarily represent a more accurate location for all enzymes than does the widespread dispersion of the dye throughout cells. There can be no serious disagreement with this statement since an overall background cytoplasmic activity of enzymes such as acid phosphatase can be clearly demonstrated. On the other hand, specious attempts should not be made to justify artefactual staining reactions consisting of random and crystalline dye deposits, as well as poorly defined dye halos surrounding sites of high activity.

# APPENDIX 4

# METAL-DYE COMPLEXES AND MOUNTING MEDIA

## 1. *Metal Chelation of Azo-Dyes (Burstone and Weisburger, 1961a; Burstone, unpublished data)*

Certain azo-dyes produced under histochemical conditions may be complexed with metals. The chelates are generally formed if the diazo component of the dye contains coordinating groups (see p. 124). Following incubation and coupling, the slide is immersed in 10% aqueous copper sulfate, cobaltous acetate, nickel chloride, or chromium acetate for a period of 30 minutes. The solutions contain several drops of ammonium hydroxide or ethanolamine. Heating the solutions to 80° may facilitate formation of a stable cyclic chelate. With reference to certain post-coupling procedures, copper and cobalt chelates of azo-dyes derived from diazotized 3-aminofluoranthene, 5*H*-aminotetrazole, and 6-amino-2-mercaptobenzothiazole are insoluble in ethanol and xylene.

Recently several new bases which contain coordinating groups for chelation have become available. These are Variogen Base I and Variogen Base II.* Variogen Salt II is also available. These compounds are primarily intended for the production of metal complexes (cobalt, copper, and nickel) of naphthol AS dyes (Gross, 1960). Their chemical composition has not been disclosed. They appear to be quite inhibitory when employed with simultaneous coupling methods, but are entirely suitable for post-coupling procedures. Coupling is carried out at room temperature or at 37° in sodium acetate solution. The chelation solution contains $CuSO_4$, $CoCl_2$, or $NiSO_4$. In addition, 1.0% of a nonionic detergent, 2.0% sodium carbonate, and a metal "complex builder" such as nitrilotriacetic acid are present. The pH of the solution should be approximately 9.0. Sections are treated with this solution for 30–60 minutes at as high a temperature as is consistent with maintenance of tissue integrity.

## 2. *Mounting Media*

The review of Lillie *et al.* (1953) should be consulted with reference to a general consideration of mounting media. In contrast to the routine hematoxylin and eosin stain, histochemically derived dyes exhibit a variable susceptibility to the action of organic solvents. Some dyes, like the formazan derived from nitro-blue tetrazolium, are insoluble in a number of organic solvents. Others, such as azo-dyes derived from $\alpha$- and $\beta$-naphthols, exhibit varying degrees of solubility in solvents and mounting media. With some azo-dye techniques only part of the dye precipitate is soluble; this is

* Carbic-Hoechst Corp., Mountainside, New Jersey.

probably the part that is least tightly bound to protein. In such cases a sharper picture is obtained when the section is dehydrated and mounted. Such sharpness may, however, be misleading, in that it is merely the result of removal of the more diffuse stained areas by the solvent. For preliminary evaluation, slides should be mounted in glycerol-gelatin or even glycerin.

*a. Glycerol-gelatin.* The following are mixed and allowed to melt in a 60° paraffin oven:

| | | |
|---|---|---|
| 20.0 | g. | gelatin |
| 100.0 | ml. | glycerin |
| 100.0 | ml. | distilled $H_2O$ |

A small crystal of thymol or a few very small crystals of merthiolate (sodium ethylmercurithiosalicytate) are added as a preservative. Phenol should be avoided in glycerol-gelatin mixtures since it may reduce certain dyes.

*b. Gum syrup modification of Lillie and Ashburn (1943).* This medium contains the following:

| | | |
|---|---|---|
| 50.0 | g. | acacia (gum arabic) |
| 50.0 | g. | cane sugar |
| 100.0 | ml. | distilled $H_2O$ |

The acacia and sugar are dissolved in the distilled water by frequent shaking at 55–60°. Distilled water should be added to restore that lost by evaporation. Fifteen mg. of merthiolate or 100.0 mg. thymol are added as a preservative. The warm mixture is placed in a vacuum oven for several minutes to remove air bubbles.

*c. Polyvinylpyrrolidone (PVP) (Burstone, 1957).* Polyvinylpyrrolidone, a product of the dyestuff industry, is the basis of a water-soluble mounting medium which is compatible with many dyes. Stained sections are washed in distilled water and mounted directly. Fifty grams of polyvinylpyrrolidone (PVP) (Plasdone C) are dissolved in 50 ml. of distilled water and permitted to stand overnight. Two ml. of glycerol are then added and the mixture stirred. (A crystal of thymol may be added as a preservative.) The medium may be conveniently applied from a 2-ounce dropper bottle. The cap may be lubricated with silicone grease in order to prevent sticking. The index of refraction of this medium, 1.43, increases after the evaporation of some of the water. The dyes from many non-enzymatic histochemical stains, including pa-S, Alcian Blue, Celestin Blue, and Aldehyde Fuchsin, are adequately preserved by this mountant. Hematoxylin is preserved but eosin is badly leached. Cover glasses become immovable within an hour after mounting and the formation of bubbles is minimal.

*d. Polyvinyl acetate (PVA) (Burstone, 1957).* This is an alcohol-water solution of low viscosity polyvinyl acetate, which enables stained sections to be mounted directly from 80% ethanol. The medium is stable, easy to use, and exhibits excellent optical qualities.

A 25% solution of polyvinyl acetate (Gelva 2.5) in 80% ethanol was prepared by allowing 5.0 g. of PVA to dissolve in 20.0 ml. 80% ethanol for approximately 24 hours. The index of refraction was found to be 1.3865 and will increase following evaporation of the ethanol and water. This mounting medium was found to be most conveniently applied from a 2-ounce dropper bottle. A large number of azo and other dyes are preserved indefinitely in this mountant.

*e. Cellulose caprate (Lillie and Henson, 1955).* A 50% solution of cellulose caprate in xylene was found to form a clear pale yellowish syrupy solution of good consistency for mounting stained and unstained sections. The refractive index of the plastic solution is 1.4860 and that of the dry resin 1.4734. Cover slips become immovable in 30 minutes and slides do not stick together after being allowed to dry for 1 day. Cobalt sulfide is considerably bleached in about 4 weeks. However, a number of histochemical and non-specific stains are well preserved.

*f. PVP/VA copolymers (Lamkie and Burstone, 1962).* Several vinylpyrrolidone/vinyl acetate copolymers have been found to be excellent mounting media for the preservation of azo, oxidation, as well as a number of non-specific histological and histochemical, stains. Plastics most suitable as mounting media are those having a vinylpyrrolidone/vinyl acetate ratio of 70:30 (E-735) and 60:40 (E-635). In general, the E-735 copolymer is used because it exhibits the highest water tolerance. The medium is prepared by pouring the syrupy PVP/VA concentrate (which comes as a 50% ethanolic or isopropanolic solution) into a 25.0 ml. beaker and then adding the contents to an equal volume of water, taking care to insure complete mixing of all of the plastic. The mountant may be stored in a dropper bottle, the cap of which is lightly coated with silicone oil to prevent sticking. Well drained slides may be mounted directly from distilled water or else from 30–50% ethanol. The mounting medium is completely compatible with higher concentrations of alcohols including absolute ethanol.

## Commercial Sources of Reagents

1. Polyvinylpyrrolidone (PVP) (Plasdone C) and PVP/VA copolymers. Antara Chemicals, Division of General Aniline and Film Corp., New York, New York
2. Polyvinyl acetate (PVA), low viscosity (Gelva 2.5). Shawinigan Resins Corp., Springfield, Massachusetts
3. Cellulose caprate (cellulose tridecanoate) Distillation Products Industries, Division of Eastman Kodak Co., Rochester, New York

## REFERENCES

Adam, N. K. (1938) "The Physics and Chemistry of Surfaces," 2nd ed., pp. 250–251. Clarendon Press, Oxford, England.

Adams, D. A. A. (1951) *J. Soc. Dyers Colourists* **67**, 223–235.

Baeyer, A. (1883) *Ber.* **16**, 2188–2204.

Baeyer, A., and Drewsen, V. (1883) *Ber.* **16**, 2205–2208.

Baker, C. L., and Johnson, L. H. (1952) *Anal. Chem.* **24**, 1625–1626.

Baker, J. R. (1958) "Principles of Biological Microtechnique. A Study of Fixation and Dyeing." Methuen, London.

Baker, J. R. (1960) "Cytological Technique." Methuen, London.

Bamford, C. H., Boulton, J., Hanby, W. E., and Ward, J. S. (1954) *Discussions Faraday Soc. No.* **16**, 222–229.

Bello, F. (1959) *Fortune* **59** (May), 144–206.

Beyer, H., and Pyl, T. (1954) *Ber.* **87**, 1505–1511.

Bhat, R. V., Forster, R. B., and Venkataraman, K. (1940) *J. Soc. Dyers Colourists* **56**, 166–178.

Brass, K., and Sommer, P. (1928) *Ber.* **61**, 993–1002.

Braude, E. A. (1949) *J. Chem. Soc.* pp. 1902–1909.

Brode, W. R. (1929) *J. Am. Chem. Soc.* **51**, 1204–1213.

Burstone, M. S. (1957) *Am. J. Clin. Pathol.* **28**, 429–430.

Burstone, M. S. (1960) *J. Natl. Cancer Inst.* **24**, 1199–1217.

Burstone, M. S., and Weisburger, E. K. (1961a) *J. Histochem. and Cytochem.* **9**, 349–355.

Burstone, M. S., and Weisburger, E. K. (1961b) *J. Histochem. and Cytochem.* **9**, 712–713.

Bury, C. R. (1935) *J. Am. Chem. Soc.* **57**, 2115–2117.

Chessick, R. D. (1953) *J. Histochem. and Cytochem.* **1**, 471–485.

Chipalkatti, H. R., Giles, C. H., and Vallance, D. G. M. (1954) *J. Chem. Soc.* pp. 4375–4390.

Conant, J. B., Lutz, R. E., and Corson, B. B. (1941) *In* "Organic Synthesis," Coll. Vol. I (H. Gilman and A. H. Blatt, eds.), 2nd ed., pp. 49–52. Wiley, New York.

Conn, H. J. (1953) "Biological Stains," (6th ed.). Biotech. Publications, Geneva, New York.

Dacey, G. C. (1962) *Science* **135**, 71–74.

Defendi, V. (1957) *J. Histochem. and Cytochem.* **5**, 1–10.

Derrien, M. (1911) Cited in *Bull. soc. chim. France* **9**(IV), 110–111.

Dewar, M. J. S. (1949) "The Electronic Theory of Organic Chemistry." Clarendon Press, Oxford, England.

Doub, L., and Vandenbelt, J. M. (1947) *J. Am. Chem. Soc.* **69**, 2714–2723.

Drew, H. D. K., and Landquist, J. K. (1938) *J. Chem. Soc.* pp. 292–304.

Eistert, B. (1938) "Tautomerie und Mesomerie." Gleichgewicht und Resonanz. Enke, Stuttgart, Germany.

Elkins, M., and Hunter, L. (1935) *J. Chem. Soc.* pp. 1598–1600.

Elöd, E., and Einsele, V. (1960) *Melliand Textilber.* **41**, 1377–1384.

Evans, J. G. (1949) *In* "Fibre Science" (J. M. Preston, ed.), pp. 248–272. Textile Institute, Manchester, England.

Feigl, F., and Casto-Neto, C. (1956) *J. Soc. Dyers Colourists* **72**, 239–240.

Forster, R. B., and Soyka, C. (1931) *J. Soc. Dyers Colourists* **47**, 99–109.

Frey-Wyssling, A. (1948) "Submicroscopic Morphology of Protoplasm and Its Derivatives." Elsevier, New York.

Frey-Wyssling, A. (1953) "Submicroscopic Morphology of Protoplasm," 2nd ed. Elsevier, New York.

Friedenwald, J. S., and Becker, B. (1948) *J. Cellular Comp. Physiol.* **31,** 303–309.

Friedländer, P. (1909) *Ber.* **42,** 765–770.

Friedländer, P. (1922) *Ber.* **55,** 1655–1658.

Funkhouser, J. A., and Brode, W. R. (1934) *J. Am. Chem. Soc.* **56,** 2172–2177.

Giles, C. H. (1944) *J. Soc. Dyers Colourists* **60,** 303–315.

Giles, C. H., and MacEwan, T. H. (1959) *J. Chem. Soc.* pp. 1791–1798.

Glenner, G. G., Burstone, M. S., and Meyer, D. B. (1959) *J. Natl. Cancer Inst.* **23,** 857–872.

Gomori, G. (1952) "Microscopic Histochemistry." Univ. Chicago Press, Chicago, Illinois.

Gomori, G., and Benditt, E. P. (1953) *J. Histochem. and Cytochem.* **1,** 114–122.

Graham, D. (1955) *In* "The Chemistry of Synthetic Dyes and Pigments" (H. A. Lubs, ed.), pp. 662–688. Reinhold, New York.

Gränacher, C., Brüngger, H., and Ackermann, F. (1941) *Helv. Chim. Acta* **24,** 40E–71E.

Green, A. G., and Wolff, S. (1913) *Ber.* **46,** 33–49.

Griess, P. (1866) *Ann.* **137,** 39–91.

Gross, Dr. (1960) *Textil-Praxis* **15,** 1046–1048.

Haendler, H. M., and Smith, G. McP. (1940) *J. Am. Chem. Soc.* **62,** 1669–1672.

Hall, T. (1961) *Science* **134,** 449–455.

Hauser, C. R., and Breslow, D. S. (1941) *J. Am. Chem. Soc.* **63,** 418–420.

Heiduschuka, A., and Goldstein, E. (1916) *Arch. Pharm.* **254,** 584–625.

Heller, G. (1912) *Ann.* **392,** 16–48.

Henn, R. W. (1953) *Photo. Soc. Am.* **19,** 131–136.

Hodgson, H. H., and Foster, C. K. (1942) *J. Chem. Soc.* pp. 30–33.

Holt, S. J. (1958) *In* "General Cytochemical Methods" (J. F. Danielli, ed.), pp. 375–397. Academic Press, New York.

Holt, S. J., and O'Sullivan, D. G. (1958) *Proc. Roy. Soc.* **B148,** 465–480.

Holt, S. J., and Sadler, P. W. (1958a) *Proc. Roy. Soc.* **B148,** 495–505.

Holt, S. J., and Sadler, P. W. (1958b) *Proc. Roy. Soc.* **B148,** 481–494.

Howitt, F. O. (1949) *In* "Fibre Science" (J. M. Preston, ed.), pp. 50–67. Textile Institute, Manchester, England.

Javan, A. (1960) *In* "Conference on Quantum Electronics; Quantum Electronics, A Symposium" (C. H. Townes, ed.), pp. 564–571. Columbia Univ. Press, New York.

Johansen, G., and Linderstrøm-Lang, K. (1952) *Acta. Med. Scand. Suppl. 266,* 601–613.

Johnson, A. J. (1955) *In* "The Chemistry of Synthetic Dyes and Pigments" (H. A. Lubs, ed.), pp. 551–576. Reinhold, New York.

Klotz, I. M., and Loh-Ming, W. C. (1954) *J. Am. Chem. Soc.* **76,** 805–814.

Knorr, L. (1887) *Ann.* **238,** 137–219.

Koltun, W. L., Dexter, R. N., Clark, R. E., and Gurd, F. R. N. (1958) *J. Am. Chem. Soc.* **80,** 4188–4194.

Koltun, W. L., Clark, R. E., Dexter, R. N., Katsoyannis, P. G., and Gurd, F. R. N. (1959) *J. Am. Chem. Soc.* **81,** 295–301.

Kramer, D. N., and Gamson, R. M. (1958) *Anal. Chem.* **30,** 251–254.

Kuhn, R., and Machemer, H. (1928) *Ber.* **61,** 118–127.
Kunz, K. (1922) *Ber.* **55,** 3688–3691.
Lamkie, N. J., and Burstone, M. S. (1962) *Stain Technol.* **37,** 109–110.
Land, E. H. (1959) *Sci. Am.* **200,** 84–99.
Lillie, R. D., and Ashburn, L. (1943) *Arch. Pathol.* **36,** 432–435.
Lillie, R. D., and Henson, J. P. (1955) *Stain Technol.* **30,** 133–134.
Lillie, R. D., Burtner, H. J., and Henson, J. P. G. (1953) *J. Histochem. and Cytochem.* **1,** 154–159.
Loveless, A., and Danielli, J. F. (1949) *Quart. J. Miscroscop. Sci.* **90,** 57–66.
Master, R. W. P. (1960) *Biochim. et Biophys. Acta* **39,** 159–160.
May, C. E., and Hunt, H. (1928) *Ind. Eng. Chem.* **20,** 384–388.
Mayer, F. (1943) "The Chemistry of Natural Coloring Matter" (Translated and revised by A. H. Cooke). Reinhold, New York.
Michaelis, L., and Hill, E. S. (1933) *J. Am. Chem. Soc.* **55,** 1481–1494.
Michaelis, L., Shubert, M. P., and Granick, S. (1939) *J. Am. Chem. Soc.* **61,** 1981–1992.
Nachlas, M. M., Crawford, D. T., and Seligman, A. M. (1957a) *J. Histochem. and Cytochem.* **5,** 264–278.
Nachlas, M. M., Young, A. C., and Seligman, A. M. (1957b) *J. Histochem. and Cytochem.* **5,** 565–583.
Nineham, A. W. (1955) *Chem. Revs.* **55,** 355–483.
O'Sullivan, D. G. (1955) *Bull. Math. Biophys.* **17,** 243–255.
O'Sullivan, D. G. (1956) *Bull. Math. Biophys.* **18,** 199–203.
Pearse, A. G. E. (1957) *J. Histochem. and Cytochem.* **5,** 515–527.
Pearse, A. G. E. (1960) "Histochemistry, Theoretical and Applied," 2nd ed. Little Brown, Boston, Massachusetts.
Person, P., and Fine, A. (1961) *J. Histochem. and Cytochem.* **9,** 190–196.
Poray-Koshits, A. E., Ginzburg, O. F., and Poray-Koshits, B. A. (1947) *J. Gen. Chem. U.S.S.R. (Eng. Transl.)* **17,** 1752–1757.
Pratt, L. S. (1947) "The Chemistry and Physics of Organic Pigments." Wiley, New York.
Ravin, H., Zacks, S., and Seligman, A. M. (1953) *J. Pharmacol. Exptl. Therap.* **107,** 37–53.
Reynolds, W. B. (1943) *Am. Dyestuff Reptr.* **32,** 455–467.
Robinson, C. (1954) *Discussions Faraday Soc.* **16,** 125–132.
Ruggli, P., and Courtin, A. (1932) *Helv. Chim. Acta* **15,** 75–108.
Russig, F. (1900) *J. prakt. chem.* **62**[2], 30–60.
Rutenburg, A. M., and Seligman, A. M. (1956) *J. Histochem. and Cytochem.* **4,** 17–22.
Sadler, P. W. (1960) *Chem. Rev.* **60,** 575–592.
Schmidt, W. A., Tulagin, V., Sprung, J. A., Gunther, R. C., Coles, R. F., and Sargent, D. E. (1953) *Ind. Eng. Chem.* **45,** 1726–1729.
Sidgwick, N. V., Taylor, T. W. I., and Baker, W. (1937) "The Organic Chemistry of Nitrogen," pp. 15, 112, 193, 224. Clarendon Press, Oxford.
Sircar, A. C. (1916) *J. Chem. Soc.* **109,** 757–776.
Smith, F. E. (1951) *Science* **113,** 751–754.
Spiegler, L. (1953) *J. Org. Chem.* **18,** 1292–1316.
Struve, W. S. (1955) *In* "The Chemistry of Synthetic Dyes and Pigments" (H. A. Lubs, ed.), pp. 607–624. Reinhold, New York.
Thomas, B. S., and Weisburger, A. (1932) *In* "The Theory of the Photographic Process" (C. E. K. Mees, ed.), p. 394. Macmillan, New York.

Thorpe, J., and Ingold, C. K. (1923) "Synthetic Colouring Matters; Vat Colours," pp. 1–169. Longmans, Green, London.

Tousimis, A. J. (1962) *I. S. A. Proc.* **8,** 53–56.

Tousimis, A. J., and Adler, I. (1963) *J. Histochem. and Cytochem.* (in press).

Valko, E. I. (1939) *J. Soc. Dyers Colourists* **55,** 173–182.

Van Alphen, J., and Drost, G. (1949) *Rec. trav. chim.* **68,** 54–56.

Venkataraman, K. (1952) "The Chemistry of Synthetic Dyes," Vol. 2. Academic Press, New York.

Vickerstaff, T. (1950) "The Physical Chemistry of Dyeing," p. 166. Interscience, New York.

Vickerstaff, T. (1953) *J. Soc. Dyers Colourists* **69,** 279–282.

Vittum, P. W., and Brown, G. H. (1946) *J. Am. Chem. Soc.* **68,** 2235–2239.

Vittum, P. W., and Brown, G. H. (1949) *J. Am. Chem. Soc.* **71,** 2287–2290.

Vittum, P. W., Sawdey, G. W., Herdle, R. A., and Scholl, M. K. (1950) *J. Am. Chem. Soc.* **72,** 1533–1536.

Weinstein, J., and Wyman, G. N. (1956) *J. Am. Chem. Soc.* **78,** 2387–2390.

Witt, O. N. (1888) *Ber.* **21,** 3468–3489.

Wizinger, R., and Biro, V. (1949) *Helv. Chim. Acta* **32,** 901–912.

Wizinger, R., and Herzog, H. (1953) *Helv. Chim. Acta* **36,** 531–536.

Wolfgram, F. (1961) *J. Histochem. and Cytochem.* **9,** 171–175.

Zollinger, H. (1958) *Textil Rundschau* **13,** 217–226.

Zollinger, H. (1959) *Textil Rundschau* **14,** 113–127.

Zollinger, H. (1961) "Azo and Diazo Chemistry." Interscience, New York.

CHAPTER 5

# PHOSPHATASES

## I. INTRODUCTION AND GENERAL CONSIDERATIONS

The fact that enzyme activity could liberate phosphoric acid from certain compounds has been known since the beginning of this century. Historical review of these earlier discoveries of the location and hydrolytic action of phosphatases has been made by Roche (1950), Roche and Thoai (1950), and Sumner and Somers (1953). By 1912 it was known that in addition to plant tissues, phosphatases could be found in kidney, spleen, pancreas, and bone. In 1922 Robison found that when phosphatase hydrolyzed hexose monophosphate in the presence of a soluble calcium salt, a precipitate of tricalcium phosphate was produced. Robison (1923) performed the first macroscopic histochemical test for the visualization of alkaline phosphatase by incubating pieces of bone from rachitic rats in a calcium hexose monophosphate substrate solution followed by treatment of the specimens with silver nitrate. An intense stain was observed in cartilage cells of the zone of provisional calcification, and not in proliferating cartilage. This, of course, is the technique which was modified by Gomori (1939) and Takamatsu (1939) for the microscopic visualization of sites of phosphatase activity.

The classification of phosphatases including pH optima and substrates modified after Sumner and Somers (1953, p. 89) is shown in Table 1.

Relatively little is known about the biological function of phosphatases despite their ubiquitous distribution. The relationship between genetic factors and the biosynthesis of phosphatases is shown in diseases such as hypophosphatasia and glycogen storage disease, which are characterized by phosphatase deficiency (Cori and Cori, 1952; Cusworth, 1958; Duncan, 1959, p. 407; Rosenthal *et al.*, 1960). Alkaline phosphatase has been associated with the formation of fibrous proteins and the passage of metabolites across cell membranes (Verzár and McDougall, 1936; Moog, 1946; Bradfield, 1950; Danielli, 1954). This hypothesis derives some support

TABLE 1

CLASSIFICATION OF PHOSPHATASES

| Type | Substrates |
|---|---|
| A. Phosphomonoesterases | |
| 1. Phosphomonoesterase I (alkaline phosphatase) pH 9.2–9.6 activated by $Mg^{++}$ | $\alpha$- or $\beta$-Glycerophosphate, dihydroxyacetone phosphate, adenine dinucleotide, phenylphosphates, and naphthyl phosphates |
| 2. Phosphomonoesterase II (acid phosphatase) pH 4.6 not activated by $Mg^{++}$ | Same as above |
| 3. Phosphomonoesterase III (acid phosphatase) pH 3.4–4.2 inhibited by $Mg^{++}$ | Same as above |
| 4. Phosphomonoesterase IV (acid phosphatase) pH 4.2–6.2 activated by $Mg^{++}$ | Same as above |
| 5. Fructose diphosphate monoesterase pH 9.3–9.5 activated by $Mg^{++}$ | Fructose diphosphate |
| 6. Glucose-6-phosphatase pH 6.5–7.0 | Glucose-6-phosphate |
| 7. Glucose-1-phosphatase pH 4.0–4.5 $Mg^{++}$ not required | Glucose-1-phosphate |
| 8. Serine phosphomonoesterase pH 7.0 requires $Mg^{++}$ Inhibited by D-serine, alanine, and glycine | Phosphoserine |
| 9. Phosphatidic acid phosphatase pH 6.0–6.5 Inhibited by $Mg^{++}$, $Ca^{++}$, and $Mn^{++}$ | L-$\alpha$-Phosphatidic acid |
| B. Phosphodiesterases | |
| 1. Nucleases (nucleophosphodiesterases) pH 8.9–9.8 activated by $Mg^{++}$ | Diglyceryl phosphoric ester, diphenyl phosphoric ester, diethyl phosphoric ester, bis(*p*-nitrophenyl)phosphate, thymidine-5′-*p*-nitrophenylphosphate |
| 2. Lipophosphodiesterases | Phospholipids (e.g. lecithin and cephalin) |
| C. Pyrophosphatases | |
| 1. Pyrophosphatase I pH 7.6–8.2 activated by $Mg^{++}$ | Sodium pyrophosphate, diphenyl and dinaphthyl pyrophosphates |
| 2. Pyrophosphatase II pH 5.0–5.5 activated by $Mg^{++}$ | Sodium pyrophosphate, diphenyl and dinaphthyl pyrophosphates |
| D. 5-Nucleotidase pH 8.5 activated by $Mn^{++}$ | Adenylic acid |
| E. Metaphosphatases pH 5.0–8.0 | Sodium metaphosphate |

TABLE 1—*Continued*

| Type | Substrates |
|---|---|
| F. Phosphoacylases pH 5.4 inhibited by $PO_4$, nucleic acid, and hyaluronic acid | Acetyl phosphate and higher homologues |
| G. Phosphamidases pH 5.5 | Phosphocreatine, phosphoarginine |
| H. Phytase pH 5.0 activated by $Mg^{++}$ | Phytin |
| I. ATPase pH 7.7 | Adenosine triphosphate and adenosine diphosphate |
| J. Ophio-ATPase pH 8.0–8.4 | Adenosine triphosphate |
| K. Apyrase pH 5.5–6.5 | Adenosine triphosphate |
| L. Phosphoprotein phosphatase pH 5.0–6.0 | Casein, tyrosine phosphate and phenylphosphate |

from the common histochemical localization of alkaline phosphatase in basement membranes of many tissues. Samorajski and McCloud (1961) have published a series of observations which suggest a relationship between alkaline phosphatase levels of the vascular endothelium of the central nervous system and changes in permeability of the blood-brain barrier. The relationship between hormonal influences and phosphatase is demonstrated by the appearance of acid phosphatase in the human prostate gland only after puberty (Gutman and Gutman, 1938). Ring (1950) has correlated changes in phosphatase activity of the vaginal mucosa with the estrous cycle in the rat. Similar changes have also been observed in the oral epithelium (Ring and Levy, 1950). The human placenta exhibits high alkaline phosphatase activity which increases during pregnancy. It has been suggested that this increase is related to the absorption of maternal blood glucose by the placenta (Anagnostopoulos and Matsudaira, 1958). The function of alkaline phosphatase in the calcification process has yet to be completely elucidated. Robison and Soames (1924) originally suggested that alkaline phosphatase effected the release of phosphate ions and thus exerted a direct effect upon the formation of bone mineral. This and similar data have been reviewed by Roche (1950, p. 486). Unfortunately, the nature of a possible *in vivo* phosphate ester has not been elucidated. Recently, definitive histochemical localizations of phosphatases in frozen-dried undecalcified tissues have been described (Burstone, 1960a, b). The fact that the *in vitro* pH optimum of alkaline phosphatase occurs at a highly alkaline pH does not preclude appreciable activity of the enzyme at physiological pH. For example, the pH optimum of the enzyme in the presence of low concentrations of glycerophosphate is near neutrality (Ross *et al.*, 1951). Kato and Moog (1958) have made the interesting observation that administration of phenylphosphate to chick embryos results in an increased alkaline phosphatase activity in the duodenum, kidney, and liver. This increase was not attributed to a non-specific effect of the substrate or products of hydrolysis, and appears to represent an adaptive enzyme response. Moog (1962) has further reported on developmental adaptation of alkaline phosphatases in the small intestine, and notes that the high alkaline phosphatase activity of the mature intestine reaches this level by intensive accumulation during certain critical periods that anticipate the onset of absorptive activity. Moog (1962) also has reviewed literature with reference to the possible functional role of alkaline phosphatase activity and notes certain implications which relate phosphatase to the "sodium pump" which may be coupled to the active intake of sugar molecules.

From a diagnostic standpoint leukocytic alkaline phosphatase activity in normal individuals and those with certain types of leukemia has been reported to exhibit characteristic variations (Valentine and Beck, 1951; Valentine *et al.*, 1952). Of interest are histochemical techniques for the demonstration of alkaline phosphatase in leukocytes (Wachstein, 1946; Kaplow, 1955; Burstone, 1958a) (Appendix 5B, Section 15).

Specific phosphomonoesterases are believed to play a significant role in intermediary metabolism (Staff of *Chem. Eng. News*, 1961; Fraenkel-Conrat and Stanley, 1961; Calvin, 1961). Glucose-6-phosphatase of mammalian liver catalyzes the reaction resulting in the formation of blood glucose. The activity of phosphatidic acid phosphatase is a significant pathway for the formation of diglycerides from glycerophosphate. With reference to ATPase, its substrate, adenosine triphosphate, which was originally isolated from muscle, is known to play an important role in many biological processes involving energy transfer, including muscle contraction and oxidative phosphorylation. Its hydrolysis to adenosine diphosphate (ADP, adenosine-5′-pyrophosphate) and adenylic acid (adenosine-5′-phosphate, AMP) is well documented in the biochemical literature. The synthesis of these compounds, including ATP, is described in the classical papers of Baddiley and Todd (1947) and Baddiley *et al.* (1949). Mandl *et al.* (1952) have shown that salts of ATP are capable of solubilizing biological inorganic and organic compounds at neutral pH, and suggest that ATP may play an important role in *in vivo* disaggregation associated with biological processes.

The metabolic role of non-specific acid phosphatase is not clear. Macrophages, and giant cells associated with lytic phenomena, have long been known to exhibit relatively high acid phosphatase activity. Osteoclasts adjacent to resorbing bone and cartilage also are highly active (Burstone, 1960a, b). Recently, a study of mononuclear phagocytes of inflammatory exudates has been made with reference to both acid phosphatase and phagocytic activities (Dannenberg *et al.*, 1960). On the other hand, Norberg (1950b) demonstrated that liver regeneration in the rat is associated with an increased acid phosphatase activity which suggests a relationship between the enzyme and protein synthesis. It is conceivable that acid phosphatase may be associated with either synthesis or differentiation as well as with phagocytosis and dissolution of tissue components, depending upon the specific tissue involved and its physiological state. Barka (1962), in an evaluation of acid phosphatase staining reactions, concluded that pinocytotic and reverse pinocytotic processes are universally visualized. In addition, the acid phosphatase activity of various cells is related to the development and actual activity of the pinocytotic and reverse pinocytotic apparatus including secretion vacuoles. Variations of acid phosphatase in the intestinal epithelium in relation to fat absorption were also noted. Increase of acid phosphatase activity in various experimental conditions resulting in hypoxia and cell damage was observed to cause compensatory increase of pinocytotic activity. Nigam *et al.* (1959) suggest that prostatic acid phosphatase functions in part as a transferase. Roche (1950, p. 486) comments that most tissues endowed with intense alkaline phosphatase activity are practically devoid of acid phosphatase activity. This view is confirmed by the mutually exclusive distribution of alkaline phosphatase and acid phosphatase in osteoblasts and osteoclasts, respectively (Burstone, 1960a, b).

A phosphatase with a pH optimum in the acid range, phosphamidase, is of interest because it hydrolyzes the N—P high energy bond. Based upon the original findings of Gomori (1948) that many neoplasms exhibit a high phosphamidase activity, a number of potential carcinostatic compounds containing an N—P bond have been prepared and studied clinically (Chapter 15).

In addition to their hydrolytic activity, non-specific phosphatases are also capable of catalyzing the synthesis of phosphates:

$$\text{Alcohol} + \text{phosphoric acid} \rightleftharpoons \text{ester} + \text{water}$$

Kay (1928) found that in the presence of an excess of alcohol, inorganic phosphate could be esterified using phosphatases derived from various tissues as catalysts. The reaction was usually carried out by mixing sodium phosphate solutions at pH 8.4 with duodenal or kidney extracts, and adding alcohols to the mixture. Glycerol, ethylene glycol, methanol, and ethanol were esterified in this manner. Phosphatases are also involved in transphosphorylation reactions (Morton, 1955d, p. 556).

a. Phenylphosphate + glucose → glucose-6-phosphate + phenol

b. Phenylphosphate + glycerol → glycerophosphate + phenol

Green and Meyerhof (1952) studied the rates of transphosphorylation to glycerol from acetyl phosphate, phosphocreatine, and *p*-nitrophenylphosphate, and found that the highest rate of transphosphorylation was given by acetyl phosphate. As yet, histochemical demonstration of this type of phosphate synthesis or transfer has not been reported. An excellent review including more recent aspects of phosphate transfer has been discussed by Axelrod (1956).

Although different phosphatases may be characterized by biochemical techniques, their histochemical differentiation is a complex problem about which there has not been unanimity of opinion. In addition to the well known phosphatases, several biochemical studies have suggested that certain substrates are capable of characterizing other types of phosphatases. Roche and Latreille (1937) have differentiated between a "glycerophosphatase" and a "phenylphosphatase," while Bowers *et al.* (1940) believe that aminoethyl(ethanolamine)phosphate is hydrolyzed by a specific enzyme. Belfanti *et al.* (1935) suggest that the alkaline phosphatase of bone is different from that of kidney and liver. Hotta (1934) studied the enzymatic hydrolysis of a number of aliphatic and aryl phosphate esters (e.g. propylphosphate, *sec*-butylphosphate, cyclohexanol phosphate, and menthol phosphate) over a wide pH range and concluded that there exists a wide spectrum of different phosphatases. Fleury *et al.* (1950) noted substrate specificities of phosphatases from kidney, bone, intestine, prostate, and urine and concluded that the enzymes were broad in their action upon various substrates but could be characterized to some degree according to pH optimum and effect of inhibitors. The substrates employed included: ethanolamine phosphate and analogues, $\beta$-glycerophosphate, choline phosphate, morpholine ethanol phosphate, benzyl alcohol phosphate, glucose-1-phosphate, and methyl naphthoquinone phosphate. More recently, Landau and Schlamowitz (1961) have differentiated alkaline phosphatases of liver, kidney, spleen, and bone from the intestinal enzyme by chromatographic and inhibitory studies. The evidence derived from these indicates that the activity of alkaline phosphatase against a variety of phosphomonoesters is due to an enzyme with a broad specificity rather than to a family of separate enzymes. They point out that the method of extraction and preparation affects the chromatographic behavior of phosphatases and makes it clear that the multiplicity of phosphatase peaks observed with the tissue such as kidney does not necessarily establish the existence of several enzymes in the *native* tissue. Multiple peaks may indeed represent etiologically distinct enzymes or they may represent a single enzyme complexed with other protein or active products of autolysis. The histochemical demonstration of phosphatases not only offers possibilities of confirming some of the aforementioned findings, but also may shed light on variations in cellular and intracellular sites of activity of the presumably different enzymes. The problem is complicated by the fact that non-specific alkaline phosphatase is known to catalyze the hydrolysis of phosphamides and compounds other than simple orthophosphoric acid monoesters. Morton (1955b) has found that purified alkaline phosphatase preparations from cow's milk and calf intestinal mucosa were capable of rapidly hydrolyzing all true orthophosphate monoesters, as well as the P—N bond of phosphocreatine, and the enol phosphate, phosphopyruvic acid. Sodium pyrophosphate, sodium hexametaphosphate, diphenylpyrophosphate, ATP, ADP, and DPN were not hydrolyzed. Whether alkaline phosphatase as it exists in tissue sections shows the same clear cut specificity is questionable. Morton (1955c) also found that purified phosphatase preparations could quantitatively convert TPN into DPN. DPN was not hydrolyzed.

Gomori (1949a) tested the histochemical specificity of a large number of phosphate esters by employing his metal-salt method for the demonstration of phosphatases at pH 5.0, 7.0, and 9.0. Among the substrates tested were: metaphosphate, pyrophosphate, methylphosphate, glycerophosphate, aminoethylphosphate, phenylphosphate, *o*-chlorophenylphosphate, $\alpha$-naphthyl phosphate, rescorcinol phosphate, phenolphthalein phosphate, glucose-1-phosphate, hexose diphosphate, adenosine triphosphate, yeast nucleate and thymonucleate, phosphorylcholine, lecithin, octanoylphosphate and *p*-chloroanilidophosphate. It was found that pyrophosphate did not give a positive reaction at any pH. With metaphosphate, a minimal reaction was obtained on the surface of intestinal villi at pH 9.0. Gomori also observed that the reaction in the liver with hexose diphosphate, or in muscle with adenosine triphosphate, was identical to that seen with glycerophosphate, and assumed that hexosediphosphatase and ATPase apparently do not survive the paraffin embedding procedure. With the last substrate, *p*-chloroanilidophosphoric acid, a strikingly different picture of enzymatic distribution was observed in the acid pH range, and this presumably delineated sites of phosphamidase activity. Gomori (1949b), in a continuation of the aforementioned study, investigated ten additional substrates. These included catechol and hydroquinone monophosphates, glucose-6-phosphate, fructose-6-phosphate, phosphoglyceric acid, yeast adenylic acid (3-nucleotide), muscle adenylic acid (5-nucleotide), dimethyl and diphenyl acid pyrophosphates, and casein. The substrates were used at two pH values, 5.0 and 9.2. Slides incubated in glycerophosphate were used as controls. No reaction was obtained with casein at either pH level. The pattern of distribution of activity was essentially the same with all substrates with the exception of 5-nucleotide. With the latter substrate, an apparently specific reaction delineated axons in the gray cortex of rat and mouse brain. In the human spleen, the centers of the Malpighian bodies are intensely stained as contrasted to the non-specific activity which is limited to the periphery of these structures. The testes of man, mouse, and rat showed an intense activity of nuclei of spermatogenic elements. The non-specific activity is much weaker. Smooth muscle of vessel walls and of some viscera (urinary bladder, stomach) exhibited sites of intense staining with 5-nucleotide as contrasted with a negligible reaction in the same sites with glycerophosphate. These histochemical findings tend to confirm previous biochemical observations that 5-nucleotidase is a specific enzyme differing from non-specific alkaline phosphatase. Gomori obtained useful histochemical localizations with catechol monophosphate and with glucose- and fructose-6-phosphates, and with phenylpyrophosphate. However, no significant differences in localization of activity were seen with the aforementioned substrates when compared with the glycerophosphate control.

Gomori (1949a) strongly cautioned against the uncritical interpretations of "specific" phosphatase reactions based upon some variations in the histological patterns seen when different substrates were used. In a similar vein, Moog and Steinbach (1946) rejected certain previous claims for the demonstration of specific enzymes and indicated the necessity of obtaining purified enzyme preparations in order to substantiate the existence of distinct phosphatases. Gomori (1949a) also pointed out that the low solubility of many substrates, as well as variations in pH optima, may produce variations in staining. Owing to these and other factors, criteria for the identity or non-identity of phosphatases pose a complex problem. Only marked and uniform differences in the histological distribution of sites of activity, obtained by changing conditions such as substrate, pH, and inhibitors and activators, can be accepted as presumptive evidence for the presence of different enzymes. Gomori questioned the work of Glick and Fisher (1946) with reference to the specific demonstration of ATPase, and also claims of other workers to the effect that they could demonstrate a number of specific phosphatases with different substrates. Despite the caution of Gomori, a number of workers have subsequently reported the demonstration of specific alkaline phosphatases. Maengwyn-Davies and Friedenwald (1950) claimed that they could distinguish a number of different types of phosphatases, without the interfering influence of alkaline phosphatase. They employed excessively long incubations of from 2 to 72 hours with unfixed frozen sections. The incubating medium contained a high concentration of sodium acetate which was stated to enhance the quality of localization. The metal-salt method of Gomori was employed, using a number of phosphate esters. In addition, $\beta$-naphthyl phosphate with Diazo Blue B was utilized in an azo-dye procedure. $\beta$-Naphthyl phosphate (as well as other $\beta$-naphthyl esters) is unsatisfactory for definitive enzyme localizations and would scarcely be expected to be of value in differentiating specific enzymes on the basis of microscopic observation.

Maengwyn-Davies and Friedenwald (1950) also reported that glycerophosphate, glucose-1-phosphate, glucose-6-phosphate, and fructose-6-phosphate are dephosphorylated by two different alkaline phosphatase enzyme systems, one of which is sharply localized in a limited number of organs and tissues and is substrate specific, while the other is present in all cells and requires magnesium for its activation. They further stated that no evidence was found for the existence in native tissue of a nonspecific alkaline phosphatase capable of hydrolyzing a wide variety of substrates. There appears to be no support for this contention in view of the numerous biochemical findings relating to the hydrolysis of a wide variety of substrates by alkaline phosphatase. Burgos *et al.* (1955) undertook similar studies employing improved histochemical procedures. However, among the substrates employed were β-naphthyl phosphate, 6-benzoyl-β-naphthyl phosphate, and indoxyl phosphate: compounds which may yield diffusion artefacts. They did find that certain inhibitors such as L-cysteine or glutathione depressed the reaction with naphthyl phosphate more than with glycerophosphate. On the other hand, the glycerophosphate reaction was more sensitive to potassium cyanide than was the naphthyl phosphate one. It is possible that careful utilization of the starch gel electrophoresis technique (zymogram) may elucidate the possible nature of different types of alkaline phosphatase. On the other hand, the finding of different "zymogram" patterns when different substrates are employed should not be construed as *ipso facto* evidence that different enzymes are invariably being distinguished. Another inherent difficulty in the attempt to characterize different phosphatases based upon variations in localization of stain relates to the fact that biochemically different phosphatases may have the same histological localization (Naidoo and Pratt, 1953).

With reference to differential staining of tissue elements, Nadel and Wodinsky (1955) found that keratohyaline granules reacted with the Gomori metal-salt technique, but not with the α-naphthyl phosphate procedure. They discuss these differences in the light of artefactual staining and do not suggest that different types of phosphatases are being demonstrated. Also of interest is the original study of Danielli (1946) in which he employed β-naphthyl phosphate for the demonstration of alkaline phosphatase by the routine azo-dye method as well as by visualization of the released phosphate by the metal-salt procedure. Phenolphthalein phosphate was employed in an analogous procedure. Differences in distribution of staining were attributed by Danielli to artefactual characteristics of each method and not to the demonstration of different phosphatases.

Bourne (1957, 1958) has employed a number of phosphate esters in conjunction with the metal-salt technique. Included were cortisone, estrone, galactose, riboflavin-5-, pyridoxal-, carbamyl-, ethanolamine, and carboxyphenyl- phosphates. Interesting variations in types of structures stained as well as intensity of reaction were reported in the central nervous system as well as in other tissues. Golarz *et al.* (1961) employed glycolytic intermediates, high energy phosphates, and related compounds including phosphorylated coenzymes as substrates in a histochemical study of human muscular dystrophy. Included among the compounds were phosphoglyceric acid, AMP, ADP, ATP, creatine phosphate, UTP, IMP, acetyl phosphate, GTP, thiamine pyrophosphate, and triphosphopyridine nucleotide (TPN). The work of this group is significant in that they have employed more phosphorylated metabolic intermediates as substrates than any other investigators. Further study should elucidate the exact specificity of these substrates when employed under histochemical conditions.

With reference to specific phosphatases, there is little question that enzymes such as ATPase, glucose-6-phosphatase, phosphamidase, and polyphosphatases can be demonstrated by histochemical techniques in conjunction with appropriate controls. Gomori (1952) has found by biochemical techniques three types of alkaline phosphatases of dog and calf intestine. As yet there does not appear to be any differential histochemical technique for the demonstration of these.

An interesting approach to the delineation of specific phosphatases has been reported by Bonting *et al.* (1960) who employed the quantitative histochemical technique developed by Lowry (1953) in a study of the properties and optimum assay conditions of four phosphomonoesterases (alkaline and acid phosphatases, ATPase, and glucose-6-phosphatase) in the human kidney. The quantitative distribution of the enzymes was determined by the technique of microdissection, weighing, and assay of anatomical units of the nephron.

## A. Substrate Types and Synthetic Approaches

The general types of substrates for phosphatases include esters and amides of phosphoric acid. Of these, monoesters of alcohols and phenols $(R{-}O)OP(OH)_2$, and diesters $(R{-}O)_2OP(OH)$, have been utilized for the most part. In general, the nature of the radical R does not affect the specificity of the reactions. However, in certain instances (e.g. ATP) the nature of the radical may confer distinct specificity upon the reaction. Esters of pyrophosphoric acid and its analogues, as well as esters of metaphosphoric acid, are also phosphatase substrates. The characteristic groupings of some of the aforementioned types are shown in Table 2. With reference to secondary esters, Neuberg and Wagner (1926) found that kidney extracts hydrolyzed diethylphosphate. More recently, however, Bamann and Riehl (1955) reported that hog and beef liver phosphatases do not hydrolyze either diethyl- or dimethylphosphates, but are capable of hy-

TABLE 2

GROUPINGS OF SPECIFIC PHOSPHATASE SUBSTRATE TYPES

| | |
|---|---|
| 1. C—C—O*P[a] | Phosphate esters of alkyl alcohols |
| 2. R—O*P | Aryl phosphate esters (R = aryl grouping) |
| 3. C═C—O*P | Enol phosphates (e.g. phosphopyruvic acid) |
| 4. R—CO—O*P | Acyl phosphates (e.g. acetyl phosphate) |
| 5. $(RO)_2$*P | Diaryl phosphates (R = aryl grouping) |
| 6. P*OP | Inorganic pyrophosphate |
| 7. R—N*P | Phosphamide (R = aryl grouping, or creatine or arginine) |

[a] Asterisk indicates point of hydrolysis.

drolyzing diphenyl- and *n*-dibutylphosphates in both alkaline and acid pH ranges. Phosphatases also hydrolyze the S—P bond of thioesters, including symmetrical anhydrides (O'Brien, 1960, p. 116). Studies are presently in progress in the laboratory of the author which deal with the application of naphthol thiophosphates as phosphatase substrates.

More substrates are available for the histochemical demonstration of phosphatases than for any other enzyme technique. This is due to the fact that phosphatases will hydrolyze a large number of diverse phosphate esters, and also that two different types of histochemical methods, the azo-dye and metal-salt procedures, are available. Substrates which are suitable for the azo-dye method may also be employed with the metal-salt technique, but the converse of this does not always apply. Among the first biochemical and histochemical substrates used for the demonstration of phosphatases were naturally occurring compounds or alkyl or aryl phosphates prepared by phosphorylating alcohols (e.g. glycerol or phenol). Subsequently a large number of aromatic and heterocyclic compounds were employed. The synthesis of a number of biochemical phosphatase substrates is described in "Methods in Enzymology" (Colowick and Kaplan, 1957). Potential histochemical substrates are found in Tables 3 and 4.

A number of synthetic approaches may be employed in the preparation of phosphatase substrates (Appendix 5A). A time-honored procedure is the phosphorylation technique employing phosphorus oxychloride ($POCl_3$) or phosphorus pentachloride ($PCl_5$). Both α- and β-naphthyl phosphates, for example, were first synthesized in 1894 by Kunz who treated the appropriate naphthol with $POCl_3$. Wolffenstein (1888) and Anschütz *et al.* (1906) prepared naphthyl phosphates by treating a suspension of hydroxynaphthoic acid in ether with $PCl_5$ while Anschütz (1927) has discussed in detail many aspects of phosphorylation employing $PCl_5$. More recently, Chanley and Gindler (1953) and Chanley and Feageson (1955) have employed $PCl_5$ in the synthesis of substituted α- and β-naphthyl phosphates.

TABLE 3

INORGANIC, ALIPHATIC, AND NUCLEOTIDE PHOSPHATE POTENTIAL PHOSPHATASE SUBSTRATES

| Chemical name and formula | Mol. wt. | Reference | Melting point (° C.) |
|---|---|---|---|
| 1. Sodium pyrophosphate, $Na_4P_2O_7$ | 265.914 | — | — |
| 2. Sodium tripolyphosphate (STPP), $Na_5P_3O_{10}$ | 367.88 | — | — |
| 3. Sodium trimetaphosphate, $(NaPO_3)_3$ | 305.898 | — | — |
| 4. Dihydroxyacetone phosphate, $C_3H_8O_7P$ | 187.07 | Kiessling, 1934; Ballou and Fischer, 1954 | — |
| 5. Acetyl phosphate (Ag and Li salts), $C_2H_3O_5PAg$ and $C_2H_3O_5PLi_2$ | — | — | 55 (dec.) |
| 6. Phosphopyruvic acid (phosphoenolpyruvic acid) (silver-barium salt), $C_3H_2O_6$-$PBaAg \cdot H_2O$ | 428.28 | Schmidt, 1957, p. 226 | — |
| 7. Ethylphosphopyruvic acid (Ba salt), $C_4H_5O_6PBa \cdot H_2O$ | 335.435 | Cherbuliez and Rabinowitz, 1956 | — |
| 8. *o*-Phosphorylethanolamine (ethanolamine phosphate) (aminoethyl-phosphate), $C_2H_8O_4NP$ | 141.069 | Outhouse, 1937 | 244 |
| 9. *o*-(L-$\alpha$-Glycerylphosphoryl)ethanolamine (L-$\alpha$-GPE), $C_5H_{14}O_6NP \cdot H_2O$ | 233.166 | Schmidt, 1957, p. 349 | 86–87 |
| 10. D-3-Phosphoglyceric acid (Ba salt), $C_3H_5O_7PBa \cdot 2H_2O$ | 357.44 | Neuberg, 1943 | — |
| 11. D-2-Phosphoglyceric acid (Na salt), $C_3H_4O_7PNa_3$ | 252.013 | Ballou and Fischer, 1954 | — |
| 12. D-Erythrose-4-phosphate, $C_4H_4O_9P$ | 227.05 | Ballou *et al.*, 1955 | — |
| 13. DL-Glyceraldehyde-3-phosphate (Ca salt), $C_3H_5O_6$-$PCa \cdot 2H_2O$ | 244.16 | Baer and Fischer, 1943b | — |
| 14. Glyceraldehyde-1,3-diphosphate (Ca salt), $C_6H_{12}O_{18}P_4$-$Ca_2 \cdot 2H_2O$ | 612.254 | Baer and Fischer, 1943a | — |
| 15. Glycerophosphate, $C_3H_9$-$O_6P$ | 172.08 | — | syrupy liquid |
| Glycerophosphate (Na salt) ($\alpha$ and $\beta$ forms), $C_3H_7O_6$-$PNa_2 \cdot 5\frac{1}{2}H_2O$ | 315.134 | — | 130 (dec.) |
| $\alpha$-Glycerophosphate (Ba salt), $C_3H_7O_6PBa$ | 307.424 | Zetzsche and Aeschlimann, 1926 | — |

TABLE 3—*Continued*

| Chemical name and formula | Mol. wt. | Reference | Melting point (° C.) |
|---|---|---|---|
| 16. Acetoneglycerophosphate (Ba salt), $C_6H_{11}O_6PBa$ | 347.49 | Asakawa, 1929 | — |
| 17. Glycolphosphate (Ba salt), $C_2H_5O_5PBa$ | 277.397 | Asakawa, 1929 | — |
| 18. 1-Chloro-2-propandiol-2,3-phosphate (Ba salt), $C_3H_6O_5ClPBa$ | 325.873 | Zetzsche and Aeschlimann, 1926 | — |
| 19. Ethyl lactic phosphate (Ba salt), $C_5H_9O_6PBa \cdot 2H_2O$ | 369.494 | Cherbuliez and Rabinowitz, 1956 | — |
| 20. Propylphosphate (Ba salt), $C_3H_7O_4PBa$ | 275.424 | Asakawa, 1929 | — |
| 21. Isopropylphosphate (Ba salt), $C_3H_7O_4PBa$ | 275.424 | Asakawa, 1929 | — |
| 22. Benzylphosphate (Ba salt), $C_7H_7O_4PBa$ | 323.468 | Zetzsche and Nachmann, 1925 | — |
| 23. Glucose-1-phosphate, $C_6H_{13}O_9P$ | 260.145 | — | — |
| Glucose-1-phosphate (K salt), $C_6H_{11}O_9PK_2 \cdot 2H_2O$ | 372.361 | — | — |
| 24. Glucose-6-phosphate, $C_6H_{13}O_9P$ | 260.145 | — | — |
| Glucose-6-phosphate (K salt), $C_6H_{11}O_9PK_2$ | 336.329 | Lardy and Fischer, 1946, Seegmiller and Horecker, 1951 | — |
| 25. Fructose-1-phosphate (F-1-P), $C_6H_{13}O_9P$ | 260.145 | — | — |
| 26. Fructose-6-phosphate (Ca, Ba, and K salts) (F-6-P), $C_6H_{13}O_9P$ | 260.145 | — | — |
| 27. Fructose-1,6-diphosphate (Ba, Ca, and Na salts), $C_6H_{14}O_{12}P_2$ | 340.13 | — | — |
| 28. Adenosine-5′-phosphate (AMP) (muscle adenylic acid), $C_{10}H_{14}O_7N_5P$ | 347.237 | — | 190 |
| 29. Adenosine-3′-phosphate (yeast adenylic acid) $C_{10}H_{14}N_5O_7P$ | 347.237 | Michelson and Todd (1949) | 195 (dec.) |
| 30. Adenosine-5′-pyrophosphate (ADP), $C_{10}H_{15}N_5O_{10}P_2$ | 427.22 | — | — |
| Adenosine-5′-pyrophosphate (acridine salt), $C_{10}H_{15}O_{10}N_5P_2 \cdot C_{13}H_9N$ | 606.443 | Baddiley and Todd, 1947 | 215 |

TABLE 3—*Continued*

| Chemical name and formula | Mol. wt. | Reference | Melting point (° C.) |
|---|---|---|---|
| 31. Adenosine-5′-triphosphate (ATP), $C_{10}H_{16}N_5O_{13}P_3$ | 507.203 | — | — |
| Adenosine-5′-triphosphate (acridine salt), $C_{10}H_{16}O_{13}N_5P_3 \cdot 3C_{13}H_9N \cdot H_2O$ | 1062.888 | Baddiley *et al.*, 1949 | 218 |
| 32. Uridine-5-phosphate (Ba salt), $C_9H_{11}O_9N_2PBa$ | 459.538 | Levene and Tipson, 1934 | — |
| 33. Uridine-5′-diphosphate (Ba salt), $C_9H_{12}O_{12}N_2PBa \cdot 3H_2O$ | 562.594 | Hall and Khorana, 1954 | — |
| 34. Inosine-5-phosphate (Ba salt), $C_{13}H_{15}O_8NPBa$ | 481.606 | Levene and Tipson, 1935 | — |
| 35. Guanosine-5′-phosphate (Ba salt), $C_9H_{11}O_9N_2PBa \cdot H_2O$ | 477.554 | Chambers *et al.*, 1955 | 264 (dec.) |
| 36. Creatine phosphate (phosphocreatine) (PC), $C_4H_{10}N_3O_5P$ (Na and Ca salt hexahydrates) | 211.123 | — | — |
| 37. Thiamine pyrophosphoric acid ester chloride (thiamine diphosphate ester chloride) (cocarboxylase), $C_{12}H_{19}ClN_4O_7P_2S \cdot H_2O$ | 478.81 | — | 240–244 (dec.) |
| 38. 2-Methyl-3-hydroxy-4-formyl-5-pyridylmethylphosphoric acid (pyridoxyl phosphate) (codecarboxylase), $C_8H_{10}NO_6P$ | 247.15 | — | 229–230 (oxime) |
| 39. Triphosphoropyridine nucleotide (TPN), $C_{21}H_{28}N_7O_{17}P_3$ | 743.436 | — | — |

Modifications of phosphorylation techniques in the preparation of alkyl and aryl phosphate substrates have been described by Neuberg and Wagner (1926), Neuberg and Jacobsohn (1928), Plimmer and Burch (1929), and Asakawa (1929). These procedures employed inert solvents including chloroform and benzene in conjunction with pyridine. Another older technique for the preparation of phosphates involves the treatment of an alcohol with polyphosphoric acid ($P_2O_5$ + $H_3PO_4$). In the synthesis of primary phosphoric acid esters, side reactions may lead to the formation of secondary and tertiary esters. The latter may be separated by treatment of the mixture with alkali to form a soluble salt of the primary ester which can then be easily separated from the other insoluble esters.

TABLE 4
ARYL PHOSPHATE POTENTIAL PHOSPHATASE SUBSTRATES

| Chemical name and formula | Mol. wt. | Reference | Melting point (° C.) |
|---|---|---|---|
| 1. Phenylphosphate, $C_6H_7O_4P$ | 174.097 | Iwatsuru, 1926; Asakawa, 1929 | 98 |
| Phenylphosphate (K salt), $C_6H_5PO_4K_2$ | 250.281 | Iwatsuru, 1926 | — |
| Phenylphosphate (Na salt), $C_6H_5PO_4Na_2$ | 218.063 | King and Nicholson, 1939, Asakawa, 1929 | — |
| Phenylphosphate (Ba salt), $C_6H_5PO_4Ba \cdot 2H_2O$ | 345.473 | King and Nicholson, 1939 | — |
| 2. *o*-Carboxyphenylphosphate (Ba salt), $C_7H_4O_6PBa_{1.5}$ | 421.124 | Walker and King, 1950 | — |
| 3. *p*-Carboxyphenylphosphate, $C_7H_4O_6PBa_{1.5}$ | 421.124 | Walker and King, 1950 | — |
| 4. *o*-Methylphenylphosphate (Ba salt), $C_7H_7O_4PBa \cdot H_2O$ | 341.484 | King and Nicholson, 1939 | — |
| 5. *p*-Nitrophenylphosphate (Ba salt), $C_6H_4O_6NPBa \cdot 2H_2O$ | 390.473 | King and Nicholson, 1939 | — |
| *p*-Nitrophenylphosphate (Na salt), $C_6H_4O_6NP \cdot Na_2$ | 263.063 | Bessey and Love, 1952 | — |
| 6. *p*-Chlorophenylphosphate, $C_6H_6O_4ClP$ | 208.546 | Rosenmund and Vogt, 1943 | 93 |
| 7. *p*-Bromophenylphosphate (Ba salt), $C_6H_4O_4PBrBa$ | 388.349 | King and Nicholson, 1939 | — |
| 8. Cyclohexanol phosphate (ammonium salt), $C_6H_{10}O_4PNH_4$ | 195.161 | Hotta, 1934 | — |
| Cyclohexanol phosphate (K salt), $C_6H_{10}PO_4K_2$ | 255.321 | King and Nicholson, 1939 | — |
| 9. 1-Methyl-2-cyclohexanol phosphate (Ba salt), $C_7H_{13}O_4PBa$ | 329.516 | Hotta, 1934 | — |
| 10. 4-Chlorothymol phosphate, $C_{10}H_{14}O_4ClP$ | 264.654 | Rosenmund and Vogt, 1943 | 142 |
| 11. Borneol phosphate, $C_{10}H_{17}PO_4H_2$, D, L, and DL forms. | 234.237 | Neuberg *et al.*, 1927 | 154–156 |
| 12. Menthol phosphate, $C_{10}H_{21}O_4P$ | 236.253 | Jacobsohn and Tapadinhas, 1931 | 100 |
| 13. *o*-Cresylphosphate, $C_7H_9O_4P$ | 188.124 | Asakawa, 1929 | 56 |
| 14. Phenylethylphosphate (K salt), $C_8H_9O_4PK_2$ | 278.335 | Asakawa, 1929 | — |

TABLE 4—*Continued*

| Chemical name and formula | Mol. wt. | Reference | Melting point (° C.) |
|---|---|---|---|
| 15. Benzylphosphate (K salt), $C_7H_7O_4PK_2$ | 264.308 | Zetzsche and Nachmann, 1925 | — |
| 16. Diphenylphosphate (K salt), $C_{12}H_{10}PO_4K$ | 288.287 | Asakawa, 1929 | — |
| 17. Di-*p*-chlorophenylphosphate (K salt), $C_{12}H_8O_4$-$Cl_2PK$ | 357.185 | Rosenmund and Vogt, 1943 | — |
| 18. Di-(2-nitro-1-naphthyl) phosphate, $C_{20}H_{13}O_8N_2P$ | 440.315 | Boyland *et al.*, 1961 | 187–189 (dec.) |
| 19. α-Naphthyl phosphate, $C_{10}H_9PO_4$ | 224.157 | Kunz, 1894 | 142 |
| α-Naphthyl phosphate (Na salt), $C_{10}H_7PO_4Na_2$ | 268.123 | Friedman and Seligman, 1950 | 155–157 |
| 20. β-Naphthyl phosphate, $C_{10}H_9PO_4$ | 224.157 | Kunz, 1894; Menten *et al.*, 1944a | 167, 172–173, 176–177 |
| β-Naphthyl phosphate (Ca salt), $C_{10}H_8PO_4Ca$ | 263.23 | Friedman and Seligman, 1950 | — |
| β-Naphthyl phosphate (Na salt), $C_{10}H_7PO_4Na_2$ | 268.123 | Menten *et al.*, 1944a | — |
| 21. 6-Bromo-2-naphthyl phosphate, $C_{10}H_8O_4BrP$ | 303.065 | Seligman *et al.*, 1949 | — |
| 22. 6-Benzoyl-2-naphthyl phosphate (Na salt), $C_{17}H_{11}O_5PNa_2$ | 372.232 | Rutenburg and Seligman, 1955 | — |
| 23. Naphthol AS-phosphates, AS-MX, AS-BI, AS-AN, AS-CL, etc. | — | Burstone, 1958a, b, c; 1961, 1962, Chapter 5, Chart 2 | — |
| 24. Fluorescein phosphate, $C_{20}H_{13}O_8P$ | 412.299 | Neumann, 1948a, b | — |
| 25. Eosin phosphate | ? | Neumann, 1948a, b | — |
| 26. 4-Methyl-7-hydroxycoumarin phosphate, $C_{10}H_9O_7P$ | 272.157 | Neumann, 1948a, b | 185 |
| 27. Phenolphthalein phosphate (Na salt), $C_{20}H_{13}$-$O_{11}P_2Na_5$ | 606.23 | Huggins and Talalay, 1945 | — |
| 28. Cholesteryl phosphate, $C_{27}H_{47}O_4P$ | 466.648 | Neuberg and Jacobsohn, 1928 | 187 |
| 29. Indoxyl phosphate, $C_8H_8O_4NP$ | 213.135 | Holt, 1954; Seligman *et al.*, 1954 | — |
| 30. 5,5′-Dibromoindoxyl phosphate, $BrC_8H_7O_4NP$ | 292.043 | Holt, 1954 | — |
| 31. Diphenylpyrophosphate (K salt), $C_{12}H_{10}K_2P_2O_7$ | 406.362 | Neuberg and Wagner, 1926 | — |

TABLE 4—*Continued*

| Chemical name and formula | Mol. wt. | Reference | Melting point (° C.) |
|---|---|---|---|
| 32. Di-*m*-cresol pyrophosphate (K salt), $C_{14}H_{14}K_2P_2O_7$ | 434.416 | Neuberg and Jacobsohn, 1928 | — |
| 33. Di-*o*-carbomethoxyphenyl pyrophosphate (Ba salt), $C_{16}H_{14}O_{11}P_2Ba$ | 581.598 | Walker and King, 1950 | — |
| 34. Dicarbethoxyphenylpyrophosphate (Ba salt), $C_{18}H_{18}O_{11}P_2Ba$ | 609.652 | Walker and King, 1950 | — |
| 35. Di-$\alpha$-naphthyl pyrophosphate, $C_{20}H_{16}O_7P_2$ | 430.298 | Neuberg and Jacobsohn, 1928 | 78 |
| Di-$\alpha$-naphthyl pyrophosphate (K salt), $C_{20}H_{14}K_2P_2O_7$ | 506.482 | Neuberg and Jacobsohn, 1928 | — |
| 36. Phenyldiamidophosphoric acid, $C_6H_9O_2N_2P$ | 172.13 | Stokes, 1894; Audrieth and Toy, 1941 | 185–190<br>183–185 |
| 37. Diphenylamidophosphoric acid, $C_{12}H_{12}O_3PN$ | 249.211 | Stokes, 1893; Audrieth and Toy, 1941 | 148, 145–146 |
| 38. Phenyldi(methylamido)-phosphoric acid, $C_8H_{13}O_2N_2P$ | 200.183 | Audrieth and Toy, 1942 | 103–105 |
| 39. *p*-Chloranilidophosphonic acid,[a] *N*-(*p*-chlorophenol)amidophosphoric acid, $C_6H_7O_3PNCl$ | 207.562 | Otto, 1895; Ichihara, 1933; Gomori, 1948 | 150–155 |
| 40. *N*-(*p*-chlorophenyl)diamidophosphoric acid,[a] $C_6H_8ClN_2O_2P$ | 206.578 | Rorig, 1949 | 156–157 |
| 41. Diphenylpyrophosphorodiamidic acid (dicyclohexylammonium salt), $C_{24}H_{40}N_4O_5P_2$ | 526.566 | Boger and Friedman, 1958 | 223–224 |
| Diphenylpyrophosphorodiamidic acid (K salt), $C_{12}H_{12}O_5P_2K_2$ | 376.378 | Boger and Friedman, 1958 | — |

[a] See discussion as to authenticity of this compound on p. 230.

In recent years, a number of new phosphates have been synthesized specifically as histochemical reagents. Menten *et al.* (1944a) prepared the first histochemical azo-dye substrate, Ca $\beta$-naphthyl phosphate, for the demonstration of alkaline phosphatase. Loveless and Danielli (1949), anticipating the value of more complex substrates, synthesized the phosphate of an azo-dye, 4-*p*-nitrobenzeneazo-1-naphthol (I).

Seligman *et al.* (1949) reported the synthesis of 6-bromo-2-naphthyl phosphate (II). Rutenburg and Seligman (1955) prepared 6-benzoyl-2-naphthyl phosphate (III) as a substrate for the post-coupling demonstration of acid phosphatase. Pearse (1954) reported 2,4-dichloro-1-naphthyl and 2,4-dibromo-1-naphthyl phosphates as substrates for the demonstration of phosphatases, but no information on the preparation of these compounds was given. Barrnett *et al.* (1956) reported in an abstract that naphthol AS phosphate could be used as a substrate for the demonstration of phosphatases in conjunction with Diazo Blue B (diazotized *o*-dianisidine). Burstone (1958a, b, c) described the synthesis and application of a number of naphthol AS phosphates for the demonstration of both alkaline and acid phosphatases. The general configuration of many of these compounds is shown in (IV), with A through F indicating the position of electronegative and/or electropositive groupings. The AS phosphates have been found to be superior to the unsubstituted α- or β-naphthyl phosphates, especially in the acid pH range. Although phosphorylation of naphthols with $POCl_3$ may yield a primary phosphate containing secondary and possibly tertiary esters, the unpurified product is still very useful for histochemical purposes. (Techniques of synthesis and purification of naphthol AS phosphates are given in Appendix 5A, Part 2.) When naphthol AS-BI phosphate is purified, the analysis indicates: calcd.: C, 47.81%; H, 3.34%; found: C, 47.68%; H, 3.59%. Infrared spectra of purified naphthol AS-BI and its corresponding phosphate are shown in Figs. 1 and 2.

New naphthol AS type substrates have been developed for the post-coupling demonstration of both alkaline and acid phosphatases (Burstone, 1960c). The success of these compounds is based upon the use of unique naphthols which were prepared by condensing 3-hydroxy-2-naphthoic acid and analogues with special arylamines.

Jeffree and Taylor (1961) reported on the synthesis and histochemical application of an impure naphthol AS-TR phosphate but no photomicrographs were published. Jeffree and Taylor gave no analysis of their compound and stated that their large amount of impurity was due to the presence of a polymer. No evidence was presented to substantiate the presence of polymer. It is a fact that polyphenol compounds such as phloroglucinol may form polymers when phosphorylated under certain conditions (Fernö *et al.*, 1953). However, no experimental data has been presented to substantiate the assumption that polymers result when naphthol AS compounds are improperly phosphorylated.

Seligman *et al.* (1954) reported indoxyl phosphate as a substrate for alkaline phosphatase. As with other unsubstituted indoxyl substrates, the final dye exhibits a grossly macrocrystalline pattern. Superior results are obtained with 5-bromoindoxyl phosphate synthesized by Holt (1954).

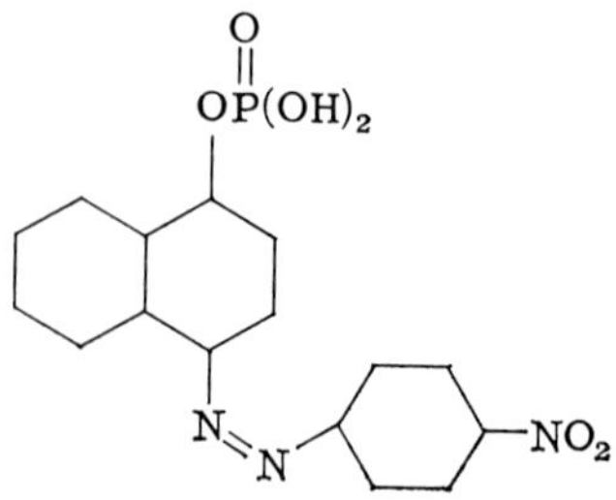


4-*p*-Nitrobenzeneazo-
1-naphthyl phosphate

(I)

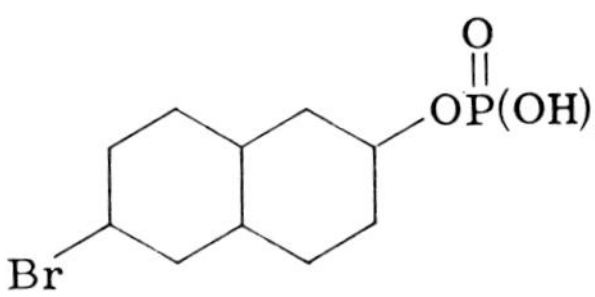


6-Bromo-2-
naphthyl phosphate

(II)

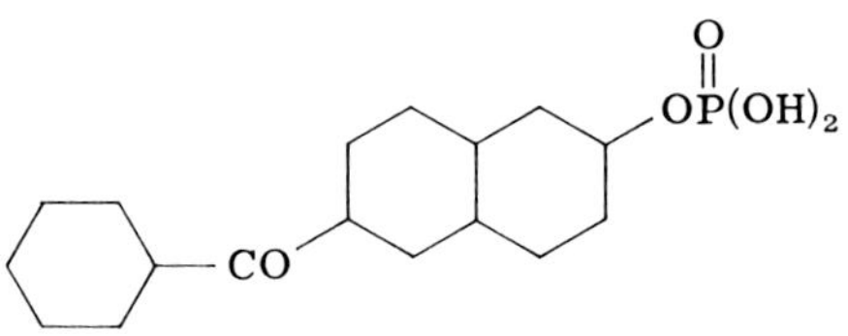


6-Benzoyl-2-
naphthyl phosphate

(III)

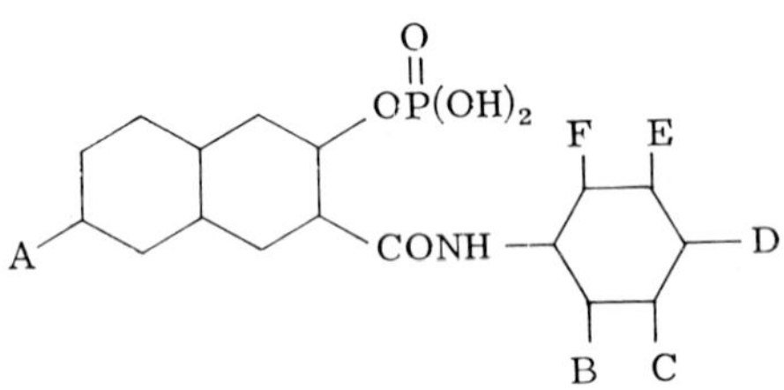


General configuration of
naphthol AS phosphates

(IV)

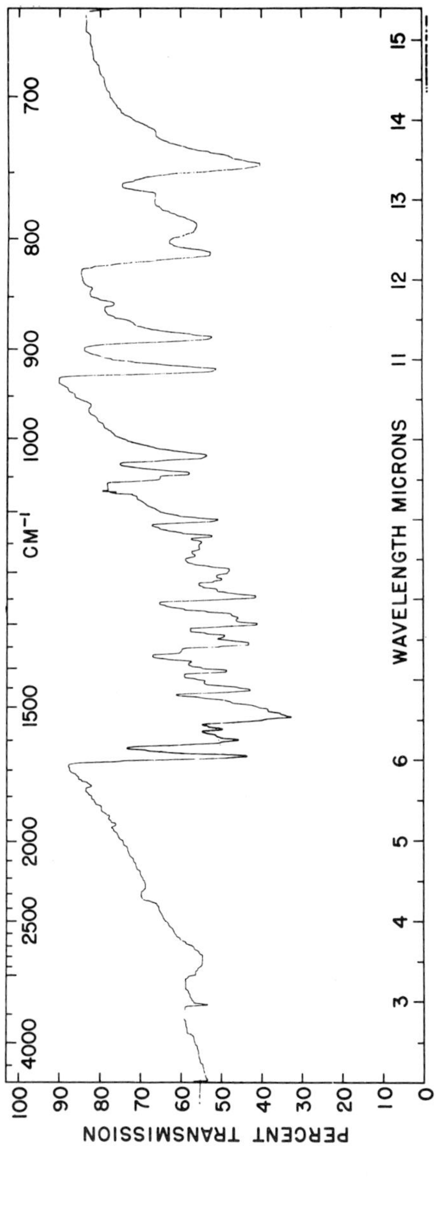


FIG. 1. Infrared spectrum of naphthol AS-BI.

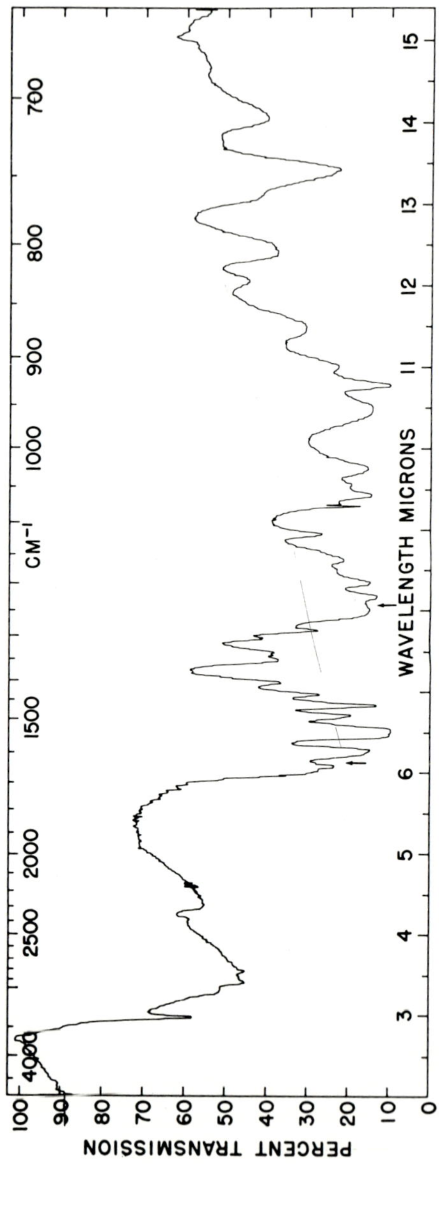


FIG. 2. Infrared spectrum of naphthol AS-BI phosphate. Arrows indicate phosphate groupings.

Baer and Fischer (1943a, b) employed diphenylchlorophosphonate (diphenylphosphorus oxychloride) in the synthesis of glyceraldehyde phosphates. The protecting phenyl groupings were removed by hydrogenation. An analogous phosphorylation technique was developed by Atherton *et al.* (1945). It employed a mild phosphorylation agent, dibenzylchlorophosphonate [$(CH_2PhO)_2POCl$] (McCombie *et al.*, 1945). The benzyl groupings are removed by catalytic hydrogenation. Atherton *et al.* (1945) prepared β-naphthyl phosphate by the aforementioned technique (yield, 80%). Dibenzylchlorophosphonate was also used to synthesize adenosine-5′-triphosphate (ATP) (Baddiley *et al.*, 1949).

Kenner and Williams (1955) prepared aryl diethylphosphates ArO—P(O)$(OEt)_2$ of α- and β-naphthols with diethylphosphite. Upon hydrogenolysis these compounds should yield primary naphthyl phosphates. When *p*-bromobenzylchlorophosphonate is employed in the synthesis of aryl phosphates, debenzylation may be accomplished with lithium chloride (Miyano, 1955).

Khorana (1953) introduced the carbodiimide technique for the synthesis of pyrophosphoric acid esters. The carbodiimide synthesis of adenosine phosphate (ADP) and adenosine triphosphate (ATP) from adenylic acid (AMP) has been described by Khorana (1954). In this procedure, AMP is simply treated at room temperature with an excess of phosphoric acid and DCC in aqueous pyridine. Tetra-*p*-nitrophenylpyrophosphate, which is prepared by treatment of di-*p*-nitrophenyl hydrogen phosphate with di-*p*-tolycarbodiimide, is also an effective phosphorylating agent for alcohols (Moffatt and Khorana, 1957). As yet the carbodiimide procedure has not been employed in the synthesis of specialized histochemical substrates.

With reference to the synthesis of nucleoside pyrophosphates, a phosphoramidic ester synthesis has also been applied (Chambers and Khorana, 1956). When the crystalline monopotassium salt of phosphoramidic acid ($HPO_3NH_2$) is allowed to react with an excess of phosphoric acid in formamide, pyrophosphates are formed. Clark *et al.* (1957) were able to convert adenosine-5′-phosphate into adenosine-5′-pyrophosphate, and thence into adenosine-5′-triphosphate through the use of benzyl hydrogen phosphoramidate and benzyl hydrogen *N*-cyclohexylphosphoramidate. Montgomery and Turnbull (1958) employed phosphoramidic halides derived from morpholine in the preparation of phenyl dihydrogen phosphate. The primary phosphate is formed by ion exchange hydrolysis of the phosphorodimorpholidates. Cyanoethylphosphate has been employed as a specific phosphorylating agent for the synthesis of phosphomonoesters (Tener, 1961). An alcohol is reacted with 2-cyanoethylphosphate in the presence of dicyclohexylcarbodiimide (DCC), and then the cyanoethyl grouping is removed from the resulting phosphodiester by mild alkaline hydrolysis to give the appropriate phosphate monoester.

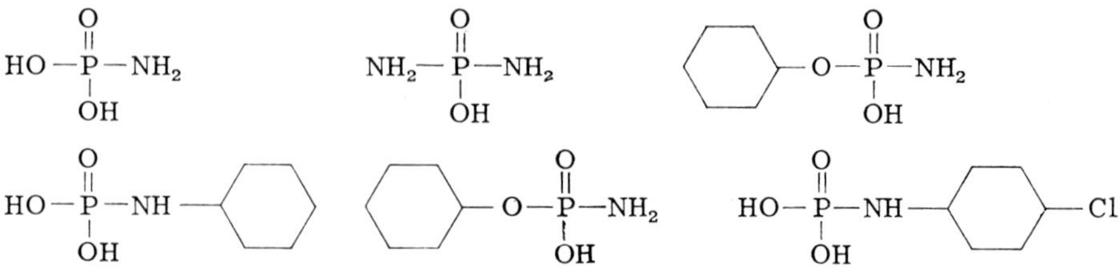

CHART 1. Potential substrates for phosphamidase.

Cramer (1961) has reviewed various aspects dealing with the chemical synthesis of pyrophosphates and nucleotides and the susceptibility of these compounds to enzymatic hydrolysis.

Although the natural substrates for phosphamidase may be phosphoarginine or phosphocreatine, histochemical as well as biochemical determinations have employed synthetic substrates prepared by phosphorylating amines. A number of these compounds have been synthesized and characterized by Ichihara (1933). They include phosphoric acid monoamide, phosphoric acid diamide, phenylphosphoric acid monoamide, phosphoric acid anilide, phenylphosphoric acid anilide, and phosphoric acid *p*-chloroanilide (Chart 1).

## B. Substrate Type, pH Optimum, and Hydrolysis Rate

As indicated in the introduction, the pH optimum of substrates varies. The concentration of the substrate in solution also influences the pH at which maximum hydrolysis occurs. Folley and Kay (1935) found that when the substrate concentration was lowered the pH optimum of alkaline phosphatase decreased. Neumann (1949) also noted the dependence of pH optimum upon the substrate concentration. Substrates also exhibit basic pH differences in optimal hydrolysis rates. According to Folley and Kay (1935) the degree of negativity of the phosphoric acid ester influences the pH at which maximum hydrolysis occurs. Asakawa (1929) found that the optimum pH for hydrolysis (kidney extracts) of β-glycero-, isopropyl-, phenylethyl-, benzyl-, and phenyl- phosphoric acid esters lay between pH 9.0 and 9.9; while that of diphenyl-, dicresyl-, diglycero- and diethylphosphates was at pH 7.8. Roche (1931) noted that the pH optimum for leukocytic and serum phosphatases was approximately 8.5 when phenylphosphate was employed as a substrate, but was higher when glycerophosphate was used. Delory and King (1943) found that both the rate of hydrolysis and the optimum pH appeared to increase with increasing acidity of the ester. Thus, bromolecithin is hydrolyzed 10 to 100 times more rapidly than lecithin, and the pH optimum is raised from 7.4 to 8.4. Nitrophenylphosphate (optimum, pH 9.80) is more rapidly hydrolyzed than phenylphosphate (optimum, pH 9.76). Of interest is the finding that *p*-nitrophenylphosphate is hydrolyzed 2.8 times more rapidly than β-glycerophosphate, while *p*-bromophenylphosphate is hydrolyzed 3.4 times more rapidly than glycerophosphate. King and Delory (1939) also studied the hydrolysis rate of secondary esters by alkaline phosphatase and found that diphenylphosphate was more rapidly hydrolyzed than dicresyl- or diethylphosphates. Walker and King (1950) found that carboxy-substituted phenylphosphates exhibited a lower rate of hydrolysis than did unsubstituted phenylphosphate. The carboxy-substituted compounds, being less acidic than phenylphosphoric acid, show a lower pH optimum (pH 8.0 for *o*-carboxyphenylphosphate). Naphthol AS phosphates exhibit a pH optimum of around 8.5 even when the substrate is in high concentration. This may be the result of the electronegative —CO— group adjacent to the phosphate ester. In any event, the lower pH optimum is advantageous in the application of AS phosphates since these compounds tend to exhibit an appreciable degree of spontaneous decomposition at pH 9.0 or above. In addition, as previously mentioned in the section on diazonium salts, a strongly alkaline pH is deleterious to the stability of diazo compounds.

Since a diverse number of phosphatase substrates have been employed for biochemical studies in the determination of alkaline and acid phosphatase activity of serum and body fluids, useful data with respect to hydrolysis rate and efficiency of the method may be gleaned from such observations. Tables 5 and 6 (Gomori, 1954) indicate the experimental conditions employed in a number of biochemical techniques and their comparative efficiency in terms of micromoles of substrate hydrolyzed per hour per milliliter.

## C. Histochemical Approaches to the Demonstration of Phosphatases

The original metal-salt phosphatase technique of Gomori (1939) depends upon the enzymatic release of phosphate ions which are precipitated as insoluble calcium phosphate. The calcium phosphate is then converted into cobaltous phosphate and finally into cobalt sulfide. This type of procedure was originally employed by Sen (1930) in the histochemical demonstration of urease. Gomori (1939) also treated incubated sections with silver nitrate followed by exposure to light according to the von Kossa procedure to form a brown-black precipitate of silver phosphate. Gomori (1941) also introduced the use of lead nitrate in place of calcium chloride when the reaction was run at an acid pH, since calcium phosphate is acid soluble. Bourne (1943) visualized calcium phosphate by the use of sodium alizarin sulfonate which formed an orange-red calcium lake. Pearse (1960, p. 388) suggested utilization of the lake formation principle with phthalocyanine dyes or else with morin (pentahydroxyflavone). The latter forms a fluorescent lake.

The calcium phosphate precipitate also may be directly visualized by means of polarized light (Bélanger, 1951) or by phase contrast microscopy

TABLE 5

BIOCHEMICAL METHODS FOR ALKALINE PHOSPHATASE (FROM GOMORI, 1954)

| Author | Substrate | pH | mg. | Definition of unit | Normal values | Efficiency |
|---|---|---|---|---|---|---|
| Kay, 1930 | Glycerophosphate | 7.6 | − | 1 mg. P/48 hr. | ±0.25/ml. | 0.16 |
| Bodansky, 1933 | Glycerophosphate | 8.6 | − | 1 mg. P/hr. | 3–5/100 ml. | 1–1.5 |
| Cayla, 1935 | Glycerophosphate | 7.6 | − | 1 mg. $H_3PO_4$/hr. | ±15/1000 ml. | 0.15 |
| Lundsteen and Vermehren, 1936 | Glycerophosphate | 8.9 | + | 1 mg. P/24 hr. | 33–50/100 ml. | 0.45–0.7 |
| Shinowara *et al.*, 1942 | Glycerophosphate | 9.3 | − | 1 mg. P/hr. | 3–8/100 ml. | 1–2.5 |
| Langemann, 1949 | Glycerophosphate | 8.4 | − | 1 mg. P/hr. | 2–5/100 ml. | 0.6–1.5 |
| de Mello, 1951 | Glycerophosphate | 10.8 | − | 1 mg. P/hr. | 1.5–4/100 ml. | 0.5–1.2 |
| Aebi, 1952 | Glycerophosphate | 9.7 | − | 1 mg. P/hr. | 2–5/100 ml. | 0.6–1.5 |
| King and Armstrong, 1934 | Phenylphosphate | 9.3 | − | 1 mg. phenol/30 min. | 5–10/100 ml. | 1–2 |
| Buch and Buch, 1939 | Phenylphosphate | 10.0 | + | 1 mg. phenol/15 min. | 2.2–6.6/100 ml. | 1–3 |
| Villela and Mello, 1948 | Phenylphosphate | 9.1 | − | 1 mg. phenol/15 min. | 1.2–8/100 ml. | 0.5–3.2 |
| Gomori, 1949c | Phenylphosphate | 9.8 | + | 1$\mu M$ phenol/hr. | 2.3–6.6/ml. | 2.3–6.6 |
| Ruppert, 1950 | Phenylphosphate | 10.0 | + | 1 mg. P/2 hr. | 8–10/100 ml. | 1.2–1.5 |
| Ohmori, 1937 | *p*-Nitrophenyl-phosphate | 9.0 | + | 40% hydrolysis/hr. | Not given | — |
| Bessey *et al.*, 1946 | *p*-Nitrophenyl-phosphate | 10.3 | + | 1m$M$ nitrophenyl-phosphate/hr. | 2–6/1000 ml. | 2–6 |
| Binkley *et al.*, 1944 | Phosphotyrosine | ±9.0 | − | 1 mg. phenol/hr. | 2–10/100 ml. | 0.6–3 |
| Huggins and Talalay, 1945 | Phenolphthalein phosphate | 9.1–9.6 | − | 0.1 mg. phenol-phthalein/hr. | 5–13/100 ml. | 0.015–0.04 |
| Seligman *et al.*, 1951a | β-Naphthyl-phosphate | 9.1 | − | 10 mg. naphthol/hr. | 1.9–3/100 ml. | 1.35–2.1 |

TABLE 6

BIOCHEMICAL METHODS FOR ACID PHOSPHATASE (FROM GOMORI, 1954)

| Author | Substrate | pH | Definition of unit | Normal values | Efficiency |
|---|---|---|---|---|---|
| Gutman and Gutman, 1938; 1940 | Phenylphosphate | 4.9 | 1 mg. phenol/hr. | 0.25–3/100 ml. | 0.3 |
| Villela and Mello, 1948 | Phenylphosphate | 4.95 | 1 mg. phenol/15 min | <3.2/100 ml. | 1.2 |
| Gomori, 1949c | Phenylphosphate | 5.0 | 1 $\mu M$ phenol/hr. | 0.6–1.5/ml. | 0.6–1.5 |
| Binkley *et al.*, 1944 | Phosphotyrosine | 4.9 | 1 mg. phenol/hr. | 0.7–2.6/100 ml. | 0.07–0.26 |
| Huggins and Talalay, 1945 | Phenolphthalein phosphate | 5.5–6.0 | 0.1 mg. phenolph-thalein/hr. | 4–8/100 ml. | 0.012–0.024 |
| Hudson *et al.*, 1947 | *p*-Nitrophenyl-phosphate | 5.4 | 1$\mu M$ nitrophenol/hr. | 1–2.3/1000 ml. | 1–2.3 |
| Altschule *et al.*, 1951 | α-Naphthyl-phosphate | 5.0 | 1 mg. naphthol/30 min. | 20–60/100 ml. (urine) | — |
| Seligman *et al.*, 1951b | β-Naphthyl-phosphate | 4.8 | 10 mg. naphthol/hr. | 0.7–1.6/100 ml. | 0.5–1.1 |
| Langemann, 1949 | Glycerophosphate | 6.0 | 1 mg. P/24 hr. | 4–7/100 ml. | 0.05–0.09 |
| de Mello, 1951 | Glycerophosphate | 5.0 | 1 mg. P/hr. | 0.04–0.64/100 ml. | 0.012–0.2 |

(Hancox and Nicholas, 1956). Danielli (1946) employed phenolphthalein phosphate in an interesting modification. Upon enzymatic hydrolysis a precipitate of calcium phosphate and colorless phenolphthalein is formed, which is changed into its red form by exposing the section to ammonia gas. Takeuchi and Nogami (1954) utilized riboflavin-5′-phosphate which resulted in formation of a fluorescent precipitate at sites of phosphatase activity. The mechanism of staining in this technique is unknown, since the enzymatically released riboflavin is known to be water soluble. In addition, riboflavin-5′-phosphate is in itself a fluorescent stain which gives a positive reaction in tissue sections. Another fluorescence procedure, the fluorescent antibody technique, also has been employed in the demonstration of a skeletal muscle myosin (Mellors, 1959, p. 45).

The development of the azo-dye technique by Menten *et al.* (1944a, b) represents a milestone in histochemical methodology. These workers synthesized β-naphthyl phosphate and employed it in a substrate solution in the presence of a diazotized amine (e.g. nitroaniline and α-naphthylamine). Upon enzymatic hydrolysis the released naphthol coupled simultaneously with the diazo component to form an insoluble dye. It is important to note that Menten, Junge, and Green specifically point out that the azo-dye principle should be applied to the histochemical demonstration of other enzymes. Subsequently modifications of the Menten, Junge, and Green azo-dye method were reported by Manheimer and Seligman (1948). Although α-naphthyl phosphate had been synthesized previously, Gomori (1951; 1952, p. 185) and Gomori and Chessick (1953) were the first to actually prove the superiority of this substrate over the β-isomer for histochemical use.

Loveless and Danielli (1949) recommended an interesting approach to the demonstration of phosphatases, namely, the application of a phosphorylated azo-dye, which upon enzymatic hydrolysis released the insoluble dye *in situ*.

Seligman *et al.* (1949) suggested the use of 6-bromo-2-naphthyl phosphate in a so-called post-coupling technique, in which incubation was carried out in the absence of a diazonium salt. 6-Benzoyl-2-naphthyl phosphate was also recommended for the post-coupling procedure (Rutenburg and Seligman, 1955). Both techniques were recommended only for the demonstration of acid phosphatase, since the enzymatically released naphthols are too soluble at an alkaline pH to be of value in the demonstration of alkaline phosphatase.

The development and application of naphthol AS phosphates have facilitated improved localizations of both alkaline and acid phosphatases by the simultaneous azo-dye technique (Burstone, 1958a, b, c, 1962). Specialized naphthol AS phosphates have also been developed for the post-coupling demonstration of both alkaline and acid phosphatases (Burstone, 1960c). These compounds are capable of giving exceedingly sharp localizations over a wide pH range.

The development of new synthetic phosphatase substrates has by no means been exhausted. Of importance is the fact that almost any primary naphthyl phosphate is susceptible to enzymatic hydrolysis, as contrasted with naphtholic fatty acid esters (Burstone, unpublished data).

## D. Quantitation of the Histochemical Reaction

The pH optimum of a substrate is difficult to assess by histochemical staining, and its evaluation requires the application of semi-quantitative or quantitative methods. Gomori (1950d) developed an interesting technique for the semi-quantitative determination of phosphatases which may be applied as well in the determination of pH optima of different substrates and enzymes. The procedure is based upon the approach of Coujard (1943), and permits approximate quantitation (±50%) of histochemical reactions for enzymes. Microscopic slides are employed upon which pen marks are drawn with enzyme dilutions of known activity. Serial dilutions by a factor of 2 are made with a diluent such as 1% gelatin or gum acacia solution. The slides are fixed and then incubated in an appropriate substrate solution. With this technique in conjunction with his metal-salt alkaline phosphatase method, Gomori was able to determine that the activity of the brush border of mouse kidney is over 10,000 $\mu M$ units (ca. 300 Bodansky units) per ml. of tissue. Activity of the cuticular border of mouse and dog intestine may exceed 20,000 $\mu M$ units. Further exploitation of this ingenious approach should be undertaken with reference to newer enzyme techniques. A potential somewhat analogous approach could employ an ion exchange resin containing enzyme. This type of model system has the advantage of containing the enzyme in an insoluble form associated with a semi-permeable membrane (Mitz, 1956).

Doyle (1950) and Doyle *et al.* (1951) were able to quantitate the alkaline phosphatase activity of tissue sections by substituting lead for cobalt in the metal-salt method, and extracting the precipitate of lead sulfide with HCl. This was subsequently converted into methylene blue according to the procedure of Fogo and Popowsky (1949), and measured spectrophotometrically. Other approaches to quantitation of the Gomori method have been made, using $P^{32}$-labeled glycerophosphate (Dalgaard, 1948) and $Pb^{212}$-labeled lead nitrate (Barka *et al.*, 1952; Shugar *et al.*, 1957).

## E. Pitfalls in Phosphatase Methods

All of the enzyme methods possess inherent defects. No enzymatically released substance is completely insoluble, and thus precipitation occurs only if the solubility of the released compound is very low. If an insoluble product is formed from two soluble components in the solution, precipitation occurs only when its insolubility is exceeded. Thus, if a sufficiently high magnification is employed it is likely that tissues stained, using present-day histochemical techniques, will reveal diffusion artefacts.

The metal-salt technique has been both overestimated and overcriticized with reference to quality and precision of localization, and there is a notably high degree of correlation between the type of evaluation and the preference of the individual authors. An excellent critique of this method is given in Chapter 15 of the recently published book by Lison (1960).

Johansen and Linderstrøm-Lang (1951, 1952) analyzed the quantitative aspects of precipitation in this histochemical procedure, and concluded that calcium phosphate exhibits a marked tendency to form supersaturated solutions. Thus, by the time the critical concentration of phosphate ions sufficient to result in precipitation at centers of enzymatic activity is reached, diffusion of phosphate from the sites of production will have caused a condition of supersaturation to prevail in a wide zone around the centers. They believe that the presence of preformed crystal nuclei within this zone will initiate the precipitation of calcium phosphate, even in the absence of any local enzymatic activity. Thus a positive reaction need not be related to the presence of the enzyme at that particular site. Gomori and Benditt (1953) concede that while under non-optimal conditions spurious staining may occur, this situation does not prevail when the metal-salt procedure is performed under optimal conditions. They further point out that calcium phosphate, when precipitated by addition of phosphate to solutions similar in composition to the histochemical substrate mixture, does not show any demonstrable tendency toward supersaturation. They also feel that the factors which may contribute to false localizations in the metal-salt technique are not well understood and not readily amenable to mathematical analysis. Gomori (1949a, 1950b) indicates that two primary factors which regulate the deposition of purely soluble phosphates in tissue sections are rates of phosphate production and rates of diffusion. Varying ratios between these two factors may result in an *all* or *none* effect, depending upon whether the rate of phosphate production will or will not overtake that of diffusion to a point where the solubility product of the phosphate in question is exceeded locally. At pH 9 the solubility of calcium phosphate is low but rapidly increases with lowering of the pH. *Thus a precipitate of calcium phosphate will form only if the production of phosphate ions is fast enough to cause local supersaturation in spite of diffusion.* Gomori (1950b) also points out that at pH 8.2 the histological picture is much less intense than it is at pH 9.0, while below pH 7.2 it is difficult to produce any precipitate of calcium phosphate despite the fact that the enzyme is known to be active at neutrality. Danielli (1954) has reported a useful technique for ascertaining diffusion of calcium phosphate as well as phosphatase. An active section is superimposed upon a similar section in which the enzyme has been inactivated, so that any diffusion artefact will be revealed by the appearance of staining in the underlying inert section. Any kind of inhibition of enzyme activity (e.g. partial inactivation by unsuitable fixation, non-optimal pH, poorly hydrolyzed substrate, or the presence of inhibitors) is likely to obliterate areas below a critical level of activity. On the other hand, areas of much higher activity may appear to be entirely uninfluenced or only slightly weakened.

The fact that the metal-salt procedure is usually performed as a multistage process, with secondary transformation of the primary precipitates, may give rise to artefacts. For example, Holter *et al.* (1951) have found that the conversion to cobalt sulfide is incomplete. Bélanger (1951) noted localized variations in the quantity and distribution of precipitate during the conversion of calcium and lead phosphates into sulfide or silver compounds, and stressed the advantage of studying the primary precipitate with polarized light.

Preformed calcifications (normal and abnormal) are essentially of the same chemical nature as the product of hydrolysis, and they thus will also be converted into CoS, PbS, or Ag and give false localizations (Gomori, 1950a). In addition, false positive reactions may also be obtained with the metal-salt acid phosphatase technique with certain pigments including hemosiderin, ceroid, and lipofuchsins (Gomori, 1952, p. 193). Although the α-naphthyl phosphate acid phosphatase method is too diffuse to permit precise localization of enzyme activity to lipofuscin granules (Goldberg *et al.*, 1960) the naphthol AS phosphate procedure reveals definitive sites of phosphatase activity associated with these granules (Spicer and Burstone, unpublished data). Bone and cartilage matrices, and also cell nuclei, readily adsorb calcium phosphate produced in other parts of the section. With reference to the latter, some investigators feel that in specific instances the staining of nuclei with the Gomori as well as with other techniques may reveal genuine localization of phosphatase. One of the most recent and comprehensive papers on this subject (including a review of the literature) has been published by Feigin and Wolf (1957). Artefactual staining occurs preferentially in the immediate vicinity of sites of high activity where transient high concentrations of phosphate ions may occur. Gomori (1951) notes that such artefacts may occur even in the optimal pH range and with short incubations. He also points out that calcium phosphate is measurably soluble even at pH 9.0, while the azo-dyes formed from α-naphthol are less soluble.

The non-enzymatic impregnation of tissue structures by Pb is well known (Palade, 1951). More recently, Clarkson and Kench (1958) have evaluated physical and physico-chemical mechanisms of lead adsorption to tissue components. Gomori (1952, p. 191) states that although unfixed frozen sections may show relatively high acid phosphatase activity, the affinity of lead salts for undenatured protein may produce artefacts. He notes that formalin fixed tissue exhibits much less affinity for lead. The danger of such impregnation is particularly great in cases where long periods of incubation resulting in spontaneous hydrolysis of substrate leads to a supersaturation with respect to Pb and possibly Ca phosphates. This type of artefact may, in many instances, be recognized by simultaneously incubating inactivated tissues as controls. On the other hand, the inactivation procedure in itself may influence the type of non-specific adsorption. The presence of lead ions also exerts an inhibitory effect upon certain enzymes such as acid phosphatase, ATPase, 5-nucleotidase, and thiamine pyrophosphatase (Pratt, 1954). Gomori (1952, p. 191) quotes Doyle to the effect that lead ions inhibit acid phosphatase very markedly (ca. 85%), although Gomori himself found the degree of inhibition to be 35%. The numerous papers of Gomori still remain as highly objective critiques of his metal-salt techniques. Contrary to what has been stated, Gomori's vast experience with his method was not limited to paraffin embedded tissues, but also included critical evaluation of results with unfixed frozen and neutral cold formalin fixed sections.

Probably the most significant finding in regard to minimizing artefacts with the acid phosphatase technique was made by Eränkö (1951, 1952). He made the highly important observation that frozen sections which had been fixed in cold formol-calcium gave improved and more consistent staining patterns with the metal-salt technique (Appendix 5B, Section 3). He still indicated that formol-calcium fixed tissues may give diffusion artefacts, presumably because of inadequate fixation.

The simultaneous coupling azo-dye methods are advantageous in that they are single-step procedures in which the dye can be seen as soon as it is formed in the tissue section. In most cases, preformed calcifications are unstained and thus the azo-dye method is especially useful in the study of calcified tissues (Burstone, 1960a, b). In general, most naphthols are not non-specifically adsorbed by tissues. Exceptions are certain brominated and benzoylated naphthols (Chapter 4, p. 139) (Rutenburg and Seligman, 1955; Burstone, 1958c). Simultaneous coupling methods employing simple naphthols ($\alpha$- and $\beta$-naphthol) are prone to give serious diffusion artefacts, especially if the sections are highly active (Gomori, 1952, p. 172). Some of the azo-dyes (e.g. those derived from $\beta$-naphthol) may remain in colloidal solution for a short time prior to precipitating (Gomori, 1952, p. 169). When diffusion artefacts (halos and streamers of dye) do occur with the azo-dye methods, they are usually more readily discernible than artefacts obtained with the metal-salt procedure. In the latter, reprecipitation phenomena may produce discrete sharp localizations including membrane staining: artefacts which are difficult to detect.

On the basis of various studies, it is apparent that useful histochemical localizations can be demonstrated even if a large percentage of phosphatase is destroyed. It is equally obvious that in considerably inactivated preparations, the histological picture cannot and does not reveal the full extent of activity because sites of low enzyme concentration are brought below the threshold of sensitivity of the method. Thus, when kidney sections are greatly inactivated with fixatives, beautiful localizations of acid phosphatase in "lysosomes" or "droplets" which stand out on a clear background are obtained. Such localizations are particularly striking because of the contrast between the droplets and the unstained background. On the other hand, there is also a genuine cytoplasmic activity of kidney tubules (Fig. 3), as well as activity of basement membranes which can be demonstrated only in tissues which are reasonably active. One cannot conclude that simply because microscopic histochemical enzyme methods are not quantitative (from a numerical standpoint), the degree of inactivation of a tissue is of no particular consequence. Furthermore, the obtaining of sharp, discrete, droplet-like precipitates does not automatically mean that the localizations are accurate or inclusive. Thus great caution must be exercised in the application of present-day phosphatase techniques to cytological localizations and electron microscopy.

The post-coupling phosphatase methods appear to represent the ideal type of procedure because of their simplicity and the avoidance of a simultaneous coupling reaction which takes a finite period of time and thus may give rise to diffusion artefacts.

## II. NON-SPECIFIC ALKALINE PHOSPHATASES

Since the initial observation that phosphate esters were hydrolyzed by animal tissues (Grosser and Husler, 1912) numerous biochemical studies have revealed the ubiquitous distribution of alkaline phosphatases in most tissues and organs (Roche, 1950, p. 473; Roche and Thoai, 1950; Sumner and Somers, 1953, p. 90; Stadtman, 1961, p. 55). Among the sources of high activity are renal cortex, intestinal epithelium, and developing bone including certain osteogenic neoplasms.

Purified intestinal phosphatase is an unconjugated protein essentially free of nucleotides and related compounds and with low carbohydrate content, and having a molecular weight of about 60,000 (Morton, 1955a). Alkaline phosphatase of animal tissues has been found to be closely associated with insoluble cellular particles (Kabat, 1941; Hers *et al.*, 1951;

Morton, 1953, 1954). Morton (1954) has found that the alkaline phosphatases of calf intestinal mucosa and of cow mammary gland are associated with microsomes.

The pH optimum of the enzyme varies from about 8.5–10.0 and is influenced by the type of substrate employed, the concentration of the substrate, the source of the enzyme and its activity, the type and concentration of activator, and the buffer employed. With reference to histochemical techniques, the metal-salt procedure is usually run at pH 9.0 while the azo-dye methods are more advantageously performed at a somewhat lower pH level.

## A. Metal-Salt Method

This type of procedure depends upon the *in situ* precipitation of the relatively insoluble calcium phosphate which may be subsequently visualized by transformation to another type of compound. Since the original report of the technique by Gomori (1939), many hundreds of papers have reported the distribution of this enzyme in tissue sections in both normal and pathological conditions. The widespread application of this technique has been primarily due to the simplicity and reproducibility of the method, coupled with the fact that the enzyme is relatively resistant to various types of fixation and embedding procedures, even though the physiological significance of the enzyme has yet to be elucidated. Gomori (1952, p. 178) reported that all fixatives cause considerable inactivation of alkaline phosphatase. In a study on the effect of various fixatives upon Coujard slides with purified intestinal phosphatase serving as the source of the enzyme, Gomori found that ethanol and acetone resulted in a relatively small loss of activity. Neutral formalin at room temperature, however, was found to inactivate over 75% of the enzyme in less than 24 hours but had very little effect under refrigeration (Chapter 1, p. 23). The work of Seligman *et al.* (1951b), Grogg and Pearse (1952), and Gomori (1952, p. 139) demonstrated the advantage of cold neutral formalin fixed tissues for the subsequent demonstration of the enzyme.

Gomori noted that formalin fixed tissues should be sectioned while frozen since subsequent embedding in paraffin usually gives poor results even if the formalin is washed out carefully. He further confirmed this observation in Coujard slide tests which revealed that paraffin embedding did not noticeably inactivate phosphatase except following formalin fixation. Gomori (1952, p. 138) specifically recommended cold acetone fixation followed by paraffin embedding (Chapter 1, p. 23). Although varying degrees of inactivation of the enzyme following fixation and paraffin embedding had been reported by a number of workers, it would appear that if small pieces of tissue are carefully fixed and dehydrated, the loss of activity should not be appreciable. Danielli (1953, p. 35) found that brief fixation in cold ethanol or acetone, as recommended by Gomori, followed by paraffin embedding did not result in an appreciable loss of enzyme activity.

The findings of Doyle (1950) with reference to frozen-dried tissues would indicate that properly dehydrated specimens are fairly resistant to the heat encountered during paraffin embedding. If, however, large pieces of tissue are inadequately fixed and dehydrated, it is apparent that the residual water present will result in a considerable (and uneven) degree of inactivation throughout the block. Freezing and drying appears to represent the best type of fixation when paraffin embedded tissues are to be employed in the demonstration of phosphatases (Chapter 1, p. 23). On the other hand, overnight fixation in cold acetone followed by rapid clearing and embedding in paraffin is also capable of producing tissue sections which exhibit good morphological integrity provided the specimen is thin (Chapter 1, p. 23). Freeze-substitution techniques may also be employed but there have been no definitive publications on the application of this technique in the demonstration of enzymes. Although unfixed frozen sections exhibit high activity, their use may result in loss of enzyme by diffiusion as well as readsorption of enzyme on normally non-active tissues (Gomori, 1950a). A simple procedure for the utilization of frozen sections in the phosphatase as well as other enzyme techniques employs brief acetone fixation of frozen sections mounted on glass slides (Burstone, 1961) (Appendix 1, Section 3).

The routine decalcification of tissues will inactivate phosphatases, although some degree of activity is retained following careful decalcification in formic acid or citrate buffer (Greep *et al.*, 1948; Lorch, 1949). Gomori (1950a) suggests that preformed calcifications in tissues may be removed by treating the sections for 5–10 minutes with a 0.1 *M* citrate buffer pH 4.5–5.5. However, for definitive studies of phosphatases associated with calcified tissues, the newer azo-dye techniques are to be preferred since the dyes formed are unaffected by the presence of calcium salts (Burstone, 1960a, b).

Following incubation the resulting precipitate of calcium phosphate may be visualized by treatment with silver nitrate to form photosensitive silver phosphate which produced a golden-brown stain, or by exposure of the sections to cobalt nitrate and subsequent transformation of the cobalt phosphate into black cobalt sulfide under the influence of ammonium sulfide (Gomori, 1939). Alternatively the calcium salt may be treated with ferrous sulfate followed by ferricyanide to form the so-called Turnbull blue (Dorfman and Epshtein, 1950), or a colored calcium lake may be formed by treatment of the section with sodium alizarin sulfonate (Bourne,

1943). Potential sources of error with this technique have been enumerated by Gomori (1950a; 1952, p. 183) as follows:

a. False negative reactions (i.e. lack of staining in spite of the known presence of alkaline phosphatase) will be obtained whenever the concentration of the enzyme is below the threshold of sensitivity of the method. The primary cause is inactivation of the enzyme due to improper or prolonged fixation.

b. False positive reactions such as nuclear staining adjacent to sites of high activity, artefactual staining of pigments such as hemosiderin, and staining of preformed calcifications.

As previously mentioned in the introduction, a major cause of artefactual staining is the diffusion of the primary reaction product, calcium phosphate. This artefact may be minimized or eliminated by employing relatively high concentrations of calcium chloride (0.1 *M*) in conjunction with relatively short incubation times (Gomori, 1951; 1952, p. 182). With reference to the numerous applications of the aforementioned alkaline phosphatase technique, the papers of Gomori listed in the bibliography of this chapter as well as the more recent excellent review of Arvy (1957–1958) should be consulted.

### B. Azo-Dye Methods

Menten *et al.* (1944a, b) devised the first azo-dye method by which alkaline phosphatase could be demonstrated by the incubation of a tissue section in a substrate solution containing β-naphthyl phosphate (Table 4) (Appendix 5A, Part 2, Section 14) and diazotized α-naphthylamine. In addition to testing other diazotized amines for use with their histochemical system, they also suggested that the azo-dye technique was applicable to the histochemical demonstration of other enzyme systems in tissue sections. In 1946 Danielli modified the aforementioned technique of Menten, Junge, and Green and used sodium β-naphthyl phosphate as a substrate in conjunction with a number of diazotized amines including β-naphthylamine, *p*-nitroaniline, α-aminoanthraquinone, and benzidine. Manheimer and Seligman (1948) modified the Menten, Junge, and Green method by employing a stabilized diazotate of α-naphthylamine. They also tested a commercially available diazotate "Naphthanil Diazo Blue B" (tetrazotized *o*-dianisidine). The preparation of a phosphorylated azo-dye by Loveless and Danielli (1949) represented an interesting innovation in substrate design in that it avoided the necessity of employing potentially inhibitory diazonium compounds. Unfortunately, the substrate in itself is a dye and potentially capable of producing non-specific staining.

Although α-naphthyl phosphate had been synthesized as early as 1894 (Table 4), Gomori (1951; 1952, p. 185) was the first to show the distinct superiority of this compound in an improved azo-dye technique for the demonstration of alkaline phosphatase. He noted that although dyes derived from $\beta$-naphthyl phosphate may exhibit brilliant shades, the resulting histological picture is blurred due to diffusion of the dye, and thus precludes the use of this substrate for definitive histological studies. On the other hand, the histological localizations obtained with the $\alpha$-isomer are considerably more detailed. Nevertheless, with highly active tissues diffusion artefacts may occur as a result of sluggish coupling of rapidly released $\alpha$-naphthol (Appendix 5B, Section 13). Davis and Ornstein (1959) recommended the use of $\alpha$-naphthyl phosphate with diazotized pararosaniline (basic fuchsin). Pearse (1954) reported the use of 1,6-dibromo-2-naphthyl phosphate and 2,4-dichloro- and 2,4-dibromo-naphthyl phosphates in the demonstration of alkaline phosphatase, but the results did not indicate any superiority over the $\alpha$-naphthyl phosphate technique of Gomori.

In 1957 the application of naphthol AS-MX phosphate in a study of calcifying tissues was described (Burstone and Keyes, 1957). Subsequently the synthesis and histochemical application of a large series of naphthol AS phosphates was reported (Burstone, 1958a, b, c, 1961). These substrates which are phosphate esters of complex arylamides of 3-hydroxy-2-naphthoic acid (Table 5, Chapter 2) release highly insoluble naphthols upon enzymatic hydrolysis. These azo-couple promptly to form highly insoluble azo-dyes which facilitate precise microscopic localizations in properly fixed tissues. Among the suitable compounds for the demonstrations of alkaline phosphatase are phosphates of naphthols AS-MX, AS-CL, AS-KB, AS-TR, AS-BS, AS-AN, AS-E, and AS-BI (Chart 2) (Table 5, Chapter 2). An interesting variation of the AS phosphate type is naphthol AS-GR phosphate (V), which is not actually a naphthol derivative but contains the anthracene nucleus. This substrate is capable of producing highly insoluble green dyes (Appendix 5B, Section 17) (Chapter 2, Table 5). The AS phosphates, which are recommended as a general purpose replacement for the older simpler naphtholic compounds, may be employed as the acid phosphate, the sodium salt, or in the form of a stable stock solution (Burstone, 1958a) (Appendix 5B, Section 14). Probably the most desirable procedure involves the use of freshly prepared substrate solutions made from either the acid phosphate or sodium salt. Although the use of a stable stock solution simplifies the procedure somewhat, there is always a danger that the buffer of such a system may deteriorate. Although unsubstituted naphthol AS phosphate is a useful substrate, the presence of nitro or halogen groupings in the arylide ring distinctly facilitates improved histochemical localizations over a wide pH range.

The AS phosphates have proved of value in the study of alkaline phos-

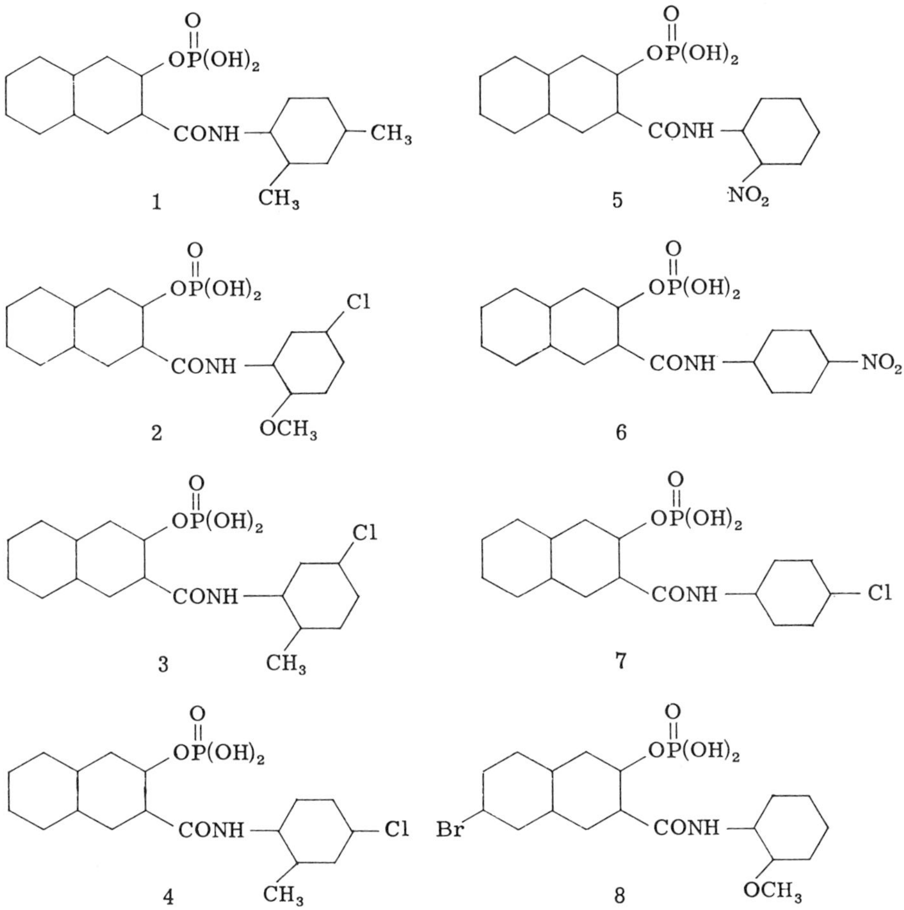

CHART 2. Formulae of naphthol AS phosphates. AS-MX phosphate (1), AS-CL phosphate (2), AS-KB phosphate (3), AS-TR phosphate (4), AS-BS phosphate (5), AS-AN phosphate (6), AS-E phosphate (7), AS-BI phosphate (8).

phatase associated with calcifying tissues (Burstone, 1960a, b), in which the metal-salt procedure may give rise to spurious staining. Samorajski and McCloud (1961) and Samorajski and Fitz (1960) have found naphthol AS phosphates to be of value in the study of alkaline as well as acid phosphatases of the central nervous system. These substrates are also very useful for the demonstration of leukocytic phosphatase in blood smears (Burstone, 1958a, 1961). Ashrafi and Fisk (1961) have advantageously

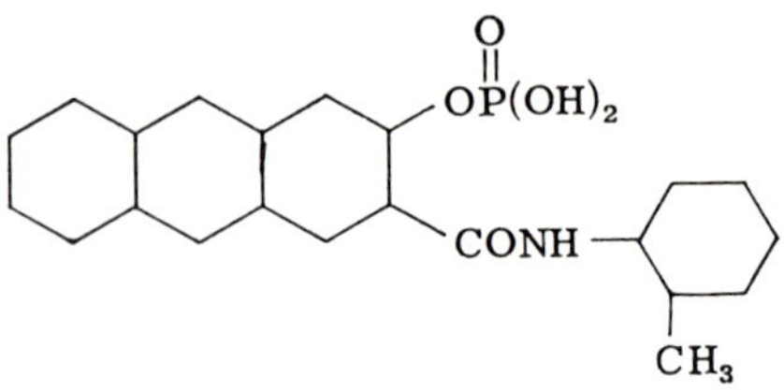


AS-GR phosphate

(V)

employed AS phosphates in the demonstration of alkaline and acid phosphatases in the stable fly (*Stomoxys calcitrans*).

Burstone (1960c) reported the successful histochemical demonstration of alkaline phosphatase by a post-coupling technique. The substrates employed were phosphorylated derivatives of complex naphthols obtained from 3-hydroxy-2-naphthoic acid (BON) and certain arylamines (Chart 3). The complex amines which were incorporated into the BON molecule included: β-aminoanthraquinone, *p*-aminoazobenzene, *p*-aminobenzophenone, *p*-aminophenyl allyl ether, *p*-dodecylaniline, dehydrothio-*p*-toluidine, and 2-nitro-3-aminodibenzofuran (Chapter 2, Table 6; Appendix 2, Section 3). Upon enzymatic hydrolysis, many of the released naphthols remained *in situ* and could be coupled with a wide variety of diazonium salts to produce exceedingly sharp histochemical localizations (Appendix 5B, Section 19). Of particular value are derivatives of β-aminoanthraquinone and aminoazobenzene [Chart 3 (1) and (2)]. The histochemical localizations obtained with the post-coupling technique exhibit the same distribution as those seen with the simultaneous naphthol AS phosphate methods, so that if there are any differences in substrate specificity they do not appear to be discernible on the basis of the staining pattern. In general, these techniques are designed for use with paraffin embedded tissues since the azo-dye may crystallize in the presence of lipid. However, in certain cases frozen sections briefly fixed in acetone may be employed.

Post-coupling techniques appear to represent an ideal type of histochemical procedure since extraneous and potentially inhibitory substances such as diazonium salts are eliminated from the incubating solution. The stability of the post-coupling substrate solutions is thus enhanced and longer incubation periods may be employed without resulting in deterioration of the substrate or objectionable non-specific background staining.

Alkaline phosphatase may also be demonstrated by a naphtholic non-coupling technique employing substrates similar to those utilized in the post-coupling method (Burstone, 1960c, 1962) [Chart 3 (2)]. The application of fluorescence techniques to the demonstration of phosphatases offers a theoretical advantage in that the visualization of a fluorochrome

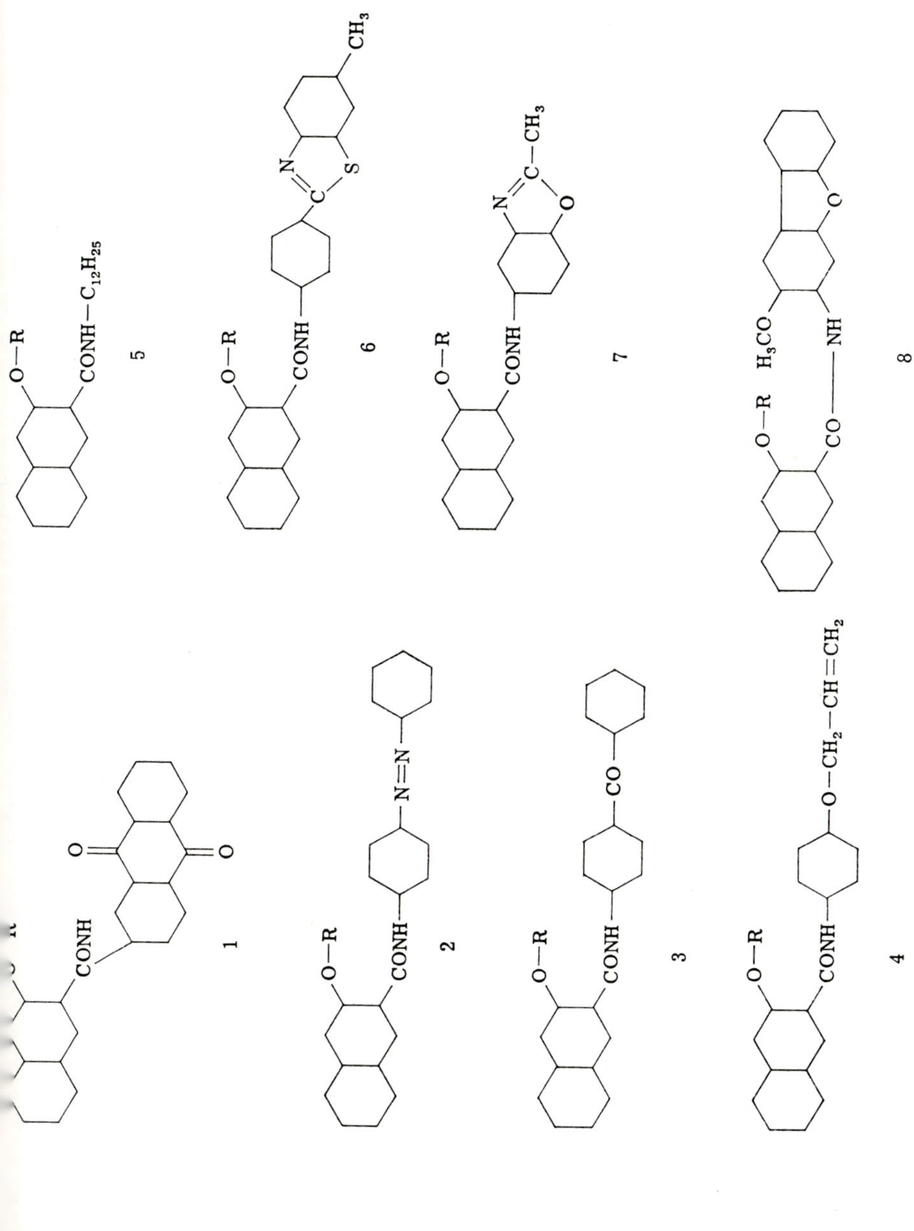


CHART 3. Post-coupling technique phosphatase substrates. R = phosphate grouping.

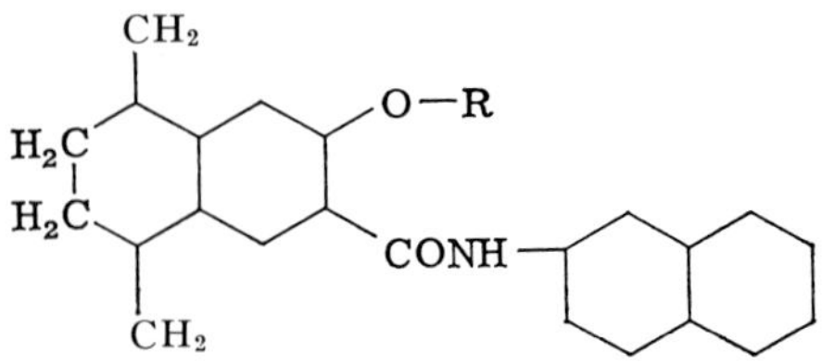

1

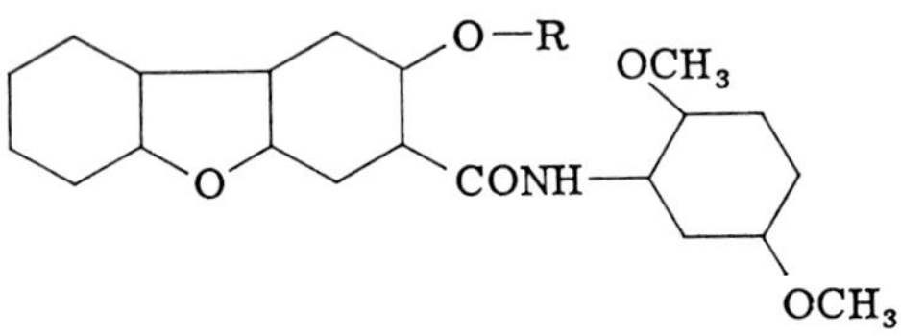

2

CHART 4. Phosphate substrates for histochemical demonstration of alkaline or acid phosphatases by a non-coupling fluorescence technique. Phosphate of 3-hydroxy-*N*-β-5,6,7,8-tetrahydro-β-naphthylamide (1), and phosphate of 2-hydroxy-2′,5′-dimethoxy-3-dibenzofurancarboxanilide (2). R = phosphate grouping

could permit detection of very low levels of activity (Chart 4) (Appendix 5B, Section 20).

The histochemical localization of alkaline phosphatase in a number of tissues as demonstrated by the naphthol AS phosphate, post-coupling, and fluorescence techniques is shown in Figs. 4 through 15.

## C. Activators and Inhibitors

The specific activator of alkaline phosphatase is $Mg^{++}$ (0.005 *M*). Other divalent cations such as $Mn^{++}$, $Zn^{++}$, and $Co^{++}$ may also behave as activators in very low concentrations. $Zn^{++}$, for example, is an activator at a concentration of 0.000001 *M* but is an inhibitor at 0.0001 *M*. Amino acids such as alanine and histidine have also been reported as activators but according to Morton (1955b) this "activation" effect is non-specific and is related to the removal of metal inhibitors. The inhibitors of alkaline phosphatase are relatively non-specific and include the following general types: metal-complexing agents such as α,α′-dipyridyldithiocarbamate and EDTA, metal salts such as beryllium and zinc chlorides, compounds which combine with amino groups, such as formaldehyde and isocyanates, competitive inhibitors such as inorganic and organic phosphates, and oxidizing agents including potassium permanganate and iodine. Cysteine and

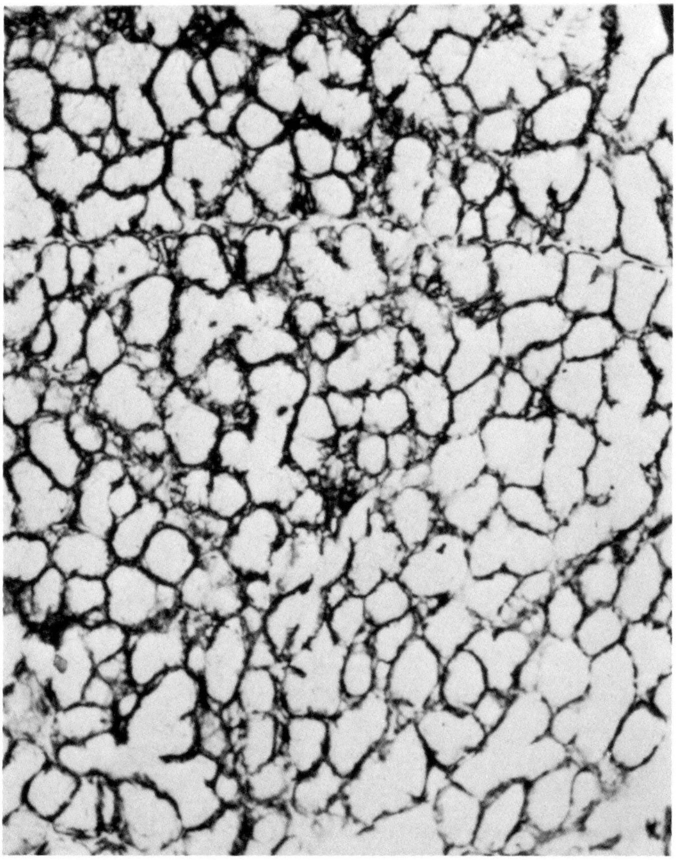

FIG. 8. Basement membrane localization of alkaline phosphatase in frozen section of rat sublingual gland. AS-CL phosphate, Blue BBN salt (× 143).

iodine (0.001 *M*) and sodium arsenate (0.01 *M*) are effective inhibitors. Polyanions such as polyestradiol phosphate, polyphloroglucinol, and polyphloretin phosphate are strong inhibitors of kidney and intestinal phosphatases (Diczfalusy *et al.*, 1953; Fernö *et al.*, 1958). Fluoride, borate, and carbonate ions are also inhibitory but to a lesser degree.

## D. Alkaline Phosphatase in Neoplasms

As might be expected, appreciable alkaline phosphatase activity is to be found in a variety of neoplasms. This activity was initially found to be reflected, in certain cases, in the sera of individuals with neoplastic disease (Franseen and McLean, 1935; Fishman, 1959, p. 732). The first significant study dealing with the histochemical distribution of the enzyme in tumors

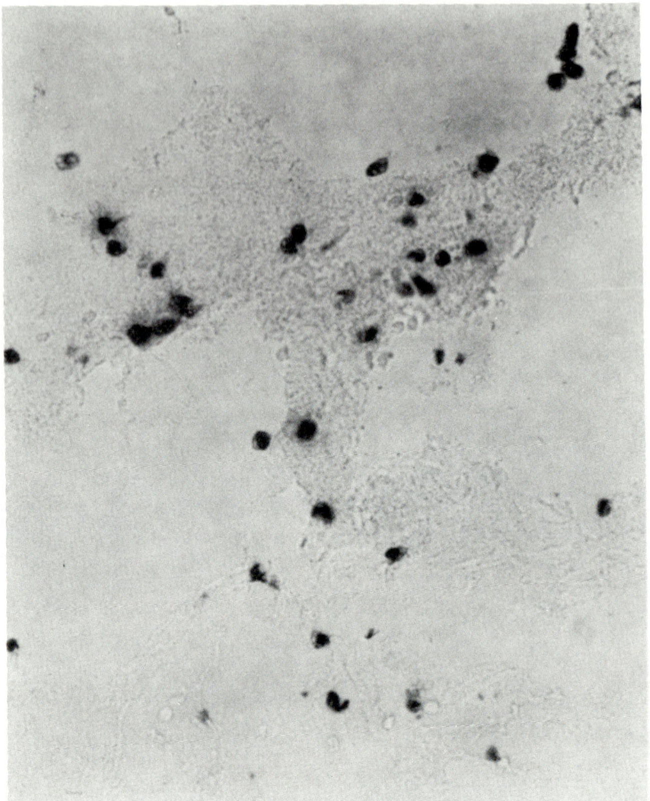

FIG. 9. Alkaline phosphatase of rat leukocytes. AS-E phosphate, Blue BBN salt (× 285).

was reported by Kabat and Furth (1941) who studied a number of human tumors including hypernephroma, fibrosarcoma, osteogenic sarcoma, hepatoma, melanocarcinoma, carcinoma of the breast, adamantoma, and carcinoma of the gastrointestinal tract. Despite the early stage of methodological development at that time, the study of Kabat and Furth is significant in that it represented one of the very first applications of modern histochemical techniques in the study of neoplasms. In 1946 Gomori noted a strong reaction of normal osteoblasts as well as of tumor cells in osteogenic sarcoma, Ewing's tumors, and synoviomas. He noted that neuroblastomas, lymphoblastomas, and reticulum cell sarcomas did not stain and suggested that the histochemical reactivity or non-reactivity of the aforementioned tumors might be useful as a diagnostic differential. More recently, Gomori (1955) documented several cases in which the alkaline phosphatase reaction was used to characterize a Ewing's tumor and a neu-

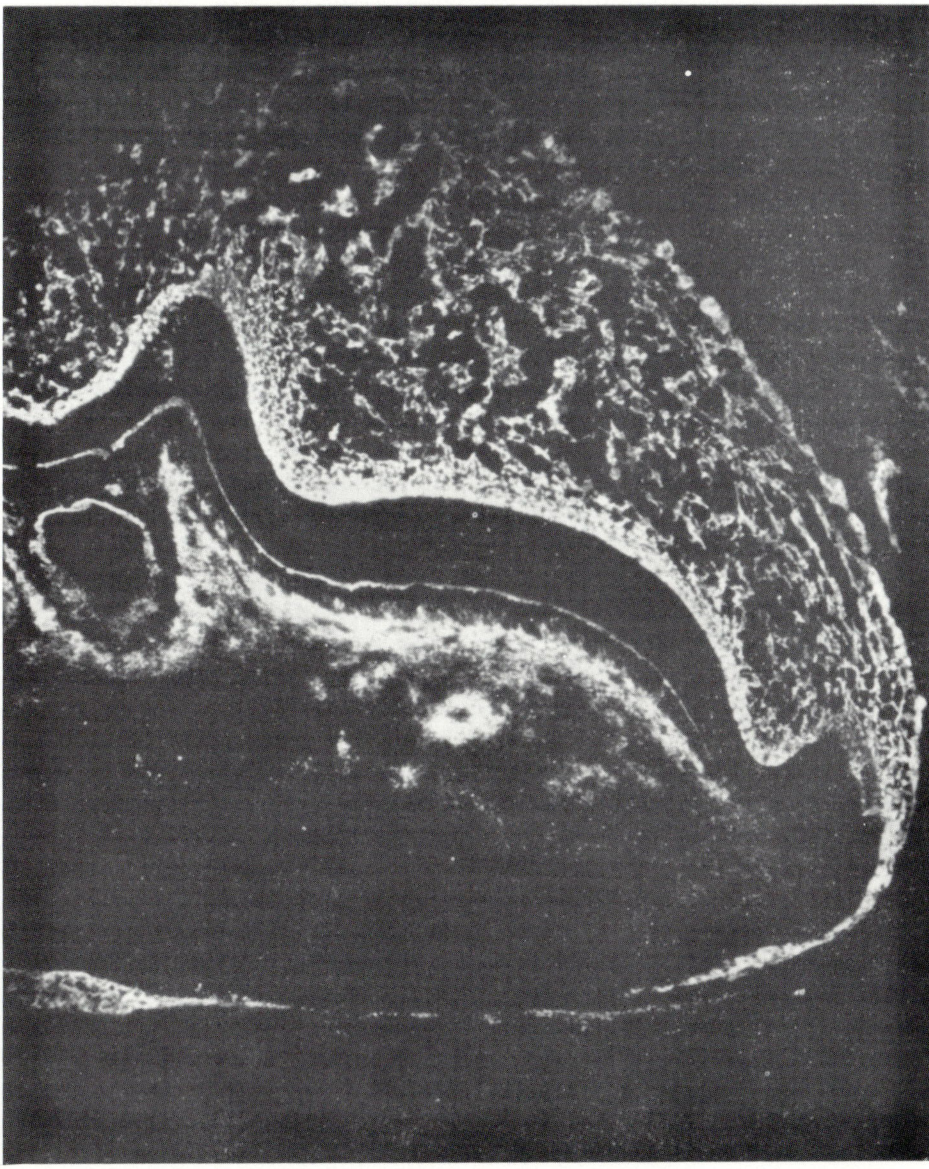

FIG. 14. Molar tooth of 3-day-old hamster. Non-coupling fluorescence method employing substrate 1 (Chart 4) (× 53).

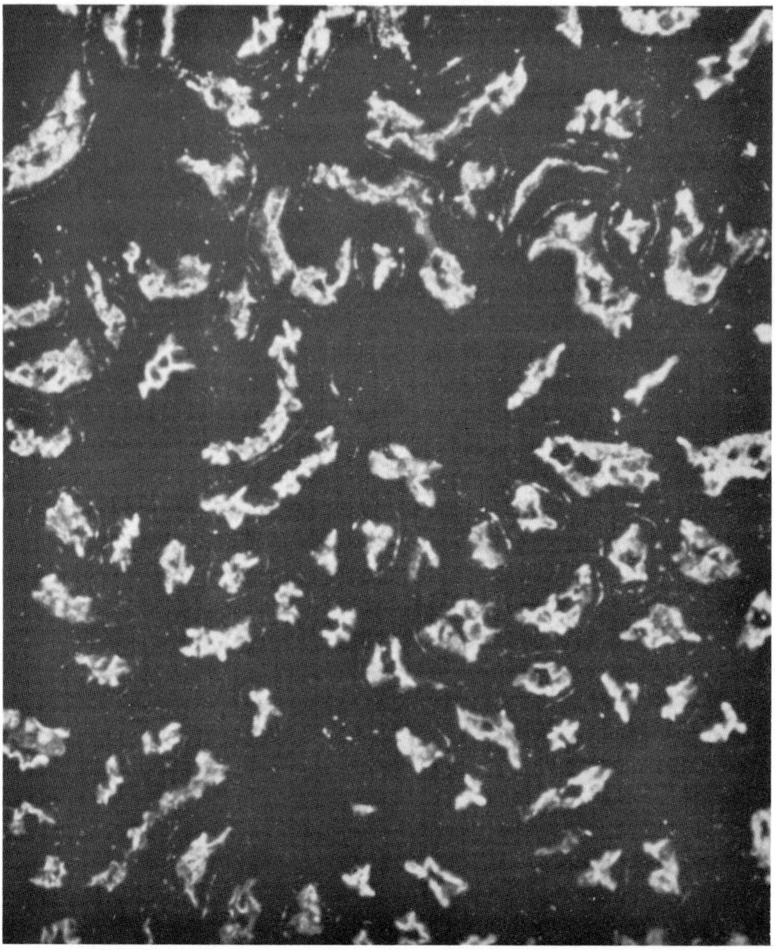

FIG. 15. Fluorescence method showing distribution of alkaline phosphatase in frozen-dried rat kidney, substrate 1 (Chart 4). Note basement membrane staining in addition to the brush border reaction (× 150).

roblastoma. Gomori also observed that intense phosphatase activity may be found in adenocarcinomas of the small intestine and in adrenal cortical tumors. Another of the significant earlier findings was the observation of Wachstein (1946, 1955) who found that polymorphonuclear leukocytes of patients with chronic myeloid leukemia exhibited a characteristic lack of phosphatase activity, as contrasted with an increased staining intensity of leukocytes in patients with infections. The studies of Valentine and Beck (1951), Kaplow (1955), Kenny and Moloney (1957), and Hayhoe (1960), as well as the more recent report by Perillie and Finch (1961), have further emphasized the diagnostic implication of alkaline phosphatase stains. The latter group (Perillie and Finch, 1961) utilized both the glycerophosphate and azo-dye techniques in the study of alkaline phosphatase activity of exudative leukocytes in acute leukemia. They found that 7 out of 11 patients with acute monocytic leukemia exhibited elevated levels of alkaline phosphatase activity, and suggest that the histochemical technique may be of assistance in classifying the stem cell forms of leukemia.

Manheimer and Seligman (1948) applied the β-naphthyl phosphate substrate of Menten *et al.* (1944a) in a study of alkaline phosphatase in over a hundred human neoplasms. As previously mentioned, the use of this particular azo-dye method is not considered to be satisfactory because of gross diffusion artefacts.

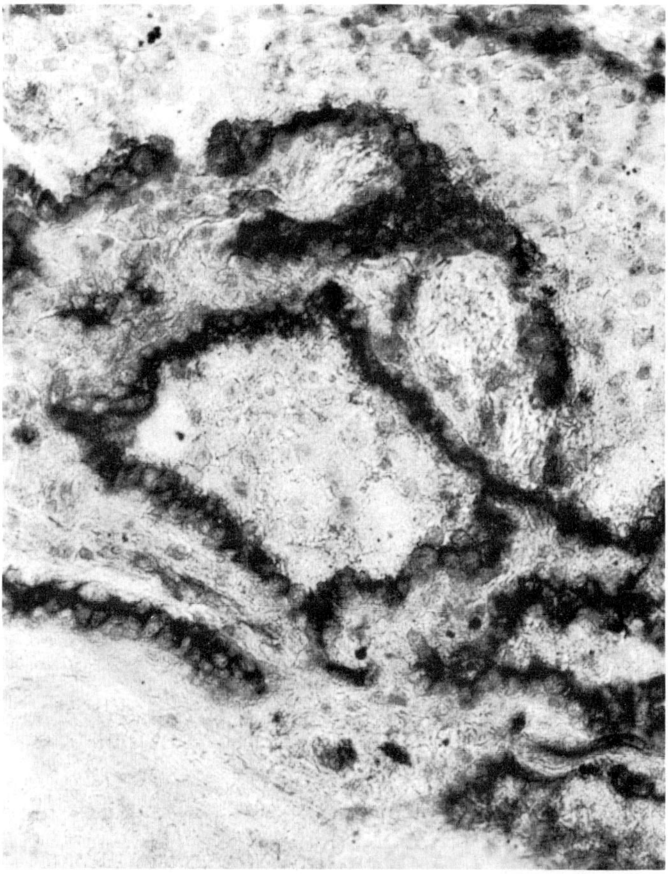

FIG. 16. Human adenocarcinoma of the lung. AS-MX phosphate, Red Violet LB salt (× 285).

Although both normal and neoplastic prostate are characterized by high acid phosphatase activity of the glandular epithelium, alkaline phosphatase of the same tissues is delineated to the capillaries (Brandes and Bourne, 1956; Butler *et al.*, 1959).

Alkaline phosphatase activity in tumors of the central nervous system has been studied by Buttuger *et al.* (1957) and Feigin and Wolf (1959). It appeared that an increasing degree of malignancy in meningiomas could be correlated with a decrease in phosphatase activity.

A number of investigators have studied phosphatase activity in carcinoma of the breast but the results have not been in complete agreement, probably due to variations in fixation and histochemical techniques. Foraker (1956a) found a decreased phosphatase activity in breast tumors as compared with normal controls. Fanger and Barker (1959, 1960) noted a

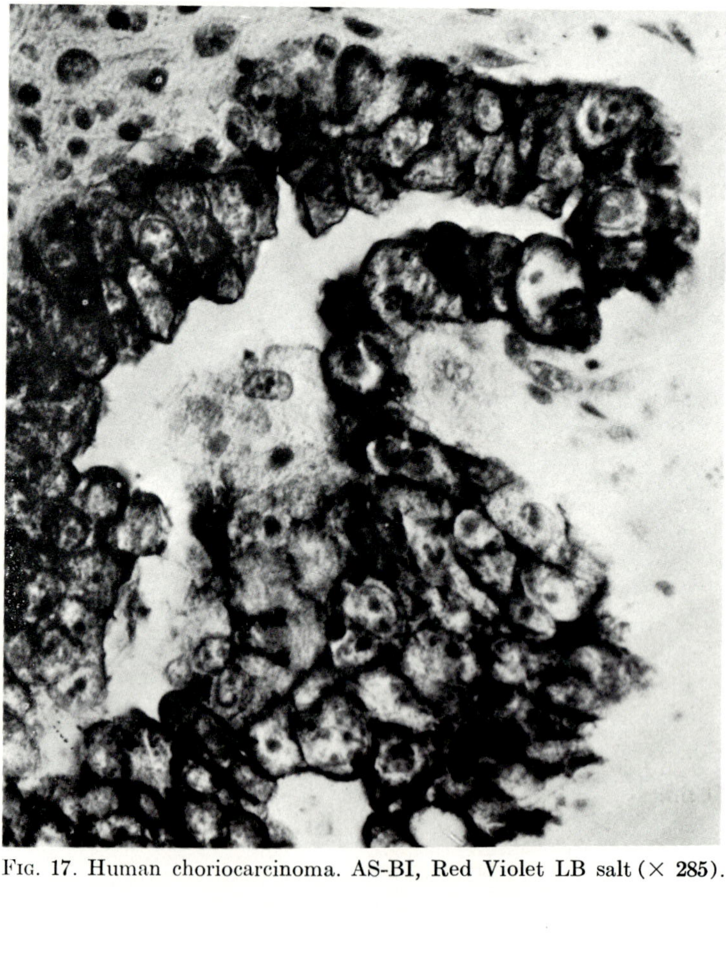

FIG. 17. Human choriocarcinoma. AS-BI, Red Violet LB salt (× 285).

characteristic staining of capillaries and myoepithelial cells in normal breast tissue as well as in adenosis. In carcinomas the enzyme was absent or else exhibited a somewhat diffuse and patchy distribution of staining. Variations in the intensity of capillary activity were also noted and indirectly correlated with the neoplastic process.

Burstone (1958a) first applied the naphthol AS phosphate method in the histochemical demonstration of alkaline phosphatase in neoplasms. Included in the study were the following neoplasms: adenocarcinoma of the lung (Fig. 16), ovarian adenocarcinoma, choriocarcinoma (Fig. 17), metastatic carcinoma of the colon (Fig. 7), and carcinoma of the liver. In addition, very high activity of a mouse osteogenic sarcoma was observed (Fig. 6). The more recent post-coupling and fluorescence techniques for the demonstration of alkaline phosphatase (Burstone, 1960c) have also been used to delineate sites of phosphatase activity in several human tumors including a carcinoma of the lung, carcinoma of the colon, and meningioma (Fig. 13).

## III. NON-SPECIFIC ACID PHOSPHATASES

From the biochemical standpoint acid phosphatases have been classified as type II, III, and IV (Roche, 1950, p. 473; Sumner and Somers, 1953, p. 89). Whether or not these types exist as such in unfixed or fixed sections as employed in enzyme histochemistry is not clear. Walker *et al.* (1954) concluded on the basis of an extensive survey of the literature that acid phosphatase is not a single enzyme in a given species but is an inclusive group including enzymes of different substrate specificities. By definition, these enzymes exhibit optimal activities at pH values below 7.0 and in the majority of instances at pH values between 3.8 and 6.0. Although the enzyme is not as widely distributed in animal tissues as its alkaline counterpart, acid phosphatase of a number of tissues including spleen, liver, kidney, and erythrocytes has been histochemically demonstrable. When a variety of substrates are employed in conjunction with acetate or tris buffers the activity curves between pH 4 and 6 are relatively flat and convex, with the maximum depending upon the type of substrate employed (Delory *et al.*, 1955; Nigam *et al.*, 1959).

Bamann and Riedel (1934) were among the first to note that certain tissues which hydrolyzed glycerophosphate at an alkaline pH also exhibited a second pH optimum in the acid range, and although the alkaline phosphatase was activated by $Mg^{++}$ the enzyme at an acid pH was unaffected by the metal. An earlier study (Bamann and Diederichs, 1934) also revealed that although the alkaline phosphatase activity of dried liver was inactivated by treatment of the tissue with dilute acetic acid, acid phosphatase activity was still retained. The most detailed biochemical studies have been made with reference to prostatic acid phosphomonoesterase (Schmidt, 1961, p. 37), presumably because of the high concentrations of the enzyme in that tissue. More recently, Goodlad and Mills (1957) have studied the acid phosphatases of rat liver and present findings which suggest the presence of two and possibly three non-specific liver acid phosphatases. The pH optima of 3.5–4.0 and 5.0–5.5 were reported for two of the enzymes, respectively. Straus (1954, 1956) reported high activity of "droplets" isolated from kidney tissues. Similar studies (Appelmans and de Duve, 1955; Appelmans *et al.*, 1955; de Duve *et al.*, 1955) indicated that acid phosphatase is attached to a special type of cytoplasmic granule which differs from both the mitochondrion and the microsome. C. de Duve (1957) termed these cellular particulates "lysosomes." The identity of lysosomes with hepatic peribiliary "dense bodies" has been suggested by the work of Rouiller (1954), Novikoff *et al.* (1956), and Novikoff (1959).

There has been some variance of opinion as to the optimal type of fixation to be employed for the preservation of acid phosphatase. Aspects of this problem have been well reviewed by Gomori (1952, p. 190), Pearse (1960, p. 431), and Novikoff (1955). Eränkö (1951, 1952) indicates that acetone fixed paraffin embedded tissues are unsatisfactory for use with the metal-salt method and recommends that the technique be employed only with unfixed frozen or formalin fixed frozen sections. On the other hand, specimens which are carefully fixed in acetone and embedded in paraffin are capable of giving excellent results with the naphthol AS phosphate azo-dye techniques (Burstone, 1958b, c, 1962). Although the enzyme has been assumed to be extremely labile, this impression probably resulted from the numerous observations relating to the extremely variable results obtained with the metal-salt technique in paraffin embedded tissues. Goetch and Reynolds (1951) claimed that two factors radically affect the method and these included the technique of mounting the sections as well as that of the de-waxing and hydration. They state that too high a water-bath temperature results in inactivation and that the trace of egg albumin used in mounting sections has an inhibitory effect upon the enzyme. It was specifically recommended that the sections not be de-waxed since this process ostensibly has an adverse effect upon the enzyme. Gomori (1956b) noted that the various modifications intended to correct the imperfections of his lead sulfide method, especially the use of cold formalin fixed sections, have resulted in more or less marked improvement of the results. On the other hand, he noted that certain artefacts such as nuclear staining were impossible to avoid. More recently, Holt (1959) reported that spurious nuclear staining could be eliminated by using freshly prepared substrate solutions. Of interest is the comment of Gomori (1956b) who states "It is certain that a significant diffusion of enzyme (acid·phosphatase) takes place from formalin-fixed frozen sections, and even more from unfixed cryostat sections." De Duve and his group (Berthet *et al.*, 1951; Appelmans and de Duve, 1955; Gianetto and de Duve, 1955) have found that acid phosphatase is readily made soluble and is released from tissues by freezing and thawing. Their work would also suggest the desirability of using fixed tissues for the histochemical demonstration of this enzyme. The optimal type of fixation depends upon the method to be employed. With the metal-salt technique the calcium-formol frozen section procedure of Eränkö (1951, 1952) (Appendix 5B, Section 3) is recommended; on the other hand, with the azo-dye techniques cold acetone fixed or frozen-dried paraffin embedded tissues yield highly reproducible results (Appendix 1, Sections 4, 5, and 11). The advantages of the freeze-dry procedure have been suggested by the biochemical studies of Pratt (1953) and Holt (1959), and by the application of the naphthol AS phosphate methods in the demonstration of acid phosphatases (Burstone, 1958b, c, Chapter 1, p. 39).

## A. Metal-Salt Method

The first histochemical method for the demonstration of acid phosphatase was the metal-salt or lead sulfide method of Gomori (1941), a procedure analogous to his original alkaline phosphatase technique. Almost from the very inception of this technique there was a considerable divergence of opinion as to the validity of the staining procedure. Among the best considerations of the many factors involved in both genuine and artefactual staining are the reviews of Gomori (1950a; 1952, p. 190), as well as the more recent evaluation of Holt (1959). Gomori (1952, p. 190) has classified the cause of partial or complete failure of the technique as follows:

1. Inactivation of the enzyme during the course of fixation, embedding, storage, and incubation. Gomori points out that inactivation may be fairly uniform or peculiarly patchy, with the boundary between active and inactive areas often being quite sharp. He also notes the reports of some authors that the enzyme is gradually inactivated in paraffin blocks, ribbons, or mounted sections which are stored over a period of time.

2. The use of unfixed or partially fixed unembedded tissues. He states that frozen sections may show a relatively high activity, but the affinity of undenatured proteins for lead salts is likely to produce false positive reactions. Although the lead adsorbed by paraffin embedded tissues may in most cases be readily removed with dilute acetic acid, frozen sections of unfixed or acetone fixed tissues will hold lead so stubbornly that even prolonged washing in strong acetic acid can not completely remove it. On the other hand, formalin fixed sections have much less of an affinity for lead.

3. Failure to use an acid rinse after incubation. Gomori points out that this is an important step since it removes protein-bound lead but does not affect enzymatically produced lead phosphate. If this step is omitted, as

it apparently has been by a number of workers, the result will be a non-enzymatic staining of many tissue components including nuclei.

4. Certain structures such as axons retain a high affinity for lead even following paraffin embedding, as is pointed out by the fact that lead impregnation is a suitable non-specific stain for these structures. Since the non-enzymatic staining of the axons may easily be mistaken for an enzymatic reaction, Gomori recommends the use of control sections which have been inactivated. However, it still may be extremely difficult or impossible to determine whether the histological localizations obtained do not represent a composite of both enzymatic and non-enzymatic staining.

5. Non-optimal composition of the incubating mixture. The effect of the composition of the substrate mixture upon the reliability of the results was originally studied by Gomori (1950c), who found that the ratio between the concentration of buffer and substrate is an important factor. An unduly high concentration of buffer will greatly reduce the sensitivity of a method, thereby giving rise to an "all or none" phenomenon in which discrete localizations may be erroneously equated with a more accurate delineation of sites of enzymatic activity.

6. Undetermined factors which include very peculiar types of failure not explainable by any of the previously enumerated factors. For example, Gomori mentioned that sections cut from a single block may stain uniformly one day but do not stain on the subsequent day, even though the incubation mixture is prepared in exactly the same way. Another type of failure is exemplified by the poor staining reaction of only one or two sections mounted on a slide, although adjacent serial sections exhibit an excellent staining reaction.

In a more recent evaluation of his lead sulfide histochemical method, Gomori (1956a, b) notes that although the use of formalin fixed frozen sections produces a more or less marked improvement of the results, nevertheless certain artefactual staining may still occur. On the other hand, other reports tend to indicate that the formol-calcium metal-salt acid phosphatase method as originally proposed by Eränkö (1951, 1952) gives unequivocally reproducible results (Holt, 1959; Novikoff, 1960b, c). It will be interesting to follow future developments with reference to this staining reaction in order to ascertain if Gomori was overly cautious with respect to his analysis of inherent pitfalls in the procedure even when cold formalin fixed sections are employed.

Several modifications of the lead sulfide method have been suggested. Takeuchi and Tanoue (1951) incubated sections at the usual pH (5.0–6.0) but visualized the lead salt by means of silver nitrate as originally suggested by Gomori (1939) for his alkaline phosphatase method. Wachstein and Meisel (1957a) also modified the method by incubating sections at neutral pH in the presence of tris buffer.

Among the first significant findings with the Gomori metal-salt acid phosphatase method was the localization of peribiliary sites of activity (Deane and Dempsey, 1945; Deane, 1947). In 1941 Gomori demonstrated by means of his method the fact that acid phosphatase is localized in the glandular epithelium of the human prostate. More recently the procedure has been employed to delineate granular or droplet-like structures in liver and kidney which are believed to correspond to the biochemically delineated lysosomes of de Duve (1959, p. 128), Holt (1959), Novikoff and Essner (1960), Novikoff (1960c), and de Man *et al.* (1960) (Figs. 18 through 20).

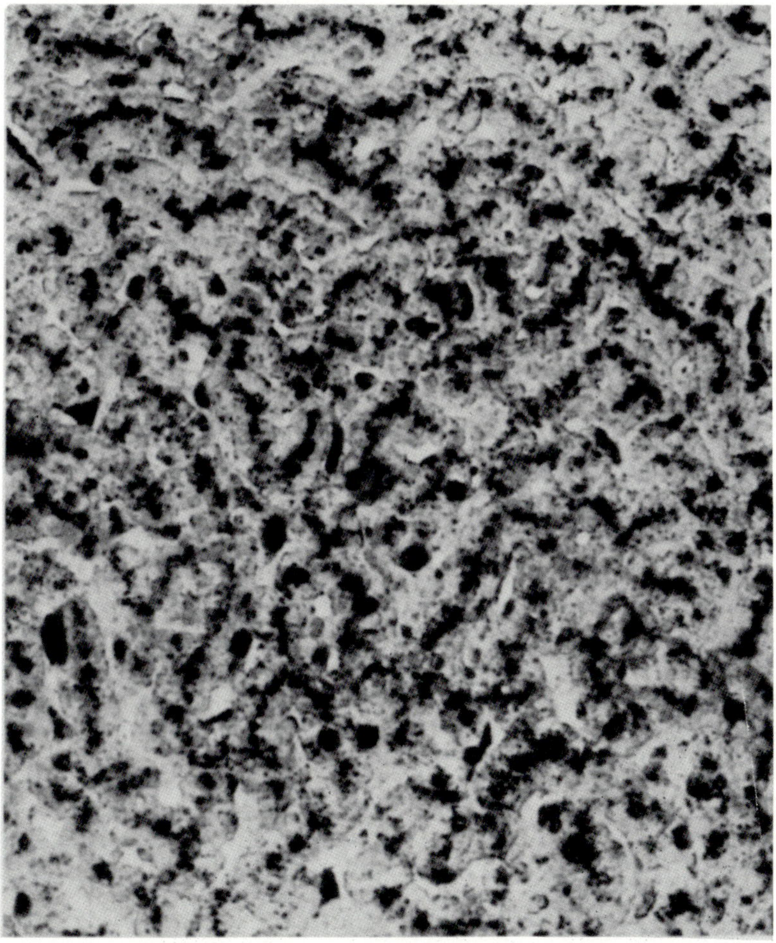

FIG. 18. Acid phosphatase of formalin fixed frozen section of rat liver, metal-salt method (× 500).

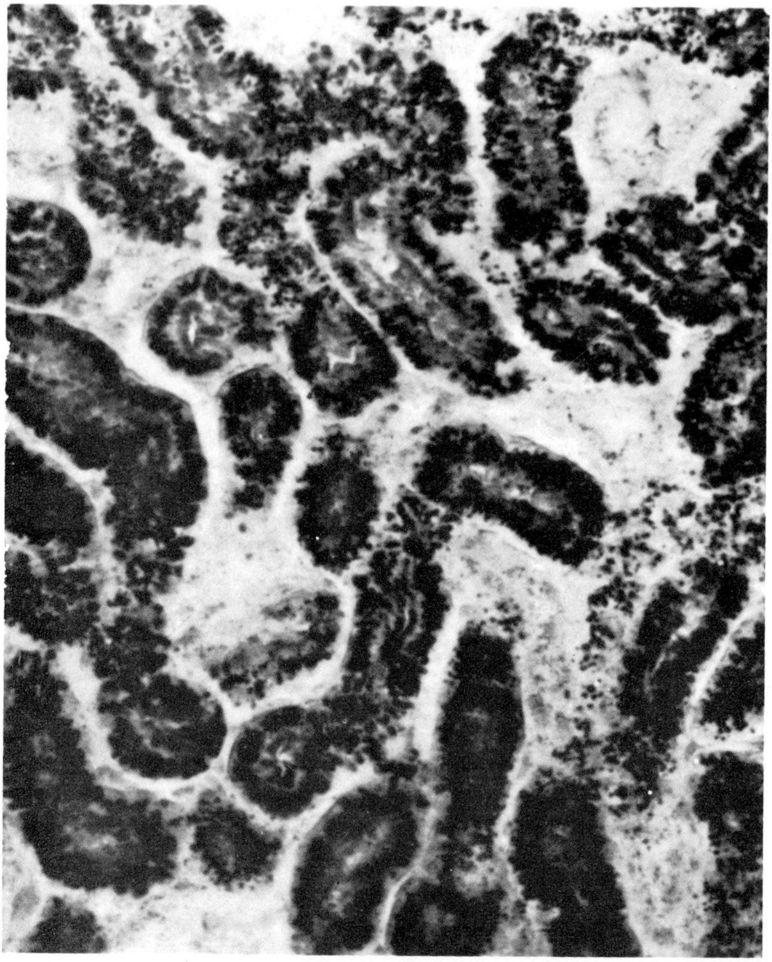

FIG. 19. Acid phosphatase of formalin fixed frozen section of rat kidney showing "droplet" staining. Metal-salt method (× 500).

## B. Azo-Dye Methods

The first azo-dye histochemical method for the demonstration of acid phosphatases was reported by Seligman and Manheimer (1949) who employed cold acetone fixed paraffin embedded tissues in conjunction with calcium α-naphthyl phosphate and diazotized α-aminoanthraquinone. A small amount of aerosol was also incorporated into the substrate solution, presumably because of its solvent action on the poorly soluble substrate. Although prostatic tissues required only a 1-hour incubation, other tissues were incubated for 12 to 24 hours. These authors noted that non-specific

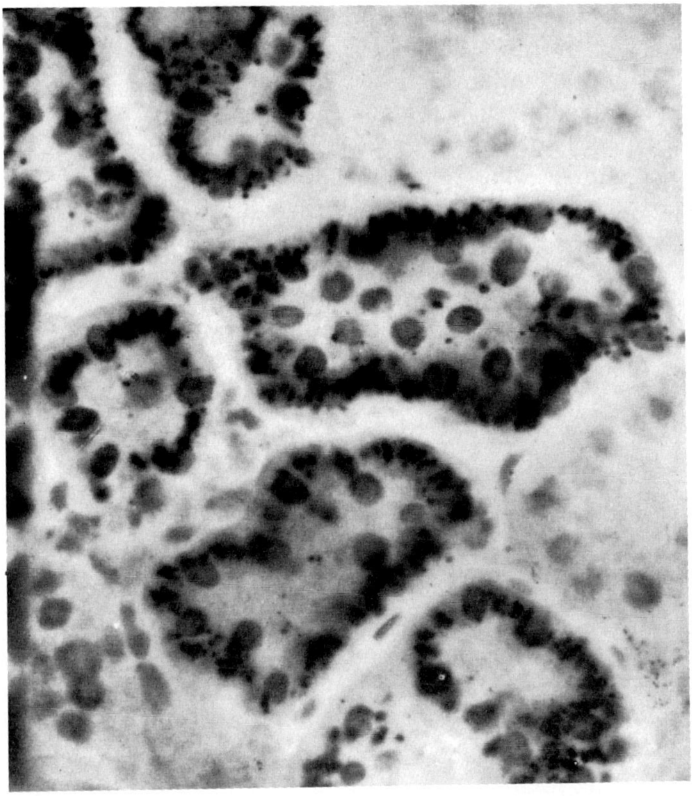

FIG. 20. Higher magnification of area similar to that seen in Fig. 18 (× 1500) (courtesy Dr. O. Eränkö).

background staining represented a serious disadvantage of the technique and recommended the study of control slides prepared by incubating sections in solutions from which the substrate had been omitted. This method was not recommended by Gomori (1952, p. 194) because of unavoidable gross diffusion artefacts. According to Pearse (1960, p. 441) the results with the Seligman-Manheimer method are far from satisfactory and compare unfavorably with the original Gomori metal-salt technique. Although definitive localization of acid phosphatases cannot be obtained with this azo-dye method, due recognition should be given to the fact that it represented a significant initial step in histochemical methodology. Grogg and Pearse (1952) and Pearse (1953, p. 255) modified the technique by employing cold formalin fixed tissues in conjunction with several diazonium salts including tetrazotized *o*-dianisidine. Although incubation times were considerably reduced, the method still suffers from obvious diffusion artefacts due to the sluggish coupling of α-naphthol with the particular diazonium salts employed. Burton (1954) attempted to improve the technique by increasing the concentration of diazonium salts and also by the addition of "coupling catalyzers" such as pyridine or 2-butoxy-*N*-(2-diethylaminoethyl)cinchoninamide hydrochloride (Percaine). Following incubation the sections were differentiated with acid alcohol and dehydrated and cleared in alcohol and xylene, respectively. Although the results of this modification appeared to be much sharper than those previously obtained, the technique is not recommended by Pearse (1960, p. 444) for several reasons. An interesting type of artefact is introduced by the differentiation process which washes out the less soluble and more diffusely distributed dye and produces what appears to be a sharper histochemical localization. Furthermore, the increased concentration of diazonium salt tends to augment the inhibitory effect of the metallic diluents contained therein. The author is in complete agreement with the original contention of Gomori (1952, p. 194) that the α-naphthyl phosphate acid phosphatase technique in conjunction with the standard diazonium salts is unsatisfactory for the histochemical localization of the enzyme because of gross diffusion artefacts. A similar view has been expressed by Rutenburg and Seligman (1955) who found that this technique revealed a lack of histological detail and diffusion artefacts exemplified by nuclear staining and furry outline of stained cells. On the other hand, Barka (1960) reported that α-naphthyl phosphate was a satisfactory acid phosphatase substrate when used with diazotized pararosaniline.

An interesting approach to the demonstration of acid phosphatase was suggested in the 6-bromo-2-naphthyl phosphate post-coupling technique of Seligman *et al.* (1949). Unfortunately, the enzymatically released naphthol is not sufficiently insoluble and diffusion artefacts occur (Gomori, 1952, p. 171; Pearse, 1960, p. 411). Another post-coupling technique was reported by Rutenburg and Seligman (1955) who employed a new substrate 6-benzoyl-2-naphthyl phosphate. They noted that upon enzymatic hydrolysis the relatively insoluble 6-benzoyl-2-naphthyl was adsorbed to cellular structures and could be subsequently coupled with a diazonium salt to form an azo-dye. Gomori (1956b) evaluated results obtained with 6-benzoyl-2-naphthyl phosphate as well as several other naphthyl phosphate esters and compared these with his original lead sulfide method. He reported that the benzoyl naphthyl post-coupling method gave useful results provided the incubation period is not extended beyond 1 hour. The substrate exhibits an appreciably spontaneous decomposition which may result in artefactual staining reactions since 6-benzoyl-2-naphthol may be partly adsorbed by tissues components (Chapter 4, p. 139). For example, kidney sections incubated in 6-benzoyl-2-naphthol and then coupled with Diazo Blue B show an intense but non-specific staining reaction of collecting tubules. Thus, when the Rutenburg-Seligman method is employed, it is always advisable to incubate a control slide in a solution of the free naphthol.

A distinctly different type of acid phosphatase substrate, the naphthol AS phosphates, was reported by Burstone (1958a, b, c, 1961, 1962). Many of these compounds had already proved to be of considerable value as improved histochemical reagents for the demonstration of alkaline phosphatases (Chart 2). The rationale for the use of arylamide derivatives of hydroxynaphthoic acid (naphthol AS) for precise localization has been previously discussed (Chapter 4). These naphthols azo-couple efficiently over a wide pH range to form brilliantly colored, highly insoluble azo-dyes of relatively non-crystalline structure at the cytological level. The most suitable naphthol AS phosphates for the demonstration of acid phosphatases are those which contain halogen or nitro substituents (e.g. AS-BI, AS-CL, AS-TR, and AS-AN phosphates). With naphthol AS-BI, as well as with similar phosphates, Burstone (1958b, c) was able to demonstrate both a granular and droplet distribution of activity which might coincide with the cytoplasmic phosphatase-containing granules described by the group of de Duve (Gianetto and de Duve, 1955; Appelmans *et al.*, 1955), (Fig. 21). Other localizations of the enzyme with the simultaneous coupling naphthol AS procedures are shown in Fig. 3, and Figs. 22 through 26.

Several of the newer of the post-coupling and fluorescence alkaline phosphatase substrates (Burstone, 1960a, b, 1962) (Charts 3 and 4) have also proved of value in the demonstration of the enzyme in the acid range (Figs. 27 and 28).

## C. Activators and Inhibitors

Review of biochemical activators and inhibitors of acid phosphatases has been made by Roche (1950, p. 486) and Walker *et al.* (1954). Acid phosphatases with the exception of phosphomonoesterase IV are not activated by $Mg^{++}$. $Mn^{++}$ appears to have a distinct activating effect when incorporated in azo-dye substrate solutions. Phosphate, arsenate, molybdate, copper, oxalate, and fluoride ions (0.001 $M$) usually exhibit a marked inhibitory effect on acid phosphatases. In higher concentrations (0.01 $M$) fluoride produces an almost complete inhibition of prostatic acid phosphatases when $\beta$-glycerophosphate and phenylphosphate are employed as substrates (Schmidt, 1961, p. 37). Phosphomolybdate, tungstate, phosphotungstate, and metavanadate ions are strong inhibitors of kidney and liver acid phosphatase (Courtois and Anagnostopoulos, 1948). Phlorizin markedly inhibits the kidney and intestinal enzyme (Beck, 1942). Compounds such as formaldehyde, isocyanates, and nitrous

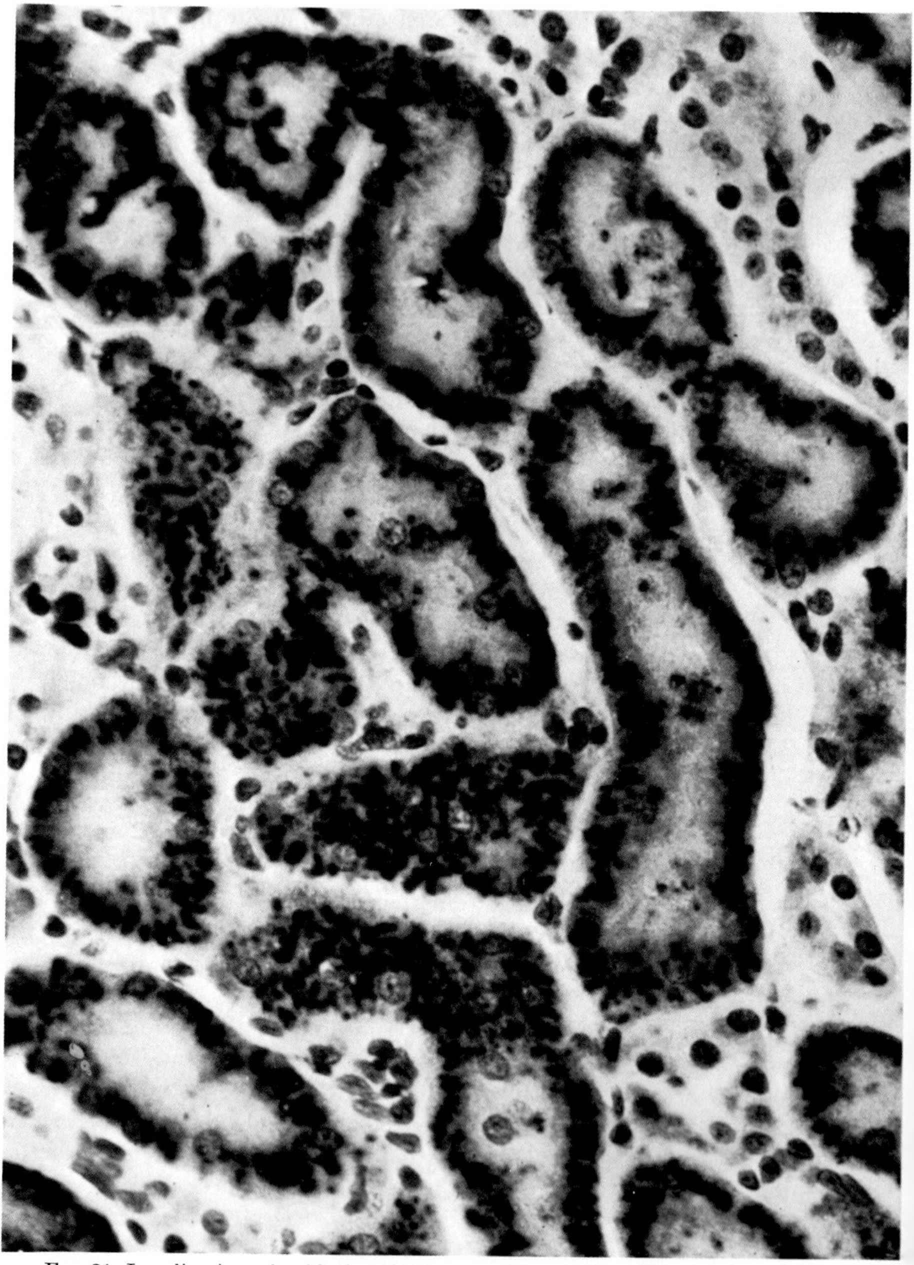

FIG. 21. Localization of acid phosphatase in "droplets" in frozen-dried paraffin embedded rat kidney. AS-TR phosphate, Red Violet LB salt (× 1425) (Burstone, 1958c).

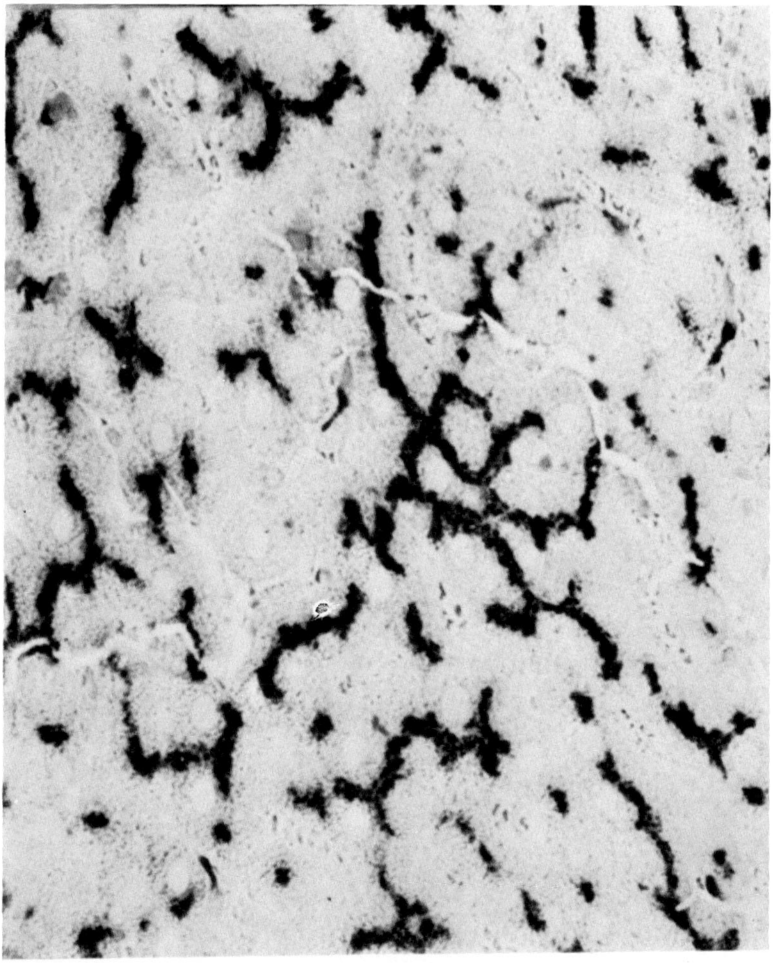

FIG. 22. Acid phosphatase of frozen-dried paraffin embedded liver showing delineation of staining pattern to peribiliary granules. AS-BI phosphate, Red Violet LB salt (× 1500) (Burstone 1958c).

acid, which react with amino groups of tissue proteins, have been found to inactivate acid phosphatases (Anagnostopoulos, 1950).

In distinct contrast to their effect on alkaline phosphatases, metal-complexing substances such as EDTA, cyanide, cysteine, and glutathione do not inhibit acid phosphatases. Alkylating reagents including iodoacetamide, iodoacetate, and iodosobenzoate have no appreciable effect upon prostatic acid phosphatase activity (Schmidt, 1961, p. 37). On the other hand, prostate phosphomonoesterase is distinctly inhibited by 0.001 *M*

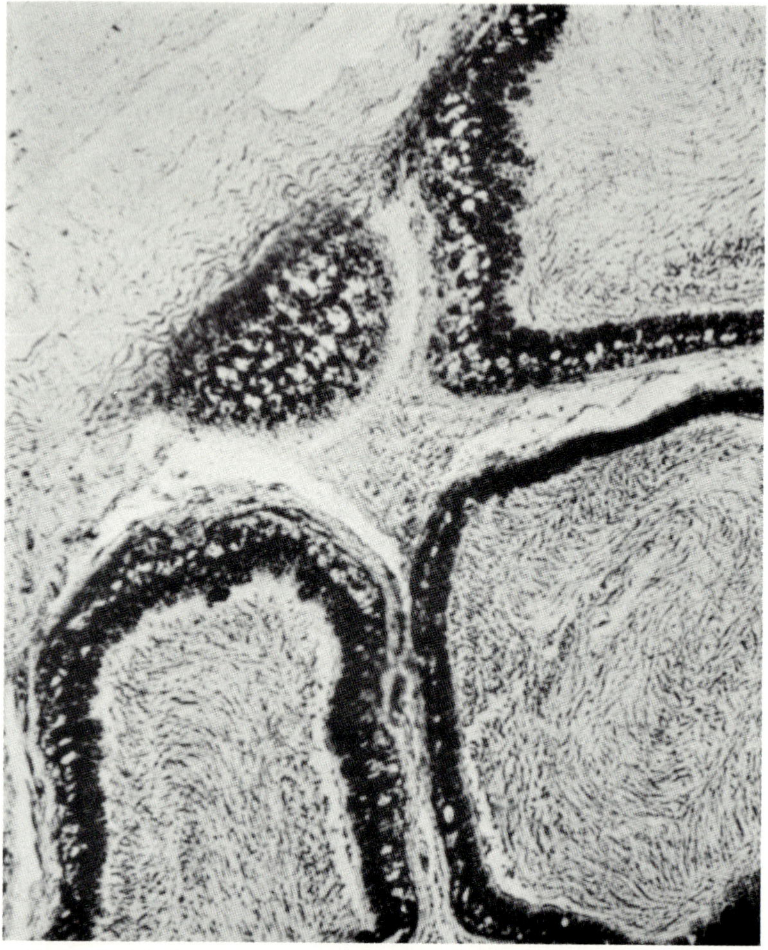

FIG. 27. Acid phosphatase of frozen-dried rat epididymis, substrate 2 (Chart 3); post-coupled with Garnet GBC salt (× 800).

mercuribenzoate (Tsuboi and Hudson, 1955). Under certain conditions, reactivation may be obtained by the addition of 0.001 $M$ cysteine. Certain $\alpha$-hydroxycarboxylic acids are distinct inhibitors of acid phosphomonoesterases (Abul-Fadl and King, 1949; Kilsheimer and Axelrod, 1957; Tsuboi and Hudson, 1955). One of the most effective hydroxycarboxylic acid inhibitors of prostate phosphomonoesterase is (+)-tartrate which has been employed in a test for the differentiation between blood serum and prostatic acid phosphatases (Fishman and Lerner, 1953). Other hydroxycarboxylic acid inhibitors include D-(+)-glyceric acid, L-(+)-threonic acid,

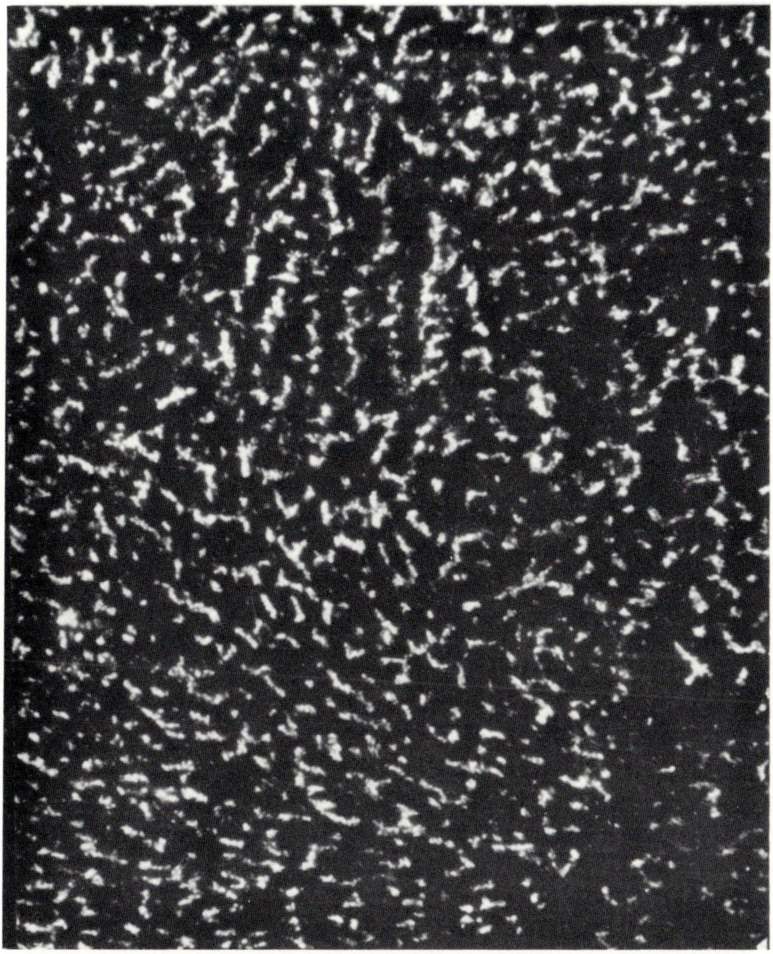

FIG. 28. Acid phosphatase of rat liver, fluorescence method employing substrate 1 (Chart 4) (× 200).

and DL-glyceric acid (Kilsheimer and Axelrod, 1957). Erythrocyte phosphomonoesterase, however, is not inhibited by (+)-tartrate but is strongly inhibited by 0.16 *M* formaldehyde (Abul-Fadl and King, 1949). The erythrocyte enzyme is also completely inhibited by $Cu^{++}$ in very low concentrations which do not affect the prostate enzyme. On the other hand, the red cell enzyme is only slightly inhibited by fluoride (Tsuboi and Hudson, 1955). Although formaldehyde in relatively low concentrations (0.5%) has little effect upon prostatic acid phosphatase, this compound does produce a distinct inhibition of enzymes of the liver, spleen, kidney and adrenal cortex (Abul-Fadl and King, 1949; Gordon, 1952). Belfanti *et al.* (1935) noted that acid phosphatases of liver and kidney are reversibly inactivated by sodium oxalate and can be reactivated by addition of calcium chloride. On the other hand, fluoride ions produced an irreversible inactivation.

### D. Acid Phosphatase in Neoplasms

From a biochemical standpoint Greenstein and Thompson (1943) reported higher acid phosphatase activity of a transplanted hepatoma as compared with adult liver, although Greenstein *et al.* (1946) indicated that the acid phosphatase activities of both normal mouse liver and mouse hepatoma were essentially similar. In a comparison of biochemical characteristics of diverse tumors as contrasted with their tissue of origin, Greenstein (1956) indicated a somewhat increased acid phosphatase activity associated with the neoplastic transformation. Lemon *et al.* (1954) found increased acid phosphatase levels in adenocarcinoma of the colon. Serum acid phosphatase is known to be elevated in carcinomas of the prostate, in primary tumors of bone, and in metastatic tumors of the liver (Fishman, 1959, p. 732).

Although the histochemical demonstration of acid phosphatase may be of potential diagnostic value as well as offering insight into the pathogenesis of the neoplastic process, the lack of reproducibility of some of the older phosphatase techniques has resulted in a very limited quantity of reliable data on this subject. In 1941 Gomori first reported the histochemical localizations of acid phosphatase in neoplasms of bladder, stomach, bronchus, prostate, bone, and colon. Dean and Woodard (1947) showed that it is possible to demonstrate whether a tumor is of prostatic origin by determining acid phosphatase activity of biopsy specimens. Histochemical techniques, of course, should be of considerable value in this type of procedure. Gomori (1955) notes that positive reaction in a section of neoplasm (1½–2 hour incubation) is practically diagnostic of a tumor of prostatic origin. He indicated that this feature could be an excellent diagnostic aid if the histochemical method (lead sulfide technique) were dependable, which he felt it was not. Lemon and Wisseman (1949) found an increased acid phosphatase activity in malignant tumors of the breast, bronchus, skin, bladder, esophagus, stomach, colon, and rectum, using both histochemical and biochemical techniques. Glück (1952) has also reported a significant enzymatic activity in carcinoma of the breast.

Reiner *et al.* (1957) noted the distribution of acid phosphatase in a number of neoplasms with the 6-benzoyl-2-naphthyl phosphate method.

Included were tumors of the prostate, breast, gastrointestinal tract, and skin. As might be expected, the strongest reaction was seen in benign and malignant prostatic tumors.

Burstone (1958b) studied frozen-dried human neoplasms with the newer naphthol AS phosphate techniques. A carcinoma of the lung showed groups and masses of intensely stained tumor cells. If this is a common characteristic of lung tumors, it could prove to be a useful diagnostic tool in the study of bronchial aspirations. An epidermoid carcinoma from the base of the tongue revealed intense staining of dyskeratotic cells comprising "pearls" (Fig. 25). The potential diagnostic implications of this finding warrant further study. A gastric carcinoma showed a granular pattern of

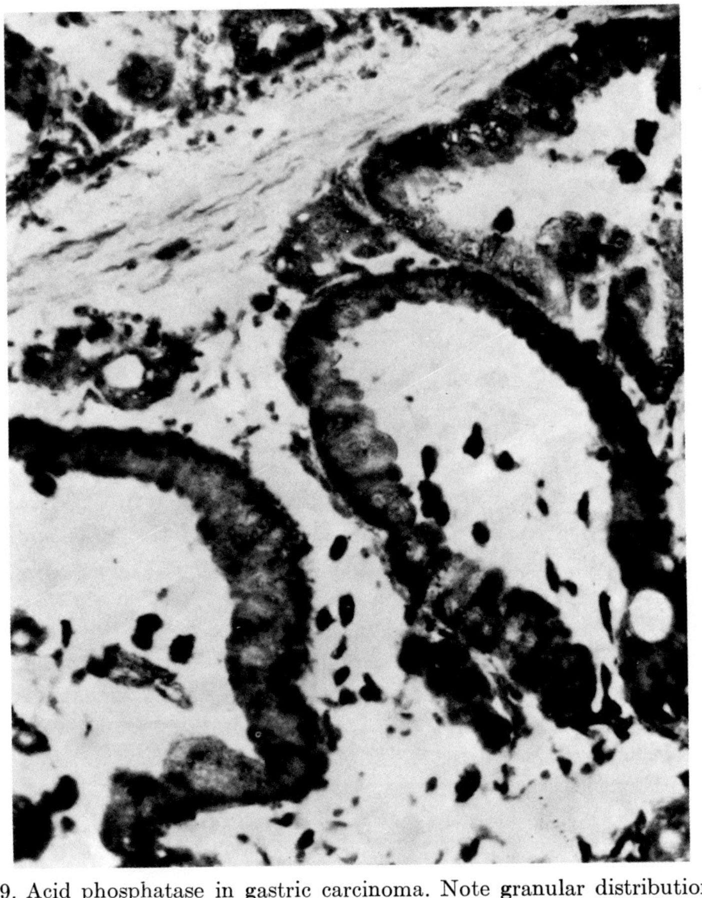

FIG. 29. Acid phosphatase in gastric carcinoma. Note granular distribution of dye which is primarily localized to basal portions of tumor cells. AS-BI phosphate, Red Violet LB salt (× 143).

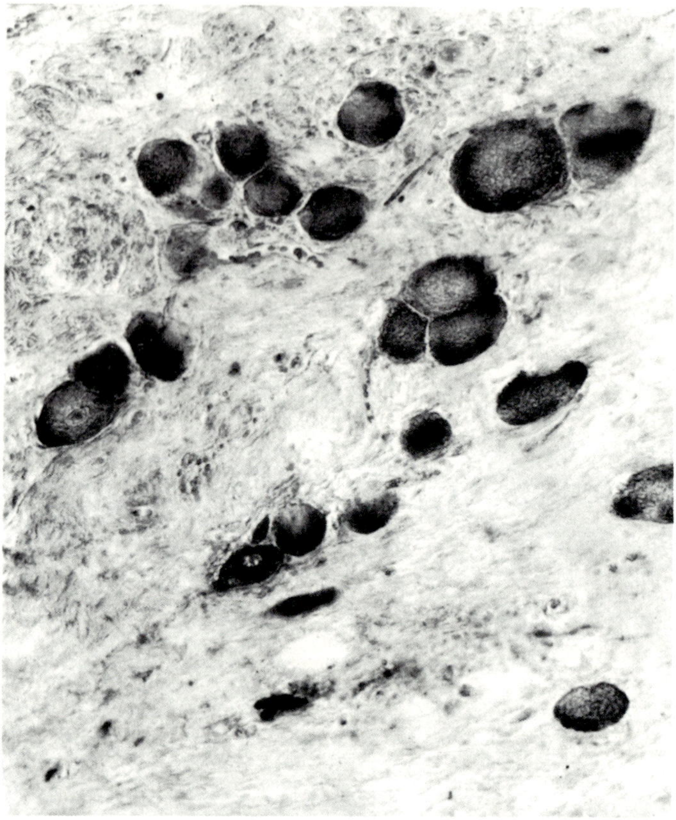

FIG. 30. Acid phosphatase of ganglioneuroma. AS-BS phosphate, Red Violet LB salt (× 143).

activity confined primarily to the basal portion of tumor cells (Fig. 29), while a ganglioneuroma contained groups of intensely staining "ganglion cells" (Fig. 30). All components of a Hodgkin's lymph node stained, with a particularly intense reaction being observed in Sternberg-Reed cells.

Fanger and Barker (1959) studied the distribution of the enzyme in a number of human breast tumors, employing unfixed frozen sections in conjunction with both azo-dye and lead sulfide methods. Many specimens from carcinoma of the breast revealed highest activity of capillaries at the infiltrating periphery of the tumor.

## IV. SPECIFIC PHOSPHATASES

### A. Adenosinetriphosphatase (ATPase)

In 1934 Lohmann discovered that muscle extracts could hydrolyze adenosine triphosphate (ATP) (VI). Subsequently, Engelhardt and Ljubimova (1939) found that muscle contained enzymatic activity in addition to ATPase by which ATP was cleaved to AMP. They also found that purified myosin exhibits ATPase activity. The work of Szent-Györgi (1947) suggested that crystalline myosin and ATPase are similar proteins. Two types of myosin are known to exist, both of which exhibit ATPase activity which is retained despite partial degradation of the macromolecule by trypsin (Kielley, 1961, p. 159). Kielley and Meyerhof (1948, 1950) and Kitiyakara and Harman (1953) have prepared an ATPase from muscle which is not associated with myosin, while a similar enzyme has been isolated from pig brain by Binkley and Olsen (1950). An ATPase in the submicroscopic fraction of rat brain has been reported by Järnefelt (1961a, b). Muscle and liver also contain an enzyme, apyrase, which hydrolyzes ATP but exhibits somewhat different characteristics from the latter (Kalckar, 1944; Meyerhof, 1945). This enzyme hydrolyzes ATP to form ADP (VII) and two molecules of phosphoric acid. Although designated as ATPase, myosin will hydrolyze a number of nucleotide triphosphates including inosine, cytidene, uridine, and guanosine triphosphates (Goldberg and Gilmour, 1954; Kielley *et al.*, 1956). In addition, ATPase will hydrolyze a number of inorganic triphosphates. These findings certainly would tend to complicate the histochemical localization of this enzyme. There has been some divergence of opinion with reference to the pH op-

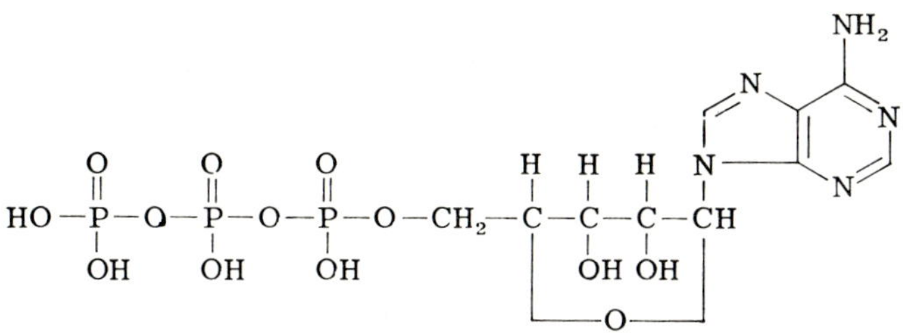

ATP

(VI)

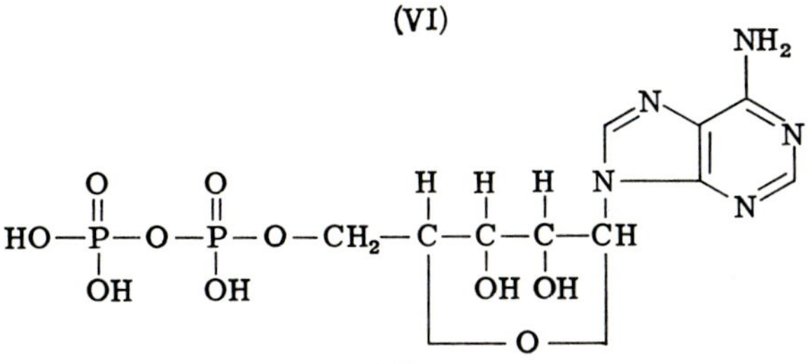

ADP

(VII)

timum of muscle ATPase. According to Kielley (1961), at one time myosin was considered to possess two pH optima, one at pH 6.4 and a more pronounced one at pH 9–10. It is now generally agreed that the optimum at the alkaline pH is due to inactivation and that there appears to be a simple linear relationship between activity and hydroxyl ion concentration. Mommaerts and Green (1954) indicate that the true pH optimum is at 6.4.

The existence of multiple ATPase systems in liver as well as in muscle has been reported (Kielley and Meyerhof, 1948; Swanson, 1951). Heppel and Hilmoe (1953) found that the enzymatic hydrolysis of ATP by bull seminal plasma exhibited a very broad pH optimum extending from pH 7.0 to 8.5. This pH curve was attributed to the net effect of one acid and two alkaline ATPases. One alkaline ATPase, which hydrolyzed ATP to form pyrophosphate and ATP, exhibited a pH optimum of 8.4–8.8 and was relatively heat stable. The second alkaline ATPase, which liberated orthophosphate from ATP, was also relatively heat stable. The acid ATPase, which hydrolyzed ATP to form orthophosphate and ADP, has a pH optimum of 5.7–6.0 and is destroyed by heating at 60° for 20 minutes. An erythrocytic ATPase has also been reported (Caffrey *et al.*, 1956). The pH optimum of the enzyme was found to lag between 7.3 and 7.6, and it catalyzed the hydrolysis of ADP, ITP, and to a lesser extent UTP. Erythrocyte ATPase has also been described by Post *et al.* (1960) and Dunham and Glynn (1961). A similar enzyme has been found in a large number of frozen-dried cat tissue homogenates (Bonting *et al.*, 1961).

As mentioned earlier in this chapter, ATPase has been associated with muscle contraction as well as with other physiological processes which may be related to the hydrolysis of high-energy bonds. The importance of ATP in the formation of energy rich compounds known as "phosphagens" is well known (Neilands and Stumpf, 1958). These substances are characterized by phosphocreatin in vertebrates and phosphoarginine in the invertebrates, and are transformed by conversion into ATP for direct metabolic utilization. Other forms of energy transfer involving ATP are also known (Strominger *et al.*, 1959). A sodium-potassium activated ATPase of cell membranes which is believed to be involved in active cation transport has been described by a number of workers (Hoffman, 1960; Bonting *et al.*, 1961).

The ATPase histochemical technique stems from the work of Glick and Fisher (1945, 1946), who employed a modified metal-salt technique for demonstration of the enzyme in plant tissues. Soulairac and Desclaux (1949) first described ATPase activity of rat muscle. Gomori (1950c) presumably reported genuine ATPase activity in animal tissues when he observed that the use of ATP as a substrate (pH 5) resulted in intense staining of muscle and erythrocytes (Appendix 5B, Section 5,*a*). Wang (1950) described the visualization of ATPase activity in several sites by means of a similar procedure. Naidoo and Pratt (1951, 1952) studied distribution of ATPase in frozen-dried paraffin embedded brain tissue, and found that the freeze-dry process did not inhibit the enzyme. They used a substrate medium at pH 6.5 which contained lead acetate (Appendix 5B, Section 5,*b*). Although Naidoo and Pratt (1951, 1952) obtained good results with frozen-dried paraffin embedded tissues, others have indicated that routine fixation and embedding procedures are not tolerated by the enzyme (Gomori, 1952, p. 173; Novikoff, 1952). Maengwyn-Davies *et al.* (1952) performed the reaction on unfixed frozen sections. Unfortunately, their incubations were excessively long and the validity of their results is seriously open to question.

Padykula and Herman (1955a, b) employed unfixed frozen sections at pH 9.4 in a modification of the Gomori Ca-Co method. Their work is highly significant in that they were the first to characterize their staining reactions by the interposition of specific activators and inhibitors. Sulfhydryl compounds such as BAL (2,3-dimercapto-1-propanol) or cysteine, which are known biochemical and histochemical inhibitors of alkaline phosphatase, enhanced sites of true ATPase activity. On the other hand, sulfhydryl inhibitors (e.g. PCMB, *p*-chloromercuribenzoate) were found to inhibit ATPase staining. The stainability could be restored by treatment of sections with SH reagents (e.g. BAL or cysteine) (Appendix 5B, Section 5,*c*). The true or SH dependent ATPase was demonstrated in cardiac and skeletal muscle (Nachmias and Padykula, 1958) (Figs. 31 and 32) and in the mitochondria of kidney tubules. Although a staining reaction was obtained in the cytoplasm of liver cells, reversible inhibition could not be demonstrated. In view of the fact that a number of enzymes can hydrolyze ATP, it would appear prudent to apply the reversible inhibitor tests of Padykula and Herman (1955b) in any definitive histochemical attempts to demonstrate ATPase. Wachstein and Meisel (1957a, b) employed a lead-salt ATPase technique similar to that of Naidoo and Pratt (1951, 1952) with the exception that the pH was adjusted to 7.2 with tris buffer (Appendix 5B, Section 5,*d*). They observed a varying degree of cytoplasmic staining as well as activity of bile canaliculi and mitochondria.

Freiman and Kaplan (1960) who utilized both biochemical and histochemical techniques found that a calcium-ADP substrate solution gave essentially the same pattern of distribution as the lead-ATP medium of Wachstein and Meisel but questioned the specificity of the latter. Novikoff *et al.* (1961), employing histochemical techniques, felt that the Wachstein-Meisel (1957a, b) or Padykula-Herman (1955a, b) methods showed similar histochemical localizations with reference to activity of

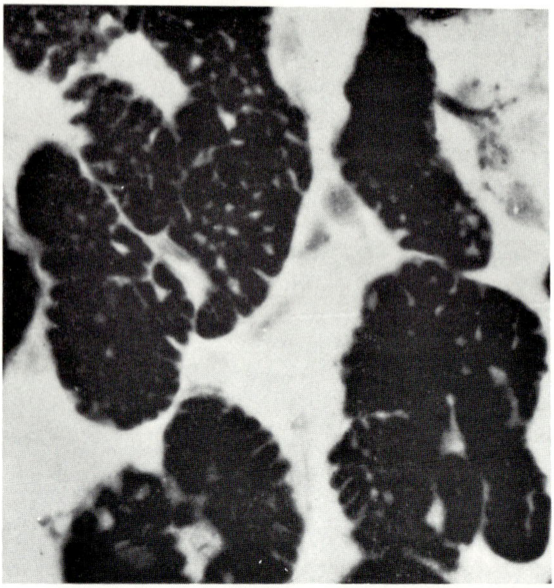

FIG. 31. ATPase of rat cardiac muscle fibers (cross-section). ATPase activity is discretely localized in branching myofibrils (× 1200) (courtesy Dr. Helen Padykula).

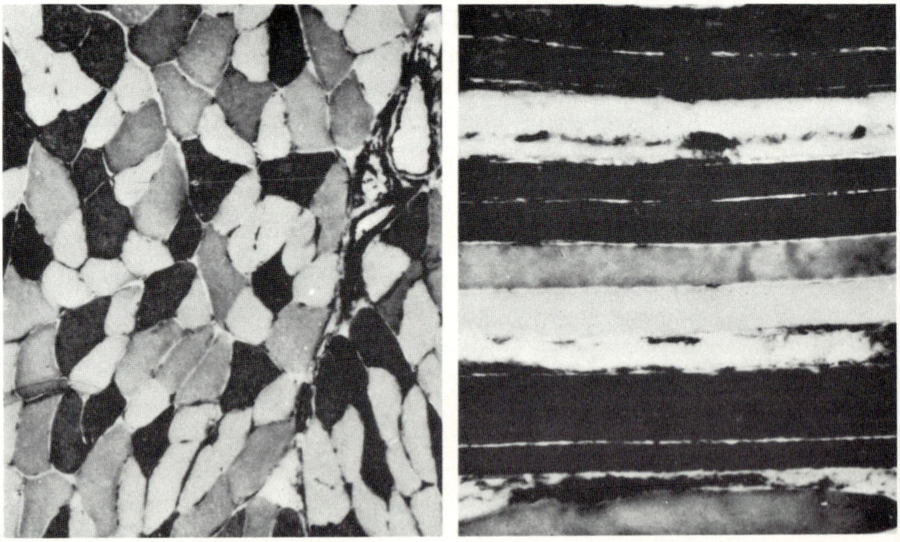

FIG. 32. (*left*) ATPase of gracilis posticus (rat) cut in cross-section (× 190) (courtesy Dr. Helen Padykula). (*right*) ATPase of gracilis posticus (rat) cut in longitudinal section (× 190) (courtesy Dr. Helen Padykula).

cell membranes. However, these authors note the well known fact that non-specific phosphatase activity is widely distributed in capillary endothelia, basement membranes, as well as in other types of cell membranes. Thus because of the ubiquitous distribution of alkaline phosphatase, methods for the delineation of sites of specific ATPase activity may necessitate some degree of revaluation. Of considerable significance with reference to the validity of the histochemical technique are recent biochemical studies of Bonting *et al.* (1962, personal communication) who found that low concentration of $Pb^{++}$ completely inhibits Na-K activated (membrane) ATPase.

### 1. *Activators and Inhibitors*

Muscle ATPase is known to be activated by calcium and potassium ions and also EDTA, while the ATPase of the central nervous system is activated by glutathione. Although myosin ATPase is inhibited by $Mg^{++}$, mitochondrial ATPase is activated by this ion. Myosin is known to be sensitive to mercurial and arsenical SH reagents but not to iodoacetate or iodoacetamide (Kielley, 1961, p. 159; Kielley and Bradley, 1956). Among the specific inhibitors are *p*-chloromercuribenzoate (PCMB) and *N*-ethylmaleimide (NEM).

With reference to the heat-stable ATPase of Heppel and Hilmoe (1953), neither $Ca^{++}$ or $Mg^{++}$ are required, as contrasted with the acid ATPase which requires $Mg^{++}$ and is inhibited by $Ca^{++}$.

Erythrocyte ATPase is activated by $Mg^{++}$, $Mn^{++}$, $Ni^{++}$, $Zn^{++}$, and $Ca^{++}$. However, in the presence of $Mg^{++}$ all other ions become inhibitory. Cysteine is an activator, while glutathione has no effect (Caffrey *et al.*, 1956).

Various potential activators and inhibitors have been employed in attempts to characterize and delineate ATPase from other phosphatases. Pratt (1953, 1954) found that neither $Ca^{++}$ or $Mg^{++}$ exerted any influence when employed with frozen-dried tissues. Padykula and Herman (1955b) felt that they could distinguish between ATPase and other phosphatases through the use of an inhibitor, PCMB (*p*-chloromercuribenzoate, $2.5 \times 10^{-3}$ *M*). True ATPase activity could be restored by subsequent treatment of the section in BAL (2,3-dimercapto-1-propanol, $5 \times 10^{-3}$ *M*) or L-cysteine ($2.5 \times 10^{-3}$ *M*).

The type of metal-salt procedure employed may influence the activity of the enzyme, particularly in the case of $Pb^{++}$ which is used to precipitate the enzymatically released phosphate. Freiman and Kaplan (1960) specifically comment on the high degree of ATPase sensitivity to lead ions and state that true ATPase cannot be demonstrated with a lead-ATP substrate medium. They state that the results observed following incubation in a lead-ATP medium at pH 7.2 is due to a less specific polyphosphatase. Wachstein *et al.* (1960b) in a partial reply to the aforementioned criticism pointed out that although lead produces considerable inactivation of ATPase, the residual activity would still be adequate for the histochemical demonstration of the enzyme.

### 2. *ATPase in Neoplasms*

Schmidt-Matthiesen (1958) reported stromal activity in a number of tumors which differed from that seen when glycerophosphate was employed as a substrate. The peripheral cells of well differentiated and poorly differentiated intestinal carcinomas showed considerable activity. Wachstein and Meisel (1959) studied the distribution of ATPase in cirrhotic and neoplastic rat liver. In frankly malignant areas the bile canalicular staining was not visible. This type of alteration may simply represent a non-specific effect due to liver damage. Novikoff (1960a) studied the effect of administration of a carcinogenic agent (3-methyl-4-dimethylaminoazobenzene) upon the distribution of ATPase in rat liver, and made the interesting observation that there was a decrease of the enzyme in bile canaliculi even though alkaline phosphatase activity was increased. Novikoff also observed staining of cell membranes in a number of lung carcinomas and two transitional cell carcinomas of the bladder. Membrane staining has also been observed in carcinomas grown in tissue culture including HeLa cells (Gropp, 1959; Novikoff, 1960b).

## B. 5′-Nucleotidase

Nucleotidases are phosphatases which hydrolyze both pyrimidine and purine nucleotides to form phosphoric acid and nucleosides. 5′-Nucleotidase was first described and characterized biochemically by Reis (1934, 1937, 1938, 1951). In addition to adenosine-5′-phosphate (VIII), inosine-5′-phosphate is also hydrolyzed while 3-adenylic acid is hydrolyzed at a

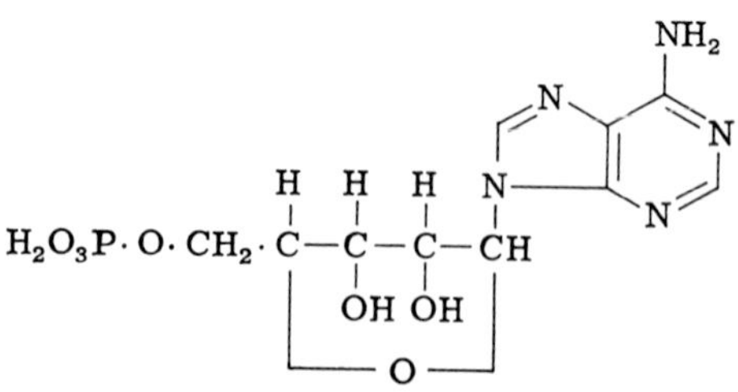


Adenosine-5′-phosphate
(5-adenylic acid)

(VIII)

very low rate. The pH optimum is approximately 7.8 but activity is quite high even at pH 9.

Gomori (1949b) was the first to demonstrate 5′-nucleotidase by a histochemical technique. The method was similar to his alkaline phosphatase technique with the exception that the incubation was carried out at pH 8.3. At this pH, the activity of the enzyme is quite high while that of alkaline phosphatase is reduced to about one-third of its optimal level. A disadvantage of the lower pH is a tendency for diffusion artefacts to occur which, according to Gomori, can be offset by a sufficiently high concentration of calcium ions. The Gomori findings and technique have been both confirmed and modified slightly by a number of investigators (Newman *et al.*, 1950; Antonini and Weber, 1951; Pearse and Reis, 1952; McManus *et al.*, 1952; Wachstein and Meisel, 1954; Padykula and Herman, 1955b; Barron and Boshes, 1961). The original procedure of Gomori (1952, p. 187) still appears to be a highly desirable technique (Appendix 5B, Section 10,*a*). Pratt (1953, 1954) incubated at a rather low pH (6.5) and employed lead ions to trap the enzymatically released phosphate. According to his data, high activity was found at this acid pH. Wachstein and Meisel (1957b) also employed a lead nitrate modification at pH 7.2 (Appendix 5B, Section 10,*b*).

Most investigators have observed a reaction in the medial layer of arterioles as well as certain other specific sites of activity (Fig. 33).

### 1. *Activators and Inhibitors*

Nucleotidase is only slightly activated by $Mg^{++}$ (Reis, 1937; Ahmed and Reis, 1958). On the other hand, $Mn^{++}$ is a strong activator, while $Ni^{++}$ and $Zn^{++}$ are strong inhibitors. $Ni^{++}$ in millimolar concentrations almost completely inhibits 5′-nucleotidase while non-specific phosphatase is relatively unaffected. A complete inhibition of activity is obtained with 0.1 *M* NaF and 0.08 *M* sodium borate (Heppel and Hilmoe, 1951).

### 2. *5′-Nucleotidase in Neoplasms*

Wachstein *et al.* (1960a) observed an occasional positive reaction in human carcinomas of the breast and adenocarcinomas of the large intestine and stomach. Wachstein and Meisel (1959) found that in cirrhosis of the rat liver, bile canaliculi often stained intensely, while in frank liver-cell carcinoma the canaliculi were unstained. This difference is probably a non-specific change related to hepatic damage. These authors (1954) also studied two seminomas and observed a distinct nuclear as well as a moderate cytoplasmic staining reaction, which was qualitatively similar to that observed in the normal tissue. Brandes and Bourne (1956) could not detect any alterations in nucleotidase in fibroadenomatous hyperplasia or carci-

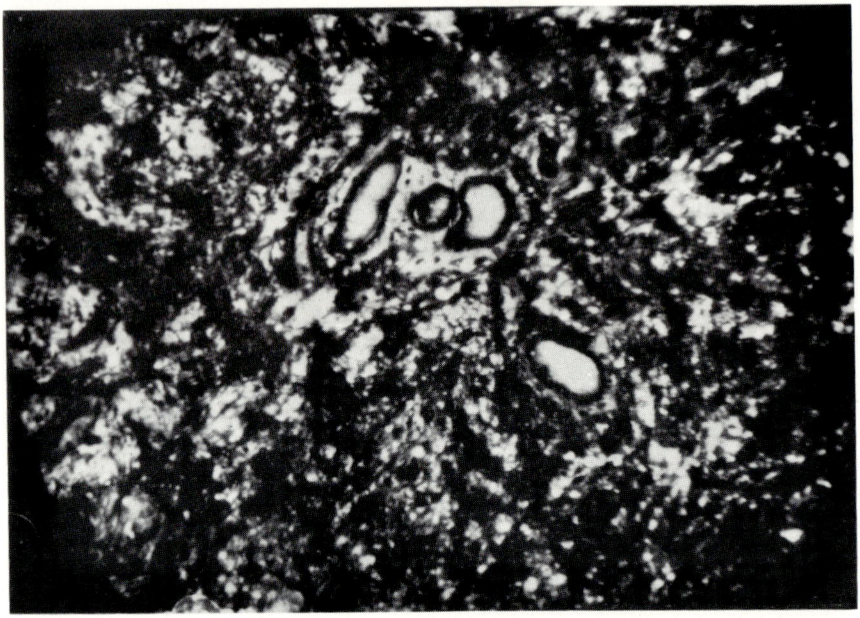

FIG. 33. 5-Nucleotidase. Posterior lobe of hypophysis. In addition to capillary staining there is a reaction of fibrillar matrix (× 150) (courtesy Dr. Sally B. Fand).

noma of the prostate. With reference to the latter, Butler *et al.* (1959) were unable to discern any activity in both normal and neoplastic prostates. Feigin and Wolf (1959) noted that the nucleotidase activity of certain tumors of the central nervous system (e.g. ependymoma, astrocytoma, and pituitary tumors) exhibited an increased nucleotidase activity associated with an increasing malignancy.

## C. Phosphamidase (Phosphoamidase)

The enzyme phosphamidase hydrolyzes *N*-substituted amides of phosphoric acids such as phosphocreatine and phosphoarginine as well as a number of synthetic substrates. Ichihara (1933) studied the enzymatic hydrolysis of simple phosphoric acid monoamides and diamides which were originally synthesized by Stokes (1894). Ichihara also employed anilides of phosphoric acid as substrates (Chart 1). He reported that kidney extracts hydrolyzed phosphamides in the alkaline range at an optimal pH of 9.0. In the acid range, the optimum was pH 3.3, but the hydrolysis rate at pH 6.1 was also considerable. He concluded that in both animal and vegetable tissue there is a specific enzyme, phosphamidase, which is capable of hydrolyzing the N—P bond. Walschmidt-Leitz and Koehler

(1933) have also indicated the existence of a phosphamidase. More recently, Holter and Li (1951) studied phosphamidase activity of rat tissue homogenates and found the pH optimum of the enzyme to be around 4.6. Of interest is the observation of Perlmann (1952, p. 179) who found that purified intestinal phosphatase will not hydrolyze *p*-chloroanilidophosphonic acid or phenyldiamidophosphonic acid at pH 5.3, 7.0, and 9.0. However, it is known that the naturally occurring substrates such as phosphocreatine are rapidly hydrolyzed by both alkaline and acid phosphatases (Winnick, 1947; Morton, 1955b). Morton suggests that it seems not unlikely that phosphocreatine (PC) is attacked at an acid pH by an enzyme of wide substrate specificity rather than by specific phosphamidase. These findings do not preclude the possibility that a true phosphamidase exists in certain tissues but they do suggest that the present classification of phosphamidases as different from phosphomonoesterases may require further clarification. Møller (1955) states that the idea that the phosphamidases form a group of enzymes distinct from phosphatases seems no longer valid. He points out that there do exist, perhaps, specific phosphatases that attack the N—P bond, and specific phosphamidases that do not attack the O—P bond, but thus far the so-called non-specific phosphomonoesterases have also been found to display phosphamidase activity.

On the other hand, the histochemical technique of Gomori (1952, p. 195) clearly delineates histological sites which are capable of hydrolyzing a phosphamide substrate but not the usual substrates for acid and alkaline phosphatase. It is important to note that Gomori obtained the specific phosphamidase picture only in the pH range of 5–6; in the alkaline range, his substrate gave patterns which were indistinguishable from the routine alkaline phosphatase reaction. A characteristic and intense reaction was found in two tissues, the gray matter of the central nervous system (especially that of the cerebellum) and in malignant tumors. He found that very high concentrations could be demonstrated in the vast majority of malignant epithelial tumors. As a rule, the intensity of the reaction roughly paralleled the degree of malignancy.

The N—P bond is a labile one and this offers certain difficulties with reference to the histochemical procedure since the substrate exhibits an appreciable degree of spontaneous decomposition. Gomori (1952, p. 196) recommends that the Coplin jar be supported in an inclined position, with the tissues facing downward in order to minimize the indiscriminate precipitation of lead phosphate over the tissues. Even then, non-specific precipitate accumulates on the section and this must be removed or differentiated in a somewhat arbitrary manner. Gomori further indicates that the method is not entirely dependable and that the difficulties are essentially the same as with the acid phosphatase technique (e.g. lack of uniformity in staining, and unexpected failures to get a positive reaction). The original technique as recommended by Gomori (1949a; 1952, p. 195) as well as the modification of Meyer and Weinmann (1955) are given in Appendix 5B, Sections 7 and 8.

A question has been raised as to the exact structure of the substrate. According to Otto (1895), Ichihara (1933), and Gomori (1948), the compound employed was *p*-chloroanilidophosphonic acid [*N*-(*p*-chlorophenyl)-amidophosphoric acid]. On the other hand, Rorig (1949) found that a compound synthesized by the method of Otto was actually *N*-(*p*-chlorophenyl)diamidophosphoric acid. Li (1950) and Holter and Li (1951) synthesized the genuine *N*-(*p*-chlorophenyl)amidophosphoric acid by two different synthetic methods and found that it was a suitable substrate which appeared to be more stable in acid solution than the compound which Ichihara and Gomori had used. The most recent consideration of *N*-(*p*-chlorophenyl)amidophosphoric acid and analogues has been made by Chanley and Feageson (1958) who were unable to confirm the previous reports dealing with the synthesis and characteristics of the aforementioned amides. On the other hand, Gomori (1952, p. 195) points out that different batches of the reagent prepared by the same method are not necessarily identical in composition. In any case, from the standpoint of enzymatic hydrolysis, both the monoamido and diamido phosphoric acids are susceptible to enzymatic hydrolysis and should give essentially the same histological picture.

The author has prepared a number of potential phosphamidase substrates which may be employed in an azo-dye procedure. Included among these was a naphthol AS phosphoric acid diamide (Burstone, 1958b). Phenyl and naphthyl phosphoric acid diamides $R{-}OPO(NH_2)_2$ are synthesized by ammonolysis of the appropriate phenyl or naphthyl phosphochloridate. More complex *N*-substituted derivatives may be prepared by the use of amines such as methylamine, butylamine, triethylamine, aniline, and morpholine (Appendix 5A, Part 2, Section 25). In certain tissues a different distribution of activity was seen, as compared with that observed with the analogous AS phosphates. However, differences were not sufficiently clear cut to unequivocally indicate that a specific phosphamidase was being demonstrated. The author has also prepared phosphamides derived from α- and β-naphthylamine, and also from substituted naphthylamines including 3-amino-2-naphthoic acid. These compounds which are prepared by treating the appropriate arylamine with phosphorus oxychloride exhibit an appreciable degree of spontaneous hydrolysis. Furthermore, azo-dyes derived from them are non-specifically adsorbed to tissue sections giving spurious staining patterns. A modified azo-dye procedure probably offers the best possibilities for the eventual accurate localization of phosphamidase.

## 1. *Activators and Inhibitors*

Holter and Li (1951) studied the effect of a number of known phosphatase activators and inhibitors upon phosphamidase activity. Comparison was made with phenylphosphate. In general, their studies did not reveal any characteristic difference between the effect of activators and inhibitors upon the phosphamide and the ester phosphate.

## 2. *Phosphamidase in Neoplasms*

As previously mentioned, Gomori (1948) found an increased phosphamidase activity in a number of neoplasms. Included were 24 adenocarcinomas of the gastrointestinal tract, 4 epidermoid carcinomas, 2 hypernephromas, and a number of other epithelial tumors, including embryonic carcinoma of the testis, and tumors of the prostate, breast, and bronchi. A Walker rat carcinoma 256 and a butter yellow tumor also reacted intensely and selectively. Gomori noted that the staining reaction may start quite abruptly at the border of a malignant change, or the apparently normal tissue may show a slight to moderate staining up to a distance of 1 or 2 mm. from the edge of the neoplasm. He reported that of a total of 64 cases, not a single carcinoma was found which did not show a characteristic reaction which delineated it from the adjacent normal tissue. The findings with sarcomas were less constant. Some of them did not stain at all while others revealed an uneven, patchy reaction. Granulomatous lesions, including those of sarcoidosis and tuberculosis, usually did not stain or else showed a faint to moderate reaction in giant cells.

Subsequently, Gomori (1955) reported that most renal cell carcinomas were inactive. On the other hand, two essentially benign lesions were found to be intensely reactive; polyps of the colon and the so-called pale cell cyst of the breast. He concluded that, while the phosphamidase technique may be valuable in the detection of invasion by malignant cells, its use in diagnosis is of limited value. The intensity of the reaction usually parallels morphological criteria of malignancy but adds little to the diagnosis of the latter.

Foraker (1956a) noted a distinct phosphamidase reaction in some carcinomas of the breast. Tomita and Takeuchi (1955) confirmed the original Gomori findings and reported a distinct reaction in certain adenocarcinomas and undifferentiated neoplasms of the stomach, pancreas, liver, urinary bladder, and lung. Braunstein *et al.* (1958) reported a positive reaction could be found only in histiocytes of normal and hyperplastic lymph nodes. Meyer and Weinmann (1953) observed high activity of intraoral epidermoid carcinoma, leukoplakia, and in a pleomorphic salivary gland tumor. A particularly intense reaction was observed in areas of proliferating connective tissue and in giant cells of a giant cell tumor. By contrast, the connective tissue stroma of normal oral mucuous membrane did not stain. Similar findings were reported by Foraker (1956b) with reference to cutaneous epidermoid carcinoma. Ober (1952) also observed the characteristic phosphamidase activity of a tumor which disappeared following x-ray treatment. Stoll *et al.* (1951) associated an increased staining reaction with infiltrating carcinomas as well as with precancerous changes. These findings were not entirely corroborated by Gross and Danzinger (1957). Phosphamidase activity of a large series of tumors has been studied by Winter (1955) and high activity was found in both invasive carcinoma as well as carcinoma *in situ*. Recently, Fand (1961) has reported on the phosphamidase activity of granular cell tumors (Fig. 34).

Although Gomori (1955) was unable to obtain satisfactory results with smears, other investigators have reported a distinct reaction of malignant cells in Papanicolaou smears (Neumann *et al.*, 1954). However, their staining reaction was primarily localized to nuclei, a finding which is considered by many to be artefactual with the metal-salt technique. On the other hand, Ober (1952) reported cytoplasmic activity.

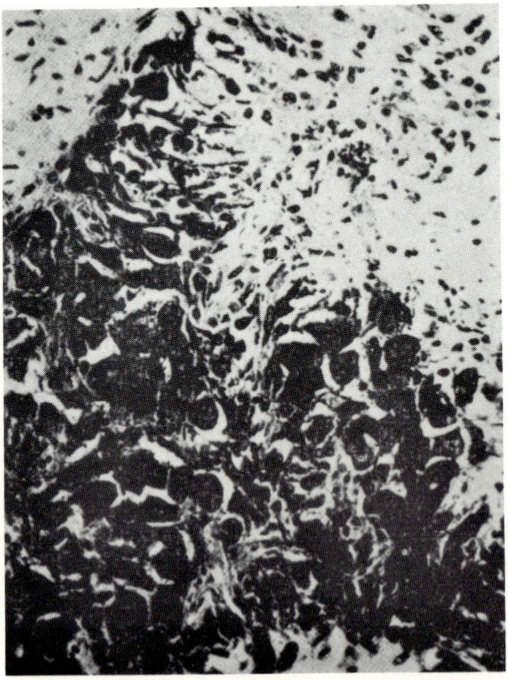

FIG. 34. Phosphamidase staining of granular cell tumor, posterior lobe of hypophysis (× 325) (courtesy Dr. Sally B. Fand).

The high phosphamidase activity of tumors is interesting in the light of the fact that there are certain indications which point to a relationship between proteolytic and phosphamidase activity (Holter and Li, 1951). According to Holter and Li (1950) certain crystalline proteolytic enzymes, namely pepsin, trypsin, chymotrypsin, and rennin, exhibit phosphamidase activity.

## D. Glucose-6-phosphatase

Glucose-6-phosphatase was first characterized as a liver enzyme by Fantl and Rome (1945). The enzyme which is associated with microsomes (Hers *et al.*, 1951) is known to be heat labile and readily inactivated below pH 5. Because of its thermal sensitivity and its relative sensitivity to acids, it may be distinguished to some degree from non-specific acid phosphatase. The enzyme preferentially hydrolyzes glucose-6-phosphate (IX) but other hexose and pentose phosphates are also dephosphorylated. Glycerophosphate, glucose-1-phosphate, or fructose-6-phosphate are not appreciably hydrolyzed. The necessity of a cation for activation has not been reported. The hydrolytic activity of the enzyme is inhibited by glucose and similar compounds. As with non-specific phosphatases the enzyme is inhibited by arsenate and fluoride ions (Byrne, 1961, p. 73). Glucose-6-phosphatase exhibits a specific transferase activity in contrast to non-specific phosphatases, in that the enzyme will catalyze the transfer of a phosphoryl group from glucose-6-phosphate to glucose or fructose.

Gomori (1949b) employed glucose-6-phosphate as a substrate in conjunction with paraffin embedded tissues and obtained no differential staining when compared with glycerophosphate and other non-specific substrates. Chiquoine (1953) employed unfixed cryostat sections and reported distinctive staining patterns which differed from those seen when sodium glycerophosphate was used either in the acid or alkaline range. Following fixation in cold 80% ethanol and paraffin embedding, glucose-6-phosphatase could not be demonstrated even after prolonged incubation. On the

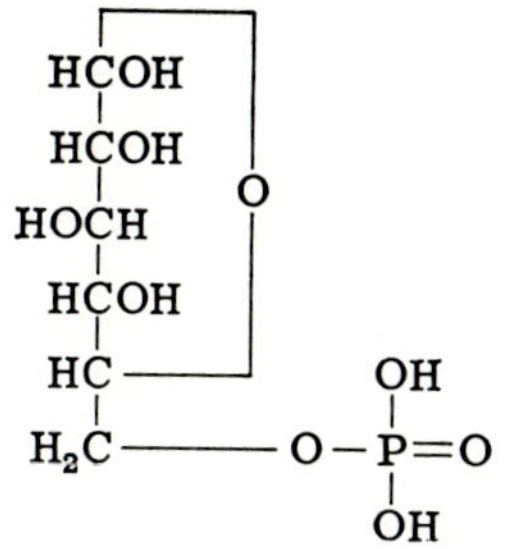


Glucose-6-phosphate
(IX)

other hand, alkaline, and to a lesser extent, acid phosphatase reactions could be delineated. Glucose-6-phosphatase activity was completely inactivated by cold formalin fixation, as contrasted with the other enzymes. Chiquoine also observed characteristic localizations when glucose-6-phosphate was employed as a substrate. In the liver, a staining reaction occurred only in hepatic cells. No activity could be demonstrated in blood vessels or bile canaliculi. A gradient of activity was discernible in the liver lobule, with darker staining of the peripheral one-third and a lighter staining of the inner two-thirds. In the kidney, the distribution of activity did not correspond to that seen for either acid or alkaline phosphatases. Furthermore, the pH optimum for glucose-6-phosphatase was 6.7 as contrasted to 5.0 for acid phosphatase and 9.4 for alkaline phosphatase. Further studies on the histochemistry and distribution of glucose-6-phosphatase were also reported by Chiquoine (1955). Appreciable activity was found only in mouse liver, kidney, small intestine, and rectum. Fluoride, zinc, and cyanide ions were inhibitory as was alloxan: findings which corroborate earlier biochemical studies.

*Glucose-6-phosphatase in Neoplasms*

Weber and Cantero (1955) found that homogenates of induced rat hepatoma exhibited no glucose-6-phosphatase activity.

## E. Thiamine Pyrophosphatase (TDPase)

This enzyme hydrolyzes the terminal phosphate bond of thiamine pyrophosphate (cocarboxylase). Thiamine coenzymes are widely distributed in plant and animal tissues and are involved in a number of metabolic transformations (Metzler, 1960, p. 295). A thiamine triphosphate (TTP) has also been found in living tissues, including liver and brain (Kiessling, 1961). Glick and Fisher (1945, 1946) in a study of wheat germs were the first to employ thiamine pyrophosphate as a histochemical substrate. They observed the distribution of enzyme activity at an acid pH in contrast to the studies of Newman *et al.* (1950) who employed thiamine pyrophosphate in the alkaline pH range. Naidoo and Pratt (1953) reported that histological sites of thiamine pyrophosphate hydrolysis were the same at pH 6.9 and 9.1. However, in the alkaline range the distribution of activity differed from that observed with glycerophosphate. These authors (1955) indicate that there is an $Mg^{++}$ activated thiamine pyrophosphatase with an optimal activity at about pH 8.4 in the chicken and 9.1 in the rat. They further suggest that the activity of this enzyme may alter with maturation of the nervous system.

Of interest are the recent observations of Kiessling and Tilander (1960) who studied the hydrolysis of TDP and TMP (thiamine monophosphate) by preparations from rat liver. It was found that TDP and TMP are hydrolyzed at a maximum rate at widely different pH levels, with the optimum of the TDP being around pH 9. Although two types of crude enzyme preparations hydrolyzed TDP into TMP, only one transformed TMP into thiamine and exhibited a pH optimum of 6.

Eränkö and Hasan (1954) found that when thiamine pyrophosphate and glycerophosphate were used in the alkaline pH range the sites of localization were identical. On the other hand, in formol-calcium fixed frozen sections incubated with thiamine pyrophosphate at an acid pH, a localization entirely different from that obtained with the glycerophosphate reagent was observed. In the kidney the glomeruli which were negative with glycerophosphate reacted strongly with thiamine pyrophosphate, while the convoluted tubules which were strongly positive with glycerophosphate gave a weak or negative reaction. They further observed a very strong nuclear reaction, particularly in connective tissue and Kupffer cells. Their findings are certainly suggestive that a different type of enzyme is being studied, but there is little question that the problem of histochemical localization of thiamine pyrophosphatase requires a critical revaluation. More recently Allen and Slater (1961) have reported the histochemical demonstration of thiamine pyrophosphatase in mouse epididymis employing either thiamine pyrophosphate or thiamine monophosphate at pH 9.5. The reaction was characterized by inclusion of cysteine in the substrate mixture as an inhibitor of non-specific alkaline phosphatase (Fig. 35). All of the histochemical techniques for this enzyme are variations of the Gomori metal-salt procedure and require either the presence of lead or calcium ions in the incubating medium (Appendix 5B, Section 9).

### F. Pyrophosphatases and Metaphosphatases

These are enzymes which catalyze the hydrolysis of the anhydride bond of organic and inorganic pyrophosphates and metaphosphates. Relatively little is known about the metabolic role of these enzymes. It is conceivable that pyrophosphatase may be involved in prevention of reversal of coenzyme synthesis as well as in activation processes for amino and fatty acids (Kielley, 1961, p. 149). Trimetaphosphatase is associated with digestive organs (Berg, 1956) and may be an adaptive enzyme of the intestinal epithelium (Berg and Shupack, 1960).

Neuberg and Wagner (1926) and Neuberg and Jacobsohn (1928) demonstrated that phosphatase preparations were capable of hydrolyzing diphenylpyrophosphate and di-*m*-cresol pyrophosphate. Lohmann (1928) reported that pyrophosphates are present in muscle and are hydrolyzed to orthophosphates under certain conditions. Kay (1928) studied the hydrolysis of sodium pyrophosphate by a number of tissue extracts at various

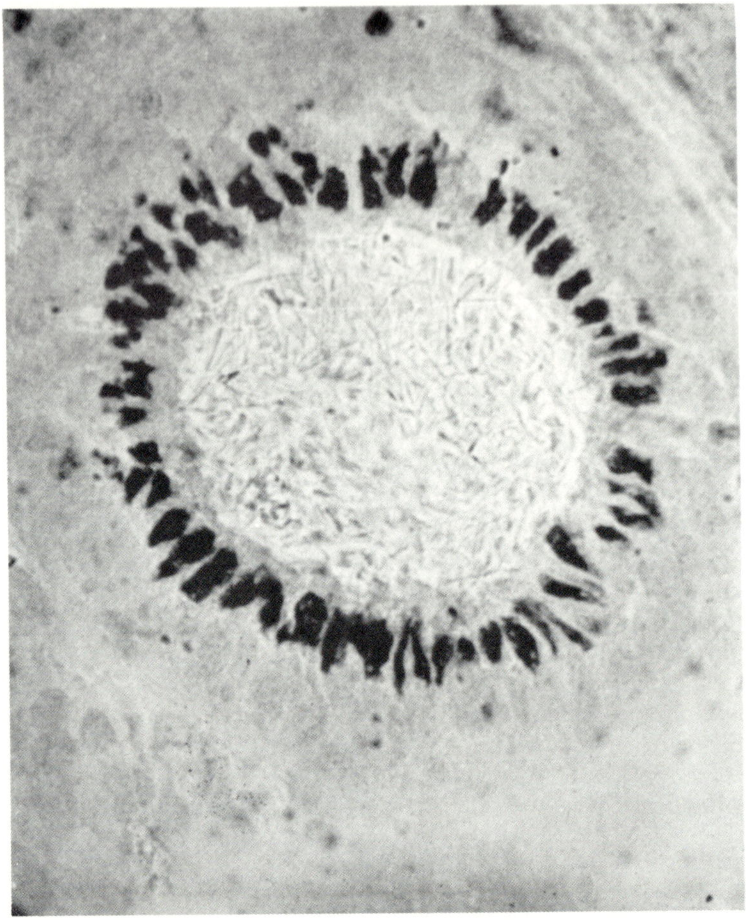

FIG. 35. Distribution of thiamine pyrophosphatase activity at pH 9.5 in lobe 4 of head portion of epididymal canal of mouse. Frozen section post-fixed for 1 hour in acetone and incubated for 2 minutes at 37° in the presence of 0.01 *M* cysteine (× 740) (courtesy Dr. John M. Allen).

pH levels and found that the pH optimum for a number of tissues lay between 7.2 and 7.8. Kay found that the relative activities of a number of tissues (in decreasing order of activity) were duodenal mucosa, growing bone, kidney, lung, and liver. Roche (1931) suggested that blood contains a specific pyrophosphatase. Bamann and Gall (1937) characterized the existence of three isodynamic pyrophosphatases in human and animal tissues while Norberg (1950b) studied pyrophosphatase activity of rat liver employing sodium pyrophosphate and found four distinct optima at pH 4.0, 5.0, 5.8, 7.0 ("neutral pyrophosphatase"), and about 8.0 ("alkaline pyrophosphatase"). The pH 4.0 pyrophosphatase was unaffected by $Mg^{++}$ while the pH 5.0 enzyme was considerably activated in certain instances. The "neutral" and "alkaline" pyrophosphatases were consistently inactivated by $Mg^{++}$. Diethylbarbituric acid was found to be inhibitory to most pyrophosphatases.

Roche (1950, p. 493) points out that the identity and characterization of pyrophosphatases depends upon their separation from the phosphomonoesterases which almost always accompany them. This may be accomplished by selective inactivation of the latter (Munemura, 1933a). In 1950 Gordon demonstrated inorganic pyrophosphatase activity of mammalian nervous tissue. More recently Seal and Binkley (1957) obtained a purified pyrophosphatase from swine brain which had a pH optimum of 7.6–7.8 and required $Mg^{++}$. It was activated by EDTA, cysteine, glutathione, and thioglycolic acid, and reversibly inhibited by polyvalent metal ions and *p*-chloromercuribenzoate. Brain pyrophosphatase is readily inactivated at 0° in a slightly acid solution (Gordon, 1957). A pyrophosphatase associated with mouse liver mitochondria and having a pH optimum of 5.3 has been reported by Rafter (1958). No requirement for $Mg^{++}$ was demonstrated, presumably because metal ions were bound to the cellular particles. Fluoride, orthophosphate, arsenate, and sulfhydryl reagents were inhibitory. Activity was rapidly lost when the particles were incubated at pH 5.0 without the presence of substrate.

With reference to the histochemical demonstration of pyrophosphatase, Gomori (1949a) could not obtain a positive reaction with inorganic pyrophosphate. The reaction which was obtained with dimethyl- and diphenylpyrophosphates revealed the same distribution as that observed with the usual glycerophosphate substrate (Gomori, 1949b). Kurata and Maeda (1956), however, have described a histochemical technique for the demonstration of inorganic pyrophosphatase (Appendix 5B, Section 12). Preliminary findings of the author indicate that naphthyl pyrophosphates may be employed in an azo-dye technique to demonstrate pyrophosphatases (Burstone, unpublished data).

Metaphosphatases are polyphosphatases which hydrolyze salts of metaphosphoric acid $(NaPO_3)_n$ to orthophosphates. Low molecular weight as well as complex colloidal metaphosphates are known to exist (Neuberg and Fischer, 1937, 1938a, b; Frankenthal *et al.*, 1943; Ingelman, 1950, p. 511; Thilo, 1958; Berg, 1960). Until recently, all polyphosphates of a polymer configuration greater than three were termed metaphosphates, but two classes of polyphosphates are presently delineated in biochemical and histochemical studies. The inorganic polymers of phosphoric acid form a series of polyphosphate linear polymers with the exception of two types. These are the ring polymers (metaphosphates) which include a trimetaphosphate and a tetrametaphosphate. Hoffmann-Ostenhof and Šlechta (1958, p. 180) note that most of the so-called metaphosphates observed in biological material are chain-like polymers, and indicate that inorganic pyrophosphate (PPi) must be considered the dimer of this class. The relationships between PPi and many important metabolic processes including transfer reactions have been reviewed by Kornberg (1957) and Hoffmann-Ostenhof and Šlechta (1958, p. 180). Berg (1955, 1959, 1960) has evolved interesting techniques for the demonstration of trimetaphosphatase, tetraphosphatase, and polyphosphatase. The metaphosphatase technique employs either trimetaphosphate or tetrametaphosphate, while the polyphosphatase procedure utilizes a stable lead polyphosphate (Figs. 36 and 37) (Appendix 5B, Section 11).

## G. Ribonuclease and Deoxyribonuclease (RNase and DNase)

Nucleodepolymerases which occur in both plant and animal tissues depolymerize high molecular weight nucleic acids to form oligonucleotides and mononucleotides. Ribonuclease and deoxyribonuclease are classified as nucleodepolymerases; phosphodiesterases which hydrolyze nucleic acids without liberating free phosphates (Laskowski, 1951, p. 956).

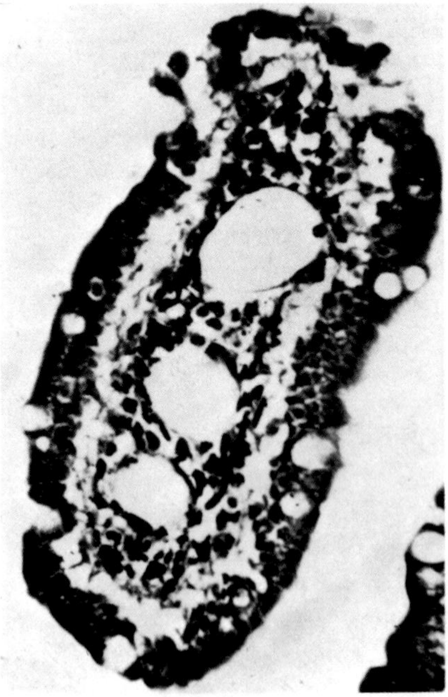

FIG. 36. Trimetaphosphatase staining of rat duodenum. Reaction products localized to lamina propria (not nuclear) (× 800) (courtesy Dr. George G. Berg).

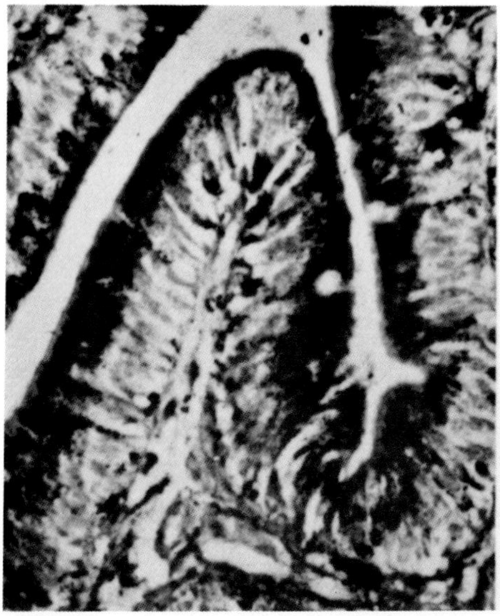

FIG. 37. Trimetaphosphatase staining. Tip of villus of frog duodenum. The stain is localized between two zones of the cell where alkaline phosphatase is usually found (× 800) (courtesy Dr. George G. Berg).

The term ribonuclease was coined by Dubos and Thompson (1938) to indicate a thermostable pancreatic enzyme which was found to depolymerize ribonucleic acid. Since that time a variety of ribonucleases which are widely distributed in nature and which catalyze the hydrolysis of RNA through the cleavage of phosphodiester bonds have been described (Anfinsen and White, 1961, p. 95). Although animal RNase has been primarily studied with reference to the pancreas, it has also been isolated from liver, spleen, lung, and polymorphonuclear leukocytes (McDonald, 1955a, p. 427). With specific reference to human sources of the enzyme, Houck (1958) has studied activity of various human body fluids including serum, saliva, spinal fluid, and urine. Most RNases exhibit a specificity for phosphodiester bonds that involve the nucleosides of cytosine and uracil, and do not attack purine nucleoside phosphate bonds. Although plant RNases exhibit a pH optimum in the neighborhood of 5, that of their animal counterpart is usually to be found near neutrality or else in the more alkaline range. RNase derived from bovine and mouse pancreas exhibits a pH optimum of approximately 7. On the other hand, a rat hepatoma enzyme was found to have a pH optimum of 7.8 as contrasted with that of a rat lymphosarcoma whose optimum was pH 5.5 (Maver and Greco, 1956). Pancreatic ribonuclease was first obtained in crystalline form by Kunitz (1940).

The histochemical localization of RNase activity in tissue sections has been reported by Daoust and Amano (1960), and Steigleder and Raab (1961). The method involved the placing of tissue sections in contact with a film composed of gelatin and RNA, and allowing the tissue enzyme to act upon the layer of substrate. The unhydrolyzed RNA remaining in the film is subsequently stained and a comparison is made between the negative pattern left in the film and corresponding untreated tissue sections (Fig. 38) (Appendix 5B, Section 21). With this substrate-film method, histological sites of RNase activity were delineated in small and large intestine, pancreas, thyroid, liver, and skin of the rat.

Sites of RNase (as well as DNase) activity have also been visualized in tissue sections by the fluorescent antibody technique (Coons and Kaplan, 1950; Marshall, 1954).

Deoxyribonucleases are phosphodiesterases which hydrolyze polymerized deoxyribonucleic acid (DNA). These enzymes have also been referred to under the names of thymonucleodepolymerase, thymonuclease, poly-

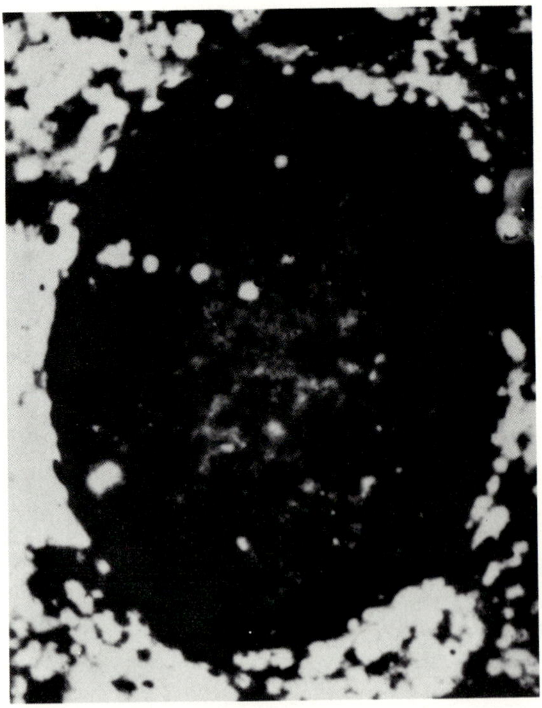

FIG. 38. RNase activity of tumor cell from human carcinoma of stomach (× 35) (courtesy Dr. Roger Daoust).

nucleotidase, and deoxyribonucleodepolymerase (Laskowski, 1951, p. 965; Sumner and Somers, 1953, p. 137; McDonald, 1955b, p. 437). The discovery of the enzyme dates to 1903 when Araki found that extracts of pancreas, thymus, liver, and spleen were capable of liquefying gels of thymonucleates. It was subsequently shown that DNase exhibits different characteristics than proteolytic enzymes. In 1948 Kunitz reported the preparation of the crystalline enzyme from beef pancreas. As with RNase the enzyme exhibits a characterisitc thermostability. In the past it was thought that the nucleases exhibited a distinct specificity toward either DNA or RNA but more recently this interpretation has been viewed with less certainty since preparations of purified DNase have been shown to exhibit some RNase activity (Laskowski, 1961, p. 123). Three types of DNase have been fairly well characterized from a biochemical standpoint. Pancreatic DNase I exhibits a fairly broad pH range with the optimum around pH 7.0. Although pancreas is the best source of the enzyme, activity has been found in several tissues including salivary glands and liver Allfrey and Mirsky (1952). Streptococcal DNase (streptodornase) which has been obtained in highly purified form also exhibits a pH optimum of around 7.0. Splenic DNase II exhibits an optimum pH in the range of 4.5–5.5 Koszalka (1959). This type of enzyme is also found in the thymus, liver, intestinal mucosa, heart, and mouse lymphosarcoma. Other enzymes, presumably DNase type II, have been found in several animal sources including snake venom, cyclostome tissues, and culture filtrate of a protozoan.

Kurnick (1953) studied DNase activity of sera of normal individuals as well as those with infectious, malignant, and "metabolic" diseases and found no significant variations in enzyme levels. Observations in hepatic disease and lupus erythematosus were not included in the study.

The substrate-film technique has also been used by Daoust and his group (Daoust, 1957, 1961; Daoust and Cantero, 1959; Daoust and Amano, 1961) for the histochemical demonstration of DNase in tissue sections (Fig. 39). The films which contain DNA dispersed in gelatin are incubated in contact with tissue sections for periods of 10–60 minutes. The section and film are then separated and the latter stained with toluidine blue or with the Feulgen method to reveal sites of unattacked DNA (Appendix 5B, Section 21,*f*).

An alternative metal-salt technique for the histochemical demonstration of this enzyme was suggested by Aronson *et al.* (1958). This approach necessitated the utilization of an exogenous source of acid phosphatase which presumably acted upon the initial products of hydrolysis to produce phosphate ions.

Vorbrodt (1961) employed the general technique of Aronson *et al.* (1958) and felt that the best results were obtained by use of formol-calcium

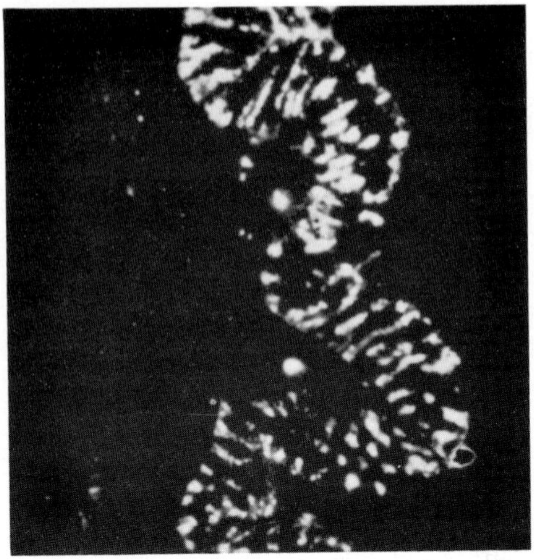

FIG. 39. DNase of rat large intestine (× 30) (courtesy Dr. Roger Daoust).

fixed frozen sections. They reported localization of DNase II in both lysozymes and nuclei of a variety of cells. Tumor cells were characterized by low enzymatic activity.

## 1. *Activators and Inhibitors of RNase*

The enzyme is not affected by trypsin or chymotrypsin but is destroyed by pepsin. Although no enzyme activity is exhibited at 85°, full activity returns upon cooling. There is considerable variation with reference to the effects of potential activators and inhibitors upon ribonucleases derived from different sources (Anfinsen and White, 1961, p. 95). It has been reported that molybdate ions inhibit pea leaf ribonuclease; but azide, fluoride, and arsenate ions have no effect. Pancreatic ribonuclease is not inhibited by arsenic, selenium, or chloride ions. Anionic polymers in relatively high concentrations have been reported to inhibit ribonuclease. Among these are polyacid substances such as heparin, poly-*p*,*p'*-dioxydibenzylphosphate, sulfonated pectin, and poly-L-aspartic acid (Vandendriessche, 1956). Acridine and other basic dyes also inhibit the enzyme. Roth (1953) noted that although 2-aminobenzimidazole inhibited ribonuclease activity, benzimidazole and 5,6-dimethylbenzimidazole were activators. EDTA, glycine, and histidine have also been reported as activators. Iodoacetate, *p*-mercuribenzoate, and other compounds which react with the SH groups of proteins also inactivate certain ribonucleases, as do reducing agents such as borohydride, thioglycolate, or cyanide.

2. *Activators and Inhibitors of DNase*

DNase I is activated by bivalent cations such as $Mg^{++}$, $Mn^{++}$, and $Co^{++}$. The enzyme is inhibited by compounds which complex with $Mg^{++}$. These include citrate, arsenate, EDTA, and fluoride. Streptococcal DNase is also activated by bivalent cations and inhibited by EDTA and similar compounds. DNase II is activated by low concentrations of $Mg^{++}$ while higher concentrations are inhibitory (Shack, 1957).

3. *DNase and RNase in Neoplasms*

De Lamirande *et al.* (1961, p. 43) has employed the substrate-film histochemical technique for the visualization of DNase activity in rat liver tumors and has found that a loss of enzyme activity is closely associated with the neoplastic transformation of normal hepatic cells. With reference to RNase, a decreased enzyme activity was found to occur at early stages following administration of the carcinogen before any perceptible change could be noted in hyperplastic cells (Amano and Daoust, 1961).

## H. Miscellaneous Phosphatases

1. *Aldolase*

This enzyme catalyzes the reversible hydrolysis of fructose-1-diphosphate (FDP) into glyceraldehyde phosphate and dihydroxyacetone phosphate (Sumner and Somers, 1953, p. 359; Rutter, 1961, p. 341). Aldolase is widely distributed in animal and plant tissues and is correlated with the Embden-Meyerhof (E-M) glycolytic pathway. Although the terms aldolase and zymohexase were originally employed synonymously, the latter term has more recently been used to characterize the aldolase-isomerase system in which dihydroxyacetone phosphate is converted to glyceraldehyde phosphate. Aldolase is a cytoplasmic enzyme which has been associated with both the soluble and non-particulate fraction (Roodyn, 1957; Brunngraber and Abood, 1960). Serum activity is elevated in a variety of pathological conditions including myocardial infarction, hepatitis, and in cancer (Sibley and Lehninger, 1949; Hanson and Biörck, 1957). Skeletal muscle of a number of species has been found to exhibit the highest activity while liver also contains appreciable amounts of the enzyme. The pH optimum is somewhat dependent upon the assay system (Rutter, 1961, p. 341). Optima of 8.5–9.5, 7.2–7.5, and 7.5–10 have been reported. Muscle aldolase is not inactivated by chelating agents and is not associated with divalent metal cations. A histochemical method for the demonstration of this enzyme has been suggested by Allen and Bourne (1943). It is based upon the fact that the enzymatically produced dihydroxyacetone phosphate and glyceraldehyde phosphate spontaneously hydrolyze at an alkaline pH and the released phosphate ions may thus be demonstrated by a metal-salt technique. The method was criticized by Gomori (1952, p. 198) on the grounds that the spontaneous dephosphorylation of the released triose phosphates occurred far too slowly to permit localization, particularly in view of the fact that these phosphates are highly diffusible. On the other hand, the approach of Allen and Bourne presents an imaginative and interesting suggestion which should offer possibilities for the demonstration of enzymes hydrolyzing phosphorylated metabolic intermediates.

### 2. *Phospholipases (Lecithinases)*

Phosphodiesterases are capable of hydrolyzing the phosphate bonds of lecithin and cephalin. A similar enzyme found in the nervous system hydrolyzes L-$\alpha$-glycerophosphocholine. Gomori (1952, p. 188) attempted to use lecithin or cephalin as substrates in the presence of $Ca^{++}$ around pH 7.0 and obtained no reaction with incubations up to 12 hours. He criticized the positive reactions obtained by Dempsey and Deane (1946) on the grounds that their prolonged incubation times resulted in artefactual staining. More recently Takeuchi (1956) employed a similar procedure and claimed that lecithinase A could be demonstrated in rabbit tissues.

### 3. *Phosphoprotein Phosphatase*

This enzyme liberates inorganic phosphate from casein and other phosphoproteins such as ovalbumin and phosvitin (Norberg, 1950b; Perlmann, 1952, p. 167; Schmidt and Laskowski, 1961, p. 3). Aromatic phosphate esters such as phenylphosphate are hydrolyzed as rapidly as casein, while the hydrolysis rate of aliphatic phosphates of low molecular weight is relatively low. The enzyme which is found in most tissues exhibits a pH optimum of 5.5 (Norberg, 1950b). Brain, adrenals, and kidneys exhibit high activity, while cardiac and skeletal muscle and erythrocytes show a low activity. The PR-enzyme (phosphorylase-rupturing) which catalyzes the transformation of muscle phosphorlyase A to phosphorylase B with the liberation of inorganic phosphate is a phosphoprotein phosphatase. The histochemical demonstration of this enzyme has not as yet been described.

### 4. *Phosphatidic Acid Phosphatases*

Mammalian enzyme systems which dephosphorylate phosphatidic acids have been described by Smith *et al.* (1957). The enzyme obtained from a number of rat tissues hydrolyzes phosphatidic acid to form D-1,2-glyceride and orthophosphate. Lecithin was not found to be hydrolyzed by the enzyme while glycerophosphate and glucose-6-phosphate were hydrolyzed at a relatively low rate. Phosphatidic acid phosphatase has not been demonstrated by histochemical techniques.

# APPENDIX 5A

## Synthesis of Phosphates

### Part 1. Laboratory Synthesis of Alcohol and Nucleotide Phosphates

This summary describes some of the characteristic syntheses of substrates which may have histochemical application, or which are of interest from a historical standpoint (e.g. synthesis of ATP).

1. *Acetyl Phosphate* (*Avison, 1955*)

This compound is of interest because of the fact that an acetyl phosphate has been isolated as one of the intermediates of glycolysis. In addition, this type of ester contains a high-energy phosphate bond.

Ninety-five grams of pyridine (1.2 moles) and 200.0 ml of dipotassium hydrogen phosphate (0.25 *M*) are stirred together at 0°. Acetic anhydride (9.6 ml., 0.1 *M*) is then added to the vigorously stirred solution during a 5 minute period causing the temperature to rise approximately 3.5°. After 35 minutes, 37.0 ml. of 4 *N* lithium hydroxide are added. After a tenfold dilution with water, a practically clear solution at pH 7.6 results. A 90% acetylation is observed at this stage. Cold ethanol (2300 ml.) is then added slowly, and the precipitate allowed to settle for 2 hours at 0° before being filtered off. Crude dilithium acetyl phosphate (6.85 g. yield) is obtained. The product may be further purified by fractional precipitation from a 10% solution by addition of ethanol. The precipitate formed with 0.5 volumes of ethanol contains most of the impurities and is discarded. The pure compound is then precipitated by the slow further addition of 5.0 volumes of cold ethanol and is collected and dried.

2. *o-Phosphorylethanolamine* (*Schmidt, 1957, p. 357*)

A mixture of 10.0 ml. of ethanolamine and 5.5 ml. of water is added dropwise from a buret to 15.0 ml. of $POCl_3$ in a 500 ml. suction flask immersed in cold water. After completion of the reaction, the mixture is evacuated on a water pump for 1 hour to remove the HCl. At this stage about 80% of the phosphorus is esterified. The mixture is taken up in 500.0 ml. of water and the solution is brought to an alkaline pH by the addition of a concentrated solution of $Ba(OH)_2$. The precipitated barium phosphate is removed by centrifugation, and the barium salt of ethanolamine phosphoric acid is precipitated from the combined supernatant solution and washings by the addition of two volumes of ethanol. The precipitate is filtered, washed with ethanol and ether, dried at room temperature, and redissolved in 200.0 ml. of water. The solution is again precipitated with alcohol. After three re-precipitations, the barium salt is crystalline (yield, 50%) ($NH_2CH_2CH_2 \cdot OPO_3H_2$).

3. *Phosphopyruvic Acid (Phosphoenolpyruvic Acid) (Schmidt, 1957, p. 226)*

To an ice-cold solution of 30.0 g. of freshly distilled pyruvic acid in 90.0 g. of freshly distilled quinoline is added an ice-cold mixture of 30.0 g. of $POCl_3$ and 30.0 g. of freshly distilled quinoline. Following completion of the reaction, the sticky phosphorylation product is mixed under constant stirring with 150.0 ml. of 25% solution of NaOH. The latter is added dropwise and simultaneously with 150.0 ml. of ice water. After all of the solid material has been dissolved, the mixture is brought to pH 8 by further additions of a 25% solution of NaOH. The mixture is centrifuged, and the dark aqueous solution (approximately 320.0 ml.) is separated from the dark layer of quinoline. The solution of the phosphate compounds is mixed with 600.0 ml. of methanol, centrifuged, and a supernatant solution precipitated by the addition of 40.0 ml. of an aqueous saturated solution of barium acetate. The centrifuged precipitate is washed twice with 600.0 ml. of a 2:1 mixture of methanol and water and dissolved in the minimal amount of 0.3 *N* nitric acid at a pH of between 2.8 and 3.2. Twenty-five ml. of a 25% solution of silver nitrate are added. After removal of the precipitate by centrifugation, the pH of the supernatant liquid is adjusted to pH 5.0–5.2 by the addition of about 450.0 ml. of 0.2 *N* solution of ammonia. The suspension is refrigerated overnight. The precipitate, which may already contain crystals of silver barium phosphopyruvate, is separated by centrifugation, and the supernatant is kept in a refrigerator for further crystallization. The precipitate is dissolved in 0.3 *N* nitric acid sufficient to bring the pH to 3 (approximately 70.0 ml.). Any insoluble material is renewed by centrifugation. The solution is brought to pH 5 with ammonia and the precipitate is centrifuged off, and the supernatant solution is placed in the refrigerator. This reprecipitation is repeated three times and finally the crystals growing from all supernatant solutions are collected and recrystallized as follows: the precipitate is dissolved in 0.3 *N* nitric acid, and the solution is diluted with an equal amount of water. After addition of 0.3 ml. of saturated barium acetate solution, the mixture is filtered and the precipitate is washed with water. Ammonia is added to the filtrate until a turbidity appears. When the suspension is shaken, crystals appear. The mixture is brought to pH 6 with ammonia, placed in the refrigerator for an hour, and filtered. The crystals are separated from the supernatant solution, washed once with a 67% aqueous methanol solution, twice with methanol, and twice with ether. The supernatant solution is mixed with 5.0 ml. of a saturated solution of barium acetate and placed in the refrigerator for several days. An additional crop of approximately 1.0 g. of crystallized silver barium phosphopyruvate can be recovered. The total yield was 4.9 g. of silver barium salt.

4. *L-α-Glycerophosphate* (*Baer, 1952; Lardy, 1957, p. 207*)

This is the natural isomer of α-glycerophosphate. The reaction is carried out in a two-necked flask fitted with a dropping funnel and a calcium chloride tube open to the atmosphere. A solution of 8.75 g. of D-isopropylidine-glycerol in 24.0 ml. of dry quinoline is added dropwise with shaking during a 20-minute period to a cold solution of 10.5 g. of $POCl_3$ in 30.0 ml. of quinoline. After 1 hour, the flask may be allowed to warm to room temperature to complete the reaction. The reaction mixture is poured slowly into a vigorously stirred mixture of 30.0 ml. of $H_2SO_4$, 28.0 g. of $Ag_2CO_3$, and 700 ml. of ice chips and water. After 30 minutes of stirring, the solution is filtered through a heavy layer of charcoal on a precoated funnel, and silver ion is removed from the solution with $H_2S$. The clear filtrate is heated in a boiling water bath for 30 minutes to free it of $H_2S$ and to remove the isopropylidine group. The solution is cooled to 0° and finely powdered $Ba(OH)_2 \cdot 8H_2O$ is slowly added with stirring until the solution becomes alkaline to phenolphthalein. Excess barium ion is removed with $CO_2$ and after filtration, the aqueous solution is concentrated by vacuum distillation (water pump) at 40°. When the volume of the solution has been reduced by about 60.0 ml. a slight turbidity is removed by filtration. The barium salt of α-glycerophosphate is precipitated from solution by heating in a boiling water bath for about 20 minutes and collected, while hot, on hard filter paper or sintered glass. It is then dried *in vacuo* over $P_2O_5$ (m.w. 307.5). For use in substrate solutions the barium salt is dissolved in $H_2O$ with the aid of a little acid. Sodium or potassium persulfate is added and the solution is filtered in order to remove the precipitate of $BaSO_4$. The pH and volume are then adjusted with suitable buffers.

5. *L-Histidinol Phosphate* (*Ames and Horecker, 1956*)

This phosphate was prepared from L-histidinol and polyphosphoric acid by the procedure used by Cherbuliez and Weniger (1946) for the synthesis of ethanolamine phosphate. Ten grams of L-histidinol dihydrochloride and 35.0 g. of polyphosphoric acid are mixed in an ice bath and 4.0 ml. of water are slowly added to effect solution. The mixture is then heated on a steam bath for 1 hour. An excess of barium hydroxide is used to remove the inorganic phosphate. Sulfuric acid is then added to adjust the pH to 4.5. Four volumes of ethanol are then added to this final supernatant solution to precipitate the phosphate ester. The histidinol phosphate comes out of solution as a highly viscous syrup, even after repeated alcohol precipitations. The product is further purified by graded elution from a 2 × 34 cm. column of Dowex 1 (X8) acetate (100–200 mesh). The mixing chamber contains 1.0 liter of water, and the reservoir 0.2 *N* acetic acid. A 500.0 mg. batch of histidinol phosphate was reported to have started to come off the column after 108.0 ml. of eluate had been collected. The fractions containing the phosphate are lyophilized to give a stable white powder of L-histidinol phosphate ester monohydrate ($C_6H_{12}O_4N_3P \cdot H_2O$). Anal., calcd.: C, 30.13%; N, 17.57%; P, 12.95%. Found: C, 30.15%; N, 17.35%; P, 13.01%.

6. *Glucose-6-phosphate* (*Seegmiller and Horecker, 1951*)

Forty ml. of water are added to 400.0 g. of anhydrous polyphosphoric acid and the mixture is stirred mechanically in an ice bath until the temperature falls to 5–10°. One hundred seventy-four grams of anhydrous glucose (0.97 moles) are added and the mixture is stirred slowly for 16 hours at room temperature. The reaction is then stopped by the addition of 1 liter of water. The solution contained 132.0 mM of glucose esterified at the 6-position. Anhydrous sodium carbonate (520.0 g.) is added to the reaction mixture with mechanical stirring. To aid in the removal of $CO_2$, the warm mixture (about 60°) is evacuated on a water pump for several minutes. The neutralized solution is diluted with 2.6 liters of water, and cooled overnight (with constant stirring) in a cold room at 3°. To reduce the rate of cooling, the reaction mixture is placed in a 10 liter water bath which is initially at a temperature of about 40°. The dense slurry of sodium polyphosphate crystals which is produced is cooled in an ice bath with stirring for 4–6 hours and filtered with suction. The crystals are permitted to drain thoroughly. By this technique, 80–90% of the inorganic phosphate may be removed. The filtrate contained 112.0 mM of ester. The ester solution was hydrolyzed by treating it with 554.0 ml. of concentrated HBr (8.7 *N*), refluxed for 16 hours, and cooled. Ninety-three mM of glucose-6-phosphate was obtained.

7. *Adenosine-2′,3′-phosphate, Trifluoroacetic Anhydride Synthesis* (*Brown et al., 1952*)

Trifluoroacetic anhydride (5.0 ml., 12 moles) is added to 1.0 g. of anhydrous yeast adenylic acid (mixture of α- and β-isomers) and the resulting solution is stored overnight in a stoppered flask at room temperature. The solution is then evaporated under reduced pressure and the residual pale yellow resin is triturated with dry ether. The resulting white powder is added rapidly to a stirred ice-cold saturated solution of ammonia in ethanol (25.0 ml.) and the mixture is set aside for 30 minutes. Solid material (primarily uncombined adenylic acid) is removed by centrifugation and washed with ethanol. The combined washings are evaporated under reduced pressure and the residual resin is thoroughly washed with acetone, and then redissolved in methanol. Addition of ether gives the ammonium salt of the cyclic diphosphate as a white deliquiscent powder which then is dissolved in water and heated in the presence of a slight excess of barium carbonate. The mixture is then filtered, the filtrate concentrated to a small bulk under reduced pressure, and an excess of ethanol is added. The microcrystalline precipitate of barium adenosine-2′,3′-phosphate is purified by reprecipitation from aqueous solution with ethanol $[(C_{10}H_{11}O_6N_5P)_2Ba \cdot 4H_2O]$. Anal., calcd.: C, 27.7%; H, 3.5%. Found: C, 27.2%; H, 4.0%.

### 8. *ATP (Phosphoramidic Ester Synthesis) (Clark et al., 1957)*

A solution of adenosine-5′-pyrophosphate (150.0 mg.) is obtained by decomposing an aqueous suspension of its barium salt with an equivalent amount of $H_2SO_4$ (N/50). Following removal of the $BaSO_4$ by centrifugation, the aqueous solution is treated with a slight excess of tri-*n*-butylamine in ether, and the resulting aqueous layer evaporated under pressure at 40°. The residue is dried over $P_2O_5$. The glassy salt so obtained is dissolved in 35.0 ml. of dry DMF containing 190.0 mg. of benzyl hydrogen *N*-cyclohexylphosphoramidate (ca. 5.0 moles), and the solution heated at 70° for 2 hours. After removal of the solvent at 40° (1.0 mm. Hg) the resulting gum is suspended in water and brought into solution by dropwise addition of dilute aqueous ammonia. The pH is adjusted to 4.0 with glacial acetic acid, and this is followed by hydrogenation at room temperature and pressure with a palladous oxide catalyst (46.0 mg.). Hydrogen uptake was complete in approximately 13 hours. The catalyst is filtered off, the pH adjusted to about 9.0 with aqueous ammonia, and the solution subjected to ion exchange chromatography on a column (7.5 × 1.8 cm.) of Dowex-2 resin ($Cl^-$). Elution with 0.003 *M* HCl gave 5.0 mg. of ATP together with orthophosphate. The ATP may be isolated as the lithium salt. Addition of barium acetate at pH 6.0 to an aqueous solution of the lithium salt precipitates the barium salt of ATP ($C_{10}H_{12}O_{13}N_5P_3Ba \cdot 6H_2O$).

### 9. *Sodium* D*-Riboflavin 5′-Acetyl Phosphate (Avison, 1955)*

Monosodium D-riboflavin 5′-phosphate dihydrate (340.0 mg., 0.66 *mM*) is dissolved in 8.0 ml. of aqueous pyridine and the solution cooled to 0°. Acetic anhydride (0.33 ml., 3.4 m*M*) is added with vigorous stirring. After 5 minutes, 50 ml. of ice-cold acetone is added and the precipitated product rapidly filtered off, washed with cold acetone and then dry ether, and then dried rapidly *in vacuo*. The sodium D-riboflavin 5′-acetyl phosphate is obtained as an orange solid (330.0 mg.) of about 90% purity ($C_{19}H_{22}O_{10} \cdot N_4PNa \cdot 4H_2O$).

### 10. *Sodium* D*-Riboflavin 5′-Chloroacetyl Phosphate (Avison, 1955)*

One hundred seventy mg. (0.33 m*M*) of monosodium D-riboflavin 5′-phosphate dihydrate is dissolved in 4.0 ml. of 40% aqueous pyridine, the solution cooled to 0°, and with vigorous stirring, a solution of chloroacetic anhydride (560.0 mg., 3.3 m*M*) in dry acetonitrile (3.0 ml.) was added, half at once and the remainder dropwise during the next 5 minutes. Then 25.0 ml. of cold acetone were added at once, and the product isolated as for the corresponding acetyl compound. The orange solid (137.0 mg.) so obtained contains about 80% of sodium D-riboflavin 5′-chloroacetyl phosphate. The chloroacetyl compound is less stable than the acetyl compound, but can be stored at $-10°$ for several days ($C_{19}H_{21}O_{10}N_4ClPNa \cdot 2H_2O$).

## APPENDIX 5A (continued)

### Synthesis of Phosphates

### Part 2. Laboratory Synthesis of Aryl Phosphates Including Phosphoamides

1. *Phenylphosphate* (*King and Nicholson, 1939*)

A solution of 10.5 g. of phenol in 50.0 ml. of dry pyridine is added slowly and with stirring to 10.0 ml. of $POCl_3$ (17.1 g.) in a 1 liter beaker. The mixture becomes hot and a crystalline precipitate of pyridine hydrochloride separates. The beaker is covered with a watch glass and the mixture is boiled gently for 10 minutes. The hot solution solidifies upon cooling. The phenylphosphoryl dichloride is decomposed with 5.0 ml. of water which is added dropwise and slowly. It is advisable to cool the beaker under the cold water tap during this process as considerable heat may be developed. Hot, saturated aqueous barium hydroxide is added until the mixture is pink to phenolphthalein. The barium salt of phenylphosphoric ester separates as a heavy flocculent precipitate. This is increased by adding alcohol, with stirring, to the capacity of the beaker and leaving it in the refrigerator overnight. The barium phenylphosphate is filtered under suction, washed first with 50%, and then with absolute ethanol and is dried *in vacuo* over $H_2SO_4$. The yield of crude barium salt was 29.4 and 30.2 g., respectively, in two separate experiments (85 and 87% of theory). The barium phenylphosphate ($C_6H_5PO_4Ba \cdot 2H_2O$) is crystallized from hot aqueous ethanol and separates as small glancing platelets. Anal., calcd.: P, 8.98%; Ba, 39.9%. Found: P, 8.95%; Ba, 40.4%.

In order to prepare the disodium compound, 10.0 g. of the barium salt are dissolved in 300.0 ml. of 0.2 *N* HCl, and the solution shaken out with six changes of 100.0 ml. of ether. The combined ether extracts are diluted with half their volume of ethanol and 40% aqueous NaOH added with stirring until the mixture is alkaline. The precipitate of disodium phenylphosphate is filtered, washed with ethanol, and dried *in vacuo* over $H_2SO_4$.

2. *p-Bromophenylphosphate* (*King and Nicholson, 1939*)

A solution of 17.3 g. of *p*-bromophenol in 50.0 ml. of pyridine is slowly added with stirring to 9.0 ml. of $POCl_3$. As the mixture cools, crystals of pyridine hydrochloride are precipitated. Water is then added dropwise and after the initial reaction has subsided, additional water is added to make a total volume of about 150.0 ml. Hot saturated aqueous barium hydroxide is then added until the mixture becomes alkaline, followed by an equal volume of ethanol. The barium salt of *p*-bromophenylphosphate separates as a white precipitate ($C_6H_4O_4PBrBa$). Anal., calcd.: P, 7.99%; Ba, 35.3%. Found: P, 7.69%; Ba, 36.6%.

3. *p-Nitrophenylphosphate* (*King and Nicholson, 1939*)

*p*-Nitrophenol (13.9 g.) is dissolved in 50.0 ml. of pyridine and added to 9.0 ml. of $POCl_3$. The mixture is diluted with water to a final volume of 150.0 ml. Hot saturated barium hydroxide and ethanol are added as in the case of the bromophenylphosphate. The barium salt is obtained as pale yellow plate-like crystals ($C_6H_4O_6NPBa \cdot 2H_2O$). Anal., calcd.: P, 7.95%. Found: P, 7.97%.

4. *Disodium Phenylphosphate, $POCl_3$ Method* (*Freeman and Colver, 1938*)

Four hundred and seventy-five grams (5 moles) of phenol and 767.0 g. (5 moles) of $POCl_3$ are placed in a 2 liter round-bottomed flask, fitted with a reflux condenser and heated to a gentle boil for approximately 10 hours. The reaction mixture, which contains mono- and diphenylphosphoryl chlorides, and possibly triphenylphosphate, is transferred to a Claisen distilling flask and fractionally distilled under reduced pressure. The monophenylphosphoryl chloride distills over at 130–134° (21 mm.) and the diphenylphosphoryl chloride at 212–215° (21 mm.). The yield of the crude monophenylphosphoryl chloride was 785 g. (74.4%), while the yield of the crude diphenylphosphoryl chloride was 110.0 g. (8.2%). Redistillation of the crude monophenylphosphoryl chloride gave a yield of 744.0 g. (70.4%) boiling at 240° (760 mm.). Careful redistillation of the impure monophenylphosphoryl chloride simplifies purification of the final product.

Phenyl dihydrogen phosphate is prepared as follows: 2 liters of water are placed in a 5 liter 3-necked flask, fitted with a mechanical stirrer, condenser, and a separatory funnel. The upper end of the condenser is fitted with a tube to carry away the HCl vapors. The monophenylphosphoryl chloride obtained in the previous reaction is added slowly to the water, with vigorous stirring. The hydrolysis initially is slow but later becomes violent if the addition is made too rapidly. Approximately 3 hours are required for the hydrolysis of 3.5 moles. The solution is then filtered and most of the HCl is removed by warming it under a stream of air at reduced pressure. The solution is then evaporated further under a vacuum desiccator containing NaOH. Some of the phenyl dihydrogen phosphate separates as large crystals and may be filtered off. The filtrate is further concentrated until free from HCl. This solution and the formed crystals are used for making the disodium salt. The latter is formed by adding solid sodium carbonate to a concentrated solution of phenyl dihydrogen phosphate which has been freed of HCl. The crystalline disodium phenylphosphate is obtained from solution by adding a mixture of ethanol and diethyl ether until there is no further precipitation. The alcohol-ether mixture tends to hold any free phenol in solution.

5. *Phenylphosphate, Phosphite Synthesis* (*Miyano, 1955*)

The starting point of this synthesis is di-*p*-bromobenzylchlorophosphonate, which is derived from the phosphite (Miyano and Funahashi, 1955). The chlorophosphonate (from 32.8 g. of the phosphite) in 200.0 ml. of benzene is added in one portion to sodium phenoxide (8.8 g.) and the mixture vigorously shaken for 30 minutes. The mixture is washed with diluted HCl, water, and aqueous potassium carbonate. It is then dried with calcium chloride and concentrated under reduced pressure. Petroleum ether is added to make the solution opalescent. Two ml. of cyclohexane are then added and the solution placed in a refrigerator. The crystalline mass is filtered under suction and washed with petroleum ether, yielding 17.4 g. of di-*p*-bromobenzylphenylphosphate. Eleven grams of the latter, together with 2.1 g. of lithium chloride, are dissolved in 50.0 ml. of ethoxyethanol and heated at 100° for 2 hours. The mixture is diluted with water, extracted with ether, acidified with sulfuric acid, and extracted with chloroform. The chloroform extract is dried with sodium bromide and concentrated to a thick syrup. This is dissolved in ethanol and excess cyclohexylamine is then added. The clear ethanolic solution is then concentrated under reduced pressure and water is added. The separated cyclohexylamine salt of *p*-bromobenzylphenyl hydrogen phosphate is filtered under suction and washed with water. It is then suspended in 10.0 ml. of dilute sulfuric acid and extracted with 30.0 ml. of chloroform. The organic layer is dried with calcium chloride and concentrated to a thick syrup. Sodium bromide (1.2 g.) and ethoxyethanol (30.0 ml.) are added and the clear solution is heated for 50 minutes at 100° and for an additional hour at 130°. The cooled solution is diluted with 50.0 ml. of water and extracted with chloroform, acidified with sulfuric acid, and extracted with ether. The ethereal extract is then concentrated to a small volume. Excess cyclohexylamine and 40.0 ml. of acetone are then added, and the solid is filtered under suction and dried. The phenylphosphate (cyclohexylamine salt) is recrystallized from aqueous ethanol (m.p. 214–215°).

6. *Phenylphosphoric Acid, Phosphoramide Halide Synthesis* (*Montgomery and Turnbull, 1958*)

One mole of phenylphosphorodichloridate ($C_6H_5OPOCl_2$) is treated with 4 moles of morpholine in benzene at 20–30°. The phenylphosphorodimorpholidate is recrystallized from cyclohexane (m.p. 84°).

An aqueous solution of the aforementioned compound (500.0 mg. in 5.0 ml. of $H_2O$) is slowly percolated (1–2 hours) through Amberlite IR-120 resin ($H^+$ form) at 60°. The dihydrogen phosphate is obtained upon evaporation, and subsequent recrystallization from chloroform (m.p. 94°) [bis-(cyclohexylammonium) salt, m.p. 211°].

7. *Borneol Phosphoric Acid* (*Neuberg et al., 1927*)

Eighty-eight grams of $POCl_3$ are added dropwise into a flask containing 350.0 ml. of dry pyridine. Sixty grams of borneol dissolved in a minimum of pyridine are added dropwise to the aforementioned solution at low temperature and with frequent shaking. The reaction product which consists partly of a mass of needle-shaped crystals is allowed to stand for about 3 hours at room temperature and is then made alkaline with 40% aqueous KOH. It is important that the reaction mixture be adequately cooled and, in addition, it is advisable that the initial amount of KOH added be relatively dilute. After the final addition of KOH, a yellow or light brown solution is produced. The solution is then acidified with 25% $H_2SO_4$, resulting in a white flocculent precipitate of the free borneol phosphoric acid. The acid is dissolved in ether, washed several times with water, and then the ether solution is dried with sodium sulfate. Upon evaporation of the ether, a very viscous oily residue is left which frequently hardens into a white mass. The crude borneol phosphate is purified as follows: the material is slowly mixed with *N* KOH while stirring, until slightly alkaline as tested with phenolphthalein. The weakly alkaline solution is then neutralized by the introduction of $CO_2$. The mixture is shaken with ether three or four times in order to remove traces of borneol or pyridine. All traces of ether must be removed from the solution which is clear to faintly yellow in appearance. The borneol phosphoric acid is again precipitated with 25% $H_2SO_4$. In some instances, this purification procedure must be carried out for a second time. In an evacuated desiccator, the purified product becomes rapidly transformed into a white mass of crystals which is now recrystallized with a mixture of benzene and petroleum ether. The dry acid is pulverized and dissolved in a minimum of dry benzene. The solution is then cooled and about ten times its volume of petroleum ether added. The borneol phosphoric acid rapidly separates in needle-shaped crystals which are dried under suction and washed several times with petroleum ether. The acid is completely insoluble in water. It is soluble in acetone, chloroform, glacial acetic acid, methanol, ethanol, benzene, toluene, and ether. Borneol phosphate exists in L, D, and DL forms. The melting points of these are 156, 154–155, and 155–156°, respectively.

8. *o-Carbomethoxyphenylpyrophosphate and p-Carbethoxyphenylpyrophosphate* (*Walker and King, 1950*)

The technique followed was essentially that of King and Nicholson (1939). The reaction mixture is cooled in an ice bath throughout the addition of $POCl_3$ to a pyridine solution of the carboxyl ester. The mixture is placed in a refrigerator for approximately 24 hours and is again cooled during the formation of the barium salt with $Ba(OH)_2$. Ba salts are recrystallized from hot aqueous ethanol. The di-*o*-carbomethoxyphenylpyrophosphate forms fine needle-like crystals ($C_{16}H_{14}O_{11}P_2Ba$). Anal., calcd.: C, 33.0%; H, 2.4%; P, 10.3%; Ba, 23.5%. Found: C, 32.0%; H, 2.9%; P, 10.4%; Ba, 24.0%. The di-*p*-carbethoxyphenylpyrophosphate forms glancing platelets ($C_{18}H_{18}O_{11}P_2Ba$). Anal., calcd.: C, 35.4%; H, 3.0%; P, 10.1%; Ba, 22.5%. Found: C, 34.8%; H, 2.9%; P, 10.3%; Ba, 22.6%.

9. *Barium o-Carboxyphenylphosphate and Barium p-Carboxyphenylphosphate (Walker and King, 1950)*

The above-mentioned pyrophosphate substrates are dissolved in water and the Ba removed by addition of a calculated amount of $H_2SO_4$. The hydrolysis of the pyrophosphate is accomplished by the addition of 5 equivalents of NaOH. It was found that after 18 hours at 37° only 2 equivalents of NaOH remained (titration with HCl to phenolphthalein end point). No inorganic phosphate was found to be liberated by this hydrolysis. The isomeric carboxyphenylphosphates are isolated by the addition of $BaCl_2$ either as the $Ba_{1.5}$ salts from alkaline solution or as the half Ba salts from acid solution (pH 3.0) ($C_7H_4O_6PBa_{1.5}$). Anal., calcd.: P, 7.4%; Ba, 48.7%. Found: (Ba *o*-carboxyphenylphosphate): P, 7.7%; Ba, 49.0%. (Ba *p*-carboxyphenylphosphate): P, 7.6%; Ba, 47.8%.

The rates of hydrolysis of these compounds are less than that of phenylphosphate. The optimum pH is also less, ranging from pH 8.0–9.1.

10. *Di-o-cresol Pyrophosphate (Neuberg and Jacobsohn, 1928)*

One hundred seven grams of phosphorus oxychloride are added to 350.0 ml. of dry pyridine and this mixture is added to an ice-cold solution containing 54.0 g. of *o*-cresol dissolved in a minimal amount of pyridine. The reaction mixture is allowed to stand for several hours at room temperature and the excess pyridine is distilled off. The potassium salt is then prepared by addition of an aqueous solution of KOH (1:1) to the well-cooled mixture. After removal of the excess moisture under vacuum, the colorless product is extracted with 500.0 ml. of 90% ethanol. The alcohol solution is evaporated revealing the formation of crystals of the pyrophosphate ester (47.0 g., 43% yield). The salt is recrystallized several times from 90% ethanol [$K_2(C_7H_7)_2P_2O_7$]. Anal., calcd.: P, 14.29%; K, 18.00%. Found: P, 14.47%; K, 17.74%.

11. *Di-α-naphthylpyrophosphate (Neuberg and Jacobsohn, 1928)*

This is prepared in a similar fashion to the aforementioned compound. Seventy-two grams of α-naphthol were found to yield 63.0 g. of di-α-naphthylpyrophosphoric acid (K salt) which constitutes a 50% yield. The melting point of the free acid was found to be 78° (the calcium salt crystallized as colorless rhombohedra) ($C_{20}H_{14}K_2P_2O_7$). Anal., calcd.: P, 12.26%; K, 15.44%. Found: P, 12.29%; K, 15.21%.

12. *α-Naphthyl Phosphate* (*Kunz, 1894*)

Fifty grams of α-naphthol and 53 g. of $POCl_3$ (equimolar quantities) are heated in a flask on an oil bath at 120° for approximately 20 hours until all of the HCl is evolved. The solution is then distilled (20 mm. Hg) given the dichloride at 198–200°. The dichloride ($C_{10}H_7O_2Cl_2P$) is a viscous liquid which is transformed into α-naphthyl acid phosphate by treatment with cold water to form colorless crystals [$C_{10}H_7OPO(OH)_2$] m.p. 142°. Anal., calcd.: C, 53.57%; H, 4.02%. Found: C, 53.04%; H, 4.22%.

13. *β-Naphthyl Phosphate* (*Kunz, 1894*)

β-Naphthyl phosphate is prepared in the same way as the aforementioned α-isomer. The dichloride is crystalline and melts at 39°. The phosphate is formed by treating the aforementioned compound with water [$C_{10}H_7OPO(OH)_2$]. Anal., calcd.: C, 53.57%; H, 4.02%. Found: C, 53.55%; H, 4.34%.

14. *α- and β-Naphthyl Phosphates* (*Menten et al., 1944a; Manheimer and Seligman, 1948; Friedman and Seligman, 1950*)

These techniques are modifications of the procedures of Kunz (1894) and King and Nicholson (1939). Menten *et al.* prepared Ba β-naphthyl phosphate as follows: to 10.0 ml. of $POCl_3$, a solution of 15.0 g. of β-naphthol and 50.0 ml. of dry pyridine is added with stirring. The reaction is allowed to proceed for 20 minutes and the resulting β-naphthyl phosphoryl chloride is decomposed with 5.0 ml. of water. A saturated aqueous solution of $Ba(OH)_2$ is added until the mixture turns pink to phenolphthalein. The barium β-naphthyl phosphate is precipitated and after the mixture is chilled in a refrigerator it is filtered with suction, washed with 50% and then with absolute ethanol, and dried. These authors attempted to prepare the free acid phosphate from the barium salt but the yield was poor.

Menten *et al.* also prepared calcium β-naphthyl phosphate. The preparation of this compound is the same as that described for the aforementioned barium β-naphthyl phosphate up to the decomposition of the phosphoryl chloride. At this point, saturated $Ca(OH)_2$ is added until the reaction mixture becomes pink to phenolphthalein (ca. 18.0 g.). The precipitate is filtered after being chilled, and is washed with 50% ethanol and then with absolute ethanol. The calcium salt of this ester was found to be more soluble than the barium salt and gave a larger yield. An additional crop of calcium naphthyl phosphate may be obtained by adding ethanol to the combined filtrate and washings. The Ca β-naphthyl phosphate was found to be readily hydrolyzed and a reaction employing an azo-dye procedure revealed development of dye in a few minutes.

Manheimer and Seligman (1948) employed essentially the same technique for the preparation of Ca β-naphthyl phosphate.

Friedman and Seligman (1950) prepared α- and β-naphthyl phosphates and their corresponding sodium salts. The α-isomer is prepared by treating a heated (reflux) solution of 25.0 g. of α-naphthol and 26.5 g. (15.5 ml.) of $POCl_3$ in 90.0 ml. of dry benzene with 13.7 g. of dry pyridine. The latter is slowly added over a period of 30 minutes, during which a precipitate of pyridine hydrochloride forms. The mixture is heated for an additional 15 minutes, and after cooling, a voluminous precipitate of pyridine hydrochloride is removed by filtration. The solvent is distilled and the residue on distillation gives a clear, colorless syrup of α-naphthyl phosphoryl dichloride (b.p. 199–201°). The dichloride is poured into Petri dishes to form a thin layer (2–3 mm.), and these are stored in a partially evacuated desiccator over aqueous KOH for several days. During this time there occurs a quantitative conversion to α-naphthyl acid phosphate which is obtained as a white crystalline solid ($C_{10}H_9PO_4$) (m.p. 155–157°). Anal., calcd.: C, 53.56%; H, 4.05%. Found: C, 53.72%; H, 3.96%.

The procedure for preparation of β-naphthyl acid phosphate ($C_{10}H_9PO_4$) is the same as that described for the α-isomer.

### 15. *β-Naphthyl Phosphate, Dibenzyl Chlorophosphonate Synthesis (Atherton et al., 1945)*

A suspension of sodium β-naphthoxide (from 8.64 g. of β-naphthol) in 300.0 ml. of toluene is cooled to −5° and a solution of dibenzyl chlorophosphonate (from 7.86 g. of dibenzyl hydrogen phosphite) in 96.0 ml. of carbon tetrachloride is added slowly with stirring so that the temperature is maintained at −5°. After being stirred for 2 hours, the mixture is left overnight at room temperature. Water is added and the phases separated by the addition of more carbon tetrachloride (200.0 ml.). The lower layer is tapped off, washed with sodium hydroxide solution and water, dried over anhydrous sodium sulfate, and evaporated. The residual oil is dissolved in chloroform and chromatographed on activated alumina. On development with chloroform, dark colored impurities are retained on the adsorbent and dibenzyl β-naphthyl phosphate passes through and is isolated as an orange-colored oil (11.0 g.) on evaporation of the solvent. The oil is dissolved in 300.0 ml. of methanol and refluxed with Raney nickel, filtered, and hydrogenated at atmospheric pressure in the presence of *N*-methylmorpholine (10.0 ml.) and a palladised charcoal catalyst (10% Pd). The solution is filtered and evaporated; the residue is evaporated twice with water, dissolved in water again, and traces of oily impurities removed by extraction with ether. The solution is acidified and a gummy precipitate is obtained which is collected and dissolved in absolute ethanol. The solution is again evaporated and the residue brought to crystallization with a small amount of chloroform. The crude product (yield, 80%) is recrystallized from chloroform-ethanol to give colorless plates of β-naphthyl phosphate (m.p. 172–173°).

16. *Tetrabromo-2-naphthyl Phosphate* (*Munemura, 1933b*)

This compound is prepared by brominating β-naphthyl phosphate. Eleven and two-tenths grams of β-naphthyl phosphate are mixed in a flask with 5.0 g. of glacial acetic acid and 2.0 g. of pyridine hydrochloride. Forty grams of bromine are slowly added dropwise into the mixture with constant stirring and cooling. The mixture is then allowed to stand overnight. It is subsequently warmed on a water bath in order to remove the excess bromine. The reaction mixture is then cooled, resulting in the formation of a crystalline precipitate which is purified by recrystallization from glacial acetic acid-benzene. Compound forms needle-shaped yellow crystals melting at 140° ($Br_4C_{10}H_5O_4P$). Anal., calcd.: P, 5.74%. Found: P, 5.65%.

17. *6-Benzoyl-2-naphthyl Phosphate* (*Rutenburg and Seligman, 1955*)

A solution containing 14.0 g. (0.056 *M*) of 6-benzoyl-2-naphthol and 8.6 g. (0.056 *M*) of $POCl_3$ and 100.0 ml. of dry benzene is heated under reflux for a few minutes. Dry pyridine (5.0 ml.) is added at once and the refluxing is continued for another 30 minutes. During the addition of pyridine, a precipitate of pyridine hydrochloride is formed. After cooling, the hydrochloride is removed by filtration. The solvent is rapidly removed by distillation (water pump) until a syrup is obtained. The temperature is not allowed to exceed 70–80°. The warm syrup is poured into a Petri dish which is placed upon a support over a saturated solution of potassium hydroxide in a desiccator jar, and evacuated by a water pump. In the course of several days, a conversion to the 6-benzoyl-2-naphthyl acid phosphate is indicated by crystallization of the syrup. The monosodium salt is precipitated from a solution of the free acid phosphate and methanol by an addition of an equivalent of sodium methoxide in methanol. The white product is collected in a Buchner funnel and thoroughly washed with a little methanol and ether. The yield was 11.2 g. (60%). In order to remove sodium chloride and sodium phosphate, the solid is suspended in a few ml. of cold distilled water and allowed to stand for 10 minutes. It is filtered cold and the insoluble salt again washed with methanol and ether.

18. *Synthesis of Napthol AS Phosphates* (*Burstone, 1958a, b*)

A large number of naphthols of the AS series as indicated in Table 5 of Chapter 2 may be phosphorylated by the following technique. Initially it is always advisable to recrystallize commerical samples as described in Appendix 2, Section 4. Four to five grams (0.011 to 0.02 *M*) of naphthol are suspended in 50.0 ml. of tetrahydrofuran (hydroquinone free) together with approximately 300.0 mg. Versene (EDTA, acid), 50–100 mg. of sodium lauryl sulfate, 2.0 g. of NaCl, and an excess of $POCl_3$ (5.0 ml.). Two to three ml. of pyridine are then added with constant stirring. The mixture is then decanted into a 1000.0 ml. beaker and partly evaporated under a jet of air until a fairly clear syrup is formed. Three or four cubes of ice (made from distilled water) are then added with constant stirring to the syrup. After 5 minutes approximately 500.0 ml. of cold water are added with stirring. In many instances a voluminous precipitate is formed. In other cases a rubbery or tarry material is precipitated which eventually hardens in the presence of water. If the precipitate is fine, it is collected under suction and washed repeatedly with cold distilled water. If the phosphate is formed as solid masses, the pieces are triturated in the presence of water and then collected and washed as previously described. Although the crude product may contain impurities in the form of secondary and other phosphates, these do not adversely affect their histochemical application.

A modified improved synthesis (Burstone, unpublished data) involves the treatment of a suspension of the naphthol in bis(2-ethoxyethyl ether) with $POCl_3$, following which the reaction mixture is treated with a large volume of ice water.

The acid phosphates may be further purified according to the following technique. Two hundred to four hundred mg. of substrate are dissolved in a minimum amount of tetrahydrofuran at room temperature and the solution is filtered through very porous filtering paper into several hundred ml. of petroleum ether. The solution is stirred and allowed to stand at room temperature until a precipitate is formed. The mixture is stirred for approximately 30 seconds and the precipitate collected under suction. This procedure is usually repeated two or three times. The disodium salt may be prepared by treating a methanolic solution of the naphthol with sodium methylate, or by addition of a calculated amount of aqueous sodium carbonate to a DMF-water solution of the naphthol.

The aforementioned techniques may be used to prepare phosphates of naphthols AS, AS-MX, AS-BI, AS-BS, AS-AN, and AS-TR, which have proved to be useful substrates in the histochemical demonstration of both alkaline and acid phosphatases.

19. *Phosphates of Fluorescein, Eosin, and 4-Methyl-7-hydroxycoumarin (Neumann, 1948b)*

Fluorescein is dissolved in pyridine. The solution is then dropped into an excess of $POCl_3$. It is hydrolyzed in the presence of water for some hours and then treated with sodium chloride. The ester is dissolved in alkali and freed from pyridine by shaking it with ether. The unchanged fluorescein is eliminated by shaking the solution at pH 3.0 with isobutanol. The phosphoric acid ester of eosin is prepared in an analogous fashion. The 4-methyl-7-hydroxycoumarin is dissolved in pyridine and this solution is dropped into a mixture of $CHCl_3$-$POCl_3$ (1:1). After hydrolysis with water, the ester is precipitated. Further purification is carried out in the same manner as for the fluorescein phosphate. The above-mentioned phosphates exhibit little fluorescence in solution but following enzymatic hydrolysis liberate highly fluorescent products. These techniques were designed for quantitative determinations of phosphatase in blood serum.

20. *p-Nitrobenzeneazo-4-α-naphthyl Phosphate (Loveless and Danielli, 1949)*

These authors employed the procedure of King and Nicholson (1939), and treated 4-nitrobenzeneazonaphthol in pyridine with a slight excess of pyridine. Phosphorylation is allowed to proceed for 1 hour with periodic shaking, and the resulting phosphoryl dichloride is decomposed with water. Two techniques were employed to obtain sodium salts of the ester. Only the first, however, yielded a substrate which was hydrolyzed by phosphatase. The phosphorylated material is diluted with 5 times its volume of water. The resulting precipitate is washed and dried. It is then digested overnight at 60° with 0.4% sodium veronal adjusted to pH 9.3 with 5 *N* NaOH. The material is extracted with successive portions of sodium veronal in this way until nothing further is dissolved (as indicated by color), and the extracts combined and filtered. A solution of this material exhibits a moderately intense orange color. The pH of the incubating solution is adjusted by addition of 5 *N* NaOH or HCl.

The substrate is hydrolyzed under histochemical conditions over a pH range of 7.0–9.5 with an optimum of pH 8.7.

21. *Sodium Phenolphthalein Phosphate (Huggins and Talalay, 1945)*

A mixture of 50.0 g. (30.0 ml.) of $POCl_3$ in about 50.0 ml. of chloroform is added fairly rapidly and with mechanical stirring to 32 g. of phenolphthalein. Twenty-five ml. of pyridine are then added slowly with rapid stirring to the ice-cooled mixture. The addition of pyridine should be regulated so that the reaction mixture does not become hot. The mixture is stirred for several hours and allowed to stand overnight. The degree of completion of the reaction may be determined by withdrawing samples of the reaction mixture and adding excess alkali. A red color indicates the presence of unreacted phenolphthalein. The mixture is evaporated *in vacuo* at room temperature or with slight warming in order to remove most of the chloroform. Approximately 100.0 ml. of water are then added slowly with stirring and cooling. There is a copious evolution of HCl and a white precipitate forms. Strong NaOH (40%) is added until all the precipitate redissolves and the solution is alkaline to phenolphthalein. The small amount of phenolphthalein in the solution serves as its own indicator. The alkaline solution is extracted twice with ether in order to remove pyridine. If the solution is too viscous, some water may be added, although it is desirable to keep the volume as small as possible in order to facilitate the later precipitation of phenolphthalein diphosphoric acid. The aqueous layer is acidified with concentrated HCl until blue to Congo red. This precipitates phenolphthalein diphosphoric acid which is a glistening, viscid mass. When precipitation is approximately complete, the supernatant solution is decanted and the free acid dried *in vacuo* over calcium chloride. The dry, crude phenolphthalein diphosphoric acid is ground to a white powder in a mortar. The powdered acid is dissolved in methanol to which some pyridine has been added. A solution of sodium ethoxide or sodium methoxide in ethanol or methanol, respectively, is added until no further precipitation occurs. The sodium phenolphthalein phosphate separates as a copious white precipitate which is filtered off and washed with repeated changes of alcohol and ether until dry. The preparation at this stage should be white and gives almost no color in alkaline solution. It is repurified by dissolving in a mixture of methanol-formamide (80:20), reprecipitating with absolute ethanol, and washing the precipitate with ethanol and ether. The sodium phenolphthalein phosphate is stable when stored under refrigeration.

Action of phosphatase liberates from the colorless soluble ester the quinonoid structure which produces a red shade at an alkaline pH. The pH optimum of the substrate in the acid phosphatase range is between 5.5–6.0 while that in the alkaline range lies between pH 9.1–9.6. The rate of hydrolysis is independent of substrate concentration between 0.0005 and 0.002 *M*.

22. *Cholesteryl Phosphate* (*Plimmer and Burch, 1929*)

One mole (4.2 g.) of $POCl_3$ is slowly added to a mechanically stirred solution of 12.0 g. of cholesterol and chloroform. After 24 hours, the mixture, which has become red, is shaken with a large volume of water and kept for an additional 24 hours. The water is separated and made neutral to phenolphthalein with barium carbonate and baryta. Barium phosphate is filtered off, and the filtrate concentrated and treated with an equal volume of ethanol. A gelatinous precipitate of barium cholesteryl phosphate, barium phosphate, and barium chloride is produced. Repeated crystallization from aqueous ethanol gives a small yield of barium monocholesteryl phosphate, which crystallizes in fine white platelets from ethanol and water. Monocholesteryl dihydrogen phosphate may be prepared from this salt by precipitating the barium with sulfuric acid. The opalescent filtrate which is not clarified even upon prolonged boiling is evaporated to dryness and the residue crystallized from chloroform.

23. *2-Methyl-3-bromo-1,4-naphthol Hydroquinone Diphosphate* (*Friedmann et al., 1952*)

Four grams of 2-methyl-3-bromo-1,4-naphthoquinone in 200.0 ml. of ether are reduced by shaking with a solution of sodium hydrosulfite (8.0 g. in 60.0 ml. of water) until no further color change is observed. The ether solution is then washed twice with a saturated sodium chloride solution containing a little sodium hydrosulfite solution and filtered through a funnel containing sodium sulfate in an atmosphere of nitrogen. The dried ethereal solution is then evaporated *in vacuo* at 50° using a nitrogen bubbler. The residual hydroquinone is not purified further but is dissolved in 90.0 ml. of pyridine kept under nitrogen, and added over a 30 minute period to an ice-cold stirred solution of 9.0 ml. of $POCl_3$ in 60.0 ml. of pyridine. After final addition, the mixture is stirred for 30 minutes at low temperature and then evaporated *in vacuo* under nitrogen at 50°. When almost dry, 50.0 ml. of pyridine are added and the solution is evaporated to dryness. The residue is left *in vacuo* overnight over sulfuric acid. The residue is then cooled in ice and treated with 200.0 ml. of ice water, shaken until dissolved and left at room temperature for 3 hours. The solution is filtered after addition of Hyflo Supercel and made slightly alkaline to phenolphthalein with saturated barium hydroxide solution (ca. 1200.0 ml.). The solution is then filtered with the aid of Hyflo, evaporated *in vacuo* at 50° to about half its bulk, and poured slowly into an equal volume of ethanol. The next day the barium salt is filtered off, washed with 50% ethanol and finally with ether. The salt is dried in a desiccator giving 4.6 g. of a colorless amorphous solid (42% crude yield). The salt is dissolved in 100.0 ml. of water and treated with *N* sulfuric acid until no free barium ions remain in solution (tested by 20–27 ml. of sodium rhodizonate solution). Evaporation of the filtered and clarified solution gives a partly crystalline solid (m.p. 237°, dec.) which is dissolved in water and neutralized by the addition of *N* sodium hydroxide solution. The solution is then put through an ion exchange column (Dowex 50) and loaded with hydrogen ions. The eluate is evaporated *in vacuo* at 50° to a small bulk, clarified by filtration through Hyflo, and allowed to evaporate to dryness in a desiccator. The diphosphate crystallizes as feathery needles, m.p. 239–240° (dec.) ($C_{11}H_{11}O_8P_2Br \cdot \frac{1}{2}H_2O$). Anal., calcd.: C, 31.3%; H, 2.87%; Br, 19.0%. Found: C, 31.2%; H, 3.26%; Br, 19.75%.

### 24. *Phenyl Esters of Amido- and Diamidophosphoric Acids (Audrieth and Toy, 1941)*

*Phenyldiamidophosphate.* Seventeen and one-tenth grams (0.111 *M*) of $POCl_3$ are dissolved in 75.0 ml. of chloroform and the mixture cooled in an ice-salt bath. A solution containing pyridine-phosphorus oxychloride (2:1 or 3:1) and 10.5 g. (0.111 *M*) of phenol is then added dropwise to the chloroform solution, the rate of addition being controlled in order to maintain the temperature at ca. 2°. The reaction product is then stirred for 15 minutes, transferred to a dropping funnel, and added slowly to an excess of liquid ammonia. Upon evaporation of the excess ammonia, the residual solution is filtered and the precipitate is washed with water to remove soluble components. The filtrate is evaporated to dryness. The water-insoluble portion is recrystallized from 95% ethanol to give phenyldiamidophosphate [$C_6H_5OPO(NH_2)_2$] (m.p. 183–185°). The residue obtained by evaporation of the chloroform-pyridine solution is combined with the product remaining after removal of the ethanol from the alcoholic mother liquor. This mixture is extracted with hot chloroform which easily dissolves the diphenylamidophosphate while the diamidophosphate is only slightly soluble. The m.p. of diphenylamidophosphate [$(C_6H_5O)_2PONH_2$] is 145–146°.

### 25. *N-Substituted Phenyl Esters of Amido- and Diamidophosphoric Acids (Audrieth and Toy, 1942)*

*a. Phenyldi(methylamido)phosphate.* An excess of an aqueous methylamine solution is placed in a 500.0 ml. 3-necked flask equipped with a thermometer, a stirrer, and a dropping funnel. Forty-two grams (0.2 *M*) of phenyldichlorophosphate ($C_6H_5OPOCl_2$) are added slowly with stirring, the temperature being maintained at ca. 2°. Following addition, the reaction mixture is allowed to come to room temperature and the precipitate removed by filtration and washed free of chloride with ice water. The filtrate is then evaporated to dryness and the residue is extracted with absolute ether. An additional quantity of product is obtained upon evaporation of the ether extract. The compound is insoluble in cold $CCl_4$, hot or cold high-boiling-point petroleum ether; and is slightly soluble in cold water, ethyl acetate, benzene, and hot $CCl_4$. It is very soluble in ethanol, chloroform, acetone, ether, and hot benzene. Upon recrystallization from

# APPENDIX 5B

## Histochemical Methods

1. *Semi-quantitative Determination of Histochemical Phosphatase Reactions (Gomori, 1950d)*

The following method permits approximate quantitation of histochemical reactions for phosphatases as well as other enzymes.

A highly active enzyme solution is made according to accepted biochemical procedures and its activity is assayed. Serial dilutions by a factor of 2 are made with a suitable diluent, such as 1% gelatin or gum acacia solution. Standard microscopic slides are prepared as follows: carefully cleaned microscopic slides are thinly coated with egg albumin-glycerol as used for histological purposes. The slides are heated over a Bunsen burner until completely dry. Using a clean steel pen, marks are made on the slides employing the serial dilutions as "ink." To avoid confusion, it is best to write with each dilution its strength as a mark. The slides are dried, fixed briefly in ethanol or acetone, coated with thin (ca. 0.1%) collodion, washed, and incubated in the substrate solution as routine histological slides would be. Following incubation, the slides may be briefly washed in water and mounted in glycerol-gelatin. With the metal-salt procedure for phosphatases, the marks will show up in gradations from colorless to black or brown. With these techniques, it was found that a slight but distinct gray shade is produced with about 20–25 $\mu M$ units ml. of alkaline phosphatase activity, and a distinctly black shade with about 200 $\mu M$ units with an incubation period of approximately 1 hour. The intensity of the reaction was found to be proportional between 5 minutes and 24 hours (i.e. blackening was obtained after 5 minutes with an activity of 3,000 $\mu M$ or over, and after 24 hours, with an activity of about 8–10 $M$ or over). With acid phosphatase, the results were found to be similar although often irregular.

The approximate estimation of enzymatic activity is done by incubating a number of tissue sections together with the standard marked slides in the same Coplin jar. At intervals (e.g. 5, 10, 20, 40 minutes, etc.) one tissue slide and a corresponding standard marked slide are removed from the incubating mixture and the color is developed. The pair of slides in which the tissue structure in question first shows up in the shade chosen (preferably black) is used for comparison. The mark made with the lowest dilution and showing black will have approximately the same activity per unit area as the histological structure in question. This technique may be used in the study of the effects of various histological procedures such as fixation and embedding on subsequent enzymatic activities since the standard slides can be exposed to a variety of physical and chemical influences.

2. *Alkaline Phosphatase, Metal-Salt Method* (*Gomori, 1939; 1952, p. 184*)

Since the enzyme is fairly stable, a wide variety of fixation and embedding procedures may be employed. Formalin fixed frozen or cryostat sections briefly fixed in formalin or other fixatives, acetone fixed and frozen-dried paraffin embedded tissues may be successfully utilized.

The incubating mixture contains the following:

| | |
|---|---|
| 3% Sodium glycerophosphate (ca. 0.1 *M*) | 5.0–10.0 ml. |
| 2% $CaCl_2$ (ca. 0.2 *M*) | 20.0–25.0 ml. |
| 10% $MgCl_2$ or $MgSO_4$ (ca. 0.5 *M*) | ca. 10 drops |
| Sodium barbital | 0.5–1.0 g. |
| Distilled water to make a total volume of 50.0 ml. | |

Instead of barbital other buffers such as tris or ammediol may be employed (Chapter 16). If the substrate solution becomes turbid it should be filtered before use. Sections are incubated at 37° for 5 minutes to 4 or 5 hours. They are then washed in tap water for approximately 1 minute and immersed in a 1–2% solution of any soluble cobalt salt (acetate, chloride, sulfate, or nitrate) for approximately 5 minutes. The sections are then washed in tap water for 1 to 2 minutes and immersed in a dilute solution of colorless or light-yellow ammonium sulfide (a few drops to a Coplin jar of distilled water) for approximately 5 minutes. The sections are washed in tap water, counterstained if desired, and mounted in a water-soluble mounting medium or dehydrated and mounted in a suitable synthetic medium. Sites of alkaline phosphatase activity are stained black.

3. *Formol-Calcium Metal-Salt Acid Phosphatase Method* (*Eränkö, 1951, 1952*)

Thin slices of tissue are fixed in formol-calcium (Appendix 1, Section 1) for 12–24 hours at 0–5°. The specimens are sectioned with a freezing microtome or in a cryostat, rinsed in two changes of distilled water and immersed in the glycerophosphate substrate solution. Incubation times may range from minutes to hours. Following incubation, the sections are washed in several changes of distilled water and treated with ammonium sulfide as in the original Gomori technique (Appendix 5B, Section 4). Alternatively, the sections may be first mounted on glass slides prior to the incubation.

4. *Metal-Salt (Pb) Acid Phosphatase Method* (*Gomori, 1950c; 1952, p. 193*)

Unfixed frozen, formalin fixed frozen, and various types of cold acetone and frozen-dried paraffin embedded tissues may be employed. In general, the formol-calcium technique of Eränkö (1951, 1952) (Appendix 5B,

Section 3) is most suitable for demonstration of acid phosphatase by the metal-salt procedure. When paraffin sections are employed, collodion coating of the sections is advisable since it appears to prevent loss of enzyme by diffusion (Goromi, 1952, p. 193).

The substrate mixture is prepared as follows: in 500.0 ml. of 0.05 *M* acetate buffer pH 5.0 dissolve 0.6 g. of $Pb(NO_3)_2$ (ca. 0.003 *M*) and add 50.0 ml. of 3% (ca. 0.1 *M*) solution of sodium glycerophosphate. The mixture will become turbid, the degree of turbidity depending upon the percentage of the β-isomer present in the glycerophosphate. Gomori (1952, p. 193) notes that a mixture containing approximately equal amounts of isomers will cause much less turbidity than the more commonly sold mixtures containing about 75% of the β-salt. The solution is placed in a 37° incubator for 24 hours and filtered. A small amount of distilled water is added to the filtrate to replace that lost by evaporation. The substrate solution is stable at icebox temperature for months. If it becomes turbid, it should be discarded.

Sections are incubated in the substrate solution at 37° for 15 minutes to 24 hours. Sites of activity become a chalky white from the deposition of lead phosphate. Following incubation, the slides are rinsed in distilled water and then briefly in 1–2% acetic acid, and finally in distilled water. The slides are immersed in a dilute solution of ammonium sulfide (a few drops to a Coplin jar of distilled water) for about 2 minutes. The slides are washed in tap water and counterstained if desired. They may then be mounted in a water-soluble medium or dehydrated in alcohols, cleared in gasoline or tetrachloroethylene, and mounted in a synthetic resin dissolved in the aforementioned solvents. Toluene or xylene should not be used since they cause some fading of the stain. Sites of activity are indicated by a dark brown-black precipitation of lead sulfide.

### 5. *ATPase*

Unfixed frozen, formalin fixed frozen, and frozen-dried paraffin embedded tissues may be employed. The various ATPase procedures are modifications of the original Gomori method (1950c; 1952, p. 193).

*a. Gomori procedure (1950c; 1952, p. 194)*. This is a metal-salt method involving the use of lead nitrate. The substrate is prepared as follows: In 500.0 ml. of a 0.5 *M*, acetate buffer pH 5.0, dissolve 0.6 g., $Pb(NO_3)_2$ (ca. 0.003 *M*) and add 50.0 ml. of a 3% (ca. 0.1 *M*) solution of ATP. An acetate buffer-lead nitrate stock solution consisting of 0.05 *M* acetate buffer pH 5.0–6.0 containing 1.2 g. of $Pb(NO_3)_2$ may be prepared in advance and stored under refrigeration. Sections are incubated at 37° for 15 minutes to several hours depending upon the initial activity of the tissue and the type of fixation employed. Following incubation the sections are rinsed in distilled water and treated in the same fashion as with the lead-glycerophosphate procedure (Appendix 5B, Section 4).

*b. Modification of Naidoo and Pratt (1951, 1952) and Pratt (1954).* Their incubation medium contains 0.05 *M* succinate buffer pH 6.5, 0.001 *M* $Pb(NO_3)_2$, and 0.0005 *M* ATP. With frozen-dried specimens 10 to 80 minute incubation periods were employed at 37°.

*c. Modification of Padykula and Herman (1955a, b).* Their substrate medium contains 20.0 ml. of 0.1 *M* sodium barbital, 10.0 ml. of 0.18 *M* $CaCl_2$, and 30.0 ml. of distilled water. One hundred fifty-two mg. of ATP (0.0005 *M* final concentration) is added to the aforementioned solution in dry form. If necessary, the pH is adjusted to 9.4 with 0.1 *M* NaOH. Distilled water is then added to make a total volume of 100.0 ml. and the solution is then filtered if cloudy. The sections are incubated at 37° for 5 minutes to 3 hours. They are then washed in three changes of 1% $CaCl_2$ for 10 minutes. The slides are transferred to 2% $CoCl_2$ for 3 minutes and washed well in several changes of distilled water. They are then immersed in 1% yellow ammonium sulfide for 2 minutes, washed in tap water, dehydrated in graded changes of ethanol, cleared in xylene, and mounted in Permount. Glycerol-gelatin, as well as a number of water-soluble and special synthetic mounting media, may also be employed.

The criterion of specificity of the staining reaction is ascertained by noting a reversible inhibition of the tissue enzyme toward ATP. Sections are incubated in the ATP substrate solution containing 0.0025 *M* PCMB (*p*-chloromercuribenzoate) for 15 minutes. If longer incubations are required, the sections are rinsed in distilled water and transferred to a fresh substrate solution which is free of inhibitor, since prolonged contact with PCMB will result in an irreversible inactivation of ATPase. The sections are washed in two changes of 2% $CaCl_2$ for 10 minutes, and then incubated for 30 minutes in an ATP substrate solution containing $4.9 \times 10^{-3}$ *M* BAL (2,3-dimercapto-1-propanol).

*d. Modification of Wachstein and Meisel (1957a).* The incubating mixture consists of 20.0 ml. of a 125.0 mg. per 100.0 ml. solution of ATP, 20.0 ml. of 0.2 *M* tris-maleate buffer pH 7.2, 3.0 ml. of 2% $Pb(NO_3)_2$, 5.0 ml. of 0.1 *M* $MgSO_4$, and 2.0 ml. of distilled water. Incubation times varied from 5 to 180 minutes at 37°. In general, human specimens required longer incubations than did animal tissues. Formalin fixation likewise necessitated the use of longer incubation periods. Following incubation the sections are washed in tap water and immersed in dilute ammonium sulfide solution for several minutes. Unfixed frozen sections are post-fixed in neutral formalin for several hours.

6. *Glucose-6-phosphatase* (*Chiquione, 1953*)

Unfixed sections cut on a freezing microtome or in a cryostat and mounted on glass slides are employed. The incubating solution contains one part of a 1% solution of sodium or potassium glucose-6-phosphate. The barium salt may be converted to the potassium salt as follows: 250.0 mg. of the barium salt are dissolved in 10.0 ml. of distilled water by the addition of a drop or two of 2 *N* HCl. The barium ions are then precipitated from solution by the addition of approximately 120.0 mg. of potassium sulfate. To insure complete precipitation the solution is allowed to stand for 2 hours at room temperature with frequent stirring. The solution is then centrifuged and the supernatant fluid tested for the presence of barium by adding a pinch of potassium sulfate. If no precipitate occurs the supernatant fluid is diluted to 30.0 ml. with distilled water and the pH adjusted to 6.7 with *N* KOH.

The incubating mixture consists of one part of the aforementioned substrate solution to which is added two parts of 0.006 *M* $Pb(NO_3)_2$ (0.2%). The addition of the lead salts to the substrate solution results in a slight precipitation which may be removed by filtering the mixture. Buffer is not employed since the glucose-6-phosphate serves as a buffer.

As with other frozen-section techniques, the slides are removed from the cryostat or deep freeze, allowed to thaw at room temperature for a minute or two, and then placed in the incubating solution in an oven adjusted to 32°. Incubation periods range from 5–15 minutes. Longer incubation times were found to result in the formation of heavy precipitates. Following incubation the lead phosphate at the sites of enzyme activity is converted to lead sulfide in the same manner as with the original Gomori acid phosphatase procedure (Appendix 5B, Section 4). The slides are then dehydrated, cleared in xylene, and mounted in a synthetic mounting medium. Alternatively, the slides may be mounted in glycerol-gelatin, water-soluble synthetic media, or alcohol-soluble plastics (Appendix 4, Section 2).

7. *Phosphamidase* (*Gomori, 1949a; 1952, p. 195*)

Dissolve 1.0 g. of *p*-chloroanilidophosphonic acid in an ice-cold mixture of 1.0 ml. of concentrated ammonia and 40.0 ml. of distilled water with vigorous shaking. The substrate usually does not dissolve completely. Filter and add two drops of 0.5% of phenolphthalein solution. The solution is then adjusted with *N* acetic acid until a barely perceptible pink shade is discernible (ca. 3.0 ml.). The solution is brought up to 50.0 ml. with distilled water and is then refrigerated until used. At the present time, the purified ammonium salt of *p*-chloroanilidophosphonic acid (My-21) is commercially available (Farbenfabriken Bayer AG, Leverkusen, Germany). This product is directly soluble in water and its use obviates the necessity of the aforementioned solubility and titration procedures.

One vial (0.22 g.) of My-21 is dissolved in 10.0 ml. of distilled water. One ml. of a 5% solution of lead nitrate is added to 50.0 ml. of a 0.05 *M* maleate buffer of pH 5.5–5.6 and the mixture is shaken until the initial white precipitate dissolves. A few drops of 0.2 *M* solution of $MnCl_2$ and 1.5–2.0 ml. of the phosphonate (My) stock solution are added. If the mixture becomes turbid, it is placed in a paraffin oven and kept for 10–30 minutes or until the turbidity settles, and then filtered into a Coplin jar. Sections are incubated for 12–24 hours at 37°. The Coplin jar must be supported in an inclined position with the tissues facing downward, in order to avoid indiscriminate precipitation of lead phosphate over the section. A wooden block cut at 45° and containing grooves or clamps to hold the base of the Coplin jar may be employed. Following incubation, the precipitate is wiped from the back surface of the slide and the slides are thoroughly rinsed in distilled water. The superficial precipitate is removed by agitating the slides in a 0.1 *M* citrate buffer pH 5.0. As soon as the glass appears completely clear around the tissues, the slides are rinsed in tap water. Gomori points out that this is the most delicate step in the entire procedure. Insufficient differentiation will leave a coarse, black precipitate in the finished section, while overdifferentiation may remove part or all of the lead phosphate deposited by enzymatic action. The slides are then treated with ammonium sulfite, etc., as with the Gomori method for acid phosphatase.

8. *Phosphamidase, Modification of Gomori Technique* (*Meyer and Weinmann, 1955*)

Dissolve 2.0 mg. of substrate in 15.0 ml. of *N* NaOH and add distilled water to make the total volume 100.0 ml. Dissolve 534.0 mg. maleic acid in 5.0 ml. *N* NaOH and add distilled water to make a total volume of 100.0 ml. and then add 175.0 mg. NaCl and 94.0 mg. of $Pb(NO_3)_2$. The mixture will become turbid and is gently heated until it clears. To this solution is added 4.5 ml. of the substrate solution. The final solution is then warmed to approximately 42° and filtered. The slides are incubated in a horizontal position with the sections facing downward. Weinmann and Meyer recommend omitting the differentiation step with the citrate for incubations less than 3 hours. They also suggest the advisability of employing inactivated control slides which have been treated with 10% nitric acid for 90 minutes.

9. *Thiamine Pyrophosphatase* (*Naidoo and Pratt, 1955; Eränkö and Hasan, 1954; Allen and Slater, 1961*)

Various types of unfixed frozen, formalin fixed frozen, and acetone fixed and frozen-dried paraffin embedded tissues may be employed.

Naidoo and Pratt (1953) employed a substrate solution containing 15.0

ml. 0.1 *M* sodium maleate, 0.75 ml. 0.04 *M* lead acetate, 3.0 ml. 0.005 TDP, and 6.0 ml. 0.1 *M* $MgCl_2$ in distilled water to make a total volume of 30.0 ml. The pH of this solution is 6.9. Following incubation the slides are briefly rinsed in distilled water and treated with ammonium sulfide as in the Gomori acid phosphatase technique.

An essentially similar technique was reported by Eränkö and Hasan (1954). Their substrate concentration with reference to TDP was .01 *M*. A pH of 5.0 was obtained by the use of acetate buffer.

A modified substrate medium was reported by Naidoo and Pratt (1955). It contains 0.025 *M* veronal buffer pH 8.7, 0.02 *M* $MgCl_2$, 0.1 *M* $CaCl_2$, and 0.008 *M* TDP. The substrate is added immediately before use and the final pH is ca. 8.4.

Allen and Slater (1961) employed the following technique. Their incubating solution contained the following final concentrations of substrate, activator, and buffer: 0.033 *M* sodium veronal pH 9.5, 0.015 *M* $CaCl_2$, and 0.005 *M* thiamine pyrophosphate chloride or thiamine monophosphate chloride. A pH meter is employed to adjust the pH of the substrate solution after addition of all components but before final dilution. The substrate solution is filtered prior to use to remove any precipitate due to the presence of free phosphate or other impurities.

Cysteine hydrochloride (final concentration 0.004 *M*) is added to the substrate solution prior to adjustment of the pH in order to inhibit nonspecific alkaline phosphatase. A comparison slide should be incubated in the same substrate solution from which the cysteine is omitted.

10. *5′-Nucleotidase*

Satisfactory results may be obtained with various types of fixed and unfixed tissues including frozen-dried, cold acetone fixed double embedded specimens, cold formalin fixed tissues, and unfixed frozen sections.

*a. Ca method (Gomori, 1952, p. 187)*. Dissolve approximately 20.0 mg. of muscle adenylic acid (adenosine-5′-phosphate, AMP) in 20.0 ml. of 0.1–0.2 *M* tris(hydroxymethyl)aminomethane buffer pH 8.3. Add 20.0 ml. of 2% $CaCl_2$ solution and a few drops of 10% $MgCl_2$. The sections are incubated for 2 to 5 hours at 37°. The calcium phosphate precipitate is converted into cobalt sulfide as in the alkaline phosphatase procedure. A control slide should always be incubated in a solution in which the AMP is substituted by glycerophosphate.

*b. Pb method (Naidoo and Pratt, 1954)*. The incubating solution contains:

| | |
|---|---|
| 0.1 *M* Sodium succinate buffer, pH 6.5 | 15.0 ml. |
| Distilled water | 10.5 ml. |

| | |
|---|---|
| 0.1 *M* AMP | 3.9 ml. |
| 0.04 *M* Lead acetate | 1.5 ml. |

Following incubation, the lead phosphate deposits are converted to lead sulfide.

Wachstein and Meisel (1957b) used a modified technique and employed tris-maleate buffer pH 7.2 with lead nitrate in the presence of magnesium sulfate as an activator.

11. *Polyphosphatases* (*Berg, 1960, and unpublished data*)

Fixation procedures for the demonstration of alkaline polyphosphatase are similar to those for the standard alkaline phosphatase procedures. According to Berg, fixation techniques for the preservation of trimetaphosphatase are relatively critical, in that much of the enzyme is lost during preparation and mounting of sections. The enzyme is inactivated by ethanol as well as by paraffin embedding temperatures over 58°. For the demonstration of acid polyphosphatase small blocks are stained following cold formalin fixation. The blocks are then dehydrated, embedded, and sectioned. Berg (unpublished data) indicates that freeze substitution techniques may also be employed.

All of the substrate mixtures are unstable and should be freshly prepared. They all exhibit a slight opalescence, but if a heavy precipitate is rapidly formed or if the solution is clear the concentration of the ingredients is not optimal and a new substrate solution should be prepared. Because of the instability of the substrates an incubation schedule must be employed on controls which have been inactivated by boiling the slides in formalin for 30 minutes. These controls when incubated up to one hour should be unstained. With longer incubation periods the controls may exhibit certain types of artefactual staining (e.g. nucleoli). With both techniques the resulting lead complex is developed with ammonium sulfide as in the Gomori metal-salt phosphatase technique.

*a. Alkaline polyphosphatase substrate.* Stock solution: $6 \times 10^{-3}$ *M* lead acetate adjusted with acetic acid to bring the pH below 6.0. Freshly prepared substrate solution: $2 \times 10^{-3}$ *M* sodium polyphosphate (for anhydrous sodium tripolyphosphate, 96.0 mg./100.0 ml. of water). To 100.0 ml. of the substrate add alternately small amounts of the lead acetate stock solution and barbital buffer keeping the pH between 8.5–9.0. The end point is reached when the mixture remains slightly opalescent. The pH is adjusted to 8.7 with NaOH or $HNO_3$. If the opalescence disappears a very small amount of the lead acetate stock solution is added to restore it.

*b. Acid polyphosphatase substrate.* Stock solution: sodium acetate (0.05 *M*) and acetic acid (0.05 *M*) to adjust the pH to 4.0. Fresh substrate solution: A fresh solution of $1 \times 10^{-3}$ *M* polyphosphate in 100.0 ml. of the stock buffer is prepared. The lead acetate is gradually added till the solution remains opalescent. Stock buffer is then added to make a total volume of 200.0 ml. and slides are incubated in the freshly prepared solution.

12. *Inorganic Pyrophosphatases* (*Kurata and Maeda, 1956*)

The substrate solution is prepared by first dissolving 1.088 g. of sodium pyrophosphate ($Na_4P_2O_7 \cdot 12H_2O$) in 20.0 ml. of distilled water. A solution of ferric chloride ($FeCl_3 \cdot 6H_2O$; 0.61 g. in 10.0 ml. of distilled water) is then added and the mixture stirred. A white precipitate forms. A concentrated solution (ca. 10%) of $Na_2CO_3$ is added cautiously drop by drop, until the precipitate is dissolved. An excess of alkali must be avoided. The pH of the mixture is then adjusted to 7.2–7.3 with a dilute solution of *N* HCl. The volume of the solution is brought to 100.0 ml. with distilled water and made hypertonic by addition of 0.9 g. of NaCl. The substrate solution should be freshly prepared, and just prior to use approximately 1.0 ml. of 10% magnesium nitrate or magnesium chloride is added drop by drop (a slight clouding of the solution occurs).

Thin slices of tissue (2–5 mm. thick) are fixed in cold, neutral 5% formalin containing 0.9% NaCl for approximately one hour. The specimens are then washed in 0.9% NaCl solution for 10 to 20 minutes and thick frozen sections (20–35 μ) are cut. The sections may be transferred directly from the knife into the substrate solution or else may be temporarily transferred to an isotonic sodium chloride solution. The sections are incubated at 37° for 3 hours and are washed twice (10–15 minutes each) in 0.9% sodium chloride, mounted on glass slides, and air-dried. The slides are incubated for a few minutes in 0.9% sodium chloride solution containing a few drops of acidic ammonium sulfide solution (Maccalum, 1932, p. 343). Pyrophosphatase activity is revealed as a green stain. The slides are then rapidly dehydrated in ethanol, cleared in xylene, and mounted in a synthetic medium. Alternatively, glycerol-gelatin or water-soluble mounting media may be employed. Unfortunately, the green dye which coincides with sites of enzyme activity is not stable.

13. *Alkaline Phosphatase, α-Naphthyl Phosphate Method* (*Gomori, 1951; 1952, p. 185*)

A variety of fixation procedures may be employed as indicated for the analogous metal-salt procedure:

The incubating solution is prepared by dissolving approximately 10.0 mg. of Na α-naphthyl phosphate in several ml. of distilled water. Two to four ml. of a 5% solution of borax ($Na_2B_4O_7 \cdot 10H_2O$) and approximately 40.0 ml. of cool distilled water and a few drops of a 10% solution of $MgCl_2$

or $MgSO_4$ are added. (It is probably advisable to use a tris or ammediol buffer since borate is a known phosphatase inhibitor.) Twenty to fifty mg. of diazonium salt (Red RC, Bordeaux GP, Blue RR, or Red-Violet LB) are added to the substrate solution which is then shaken and filtered through porous filtering paper into a Coplin jar.

Slides are incubated for 5 to 30 minutes or longer if necessary. With the azo-dye technique the slides may be removed from the incubating mixture and inspected under the microscope. Gomori advises mechanical stirring of the α-naphthyl phosphate substrate solution during incubation as a means of avoiding diffusion artefacts with highly active tissues. Following incubation the slides are washed in tap water, counterstained if desired, and mounted in a water-soluble medium or dehydrated and mounted in a synthetic medium if the dye is insoluble in organic solvents. Sites of phosphatase activity appear as red, red-brown, or dark brown shades, depending upon the type of diazonium salt used.

14. *Alkaline Phosphatase, Naphthol AS Phosphate Method* (*Burstone, 1958a, c, 1961*)

As previously indicated, a variety of fixation procedures may be employed. Although unfixed cryostat sections mounted on glass slides may be used, superior localizations are obtained if these sections are briefly fixed in acetone or diacetone alcohol for several minutes prior to incubation (Burstone, 1961). Cold acetone fixed paraffin embedded as well as frozen-dried paraffin embedded sections give particularly useful localizations (Appendix 1, Sections 4, 5, and 11).

A large number of naphthol AS substrates including AS-MX, AS-AN, AS-CL, AS-KB, AS-TR, AS-BI, AS-E phosphates may be employed. Approximately 5.0 mg. of substrate are dissolved in 0.25 ml. of DMF (*N,N*-dimethylformamide) or 0.25 ml. dimethyl sulfoxide. (If the Na salt of the acid phosphate is used the solvent may be omitted; on the other hand, the presence of small amounts of solvent appears to facilitate dispersion of the substrate and enhances coupling.) Twenty-five ml. of distilled water are added followed by 25.0 ml. of buffer (0.2 *M* tris, pH 8.3–9.3). Thirty mg. of Red-Violet LB salt are added and the mixture shaken and filtered into a Coplin jar. Other well known diazonium salts may also be employed. An activator is usually not necessary for the alkaline phosphatase reaction. Sections are incubated for 5 minutes to several hours, washed in tap water and mounted in glycerol-gelatin or dehydrated and mounted in a synthetic medium. Dehydration in 80% ethanol followed by mounting in an 80% ethanolic solution of polyvinyl acetate has proved to be very useful in obtaining thin mounts of improved optical quality.

A stable stock solution containing both buffer and substrate may also be employed. Fifty mg. of substrate are dissolved in 20 ml. (*N*,*N*-dimethylformamide) DMF. Twenty ml. water are added and enough 1 *M* sodium carbonate (usually 0.5–1.0 ml.) to adjust the pH to 8. Laboratory pH test paper is adequate to determine this point. Six hundred ml. distilled water are added and the volume is then brought up to 1 liter with 0.2 *M* tris buffer, pH 8.3–8.6. Upon addition of buffer the solution will become opalescent. This stock solution is stable at room temperature for several months and is available for immediate use. For histochemical application 50.0 ml. of stock solution plus diazonium salt are shaken and filtered into a Coplin jar.

15. *Demonstration of Alkaline Phosphatase in Leukocytes, Azo-Dye Method* (*Burstone, 1958a, 1961*)

The aforementioned simultaneous coupling naphthol AS phosphate technique is employed. Freshly prepared blood smears, on chemically clean glass slides or cover slips, may be employed. The smears may be briefly fixed in cold neutral formalin. Alternatively, smears may be prepared on Mylar film (Burstone and Fleming, 1959), and pieces of the film incubated without prior fixation. Mylar film is insoluble in most solvents and the film may be subsequently run through dehydrating solvents provided the azo-dye is sufficiently insoluble.

16. *Acid Phosphatase, Naphthol AS Phosphate Method* (*Burstone 1958b, c, 1961*)

Cold acetone fixed and frozen-dried paraffin embedded tissues are recommended. Cryostat sections mounted on glass slides and briefly fixed in acetone prior to incubation are also capable of giving reproducible reactions.

In general, the more complex halogen or nitro substituted naphthol AS phosphates are recommended for use in the acid phosphatase range. These include naphthols AS-BI, AS-CL, AS-AN, and AS-TR phosphates. Approximately 5.0 mg. of substrate are dissolved in 0.25 ml. of DMF or dimethyl sulfoxide followed by 25.0 ml. of distilled water. Twenty-five ml. of 0.2 *M* acetate buffer pH 5.2 are then added followed by 30.0 mg. Red-Violet LB salt and the mixture shaken and filtered into a Coplin jar. As previously indicated for the alkaline phosphatase AS technique a large variety of diazonium salts may be employed. Two drops of a 10% solution of $MnCl_2$ are added to the Coplin jar and the mixture stirred. Sections are incubated for periods of 30 minutes to 8 hours, depending upon the initial activity of the section and the type fixation employed. The slides are then washed in tap water, counterstained if desired, and mounted in a water-soluble medium. The sections may also be dehydrated and mounted in 80% ethanolic solution of polyvinyl acetate. Sites of acid phosphatase activity are revealed by homogeneous precipitates of azo-dyes whose shades range from red to red-brown and blue-black depending upon the type of diazonium salt used and the fixation employed.

17. *Simultaneous Demonstration of Alkaline and Acid Phosphatases in Bone* (*Burstone, 1960a*)

Cold acetone fixed or frozen-dried paraffin embedded sections are incubated in a simultaneous coupling acid phosphatase substrate solution containing naphthol AS-GR phosphate (see p. 196) and Blue BBN salt. The substrate is initially dissolved in 0.5 ml. DMF or dimethyl sulfoxide. In other respects the procedure is identical with the previously discussed AS phosphate method. Following an incubation period of 30–60 minutes at room temperature the slides are briefly rinsed in distilled water and placed in an alkaline phosphatase substrate solution containing naphthol AS-MX phosphate and Red-Violet LB salt. The incubation times in the alkaline medium are usually considerably less and range from 10–30 minutes.

Sites of acid phosphatase (osteoclasts) are stained green while those of alkaline phosphatase are stained red. The slides may be mounted in glycerol-gelatin or dehydrated and mounted in 25% polyvinyl acetate in 80% ethanol.

18. *Acid Phosphatase, Benzoyl Naphthol Post-Coupling Method* (*Rutenburg and Seligman, 1955*)

The authors specifically recommend the use of calcium-formol fixed frozen sections. Following overnight fixation thin blocks of tissue are washed in two changes of 0.85% NaCl solution for 30 minutes. The frozen sections (10–20 $\mu$ thick) are transferred directly from the knife to approximately 10.0 ml. of substrate solution in a beaker and incubated at room temperature. The substrate solution is prepared by dissolving sodium 6-benzoyl-2-naphthyl phosphate (25.0 mg.) in 80.0 ml. of distilled water and adding 20.0 ml. of acetate buffer (0.5 *M*) pH 5.0. The solution is then made hypertonic by the addition of 2.0 g. of NaCl to each 100.0 ml. of solution. The formalin fixed sections are incubated at 25° for 1–2 hours. Following incubation the sections are rinsed in several changes of cold water and transferred to a cold (4°) freshly prepared solution of tetrazotized *o*-dianisidine (1.0 mg./ml.) in water made alkaline with sodium bicarbonate. The sections are gently agitated in the diazo solution for several minutes during which time there occurs rapid formation of a deep blue azo-dye. They are then washed several times in cold 0.85% NaCl solution and water and mounted in glycerol-gelatin.

The dark blue or purplish azo-dye formed is lipid soluble to some degree and is not permanent. After some hours or days the localization of the pigment will become diffuse and the dye will also exhibit a tendency toward crystallization.

19. *Alkaline and Acid Phosphatases, Special Naphthol AS Post-Coupling Method* (*Burstone, 1960c, 1962*)

This technique is designed primarily for use with cold acetone fixed or frozen-dried paraffin embedded tissues. The substrates employed are derivatives of 3-hydroxy-2-naphthoic acid and are shown in Chart 3. Among the most suitable substrates are phosphates of 3-hydroxy-2-naphthoic acid-2-aminoanthraquinonylamide and 3-hydroxy-2-naphthoic acid-aminoazobenzanilide. Approximately 5.0 mg. of substrate (excess) is placed in a 50.0 ml. Erlenmeyer flask and 0.25 ml. DMF added. The flask is gently agitated in order to dissolve the substrate. Twenty-five ml. of distilled water are then added followed by 25.0 ml. 0.2 *M* tris buffer pH 8.3–9.2. In general, pH 8.7 tris buffer was employed. Two drops of 10% $MgCl_2$ are added and the solution is shaken several times and filtered through rapid filtering paper into a Coplin jar. (In the case of acid phosphatase the tris buffer is substituted by an equivalent of 0.2 *M* acetate buffer pH 5.2, and two drops of $MnCl_2$ are employed instead of $MgCl_2$.) Incubations may be carried out at room or elevated temperature. The Coplin jar may be placed in a 60° paraffin oven for approximately 15 minutes in order to warm the solution rapidly. The solution is then maintained at room temperature for the remainder of the incubation period which ranges from 30 minutes to several hours or longer. The slides are then washed in running tap water for several minutes and transferred to a suitable diazonium salt solution.

The diazo coupling solution consists of diazonium salt (ca. 50.0 mg.), 20.0 ml. 0.2 *M* sodium acetate, and 30.0 ml. distilled water. This mixture is shaken and filtered into a Coplin jar. The slides are placed in the diazo solution for 10–15 minutes at room temperature and are then washed in running water and mounted in glycerol-gelatin or 25% polyvinyl acetate in 80% ethanol. Among the diazonium salts which may be used are Garnet GBC, Red RL, Red B, Red Violet LB, and Blue BBN. In addition, special complex diazotates including diazo oxides (Burstone and Weisburger, 1961) may be employed (Appendix 3, Section 5). With the latter, the azo-dyes may be complexed with metals including copper, cobalt, and nickel.

20. *Alkaline and Acid Phosphatases, Naphtholic Fluorescence Method* (*Burstone, 1960c, 1962*)

This technique is specifically intended for use with cold acetone fixed or frozen-dried paraffin embedded tissues.

The substrate employed is 5,6,7,8-β-tetralol carboxylic acid-β-naphthylamide phosphate, and the substrate solution is prepared in the same manner as the analogous post-coupling compounds described in Appendix 5B, Section 19. Following incubation the slides are washed in running tap water for several minutes, followed by two changes of 30% ethanol, and transferred to distilled water. The slides are mounted in glycerol or glycerol-gelatin and examined with a fluorescence microscope.

21. *DNase and RNase* (*Daoust and Amano, 1960, 1961; Daoust, 1961, p. 159; Daoust, personal communication*)

*a. Preparation of gelatin-DNA and RNA films.* A 5% solution of gelatin (commercial sheet form) prepared by heating in a water bath is mixed in a test tube with an equal volume of a 7% solution of DNA (calf thymus sodium deoxyribonucleate) or a 10% solution of RNA (yeast sodium ribonucleate). Two or three drops of the gelatin-nucleic acid mixture are placed on a glass slide and rapidly spread with the tip of the pipette over an area measuring approximately 2.5 × 4.0 cm., and the excess allowed to drain by holding the slide vertically against a piece of filter paper. The solution covering the lower end of the slide is wiped with the filter paper and the slide is allowed to dry at room temperature while in a vertical position.

*b. Fixation of substrate films.* The gelatin DNA and RNA films are fixed overnight in neutral 20% formaldehyde (50% formalin) at room temperature (ca. 20–23°). The fixed films are washed in three changes of distilled water, 5 minutes each, and allowed to dry at room temperature.

*c. Preparation of tissues.* Freshly obtained tissues should be used. For best results the animal is exsanguinated from the abdominal aorta, while under ether anesthesia. The organs are excised with scissors, washed in distilled water, blotted on filter paper, and placed in a Petri dish kept on ice. Pieces of organs are removed and held with forceps on a layer of ice covering the stage of a microtome in a refrigerated cabinet (cryostat) (−20 to −15°). A small amount of water is added to affix the tissue specimens firmly to the stage.

*d. Mounting of tissue sections on gelatin-glycerol.* When sections are ready to be cut, a mixture of gelatin and glycerol (final concentrations 7 and 40%, respectively) is liquefied on a water bath. Ten to twelve drops of the solution (ca. 0.5 ml.) are placed on a glass slide and spread with the tip of a pipette over an area approximately 2.5 × 4.0 cm. The slide is left on a level surface at room temperature until gelation occurs in 2 or 3 minutes. Sections of unfixed frozen tissues, 15 $\mu$ thick, are cut in the refrigerated cabinet or cryostat. The sections are picked up from the knife with a fine camel's-hair brush and deposited on the slide covered with gelatin-glycerol. The latter is introduced into the cryostat chamber only for the time necessary to transfer the sections, in order to prevent freezing of the semi-solid support. The slide with tissue sections is placed on a level surface at 40° until the gelatin-glycerol liquefies and the sections spread on its surface. The slide is then transferred onto a level surface at room temperature and the gelatin-glycerol support is allowed to form a firm gel (5 minutes).

*e. Exposure of gelatin DNA and RNA films to tissue sections.* One drop of the gelatin-glycerol mixture is deposited on each end of the slide covered with a film of nucleic acid, and the slide carrying the tissue section is simply pressed against the film of substrate. During exposure the slides are allowed to remain at room temperature with the section above the substrate film. After contact for a given period of time (5–40 minutes), the two slides are separated by insertion of a blade between the slides at one end. The film of substrate is stained directly with toluidine blue. The tissue sections are fixed overnight in 10% neutral formalin and washed for 10 minutes in two successive changes of distilled water. The gelatin-glycerol support is allowed to dry completely, at room temperature, before staining the tissue section.

*f. Staining of substrate films and tissue sections.* Following exposure to tissue sections, the film of nucleic acid is stained with 0.2% toluidine blue for 5 minutes. The film is washed in two changes of distilled water, 5 minutes each, allowed to dry at room temperature, and mounted in a synthetic mounting medium. The slide carrying the tissue section is immersed into a 1:3 solution of acetic acid-ethanol for 5 minutes before staining with toluidine blue. The latter procedure prevents staining of the supporting medium. The slide is transferred from the acetic acid-alcohol to a 0.1% solution of toluidine blue for 1–2 minutes, washed in distilled water to remove excess dye, dried at room temperature, and mounted with a synthetic mounting medium.

## Commercial Sources of Reagents

1. Aldrich Chemical Company, Milwaukee, Wisconsin
2. Carbic-Hoechst Corp., Mountainside, New Jersey
3. Dajac Laboratories, The Borden Chemical Company, Philadelphia, Pennsylvania
4. Eastman Kodak, Organic Chemicals Division, Rochester, New York
5. General Aniline and Film Corp., New York, New York
6. Mann Research Laboratories, New York, New York
7. Matheson, Coleman and Bell, East Rutherford, New Jersey
8. Nutritional Biochemicals Corp., Cleveland, Ohio
9. Pfister Chemical Works, Ridgefield, New Jersey
10. Sigma Chemical Company, St. Louis, Missouri
11. Verona Dyestuffs, Union, New Jersey

## References

Abul-Fadl, M. A. M., and King, E. J. (1949) *Biochem. J.* **45**, 51–60.
Aebi, H. (1952) *Schweiz. med. Wochschr.* **82**, 135–140.

Ahmed, Z. J., and Reis, J. L. (1958) *Biochem. J.* **69,** 386–387.

Allen, J. M., and Slater, J. J. (1961) *J. Histochem. and Cytochem.* **9,** 418–423.

Allen, R. J., and Bourne, G. H. (1943) *J. Exptl. Biol.* **20,** 61–64.

Allfrey, V., and Mirsky, A. E. (1952) *J. Gen. Physiol.* **36,** 227–241.

Altschule, M. D., Parkhurst, B. H., and Zager, G. R. (1951) *Am. J. Clin. Pathol.* **21,** 480–483.

Amano, H., and Daoust, R. (1961) *J. Histochem. and Cytochem.* **9,** 161–164.

Ames, B. N., and Horecker, B. L. (1956) *J. Biol. Chem.* **220,** 113–128.

Anagnostopoulos, C. (1950) *Biochim. et Biophys. Acta* **4,** 584–595.

Anagnostopoulos, C., and Matsudaira, H. (1958) *In* "Proceedings of the International Symposium on Enzyme Chemistry, Tokyo and Kyoto, 1957" (K. Ichihara, ed.), pp. 166–172. Academic Press, New York.

Anfinsen, C. B., and White, F. H. (1961) *In* "The Enzymes" (P. D. Boyer, H. Lardy, and K. Myrbäck, eds.), 2nd ed., Vol. 5, pp. 95–122. Academic Press, New York.

Anschütz, L. (1927) *Ann.* **454,** 71–120.

Anschütz, L., Weber, E., and Runkel, K. (1906) *Ann.* **346,** 360–381.

Antonini, F. M., and Weber, G. (1951) *Arch. Vecchi. Anat. Patol. Med. Clin.* **16,** 985–1012.

Appelmans, F., and de Duve, C. (1955) *Biochem. J.* **54,** 426–433.

Appelmans, F., Wattiaux, R., and de Duve, C. (1955) *Biochem. J.* **59,** 438–445.

Araki, T. (1903) *Z. physiol. Chem. Hoppe-Seyler's* **38,** 84–97.

Aronson, J., Hempelmann, L. H., and Okada, S. (1958) *J. Histochem. and Cytochem.* **6,** 255–259.

Arvy, L. (1957–1958) "Les Techniques Actuelles D'Histoenzymologie." Collection of reprints from *Biol. med.* (*Paris*) **46–47,** (1957–1958).

Asakawa, K. (1929) *J. Biochem.* (*Tokyo*) **11,** 143–172.

Ashrafi, S. H., and Fisk, F. W. (1961) *Ohio J. Sci.* **61,** 7–13.

Atherton, F. R., Openshaw, H. T., and Todd, A. R. (1945) *J. Chem. Soc.*, pp. 382–385.

Audrieth, L. F., and Toy, A. D. F. (1941) *J. Am. Chem. Soc.* **63,** 2117–2119.

Audrieth, L. F., and Toy, A. D. F. (1942) *J. Am. Chem. Soc.* **64,** 1337–1339.

Avison, A. W. D. (1955) *J. Chem. Soc.* pp. 732–738.

Axelrod, B. (1956) *Advances in Enzymol.* **17,** 159–188.

Baddiley, J., and Todd, A. R. (1947) *J. Chem. Soc.* pp. 648–651.

Baddiley, J., Michelson, A. M., and Todd, A. R. (1949) *J. Chem. Soc.* pp. 582–586.

Baer, E. (1952) *Biochem. Preparations* **2,** 31–38.

Baer, E., and Fischer, H. O. L. (1943a) *J. Biol. Chem.* **150,** 213–222.

Baer, E., and Fischer, H. O. L. (1943b) *J. Biol. Chem.* **150,** 223–229.

Ballou, C. E., and Fischer, H. O. L. (1954) *J. Am. Chem. Soc.* **76,** 3188–3193.

Ballou, C. E., Fischer, H. O. L., and MacDonald, D. L. (1955) *J. Am. Chem. Soc.* **77,** 5967–5970.

Bamann, E., and Diederichs, K. (1934) *Ber.* **67,** 2019–2021.

Bamann, E., and Gall, H. (1937) *Biochem. Z.* **293,** 1–15.

Bamann, E., and Riedel, E. (1934) *Z. physiol. Chem. Hoppe-Seyler's* **229,** 125–150.

Bamann, E., and Riehl, J. (1955) *Biochem. Z.* **327,** 136–148.

Barka, T. (1960) *Nature* **187,** 248–249.

Barka, T. (1962) *J. Histochem. and Cytochem.* **10,** 231–232.

Barka, T., Szalay, S., Pósolaky, Z., and Kertesz, L. (1952) *Acta Anat.* **16,** 45–53.

Barrnett, R. J., Bressler, R., and Rutenburg, A. M. (1956) *Anat. Record* **124,** 255–256.

Barron, J. H., and Boshes, C. (1961) *J. Histochem. and Cytochem.* **9,** 455–457.
Beck, L. V. (1942) *Proc. Soc. Exptl. Biol. Med.* **49,** 435–439.
Bélanger, L. F. (1951) *Proc. Soc. Exptl. Biol. Med.* **77,** 266–269.
Belfanti, S., Contardi, A., and Ercoli, A. (1935) *Biochem. J.* **29,** 842–846.
Berg, G. G. (1955) *J. Histochem. and Cytochem.* **3,** 22–31.
Berg, G. G. (1956) *J. Histochem. and Cytochem.* **4,** 424.
Berg, G. G. (1959) *J. Histochem. and Cytochem.* **7,** 294–295.
Berg, G. G. (1960) *J. Histochem. and Cytochem.* **8,** 92–101.
Berg, G. G., and Shupack, B. (1960) *J. Cellular Comp. Physiol.* **56,** 165–172.
Berg, G. G., and Szekerczes, J. (1960) *J. Histochem. and Cytochem.* **8,** 322.
Berthet, J., Berthet, L., Appelmans, F., and de Duve, C. (1951) *Biochem. J.* **50,** 182–189.
Bessey, O. A., and Love, R. H. (1952) *J. Biol. Chem.* **196,** 175–178.
Bessey, O. A., Lowry, O. H., and Brock, M. J. (1946) *J. Biol. Chem.* **164,** 321–329.
Binkley, F., and Olsen, C. K. (1950) *J. Biol. Chem.* **186,** 725–730.
Binkley, F., Shank, R. E., and Hoagland, C. L. (1944) *J. Biol. Chem.* **156,** 253–256.
Bodansky, A. (1933) *J. Biol. Chem.* **101,** 93–104.
Boger, E., and Friedman, O. M. (1958) *J. Am. Chem. Soc.* **80,** 2583–2584.
Bonting, S. L., Simon, K. A., and Hawkins, N. M. (1961) *Arch. Biochem. Biophys.* **95,** 416–423.
Bonting, S. L., Caravaggio, L. L., and Hawkins, N. M. (1962) *Arch. Biochem. Biophys.* **98,** 413–419.
Bonting, S. L., Tsoodle, A. D., de Bruin, H., and Mayron, B. R. (1960) *J. Biochem. Biophys.* **91,** 130–137.
Bourne, G. H. (1943) *Quart. J. Exptl. Physiol.* **32,** 1–19.
Bourne, G. H. (1957) *Gerontologia* **1,** 50–58.
Bourne, G. H. (1958) *Exptl. Cell Research Suppl.* **5,** 101–117.
Bowers, R. V., Outhouse, E. L., and Forbes, J. C. (1940) *J. Biol. Chem.* **132,** 675–680.
Boyland, E., Kinder, C. H., and Manson, D. (1961) *Biochem. J.* **78,** 175–179.
Bradfield, J. G. R. (1950) *Biol. Revs.* **25,** 113–157.
Brandes, D., and Bourne, G. H. (1956) *J. Pathol. Bacteriol.* **71,** 33-36.
Braunstein, H., Freiman, D. G., and Gall, E. A. (1958) *Cancer* **11,** 829–837.
Brown, D. M., Magrath, D. I., and Todd, A. R. (1952) *J. Chem. Soc.* pp. 2708–2714.
Brunngraber, E. G., and Abood, L. G. (1960) *J. Biol. Chem.* **235,** 1847–1853.
Buch, I., and Buch, H. (1939) *Acta Med. Scand.* **101,** 211–236.
Burgos, M. H., Deane, H. W., and Karnovsky, M. L. (1955) *J. Histochem. and Cytochem.* **3,** 103–121.
Burstone, M. S. (1958a). *J. Natl. Cancer Inst.* **20,** 601–615.
Burstone, M. S. (1958b) *J. Natl. Cancer Inst.* **21,** 523–539.
Burstone, M. S. (1958c) *J. Histochem. and Cytochem.* **6,** 322–339.
Burstone, M. S. (1960a) *In* "Calcification in Biological Systems." (R. F. Sognnaes, ed.), pp. 217–243. Am. Assoc. Advance. Sci., Washington, D. C.
Burstone, M. S. (1960b) *Ann. N. Y. Acad. Sci.* **85,** 431–444.
Burstone, M. S. (1960c) *J. Natl. Cancer Inst.* **24,** 1199–1217.
Burstone, M. S. (1961) *J. Histochem. and Cytochem.* **9,** 146–153.
Burstone, M. S. (1962) "Proceedings of the 1st International Congress of Histochemistry and Cytochemistry." Pergamon Press, Oxford, England.
Burstone M. S., and Fleming, T. J. (1959) *J. Histochem. and Cytochem.* **7,** 203.
Burstone, M. S., and Keyes, P. H. (1957) *Am. J. Pathol.* **13,** 1229–1235.

BURSTONE, M. S., AND WEISBURGER, E. K. (1961) *J. Histochem. and Cytochem.* **9,** 301–303.
BURTON, J. F. (1954) *J. Histochem. and Cytochem.* **2,** 88–94.
BUTLER, J., BRAUNSTEIN, H., FREIMAN, D. G., AND GALL, E. A. (1959) *Arch. Pathol.* **68,** 243–251.
BUTTUGER, H. W., SCARLATO, G., MÜLLER, W., AND KEMALI, D. (1957) *Deut. Z. Nervenheilk.* **176,** 67–76.
BYRNE, W. L. (1961) *In* "The Enzymes" (P. D. Boyer, H. Lardy, and K. Myrbäck, eds.), 2nd ed., Vol. 5, pp. 73–78. Academic Press, New York.
CAFFREY, R. W., TREMBLAY, R., GABRIO, B. W., AND HUENNEKENS, F. M. (1956) *J. Biol. Chem.* **223,** 1–8.
CALVIN, B. (1961) *Chem. Eng. News* **39,** 96–104.
CAYLA, J. (1935) *Bull. soc. chim. biol.* **17,** 1707–1714.
CHAMBERS, R. W., AND KHORANA, H. G. (1956) *Chem. & Ind. (London)* 1022–1023.
CHAMBERS, R. W., MOFFATT, J. G., AND KHORANA, H. G. (1955) *J. Am. Chem. Soc.* **77,** 3416–3417.
CHANLEY, J. D., AND FEAGESON, E. (1955) *J. Am. Chem. Soc.* **77,** 4002–4007.
CHANLEY, J. D., AND FEAGESON, E. (1958) *J. Am. Chem. Soc.* **80,** 2686–2691.
CHANLEY, J. D., AND GINDLER, E. M. (1953) *J. Am. Chem. Soc.* **75,** 4035–4040.
CHERBULIEZ, E., AND RABINOWITZ, J. (1956) *Helv. Chim. Acta* **39,** 1461–1467.
CHERBULIEZ, E., AND WENIGER, H. (1946) *Helv. Chim. Acta* **29,** 2006–2017.
CHIQUOINE, A. D. (1953) *J. Histochem. and Cytochem.* **1,** 429–435.
CHIQUOINE, A. D. (1955) *J. Histochem. and Cytochem.* **3,** 471–478.
CLARK, V. M., KIRBY, G. W., AND TODD, A. (1957) *J. Chem. Soc.* pp., 1497–1501.
CLARKSON, T. W., AND KENCH, J. E. (1958) *Biochem. J.* **69,** 432–439.
COLOWICK, S. P., AND KAPLAN, N. O. (1957) "Methods in Enzymology," Vol. III. Academic Press, New York.
COONS, A. H., AND KAPLAN, M. H. (1950) *J. Exptl. Med.* **91,** 1–13.
CORI, G. T., AND CORI, C. F. (1952) *J. Biol. Chem.* **199,** 661–667.
COUJARD, R. (1943) *Bull. histol. appl. physiol. et pathol. et tech. microscop.* **20,** 161–173.
COURTOIS, J., AND ANAGNOSTOPOULOS, C. (1948) *Bull. soc. chim. biol.* **31,** 1494–1509.
CRAMER, F. (1961) *Angew. Chem.* **73,** 49–56.
CUSWORTH, D. C. (1958) *Biochem. J.* **68,** 262–264.
DALGAARD, J. B. (1948) *Nature* **162,** 811.
DANIELLI, J. F. (1946) *J. Exptl. Biol.* **22,** 110–117.
DANIELLI, J. F. (1953) "Cytochemistry, a Critical Approach." Wiley, New York.
DANIELLI, J. F. (1954) *Proc. Roy. Soc.* **B142,** 146–154.
DANNENBERG, A. M., BURSTONE, M. S., AND WILDHACK, J. A. (1960) *J. Histochem. and Cytochem.* **8,** 318–319.
DAOUST, R. (1957) *Exptl. Cell Research* **12,** 203–211.
DAOUST, R. (1961) *In* "General Cytochemical Methods" (J. F. Danielli, ed.), Vol. 2, pp. 153–170. Academic Press, New York.
DAOUST, R., AND AMANO, H. (1960) *J. Histochem. and Cytochem.* **8,** 131–134.
DAOUST, R., AND AMANO, H. (1961) *Exptl. Cell Research* **24,** 559–564.
DAOUST R., AND CANTERO, A. (1959) *J. Histochem. and Cytochem.* **7,** 139–143.
DAVIS, B. J., AND ORNSTEIN, L. (1959) *J. Histochem. and Cytochem.* **7,** 297–298.
DEAN, A. L., AND WOODARD, H. Q. (1947) *J. Urol.* **57,** 172–174.
DEANE, H. W. (1947) *Am. J. Anat.* **80,** 321–359.
DEANE, H. W., AND DEMPSEY, E. W. (1945) *Anat. Record* **93,** 401–417.
DE DUVE, C. (1957) *Symposia Soc. Exptl. Biol.* **10,** 50–61.

DE DUVE, C. (1959) *In* "Subcellular Particles" (T. Hayashi, ed.), pp. 128–158. Ronald Press, New York.

DE DUVE, C., PRESSMAN, B. C., GIANETTO, R., WATTIAUX, R., AND APPELMANS, F. (1955) *Biochem. J.*, **60**, 604–617.

DE LAMIRANDE, G., DAOUST, R., AND CANTERO, A. (1961) *Can. Cancer Conf.* **4,** 43–56.

DELORY, G. E., AND KING, E. J. (1939) *Biochem. J.* **33,** 1185–1190.

DELORY, G. E., AND KING, E. J. (1943) *Biochem. J.* **37,** 547–550.

DELORY, G. E., WILBERG, G. S., AND HETHERINGTON, M. (1955) *Can. J. Biochem. and Physiol.* **33,** 539–544.

DE MAN, J. C. H., DAEMS, W. TH., WILLIGHAGEN, R. G. J., AND VAN RIJSSEL, TH. G. (1960). *J. Ultrastruct. Research* **4,** 43–57.

DE MELLO, J. B. (1951) *Hospital, Rio de Janeiro* **40,** 857–861.

DEMPSEY, E. W. AND DEANE, H. W. (1946) *J. Cell. and Comp. Physiol.* **27,** 159–179.

DICZFALUSY, E., FERNÖ, O., FEX, H., HÖGBERG, B., LINDEROT, T., AND ROSENBERG, TH. (1953) *Acta Chem. Scand.* **7,** 913–920.

DORFMAN, V. A., AND EPSHTEIN, S. M. (1950) *Doklady Akad. Nauk S.S.S.R.* **72,** 977–978.

DOYLE, W. L. (1950) *Science* **111,** 64–65.

DOYLE, W. L., OMOTO, J., AND DOYLE, M. E. (1951) *Exptl. Cell Research* **2,** 20–38.

DUBOS, R. J., AND THOMPSON, R. H. S. (1938) *J. Biol. Chem.* **124,** 501–510.

DUNCAN, G. G. (1959) "Diseases of Metabolism." Saunders, Philadelphia, Pennsylvania.

DUNHAM, E. T., AND GLYNN, I. M. (1961) *J. Physiol.* **156,** 274–293.

ENGELHARDT, W. A., AND LJUBIMOVA, M. N. (1939) *Nature* **144,** 668–669.

ERÄNKÖ, O. (1951) *Ann. Med. Exptl. et Biol. Fenniae (Helsinki)* **29,** 287–289.

ERÄNKÖ, O. (1952) *Acta Anat.* **16,** Suppl. 17, 1–60.

ERÄNKÖ, O., AND HASAN, J. (1954) *Acta Pathol. Microbiol. Scand.* **35,** 563–572.

FAND, S. B. (1961) *Am. J. Clin. Pathol.* **35,** 210–221.

FANGER, H., AND BARKER, B. (1959) *Arch. Pathol.* **67,** 293–305.

FANGER, H., AND BARKER, B. (1960) *Arch. Pathol.* **69,** 67–76.

FANTL, P., AND ROME, M. N. (1945) *Australian J. Exptl. Biol. Med. Sci.* **23,** 21–27.

FEIGIN, I., AND WOLF, A. (1957) *J. Histochem. and Cytochem.* **5,** 53–64.

FEIGIN, I., AND WOLF, A.(1959) *Arch. Pathol.* **76,** 670–678.

FERNÖ, O., FEX, H., HÖGBERG, B., LINDEROT, T., AND ROSENBERG, TH. (1953) *Acta Chem. Scand.* **7,** 921–927.

FERNÖ, O., FEX, H., HÖGBERG, B., LINDEROT, T., VEIGE, S., AND DICZFALUSY, E. (1958) *Acta Chem. Scand.* **12,** 1675–1689.

FISHMAN, W. H. (1959) *In* "Physiopathology of Cancer" (F. Homburger, ed.), pp. 732–760. Hoeber, New York.

FISHMAN, W. H., AND LERNER, F. (1953) *J. Biol. Chem.* **200,** 89–97.

FLEURY, P., COURTOIS, J., ANAGNOSTOPOULOS, C., AND DESJOBERT, A. (1950) *Bull. soc. chim. biol.* **32,** 771–783.

FOGO, J. K., AND POPOWSKY, M. (1949) *Anal. Chem.* **21,** 732–734.

FOLLEY, S. J., AND KAY, H. (1935) *Biochem. J.* **29,** 1837–1850.

FORAKER, A. (1956a) *Surg. Gynecol. Obstet.* **102,** 1–8.

FORAKER, A. (1956b) *Cancer* **9,** 367–373.

FRAENKEL-CONRAT, H., AND STANLEY, W. M. (1961) *Chem. Eng. News* **39,** 136–144.

FRANKENTHAL, L., NEUBERG, C., AND ROBERTS, I. S. (1943) *Exptl. Med. Surg.* **1,** 386–401.

FRANSEEN, C. C., AND MCLEAN, R. (1935) *Am. J. Cancer* **24,** 299–317.

FREEMAN, H. F., AND COLVER, C. W. (1938) *J. Am. Chem. Soc.* **60,** 750–751.

FREIMAN, D. G., AND KAPLAN, N. (1960) *J. Histochem. and Cytochem.* **8,** 159–170.
FRIEDMAN, O. M., AND SELIGMAN, A. M. (1950) *J. Am. Chem. Soc.* **72,** 624–625.
FRIEDMANN, E., MARRIAN, D. H., AND SIMON-REUSS, I. (1952) *Biochim. et Biophys. Acta* **8,** 680–686.
GIANETTO, R., AND DE DUVE, C. (1955) *Biochem. J.* **59,** 433–438.
GLICK, D., AND FISHER, E. E. (1945) *Arch. Biochem.* **8,** 91–96.
GLICK, D., AND FISHER, E. E. (1946) *Arch. Biochem.* **11,** 65–79.
GLÜCK, E. (1952) *Acta Pathol. Microbiol. Scand. Suppl.* **93,** 106–120.
GOETCH, J. B., AND REYNOLDS, P. N. (1951) *Stain Technol.* **26,** 145–151.
GOLARZ, M. N., BOURNE, G. H., AND RICHARDSON, H. D. (1961) *J. Histochem. and Cytochem.* **9,** 132–140.
GOLDBERG, L., MARTIN, L. E., AND BATCHELOR, A. (1960) *Biochem. J.* **77,** 252–262.
GOLDBERG, M., AND GILMOUR, D. (1954) *Arch. Biochem. Biophys.* **51,** 411–418.
GOMORI, G. (1939) *Proc. Soc. Exptl. Biol. Med.* **42,** 23–26.
GOMORI, G. (1941) *Arch. Pathol.* **32,** 189–199.
GOMORI, G. (1946) *Am. J. Clin. Pathol.* **16,** 347–352.
GOMORI, G. (1948) *Proc. Soc. Exptl. Biol. Med.* **69,** 407–409.
GOMORI, G. (1949a) *Proc. Soc. Exptl. Biol. Med.* **70,** 7–11.
GOMORI, G. (1949b) *Proc. Soc. Exptl. Biol. Med.* **72,** 449–450.
GOMORI, G. (1949c) *J. Lab. Clin. Med.* **34,** 275–281.
GOMORI, G. (1950a) *J. Lab. Clin. Med.* **35,** 802–809.
GOMORI, G. (1950b) *Ann. N. Y. Acad. Sci.* **50,** 968–981.
GOMORI, G. (1950c) *Stain Technol.* **25,** 81–85.
GOMORI, G. (1950d) *Exptl. Cell Research* **1,** 33–36.
GOMORI, G. (1951) *J. Lab. Clin. Med.* **37,** 526–531.
GOMORI, G. (1952) "Microscopic Histochemistry. Principles and Practice." Univ. Chicago Press, Chicago, Illinois.
GOMORI, G. (1954) *Am. J. Clin. Pathol.* **24,** 99–110.
GOMORI, G. (1955) *Texas Repts. Biol. and Med.* **13,** 636–640.
GOMORI, G. (1956a) *J. Histochem. and Cytochem.* **4,** 425–427.
GOMORI, G. (1956b) *J. Histochem. and Cytochem.* **4,** 453–461.
GOMORI, G., AND BENDITT, E. P. (1953) *J. Histochem. and Cytochem.* **1,** 114–122.
GOMORI, G., AND CHESSICK, R. D. (1953) *J. Neuropathol. Exptl. Neurol.* **12,** 387-396.
GOODLAD, G. A. J., AND MILLS, G. T. (1957) *Biochem. J.* **66,** 346–354.
GORDON, J. J. (1950) *Biochem. J.* **46,** 96–99.
GORDON, J. J. (1952) *Biochem. J.* **51,** 97–103.
GORDON, J. J. (1957) *Biochem. J.* **66,** 50–53.
GREEN, H., AND MEYERHOF, O. (1952) *J. Biol. Chem.* **197,** 347–364.
GREENSTEIN, J. P. (1956) *Cancer Research* **16,** 641–653.
GREENSTEIN, J. P., AND THOMPSON, J. W. (1943) *J. Natl. Cancer Inst.* **4,** 271-274.
GREENSTEIN, J. P., CARTER, C. E., AND LEUTHARDT, F. M. (1946) *J. Natl. Cancer Inst.* **7,** 47–49.
GREEP, R. O., FISCHER, C. J., AND MORSE, A. (1948) *J. Am. Dental Assoc.* **36,** 427–442.
GROGG, E., AND PEARSE, A. G. E. (1952) *Brit. J. Exptl. Pathol.* **33,** 567–576.
GROPP, A. Z. (1959) *Z. Krebsforsch.* **63,** 156-167.
GROSS, S. J., AND DANZINGER, S. (1957) *Am. J. Obstet.* **73,** 94–119.
GROSSER, P., AND HUSLER, J. (1912) *Biochem. Z.* **39,** 1-5.
GUTMAN, A. B., AND GUTMAN, E. B. (1938) *J. Clin. Invest.* **17,** 473–478.
GUTMAN, E. B., AND GUTMAN, A. B. (1940) *J. Biol. Chem.* **136,** 201–209.
HALL, R. H., AND KHORANA, H. G. (1954) *J. Am. Chem. Soc.* **76,** 5056–5060.
HANCOX, N. M., AND NICHOLAS, E. (1956) *Acta Anat.* **26,** 302–317.

Hanson, A., and Biörck, G. (1957) *Acta Med. Scand.* **157,** 493–502.
Hayhoe, F. G. J. (1960) "Leukemia: Research and Clinical Practice." Little, Brown, Boston, Massachusetts.
Heppel, L. A., and Hilmoe, R. J. (1951) *J. Biol. Chem.* **188,** 665–676.
Heppel, L. A., and Hilmoe, R. J. (1953) *J. Biol. Chem.* **202,** 217–226.
Hers, H. G., Berthet, J., Berthet, L., and de Duve, C. (1951) *Bull. soc. chim. biol.* **33,** 21–41.
Hoffman, J. F. (1960) *Federation Proc.* **19,** 127.
Hoffmann-Ostenhof, O., and Šlechta, L. (1958) "Proceedings of the International Symposium on Enzyme Chemistry, Tokyo and Kyoto, 1957" (K. Ichihara, ed.), pp. 180–189. Academic Press, New York.
Holt, S. J. (1954) *Proc. Roy. Soc.* **B142,** 160–169.
Holt, S. J. (1959) *Exptl. Cell Research Suppl.* **7,** 1–27.
Holter, H., and Li, S.-O. (1950) *Acta Chem. Scand.* **4,** 1321–1322.
Holter, H., and Li, S.-O. (1951) *Compt. rend. trav. lab. Carlsberg* [*Sér. chim.*] **27,** 393–407.
Holter, H., Løvtrup, L., and Rubin, I. (1951) *Acta Chem. Scand.* **5,** 194–195.
Hotta, R. (1934) *J. Biochem.* (*Tokyo*) **20,** 343–360.
Houck, J. C. (1958) *Arch Biochem. Biophys.* **73,** 384–390.
Hudson, P. B., Brendler, H., and Scott, W. W. (1947) *J. Urol.* **58,** 89–92.
Huggins, C., and Talalay, P. (1945) *J. Biol. Chem.* **159,** 399–410.
Ichihara, M. (1933) *J. Biochem.* (*Tokyo*) **18,** 87–106.
Ingelman, B. (1950) *In* "The Enzymes" (J. B. Sumner and K. Myrbäck, eds.), Vol. I, Part 1, pp. 511–516. Academic Press, New York.
Iwatsuru, R. (1926) *Biochem. Z.* **173,** 348–357.
Jacobsohn, K. P., and Tapadinhas, J. (1931) *Biochem. Z.* **230,** 304–311.
Järnefelt, J. (1961a) *Biochim. et Biophys. Acta* **48,** 104–110.
Järnefelt, J. (1961b) *Biochim. et Biophys. Acta* **48,** 111–116.
Jeffree, G. M., and Taylor, K. B. (1961) *J. Histochem. and Cytochem.* **9,** 93–96.
Johansen, G., and Linderstrøm-Lang, K. (1951) *Acta Chem. Scand.* **5,** 965–967.
Johansen, G., and Linderstrøm-Lang, K. (1952) *Acta Med. Scand.* **142,** Suppl. 266, 601–613.
Kabat, E. A. (1941) *Science* **93,** 43–44.
Kabat, E. A., and Furth, J. (1941) *Am. J. Pathol.* **17,** 303–318.
Kalckar, H. M. (1944) *J. Biol. Chem.* **153,** 355-367.
Kaplow, L. S. (1955) *Blood* **10,** 1023–1029.
Kato, Y., and Moog, F. (1958) *Science* **127,** 812–813.
Kay, H. D. (1928) *Biochem. J.* **22,** 855–866.
Kay, H. D. (1930) *J. Biol. Chem.* **89,** 235–247.
Kenner, G. W., and Williams, N. R. (1955) *J. Chem. Soc.* pp. 522–525.
Kenny, J. J., and Moloney, W. C. (1957) *Blood* **12,** 295–302.
Khorana, H. G. (1953) *Chem. Revs.* **53,** 145–166.
Khorana, H. G. (1954) *J. Am. Chem. Soc.* **76,** 3517-3522.
Kielley, W. W. (1961) *In* "The Enzymes" (P. D. Boyer, H. Lardy, and K. Myrbäck, eds.), 2nd ed., Vol. 5, pp. 149–168. Academic Press, New York.
Kielley, W. W., and Bradley, L. B. (1956) *J. Biol. Chem.* **218,** 653–659.
Kielley, W. W., and Meyerhof, O. (1948) *J. Biol. Chem.* **176,** 591–601.
Kielley, W. W., and Meyerhof, O. (1950) *J. Biol. Chem.* **183,** 391–401.
Kielley, W. W., Kalckar, H. M., and Bradley, L. B. (1956) *J. Biol. Chem.* **219,** 95–101.
Kiessling, W. (1934) *Ber.* **67,** 869–874.

Kiessling, K. H. (1961) *Biochim. et Biophys. Acta* **46**, 603-604.

Kiessling, K. H., and Tilander, K. (1960) *Biochim. et Biophys. Acta* **43**, 335-336.

Kilsheimer, G. S., and Axelrod, B. (1957) *J. Biol. Chem.* **227**, 879-890.

King, E. J., and Armstrong, A. R. (1934) *Can. Med. Assoc. J.* **31**, 376-381.

King, E. J., and Delory, G. E. (1939) *Biochem. J.* **33**, 1185-1190.

King, E. J., and Nicholson, T. F. (1939) *Biochem. J.* **33**, 1182-1184.

Kitiyakara, A., and Harman, J. W. (1953) *J. Exptl. Med.* **97**, 553-572.

Kornberg, A. (1957) *Advances in Enzymol.* **18**, 191-240.

Koszalka, T. R., Horne, S., Schlegel, B., and Altman, K. I. (1959) *Biochim. et Biophys. Acta* **35**, 197-202.

Kunitz, M. (1940) *J. Gen. Physiol.* **24**, 15-32.

Kunitz, M. (1948) *Science* **108**, 19-20.

Kunz, Ph. (1894) *Ber.* **27**, 2559-2565.

Kurata, Y., and Maeda, S. (1956) *Stain Technol.* **31**, 13-16.

Kurnick, N. B. (1953) *Arch. Biochem. Biophys.* **43**, 97-107.

Landau, W., and Schlamowitz, M. (1961) *Arch. Biochem. Biophys.* **95**, 474-482.

Langemann, H. (1949) *Schweiz. med. Wochschr.* **79**, 138-142.

Lardy, H. (1957) *In* "Methods in Enzymology" (S. P. Colowick, and N. O. Kaplan, eds.), Vol. III, pp. 198-201; 207-208. Academic Press, New York.

Lardy, H. A., and Fischer, H. O. L. (1946) *J. Biol. Chem.* **164**, 513-519.

Laskowski, M. (1951) *In* "The Enzymes" (J. B. Sumner and K. Myrbäck, eds.), Vol. I, Part 2, pp. 956-985. Academic Press, New York.

Laskowski, M. (1961) *In* "The Enzymes" (P. D. Boyer, H. Lardy, and K. Myrbäck, eds.), 2nd ed., Vol. 5, pp. 123-147. Academic Press, New York.

Lemon, H. M., and Wisseman, C. L. (1949) *Science* **109**, 233-235.

Lemon, H. M., Davidson, M. M., and Asimov, I. (1954) *Cancer* **7**, 92-99.

Levene, P. A., and Tipson, R. S. (1934) *J. Biol. Chem.* **106**, 113-124.

Levene, P. A., and Tipson, R. S. (1935) *J. Biol. Chem.* **111**, 313-323.

Li, S.-O. (1950) *Acta Chem. Scand.* **4**, 610-612.

Lison, L. (1960) "Histochimie et Cytochimie Animales," 3rd ed., Vol. 2. Gauthier Villars, Paris.

Lohmann, K. (1928) *Naturwissenschaften* **16**, 298.

Lohmann, K. (1934) *Biochem. Z.* **271**, 264-277.

Lorch, I. J. (1949) *Quart. J. Microscop. Sci.* **90**, 183-208.

Loveless, A., and Danielli, J. F. (1949) *Quart J. Microscop. Sci.* **90**, 57-66.

Lowry, O. H. (1953) *J. Histochem. and Cytochem.* **1**, 420-428.

Lundsteen, E., and Vermehren, E. (1936) *Enzymologia* **1**, 273-279.

Maccalum, A. B. (1932) *In* "Taschenbuch der mikroskopischen Technik" (B. Romeis, ed.), 13th ed., pp. 343-344. Oldenbourg, Munich and Berlin.

McCombie, H., Saunders, B. C., and Stacey, G. J. (1945) *J. Chem. Soc.*, pp. 380-382.

McDonald, M. R. (1955a) *In* "Methods in Enzymology" (S. P. Colowick and N. O. Kaplan, eds.), Vol. II, pp. 427-436. Academic Press, New York.

McDonald, M. R. (1955b) *In* "Methods in Enzymology" (S. P. Colowick and N. O. Kaplan, eds.), Vol. II, pp. 437-447. Academic Press, New York.

McManus, J. F. A., Lupton, C. H., and Harden, G. (1952) *Lab. Invest.* **1**, 480-487.

Maengwyn-Davies, G. D., and Friedenwald, J. S. (1950) *J. Cellular Comp. Physiol.* **36**, 421-459.

Maengwyn-Davies, G. D., Friedenwald, J. S., and White, R. T. (1952) *J. Cellular Comp. Physiol.* **39**, 395-447.

Mandl, I., Grauer, A., and Neuberg, C. (1952) *Biochim. et Biophys. Acta* **8**, 654-663.

MANHEIMER, L. H., AND SELIGMAN, A. M. (1948) *J. Natl. Cancer Inst.* **9**, 181–199.
MARSHALL, J. M. (1954) *Exptl. Cell Research* **6**, 240–242.
MAVER, M. E., AND GRECO, A. (1956) *J. Natl. Cancer Inst.* **17**, 503–516.
MELLORS, R. C., ed. (1959) "Analytic Cytology." McGraw-Hill, New York.
MENTEN, M. L., JUNGE, J., AND GREEN, M. H. (1944a) *Proc. Soc. Exptl. Biol. Med.* **57**, 82–86.
MENTEN, M. L., JUNGE, J., AND GREEN, M. H. (1944b) *J. Biol. Chem.* **153**, 471–477.
METZLER, D. E. (1960) *In* "The Enzymes" (P. D. Boyer, H. Lardy, and K. Myrbäck, eds.), 2nd ed., Vol. 2, pp. 295–337. Academic Press, New York.
MEYER, J., AND WEINMANN, J. P. (1953) *J. Histochem. and Cytochem.* **1**, 305–314.
MEYER, J., AND WEINMANN, J. P. (1955) *J. Histochem. and Cytochem.* **3**, 134–140.
MEYERHOF, O. (1945) *J. Biol. Chem.* **157**, 105–119.
MICHELSON, A. M., AND TODD, A. R. (1949) *J. Chem. Soc.*, pp. 2476–2486.
MITZ, M. A. (1956) *Science* **123**, 1076–1077.
MIYANO, M. (1955) *J. Am. Chem. Soc.* **77**, 3524–3526.
MIYANO, M., AND FUNAHASHI, S. (1955) *J. Am. Chem. Soc.* **77**, 3522–3524.
MOFFATT, J. G., AND KHORANA, H. G. (1957) *J. Am. Chem. Soc.* **79**, 3741–3746.
MØLLER, K. M. (1955) *Biochim. et Biophys. Acta* **16**, 162–165.
MOMMAERTS, W. S. H. M., AND GREEN, I. (1954) *J. Biol. Chem.* **208**, 833–843.
MONTGOMERY, H. A. C., AND TURNBULL, J. H. (1958) *J. Chem. Soc.* pp. 1963–1967.
MOOG, F. (1946) *Biol. Revs.* **21**, 41–59.
MOOG, F. (1962) *Federation Proc.* **21**, 51–56.
MOOG, F., AND STEINBACH, H. B. (1946) *Science* **103**, 144.
MORTON, R. K. (1953) *Biochem. J.* **55**, 786–795.
MORTON, R. K. (1954) *Biochem. J.* **57**, 595–603.
MORTON, R. K. (1955a) *Biochem. J.* **60**, 573–582.
MORTON, R. K. (1955b) *Biochem. J.* **61**, 232–240.
MORTON, R. K. (1955c) *Biochem. J.* **61**, 240–244.
MORTON, R. K. (1955d) *In* "Methods in Enzymology" (S. P. Colowick and N. O. Kaplan, eds.), Vol. II, pp. 533–539, 556–561. Academic Press, New York.
MUNEMURA, S. (1933a) *J. Biochem. (Tokyo)* **17**, 343–366.
MUNEMURA, S. (1933b) *J. Biochem. (Tokyo)* **18**, 23–32.
NACHMIAS, V. T., AND PADYKULA, H. A. (1958) *J. Biophys. Biochem. Cytol.* **4**, 47–54.
NADEL, E. M., AND WODINSKY, I. (1955) *J. Histochem. and Cytochem.* **3**, 339–341.
NAIDOO, D., AND PRATT, O. E. (1951) *J. Neurol. Neurosurg. Psychiat.* **14**, 287–294.
NAIDOO, D., AND PRATT, O. E. (1952) *J. Neurol. Neurosurg. Psychiat.* **15**, 164–168.
NAIDOO, D., AND PRATT, O. E. (1953) *Enzymologia* **16**, 91–98.
NAIDOO, D., AND PRATT, O. E. (1954) *Enzymologia* **16**, 298–304.
NAIDOO, D., AND PRATT, O. E. (1955) *Biochim. et Biophys. Acta* **16**, 75–81.
NEILANDS, J. B., AND STUMPF, P. K. (1958) "Outlines of Enzyme Chemistry," 2nd ed. Wiley, New York.
NEUBERG, C. (1943) *Arch. Biochem.* **3**, 105–112.
NEUBERG, C., AND FISCHER, H. A. (1937) *Enzymologia* **2**, 191.
NEUBERG, C., AND FISCHER, H. A. (1938a) *Enzymologia* **2**, 241–257.
NEUBERG, C., AND FISCHER, H. A. (1938b) *Enzymologia* **2**, 360–366.
NEUBERG, C., AND JACOBSOHN, K. P. (1928) *Biochem. Z.* **199**, 498–517.
NEUBERG, C., AND WAGNER, J. (1926) *Biochem. Z.* **171**, 485–500.
NEUBERG, C., WAGNER, J., AND JACOBSOHN, K. P. (1927) *Biochem. Z.* **188**, 227–240.
NEUMANN, H. (1948a) Dissertation, University of Amsterdam, "Onderzoekingen over Phosphomonoesterase en Pyrophosphatasen van het dierlijk Organisme." Drukkerij Kinsbergen, Amsterdam.

NEUMANN, H. (1948b) *Experientia* **4,** 74–75.
NEUMANN, H. (1949) *Biochim. et Biophys. Acta* **3,** 117-125.
NEUMANN, K., OEHLERT, G., AND HANSMANN, H. (1954) *Z. Geburtshilfe u. Gynäkol.* **141,** 109-130.
NEWMAN, W., FEIGIN, I., WOLF, A., AND KABAT, E. A. (1950) *Am. J. Pathol.* **26,** 257–305.
NIGAM, V. N., DAVIDSON, H. M., AND FISHMAN, W. H. (1959) *J. Biol. Chem.* **234,** 1550–1554.
NORBERG, B. (1950a) *Acta Chem. Scand.* **4,** 601–609.
NORBERG, B. (1950b) *Acta Chem. Scand.* **4,** 1206–1215.
NOVIKOFF, A. B. (1952) *Exptl. Cell Research Suppl.* **2,** 123–143.
NOVIKOFF, A. B. (1955) *In* "Analytical Cytology" (R. C. Mellors, ed.), pp. 69–168. McGraw-Hill, New York.
NOVIKOFF, A. B. (1959) *Biol. Bull.* **117,** 385.
NOVIKOFF, A. B. (1960a) *Acta Unio Intern. contra Cancrum* **16,** 966–972.
NOVIKOFF, A. B. (1960b) *In* "Cell Physiology of Neoplasia" (J. C. Hsu, ed.), pp. 219–268. Univ. Texas Press, Austin, Texas.
NOVIKOFF, A. B. (1960c) *In* "Developing Cell Systems and Their Control." (D. Rudnick, ed.), pp. 167–203. Ronald Press, New York.
NOVIKOFF, A. B., AND ESSNER, E. (1960) *Am. J. Med.* **29,** 102–131.
NOVIKOFF, A. B., BEAUFAY, H., AND DE DUVE, C. (1956) *J. Biophys. Biochem. Cytol. Suppl.* **2,** 179–184.
NOVIKOFF, A. B., DRUCKER, J., SHIN, W.-Y., AND GOLDFISCHER, S. (1961) *J. Histochem. and Cytochem.* **9,** 434–451.
OBER, K. G. (1952) *Zbl. Gynaek.* **74,** 115-117.
O'BRIEN, R. D. (1960) "Toxic Phosphorus Esters." Academic Press, New York.
OHMORI, Y. (1937) *Enzymologia* **4,** 217–231.
OTTO, P. (1895) *Ber.* **28,** 616–620.
OUTHOUSE, E. L. (1937) *Biochem. J.* **31,** 1459–1463.
PADYKULA, H. A., AND HERMAN, E. (1955a) *J. Histochem. and Cytochem.* **3,** 161–169.
PADYKULA, H. A., AND HERMAN, E. (1955b) *J. Histochem. and Cytochem.* **3,** 170–195.
PALADE, G. E. (1951) *J. Exptl. Med.* **94,** 535–547.
PEARSE, A. G. E. (1953) "Histochemistry, Theoretical and Applied." Little, Brown, Boston, Massachusetts.
PEARSE, A. G. E. (1954) *Intern. Rev. Cytol.* **3,** 329–358.
PEARSE, A. G. E. (1960) "Histochemistry, Theoretical and Applied," 2nd ed. Little Brown, Boston, Massachusetts.
PEARSE, A. G. E., AND REIS, J. L. (1952) *Biochem. J.* **50,** 534–536.
PERILLIE, P. E., AND FINCH, S. C. (1961) *Blood* **18,** 572–580.
PERLMANN, G. E. (1952) *In* "Phosphorus Metabolism" (W. D. McElroy and B. Glass, eds.), Vol. II, pp. 167–185. Johns Hopkins Press, Baltimore, Maryland.
PLIMMER, R. H. A., AND BURCH, W. J. N. (1929) *J. Chem. Soc.* pp. 279–300.
POST, R. L., MERRITT, C. R., KINSOLVING, C. R., AND ALBRIGHT, C. D. (1960) *J. Biol. Chem.* **235,** 1796–1802.
PRATT, O. E. (1953) *Biochem. J.* **55,** 140–145.
PRATT, O. E. (1954) *Biochim. et Biophys. Acta* **14,** 380–389.
RAFTER, G. W. (1958) *J. Biol. Chem.* **230,** 643–648.
REINER, L., RUTENBURG, A. M., AND SELIGMAN, A. M. (1957) *Cancer* **10,** 563–576.
REIS, J. L. (1934) *Bull. soc. chim. biol.* **16,** 385–399.
REIS, J. L. (1937) *Enzymologia* **2,** 183–190.
REIS, J. L. (1938) *Enzymologia* **5,** 251–253.

Reis, J. L. (1951) *Biochem. J.* **48,** 548–551.
Ring, J. R. (1950) *Anat. Record* **107,** 121–132.
Ring, J. R., and Levy, B. (1950) *J. Dental Research* **29,** 817–824.
Robison, R. (1922) *Biochem. J.* **16,** 809–824.
Robison, R. (1923) *Biochem. J.* **17,** 286–293.
Robison, R., and Soames, K. M. (1924) *Biochem. J.* **18,** 740–754.
Roche, J. (1931) *Biochem. J.* **25,** 1724–1733.
Roche, J. (1950) *In* "The Enzymes," (J. B. Sumner and K. Myrbäck, eds.), Vol. I, Part 1. pp. 473–510. Academic Press. New York.
Roche, J., and Latreille, M. (1937) *Enzymologia* **3,** 75–81.
Roche, J., and Thoai, N. (1950) *Advances in Enzymol.* **10,** 83–122.
Roodyn, D. B. (1957) *Biochim. et Biophys. Acta* **25,** 129–131.
Rorig, K. (1949) *J. Am. Chem. Soc.* **71,** 3561–3562.
Rosenmund, K. W., and Vogt, H. (1943) *Arch. Pharm.* **281,** 317–327.
Rosenthal, I. M., Bonting, S. L., Hogan, W., and Pirani, C. L. (1960) *A.M.A.J. Diseases Children* **99,** 185–192.
Ross, M. H., Ely, J. O. and Archer, J. G. (1951) *J. Biol. Chem.* **192,** 561–568.
Roth, J. S. (1953) *Arch. Biochem. Biophys.* **44,** 265–270.
Rouiller, C. (1954) *Compt. rend. soc. biol.* **148,** 2008–2011.
Ruppert, F. (1950). *Z. ges. exptl. Med.* **116,** 378–383.
Rutenburg, A. M., and Seligman, A. M. (1955) *J. Histochem. and Cytochem.* **3,** 455–469.
Rutter, W. J. (1961) *In* "The Enzymes" (P. D. Boyer, H. Lardy, and K. Myrbäck, eds.), 2nd ed., Vol. 5, pp. 341–366. Academic Press, New York.
Samorajski, T., and Fitz, G. R. (1960) *Lab. Invest.* **9,** 517–534.
Samorajski, T., and McCloud, O. (1961) *Lab. Invest.* **10,** 492–501.
Schmidt, G. (1957) *In* "Methods in Enzymology" (S. P. Colowick and N. O. Kaplan, eds.), Vol. III, pp. 223–228, 346–358, 671–679, 687–691, 747–781. Academic Press, New York.
Schmidt, G. (1961) *In* "The Enzymes" (P. D. Boyer, H. Lardy, and K. Myrbäck, eds.), 2nd ed., Vol. 5, pp. 37–47. Academic Press, New York.
Schmidt, G., and Laskowski, M. (1961) *In* "The Enzymes" (P. D. Boyer, H. Lardy, and K. Myrbäck, eds.), 2nd ed., Vol. 5, pp. 3–35. Academic Press, New York.
Schmidt-Matthiesen, H. (1958) *Arch. Geschwulstforsch.* **14,** 265–276.
Seal, U. S., and Binkley, F. (1957) *J. Biol. Chem.* **228,** 193–199.
Seegmiller, J. E., and Horecker, B. L. (1951) *J. Biol. Chem.* **192,** 175–180.
Seligman, A. M., and Manheimer, L. H. (1949) *J. Natl. Cancer Inst.* **9,** 427–434.
Seligman, A. M., Nachlas, M. M., Manheimer, L. H., Friedman, O. M., and Wolf, G. (1949) *Ann. Surg.* **130,** 333–341.
Seligman, A. M., Chauncey, H. H., Nachlas, M. M., Manheimer, L. H., and Ravin, H. A. (1951a) *J. Biol. Chem.* **190,** 7–15.
Seligman, A. M., Chauncey, H. H., and Nachlas, M. M. (1951b) *Stain Technol.* **26,** 19–23.
Seligman, A. M., Heymann, H., and Barrnett, R. J. (1954) *J. Histochem. and Cytochem.* **2,** 441–442.
Sen, P. B. (1930) *Indian J. Med. Research* **18,** 79–82.
Shack, J. (1957) *J. Biol. Chem.* **226,** 573–581.
Shinowara, G. Y., Jones, L. M., and Reinhart, H. L. (1942) *J. Biol. Chem.* **142,** 921–933.

SHUGAR, D., SZENBERG, A., AND SIERAKOWSKA, H. (1957) *Exptl. Cell Research* **13,** 424–426.

SIBLEY, J. A., AND LEHNINGER, A. L. (1949) *J. Biol. Chem.* **177,** 859–872.

SMITH, S. W., WEISS, S. B., AND KENNEDY, E. P. (1957) *J. Biol. Chem.* **228,** 915–922.

SOULAIRAC, A., AND DESCLAUX, P. (1949) *Compt. rend. soc. biol.* **143,** 470.

STADTMAN, T. C. (1961) *In* "The Enzymes" (P. D. Boyer, H. Lardy, and K. Myrbäck, eds.), 2nd ed., Vol. 5, pp. 55–71. Academic Press, New York.

STAFF, Chem. Eng. News (1961) *Chem. Eng. News* **39,** 80–89, May 8.

STEIGLEDER, G. K., AND RAAB, W. (1961) *Klin. Wochschr.* **39,** 598–599.

STOKES, H. N. (1893) *Am. Chem. J.* **15,** 198–214.

STOKES, H. N. (1894) *Am. Chem. J.* **16,** 123–155.

STOLL, P., EBNER, H., AND STRECKER-HEIDELBERG, H. (1951) *Arch. Gynäkol.* **180,** 76–84.

STRAUS, W. (1954) *J. Biol. Chem.* **207,** 745–755.

STRAUS, W. (1956) *J. Biophys. Biochem. Cytol.* **2,** 513–521.

STROMINGER, J. L., HEPPEL, L. A., AND MAXWELL, E. S. (1959) *Biochim. et Biophys. Acta* **32,** 412–421.

SUMNER, J. B., AND SOMERS, G. F. (1953) "Chemistry and Methods of Enzymes," 3rd ed. Academic Press, New York.

SWANSON, M. A. (1951) *J. Biol. Chem.* **191,** 577–590.

SZENT-GYÖRGI, A. (1947) "Chemistry of Muscular Contractions." Academic Press, New York.

TAKAMATSU, H. (1939) *Trans. Soc. Pathol. Japon.* **29,** 492–498.

TAKEUCHI, T. (1956) *Acta Pathol. Japon.* **6,** 13–18.

TAKEUCHI, T., AND NOGAMI, S. (1954) *Acta Pathol. Japon.* **4,** 277–283

TAKEUCHI, T., AND TANOUE, M. (1951) *Kumamoto Med. J.* **4,** 41–49.

TENER, G. M. (1961) *J. Am. Chem. Soc.* **83,** 159–168.

THILO, E. (1958) *Allg. Chem.* **67,** 141–145.

TOMITA, K., AND TAKEUCHI, T. (1955) *Gann* **46,** 333–334.

TSUBOI, K. K., AND HUDSON, P. B. (1955) *Arch. Biochem. Biophys.* **55,** 191–205.

VALENTINE, W. N., AND BECK, W. S. (1951) *J. Lab. Clin. Med.* **38,** 39–55.

VALENTINE, W. N., BECK, W. S., FOLLETTE, J. H., MILLS, H., AND LAWRENCE, J. S. (1952) *Blood* **7,** 959–977.

VANDENDRIESSCHE, L. (1956) *Arch. Biochem. Biophys.* **65,** 347–353.

VERZÁR, F., AND MCDOUGALL, E. J. (1936) "Absorption from the Intestine." Longmans, Green, London.

VILLELA, G. G., AND MELLO, M. I. (1948) *Hospital, Rio de Janeiro* **34,** 433–438.

VORBRODT, A. (1961) *J. Histochem. and Cytochem.* **9,** 647–655.

WACHSTEIN, M. (1946) *J. Lab. Clin. Med.* **31,** 1–17.

WACHSTEIN, M. (1955) *Ann. N. Y. Acad. Sci.* **59,** 1052–1065.

WACHSTEIN, M., AND MEISEL, E. (1954) *J. Histochem. and Cytochem.* **2,** 137–148.

WACHSTEIN, M., AND MEISEL, E. (1957a) *Am. J. Clin. Pathol.* **27,** 13–23.

WACHSTEIN, M., AND MEISEL, E. (1957b) *J. Histochem. and Cytochem.* **5,** 204–220.

WACHSTEIN M., AND MEISEL, E. (1959) *J. Histochem. and Cytochem.* **7,** 189–201.

WACHSTEIN, M., MEISEL, E., AND FALCON, E. (1960a) *J. Histochem. and Cytochem.* **8,** 353.

WACHSTEIN, M., MEISEL, E., AND NIEDZWIEDZ, A. (1960b) *J. Histochem. and Cytochem.* **8,** 387–388.

WALKER, B. S., LEMON, H. M., DAVIDSON, M. M., AND SCHWARTZ, M. K. (1954) *Am. J. Clin. Pathol.* **24,** 807–837.

Walker, P. G., and King, E. J. (1950) *Biochem. J.* **47,** 93-95.
Walschmidt-Leitz, E., and Koehler, E. (1933) *Biochem. Z.* **258,** 360–364.
Wang, K. J. (1950) *Chinese J. Physiol.* **17,** 317–324.
Weber, G., and Cantero, A. (1955) *Cancer Research* **15,** 105–108.
Winnick, T. (1947) *Arch. Biochem.* **12,** 209-211.
Winter, G. F. (1955) *Acta Histochem.* **1,** 303–318.
Wolffenstein, R. (1888) *Ber.* **21,** 1186–1192.
Zetzsche, F., and Aeschlimann, F. (1926) *Helv. Chim. Acta* **9,** 708–714.
Zetzsche, F., and Nachmann, M. (1925) *Helv. Chim. Acta* **8,** 943–945.

CHAPTER 6

# ESTERASES

## I. INTRODUCTION

By histochemical definition esterases or carboxylic acid esterases which are widely distributed in nature are enzymes which catalyze the following reversible reaction:

$$R—COOR' + HOH \rightleftharpoons R—COOH + R'OH$$

Although the aforementioned reaction is fairly characteristic, these enzymes are also known to hydrolyze amide derivatives (Myers *et al.*, 1957). Apparently this overlap in substrate specificity between the different classes of hydrolytic enzymes may be due, in some cases, to an indiscriminate bond-breaking mechanism which involves any ester which can approach the active enzyme center (Myers, 1960, p. 475).

The physiological role of esterases involves regulatory processes such as the regulation of the level and action of acetylcholine (Volle and Koelle, 1961). The enzymes may also be related to phagocytosis and related phenomena. The finding that liver aliesterase is capable of hydrolyzing amides and amino acid esters suggests that esterases may play a role in protein metabolism (Myers *et al.*, 1957).

In the past, esterases have been divided into three groups; simple esterases or aliesterases, lipases, and cholinesterases. Most authors have defined simple esterases as those which hydrolyze short-chain aliphatic esters, while lipases attack fatty acid esters with a long carbon chain. An excellent review of the literature until 1952 with orientation toward the histochemistry of these enzymes has been given by Gomori (1952a). In general, aliesterases have been reported to preferentially attack monohydric alcohol esters of aromatic acids which are characterized by an iso-chain of relatively short length containing less than twelve carbon atoms. There are, however, numerous reports relating to the overlapping specificities of these enzymes. For example, liver esterase, which does not appear to hydrolyze true fats, will nevertheless attack solutions of aliphatic or aromatic esters with a long carbon chain at a much higher rate than ones with a short carbon chain (Hofstee, 1960, p. 485). Such observations, together with the finding that the activity of purified lipase from pancreas becomes marked only when the saturation point of the substrate (e.g. methyl butyrate) is reached, suggest that esterases may be classified as those acting on substrates in solution (esterases proper) and those acting predominantly on undissolved substrates (lipase-type esterases).

Desnuelle (1961) points out that the term "lipase" usually covers a large series of enzymes hydrolyzing various esters which may be water soluble or insoluble mono- and polyesters. The acyl moiety may be aliphatic or aromatic and the second moiety, alcoholic or phenolic. Thus the term "lipase," as commonly used, is not very different from the term "esterase."

The different types of esterases may be characterized by their susceptibility to certain inhibitors. For example, aliesterases are little affected even by $10^{-3}$ *M* eserine, while cholinesterases are completely inhibited by $10^{-6}$ *M* of this compound (Richter and Croft, 1942). When $\beta$-naphthyl esters (acetate, laurate, and palmitate-stearate) are employed as substrates, all are hydrolyzed by pancreatic lipase and esterase while kidney and liver esterase are found to hydrolyze the acetate and laurate. Eserine was found to inhibit esterase and lipase, arsanilate and fluoride inhibited esterase, quinine inhibited lipase, while taurocholate activated lipase activity and slightly inhibited esterase activity (Nachlas and Seligman, 1949b).

A fairly large number of presumably specific esterases have been reported. These include lipoprotein lipase which hydrolyzes neutral triglycerides associated with proteins; lecithinase A which hydrolyzes lecithin giving lysolecithin plus one molecule of unsaturated fatty acid; acetyl esterase which hydrolyzes acetyl esters; tropinesterases which hydrolyze atropine, cocaine, and other tropine alkaloids; acetylcholinesterase or true cholinesterase which hydrolyze acetylcholine; pseudocholinesterase which hydrolyzes choline esters; benzoylcholinesterase which hydrolyzes benzoylcholine; vitamin A esterase which hydrolyzes vitamin A acetate; and cholesterol esterase which hydrolyzes steroid esters, especially cholestanol and cholesterol (Dixon and Webb, 1958, p. 190).

All esterases which are readily inhibited by DFP may be closely related with reference to mechanism of hydrolytic action (Myers, 1960, p. 477). Aldridge (1953a,b,c, 1954) has reported an enzyme or group of enzymes which are designated as A-esterases (aromesterases), which like aliesterases (B-esterases) catalyze the hydrolysis of carboxylic acid esters such as *p*-nitrophenyl acetate and phenyl acetate. However, unlike the aliesterases, serum A esterase exhibits little activity toward tributyrin and is not inhibited by DFP, although the latter compound is actually hydrolyzed by the enzyme as contrasted with the cholinesterases and aliesterases. A-esterase is inhibited by chelating agents and activated by calcium ions. A third type of enzyme, C-esterase, which also hydrolyzes simple carboxylic acid esters, is neither inhibited by DFP nor does it hydrolyze the compound (Bergmann *et al.*, 1957; Bergmann and Rimon, 1958). As contrasted with A-esterase, aliesterases, and cholinesterases, the C-esterase is activated by low concentrations of *p*-mercuribenzoate and other sulfhydryl reagents.

Esterases hydrolyzing naphtholic and indoxylic esters are believed to be localized in microsomes (de Duve, 1959). There exists, however, a discrepancy between biochemical findings and cytochemical staining reactions in that the latter primarily reveal enzyme activity of "lysosomes" with the indoxyl procedures (de Duve, 1960). On the other hand, Burstone (1957a) (Fig. 1) has shown that esterase activity in the kidney can be localized with an azo-dye method to "droplets" (lysosomes?) as well as in very fine granules.

When histochemical methods (e.g. azo-dye procedures) are employed in the demonstration of esterases, there is little question as to the general specificity of the enzyme type being demonstrated. However, the studies of Gomori and Chessick (1953a,b), Chessick (1954), and Gomori (1955a) appear to indicate that the naphtholic substrates are capable of demonstrating a spectrum of simple esterase and cholinesterase type enzymes with considerable overlapping of substrate specificity and influence of inhibitors.

## II. SUBSTRATE TYPES AND SYNTHETIC APPROACHES

The development of the simultaneous azo-dye coupling method for alkaline phosphatase by Menten *et al.* (1944a,b), which employed β-naphthyl phosphate, stimulated the subsequent development of analogous procedures for the demonstration of other esterases. Fortunately, the synthesis of naphthol acetates and other naphtholic fatty acid esters had been described earlier in the chemical literature (Table 1). For example, β-naphthyl acetate was originally synthesized in 1869 by Schaeffer, while α-naphthyl acetate (I) was prepared by Miller in 1881.

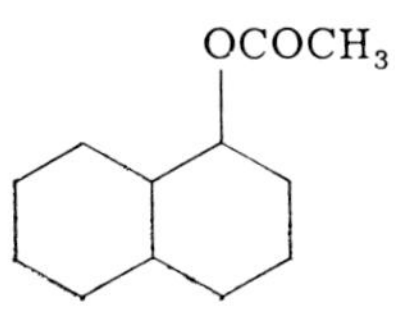


α-Naphthyl acetate

(I)

TABLE 1

ESTERASE SUBSTRATES

| Chemical name and formula | Mol. wt. | Reference | Melting point (° C.) |
|---|---|---|---|
| 1. α-Naphthyl acetate, 1-acetoxynaphthalene, $C_{12}H_{10}O_2$ | 186.21 | Miller, 1881; Chattaway, 1931 | 48–49 |
| 2. α-Naphthyl propionate, $C_{13}H_{12}O_2$ | 200.24 | — | Oil |
| 3. α-Naphthyl butyrate, $C_{14}H_{14}O_2$ | 214.27 | — | Liquid |
| 4. α-Naphthyl isobutyrate, $C_{14}H_{14}O_2$ | 214.27 | — | Liquid |
| 5. α-Naphthyl valerate, $C_{15}H_{16}O_2$ | 228.29 | — | Oil |
| 6. α-Naphthyl caprylate, $C_{18}H_{22}O_2$ | 270.37 | Price and May, 1943 | Liquid |
| 7. α-Naphthyl stearate, $C_{28}H_{42}O_2$ | 410.64 | Seidel and Engelfried, 1936 | 67.5–69 |
| 8. α-Naphthyl chloroacetate, $C_{12}H_9O_2Cl$ | 220.66 | Ullmann, 1897 | 48 |
| 9. α-Naphthyl α-bromopropionate, $C_{13}H_{11}O_2Br$ | 279.15 | — | Oil |
| 10. α-Naphthyl benzoate, $C_{17}H_{12}O_2$ | 248.28 | — | 56.5–57 |
| 11. α-Naphthyl phenylacetate, $C_{18}H_{14}O_2$ | 262.31 | — | 48 |
| 12. β-Naphthyl acetate (2-naphthyl acetate), $C_{12}H_{10}O_2$ | 186.21 | Schaeffer, 1869; Chattaway, 1931 | 71–72 |
| 13. β-Naphthyl propionate, $C_{13}H_{12}O_2$ | 200.24 | — | 51 |
| 14. β-Naphthyl butyrate, $C_{14}H_{14}O_2$ | 214.27 | — | Oil |
| 15. β-Naphthyl valerate, $C_{15}H_{16}O_2$ | 228.29 | — | Oil |
| 16. β-Naphthyl caproate, $C_{16}H_{18}O_2$ | 242.32 | — | Oil |
| 17. β-Naphthyl stearate, $C_{28}H_{42}O_2$ | 410.64 | Seidel and Engelfried, 1936 | 73–75 |
| 18. β-Naphthyl chloroacetate, $C_{12}H_9O_2Cl$ | 220.66 | Fries and Frellstedt, 1921 | 97 |
| 19. β-Naphthyl benzoate, $C_{17}H_{12}O_2$ | 248.28 | — | 97 and 110 |
| 20. 6-Bromo-2-naphthyl acetate, $C_{12}H_9O_2Br$ | 265.12 | Tilden and Armstrong, 1901 | 103 |
| 21. 6-Benzoyl-2-naphthyl acetate, $C_{19}H_{14}O_3$ | 290.32 | Ravin *et al.*, 1953 | — |
| 22. Naphthol AS acetate, $C_{19}H_{15}O_3N$ | 305.34 | Graff, 1909; Brass and Sommer, 1928; Gomori, 1952a | 152, 160 |
| 23. Naphthol AS-D acetate, $C_{20}H_{17}O_3N$ | 319.36 | Burstone, 1957b | 165 |
| 24. Naphthol AS-OL acetate, $C_{20}H_{17}O_4N$ | 335.36 | Burstone, 1957b | 128 |
| 25. Naphthol AS-MX acetate, $C_{21}H_{19}O_3N$ | 333.39 | Burstone, 1957b | 112 |
| 26. Naphthol AS-E acetate, $C_{19}H_{14}O_3NCl$ | 339.79 | Burstone, 1957a | 178 |
| 27. Naphthol AS-LC acetate, $C_{21}H_{18}O_5NCl$ | 399.84 | Burstone, 1957a | 138 |
| 28. Naphthol AS-LC propionate, $C_{22}H_{20}O_5NCl$ | 413.87 | Burstone, 1957a | 135 |

TABLE 1—*Continued*

| Chemical name and formula | Mol. wt. | Reference | Melting point (° C.) |
|---|---|---|---|
| 29. Naphthol AS-LC butyrate, $C_{23}H_{22}O_5NCl$ | 427.89 | Burstone, 1957a | 133 |
| 30. Naphthol AS-chloroacetate, $C_{19}H_{14}O_2NCl$ | 323.79 | Gomori, 1953b | — |
| 31. Naphthol AS-D chloroacetate, $C_{20}H_{16}O_3NCl$ | 353.81 | Burstone, 1957b | 162 |
| 32. Indoxyl acetate (*o*-acetylindoxyl), $C_{12}H_{13}NO_2$ | 203.24 | Vorländer and Drescher, 1901; Spencer, 1931; Barrnett and Seligman, 1951 | 126–128 |
| 33. 5-Bromoindoxyl acetate, $C_{10}H_8O_2NBr$ | 254.10 | Holt and Sadler, 1958 | 134 |
| 34. 5-Bromo-6-chloroindoxyl acetate $C_{10}H_7O_2NBrCl$ | 288.55 | — | 130 |
| 35. 5-Bromo-4-chloroindoxyl acetate, $C_{10}H_7O_2NBrCl$ | 288.55 | Holt and Sadler, 1958 | 107 |
| 36. Thiolacetic acid, $C_3H_6O_2S$ | 106.15 | — | Oil |
| 37. Acetylcholine chloride, $C_7H_{16}ClNO_2$ | 181.67 | — | — |
| 38. Acetylthiocholine iodide [$\beta$-(acetthio)ethyltrimethylammonium iodide], $C_7H_{16}INOS$ | 289.19 | Renshaw *et al.*, 1938 | 203–204 |
| 39. 6-Bromo-$\beta$-carbonaphthoxycholine iodide, $C_{16}H_{19}NO_3BrI$ | 480.16 | Ravin *et al.*, 1951 | 179–180 |

One of the classical synthetic methods for the preparation of esters is the Schotten-Baumann reaction, in which acetylation of a phenol or naphthol is accomplished by means of an acid chloride in the presence of dilute alkali;

$$ROH + R'COCl + NaOH \rightarrow R'COOR + NaCl + H_2O$$

The replacement of the hydrogen of the hydroxyl group by acyl groups to yield esters can also be effected with the appropriate acid anhydride under relatively mild conditions. A brief but excellent review of acetylation techniques is discussed by Hickinbottom (1957, p. 118). In recent years, certain newer procedures for the synthesis of esters have been described, such as the acetylation of hydroxyl compounds in the presence of trifluoroacetic anhydride (Bourne *et al.*, 1949) (Appendix 6A, Section 3). The use of boron trifluoride in the presence of acetic anhydride has also found application, especially in the synthesis of compounds which are not readily acetylated (Campbell and Coppinger, 1951; Fieser and Rajagopalan, 1949). Isopropenyl acetate has proved to be of value in the acetylation of compounds under mild and readily controlled conditions (Hagemeyer and Hull, 1949; Anonymous, 1950).

The synthesis of a large number of aromatic fatty acid esters according to standard techniques was reported by Gomori (1953a). Among the compounds were a series of phenyl and naphthyl fatty acid esters including acetates of substituted benzoic acid anilides.

Ravin and Seligman (1953) reported the synthesis of longer-chain fatty acid esters of dinaphtholic compounds including 2,7-naphthol dicaprate, dilaurate, dimyristate, and diundecylenate. In the same year, Ravin *et al.* reported the preparation of several complex naphthyl acetates including 5-acetamido-α-naphthyl acetate, 3-phenanthryl acetate, and 2-anthranyl acetate. Details of the synthesis were not given.

Of interest as potential substrates are acetylated azo-dyes derived from naphthol AS and other naphthols (Gränacher *et al.*, 1941). Such compounds have not as yet received histochemical application.

Gomori (1952a) reported that naphthol AS acetate could be employed as a histochemical substrate in the demonstration of esterase. Naphthol AS acetate (3-acetoxy-2-naphthanilide) (Table 1) (see Chapter 2, Table 5, for structure of naphthol AS type compounds) was first synthesized by Graff (1909) who treated 3-acetoxy-2-naphthoyl chloride with aniline. Subsequently, this compound was also prepared by Brass and Sommer (1928) who refluxed naphthol AS with acetic anhydride. This compound was similarly synthesized by Gomori (1952a) who did not report analytical data with the exception of the melting point. The latter closely corresponded with previously described values. The author (Burstone, unpublished data) has synthesized pure naphthol AS acetate by treating an acetone suspension of naphthol AS containing acetic anhydride with an excess of NaOH (Appendix 6A, Section 4). The melting points of the compounds so prepared were 155–158° (Kofler) and 158–160° (capillary tube). Higgins (1927) reported that naphthol AS acetate couples rapidly with diazonium salts, but this finding cannot be confirmed with the substrate as synthesized for histochemical use. In fact, the rationale of the application of the histochemical substrate depends upon its inability to azo-couple until the acetoxy group is hydrolyzed. Gomori (1953b) reported the preparation of naphthol AS chloroacetate. This compound was found always to contain 7–12% free naphthol which persisted as an impurity despite repeated recrystallizations from dilute ethanol or acetone.

In 1954 Pearse indicated that acetates of several additional naphthol AS type compounds could be employed as histochemical substrates. Among these were AS-OT, AS-FR, and AS-BG acetates. These correspond to AS-D, AS-OL, and AS-BG acetates according to the American and German system of naphthol nomenclature (Chapter 2, Table 1). Pearse, however, did not describe the synthetic procedure and gave no analytical data or physical characteristics of the substrates. Burstone (1957a,b) reported the synthesis and histochemical application of a number of naphthol AS acetates and related compounds. A different synthetic approach was employed in that the substrates were prepared at room temperature under relatively mild conditions by treatment of a suspension of the naphthol in acetone or tetrahydrofuran and acetic anhydride with aqueous alkali (Appendix 6A, Section 4). Among the compounds prepared and characterized (Table 1) were naphthol AS-D (II), AS-OL, AS-RT, AS-

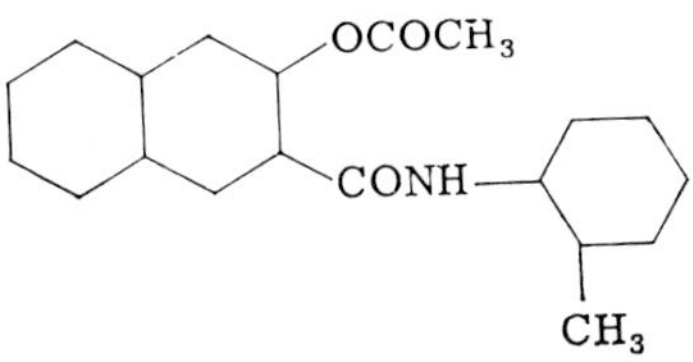

Naphthol AS-D acetate

(II)

MX, AS-E, and AS-LC acetates. Naphthol AS-LC propionate and butyrate were also described. An analogous procedure was employed to prepare an analytically pure naphthol AS-D chloroacetate (Appendix 6A, Section 5).

The ability of indoxyl and substituted indoxyls to form oxidation dyes (indigos) (Chapter 4, Section III, D) has been utilized as a principle in the histochemical demonstration of esterases. A number of indoxyl fatty acid esters have been employed as substrates. The first of these, indoxyl acetate (III), (Table I) was first synthesized by Vorländer and Drescher

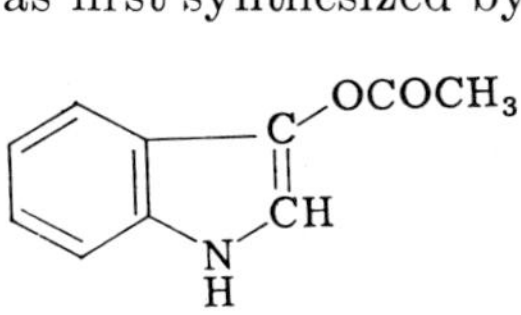

Indoxyl acetate

(III)

(1901). More recently the synthesis of indoxyl acetate has been described by Spencer (1931) and Barrnett and Seligman (1951) (Appendix 6A, Section 6). Subsequently the preparation of a large number of indoxyl acetates has been reported by Holt and Sadler (1958) and Holt (1958, p. 375). The synthetic procedure utilized involved the formation of an indoxyl ring system, by ring closure of phenylglycine-*o*-carboxylic acid to a diacetylindoxyl, by heating with acetic anhydride in the presence of sodium acetate (Chapter 4, Chart 3). The diacetylindoxyl was subsequently hydrolyzed to the free indoxyl which was then converted, by controlled acetylation, to the acetate. Of interest from a histochemical standpoint are the more recently synthesized 5-bromo-4-chloroindoxyl and 5-bromo-6-chloroindoxyl acetates (Holt, 1958, p. 375) (Appendix 6A, Section 7).

The cholinesterase group of enzymes are capable of hydrolyzing acetylcholine and analogous compounds (Wilson, 1960, p. 501; Augustinsson, 1960, p. 521). From the standpoint of enzyme histochemistry it has been shown that acetylthiocholine (AThCh) is hydrolyzed at a much more rapid rate than acetylcholine (ACh) by both specific cholinesterases as well as non-specific esterases, presumably because of the weaker linkage of the —C(=O)—S— than of the —C(=O)—O— bond (Koelle and Friedenwald, 1949). Koelle and Friedenwald (1949) and Koelle (1950) first employed as histochemical substrates acetylthiocholine iodide (Table 1) and butyrylthiocholine iodide which were synthesized according to the procedure described by Renshaw *et al.* (1938). According to Gomori (1952b, p. 210) butyrylthiocholine is hydrolyzed by a non-specific cholinesterase.

The work of Glick (1941) demonstrated that choline esters of higher fatty acids are appreciably hydrolyzed by cholinesterases. It was specifically observed that the enzymatic hydrolysis of *n*-acyl choline esters increases with the lengthening of the hydrocarbon chain to the butyryl compound, and decreases thereafter. Gomori (1948), cognizant of these observations, synthesized lauroyl-, myristoyl-, palmitoyl-, and stearoylcholine chlorides for use as histochemical substrates. These compounds have been previously reported by Fourneau and Page (1914). The method of synthesis of Gomori differed from the older procedure in that the compounds were prepared by direct esterification of choline chloride with the corresponding acyl chlorides.

Benzoylcholine and analogues are also known to be hydrolyzed by cholinesterase. Among the specific substrates are benzoylcholine iodide (Ford-Moore, 1950) and iodides of *p*-toluoyl-, *m*-toluoyl-, *p*-methoxybenzoyl-, *m*-nitrobenzoyl-, *p*-fluorobenzoyl-, and *m*-chlorobenzoylcholine (Ormerod, 1953). Ravin *et al.* (1951) reported the synthesis of analogous compounds, β-carbonaphthoxycholine iodide and 6-bromo-β-carbonaphthoxycholine iodide. Upon enzymatic hydrolysis the ester linkage between choline and the carboxylic acid is split to release β-naphthylcarbonic acid, which spontaneously decarboxylates to yield β-naphthol. The latter, of course, may be coupled with a suitable diazonium salt to produce an insoluble azo-dye.

Wolf *et al.* (1950) reported the synthesis of naphthalene and quinoline derivatives related to acetylcholine, including 2-acetoxy-3-dimethylaminonaphthalene methiodide, 8-acetoxy-1-methylquinolinium chloride, 2-propionoxy-3-dimethylaminonaphthalene methiodide, and 2-acetoxyl-3-dimethylaminonaphthalene methiodide. Unfortunately, none of these compounds proved to be of histochemical value either because of extensive spontaneous hydrolysis or else inability of the naphthalene compound to couple with diazonium salts (Ravin *et al.*, 1953).

An interesting colorimetric technique for the determination of acetylcholinesterase activity employs indophenyl acetate, which upon enzymatic hydrolysis produces a blue indophenol (Kramer and Gamson, 1958). This type of chromogenic reaction may have application in the histochemical visualization of the enzyme.

## III. HISTOCHEMICAL METHODS

In general, esterases are relatively resistant to various types of fixatives when employed in the cold. Gomori (1952b, p. 203) specifically recommends cold acetone as the best fixative for esterase and lipase. Following fixation the tissue may be double embedded, employing celloidin and paraffin (Appendix 1, Section 4). Freeze-drying, in conjunction either with single or double embedding, results in good preservation of the enzyme. Overnight fixation in 10% neutral formalin or formol-calcium is also a useful fixation procedure when frozen sections are to be subsequently employed (Seligman *et al.*, 1951). Gomori and Coutinho (1953) found that in dog tissues, 95% ethanol selectively inactivates aliesterase while cholinesterase is preserved reasonably well.

The histochemical procedures for cholinesterase have generally been performed in conjunction with fresh teased tissues or with unfixed frozen sections (Koelle and Friedenwald, 1949; Koelle, 1950; Ravin *et al.*, 1951, 1953; Holmstedt, 1957b). The advisability of employing unfixed tissues is indicated by the manometric studies of Taxi (1952) who demonstrated that formalin is a strong inhibitor of cholinesterases. On the other hand, a number of reports have indicated that the enzyme may be readily demonstrated in formalin and acetone fixed tissues (Chessick, 1954; Coupland and Holmes, 1957; Gerebtzoff, 1959, p. 4). In the case of cholinesterase or cholinesterase-like enzymes in motor end plates, the procedure may be used to advantage in tissues fixed in cold acetone and embedded in paraffin (Fig. 2).

According to Gomori (1952b, p. 202) the first histochemical method for esterases (aliesterases) was described by Kurata and Hoso who employed a suspension of olive oil as a substrate. This method, however, was very insensitive and revealed only sites of extremely high activity. In 1945 Gomori reported the first workable technique for the histochemical demonstration of esterases employing Tweens as substrates. These commercially available compounds are partly esterified common fatty acids (e.g. lauric, palmitic, stearic, and oleic) and hexitol anhydrides derived

FIG. 2. Cholinesterase or cholinesterase-like enzyme in motor end plates from rat skeletal muscle. Naphthol AS acetate (× 400) (courtesy Dr. R. D. Chessick).

from sorbitol. Polyoxyethylene chains are condensed with the non-esterified hydroxyls. These water-soluble substrates are readily hydrolyzed by esterases, and the fatty acid liberated may be precipitated in the form of an insoluble calcium soap. Esters of polyethylene glycols may also be employed as substrates. Most of the lauric, palmitic, stearic, oleic, and ricinoleic compounds are mixed esters and may be satisfactorily employed provided that they are water soluble. Gomori (1950) notes that the rate of hydrolysis decreases with increasing chain length (e.g. laurates being hydrolyzed more rapidly that stearates). On the other hand, the crystal size of the precipitate decreases with increasing chain length, and the localization is far more accurate with the more slowly hydrolyzed substrates. The saturated compounds are general substrates for lipase-esterase and are hydrolyzed by the enzymes of many tissues, while the unsaturated compounds are attacked only by the lipase of the pancreas and of the chief cells of the mouse stomach (Gomori, 1949) (Appendix 6B, Section 1).

Gomori (1952a) compared the histochemical localization of esterases, as influenced by potential activators and inhibitors, with a number of techniques including the Tween procedure. With the latter, some of the histochemical results were found to be in complete agreement with biochemical findings, while other methods indicated an overlapping between enzymes. Bile acids, for example, differentiate sharply between pancreatic lipase which is greatly activated, and esterases of other organs which are markedly inhibited. Gomori reported that the only other lipolytic enzyme besides pancreatic lipase to be activated by bile acids was found in the chief cells in a limited area of the mouse stomach. The distribution of staining with the Tween and naphtholic methods was similar although many structures which are completely negative with the Tween method are more or less intensely positive with the naphthol method (Brunner's glands, macrophages, and kidney tubules). Gomori notes that this difference is not merely an expression of the superior sensitivity of the azo-dye method, but indicates that different esterases are being demonstrated. As previously indicated, the Gomori Tween technique results in the formation of insoluble calcium soaps, which can subsequently be converted to lead soaps and then into brown-black lead sulfide. An interesting modification of the Tween procedure was suggested by Takeuchi and Furuta (1956) who stained the fatty soaps liberated from Tween with Nile Blue.

The first azo-dye method for the histochemical demonstration of esterase was described by Nachlas and Seligman (1949a) who employed the same type of procedure previously utilized by Menten *et al.* (1944a, b) for the demonstration of phosphatase. Nachlas and Seligman employed β-naphthyl acetate as a substrate in the presence of diazotized α-naphthylamine. Their method was considered to be unsatisfactory by Gomori (1952b, p. 206), Chessick (1953), and Pearse (1960, p. 464) because the final azo-dye product is readily diffusible and is also lipid soluble. Nevertheless, the contribution of the technique of Nachlas and Seligman should not be minimized since it represented the first histochemical demonstration of esterase, if only at a rather gross level.

Gomori (1952a; 1952b, p. 206) recommended that β-naphthyl acetate be replaced by the α-isomer (I) or by naphthol AS acetate in order to minimize diffusion artefacts (Appendix 6B, Sections 2 and 3). Nevertheless, the α-naphthyl acetate modification is not without defects. Gomori (1952b, p. 206) notes that the procedure must be run at a non-optimum pH (ca. 8) because of the more efficient azo-coupling of α-naphthol in the alkaline range. Despite this, diffusion artefacts are frequently seen in the case of highly active tissues which liberate naphthol so rapidly that the diazonium salt cannot couple with it to precipitate promptly the product of hydrolysis. In the case of the AS acetate, Gomori (1952a; 1952b, p. 207) specifically recommended that the substrate solution contain at least 20% propylene glycol in addition to the initial solvent (acetone) (Appendix 6B, Section 3). This type of substrate solution may result in diffusion artefacts because of the high percentage of solvent. Pearse (1954) also indicated that α-naphthyl acetate and naphthol AS acetate were advantageous in the histochemical demonstration of esterase and specifically stated that free-floating and not mounted sections should be employed. Pearse (1954; 1960, p. 466) indicated that naphthol AS-OT (AS-D) acetate was superior to the original AS acetate of Gomori. Since the only difference between AS acetate and AS-D acetate is the presence of a methyl group, this minor alteration would scarcely be expected to improve the quality of localizations. In the experience of the author, naphthol AS-D acetate is preferable to the original naphthol AS acetate primarily because it is somewhat more resistant to spontaneous hydrolysis, especially at neutral pH.

As previously mentioned, Burstone (1957a, b, 1958) synthesized and explored the applications of a number of substituted naphthol AS acetates and related fatty acid esters including naphthol AS-D (II), AS-OL, AS-MX, and AS-LC acetates. The technique was also modified in that the substrates were dissolved in a small amount (0.25–0.5 ml.) of an inert solvent (e.g. DMF), and no propylene glycol was employed (Appendix 6B, Section 4). It was found that these substrates were capable of giving consistent and exceedingly sharp staining localizations when used in conjunction with frozen-dried paraffin embedded tissues mounted on glass slides (Fig. 1), as well as with the older cold acetone fixation technique of Gomori.

Gomori (1953b) reported the preparation of an impure naphthol AS chloroacetate (containing 7–12% of free naphthol) which could be employed in a simultaneous azo-dye technique. The histological localizations obtained with this substrate in various tissues fixed in acetone or neutral formalin were different from those seen with the usual naphtholic fatty acid esters, in that mast cells of a number of species (man, dog, rat, mouse, rabbit) became intensely stained in a matter of minutes. No other tissue components exhibited more than a trace of reactivity during this short period. After longer incubations, other tissue elements became reactive, but the pattern of staining was the same as seen with the usual naphthol acetates. Burstone (1957b) reported the synthesis of a pure naphthol AS-D chloroacetate (Fig. 3) which was later employed by Moloney *et al.* (1960) in a study of human myeloid cells in normal and pathological conditions.

Barrnett and Seligman (1951) first proposed the use of indoxyl acetate (Table 1) (III) as a histochemical substrate for the demonstration of esterases. These authors similarly described the histochemical application of indoxyl butyrate. Unfixed frozen sections were incubated in a substrate mixture containing indoxyl acetate buffered at pH 8.5. The enzymatically released indoxyl auto-oxidized to form a precipitate of indigo (Chapter 4, Section III, D) (Chart 1). Gomori (1952b, p. 147) has criticized the technique in that the process of oxidation to insoluble indigo is a relatively slow process under the conditions employed by Barrnett and Seligman. Their study, however, is of considerable interest because they suggest that indoxyl acetate is hydrolyzed by lipase, non-specific esterase, and cholinesterases.

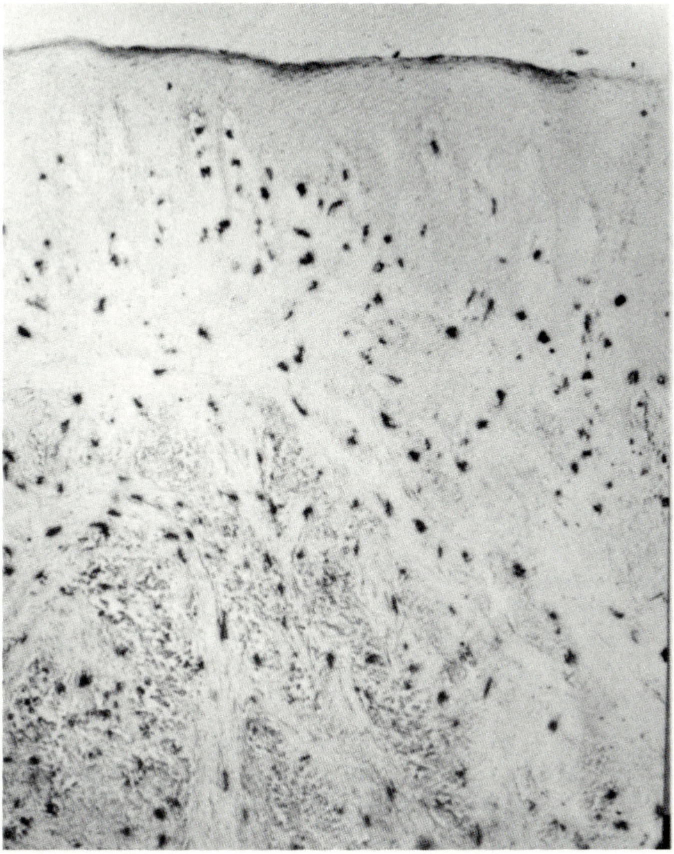

FIG. 3. Selective staining of mast cells in frozen-dried human gingiva. Naphthol AS-D chloroacetate (× 73).

Holt (1952) and Holt and Withers (1952) reported an improved indoxyl technique which employed substituted indoxyl esters including 5-bromoindoxyl acetate. In addition, the substrate medium contained an oxidizing agent (ferricyanide-ferrocyanide) which greatly accelerated the transformation of the enzymatically released indoxyl into an indigo precipitate, and thus eliminated to a great extent the initial defects of the technique as criticized by Gomori. In addition to the slow rate of oxidation and precipitation of indigo, the use of simple indoxyl acetate also suffers from another defect in that the precipitated indigo deposits in a coarse granular crystalline pattern in tissue sections and is not capable of giving accurate intracellular localizations. Further studies of Holt and Withers (1958) and Holt (1958) have described the application of a variety of halogen substituted

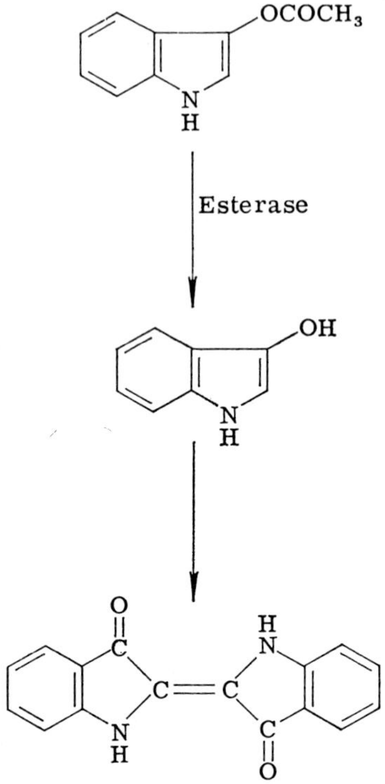


CHART 1. Enzymatic hydrolysis of indoxyl acetate resulting in the formation of an insoluble precipitate of indigo.

indoxyl acetates as histochemical substrates. Of the various substrates employed, 5-bromo-4-chloroindoxyl, 5-bromoindoxyl, and 5-bromo-6-chloroindoxyl acetates were found to be the most suitable substrates (Appendix 6B, Section 6) in that the final dye product was relatively homogeneous and lipid insoluble. Although precise staining reactions (e.g. kidney "droplets") may be facilitated with the indoxyl procedures, equally good localizations may be obtained with the more readily available and less expensive naphtholic substrates.

According to Underhay (1957), human blood contains five esterases capable of hydrolyzing unsubstituted indoxyl esters. Three of the esterases, acetylcholinesterase, and aliphatic esterase, and an aromatic esterase, are found in erythrocytes. The remaining two esterases, butyrylcholinesterase and an aromatic esterase (which is not identical with the aromatic esterase of the erythrocytes), are present in plasma. Hobbinger (1957) has found that unsubstituted indoxyl esters are hydrolyzed by rat erythrocyte preparations and plasma and liver homogenates at a considerably greater rate than any of the corresponding choline esters. The enzymes chiefly responsible for the hydrolysis of the indoxyl compounds are organophosphate-sensitive esterases.

The thiocholine method for the demonstration of cholinesterase (Koelle and Friedenwald, 1949; Koelle, 1950, 1951) depends upon the enzymatic hydrolysis of carboxylic acid esters of thiocholine sulfate in the presence of a cupric iron complex. The liberated thiocholine sulfate is precipitated as the copper mercaptide, according to Malmgren and Sylvén (1955) ($[Cu^{+}—S—CH_2—N^{+}(CH_3)_3]\ SO_4^{--}$). The precipitated thiocholine is converted into copper sulfide by subsequent treatment with ammonium sulfide (Appendix 6B, Section 7). It was reported that the use of acetylthiocholine iodide rendered the method specific for acetylcholinesterase. When butyrylthiocholine iodide was employed as a substrate, the reaction was characterized by the use of a selective inhibitor. Modifications of the technique have been reported by Coupland and Holmes (1957) and by Holmstedt (1957a,b). The latter author employed the use of Mipafox and BW284C51 as specific inhibitors of butyrylcholinesterase and acetylcholinesterase, respectively.

In view of the known acetyl esterase function of cholinesterase, simple aromatic fatty acid esters have been found to be hydrolyzed by cholinesterase. For example, human serum cholinesterase hydrolyzes *o*-nitrophenyl butyrate about twice as rapidly as acetylcholine (Main *et al.*, 1961). Ravin *et al.* (1953) indicated that $\beta$-naphthyl acetate is susceptible to enzymatic hydrolysis by acetylcholinesterase, and that the enzyme could be histochemically delineated by the clear cut inhibitory effect of low concentrations of eserine. Chessick (1954) compared both the azo-dye and thiocholine techniques in conjunction with various inhibitors and concluded that cholinesterase of the various species and tissues should be viewed as a family of enzymes, forming a spectrum with multiple overlapping and differences with respect to physical properties, effective inhibitors, and substrate specificity. A similar study was reported by Denz (1953). In any case, it is possible to clearly visualize sites of known cholinesterase activity (e.g. motor end plates) with naphtholic acetates such as naphthol AS acetate (Fig. 2).

Naphtholic compounds similar to acetylcholine may also have application in the histochemical demonstration of cholinesterase. Ravin *et al.* (1951) reported 6-bromo-$\beta$-carbonaphthoxycholine iodide which could be employed in a simultaneous azo-dye coupling procedure for the demonstration of a plasma type cholinesterase in tissue sections.

As previously indicated, indoxyl esters may serve as substrates for cholinesterases (Barrnett and Seligman, 1951; Holt, 1952). Holt (1952) has reported on kinetic studies of the comparative rates of hydrolysis of a number of indoxyl esters by specific and non-specific cholinesterase, and notes a significant structural resemblance between *o*-acetylindoxyl and acetylcholine.

In 1955 Crevier and Bélanger first reported that thiolacetic acid ($CH_3COSH$) could be employed as a histochemical substrate for the demonstration of cholinesterase. Biochemical studies have previously shown that acetylcholinesterase catalyzed the hydrolysis of thiolacetic acid with the liberation of acetic acid and hydrogen sulfide (Wilson, 1951). The evolved $H_2S$ could be trapped *in situ* by either lead nitrate or lead acetate to form an insoluble precipitate. Minor modifications of this principle have been described by Sávay and Csillik (1959) and Wachstein *et al.* (1961). According to the latter group, although the thiolacetic acid method is clearly capable of delineating cholinesterase activity, it also reveals sites of aliesterase and possibly other types of esterases in a number of tissues.

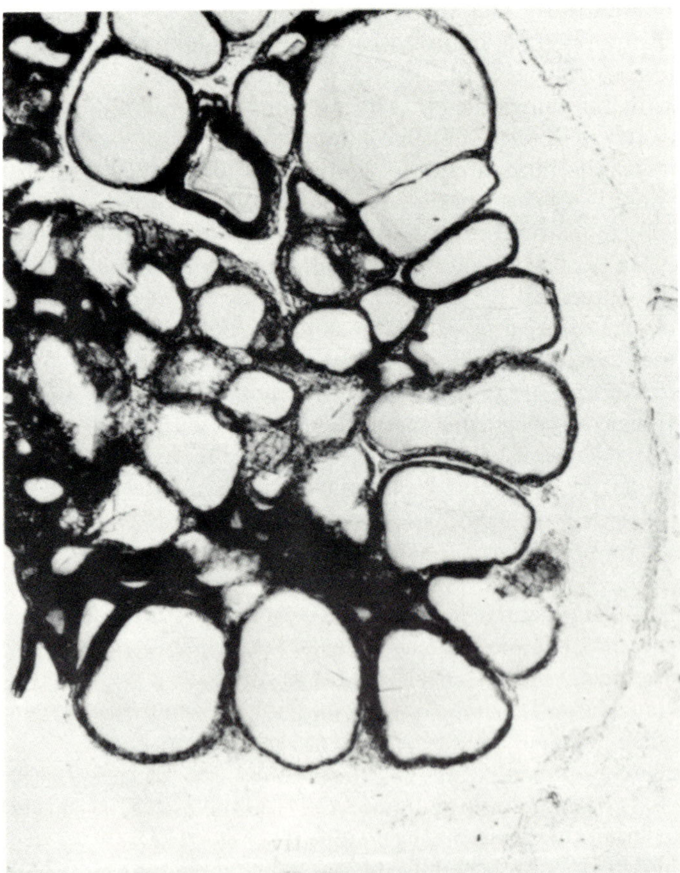

FIG. 5. Esterase activity of rat thyroid. Naphthol AS-D acetate (× 80).

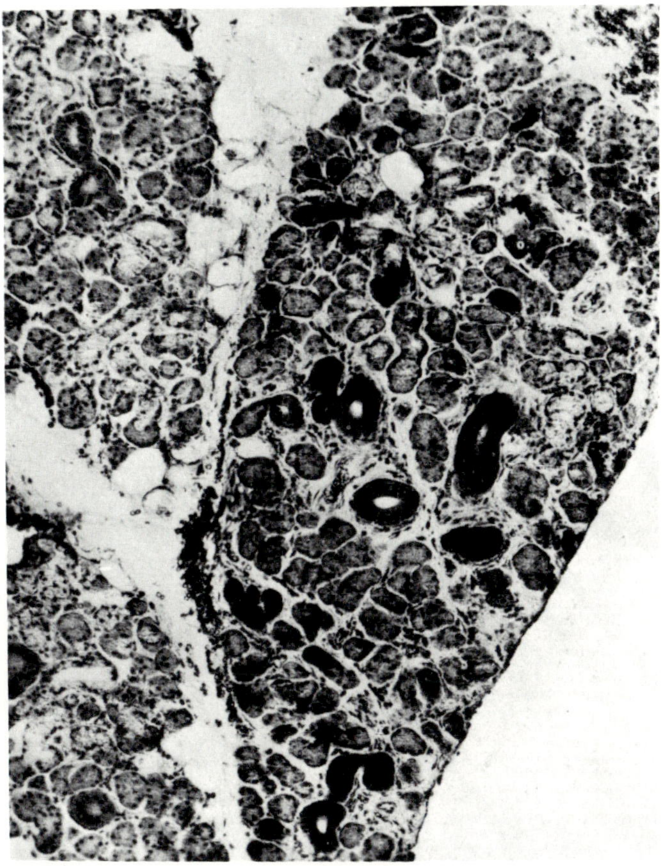

FIG. 6. Esterase activity of ducts of frozen-dried human submaxillary gland. Naphthol AS-D acetate. Hematoxylin counter stain (× 110).

Among the more comprehensive reviews dealing with the histochemical application of esterase techniques are those by Gomori (1952a), Denz (1953), Chessick (1953), Gomori and Chessick (1953b) Montagna (1955), Snell and McIntyre (1956), Pearson and Defendi (1957), Pepler and Pearse (1957), Hess and Pearse (1958), Holt (1958), Gerebtzoff (1959), Fukuda and Koelle (1959), and Eränkö *et al.* (1959).

Esterase activities of various tissues are illustrated by Figs. 1 through 13.

## IV. ACTIVATORS AND INHIBITORS

Biochemical and histochemical reports dealing with general and specific aspects of various esterase inhibitors are considered in the reviews and reports of Ammon and Jaarma (1950, p. 390), Augustinsson (1950, p. 433), Gomori (1952a), Gomori and Chessick (1953a), Aldridge (1953a, b, c,

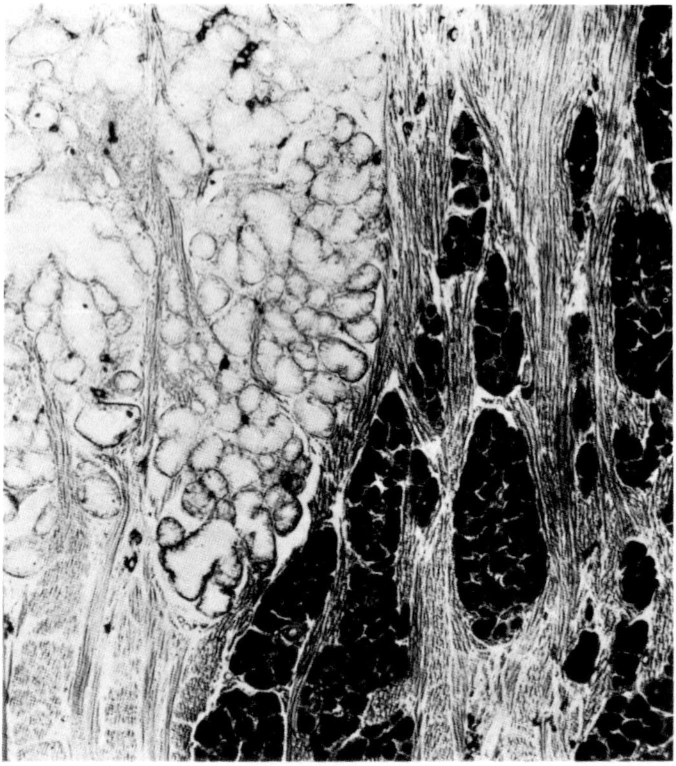

FIG. 7. Esterase activity of serous cells of frozen-dried rat tongue. α-Naphthyl acetate. The mucous glands are unstained with the exception of subjacent demilunes (× 70).

1954), Chessick (1954), Gomori (1955a), Bergmann *et al.* (1957), Bergmann and Rimon (1958), Hofstee (1960, p. 485), and Wilson (1960, p. 501).

Sodium fluoride, sodium arsanilate, and sodium taurocholate are inhibitors for simple esterases while quinine has no effect (Gomori, 1952a). On the other hand, taurocholate behaves as a lipase activator (Nachlas and Seligman, 1949b; Gomori, 1952a). Of interest is the apparent delineation of two types of esterases as demonstrated with naphthol AS acetate. One type present in the liver is inhibited by fluoride and DFP ($10^{-6}$ *M*), while a more general type found in the stomach, kidney, and other organs is insensitive to fluoride and DFP (Gomori, 1955a). Gomori and Chessick (1953a) have conducted studies on the inhibition of esterases as revealed by the azo-dye method.

The previously mentioned A, B, and C esterases may be characterized

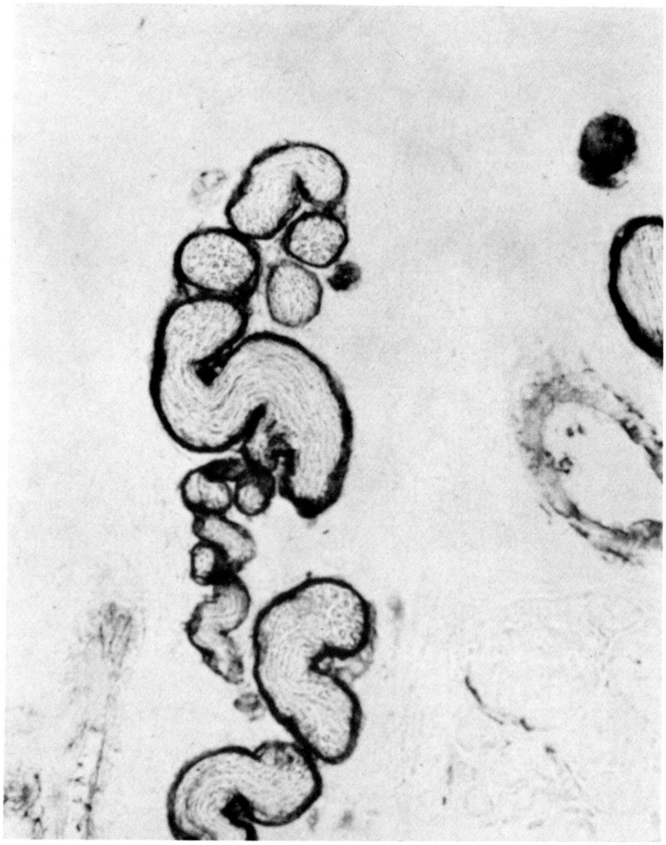

FIG. 8. Esterase activity of perineurium. Frog tongue. Naphthol AS-D acetate (× 143).

by their sensitivity to certain inhibitors (Aldridge, 1953a, b, c, 1954; Bergmann *et al.*, 1957; Bergmann and Rimon, 1958). The A-esterases which are believed to contain a sulfhydryl group are sensitive to certain metallic ions including $Hg^{++}$, $Ni^{++}$, and $Cu^{++}$. Iodoacetate is also inhibitory. A-esterases are not inhibited by organophosphorus compounds while the B-esterases, on the other hand, are inhibited by very low concentrations of organophosphorus compounds such as DFP and E-600 (diethyl-*p*-nitrophenyl phosphate). The C-esterase is unaffected by DFP but is strongly activated by aromatic mercurials such as *p*-chloromercuribenzoate (10–50 $\mu M$). The enzyme is inhibited by metallic salts including mercuric acetate and copper sulfate. Thus far, the histochemical delineation of A, B, and C esterases has not been established although preliminary

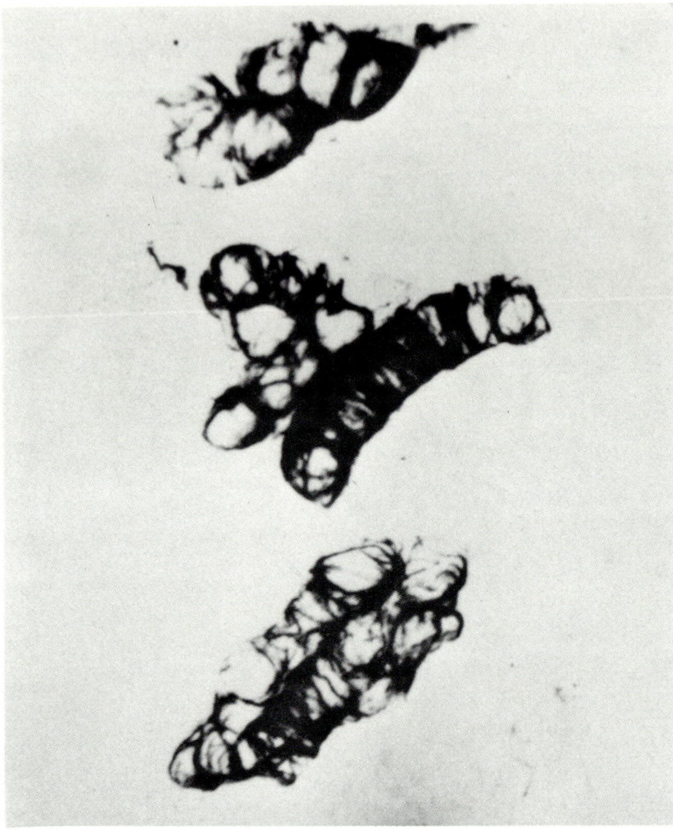

FIG. 9. Specific cholinesterase in human eccrine sweat gland. Thiocholine method (× 500) (courtesy Dr. William Montagna).

attempts in this direction have been reported by Pepler and Pearse (1957) and Hess and Pearse (1958). Koelle (1951) employed DFP for the inhibition of cholinesterase in conjunction with the thiocholine method. According to Gomori (1952b, p. 210), the non-specific enzyme is almost completely inhibited by a 30 minute incubation with $10^{-6}$ $M$ DFP while true cholinesterase is inhibited approximately 40%. It has, however, been suggested that acetylthiocholine may be hydrolyzed by enzymes other than cholinesterase even when DFP is employed as an inhibitor. Bayliss and Todrick (1956) suggested that acetylcholinesterase could be delineated more precisely through the use of selective inhibitors. Histochemical application of this general approach was reported by Pepler and Pearse (1957) who employed specific acetylcholinesterase inhibitors including 62.C.47 [1,5-bis(4-trimethylammoniumphenyl)pentan-3-one di-iodide].

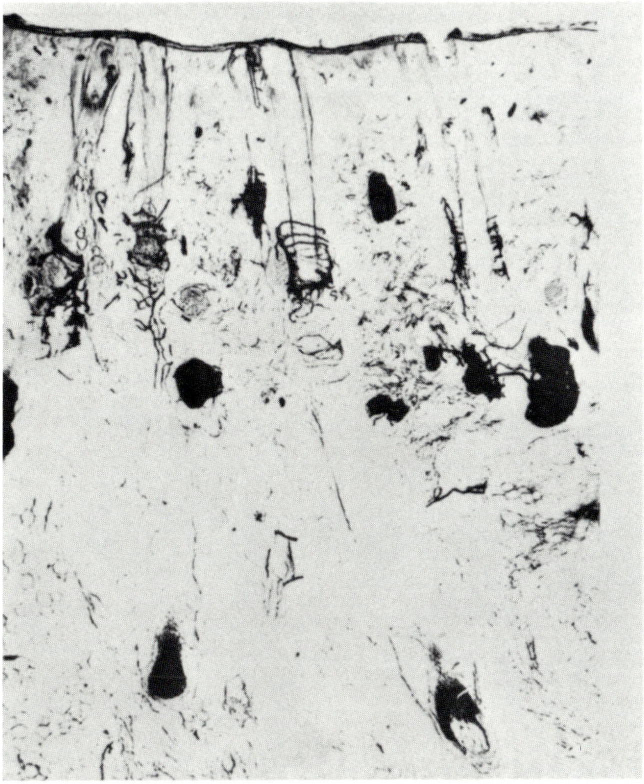

FIG. 10. Specific cholinesterase activity in human scalp. Thiocholine method (× 80) (courtesy Dr. William Montagna).

Pearse (1960, p. 485) questions the efficacy of these inhibitors and indicates that iso-OMPA and lysivane (ethopropazine methosulfate) should be more effective. Other cholinesterase inhibitors which have histochemical application are Nu-683 (2-hydroxyphenylbenzyltrimethylammonium bromide dimethyl carbamate) (Hawkins and Gunter, 1946) and Nu-1250 (*N*-*p*-chlorophenyl-*N*-methyl carbamate of *m*-hydroxyphenyltrimethylammonium bromide) (Hawkins and Mendel, 1949). Non-specific cholinesterase was inhibited by iso-OMPA (tetraisopropylpyrophosphoramide). An interesting discussion on the effects of potential cholinesterase and cholinesterase-like enzyme inhibitors when employed with the thiocholine and azo-dye techniques is given by Chessick (1954).

Koelle (1957) has reported on the histochemical demonstration of reactivated acetylcholinesterase *in vivo* by the use of pyridine-2-aldoxime methiodide (2-PAM) and certain other oxime and hydroxaminic acid de-

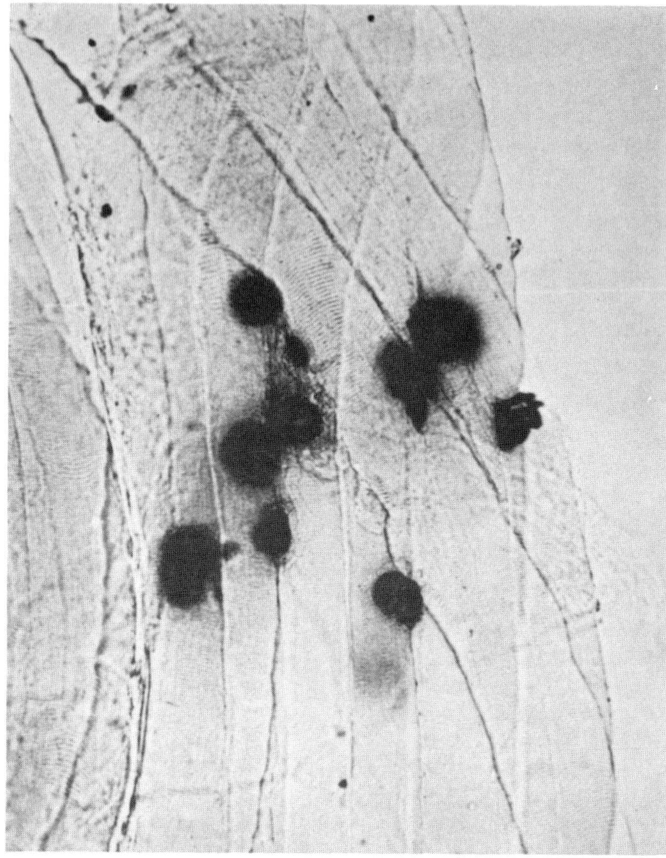

FIG. 11. Motor end plates of rat intercostal muscle. Acetylthiocholine method (× 125) (courtesy Dr. George B. Koelle).

rivatives. This reactivation could be demonstrated following inactivation by DFP and other irreversible alkyl phosphate anticholinesterase agents.

## V. ESTERASES IN NEOPLASMS

The earliest study on the histochemical distribution of esterase (lipase) activity in neoplasms was that of Gomori (1946) who employed the Tween method. Although a number of investigators have subsequently reported on the distribution of esterases employing the aforementioned procedure, Gomori (1955b) notes that the Tween technique is outmoded because of its low sensitivity. Cohen *et al.* (1951) described the histochemical demonstration of esterases in malignant neoplasms employing $\beta$-naphthyl acetate. As previously indicated (p. 303) this technique is inadequate for the

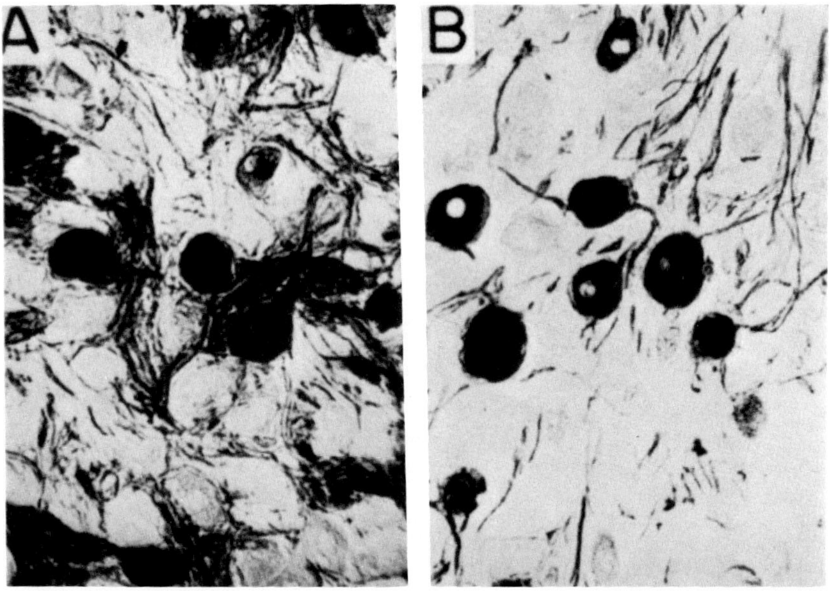

FIG. 12. Autonomic ganglia of cat. Acetylthiocholine method. A. Normal stellate ganglion. B. Stellate ganglion, preganglionically denervated (× 200) (courtesy Dr. George B. Koelle).

accurate visualization of enzyme activity because of the marked tendency for diffusion artefacts to occur. In general, the most satisfactory substrates have been naphtholic esters including $\alpha$-naphthyl acetate and naphthol AS acetate and derivatives. Although substituted indoxyl esters are capable of giving reproducible results, they have found limited application in the histochemical study of neoplasms.

As previously indicated, esterases are widely distributed in many tissues but highest activity is found in the liver, pancreatic acini, chief cells of the gastric mucosa, kidney tubules, and bronchial mucosa. The enzymatic pattern of tumors may or may not duplicate that of the parent tissue (Gomori, 1955b). Hepatomas give intensely positive reactions; cholagiomas and metastatic carcinomas of the liver are usually unstained. Gastric carcinomas are either negative or else exhibit a slightly positive reaction. Epidermoid carcinomas of the esophagus and bronchi may be intensely positive; and when a bronchiogenic carcinoma invades the pleura, the large esterase positive tumor cells in the pleural tap present a distinctive picture of diagnostic significance. Gomori (1955b) also reports that tumors of the breast behave erratically, in that some are completely negative while in rare cases an extensive staining reaction may be seen.

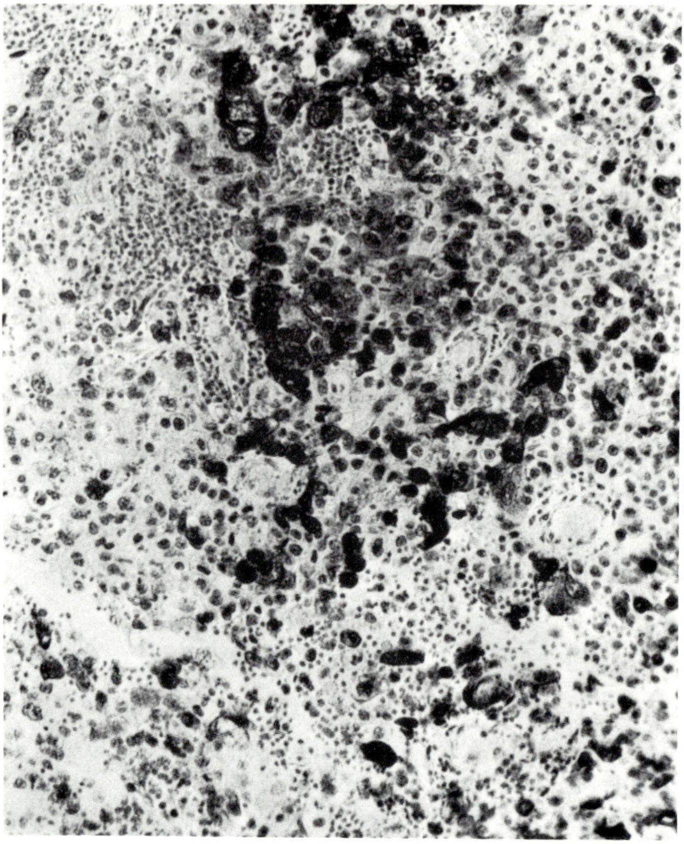

FIG. 13. Esterase activity of tumor cells from undifferentiated lung adenocarcinoma. Naphthol AS-D acetate (× 143).

Numerous cases are seen in which highly active cells are scattered among a majority of morphologically identical but inactive cells. Tumors of the thyroid often retain the typical enzymatic distribution pattern of the normal thyroid in that esterase activity is localized in the basement membrane and/or the basal portion of the cell. Renal carcinomas are negative or moderately positive.

Gomori (1955a, b) reported the case of a 5 year old girl who had a large tumor mass involving the entire root of the mesentery, the back surface of the liver, and the proximal half of the pancreas. Microscopically the tumor consisted of fairly large, polyhedral, rather clear cells which contained no glycogen. It was impossible to determine whether the tumor originated in the liver or pancreas and no other tumor was found which could be considered to be the primary. From a histochemical standpoint the tumor exhibited a very intense esterase reaction. When inhibitors were used it was found that the tumor enzyme was completely insensitive to fluoride, totally inhibited by DFP, and somewhat activated by anionic detergents. These features clearly classify it as a pancreatic esterase. On the basis of these findings the tentative diagnosis of embryonic carcinoma of the pancreas was made. In a study of over one hundred various human tumors by histochemical esterase techniques, Gomori (1955a) found that if a tumor contains any esterase at all, it is identical in type with that of the parent tissue. With reference to the effect of inhibitors, hepatomas are inhibited by fluoride whereas tumors of the thyroid, prostate, and bronchi are not. On the basis of these findings, the identification of the origin of some neoplasms may be made possible by histochemical means when routine morphological criteria are inadequate. There is little question that this type of diagnostic approach will be embodied in histopathological diagnostic procedures of the future.

Burstone (1957c) reported esterase activity of several fairly typical frozen-dried human oral epidermoid carcinomas. With the α-naphthyl acetate procedure activity was observed to be associated with sites of keratinization (e.g. "pearls" and dyskeratotic cells), as well as in groups and clusters of poorly differentiated cells which appeared to be morphologically similar to adjacent non-active tumor cells. Burstone (1956) also reported high esterase activity of certain groups of cells in a poorly differentiated lung adenocarcinoma (Fig. 13). Brandes and Bourne (1956) reported on a diminished esterase activity in acetone-fixed prostatic adenomas and carcinomas in conjunction with the α-naphthyl acetate procedure.

Gross and Danzinger (1957) found that the esterase activity of normal cervical epithelium was limited to the basal layer. A slight increase was associated with benign changes, a moderate increase with carcinoma *in situ,* and very high activity was found in infiltrating epidermoid carcinoma of the cervix. Similar observations were reported by Hopman (1960).

Pearse and Pepler (1957) studied three carcinoids employing 5-bromoindoxyl acetate and suggested that the esterase demonstrated might be related to a cathepsin, although biochemical evidence was not presented to substantiate this assumption.

Takeuchi *et al.* (1955) studied smears from ascites of patients with gastric and liver neoplasms and found esterase more frequently in cancer cells from the liver than those from the stomach.

With reference to tissue cultures, Sandritter and Schiemer (1959) have described various esterases in HeLa cells in the cytoplasm and Golgi region.

Wachstein and Meisel (1959) found distinct differences in the staining patterns between fresh and formalin fixed frozen sections of rat liver in conjunction with naphthol AS acetate. It is possible that the difference in distribution of dye particles may be related to alterations in lipids as influenced by fixation. Strong esterase activity was observed in proliferating liver cells in ethionine-induced cirrhosis. Malignant hepatomas frequently exhibited a marked staining reaction.

Gomori (1953b) and Moloney *et al.* (1960) reported that blood smears from patients with myeloid leukemia exhibited a decreased reactivity when naphthol AS or AS-D chloroacetate were employed as substrates.

Zacks (1958) studied several non-chromaffin paragangliomas with an azo-dye procedure and felt that the inhibition of the reaction by eserine indicated a cholinesterase-like reaction. Winkelmann (1960a, b) studied the distribution of acetylcholinesterase and non-specific cholinesterase using the thiocholine procedures. Neuroepithelial nevi exhibited non-specific cholinesterase activity. Positive reacting cells in junctional nevi were found in the deeper intradermal regions. Melanocarcinomas and blue nevi did not stain.

# APPENDIX 6A

# SYNTHESIS OF SUBSTRATES

## 1. *General Procedure for Synthesis of Simple Naphtholic Fatty Acid Esters*

The following technique is useful in the preparation of α-naphthyl acetate, propionate, butyrate, and analogous naphtholic esters. This procedure is also useful in the preparation of esters which form oils or liquids at room temperature.

α-Naphthol (7.0 g.) is dissolved in 150 ml. of acetone and a little more than the calculated amount of NaOH is added (3.0 g./10.0 ml. of $H_2O$). The solution is cooled in ice water and the appropriate acid chloride or anhydride is slowly added with constant stirring or shaking. If the acid chloride is employed, a colorless supernatant is observed and NaCl is precipitated. The supernatant is poured into ice water, collected, and dissolved in excess ether. The ether solution is placed in a separatory funnel and ice cold 1% NaOH solution (same volume as the ether mixture) is added and the ether solution shaken out twice with the cold alkaline solution. After the second washing the NaOH solution should appear clear and free of any residue. An equivalent volume of 1% $H_2SO_4$ is then added to the separatory funnel and the mixture shaken in order to remove any traces of alkali. The ether mixture is then added to a beaker containing anhydrous sodium sulfate, the mixture stirred, and then the solution is decanted. The resulting naphtholic ester may be purified by fractional distillation or by recrystallization at lower temperatures at which the oils tend to solidify.

## 2. *α-Naphthyl Acetate* (*Chattaway, 1931*)

This is a relatively simple technique by which phenolic and naphtholic compounds may be rapidly acetylated. The compound to be acetylated is dissolved in an aqueous solution containing a slight excess of the theoretical amount of sodium or potassium hydroxide to which has been added crushed ice. The mixture is then poured rapidly into an equivalent of acetic anhydride.

One mole of α-naphthol is dissolved in a solution containing 1.5 moles of sodium hydroxide (60.0 g.) in 100.0 ml. of water. The solution is cooled and mixed with 500.0 g. of crushed ice. Acetic anhydride (1.25 moles, 128.0 g.) is quickly added and the mixture vigorously shaken for a few minutes. Naphthyl acetate separates as a colorless liquid. It is washed with a little dilute NaOH and then with water, and dried over calcium chloride. The compound may be further purified by recrystallization from ethanol-water.

3. *β-Naphthyl Acetate, Trifluoroacetic Anhydride Synthesis* (*Bourne et al., 1949*)

Trifluoroacetic anhydride (0.080 ml.) is added to a solution of β-naphthol (0.5 g.) in glacial acetic acid (3.5 ml.). Heat is generated, and the reaction mixture is allowed to cool slowly and is then poured into sodium bicarbonate solution. The solution is exhaustively extracted with chloroform and the extract dried with magnesium sulfate and evaporated to a syrup. The product is recrystallized from aqueous ethanol (m.p. 70°).

β-Naphthyl benzoate, β-naphthyl propionate, and β-naphthyl palmitate may be similarly synthesized.

4. *Naphthol AS Acetates* (*Burstone 1957a, b*)

This procedure is advantageous in that it is performed at room temperature and without the necessity of a reflux system. Naphthol AS-D, AS-E, AS-OL, AS-RT, AS-MX, and AS-LC acetates (Table 1) may be prepared by this technique. If the purity of the starting naphthol is in question, it is always advisable to recrystallize the compound.

To a suspension of ca. 5.0 g. of naphthol in 50.0 ml. of tetrahydrofuran or acetone is added an excess of acetic anhydride (5.0 ml.). A slight excess of 5 *M* NaOH is then added with constant stirring until the solution becomes clear. The clear supernatant is carefully decanted and poured into ca. 800 ml. of petroleum ether. Upon standing and with gentle stirring, a precipitate forms which is collected under suction and washed with petroleum ether. The precipitate is then dried, dissolved in hot methanol containing charcoal, filtered, and recrystallized several times from methanol or methanol-ethanol.

5. *Naphthol AS-D Chloroacetate* (*Burstone, 1957b*)

Approximately 0.5 *M* (4.0 g.) of recrystallized naphthol AS-D is suspended in 50.0 ml. of acetone or tetrahydrofuran (hydroquinone-free) containing 8.0 g. of chloroacetic anhydride. Approximately 8.0 ml. of 5 *N* NaOH solution are added with constant stirring, and the mixture is allowed to stand for 30 minutes at room temperature. The supernatant is filtered through rapid filtering paper into several hundred ml. of petroleum ether, in which the chloroacetate crystallizes. The compound may be further purified by dissolving it in a minimum of tetrahydrofuran and recrystallizing again from petroleum ether (m.p. 162°) ($C_{20}H_{16}O_3NCl$). Anal. calcd.: C, 67.8%; H, 4.5%; N, 4.1%; Cl, 10.02%. Found: C, 67.3%; H, 4.5%; N, 4.5%; Cl, 10.61%.

6. *Indoxyl Acetate (Vorländer and Drescher, 1901; Spencer, 1931; Barrnett and Seligman, 1951)*

Indoxyl acetate may be readily prepared from the commercially available sodium indoxyl-alkali flux melt (General Dyestuff Corp., N. Y.) which is an intermediate in the industrial preparation of indigo.

Approximately 64 g. of the aforementioned compound (powder) are placed in a liter of water and ice in a 2 liter Erlenmeyer flask into which illuminating gas is constantly passed to prevent extensive auto-oxidation. As soon as the melt dissolves, 40.0 ml. of 50% sulfuric acid are rapidly added and with stirring. This is followed by the addition of 25.0 ml. of acetic anhydride. The flask is then stoppered and shaken. Illuminating gas is bubbled through the mixture for 10 minutes. A green-blue precipitate which forms is collected with suction and washed with water. The filter cake is extracted approximately 6 times by suspending it in 300.0 ml. of boiling water, followed by rapid filtration under suction. Activated charcoal is employed during the last few extractions. Upon cooling of the combined filtrates, fine grayish needles are precipitated. The product is recrystallized from hot water or dilute ethanol to form white needles (m.p. 126–128°).

Spencer (1931) employed a partly neutralized indoxylic acid melt which was boiled for 10 minutes, cooled, and then acetylated with acetic anhydride. After allowing the reaction mixtures to stand for 30 minutes the *o*-acetylindoxyl is filtered off, washed free from acetylindoxylic acid and recrystallized from water.

7. *Indoxyl Acetate (Holt, 1958, p. 389)*

The synthesis of this compound as described by Holt involves the formation of phenylglycine-*o*-carboxylic acid and diacetylindoxyl.

A suspension of anthranilic acid (13.7 g.) in 20.0 ml. of water and a solution of monochloroacetic acid (9.5 g.) in 20.0 ml. of water are separately treated with 30% KOH solution (ca. 19.0 ml.) until clear solutions, just alkaline to phenolphthalein, are obtained. The cooled solutions are mixed, treated with activated charcoal (1.0 g.) and filtered. The filtrate is then incubated in a covered vessel at 50° for 48 hours. The resulting suspension is cooled and the precipitate of the required salt collected, washed with water, and dried.

A mixture of potassium acetate (15.0 g.) and acetic anhydride (90.0 ml.) is gently boiled in a 500 ml. beaker, and finely powdered dry monopotassium salt of phenylglycine-*o*-carboxylic acid (15.0 g.) is added as rapidly as possible with stirring, allowing the reaction to subside between each addition. The resulting orange solution is boiled for an additional 15 minutes and cooled. Water (90.0 ml.) is added to hydrolyze the excess acetic anhydride. The resulting suspension of solid diacetylindoxyl is collected, washed with 30% acetic acid (20.0 ml.), 50% ethanol, and dried. The diacetylindoxyl is a cream-colored crystalline solid with a m.p. of 80–82°. The purified compound is a white crystalline solid with a m.p. of 83–84°.

The aforementioned diacetyl compound is transformed into indoxyl acetate as follows: sodium hydroxide solution (55.0 ml. of a 2 *N* solution) is added to a multi-necked flask fitted with a stirrer, reflux condenser, tap funnel, and delivery tube reaching to the bottom of the flask. A slow stream of oxygen-free nitrogen is passed through the solution and apparatus for 5 minutes to displace air, and diacetylindoxyl (2.17 g.) is added into the system through a stoppered side neck. The stopper is replaced and nitrogen is passed into the system for 5 minutes. The mixture is heated to boiling and then under reflux until the diacetylindoxyl has completely dissolved. The indoxyl solution is cooled to 0° and acetic anhydride (5.3 ml.) is slowly added from the tap funnel with rapid stirring. After allowing the mixture to stir for approximately 20 minutes, the precipitated indoxyl acetate is collected, washed with water, and crystallized from aqueous ethanol by the following procedure. The moist product is dissolved in 20.0 ml. of absolute ethanol by applying gentle heat. Activated charcoal (0.5 g.) is added, the mixture filtered, and the filtrate reheated to about 60°. An equal volume of water at the same temperature is then added. Upon cooling of the solution, pure indoxyl acetate separates as white plates (m.p. 129°).

8. *5-Bromo-4-chloroindoxyl and 5-Bromo-6-chloroindoxyl Acetates* (*Holt, 1958*).

These substrates are synthesized from *m*-chloroaniline by conversion into a mixture 4-chloroisatin and 6-chloroisatin via 3-chloroisonitrosoacetanilide. The isatins are then separated and converted independently into the appropriate substrates. Because of the detailed description of the synthesis the original paper of Holt (1958) should be consulted.

# APPENDIX 6B

# HISTOCHEMICAL METHODS

## 1. *Esterase-Lipase, Tween Method (Gomori, 1952b, p. 203)*

Cold acetone fixed as well as frozen-dried paraffin embedded tissues may be employed. A fairly large number of Tween and related compounds may be utilized as substrates. The saturated compounds are substrates for lipase-esterase, while the unsaturated ones are attacked only by lipase.

A 5% Tween 60 solution in 0.1 *M* tris buffer pH 7.0–7.4 is prepared. Calcium chloride (0.5%) and a few crystals of camphor are added. The solution is incubated at 37° for 24–48 hours and filtered through a fritted glass plate or Seitz asbestos filter. The filtered solution is stored under refrigeration. One ml. of the aforementioned stock solution is added to ca. 45.0 ml. of 0.05 *M* tris buffer pH 7.0–7.4 containing 3.0 ml. of a 2% solution of calcium chloride and a few crystals of camphor.

Slides are incubated in the substrate solution at 37° for 8–24 hours. The slides are rinsed in distilled water and immersed in a 1–2% solution of lead nitrate for 10 minutes at 55–60°. The slides are then rinsed in repeated changes of distilled water for 5–10 minutes and immersed for about 5 minutes in a dilute solution of light-yellow ammonium sulfide. The slides are washed in tap water, counterstained if desired, dehydrated, cleared in gasoline or tetrachloroethylene, and mounted in a suitable synthetic medium. Sites of lipase-esterase activity appear in gold-brown shades.

Alternatively the precipitate of calcium soap may be visualized by staining with Nile Blue. A colloidal Nile Blue solution is prepared by dissolving 20.0 mg. of the dye in 20.0 ml. of ethanol and then filtering the solution. The filtrate is added drop by drop to a colloidal solution prepared by dissolving 5.0 g. of gum arabic in 200.0 ml. of boiling distilled water. The solution is then evaporated to a volume of 100.0 ml. A 5 minute staining period in this solution is adequate for the visualization of sites of lipase activity as a blue stain (Takeuchi and Furuta, 1956).

## 2. *Esterase, α-Naphthyl Acetate Method (Gomori, 1952b, p. 207)*

This technique is of use in conjunction with a variety of fixed tissues. Unfortunately, there is a marked tendency for diffusion artefacts to occur, especially in the case of highly active tissues such as liver, intestine, and pancreas (Gomori, 1952b, p. 206).

A 1% stock solution of α-naphthyl acetate in 50% acetone is prepared and stored under refrigeration. For use, 0.5–1.0 ml. of the stock solution is blown into 50.0 ml. of distilled water and a few ml. of 0.2 *M* solution of phosphate buffer (ca. pH 8.0) and 20–50 mg. of diazonium salt are added.

Among the diazo compounds which may be employed are Fast Blue B, Fast Blue RR, and Fast Red RC salts. The mixture is stirred and filtered into a Coplin jar.

Sections or slides are incubated, preferably at a temperature not over 20° for a period of 5–20 minutes or until the desired staining intensity is obtained. In the case of highly active tissues the substrate solution is cooled and/or stirred to minimize diffusion artefacts. The slides are washed in running water, counterstained if desired, and mounted in glycerol-gelatin or a synthetic mountant.

3. *Esterase, Naphthol AS Acetate Method* (*Gomori, 1952b, p. 207*)

This technique is primarily intended for use with paraffin embedded sections mounted on glass slides. The use of frozen sections may result in diffusion artefacts because of the high percentage of solvent contained in the substrate solution.

A 1% solution of naphthol AS acetate in equal parts of acetone and propylene glycol is prepared and stored under refrigeration. For use, 0.5–1.0 ml. of the stock solution is blown into 10–15 ml. of propylene glycol and the mixture shaken. The solution is then diluted, with constant stirring, with water to form a total volume of 50.0 ml. A few ml. of 0.2 *M* phosphate buffer ca. pH 6.5 is added (the mixture may become opalescent). Then 20–50 mg. of diazonium salt (e.g. Garnet GBC salt) are added, the solution stirred and filtered into a Coplin jar.

Slides are incubated at room temperature for 30–120 minutes, rinsed in water, counterstained if desired, and mounted in glycerol-gelatin or a water-soluble synthetic mounting medium.

4. *Esterase, Naphthol AS-D Acetate Method* (*Burstone, 1957b, 1958*)

This procedure is useful for the demonstration of the enzyme in formalin fixed frozen sections, and a variety of paraffin embedded tissues including frozen-dried material.

Approximately 5.0 mg. of naphthol AS-D acetate (Table 1) are placed in a 50 ml. Erlenmeyer flask and the substrate dissolved by adding 0.5 ml. of DMF or acetone. Twenty five ml. of distilled water, 25.0 ml. of tris buffer pH 7.1, and 20–40 mg. of diazonium salt (e.g. Garnet GBC or Fast Blue RR) are added, the mixture shaken and filtered through rapid filtering paper into a Coplin jar.

Sections may be incubated for periods of minutes to several hours, the length of time depending upon the stability of the substrate solution. This stability partly depends upon the pH of the medium and under no circumstances should a pH of over 7.0–7.1 be employed.

Following incubation, the sections or slides are briefly washed in running tap water, counterstained if desired, transferred to distilled water, and mounted in glycerol-gelatin or a water miscible synthetic mountant.

Incorporation of $10^{-5}$ *M* eserine in the substrate solution will inhibit the cholinesterase or cholinesterase-like staining reaction in motor end plates.

5. *Esterase, Naphthol AS-LC Acetate Method* (*Burstone, 1957a*)

This technique is primarily designed for application with frozen-dried paraffin embedded tissues although under certain conditions formalin fixed frozen sections may be employed.

Approximately 20 mg. of naphthol AS-LC acetate (Table 1) are dissolved in 2–4 ml. of DMF or dimethyl sulfoxide, 5–10 ml. of ethylene glycol monoethyl ether are added plus 10.0 ml. of 0.2 *M* tris buffer pH 7.1, and sufficient distilled water to make 50.0 ml. of solution. Twenty to thirty mg. of diazonium salt (Garnet GBC or Corinth LB salts) are added, the mixture shaken and filtered into a Coplin jar. (Since filtration may be relatively slow, the substrate solution is usually first poured into the Coplin jar and the diazonium salt dissolved in a few ml. of buffer is then filtered into the jar.) If free-floating cold formalin fixed frozen sections are to be employed, the substrate is dissolved in 2.0 ml. of DMF and 5.0 ml. of ethylene glycol monoethyl ether plus the usual quantity of buffer and water.

Incubation periods usually range from 10–60 minutes. Following incubation the slides or sections are briefly washed in water and counterstained if desired and mounted in an 80% ethanolic solution of polyvinyl acetate (Appendix 4, Section 2, *d*) or other synthetic mountant.

6. *Esterase, Halogenated Indoxyl Methods* (*Holt, 1958*)

Tissues are fixed in formol-calcium gum sucrose (Appendix 1, Section 1, *a*) and sectioned on a freezing microtome.

The oxidizing solution contains potassium ferrocyanide and potassium ferricyanide ($5 \times 10^{-2}$ *M*) and is stable at 0–2° for 1 week.

The initial substrate solution contains 1.5 mg. of 5-bromo-4-chloroindoxyl acetate dissolved in 0.1 ml. of absolute ethanol. Alternatively 5-bromoindoxyl acetate (1.3 mg.) or 5-bromo-6-chloroindoxyl acetate (3.0 mg.) may be employed. To the aforementioned ethanolic substrate solution is added rapidly with shaking a solution containing the following:

| | |
|---|---|
| Buffer (0.1 *M* tris, pH 8.5) | 2.0 ml. |
| Oxidizing solution | 1.0 ml. |
| *M* Calcium chloride | 0.1 ml. |
| 2 *M* Sodium chloride | 5.0 ml. |

The volume of the solution is then brought up to a total of 10.0 ml. with distilled water.

Sections are incubated for a period of minutes to several hours and may be examined during incubation to ascertain the stage of dye development. Following incubation, the sections are rinsed in water and mounted in glycerol-gelatin.

7. *Cholinesterase, Thiocholine Method (Koelle and Friedendwald, 1949; Koelle, 1950; Gomori, 1952b, p. 210)*

This technique has been recommended for use with teased tissues or unfixed frozen sections, although Gerebtzoff (1959, p. 3) indicates that paraffin embedded tissues as well as smears may also be employed.

The following stock solution is prepared:

| | |
|---|---|
| $CuSO_4 \cdot H_2O$ | 0.3 g. |
| Glycine | 0.375 g. |
| $MgCl_2 \cdot 6H_2O$ | 1.0 g. |
| *N* (4%) NaOH | 30.0 ml. |
| Hot saturated solution of $Na_2SO_4$ (ca. 40%) | 170.0 ml. |

This stock solution has a pH of around 6 and will keep indefinitely, although some of the $Na_2SO_4$ may crystallize upon cooling of the solution.

For use, ca. 20.0 mg. of acetylthiocholine iodide are dissolved in a few drops of water and 10.0 ml. of the stock solution are added. Tissues are incubated for 10–60 minutes at 37° and are then rinsed in 2–3 changes of saturated $Na_2SO_4$ and immersed in a dilute solution of yellow ammonium sulfide. The sections are counterstained if desired, dehydrated, and mounted. Sites of activity are revealed as a dark-brown stain (cupric sulfide).

If diisopropyl fluorophosphate (DFP) is employed as an inhibitor, the tissues are first treated with a $10^{-6}$ *M* solution of the compound in saturated $Na_2SO_4$ for 30 minutes, and then incubated with a substrate solution containing the same concentration of DFP. According to Gomori (1952b, p. 210) "true" cholinesterase is inhibited by the DFP treatment only to the extent of 40%, while the non-specific enzyme is completely inhibited by DFP.

8. *Cholinesterases, Choline Fatty Acid Ester Method (Gomori, 1952b, p. 209)*

Acetone fixed paraffin embedded tissues may be employed.

A 0.02 *M* (ca. 0.7%) stock solution of the substrates are prepared and stored under refrigeration. The solutions are opalescent and quite viscous.

A buffered cobalt acetate solution containing ca. 0.0125 *M* cobalt acetate in 0.025–0.030 *M* tris maleate buffer pH 7.6–7.8. Approximately 0.5 ml. of 0.2 *M* solutions of each of the following salts (activators) are added to

100.0 ml. of the mixture; $MgCl_2$, $MnCl_2$, and $CaCl_2$. A few crystals of camphor are added as a preservative.

One ml. of the lauroyl- or myristoylcholine stock solution is blown into 50.0 ml. of the buffered Co solution. In the case of palmitoylcholine, 1.0 ml. of the stock solution is dissolved in 10–15 ml. of propylene glycol and the mixture diluted with the Co solution to make a total volume of 50.0 ml. All substrate mixtures are filtered into a Coplin jar.

Slides are incubated for 6–48 hours (at 37°), with the longer-chain substrates requiring increased incubation periods. The slides are then washed in tap water and immersed in a dilute solution of yellow ammonium sulfide prepared with 70% ethanol. The slides are then counterstained if desired and dehydrated and mounted. Sites of activity are dark-gray to black in color and may be coarsely granular.

### 9. *Cholinesterase, Thiolacetic Acid Method* (*Crevier and Bélanger, 1955*)

Formalin fixed frozen sections mounted on glass slides have proved to be satisfactory for use with the thiolacetic acid method.

The substrate solution is prepared by dissolving thiolacetic acid (0.12 *M*) and lead nitrate (0.001 *M*) in 83.0 ml. of 0.1 *M* $Na_2HPO_4$. Seventeen ml. of McIlvaine phosphate-citrate buffer pH 6.2 are then added to the mixture.

The slides are incubated in the substrate solution for 30–60 minutes at room temperature. They are then washed for 5 minutes in cold running tap water. Sections may be mounted in glycerol-gelatin or solvent-miscible synthetic media.

## Commercial Sources of Reagents

1. Aldrich Chemical Company, Milwaukee, Wisconsin
2. Dajac Laboratories, The Borden Chemical Company, Philadelphia, Pennsylvania
3. Eastman Kodak, Rochester, New York
4. General Aniline and Film Corp., New York, New York
5. Mann Research Laboratories, New York, New York
6. Matheson, Coleman and Bell, East Rutherford, New Jersey
7. Nutritional Biochemicals Corp., Cleveland, Ohio
8. Pfister Chemical Works, Ridgefield, New Jersey
9. Sigma Chemical Company, St. Louis, Missouri

## References


Aldridge, W. N. (1953a) *Biochem. J.* **53,** 110–117.

Aldridge, W. N. (1953b) *Biochem. J.* **53,** 117–124.

Aldridge, W. N. (1953c) *Biochem. J.* **54,** 442–448.

Aldridge, W. N. (1954) *Biochem. J.* **57,** 692–702.

Ammon, R., and Jaarma, M. (1950) *In* "The Enzymes" (J. B. Sumner and K. Myrbäck, eds.), Vol. I, Part 1, pp. 390–442. Academic Press, New York.

ANONYMOUS (1950) *Chem. Age (London)* **62,** 534.
AUGUSTINSSON, K.-B. (1950) *In* "The Enzymes" (J. B. Sumner and K. Myrbäck, eds.), Vol. I, Part 1, pp. 443–472. Academic Press, New York.
AUGUSTINSSON, K.-B. (1960) *In* "The Enzymes" (P. D. Boyer, H. Lardy, and K. Myrbäck, eds.), 2nd ed., Vol. 4, pp. 521–540. Academic Press, New York.
BARRNETT, R. J., AND SELIGMAN, A. M. (1951) *Science* **114,** 579–583.
BAYLISS, B. J., AND TODRICK, A. (1956) *Biochem. J.* **62,** 62–67.
BERGMANN, F., AND RIMON, S. (1958) *Biochem. J.* **70,** 339–344.
BERGMANN, F., SEGAL, R., AND RIMON, S. (1957) *Biochem. J.* **67,** 481–486.
BOURNE, E. G., STACEY, M., TATLOW, J. C., AND TEDDER, J. M. (1949) *J. Chem. Soc.* pp. 2976–2979.
BRANDES, D., AND BOURNE, G. H. (1956) *J. Pathol. Bacteriol.* **71,** 33–36.
BRASS, K., AND SOMMER, P. (1928) *Ber.* **1,** 993–1002.
BURSTONE, M. S. (1956) *J. Natl. Cancer Inst.* **17,** 49–63.
BURSTONE, M. S. (1957a) *J. Natl. Cancer Inst.* **18,** 167–172.
BURSTONE, M. S. (1957b) *Arch. Pathol.* **63,** 164–167.
BURSTONE, M. S. (1957c) *Oral Surg. Oral Med. Oral Pathol.* **10,** 296–303.
BURSTONE, M. S. (1958) *J. Histochem. and Cytochem.* **6,** 322–339.
CAMPBELL, T. W., AND COPPINGER, G. M. (1951) *J. Am. Chem. Soc.* **73,** 2708–2712.
CHATTAWAY, F. D. (1931) *J. Chem. Soc.* pp. 2495–2496.
CHESSICK, R. D. (1953) *J. Histochem. and Cytochem.* **1,** 471–485.
CHESSICK, R. D. (1954) *J. Histochem. and Cytochem.* **2,** 258–273.
COHEN, R. B., NACHLAS, M. M., AND SELIGMAN, A. M. (1951) *Cancer Research* **11,** 709–711.
COUPLAND, R. E., AND HOLMES, R. L. (1957) *Quart. J. Microscop. Sci.* **98,** 327–330.
CREVIER, M., AND BÉLANGER, L. F. (1955) *Science* **122,** 556.
DE DUVE, C. (1959) *Exptl. Cell Research Suppl.* **7,** 169–182.
DE DUVE, C. (1960) *Nature* **187,** 836 and 853.
DENZ, F. A. (1953) *Brit. J. Exptl. Pathol.* **34,** 329–339.
DESNUELLE, P. (1961) *Advances in Enzymol.* **23,** 129–161.
DIXON, M., AND WEBB, E. C. (1958) "Enzymes." Academic Press, New York.
ERÄNKÖ, O., HOPSU, V., AND RÄISÄNEN, L. (1959) *J. Neurochem.* **4,** 332–337.
FIESER, L. F., AND RAJAGOPALAN, S. (1949) *J. Am. Chem. Soc.* **71,** 3938–3941.
FORD-MOORE, A. H. (1950) *Org. Syntheses* **30,** 10–14.
FOURNEAU, E., AND PAGE, H. J. (1914) *Bull. soc. chim France* **15,** 544–553.
FRIES, K., AND FRELLSTEDT, R. (1921) *Ber.* **54,** 715–725.
FUKUDA, T., AND KOELLE, G. B. (1959) *Biophys. Biochem. Cytol.* **5,** 433–440.
GEREBTZOFF, M A. (1959) "Cholinesterases." Pergamon Press, New York.
GLICK, D. (1941) *J. Biol. Chem.* **137,** 357–362.
GOMORI, G. (1945) *Proc. Soc. Exptl. Biol. Med.* **58,** 362–364.
GOMORI, G. (1946) *Arch. Pathol.* **41,** 121–129.
GOMORI, G. (1948) *Proc. Soc. Exptl. Biol. Med.* **68,** 354–358.
GOMORI, G. (1949) *Proc. Soc. Exptl. Biol. Med.* **72,** 697–700.
GOMORI, G. (1950) *In* "Menstruation and Its Disorders" (E. T. Engle, ed.), pp. 52–60. Charles C Thomas, Springfield, Illinois.
GOMORI, G. (1952a) *Intern. Rev. Cytol.* **1,** 323–335.
GOMORI, G. (1952b) "Microscopic Histochemistry." Univ. Chicago Press, Chicago, Illinois.
GOMORI, G. (1953a) *J. Lab. Clin. Med.* **42,** 445–453.
GOMORI, G. (1953b) *J. Histochem. and Cytochem.* **1,** 469–470.
GOMORI, G. (1955a) *J. Histochem. and Cytochem.* **3,** 479–484.

Gomori, G. (1955b) *Texas Repts. Biol. Med.* **13,** 636–640.

Gomori, G., and Chessick, R. D. (1953a) *J. Cellular Comp. Physiol.* **41,** 51–63.

Gomori, G., and Chessick, R. D. (1953b) *J. Neuropathol. Exptl. Neurol.* **12,** 387–396.

Gomori, G., and Coutinho, H. B. (1953) *Proc. Soc. Exptl. Biol. Med.* **83,** 809–811.

Gränacher, Ch., Brünggeer, H., and Ackermann, F. (1941) *Helv. Chim. Acta* **24,** 40E–71E.

Graff, J. (1909) *Ann.* **367,** 253–263.

Gross, S. J., and Danzinger, S. (1957) *Am. J. Obstet. Gynecol.* **73,** 94–119.

Hagemeyer, H. J., and Hull, D. C. (1949) *Ind. Eng. Chem.* **41,** 2920–2924.

Hawkins, R. D., and Gunter, J. M. (1946) *Biochem. J.* **40,** 192–197.

Hawkins, R. D., and Mendel, B. (1949) *Biochem. J.* **44,** 260–264.

Hess, R., and Pearse, A. G. E. (1958) *Brit. J. Exptl. Pathol.* **39,** 292–299.

Hickinbottom, W. J. (1957) "Reactions of Organic Compounds," 3rd ed. Longmans, Green, London.

Higgins, E. B. (1927) *J. Soc. Dyers Colourists* **43,** 213–219.

Hobbinger, E. E. (1957) *Biochem. J.* **67,** 600–607.

Hofstee, H. J. (1960) *In* "The Enzymes" (P. D. Boyer, H. Lardy, and K. Myrbäck, eds.), 2nd ed. Vol. 4, pp. 485–500. Academic Press, New York.

Holmstedt, B. (1957a) *Acta Physiol. Scand.* **40,** 322–330.

Holmstedt, B. (1957b) *Acta Physiol. Scand.* **40,** 331–337.

Holt, S. J. (1952) *Nature* **169,** 271–273.

Holt, S. J. (1958) *In* "General Cytochemical Methods" (J. F. Danielli, ed.), Vol. 1, pp. 375–398. Academic Press, New York.

Holt, S. J., and Sadler, P. W. (1958) *Proc. Roy. Soc.* **B148,** 481–494.

Holt, S. J., and Withers, R. F. J. (1952) *Nature* **170,** 1012–1014.

Holt, S. J., and Withers, R. F. J. (1958) *Proc. Roy. Soc.* **B148,** 520–532.

Hopman, B. C. (1960) *Am. J. Obstet. Gynecol.* **79,** 346–369.

Koelle, G. B. (1950) *J. Pharm. Exptl. Therap.* **100,** 158–179.

Koelle, G. B. (1951) *J. Pharm. Exptl. Therap.* **103,** 153–171.

Koelle, G. B. (1957) *Science* **125,** 1195–1196.

Koelle, G. B., and Friedenwald, J. S. (1949) *Proc. Soc. Exptl. Biol. Med.* **70,** 617–622.

Kramer, D. N., and Gamson, R. M. (1958) *Anal. Chem.* **30,** 251–254.

Main, A. R., Miles, K. E., and Braid, P. E. (1961) *Biochem. J.* **78,** 769–776.

Malmgren, H., and Sylvén, B. (1955) *J. Histochem. and Cytochem.* **3,** 441–445.

Menten, M. L., Junge, J., and Green, M. H. (1944a) *Proc. Soc. Exptl. Biol. Med.* **57,** 82–86.

Menten, M. L., Junge, J., and Green, M. H. (1944b) *J. Biol Chem.* **153,** 471–477.

Miller, O. (1881) *Ann.* **208,** 223–248.

Moloney, W. C., McPherson, K., and Fliegelman, L. (1960) *J. Histochem. and Cytochem.* **8,** 200–207.

Montagna, W. (1955) *J. Biophys. Biochem. Cytol.* **1,** 13–16.

Myers, D. K. (1960) *In* "The Enzymes" (P. D. Boyer, H. Lardy, and K. Myrbäck, eds.), 2nd ed., Vol. 4, pp. 475–483. Academic Press, New York.

Myers, D. K., Tol, J. W., and De Junge, M. H. T. (1957) *Biochem. J.* **65,** 223–232.

Nachlas, M. M., and Seligman, A. M. (1949a) *J. Natl. Cancer Inst.* **9,** 415–425.

Nachlas, M. M., and Seligman, A. M. (1949b) *J. Biol. Chem.* **181,** 343–355.

Ormerod, W. E. (1953) *Biochem. J.* **54**(4), 701–704.

Pearse, A. G. E. (1954) *Intern. Rev. Cytol.* **3,** 329–358.

Pearse, A. G. E. (1960) "Histochemistry, Theoretical and Applied," 2nd ed. Little, Brown, Boston, Massachusetts.

PEARSE, A. G. E., AND PEPLER, W. J. (1957) *Nature* **179,** 589.
PEARSON, B., AND DEFENDI, V. (1957) *J. Histochem. and Cytochem.* **5,** 72–83.
PEPLER, W. J., AND PEARSE, A. G. E. (1957) *J. Neurochem.* **1,** 193–202.
PRICE, D., AND MAY, E. L. (1943) *J. Am. Chem. Soc.* **65,** 297.
RAVIN, H. A., AND SELIGMAN, A. M. (1953) *Arch. Biochem. Biophys.* **42,** 337–354.
RAVIN, H. A., TSOU, K. C., AND SELIGMAN, A. M. (1951) *J. Biol. Chem.* **191,** 843–857.
RAVIN, H. A., ZACKS, S. I., AND SELIGMAN, A. M. (1953) *J. Pharm. Exptl. Therap.* **107,** 37–53.
RENSHAW, R. R., DRIESBACK, P. F., ZIFF, M., AND GREEN, D. (1938) *J. Am. Chem. Soc.* **60,** 1765–1770.
RICHTER, D., AND CROFT, P. G. (1942) *Biochem. J.* **36,** 746–757.
SANDRITTER, W., AND SCHIEMER, H. G. (1959) *Verhandl. deut. Ges. Pathol.* **42,** 449–458.
SÁVAY, GY., AND CSILLIK, B. (1959) *Acta Histochem.* **6,** 307–314.
SCHAEFFER, L. (1869) *Ann.* **152,** 279–298.
SEIDEL, F., AND ENGELFRIED, O. (1936) *Ber.* **69,** 2567–2591.
SELIGMAN, A. M., CHAUNCEY, H. H., AND NACHLAS, M. M. (1951) *Stain Technol.* **26,** 19–23.
SNELL, R. S., AND MCINTYRE, N. (1956) *Brit. J. Exptl. Pathol.* **37,** 44–48.
SPENCER, G. (1931) *J. Soc. Chem. Ind.* (*London*) **50,** 63T–64T.
TAKEUCHI, T., AND FURUTA, M. (1956) *Kumamoto Med. J.* **9,** 70–73.
TAKEUCHI, T., MAEDA, A., AND TSUTSUMI, I. (1955) *Gann* **46,** 209–210.
TAXI, J. (1952) *J. physiol.* (*Paris*) **44,** 595–599.
TILDEN, W. A., AND ARMSTRONG, H. E. (1901) *Rept. Brit. Assoc. Advance Sci.* pp. 152–154.
ULLMANN, G. (1897) *Ber.* **30,** 1466–1470.
UNDERHAY, E. E. (1957) *Biochem. J.* **66,** 383–390.
VOLLE, R. L., AND KOELLE, G. B. (1961) *J. Pharm. and Exptl. Therap.* **133,** 223–240.
VORLÄNDER, D., AND DRESCHER, B. (1901) *Ber.* **34,** 1854–1860.
WACHSTEIN, M., AND MEISEL, E. (1959) *J. Histochem. and Cytochem.* **7,** 189–201.
WACHSTEIN, M., MEISEL, E., AND FALCON, C. (1961) *J. Histochem. and Cytochem.* **9,** 325–339.
WILSON, I. B. (1951) *Biochim. et Biophys. Acta* **7,** 520–525.
WILSON, I. B. (1960) *In* "The Enzymes" (P. D. Boyer, H. Lardy, and K. Myrbäck, eds.), 2nd ed., Vol. 4, pp. 501–520. Academic Press, New York.
WINKELMANN, R. K. (1960a) *A. M. A. Arch. Dermatol.* **82,** 17–23.
WINKELMANN, R. K. (1960b) *Cancer* **13,** 626–630.
WOLF, G., FRIEDMAN, O. M., DICKINSON, S. J., AND SELIGMAN, A. M. (1950) *J. Am. Chem. Soc.* **72,** 390–392.
ZACKS, S. I. (1958) *Am. J. Pathol.* **34,** 293–310.

*CHAPTER 7*

# SULFATASES

Sulfatases are hydrolytic enzymes which catalyze the following type of reaction:

$$ROSO_3^- + H_2O \rightleftharpoons ROH + H^+ + SO_4^{--}$$

Reviews of the biochemical aspects of sulfatases have been made by Fromageot (1950, p. 517; 1955, p. 324) and Roy (1960). From a historical standpoint Derrien (1911) discovered an enzyme derived from a mollusk (*M. trunculus*) which hydrolyzed an indoxyl sulfate with the subsequent formation of a blue dye. In 1923 Neuberg and Kurono reported the enzymatic hydrolysis of phenyl sulfate, and in 1924 Neuberg coined the term "sulfatase." The enzymes which hydrolyze sulfate type bonds may be classified as: arylsulfatases, steroid sulfatases, glucosulfatases, chondrosulfatases, myrosulfatases, and several less well defined enzymes which may be categorized as miscellaneous sulfatases.

Although sulfatases are widely distributed, their physiological role is not known. Spencer (1956, p. 135) suggests that arylsulfatases may function as transferases. Roy (1960, p. 226) notes that both sulfatase and β-glucuronidase are lysosomal enzymes, and suggests that an answer to the arylsulfatase problem will only be obtained when the significance of the lysosomes is better understood. With reference to specific aryl sulfate substrates he suggests that they should be considered not from the standpoint of sulfate metabolism but rather from that of metabolism of the phenolic part of the molecule. Of interest is the fact that estrone and stilbestrol sulfates, as well as insulin sulfate, exhibit physiological activity.

Arylsulfatases, which are widely distributed in both plants and animals, catalyze the hydrolysis of phenyl sulfate and analogues (Boyland *et al.*, 1956; Dodgson *et al.*, 1956), and indoxyl sulfate. Among the specific substrates hydrolyzed by the enzyme are *p*-nitrophenyl sulfate, nitrocatechol sulfate, *p*-acetylphenyl sulfate, β-naphthyl sulfate, and 6-benzoyl-2-naphthyl sulfate (Table 1). Of interest is the enzymatic hydrolysis by sulfatase of β-naphthyl sulfate which was first reported by Noguchi in 1924.

TABLE 1

POTENTIAL SULFATASE SUBSTRATES

| Chemical name and formula | Mol. wt. | Reference | Melting point (° C.) |
|---|---|---|---|
| 1. Glucose-6-sulfate, $C_6H_{12}O_9S \cdot 1/2H_2O$ | 269.24 | Soda, 1933; Dodgson and Spencer, 1957 | 178 |
| 2. Sinigrin (potassium myronate), $C_{10}H_{16}NO_9S_2K \cdot H_2O$ | 415.49 | Ettlinger and Lundeen, 1956 | 127–129 |
| 3. Tyrosine-*o*-sulfate (K salt), $C_9H_{10}O_6NSK$ | 299.35 | Tallan *et al.*, 1955, Dodgson *et al.*, 1959 | — |
| 4. Estrone sulfate (Na salt), $C_{18}H_{21}OSNa \cdot H_2O$ | 326.44 | Butenandt and Hofstetter, 1939 | 228–230 |
| 5. Indoxyl sulfate (K salt), $C_8H_6NSO_4K$ | 251.31 | Schwenk and Jolles, 1915 | 179–180 (dec.) |
| 6. Phenyl sulfate (K salt), $C_6H_5O_4SK$ | 212.27 | Baumann, 1878; Czapek, 1914; Burkhardt and Lapworth, 1926 | — |
| 7. *o*-Nitrophenyl sulfate (K salt), $C_6H_4O_6NSK$ | 257.27 | Burkhardt and Wood, 1929 | — |
| 8. *p*-Nitrophenyl sulfate (K salt), $C_6H_4O_6NSK$ | 257.27 | Burkhardt and Wood, 1929; Huggins and Smith, 1947 | — |
| 9. 2-Hydroxy-5-nitrophenyl sulfate (nitrocatechol sulfate) (K salt), $C_6H_4O_7NSK_2 \cdot 2H_2O$ | 348.40 | Roy, 1953 | — |
| 10. *p*-Aminophenyl sulfate (K salt), $C_6H_6O_4NSK \cdot H_2O$ | 245.30 | Burkhardt and Wood, 1929 | — |
| 11. 4-Amino-1-naphthyl sulfate, $C_{10}H_9O_4NS$ | 239.26 | Burkhardt and Wood, 1929 | — |
| 12. $\alpha$-Naphthyl sulfate (K salt), $C_{10}H_7O_4SK$ | 262.33 | Czapek, 1914; Burkhardt and Lapworth, 1926; Whitehead *et al.*, 1952 | 182 |
| 13. 4-Nitro-1-naphthyl sulfate (K salt), $C_{10}H_6O_6NSK$ | 307.33 | Burkhardt and Wood, 1929 | — |
| 14. $\beta$-Naphthyl sulfate (K salt), $C_{10}H_7O_4SK \cdot 4H_2O$ | 334.40 | Noguchi, 1924; Burkhardt and Lapworth, 1926 | — |
| 15. 8-Hydroxyquinoline sulfate (K salt), $C_9H_6NO_4SK$ | 263.32 | Czapek, 1914 | — |
| 16. 4-Methylumbelliferone sulfate (K salt), $C_{10}H_7O_6SK$ | 294.33 | Mead *et al.*, 1955 | 185–186 |
| 17. 6-Benzoyl-2-naphthyl sulfate (K salt), $C_{17}H_{11}O_5K$ | 334.37 | Rutenburg *et al.*, 1952 | — |
| 18. Naphthol AS sulfates | — | Rath and Feess, 1954 | — |
| 19. $\beta$-Hydroxynaphthaleneazophenyl sulfate (K salt), $C_{10}H_{11}O_5N_2SK \cdot 1/2H_2O$ | 319.39 | Burkhardt and Wood, 1929 | — |
| 20. Phenolphthalein sulfate (K salt), $C_{20}H_{13}O_{11}S_2K_3 \cdot 2H_2O$ | 646.79 | Whitehead *et al.*, 1952 | — |

According to Dodgson and Spencer (1957, p. 211) arylsulfatase may be divided into two groups, type I and type II, according to their relative substrate specificities and response to inhibitors. Arylsulfatases are widely distributed in mammalian tissues and relatively high activities are exhibited by liver, kidney, and adrenal. Huggins and Smith (1947) made the interesting observation that various rat neoplasms hydrolyze phenyl sulfate more rapidly than do normal tissues. The pH optimum for arylsulfatase varies with the substrate employed, the tissue, and the buffer. Huggins and Smith (1947) found the pH optimum to be 6.12 with reference to the enzyme obtained from tissues and body fluids. Roy (1953) found that ox liver sulfatase exhibited a pH optimum of 5.0 and 6.0 in the presence of acetate buffer, while Maengwyn-Davies and Friedenwald (1954) reported an optimum of 6.0–6.1 (rabbit liver and kidney). Rutenburg and Seligman (1956) employing 6-benzoyl-2-naphthyl sulfate found the pH optimum for the rat kidney enzyme to be 5.8 (acetate buffer), 7.6 (phosphate buffer), and 7.7 (phosphocitrate buffer). The rat liver sulfatase exhibited a somewhat higher pH optimum. Pulkkinen (1957) noted that the pH optimum of human embryonic tissues including liver, pancreas, and intestine was 7.3.

Biochemical studies have indicated that arylsulfatases of mammalian livers are found in both mitochondrial and microsomal fractions as well as in the so-called "lysosomes." With the development of new and definitive histochemical techniques for the visualization of sites of sulfatase activity, precise delineation of the enzyme specific subcellular components should be possible.

Steroid sulfatases which exhibit a pH optimum of 7.7–7.9 catalyze the hydrolysis of sulfate esters of steroid alcohols. These enzymes exhibit rather definitive substrate specificities and appear to hydrolyze only $3\beta$-sulfates of the $5\alpha$- and $\Delta^5$-steroids. Liver is known to contain a steroid sulfatase which hydrolyzes dehydroepiandrosterone sulfate (DHAS) (Roy, 1957).

Glucosulfatase (glycosulfatase) is known to hydrolyze mono-, di-, and trisulfuric acid esters of glucose, galactose, maltose, and other simple carbohydrates. The natural substrates of the enzyme are unknown. In certain mollusks the pH optimum with glucose-6- or glucose-3-sulfate is around 6.0.

Chondrosulfatase which has a pH optimum of around 7.0 has been found in mollusks but not in mammalian tissues, although the enzyme will possibly be found in the latter when suitably sensitive methods are developed.

Myrosulfatase liberates sulfate ions from sinigrin (Na myronate) and related mustard oil glycosides. According to Roy (1960, p. 230) it is possible that myrosulfatase does not exist as such and its apparent action may simply be related to thioglycosidase activity.

The possible existence of sulfamatases or *N*-sulfatases, enzymes which hydrolyze —NH—$SO_3$— containing compounds, should also be considered. Among the biological materials which contain this grouping is the sulfamate derivative of glucosamine which is found in heparin, and aryl sulfamates found in urine after the administration of arylamines.

Sulfate esters may be biosynthetically formed by enzymes known as sulfurylases (Roy, 1960, p. 205). For example "microsome-free" supernatant solutions of rat liver will synthesize phenyl sulfate from phenol and sulfate ions in the presence of ATP and $Mg^{++}$ (de Meio *et al.*, 1953). The synthetic or sulfate-activating enzymes are in some instances associated with sulfate transferase or sulfokinases (Roy, 1960, p. 205). Among the latter is phenol sulfokinase which will catalyze the transfer of sulfate ions from nitrophenyl sulfate to a suitable phenol, with formation of an aryl sulfate and *p*-nitrophenol. The so-called arylsulfatase may also be involved in sulfate transfer.

## I. HISTOCHEMICAL DEMONSTRATION OF SULFATASES

To date the histochemical techniques for the demonstration of sulfatases have not proved to be satisfactory. Peche (1913) in an attempt to localize myrosulfatase incubated plant tissues in a substrate solution containing potassium myronate and $BaCl_2$, and assumed that the resulting precipitate consisted of $BaSO_4$. Ohara and Kurata (1951) employed a Gomori type metal-salt technique in an attempt to demonstrate arylsulfatase in tissue sections. Among the substrates used were phenyl sulfate and 8-hydroxyquinoline sulfate. Upon enzymatic hydrolysis the sulfate ion was precipitated in the presence of lead to form lead sulfide. These authors also reported that benzidine and 1,2-naphthoquinone-4-sulfonate could also be used to precipitate the sulfate ion. Of interest is their report that the procedure is satisfactory only when employed with acetone fixed unembedded tissues. Seligman *et al.* (1949) attempted to demonstrate the enzyme by means of an azo-dye method employing α- or β-naphthyl sulfates but reported that the hydrolysis of these compounds was relatively low and that azo-coupling at the pH optimum of the enzyme was not satisfactory. Pearse (1960, p. 595) also reported that β-naphthyl sulfate was inadequate as a substrate when utilized in an azo-dye procedure. Seligman *et al.* (1949), Seligman *et al.* (1950), and Rutenburg *et al.* (1952) indicated that a post-coupling type of procedure utilizing either 6-bromo-2-naphthyl sulfate or 6-benzoyl-2-naphthyl sulfate could be employed for the demonstration of sulfatase. Gomori (1952, p. 200) notes that human tissues do not hydrolyze the benzoyl compound and furthermore the incubation periods with these techniques are exceedingly long. Pearse (1960, p. 596) considers both the 6-bromo- and 6-benzoylnaphthyl sulfate methods as unsatisfactory. Gössner (1958) reported the demonstration of sulfatase with naphthol AS sulfate in conjunction with Fast Red TR salt which was employed in a simultaneous coupling azo-dye procedure. Unfortunately, a prolonged incubation period was required and the use of suitable control slides was not indicated. Holt (1952) suggested the application of indoxyl sulfate as a substrate but no reports on the application of this technique have been forthcoming. It would, however, appear that the indoxyl techniques are suitable for the histochemical demonstration of arylsulfatase in view of the original report of Derrien (1911), as well as the more recent applications of indoxyl sulfate as a biochemical substrate. Dohlman and Friedenwald (1955) were unsuccessful in attempts to demonstrate mucoid sulfatase activity by a metal-salt technique which employed either heparin or chondroitin sulfate in the presence of lead ions. When unfixed frozen sections of rat kidney were incubated in the aforementioned medium a sharp brush border staining was obtained. However, corroborative biochemical evidence to verify the demonstration of sulfatase could not be obtained. Histochemical attempts to demonstrate steroid, gluco-, and myrosulfatases as well as a possible sulfamatase have not been reported.

Even though the azo-dye methods for the histochemical localization of sulfatases appear to be unsatisfactory, Roy (1962) has found on the basis of quantitative studies that aryl sulfates such as 6-benzoyl-2-naphthyl sulfate are a satisfactory substrate only for sulfatase C but not for sulfatases A and B. He further indicates that substrates such as naphthol AS sulfate undoubtedly would have the same limitation as the aforementioned benzoylnaphthyl sulfate when employed under histochemical conditions.

The successful histochemical demonstration of sulfatases probably will depend upon the synthesis and application of new histochemical substrates in metal-salt, azo-dye, or oxidation techniques. It is also conceivable that some of the substrates listed in Table 1 (or their analogues) may be of value as histochemical reagents in future techniques. Therefore the synthesis of a number of sulfate esters which may be used as a guide in the preparation of new compounds is given in Appendix 7. The basic type of synthetic approach to the preparation of aryl sulfates involves the treatment of a phenol with a chlorosulfonic acid in the presence of a base, the treatment of a phenol with a persulfate in the presence of a base, or the treatment of a phenol with sulfur trioxide in the presence of a base (Burkhardt and Lapworth, 1926; Burkhardt and Wood, 1929; Sobel and Spoerri, 1941; Baker and Brown, 1948). Certain industrial modifications employed in the preparation of aliphatic, cyclic, carbohydrate, as well as sulfamate compounds, may also offer suggestions for the synthesis of new sulfatase reagents (Gilbert and Jones, 1955, 1956).

## II. ACTIVATORS AND INHIBITORS

The various activators and inhibitors of sulfatases are discussed in the reviews of Sumner and Somers (1953, p. 85), Dodgson and Spencer (1957, p. 211), and Roy (1960, p. 205). Type I arylsulfatases are strongly inhibited by sulfite and by cyanide, but are relatively insensitive to sulfate or phosphate ions. Fluoride produces a moderate degree of inhibition. The type II arylsulfatase is strongly inhibited by phosphate, sulfate, and sulfite ions but not by cyanide or fluoride. Mammalian steroid sulfatase is distinctly inhibited by phosphate ions but is only slightly inhibited by sulfate. Glucosulfatase, on the other hand, is inhibited by phosphates, borates, and fluorides.

# APPENDIX 7

## SYNTHESIS OF SULFATES

### 1. *Glucose-6-sulfate* (*Soda, 1933; Dodgson and Spencer, 1957, p. 211*)

A mixture of 7.0 ml. of chlorosulfonic acid and 18.0 ml. of chloroform is added dropwise with stirring to a solution of 20.0 g. of glucose in 75.0 ml. of dry pyridine at a temperature below 10°. The mixture is stirred at room temperature for approximately 1 hour and allowed to stand overnight. The oil which forms is separated, dissolved in 300.0 ml. of water, and freed of chloroform by concentrating to about 250.0 ml. *in vacuo* at 40°. Powdered lead oxide (80.0 g.) is added and, after mixing, is distilled *in vacuo* to remove the pyridine. The filtered solution which is free of pyridine and inorganic sulfate contains in addition to the lead salt of glucose-6-sulfate: free glucose, chloride ions, and lead hydroxide. The latter is precipitated by passing carbon dioxide through the solution. After filtering, the chloride is removed by the addition of a solution of silver sulfate and the precipitate is then separated by filtration. The filtrate is diluted with about 600.0 ml. of water and the solution is then treated with hydrogen sulfide till formation of lead sulfide ceases. The precipitate of lead sulfide is filtered off and the excess hydrogen sulfide is removed by passing a stream of air through the solution. The filtrate is then treated with a slight excess of an ethanolic solution of brucine and is then concentrated *in vacuo* at 40°. The first crop of crystals (free brucine) is discarded and after further concentration a syrup remains which solidifies upon cooling. This is dissolved in a small volume of hot water, and ethanol is added until a slight turbidity appears. Upon cooling the crystalline brucine salt separates out. It is then recrystallized from aqueous ethanol approximately 5 times and is then dissolved in cold water and converted into the potassium salt by the slow addition, with stirring, of the required amount of 2 $N$ potassium hydroxide. The precipitated brucine is filtered off and the filtrate concentrated to about 50.0 ml. *in vacuo*. It is then poured slowly (with stirring) into 1 liter of ethanol. After standing a few minutes the precipitated potassium glucose-6-sulfate is filtered off, washed with ethanol, and dried *in vacuo* at 50°. The salt is very hygroscopic.

### 2. *Phenyl Sulfate* (*Burkhardt and Lapworth, 1926*)

Although aryl sulfates are frequently synthesized by the chlorosulfonic acid technique, they may also be readily made by the action of potassium persulfate on a solution of phenol (or naphthol) in dry dimethylaniline, followed by extraction of the product with aqueous alkali.

A mixture of phenol (1.0 $M$) and dimethyl- or diethylaniline (3.0 $M$)

is heated on a steam bath and stirred mechanically while finely divided potassium persulfate (2.5 *M*) is added. The mixture is heated and stirred for 4 hours and is then poured into a suspension of excess barium hydroxide in cold water. The base is removed by steam distillation and the excess barium hydroxide is precipitated with carbon dioxide. After filtering and evaporating, potassium phenyl sulfate is readily isolated (yield, 90%).

Burkhardt and Lapworth noted that this method is also satisfactory for the preparation of sulfates of α- and β-naphthols, and for eugenol and isoeugenol with the exception that potassium hydroxide was substituted for barium hydroxide in the separation procedure.

3. *p-Nitrophenyl Sulfate* (*Huggins and Smith, 1947*)

Forty-seven ml. of dimethylaniline and 50.0 ml. of carbon disulfide are mixed in a 500.0 ml. suction flask and placed in an ice bath under a hood. Chlorosulfonic acid (9.1 ml.) is added dropwise to the chilled mixture, followed by rapid addition of 13.9 g. of *p*-nitrophenol. The mixture is stirred for approximately 1 hour and then allowed to stand overnight. One hundred ml. of 0.4 *M* KOH are added, which results in the immediate formation of bright yellow crystals. After thorough stirring of the mixture the flask is placed in a water bath (ca. 80°) and most of the carbon disulfide is evaporated *in vacuo*. The excess dimethylaniline is separated by centrifugation and the residual yellow mass is dried *in vacuo*. The crude product is recrystallized 3–4 times from 80% ethanol. The purified dried compound is stored under refrigeration to minimize decomposition.

4. *2-Hydroxy-5-nitrophenol Sulfate* (*Nitrocatechol Sulfate*) (*Roy, 1953; Dodgson and Spencer, 1957, p. 211*)

Thirty grams of *p*-nitrophenol are added to a solution of 70.0 g. of KOH and 70.0 g. of potassium persulfate in 1 liter of water. The reaction mixture is allowed to stand for 48 hours at 37°, acidified with sulfuric acid to pH 4.0, and the unchanged *p*-nitrophenol extracted with ether. The aqueous solution which is made strongly alkaline to litmus with strong KOH solution is concentrated *in vacuo* at 35° to about 300.0 ml. The solution together with any precipitate is poured into 2 volumes of acetone and the mixture is then filtered. The residue is washed with aqueous acetone (1:2) and the combined filtrates are evaporated to dryness. The residue is then dissolved in a minimal amount of water and the pH is adjusted to 4.2 with glacial acetic acid. The precipitated material is filtered under suction and is then recrystallized 4 times from water. The recrystallized material is suspended in water and a concentrated solution of KOH is then added to bring the pH to approximately 12.0. The solution is poured slowly into 10 volumes of ethanol, the precipitated dipotassium salt filtered off, washed with 90% ethanol, and dried *in vacuo* (yield, ca. 3.0 g.) ($C_6H_4O_7NSK_2 \cdot 2H_2O$). Anal., calcd.: C, 20.3%; H, 2.03%; N, 4.03%; K, 22.5%. Found: C, 20.2%; H, 1.87%; N, 4.02%; K, 23.0%.

5. *α- and β-Naphthyl Sulfates* (*Burkhardt and Lapworth, 1926; Feigenbaum and Neuberg, 1941; Whitehead et al., 1952*)

Fifteen grams of chlorosulfonic acid are added drop by drop and with constant stirring to an ice cold solution of 10.0 g. of the appropriate naphthol dissolved in approximately 40.0 ml. of dimethylaniline. When the reaction has subsided the mixture is made alkaline with strong aqueous KOH (1:1). The precipitate is filtered off by suction and washed thoroughly with ether. The ester salt is extracted from the residue with 95% hot ethanol and then filtered. The filtrate is cooled and the crystals which form are filtered off and washed with cold ethanol. Recrystallization from a minimum of hot water gives the pure potassium salt of naphthol sulfate. Although pyridine may be substituted for dimethylaniline, the latter results in a higher yield of sulfate ester.

Whitehead *et al.* (1952), purified the potassium salt of α-naphthyl sulfate by converting it to the *p*-toluidine salt (Appendix 7, Section 10), recrystallizing this compound, and then converting it back to the potassium salt.

According to Burkhardt and Lapworth (1926) potassium α-naphthyl sulfate crystallizes from water as a coarse crystalline powder or in long soft needles. It is soluble in 40 parts of water at 17°, is freely soluble in hot water, and is practically insoluble in most organic media ($C_{10}H_7O_4SK$). Anal., calcd.: K, 14.9%. Found: K, 14.7%. Potassium β-naphthyl sulfate resembles the α-isomer in general appearance and characteristics, but the salt is hydrated and it dissolves in 70 parts of water at 17° ($C_{10}H_7O_4SK \cdot 4H_2O$). Anal., calcd.: K, 11.7%. Found: K, 11.9%.

6. *6-Benzoyl-2-naphthol Sulfate* (*Rutenberg et al., 1952*)

Chlorosulfonic acid (7.5 ml.) is cautiously added to anhydrous pyridine (50.0 ml.) in an Erlenmeyer flask cooled by ice. During the addition a vigorous reaction occurs with the formation of a white crystalline mass of sulfur trioxide. A solution of 6-benzoyl-2-naphthol (25.0 g.) in anhydrous pyridine (50.0 ml.) is rapidly added and the mixture is stirred and heated to boiling for 2–3 minutes. The solution is then cooled and transferred to a 3 liter beaker containing 100.0 ml. of methanol. A saturated solution of methanolic KOH is added in small portions until the solution becomes alkaline, at which point a darkening of the mixture occurs. Two liters of dry ether are added and the precipitate is filtered off under suction, washed with ether, and dried. The white powder is then suspended in 100.0 ml. of cold distilled water for 30 minutes in order to remove KCl and other inorganic salts which may inhibit arylsulfatase. The relatively insoluble potassium 6-benzoyl-2-naphthol sulfate is collected under suction, washed with a small amount of cold distilled water, acetone, and ether, and dried [yield, 26.0 g. (74.0%)].

7. *Naphthol AS Sulfates* (*Rath and Feess, 1954*)

Hydroxycarboxylic acid arylamides of the naphthol AS series may be converted to their sulfuric acid ester salts by a modification of the procedure of Feigenbaum and Neuberg (1941).

An excess of chlorosulfonic acid is slowly added to a cold suspension of naphthol in dimethylaniline. The reaction mixture is heated to 50–60°. After cooling, the potassium salt is prepared by addition of aqueous KOH in which the dimethylaniline is thrown out of solution. If another base (e.g. pyridine) is employed, the sulfonic acid of naphthol AS is produced instead of the sulfuric acid ester.

8. *Phenolphthalein Disulfate* (*Whitehead et al., 1952*)

A mixture of 12.5 ml. of dry pyridine and 12.5 ml. of chloroform in a 100.0 ml. beaker is cooled in an ice bath and 5.0 ml. of chlorosulfonic acid are slowly run down the side of the beaker with constant stirring. Eight grams of phenolphthalein in 15.0 ml. of chloroform are rapidly added to the aforementioned solution. The cooling mixture is removed and the contents of the beaker are stirred for approximately 2 hours and then allowed to stand at room temperature overnight. The beaker is then placed in an ice bath and 80% aqueous KOH solution is added (with constant stirring) until the mixture becomes barely alkaline to litmus (ca. 14.0 ml. of KOH solution). The mixture is stirred for approximately 10 minutes after the addition of alkali and then it is poured with stirring into 150.0 ml. of ethanol and is then refrigerated. The precipitate is separated by centrifugation and the solid material extracted on a water bath with three 100.0 ml. portions of 80% ethanol-water. During this process the mixture is kept just alkaline to phenolphthalein by the addition of a few drops of KOH solution. The extracts are cooled separately to room temperature and then cooled in an ice bath. The heavy precipitate which forms in the first extract is separated by filtration. The filtrate is then combined with the second and third extracts, 900.0 ml. of ethanol are added, and the precipitate which forms is filtered off. The combined precipitates are dissolved in a minimum of distilled water and the compound is salted out by the addition (with vigorous stirring) of 80% aqueous KOH. The potassium phenolphthalein disulfate is removed by filtration on a sintered glass filter, washed with ethanol, and dried (yield, 64%). The compound is then crystallized twice from water and dried over $P_2O_5$ *in vacuo* at room temperature. ($C_{20}H_{13}O_{11}S_2K_3 \cdot 2H_2O$) Anal., calcd.: C, 37.1%; H, 2.6%; S, 9.9%; K, 18.1%. Found: C, 36.7%; H, 2.9%; S, 9.8%; K, 17.0%.

9. *Dehydroepiandrosterone (3β-Hydroxyandrost-5-ene-17-one) sulfate (DHAS) (Roy, 1956; Dodgson and Spencer, 1957, p. 211)*

Dehydroepiandrosterone (50.0 mg.) is dissolved in 3.0 mg. of dry benzene and an equal weight of pyridine-sulfur trioxide is added. The mixture is refluxed on a water bath for 90 minutes, cooled, and diluted with 30.0 ml. of petroleum ether (b.p. 60–80°). After standing for 1 hour at 0° the precipitated pyridinium salt of the steroid sulfate is removed by filtration and washed with petroleum ether. The residue is extracted 5 times with 2.0 ml. portions of boiling chloroform in order to separate the soluble pyridinium salt from the excess sulfur trioxide reagent. The extracts are then combined and evaporated to dryness *in vacuo*. The dried material is dissolved in a minimum volume of boiling 70% ethanol, and an equal volume of hot potassium sulfate solution (saturated when cold) is added. Water is added dropwise to the boiling mixture until complete solution is effected. Potassium dehydroepiandrosterone sulfate crystallizes out and is then recrystallized from water. Contact of the sulfate ester with boiling water is kept to a minimum in order to avoid spontaneous hydrolysis. Other steroid sulfates may be prepared according to this procedure.

10. *p-Toluidine Salts of Monoaryl Sulfates (Barton and Young, 1943)*

These authors point out that many monoaryl sulfates are unstable in the free acid form and are thus usually prepared as alkali-metal salts. Such salts are relatively unsatisfactory for identification purposes since they do not melt sharply. In addition they often contain inorganic impurities which are not readily removed.

Potassium aryl sulfates of phenol, *o*-cresol, *m*-cresol, *p*-cresol, and *p*-bromophenol are prepared according to the method of Burkhardt and Wood (1929). These compounds crystallize in the form of colorless plates with the exception of *p*-nitrophenyl sulfate which forms bright yellow prisms.

Solutions of equimolar portions of the potassium aryl sulfate and of *p*-toluidine hydrochloride dissolved in a minimum amount of water at 60° are mixed rapidly and chilled. The *p*-toluidine salt of the aryl sulfate separates almost immediately and is removed by filtration. The crude product is washed with a small amount of ice water and recrystallized from water, using charcoal, if necessary to remove any colored impurities. The material is then recrystallized once again from water and finally from aqueous ethanol, yielding colorless needles in each case. Melting points of several *p*-toluidine salts are:

*p*-cresyl ($C_{14}H_{17}NO_4S$), m.p. 162–163°

*p*-bromophenyl ($C_{13}H_{14}NO_4SBr$), m.p. 193–194°

*p*-nitrophenyl ($C_{13}H_{14}N_2O_6S$), m.p. 167–168°

The highly purified potassium salt of the aryl sulfate may readily be reformed from the aforementioned toluidine salts.

## Commercial Sources of Reagents

1. General Aniline and Film Corp., New York, New York
2. Pfister Chemical Works, Ridgefield, New Jersey
3. Sigma Chemical Company, St. Louis, Missouri

## References

Baker, W., and Brown, N. C. (1948) *J. Chem. Soc.* pp. 2303–2307.
Barton, A. D., and Young, L. (1943) *J. Am. Chem. Soc.* **65,** 294–295.
Baumann, E. (1878) *Ber.* **11,** 1907–1916.
Boyland, E., Manson, D., Sims, P., and Williams, D. C. (1956) *Biochem. J.* **62,** 68–71.
Burkhardt, G. N., and Lapworth, A. (1926) *J. Chem. Soc.* pp. 684–690.
Burkhardt, G. N., and Wood, H. (1929) *J. Chem. Soc.* pp. 141–152.
Butenandt, A., and Hofstetter, H. (1939) *Z. physiol. Chem. Hoppe-Seyler's* **259,** 222–234.
Czapek, E. (1914) *Monatsh. Chem.* **35,** 635–642.
de Meio, R. H., Wizerkaniuk, M., and Fabiani, E. (1953) *J. Biol. Chem.* **203,** 257–263.
Derrien, M. (1911) Cited in *Bull. soc. chim. France* (IV) **9,** 110–111.
Dodgson, K. S., and Spencer, B. (1957) *In* "Methods of Biochemical Analysis" (D. Glick, ed.), Vol. 4, pp. 211–255. Interscience, New York.
Dodgson, K. S., Spencer, B., and Williams, K. (1956) *Biochem. J.* **64,** 216–221.
Dodgson, K. S., Rose, F. A., and Tudball, N. (1959) *Biochem. J.* **71,** 10–15.
Dohlman, C. H., and Freidenwald, J. S. (1955) *J. Histochem. and Cytochem.* **3,** 492–494.
Ettlinger, M. G., and Lundeen, A. J. (1956) *J. Am. Chem. Soc.* **78,** 4172–4173.
Feigenbaum, J., and Neuberg, C. A. (1941) *J. Am. Chem. Soc.* **63,** 3529–3530.
Fromageot, C. (1950) *In* "The Enzymes" (J. B. Sumner and K. Myrbäck, eds.), Vol. I, Part 1, pp. 517–526. Academic Press, New York.
Fromageot, C. (1955) *In* "Methods in Enzymology" (S. P. Colowick and N. O. Kaplan, eds.) Vol. II, pp. 324–333. Academic Press, New York.
Gilbert, E. E., and Jones, E. P. (1955) *Ind. Eng. Chem.* **47,** 1916–1925.
Gilbert, E. E., and Jones, E. P. (1956) *Ind. Eng. Chem.* **48,** 1658–1669.
Gomori, G. (1952) "Microscopic Histochemistry, Principles and Practice." Univ. Chicago Press, Chicago, Illinois.
Gössner, W. (1958) *Histochemie* **1,** 48–96.
Holt, S. J. (1952) *Nature* **169,** 271–273.
Huggins, C., and Smith, D. (1947). *J. Biol. Chem.* **170,** 391–398.

MAENGWYN-DAVIES, G. D., AND FRIEDENWALD, J. S. (1954) *Arch. Biochem. Biophys.* **53,** 29–43.

MEAD, J. A. R., SMITH, J. N., AND WILLIAMS, R. T. (1955) *Biochem. J.* **61,** 569–574.

NEUBERG, C. (1924) *Naturwissenschaften* **12,** 797–800.

NEUBERG, C., AND KURONO, K. (1923) *Biochem. Z.* **140,** 295–298.

NOGUCHI, J. (1924) *Biochem. Z.* **144,** 138–140.

OHARA, M., AND KURATA, J. (1951) *Chem. Abstr.* **45,** 1187.

PEARSE, A. G. E. (1960) "Histochemistry, Theoretical and Applied," 2nd ed. Little, Brown, Boston, Massachusetts.

PECHE, K. (1913) *Ber. deut. botan. Ges.* **31,** 358–462.

PULKKINEN, M. (1957) *Acta Physiol. Scand.* **42,** *Suppl. 145,* 115–116.

RATH, H., AND FEESS, E. (1954) *Melliand Textilber.* **35,** 267–269.

ROY, A. B. (1953) *Biochem. J.* **53,** 12–15.

ROY, A. B. (1956) *Biochem. J.* **62,** 41–50.

ROY, A. B. (1957) *Biochem. J.* **66,** 700–703.

ROY, A. B. (1960) *Advances in Enzymol.* **22,** 205–235.

ROY, A. B. (1962) *J. Histochem. and Cytochem.* **10,** 106–107.

RUTENBURG, A. M., AND SELIGMAN, A. M. (1956) *Arch. Biochem. Biophys.* **60,** 198–214.

RUTENBURG, A. M., COHEN, R. B., AND SELIGMAN, A. M. (1952) *Science* **116,** 539–543.

SCHWENK, E., AND JOLLES, A. (1915) *Biochem. Z.* **69,** 467–470.

SELIGMAN, A. M., NACHLAS, M. M., MANHEIMER, L. H., FRIEDMAN, O. M., AND WOLF, G. (1949) *Ann. Surg.* **130,** 333–341.

SELIGMAN, A. M., NACHLAS, M. M., AND COHEN, R. B. (1950) *Cancer Research* **10,** 240.

SOBEL, A. E., AND SPOERRI, P. E. (1941) *J. Am. Chem. Soc.* **63,** 1259–1261.

SODA, T. (1933) *Bull. Chem. Soc. Japan* **8,** 37–49.

SPENCER, B. (1956) *In* "Colloque sur la Biochimie du Soufre," pp. 135–142. Editions du Centre National de la Recherche Scientifique, Paris, France.

SUMNER, J. B., AND SOMERS, G. F. (1953) "Chemistry and Methods of Enzymes," 3rd ed. Academic Press, New York.

TALLAN, H. H., BELLA, S. T., STEIN, W. H., AND MOORE, S. (1955) *J. Biol. Chem.* **217,** 703–708.

WHITEHEAD, J. E. M., MORRISON, A. R., AND YOUNG, L. (1952) *Biochem. J.* **51,** 585–594.

*CHAPTER 8*

# GLYCOSIDASES

Glycosidases include hydrolases which attack glycosidic linkages, the latter consisting of a bond between the aldehyde grouping of a sugar (glycone) and the hydroxyl grouping of another compound (aglycone) (Hoffmann-Ostenhof, 1953). Among the enzyme systems which act upon glycosidic bonds are α- and β-amylases and glucoamylases which act upon α-1,4-glucose links; amylo-1,6-α-glucosidases and dextranase which hydrolyze α-1,6-glucose links; α-1,3-glucosidase which hydrolyzes α-1,3-glucosides; α-mannosidase, an enzyme widely distributed in mammalian tissues, which hydrolyzes α-D-mannosides (Conchie *et al.*, 1959); heparinase which hydrolyzes α-1,4-links between glucosamine sulfate and glucuronate; lysozyme which acts upon glycosidic links of mucopolysaccharides; neuraminidase (sialidase) which hydrolyzes β-D-acetyl-neuraminides; β-glucosidases which hydrolyze many β-D-glucopyranosides; and β-galactosidases which hydrolyze β-D-galactosides (Dixon and Webb, 1958, pp. 193–195; Fischer and Stein, 1960, p. 301; Larner, 1960, p. 369). From a histochemical standpoint glycosidases which have attracted the most attention are the β-glucuronidases and the β-glucosaminidases. The former hydrolyze β-D-glucuronides (glucosiduronic acids; glucopyranosiduronic acids; glucopyruronosides) while the latter attack alkyl and aryl *N*-acetyl-β-glucosaminides.

## I. SYNTHETIC APPROACHES AND SUBSTRATE TYPES

Table 1 illustrates various types of aryl glycosidic substrates several of which have found application in histochemistry. The others may find use in future techniques or may suggest possibilities for the development of newer substrates. The nomenclature employed is generally that indicated in the original references to the synthesis.

The fact that administration of phenolic compounds, including naphthols, to animals may result in the excretion of glucuronides in the urine has been known since the 19th century (Artz and Osman, 1950). The biosyn-

TABLE 1

GLYCOSIDE POTENTIAL SUBSTRATES

| Chemical name and formula | Mol. wt. | Reference | Melting point (° C.) |
|---|---|---|---|
| 1. Phenyl-α-D-glucopyranosiduronic acid (glucopyruronoside), $C_{12}H_{14}O_7$ | 270.24 | Bollenback *et al.*, 1955; Marsh and Levvy, 1958b | 147–149 |
| 2. Phenyl-β-D-glucopyranosiduronic acid (glucopyruronoside), $C_{12}H_{14}O_7$ | 270.24 | Garton *et al.*, 1949; Tsou and Seligman, 1953; Marsh and Levvy, 1958b | 161–162 |
| 3. *p*-Nitrophenyl-α-D-glucopyranosiduronic acid, $C_{12}H_{13}O_9N \cdot H_2O$ | 333.26 | Marsh and Levvy, 1958b | 211–212 |
| 4. L-Menthyl-β-D-glucuronide (glucopyruronoside), $C_{16}H_{28}O_7 \cdot 1\frac{1}{2}H_2O$ | 359.42 | Quick, 1924; Marsh, 1952 | 78–80 |
| 5. *p*-Biphenyl-β-D-glucuronide, $C_{18}H_{18}O_7 \cdot 1\frac{1}{2}H_2O$ | 373.37 | Dodgson *et al.*, 1948 | 183–184 |
| 6. α-Naphthyl glucuronide (naphthyl-β-D-glucopyranosiduronic acid), $C_{16}H_{16}O_7 \cdot 1\frac{1}{2}H_2O$ | 347.33 | Lesnik and Nencki, 1886; Berenbom and Young, 1951 | 202–203<br>198–199 |
| 7. β-Naphthyl glucuronide (glucopyranosiduronic acid), $C_{16}H_{16}O_7$ | 320.30 | Lesnik and Nencki, 1886; Berenbom and Young, 1951; Bollenback *et al.*, 1955 | 149–150<br>151.5–152 |
| 8. Umbilliferone (7-hydroxycoumarin) glucuronide, $C_{15}H_{14}O_9 \cdot H_2O$ | 356.29 | Mead *et al.*, 1954, 1955 | 150 |
| 9. 4-Methylumbelliferone (hydroxycoumarin) glucuronide, $C_{16}H_{16}O_9 \cdot 3\frac{1}{2}H_2O$ | 415.36 | Mead *et al.*, 1955; Marsh and Levvy, 1956 | 140 |
| 10. 8-Hydroxyquinoline glucuronide, $C_{15}H_{15}O_7N \cdot 3H_2O$ | 375.34 | Brahm, 1899; Robinson *et al.*, 1953a | 153 |
| 11. 6-Bromo-2-naphthyl-β-D-glucopyruronoside, $C_{16}H_{15}BrO_7$ | 399.21 | Seligman *et al.*, 1954 | 168–169 (dec.) |
| 12. 8-Benzoylamino-2-naphthyl-β-D-glucuronide, $C_{23}H_{20}O_8$ | 424.41 | Rutenburg and Seligman, 1953 | — |
| 13. Naphthol AS-LC glucuronide, $C_{25}H_{24}O_{10}NCl$ | 533.93 | Pugh and Walker, 1961a | — |
| 14. Phenolphthalein (mono) β-glucuronide, $C_{45}H_{44}O_{11}N_2$ | 788.86 | di Somma, 1940; Fishman, 1955, p. 55 | — |
| 15. *p*-Aminophenyl-*N*-acetyl-β-glucosaminide, $C_{14}H_{20}O_6N_2$ | 312.33 | Westphal and Schmidt, 1951 | 227–228 |
| 16. Phenyl-*N*-acetyl-β-D-glucosaminide, $C_{14}H_{19}O_6N$ | 297.31 | Helferich and Iloff; 1933; Roseman and Dorfman, 1951 | 238<br>249–250 |
| 17. α-Naphthyl-*N*-acetyl-β-glucosaminide, $C_{18}H_{21}O_6N$ | 347.37 | Leaback and Walker, 1957 | 244–246 (dec.) |

TABLE 1—*Continued*

| Chemical name and formula | Mol. wt. | Reference | Melting point (° C.) |
|---|---|---|---|
| 18. Naphthol AS-LC *N*-acetyl-β-glucosaminide, $C_{27}H_{29}O_9NCl$ | 546.99 | Pugh and Walker, 1961a | — |
| 19. Indoxyl-β-D-glucoside, $C_{14}H_{17}NO_6$ | 295.30 | Robertson and Waters, 1933 | 178–180 (dec.) |
| 20. Phenyl-β-D-glucoside, $C_{12}H_{16}O_6 \cdot 2H_2O$ | 292.29 | Robertson and Waters, 1930 | 176–177 |
| 21. *p*-Nitrophenyl-β-D-glucoside, $C_{12}H_{15}O_8N$ | 301.26 | Glaser and Wulwek, 1924; Montgomery *et al.*, 1942 | 164 |
| 22. *o*-Nitrophenyl-β-D-glucoside, $C_{12}H_{15}O_8N$ | 301.26 | Glaser and Wulwek, 1924; Conchie *et al.*, 1959 | 152 |
| 23. *o*-Diphenyl-β-D-glucoside, $C_{18}H_{20}O_6$ | 332.36 | Bembry and Powell, 1942 | 76–77 |
| 24. α-Naphthyl-β-D-glucoside, $C_{16}H_{18}O_6 \cdot H_2O$ | 324.34 | Drouin, 1895 | 147 |
| 25. β-Naphthyl-β-D-glucoside, $C_{16}H_{18}O_6 \cdot H_2O$ | 324.34 | Ryan, 1899; Robertson and Waters, 1930 | 184–186 |
| 26. Naphthol AS glucosides | — | Rath and Feess, 1954 | — |
| 27. 6-Bromo-2-naphthyl-β-D-galactopyranoside, $C_{16}H_{17}BrO_6$ | 385.23 | Cohen *et al.*, 1952b | 198–200 |
| 28. *o*-Nitrophenyl-β-D-galactoside (galactopyranoside), $C_{12}H_{15}O_8N$ | 301.26 | Seidman and Link, 1950 | 193–194 |
| 29. α-Naphthyl-D-galactoside, $C_{16}H_{18}O$ | 226.32 | Ryan and Mills, 1901 | 202–203 |
| 30. 6-Bromo-2-naphthyl-β-D-galactopyranoside, $C_{16}H_{17}O_6Br$ | 385.23 | Cohen *et al.*, 1952a | 205 |

thesis of phenyl-β-D-glucuronide was reported by Külz in 1890 and by Salkowski and Neuberg in 1906 while that of both α- and β-naphthyl glucuronides was originally described by Lesnik and Nencki in 1886. Subsequently Masamune (1933, 1934) who coined the term "glucuronosidase" found that biosynthetically prepared β-naphthol, borneol, and phloroglucinol glucuronic acids were hydrolyzed by the enzyme. Even though biosynthetic procedures are cumbersome and to a certain extent have been supplanted by direct synthetic approaches, the former techniques are still of value in the preparation and characterization of many aryl glucuronides (Kamil *et al.*, 1951; Robinson *et al.*, 1953b). Although the phenol may be directly administered to the animal, the use of either the sodium salt or else the aryl phosphate is employed when appreciable yields of some of the more insoluble compounds are desired (Talalay *et al.*, 1946; Kamil *et al.*, 1951).

The first direct synthesis of a glucuronide was reported by Neuberg and Neimann (1905) who prepared phenyl-β-D-glucuronide from phenol and diacetylbromoglucuronolactone. In 1935 Goebel and Babers synthesized α-1-bromo-2,3,4-triacetylglucuronic acid methyl ester and from it the corresponding ester of β-D-*p*-nitrobenzyl glucuronide. In this type of procedure the aglycone is coupled with a bromoglucuronic acid ester followed by deacetylation and removal of the halogen group. Another synthetic approach which has received considerable attention involves the oxidation of glucosides to glucosiduronic acids employing oxygen in the presence of a platinum catalyst. An impetus to this work was given by studies of Fernández-García *et al.* (1950), who found that isopropylidene-D-glucose could be oxidized stoichiometrically in a neutral or mildly alkaline solution to give 1,2-*o*-isopropylidene-D-glucuronic acid, which upon hydrolysis yielded D-glucuronic acid. Subsequently this type of reaction was employed in the synthesis of a variety of alkyl and aryl glucosides (Barker *et al.*, 1951; Marsh, 1951, 1952; Tsou and Seligman, 1952b, 1953; Seligman *et al.*, 1954; Bollenback *et al.*, 1955; Marsh and Levvy, 1958b). Bollenback *et al.* (1955) have carefully reviewed some of the aforementioned work and outline methods for the preparation of aryl glucosides and their subsequent transformation into glucopyruronosides:

1. Treatment of a phenol with methyl (tri-*o*-acetyl-α-D-glucopyranosyl bromide)uronate in the presence of silver carbonate.
2. Treatment of a potassium phenolate with methyl (tri-*o*-acetyl-α-D-glucopyranosyl bromide)uronate in ethanol.
3. Fusing a phenol with methyl tetra-*o*-acetyl-β-D-glucopyranuronate in the presence of either *p*-toluenesulfonic acid or zinc chloride.
4. Fusing a phenol with methyl tetra-*o*-acetyl-α-D-glucopyranuronate in the presence of zinc chloride.
5. Treatment of a phenol with methyl (tri-*o*-acetyl-α-D-glucopyranosyl bromide)uronate in the presence of quinoline.

Acetylated methyl glucuronates may be prepared from glucuronolactone which is commercially available. The corresponding aryl glucuronides are obtained by de-esterification reactions. Review of synthetic approaches to the preparation of glucuronides has also been made by Conchie *et al.* (1957).

In addition to the aforementioned papers of Tsou and Seligman (1953), Bollenback *et al.* (1955), and Marsh and Levvy (1958b), which deal with the preparation of aryl glucuronides, there have been interesting reports on the synthesis of glucuronides of phenolic amino acids (Taurog *et al.*, 1953) and of steroids (Schapiro, 1939; Wotiz *et al.*, 1959).

Mammalian glucuronidases hydrolyze alkyl, aryl, alicyclic, and aryl β-glucuronides, but have no effect upon α-glucuronides or β-D-glucopyranosides (Levvy and Marsh, 1959; 1960, p. 397). Glucuronidase also hydrolyzes β-D-galactopyranosiduronic acids (Marsh and Levvy, 1958a). In general, it is believed that biosynthetically produced glucuronides contain the pyranose ring structure. This is based on the observation that oxidation of β-D-glucopyranosides results in formation of the corresponding β-D-glucopyranosiduronic acids (pyruronosides). Also of interest is the observation that β-D-glucuronic acid-1-phosphate (Touster and Reynolds, 1952) is hydrolyzed by glucuronidase (Levvy and Marsh, 1952). Although Tsou and Seligman (1952b) claimed that substrates containing the furanose ring could not be hydrolyzed by glucuronidase, Fishman (1955) and Levvy and Marsh (1959; 1960, p. 397) indicate considerable reservation as to whether or not the enzyme can hydrolyze β-D-glucofuranosiduronic acids since many of these compounds have not as yet been isolated. Of interest is the enzymatic hydrolysis of an aminobenzoyl β-glucuronide which is believed to have a furanose structure (Dutton, 1956).

From a histochemical standpoint, only two of the substrates shown in Table 1 have thus far proved to be of value: 8-hydroxyquinoline and 6-bromo-β-naphthyl glucuronides. The general configuration of these compounds is illustrated by α-naphthyl glucuronide (I), although this compound has not as yet been utilized in a successful histochemical technique. It is reasonably certain, however, that future histochemical methods will employ a variety of complex aryl (including steroid) glucuronides.

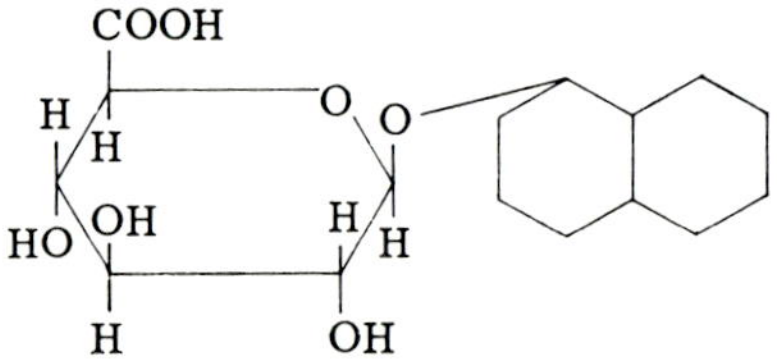


α-Naphthyl glucuronide

(I)

The first organic synthesis of an aryl glycoside, phenyl-β-D-glucopyranoside, dates back to the 19th century (Michael, 1879). In 1895 Drouin reported the preparation of α-naphthyl glucoside. The synthesis of β-naphthyl glucoside was reported by Ryan in 1899 and 1901, while that of α-naphthyl galactoside was reported by Ryan and Mills in 1901. These earlier techniques involved the treatment of a phenol or naphthol with acetochloroglucose or acetochlorogalactose which were prepared by treating D-glucose or D-galactose with acetyl chloride. An analogous procedure employing the corresponding bromide was reported by Koenigs and Knorr

(1901). Reviews dealing with the preparation of aryl glycosides and intermediates have been reported by Robertson and Waters (1930), Aizawa (1939), Montgomery *et al.* (1942), Haynes and Newth (1955), Conchie *et al.* (1957), and Hickinbottom (1957, p. 278). The paper of Montgomery *et al.* (1942) offers a succinct review of useful synthetic procedures prior to that date. In 1952 the preparation of several naphtholic β-D-glucosides and galactosides was reported by the Seligman group (Cohen *et al.*, 1952a, b; Tsou and Seligman, 1952a). These authors employed the well known synthetic technique devised by Helferich and Schmitz-Hillebrecht (1933) as modified by Montgomery *et al.* (1942). In this technique the acetylated sugar is heated with a naphthol in the presence of a catalyst such as *p*-toluenesulfonic acid. The resulting tetraacetyl-β-D-glucoside is then deacetylated to give the appropriate naphthyl-β-D-glucoside (glucopyranoside) or galactoside (galactopyranoside).

With reference to substrate specificity, β-D-glucosidase will hydrolyze aryl β-D-xyloside and mannoside but at a much lower rate than the glucoside. Various aspects of substrate specificity from a biochemical standpoint are considered in the reviews by Veibel (1950, p. 583) and Sumner and Somers (1953, p. 107). As in the case of the aforementioned glycosides the pyranose ring structure seems to be essential for the hydrolytic activity of β-galactosidase (Wallenfels and Malhotra, 1960, p. 416). Mammalian β-glucuronidase has no effect upon β-glucosides (Marsh and Levvy, 1958a; Levvy and Marsh, 1960, p. 404). From the standpoint of applied histochemistry two substrates have been suggested for these enzymes, 6-bromo-2-naphthyl-β-D-glucopyranoside and 6-bromo-2-naphthyl-β-D-galactopyranoside (Cohen *et al.*, 1952a, b) (Table 1), (see also Appendix 8A, Sections 15 and 17). It would be interesting to explore the possible application of α-naphthyl galactoside and α-naphthyl glucoside (II and III) (Table 1) in simultaneous coupling azo-dye procedures. With reference to the latter compound, Bembry and Powell (1942) have described a convenient synthesis of tetraacetyl-α-naphthyl-β-D-glucoside.

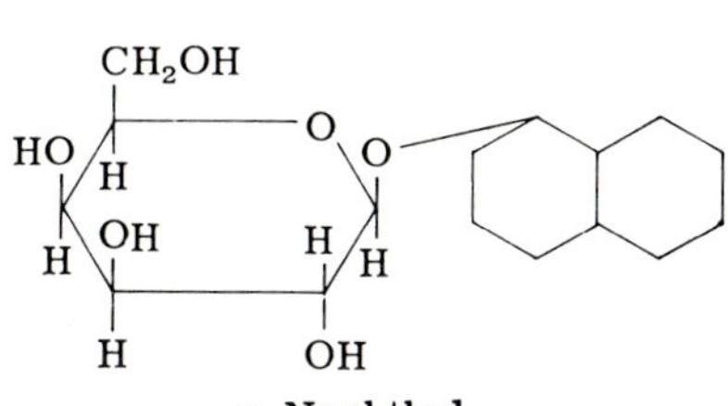


α-Naphthyl
galactoside

(II)

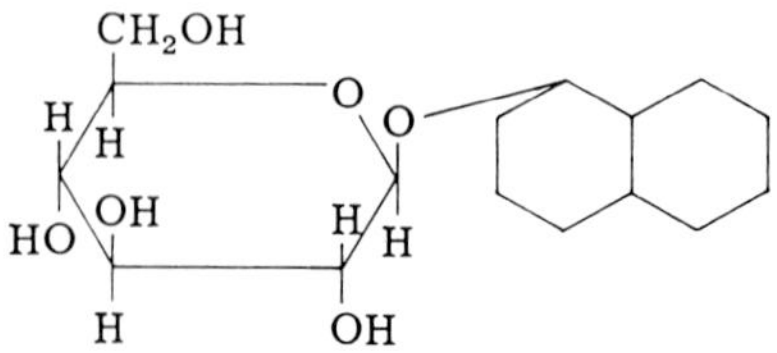


α-Naphthyl
glucoside

(III)

Both alkyl and aryl *N*-acetyl-β-D-glucosaminides (glucosaminopyranosides) have been shown to be suitable substrates for the biochemical and histochemical determination of glucosaminidase (Helferich and Iloff, 1933; Watanabe, 1936; Neuberger and Pitt Rivers, 1939; East *et al.*, 1941; Westphal and Schmidt, 1951; Kuhn and Tiedemann, 1954; Pugh *et al.*, 1957).

From a histochemical standpoint the synthesis of phenyl-*N*-acetyl-β-D-glucosaminide by Helferich and Iloff (1933) and subsequent modification by Roseman and Dorfman (1951) (see Appendix 8A, Section 10), as well as the synthesis of the *p*-nitrophenol analogue by Westphal and Schmidt (1951), suggested the possibility of other aryl glucosaminides as biochemical and histochemical substrates. The method of Helferich and Iloff (1933) involves the synthesis of phenyl-3,4,6-triacetyl-β-D-glucosaminopyranoside hydrochloride, which upon acetylation with acetic anhydride followed by partial deacetylation yields the *N*-acetyl-β-D-glucosaminopyranoside. More recently, Leaback and Walker (1957, 1961) reported the synthesis of a number of aryl β-glucosaminides. They employed acetochloroglucosamine (Baker *et al.*, 1954) which was allowed to react with a number of phenolic and naphtholic compounds including phenol, *p*-nitrophenol, *o*-nitrophenol, α-naphthol, and 4-methylumbelliferone. The acetylated glycoside was then de-*o*-acetylated with sodium methylate or other compounds to give the appropriate glucosaminide. Of these compounds only the α-naphthol derivative (IV) has been employed as a histochemical reagent (Pugh and Walker, 1961b), although brief mention has been made of the

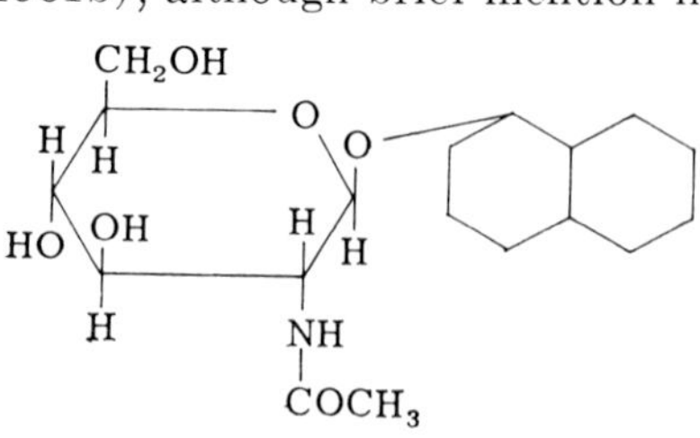


α-Naphthol *N*-acetyl-
β-D-glucosaminide

(IV)

potential value of naphthol AS-LC glucosaminide (Pugh and Walker, 1961a).

## II. β-GLUCURONIDASE

Comprehensive reviews on the biochemistry of β-glucuronidase have been given by Fishman (1950, p. 635; 1955), Bray (1953), Levvy (1956), and Levvy and Marsh (1959; 1960, p. 397).

The enzymatic hydrolysis of a conjugated glucuronide, phenylglucuronic acid, was first reported by Neuberg and Neimann (1905). Although subsequent studies demonstrated that certain aryl glucuronides could be hydrolyzed by extracts of animal tissues (Sera, 1914) it remained for Masamune and associates to clearly demonstrate the existence of a β-glucuronidase in animal tissues (Masamune, 1934; Oshima, 1934, 1936). Masamune (1934) noted that the enzyme which had a pH optimum of around 5.6 catalyzed the hydrolysis of a number of biosynthetically prepared glucuronides including β-naphthyl glucuronide. Oshima (1936) observed that dicarboxylic acids such as maleic and tartaric acids were distinct inhibitors of the enzyme. The subsequent extension of these studies has been described in the aforementioned review papers. In mammalian tissues highest activity is found in liver, kidney, spleen, epididymis, and neoplasms. According to de Duve *et al.* (1955) β-glucuronidase is localized to the so-called "lysosomes." Goldfarb and Barka (1960), employing histochemical techniques, observed a characteristic peribiliary staining of liver granules which they assumed associates the enzyme with "lysosomes." More recent studies (Fishman, 1950, p. 640; Levvy and Marsh, 1960, p. 401) indicate that mammalian β-glucuronidase preparations display pH optima at 4.5 and 5.2. With reference to histochemical procedures, tissue sections are generally incubated at pH 5.0 in the presence of acetate or phosphate-citrate buffer. Fishman (1955) has reviewed the possible *in vivo* roles of β-glucuronidase and notes a correlation of high enzymatic activity with growth and tissue proliferation. The enzyme in certain tissues responds to the action of estrogens, and is also involved in the physiological production of steroid and aryl glucuronides. It has also been shown that bilirubin excreted in the bile is an ester glucuronide (Billing *et al.*, 1957). Of particular significance is the role of β-glucuronidase in the metabolism of drugs and toxic compounds (Fishman, 1956; Williams, 1959, p. 290). *In vitro* studies indicate the enzyme may function as a transferase (Fishman and Green, 1957).

### A. Oxidation-Chelation and Azo-Dye Methods

The first histochemical methods for the demonstration of this enzyme were suggested by Friedenwald and Becker (1948). One of their methods employed the glucuronide of a water-insoluble azo-dye as the substrate

(1-*o*-hydroxyphenylazo-2-naphthol). Upon enzymatic hydrolysis the insoluble azo-dye was precipitated *in situ*. Although the unhydrolyzed substrate is in itself a dye, it produces only a pale orange-yellow stain in certain tissue components (e.g. lipids). The other ingenious approach of Friedenwald and Becker employed a substrate solution containing 8-hydroxyquinoline glucuronide in conjunction with ferric chloride. The enzymatically released 8-hydroxyquinoline formed an insoluble black precipitate of ferric hydroxyquinoline which was subsequently converted into Prussian blue by ferrocyanide. These authors reported that various metal salts including less ionized ones were employed but were abandoned in favor of the ferric iron. Campbell (1949) reported successful application of the Friedenwald-Becker 8-hydroxyquinoline glucuronide method but were unable to obtain specific staining with the phenylazo glucuronide technique. Among the criteria which they employed to ascertain the validity of the method was the inclusion in the substrate solution of a known inhibitor, potassium saccharate. Similar confirmation of the validity of the hydroxyquinoline procedure has also been reported by Campbell and Levvy (1950) and by Burton and Pearse (1952). Pearse (1960, p. 499) recommends cold formalin fixed frozen sections and substitution of ferric chloride by ferric sulfate in the substrate solution. It was claimed that these modifications considerably widened the range of tissues to which the method could be applied and also reduced the degree of non-specific precipitate formation. The most recent and best of the hydroxyquinoline glucuronide modifications is that reported by Fishman and Baker (1956, also unpublished data) who in addition to offering an excellent critique of the subject have reported a highly reproducible histochemical procedure (Chart 1). Among their modifications is the introduction of small amounts of chloral hydrate into the fixative solution, a factor which contributes greatly to the preservation of the enzyme. In addition, a substrate solution which is stable over a period of months was evolved. Improvements in the technique have resulted in a relatively homogeneous dye which permits good localizations at the histological and in some cases at the cytological level. With incubation periods of 1–4 hours, excellent results may be obtained in a variety of tissues. Particularly fine localizations are obtained in certain specialized cells and tissues (e.g. Purkinje cells, pancreatic islet cells, uterine endometrium, and cartilage (Figs. 1 through 4).

Simultaneous and post-coupling azo-dye techniques for the demonstration of glucuronidase were first reported by the Seligman group (Seligman *et al.*, 1949, 1954). Both $\alpha$- and $\beta$-naphthyl glucuronides were employed without success in a simultaneous coupling procedure. Fishman and Baker (1956) were also unsuccessful in attempts to utilize stable diazotates in conjunction with biphenyl-, *p*-hydroxyphenyl-, and 8-hydroxyquinoline glucuronides. On the other hand, Pearse (1960, p. 496) reported limited

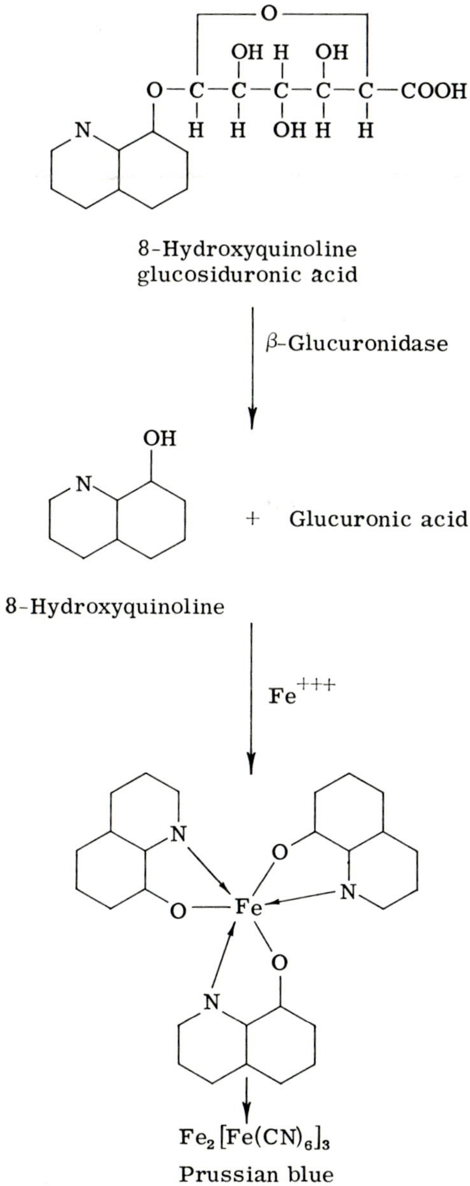


**CHART 1.** Modified Fishman-Baker (1956) oxidation-chelation technique for demonstration of β-glucuronidase.

results with a hydroxyquinoline glucuronide azo-dye procedure. A significant advance was made by Seligman *et al.* (1954) who subsequently reported the synthesis and histochemical application of 6-bromo-2-naphthyl-β-glucopyruronoside. They noted that mammalian β-D-glucuronidase

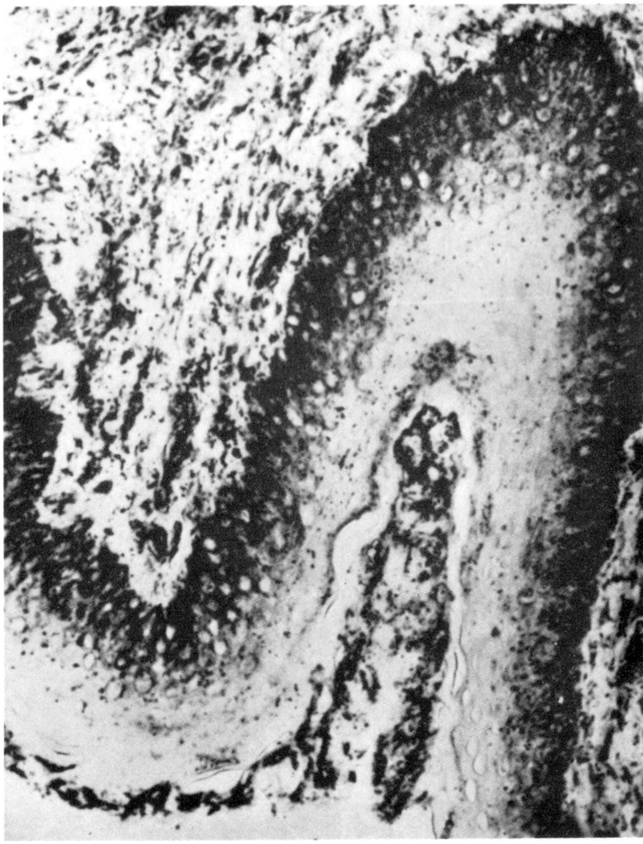

FIG. 1. Rat vagina at early estrus. β-glucuronidase. An intense reaction of β-glucuronidase is characteristically restricted to the germinal layer of the epithelium. A strong reaction is also observed in desquamated epithelial cells in the lumen. Note also enzyme-rich granules (× 150). (Courtesy Dr. W. Fishman.)

hydrolyzes glucuronides with a pyranose structure (6-membered ring) but does not hydrolyze glucuronides with a furanose structure (5-membered ring) or the corresponding glucoside. Although they indicated that the localization of glucuronidase activity with the new method was roughly similar to that obtained with the Friedenwald-Becker hydroxyquinoline procedure, Gomori (1952, p. 217) pointed out that the bromonaphthyl glucuronide technique is not likely to work out satisfactorily. According to Pearse (1960, p. 498) the overall picture resembles that given by the modified Fishman-Baker method but the localizations are not as sharp as with the latter. As previously indicated (Chapter 4, p. 139), 6-bromo-2-naphthol is capable of non-specifically staining tissue sections by virtue of its selec-

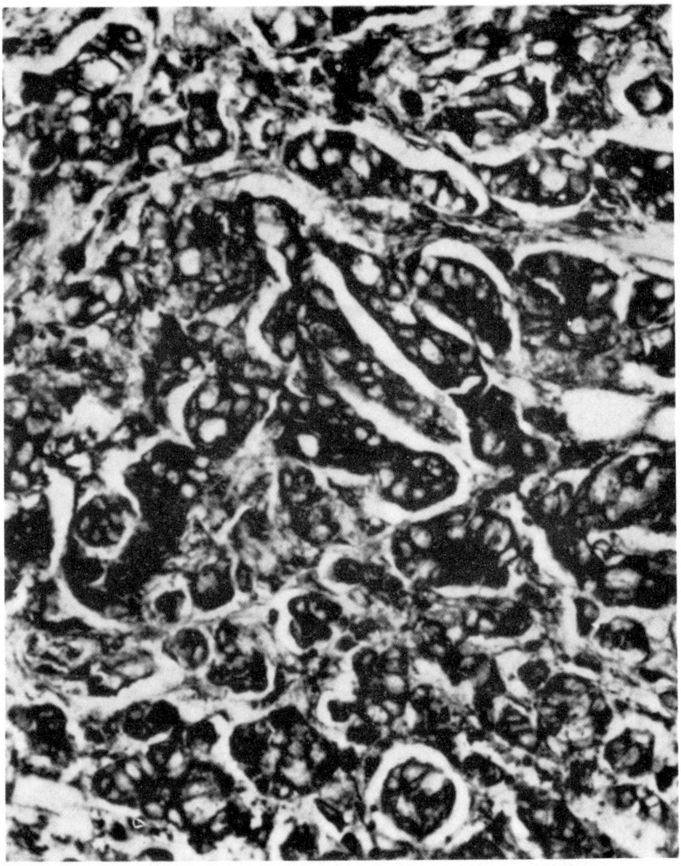

Fig. 2. Carcinoma from human jejunum. β-glucuronidase. The enzyme reaction is evident in clusters of pleomorphic cells, where it usually exhibits a cytoplasmic location (× 200). (Courtesy Dr. W. Fishman.)

tive affinity for certain structures. *Therefore, as with all other techniques employing this naphthol as a basis for a substrate, control slides should always be incubated in solutions of the free naphthol and then coupled with a diazonium salt to ascertain if the enzymatic reaction is genuine or if the staining reaction is simply due to non-specific adsorption of the naphthol released either by enzymatic hydrolysis or by spontaneous decomposition of the substrate.*

## B. Activators and Inhibitors

Fishman (1955), Levvy (1956), and Levvy and Marsh (1959) have noted a number of potential activators and inhibitors of β-glucuronidase.

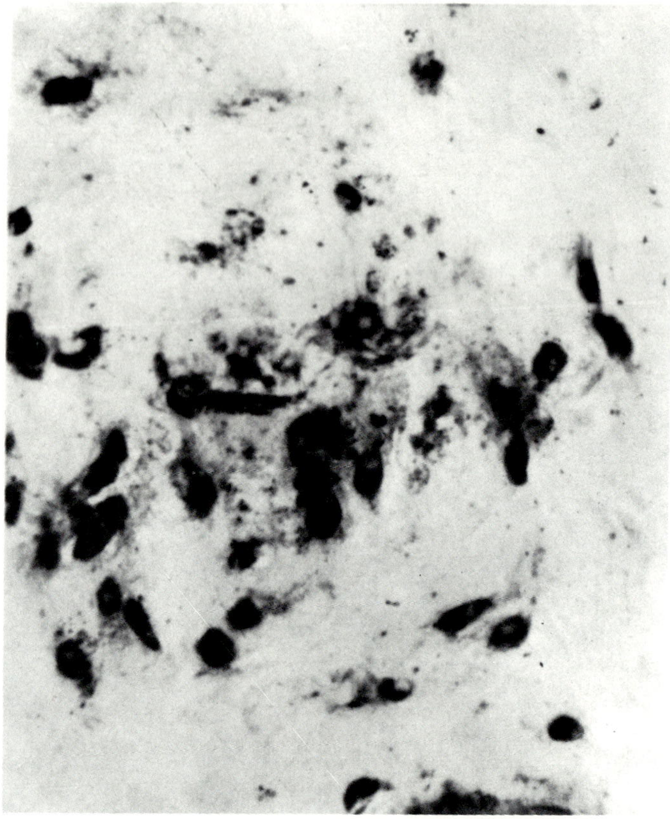

FIG. 3. Histiocytes of human vaginal stroma. β-glucuronidase. Intense reaction in histiocytes and capillary endothelium (× 400). (Courtesy Dr. W. Fishman.)

Among the activators are chitosan, protamine, crystalline bovine serum albumin, DNA, gelatin, crystalline chymotrypsin, and yeast RNA.

As previously mentioned, Oshima (1936) found that certain hydroxycarboxylic acids were inhibitors. Karunairatnam and Levvy (1949) first reported that the enzyme is markedly inhibited by saccharate solutions, and it was subsequently found that the inhibitory effect was due to the formation of saccharo-1,4-lactone (Levvy, 1952). The presence of saccharo-1,4-lactone (D-glucaro-1,4-lactone) in a concentration of $2.5 \times 10^{-6}$ *M* results in approximately a 50% inhibition of mammalian β-glucuronidase. It is important to note that this highly specific competitive inhibitor has no effect upon other glycosidases such as β-glucosaminidase, β-galactosidase, and α-mannosidase. In general, the inhibitor is prepared by boiling saccharate solutions (ca. 0.25 m*M*), which results in a conversion of one-

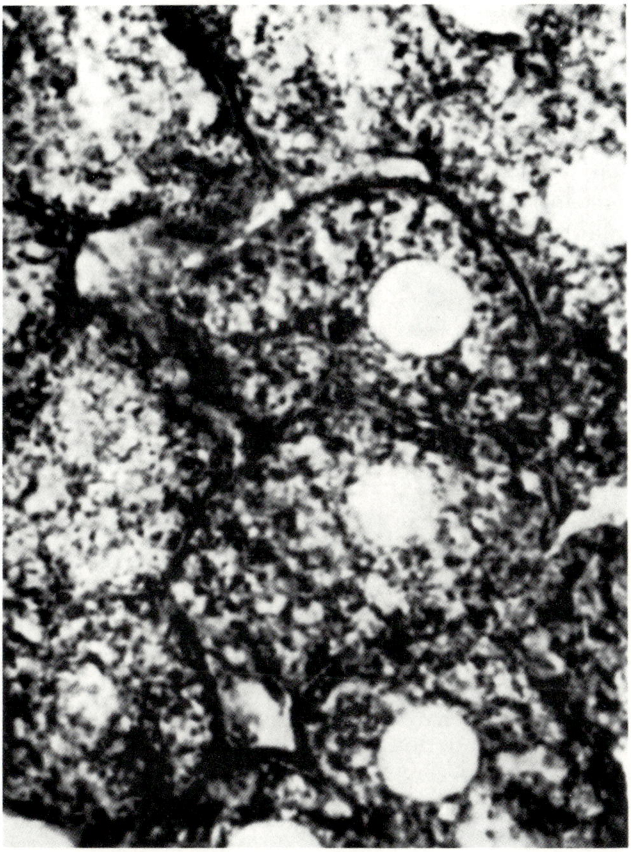

FIG. 4. Rat liver, β-glucuronidase. The enzyme activity is intense in the granules clearly visible in the cell cytoplasm bordering the bile canaliculi. A large number of small stained granules are distributed uniformly in the cytoplasm (× 1800). (Courtesy Dr. W. Fishman.)

third of the anion into the 1,4-lactone. Alternatively the pure lactone may be employed. Related compounds such as D-glucarate and galactarate have a similar inhibitory effect though not as pronounced. Non-competitive inhibitors of the enzyme include ascorbic acid, heparin, $Ag^{+}$, and $Cu^{++}$. Of interest is the observation that β-glucuronidase activity *in vitro* is unaffected by fluoride, iodoacetate, cyanide, azide, and dinitrophenol.

Animal tissues are known to contain a thermostable β-glucuronidase inhibitor which in certain cases resembles the acid mucopolysaccharides. The endogenous inhibitor is not pH-dependent and produces an inhibition which is considerably below 100%. Furthermore, there is considerable variation as to the susceptibility of glucuronidases from different sources.

### C. β-Glucuronidase in Neoplasms

The original biochemical studies of Fishman and associates (Fishman, 1947; 1950, p. 651; 1955; 1959, p. 746) showed that a number of primary human neoplasms including those from the skin, lung, breast, esophagus, stomach, colon, uterine cervix, and brain exhibited high β-glucuronidase activity as compared with the adjacent "normal" tissue. Endometrial carcinoma did not exhibit increased β-glucuronidase levels. Also of interest was was the observation that increased glucuronidase levels could be found in certain polyps of the colon which were in a premalignant state. Confirmation of the aforementioned findings of the Fishman group was made by Kerr *et al.* (1950), Cohen and Bittner (1951), and Kun *et al.* (1951) with reference to rodent neoplasms. In connection with human neoplasms, Boyland *et al.* (1954) reported that carcinoma of the bladder is associated with high urinary β-glucuronidase levels. Odell and Burt (1950) originally noted high β-glucuronidase activity of carcinoma of the cervix, and suggested that measurement of the enzyme in vaginal fluid might constitute a diagnostic aid. The more recent work of Fishman and Mitchell (1959) has evaluated the relationship between vaginal β-glucuronidase activity and certain physiological and pathological conditions including malignancies. Although they indicate that it may be possible to associate sites of enzyme activity with ovarian function, diagnostic interpretation of enzyme levels in individual samples presents certain difficulties in the absence of exact knowledge as to the source of the vaginal fluid enzyme (e.g. vaginal mucosa, uterine cervix, endometrium, bacterial flora, protozoa, benign erosions and inflammations, gestational tissues, and cancer). Rauramo (1959), in an evaluation of the aforementioned diagnostic potential suggested by Odell and Burt (1950), also indicated that there should be a *critical* evaluation with reference to diagnostic implications of vaginal β-glucuronidase levels. The buffy coat of red cells has been found to exhibit high β-glucuronidase activity and, according to Anlyan *et al.* (1950), there is a markedly increased activity of the enzyme in myelogenous leukemia.

The initial histochemical findings which demonstrated high glucuronidase activity of animal and human tumors were made by Campbell (1949) and Seligman *et al.* (1949). The first definitive and comprehensive study on the histochemical localization of the enzyme in human tumors was reported by Fishman *et al.* (1959) who employed an improved modification of the Friedenwald-Becker (1948) 8-hydroxyquinoline glucuronide procedure. Among the neoplasms studied were tumors of the jejunum, sigmoid colon, rectosigmoid colon, uterus, adrenal, prostate, and thymus. It was consistently observed that the cytoplasm of neoplastic cells exhibited β-glucuronidase activity. Nuclei for the most part did not stain or else exhibited a weakly positive reaction. Fishman *et al.* (1959) specifically noted that the cancer cell contains β-glucuronidase activity which exceeds that seen in non-neoplastic cells. They felt that this characterized the enzyme as one of the few which do not exhibit decreased activity concomitant with the neoplastic transformation. They also observed that the stromal activity of cancer tissue is variable. In some neoplasms stromal activity is absent while in others (adenocarcinoma) activity is quite high. The latter may be due to disintegrated β-glucuronidase-rich cells, to phagocytized fragments of cancer cells, or to β-glucuronidase-containing secretions released by cancer cells. They also point out the known activity of leukocytes and fibroblasts. Accordingly, the β-glucuronidase activity of a tumor may arise from several different sources, but nevertheless, it is felt that the neoplastic cell constitutes the major contributor of the enzyme. Monis *et al.* (1960) confirmed the aforementioned findings of Fishman with reference to high β-glucuronidase activity of human neoplasms. In addition, they found that malignant epithelial tumors generally exhibited an intense staining reaction as contrasted with mesenchymal tumors which showed little or no activity.

## III. *N*-ACETYL-β-GLUCOSAMINIDASE

This enzyme which hydrolyzes β-glycosides of *N*-acetyl-D-glucosaminopyranosides was discovered by Helferich and Iloff (1933). East *et al.* (1941) demonstrated that β-glucosaminidase is a distinct enzyme from hyaluronidase. Since amino sugars including glucosamine are widely distributed in tissues, and in particular in the so-called "mucoid" substances, there has been renewed interest in the study of enzymes which hydrolyze glucosamine complexes. According to Sellinger *et al.* (1960) β-glucosaminidase is associated with lysosomes.

The first comprehensive biochemical survey on the enzyme which dealt with the distribution, specificities, as well as synthesis and application of an aryl glucosaminide as a substrate, was reported by Roseman and Dorfman (1951). High β-glucosaminidase activity was found in a variety of rat tissues including testis, spleen, liver, lung, heart, brain, and muscle as measured by the hydrolysis of phenyl-*N*-acetyl-β-D-glucosaminide. Findlay and Levvy (1960) note that the epididymis is the richest source of the enzyme in mammals. According to East *et al.* (1941) and Roseman and Dorfman (1951) the pH optimum of the enzyme is about 4.5 in the presence of phosphate-citrate buffer. Of interest is their observation that activity of glucosaminidase is diminished in the presence of acetate buffer. The studies of Roseman and Dorfman (1951) also indicate the enzyme differs from α- and β-glucosidases and from hyaluronidase. Of interest is their evidence for the existence of a specific α-glucosaminidase which is widely distributed in tissues. More recently Roseman (1957) has reported an *N*-acetylglucosamine deacetylase which is found in bacteria but not in animal tissues.

Conchie *et al.* (1959) have studied the biochemical distribution of several glucosaminidases in mammalian tissue and characterized them by the use of selective inhibitors. It was found, for example, that *N*-acetylglucosaminonolactone (Findlay *et al.*, 1958) in concentrations of 0.1 m*M* was a distinct inhibitor of β-glucosaminidase but had no effect upon β-glucuronidase, β-galactosidase, or α-mannosidase. Borooah *et al.* (1961) studied the hydrolytic activity of the enzyme and found that the *p*-nitrophenyl-*N*-acetyl-β-D-glucosaminide substrate as originally employed by Westphal and Schmidt (1951) was the most sensitive substrate for estimation of enzyme activity in crude or purified preparations.

The function of β-glucosaminidase is not known. Conchie *et al.* (1959) point out that the enzyme acts upon degradation products of hyaluronic acid and this activity may also include chondroitin. Since glucosamine as well as galactosamine constitute in part the sugar component of mucoproteins, related glycosides may have a role in "mucoid" metabolism.

## A. Azo-Dye Methods

Pugh and Walker (1961b) reported a histochemical technique for the demonstration of this enzyme in which formalin fixed frozen sections were incubated in a solution of α-naphthyl-*N*-acetyl-β-glucosaminide (IV) in the presence of a diazonium salt at pH 5.5. The tissues were fixed in a cold formol-calcium solution and either impregnated with gum-sucrose solution or embedded in gelatin prior to sectioning. Frozen sections 5–15 μ thick were transferred directly to the substrate solution, and incubated at room temperature for periods of 10–60 minutes. Control slides consisted of heat inactivated sections as well as sections incubated in substrate solutions containing *N*-acetylglucosaminonolactone, a known powerful competitive inhibitor of the enzyme (Findlay *et al.*, 1958). These authors reported that the quality of fixation with formol-calcium was superior to that obtained with either cold acetone or ethanol, even though a high degree of preservation of the enzyme was obtained with the latter fixatives. It seems reasonable that improved results could be obtained by pre-chilling the specimens prior to immersion in cold solvents (Appendix 8B, Section 4). The relative resistance of the enzyme to acetone or ethanol also suggests the feasibility of employing paraffin embedded tissues including frozen-dried specimens.

They employed several diazonium salts including Fast Blue B, Fast Reds TR and RC, as well as Fast Garnet GBC, and found the latter to be the most suitable. It was found that Garnet GBC produced a minimal inhibition of the enzyme, a finding which confirms the previous work of Burstone (1958) who first emphasized the value of Garnet GBC in enzyme histochemistry because it is stabilized as a diazonium sulfate and industrial samples are usually specified to be relatively free of diluents. Pugh and Walker (1961b) were able to obtain satisfactory localizations of *N*-acetyl-β-glucosaminidase in respiratory and intestinal epithelia, mucous glands, synovial membranes, thyroid, and epididymis. Walker *et al.* (1960) also noted high activity of decidual cells of the human placenta. Although the histochemical localizations revealed certain diffusion artefacts, there is little question as to the specificity of the method and its value as a guide for improved techniques.

Pugh and Walker (1961a) in a brief report noted that *N*-acetyl-β-glucosaminidase could be demonstrated by an azo-dye procedure employing naphthol AS-LC *N*-acetyl-β-glucosaminide. These authors reported this to be an improved substrate and noted that advantage had been taken of the relative insolubility, high substantivity, and rapid coupling of naphthol AS compounds as reported by Burstone (1957, 1958). As with the aforementioned α-naphthol procedure, Garnet GBC salt was employed.

### B. Activators and Inhibitors

From a histochemical standpoint the most selective inhibitor for the characterization of β-D-glucosaminidase is *N*-acetylglucosaminonolactone (Findlay *et al.*, 1958; Pugh and Walker, 1961b) (Appendix 8B, Section 4).

### C. β-D-Glucosaminidase in Neoplasms

Biochemical assay of the enzyme in neoplastic tissues has been reported by Conchie and Levvy (1957) and Conchie *et al.* (1959). Spontaneous mammary tumors from mice exhibited a high activity, although in other neoplasms the normal parent tissue was often observed to exhibit the highest activity. There have been no reports on the histochemical distribution of the enzyme in tumors.

## IV. β-D-GLUCOSIDASE AND β-D-GALACTOSIDASE

The enzyme biochemistry of the two enzymes has been reviewed by Aizawa (1939), Veibel (1950, pp. 583 and 623), Heilskov (1951), Sumner and Somers (1953, pp. 107 and 111), Conchie *et al.* (1959), Larner (1960, p. 369), and Maio and Rickenberg (1960). According to Sellinger *et al.* (1960) galactosidase activity is associated with the lysosomes. Aizawa (1939) found that the pH optimum of mammalian glucosidase is around

7.0. Sumner and Somers (1953, p. 108) note that $Ag^{+}$, $Cu^{++}$, and $Hg^{++}$ ions, as well as oxidizing agents, inactivate the enzyme. Formaldehyde has also been reported to rapidly inactivate β-glucosidase (Veibel, 1950, p. 617).

High activity of β-galactosidase is associated with the gastrointestinal tract of a number of species (Wallenfels and Malhotra, 1960, p. 409; Heilskov, 1951). The β-galactosidase of mouse strain L-cells and mouse intestine, liver, skeletal muscle, stomach, kidney, pancreas, heart, brain, placenta, and lymph nodes exhibit a pH optimum of 2.8 (Maio and Rickenberg, 1960). Cohen *et al.* (1952a) assayed β-galactosidase activity at pH 5.0, although Maio and Rickenberg note that the activity at that pH is only 55% of that observed at the optimum of 2.8. Maio and Rickenberg (1960) also reported that galactose and lactose competitively inhibited the hydrolysis of their aryl galactoside substrate. With reference to specific inhibitors, Conchie *et al.* (1959) reported that galactosidase is selectively inhibited by galactosolactone (10.0 m*M*), but this compound had no effect upon β-glucosaminidase or β-glucuronidase.

## A. Azo-Dye Methods for α- and β-D-Glucosidases

Rutenburg *et al.* (1960) reported that α-glucosidase activity could be demonstrated by a post-coupling procedure employing 6-bromo-2-naphthyl-α-D-glucopyranoside in conjunction with formalin fixed frozen sections. Alternatively, the substrate was employed in a simultaneous coupling procedure. The enzyme was reported to be demonstrable only in the small intestine and in the kidney of the rat. With reference to the demonstration of β-D-glucosidase, Cohen *et al.* (1952b) stated that although the technique was adequate for biochemical determinations, it did not prove satisfactory from a histochemical standpoint. On the other hand, Pearse (1960, p. 502) reported that the technique may successfully reveal sites of high activity in certain non-mammalian species.

The most significant advance in the demonstration of β-glucosidases was reported by Pearson *et al.* (1961) who employed a new substrate, 3-(5-bromoindolyl)-β-D-glucopyranoside ($C_{14}H_{16}BrNO_6$). Frozen sections mounted on glass slides were incubated in a solution containing 4.0 mg. of the aforementioned compound dissolved in 1.0 ml. of DMF, 5.0 mg. of PVP in 1.0 ml. of water, and 14.0 ml. of 0.2 *M* acetate buffer pH 5.4. In general, the enzyme activity was greatest in the preputial glands, duodenum, and small intestine. Slight activity was observed in adrenals, liver, kidney, parotid, pancreas, stomach, spleen, lymph nodes, epididymis, and skin.

## B. Azo-Dye Method for β-Galactosidase

Attempts to demonstrate this enzyme in tissue sections have been reported by the Seligman group (Cohen *et al.*, 1952a; Rutenburg *et al.*, 1958). Unfixed frozen and formalin fixed frozen sections were employed in a post-coupling technique utilizing 6-bromo-2-naphthyl-β-D-galactopyranoside. The substrate solution contained a fairly high concentration of sodium chloride (ca. 5%) which was added in an attempt to minimize diffusion of the enzyme. In general, free floating formalin fixed frozen sections were found to be superior to unfixed cryostat sections mounted on glass slides. Both unfixed and formalin fixed tissues were usually incubated for 4 hours at 37°. Following incubation the sections were treated with Diazo Blue B to develop a dye at sites of deposition of 6-bromo-2-naphthol. Reactive sites were seen in rat kidney, liver, gastrointestinal tract, testis, and epididymis. An intense staining reaction was observed in leukocytes. Rutenburg *et al.* (1958) noted that when tissue sections were treated with 6-bromo-2-naphthol and then coupled with the diazonium salt, the distribution of staining resembled the general picture as seen with their presumed method for the histochemical demonstration of galactosidase. On the other hand, they observed that when sections were incubated in very dilute solutions of 6-bromo-2-naphthol a spurious staining reaction was not obtained. They therefore feel that the latter observation in conjunction with lack of staining of heat inactivated sections adequately demonstrated the validity of their technique.

## APPENDIX 8A

# SYNTHESIS OF GLYCOSIDES

### 1. *Aminophenol Glucuronide* (*Williams, 1943*)

Aminophenol glucuronides are of interest from a histochemical standpoint because they may be diazotized and coupled to various phenolic compounds to form interesting histochemical substrates of the type described by Friedenwald and Becker (1948). They may be biosynthetically prepared by administration of *o*-, *m*-, or *p*-aminophenol. The synthesis of *o*-aminophenol glucuronide will be described.

Six rabbits (ca. 2.5 kg.) are given 1.0 g. of *o*-aminophenol in water by stomach tube. The urine is collected for 1–4 hours following administration of the compound and it is then acidified with a few drops of glacial acetic acid and treated with saturated normal lead acetate until no further precipitate forms. The filtrate is neutralized with ammonia and is treated with excess basic lead acetate solution. The precipitate is centrifuged, washed twice in the centrifuge with water, and the lead salt suspended in water and decomposed with $H_2S$. The filtrate from the PbS is concentrated *in vacuo* at 40–50°, during which *o*-aminophenol glucuronide begins to separate out as a white crystalline solvent. The white residue is shaken with ethanol and the entire mixture filtered. The glucuronide is then washed with ethanol and ether (yield 3.3 g.). The compound is then recrystallized from a large volume of hot water to form cigar-shaped platelets which blacken without melting at 300°. The compound is insoluble in cold water, ethanol, ether, and in most organic solvents; it is fairly soluble in hot water and quite soluble in acid and alkali media ($C_{12}H_{15}O_7N$). Anal., calcd.: C, 50.5%; H, 5.3%; N, 4.9%. Found: C, 50.7%; H, 5.35%; N, 4.9%.

### 2. *Phenyl-α- and Phenyl-β-D-glucopyranosiduronic Acids* (*glucopyruronosides*) (*Tsou and Seligman, 1953; Marsh and Levvy, 1958b*)

The synthesis of the following phenolic glucuronides is given in order to outline the general type of procedure which may be employed in the preparation of a number of aryl glucopyruronosides.

The α-isomer is prepared by oxidizing at 90° a solution of phenyl-α-D-glucopyranoside monohydrate (2.16 g.) in 50.0 ml. of water (pH 8–10) containing 1.0 g. of platinum catalyst (Marsh, 1952). The pH is maintained during the oxidation period by frequent addition of 0.5 *N* $NaHCO_3$ solution (ca. 15.5 ml.). After completion of the reaction in approximately 1½ hours the solution is filtered, evaporated to 10.0 ml., and 100.0 ml. of hot ethanol added. The resulting precipitate is removed, and the sodium salt of phenyl-α-D-glucopyranosiduronic acid separates upon cooling. This is dissolved in 18.0 ml. of water and the pH brought to 2.2 with $H_2SO_4$. The solution is extracted with ethyl acetate for several hours followed by concentration of the ethyl acetate solution which gives phenyl-α-D-glucopyranosiduronic acid hemihydrate. After recrystallization from moist ethyl acetate the m.p. was found to be 149–150°.

The β-isomer is prepared in a similar fashion. Two grams of phenyl-β-D-glucopyranoside dihydrate in 25.0 ml. of water with 0.8 g. of catalyst are oxidized at 90°, with the pH kept at 8–10 by 0.5 $N$ $NaHCO_3$ solution (12.6 ml.). After completion of the reaction in about 3 hours the solution is filtered, concentrated to a volume of 10.0 ml., treated with charcoal, and evaporated to yield a gum. Evaporation with methanol, followed by addition of 50.0 ml. of ethanol, yields the insoluble crude sodium salt of phenyl-β-D-glucopyranosiduronic acid. This is dissolved in 5.0 ml. of water, the pH adjusted to 2.4 with $H_2SO_4$, and extracted repeatedly with ethyl acetate (total 100.0 ml.). Evaporation of the dried ethyl acetate solution yields a gummy solid which is recrystallized from moist ethyl acetate yielding phenyl-β-D-glucopyranosiduronic acid dihydrate. After recrystallization, an m.p. of 160–161° has been reported ($C_{12}H_{14}O_7$). Anal., calcd.: C, 53.33%; H, 5.22%. Found: C, 53.35%; H, 5.38%.

### 3. *β-Naphthyl-β-D-glucopyruronoside* (*Tsou and Seligman, 1952b*)

Oxygen is bubbled through an aqueous solution (200.0 ml.) of β-naphthyl-β-D-glucopyranoside (2.0 g.) (Tsou and Seligman, 1952a) in the presence of 0.5 g. of platinum black and 0.5 g. of sodium bicarbonate on a steam bath with vigorous mechanical stirring. The oxygen is introduced through two gas bubblers, each fitted with a fritted glass disc, and the sodium bicarbonate is added in two portions at 30 minute intervals. At the end of an hour the amber colored solution is filtered and cooled. The insoluble glucopyranoside, if any, is removed by filtration and basic lead acetate is added to precipitate the lead salts of the glucopyruronoside. The lead salt is collected and converted to the acid with hydrogen sulfide. Lead sulfide is removed and a filtrate is extracted with ethyl acetate. Upon recrystallization from ethyl acetate the sample was found to have an m.p. of 150–151°.

### 4. *8-Hydroxyquinoline Glucuronide* (*Brahm, 1899; Robinson et al., 1953a*)

A total of 18.0 g. of 8-hydroxyquinoline in water is fed by stomach tube to six rabbits. The 24 hour urine (ca. 750.0 ml.) is filtered through glass wool and brought to pH 4.0 with glacial acetic acid (ca. 3.0 ml.). The solution is allowed to stand under refrigeration for several days during which time β-hydroxyquinoline glucuronide slowly separates as large greenish crystals which are recrystallized several times from hot water (yield, 12.0 g.). The compound forms pale yellow-green prisms with an m.p. of 153° (dec.) ($C_{15}H_{15}O_7N \cdot 3H_2O$). Anal., calcd.: C, 48.0%; H, 5.6%; N, 3.7%. Found: C, 47.4%; H, 5.6%; N, 3.7%.

5. *Umbelliferone (7-Hydroxycoumarin) Glucuronide (Mead et al., 1955)*

This type of substrate is of interest because the product of hydrolysis, umbelliferone, is highly fluorescent and suggests the possibility of a histochemical fluorescent technique for the demonstration of glucuronidase.

Six rabbits are each given by stomach tube 2.0 g. of umbelliferone suspended in water. The urine (ca. 750.0 ml.) is collected during the following 24 hours and the glucuronide fraction separated by a systematic lead acetate precipitation (Kamil *et al.*, 1951). The filtrate obtained after removal of lead by $H_2S$ is evaporated under reduced pressure at 40° to a volume of 150.0 ml. On standing overnight a crystalline precipitate is formed. Upon recrystallization from 70.0 ml. of hot water umbelliferone glucuronide (7-β-D-glucuronosidocoumarin) is formed (yield 0.91 g.). The compound which is recrystallized from ethanol has an m.p. of 149–150° ($C_{15}H_{14}O_9 \cdot H_2O$). Anal., calcd.: C, 50.6%; H, 4.5%. Found: C, 50.5%; H, 4.7%.

6. *4-Methylumbelliferone-β-D-glucuronide (Marsh and Levvy, 1956)*

4-Methylumbelliferone-β-D-glucoside hemihydrate (1.5 g.) in 60.0 ml. of water is maintained at 90° while being vigorously stirred. Oxygen is passed into the solution through a sintered glass disc, and freshly reduced powdered platinum catalyst, prepared from Adam's platinum oxide catalyst by hydrogenation at atmospheric pressure, is added. The pH is maintained between 7.0–9.0 throughout the oxidation by addition of 0.5 *N* sodium bicarbonate. After 1 hour additional catalyst (0.2 g.) is added. The reaction is complete in approximately 30 minutes with an almost theoretical uptake of alkali. The filtered solution is reduced to half volume, treated with charcoal, and made acid to Congo red with HCl. The hydrated glucuronide is filtered off and recrystallized from 50% aqueous ethanol and then from water. After drying over phosphorus pentoxide the m.p. was found to be 139–140° ($C_{16}H_{16}O_9 \cdot 3\frac{1}{2}H_2O$).

7. *Phenolphthalein Glucuronic Acid (Talalay et al., 1946; Fishman et al., 1948; Fishman, 1957, p. 55)*

This is a biosynthetic procedure in which rabbits injected with sodium phenolphthalein phosphate excrete phenolphthalein glucuronic acid in the urine. The synthesis is divided into three parts: the preparation of sodium phenolphthalein phosphate, preparation of the cinchonidine derivative of phenolphthalein glucuronic acid, and preparation of phenolphthalein glucuronic acid.

Forty ml. of chloroform and 50.0 ml. of $POCl_3$ are added consecutively and with mechanical stirring to 50.0 g. of phenolphthalein in an ice-cooled 1 liter flask. Forty ml. of pyridine are added dropwise and with constant stirring for 3–5 hours. The next day 100.0 ml. of distilled water are added slowly and in small quantities to the ice-cooled reaction flask followed by 40% aqueous NaOH (ca. 300.0 ml.). Upon cooling, needle-like crystals of sodium phosphate appear and are removed by filtration. An excess of concentrated HCl is then added to the aqueous phase and the phenolphthalein phosphoric acid precipitates as a gum which may be readily separated from the mixture by a glass rod. The gum is then warmed in a porcelain dish (on a boiling water bath) and 20.0 ml. of cencentrated NaOH solution (1:1) are slowly added with stirring. Distilled water is then added until complete solution is effected (final volume ca. 200.0 ml.).

Six ml. of phenolphthalein phosphate are adjusted to neutral pH with weak alkali and distilled water is added to make a total volume of 20.0 ml. This amount is injected subcutaneously daily at two widely separated sites on the skin of rabbits for a period of 6 days. Water intake of the animal should be limited. The daily urine collections (under toluene) are completed on the morning of the 8th day. The filtered toluene-free urine is acidified to Congo red paper with 6 *N* HCl and shaken vigorously. Then 600.0 ml. portions of urine are extracted with four successive 170.0 ml. portions of ethyl acetate, and the ethyl acetate phase is centrifuged and dried by decantation through cotton. The ethyl acetate extract is then slowly saturated with cinchonidine by agitating a gram or two of the material contained in the cotton bag in the solution. This procedure is repeated several times. A fine white crystalline precipitate consisting of the cinchonidine derivative of phenolphthalein glucuronic acid appears. The separation of the product is completed overnight under refrigeration. The precipitate is collected on filter paper, dried, and dissolved in a minimum of hot methanol to which are added 4 volumes of hot ethyl acetate. A relatively pure white crystalline product is thus obtained. Additional amounts of the glucuronic acid may be obtained by extracting with methanol the gum which forms on the outside of the gauze bag and crystallizing the product by addition of ethyl acetate and also by concentrating the mother liquors to a small volume at 50°. The recrystallized cinchonidine salt exhibits an m.p. of 200–211°. However, products with an m.p. of 200° have been found to be satisfactory for assay purposes. The yield varies between 0.5 g. and 1.0 g. per rabbit per day.

The recrystallized cinchonidine salt of phenolphthalein monoglucuronide (m.w. 788) (0.788 g.) is mixed with an excess (20.0 ml.) of 2 *N* HCl and 20.0 ml. of ethyl acetate. The mixture is transferred in its entirety to a separatory funnel with several washings of ethyl acetate. The ethyl acetate is collected and decanted through a cotton plug in a funnel into a 500.0 ml. Erlenmeyer flask. The aqueous layer is then extracted with six successive portions (20.0 ml.) of ethyl acetate, the ethyl acetate layers being collected in the Erlenmeyer flask after decantation through the same cotton plug. The solvent is evaporated by a current of dry air which is introduced into the flask. Phenolphthalein glucuronic acid forms as a gummy colorless residue to which is added 20.0 ml. of hot distilled water and sufficient 0.1 *N* NaOH solution to bring the solution to pH 5.0. The mixture is transferred in its entirety with water washings to a 100.0 ml. volumetric flask and is well shaken. If the solution becomes turbid a small amount of charcoal is added and the mixture is then filtered.

8. *6-Bromo-2-naphthyl-β-*D*-glucopyruronoside* (*Pyranosiduronic Acid*) (*Seligman et al., 1954*)

The synthesis is a three-step procedure involving preparation of 6-bromo-2-naphthyl tetraacetyl-β-D-glucopyranoside and 6-bromo-2-naphthol-β-D-glucopyranoside, and oxidation of the latter to form the naphthyl glucopyruronoside (pyranosiduronic acid).

An intimate mixture of 108.0 g. (0.28 *M*) of glucose β-D-pentaacetate, 130.0 g. (0.68 *M*) of 6-bromonaphthol, and 5.4 g. of *p*-toluenesulfonic acid is prepared with a mortar and pestle. The mixture is heated on a steam bath in a 1 liter Claisen distillation flask connected through an ice-chilled trap to a high vacuum pump. Melting and bubbling occur during the first 5 minutes. The flask is shaken at 5 minute intervals during the 30 minutes required for the completion of the reaction. The melt is dissolved in 300.0 ml. of chloroform and washed 3–4 times with 100.0 ml. portions of cold water or until the washings are free of the sulfonic acid. The chloroform solution is then washed 10 times with 100.0 ml. portions of 2% NaOH and ice to remove unreactive naphthol, and finally 3–4 times with ice water (100.0 ml.) to remove alkali. The chloroform solution is dried with anhydrous calcium sulfate, treated with 2.0 g. of activated charcoal, and filtered with suction. The chloroform is evaporated to dryness *in vacuo* or overnight with an air jet. The residue is dissolved in 250.0 ml. of methanol from which a pale yellow product crystallizes. The compound is recrystallized from hot ethanol to form white needles (m.p. 146–146.5°).

The aforementioned tetraacetate (7.1 g.) is suspended in 4 liters of methanol cooled by an ice bath. Gaseous ammonia is introduced into the stirred solution for approximately 1 hour and the stoppered flask is then allowed to stand overnight under refrigeration. A small amount of the tetraacetate may be recovered by filtration. The methanolic solution is evaporated *in vacuo* at 40° until the product begins to crystallize; the mixture is chilled and the product is collected by filtration. The filtrate is further evaporated to give another crop of compound which forms short white needles, m.p. 211–212° (yield 45.0 g.).

A milky aqueous suspension of finely ground 6-bromo-2-naphthyl-D-glucopyranoside (10.0 g. in 3 liters) is placed in a 5 liter 3-necked flask equipped with a mechanical stirrer and 4 gas bubblers. Two grams of platinum black are added and the mixture vigorously stirred on a steam bath (90–100°) while oxygen is introduced. The mixture is kept slightly alkaline by addition of 2.4 g. of sodium bicarbonate in small portions at 15 minute intervals. At the end of 3½ hours the amber colored solution is filtered and the clear filtrate is evaporated to a volume of 50.0 ml. *in vacuo* at 50–60°. The brownish solution is decolorized with charcoal and the solution cooled to 0° and acidified with cold 6 *N* HCl until complete precipitation is effective (pH 2.0). The precipitated glucuronide is collected with suction, washed with cold water and 5.0 ml. of ether. The compound which is dried in a vacuum oven at 70° is recrystallized from ethanol-water to give lusterous pink flakes. The compound is further purified by conversion to the *p*-toluidine salt by addition of *p*-toluidine hydrochloride to a solution of the sodium salt of the glucuronide. The salt is recrystallized from water and a few drops of ethanol and is reconverted to the glucuronide by suspension in 3 *N* HCl and subsequent recrystallization under refrigeration ($C_{16}H_{15}BrO_7$). Anal., calcd.: C, 48.14%; H, 3.79%. Found: C, 48–23%; H, 4.06%.

9. *Naphthol AS-LC Glucuronide* (*Pugh and Walker, 1961a*)

Only the general synthetic approach to this new substrate for the demonstration of glucuronidase has been given. Naphthol AS-LC is coupled in aqueous alkaline acetone with acetobromomethyl glucuronate (Bollenback *et al.*, 1955) or acetochloroglucosamine (Leaback and Walker, 1957). The β-D-glucuronide is isolated after removal of protecting groups (Leaback and Walker, 1957).

10. *Phenyl-N-acetyl-β-D-glucosaminide* (*Roseman and Dorfman, 1951*)

This procedure is applicable to the preparation of a variety of phenolic (or naphtholic) *N*-acetylglucosaminides.

The initial compound prepared in this synthesis is phenyl-2,3,4,6-tetraacetyl-β-D-glucosaminide. One hundred twenty-five grams of pentaacetylglucosamine (syrup) are added to 175.0 g. of phenol and 2.3 g. of toluenesulfonic acid. The mixture is heated on a water bath for exactly 20 minutes and is then poured into 1 liter of chloroform and 1 liter of ice water. The chloroform layer is washed with cold 5% NaOH solution until the alkali is free of color. The chloroform layer is then washed with ice water, dried over calcium chloride, and concentrated. A small amount of anhydrous ether is added to the syrup in order to initiate crystallization. Crystals formed consist of a mixture of the anomeric phenol tetraacetylglucosaminides and of glucosamine pentaacetate. The mixture is fractionally crystallized by dissolving in hot isopropyl alcohol and cooling immediately by scratching and seeding with the phenyl-β-glycoside. Shortly after crystallization begins, the mixture is filtered, and the filtrate treated with increasing amounts of anhydrous ether to yield first the crude phenol α-glycoside and then the glucosamine pentaacetate. The crude β-glycoside is recrystallized several times from isopropanol.

Phenyl-*N*-acetyl-β-D-glucosaminide is obtained by alkaline hydrolysis of the aforementioned compound. Two hundred forty ml. of *N* NaOH are added to a cool solution of 3.1 g. of the tetraacetate in 270.0 ml. of acetone. The mixture is refrigerated for 16 hours and is then treated with the acid form of Amberlite IR-120 by a shaking technique in which the resin is initially prepared by shaking vigorously with a 10% solution of NaOH until the washings are colorless. The resin is then treated with 15% HCl until converted to the acid form, and it is finally washed with water and acetone. The resulting acid, salt-free solution is concentrated and crystallizes to form needles. The compound is recrystallized several times from water to give a product with an m.p. of 249–250° ($C_{14}H_{19}O_6N$). Anal., calcd.: C, 56.56%; H, 6.40%; N, 4.71%. Found: C, 56.68%; H, 6.43%; N, 4.73%.

11. *α-Naphthyl-N-acetyl-β-*D*-glucosaminide* (*Leaback and Walker, 1957*)

This synthesis involves the coupling of α-naphthol and acetochloroglucosamine (Baker *et al.*, 1954; Leaback and Walker, 1957) in aqueous alkaline acetone.

Acetochloroglucosamine is prepared by shaking dry crystalline 2-acetamidotetra-*o*-acetyl-2-deoxy-β-D-glucose with acetic anhydride (30.0 ml.) saturated at 0° with dry hydrogen chloride. After complete solution is effected, the mixture is allowed to stand for 16 hours at room temperature. It is subsequently added to 125.0 ml. of chloroform, cooled, and extracted twice with cold saturated sodium hydrogen carbonate solution, and twice with cold water. The chloroform layer is dried ($MgSO_4$) and evaporated to a syrup which subsequently crystallizes. The residue is triturated with dry ether and the solid filtered off to give acetochloroglucosamine, m.p. 126–

127°. The compound is recrystallized from ethyl acetate by addition of petroleum ether followed by cooling overnight under refrigeration.

The details of synthesis of the naphthyl glucosaminide were not given but it was stated that the method was similar to that employed in the preparation of the phenol and nitrophenol analogues. The technique involves the treatment of the phenol and acetochloroglucosamine in approximately 100.0 ml. of acetone and 45.0 ml. of 3.3% aqueous NaOH for 6 hours at room temperature and overnight at 5°. Following evaporation of the solvent the product is treated with 50.0 ml. of ether. The solid is filtered off, washed with water, ether, and then recrystallized from propylene glycol ($C_{24}H_{27}O_9N$) (m.p. 210–211°).

The *o*-acetyl group is removed according to the method of Westphal and Schmidt (1951) to form α-naphthyl-*N*-acetyl-β-D-glucosaminide.

12. *Methylumbelliferyl-(Methylhydroxycoumarin) N-acetyl-β-D-glucosaminide (Leaback and Walker, 1961)*

Methylumbelliferone (1.9 g.) and 5.0 g. of acetochloroglucosamine (Leaback and Walker, 1957) are dissolved in 70.0 ml. of cold acetone, and 16.5 ml. of *N* NaOH are added. After standing for 6 hours at room temperature followed by several days at 5°, the acetylated glycoside crystallizes (only a small amount remains in solution). The glycoside is recrystallized from ethanol to give methylumbelliferyltetraacetyl-β-glucosaminide ($C_{24}H_{27}O_{11}N$). One gram of the aforementioned acetylated glycoside is suspended in 50.0 ml. anhydrous methanol. Upon addition of 1.0 ml. of *N* sodium methoxide the glycoside dissolves and after several minutes the deacetylated compound starts to crystallize. The mixture is allowed to stand overnight and the solid filtered off. It is then recrystallized once from hot water, and traces of methylumbelliferone are removed by stirring a suspension of the compound in 100.0 ml. of acetone for 16 hours. The recrystallized methylumbelliferyl-*N*-acetyl-β-glucosaminide has an m.p. of 212–214° ($C_{18}H_{21}O_8N$). Anal., calcd.: C, 57.0%; H, 5.5%; N, 3.7%. Found: C, 57.1%; H, 5.6%; N, 3.6%.

13. *β-2-Naphthyl Glucoside (Robertson and Waters, 1930)*

The synthetic approach involves the interaction of *o*-tetraacetyl-α-glucosidyl bromide and a phenolic compound (Takahashi, 1925; Zemplen and Müller, 1929; Müller, 1929) in the presence of silver oxide and a base such as pyridine or quinoline.

β-Naphthol (4.0 g.) and *o*-tetraacetyl-α-glucosidyl bromide (10.0 g.) are allowed to react in the presence of 5.0 g. of silver oxide and 10.0 ml. of quinoline under moderate cooling for approximately 10 minutes. The mixture is placed in a desiccator for 30 minutes, extracted with acetic acid, and the filtered extract poured into approximately 400.0 ml. of ice water. The light-brown crystalline tetraacetate on recrystallization from methanol (charcoal), and subsequently from petroleum ether, forms colorless prisms with an m.p. of 135° ($C_{24}H_{26}O_{10}$). The tetraacetate is deacetylated with methanolic-ammonia to give an almost theoretical yield of β-2-naphthyl glucoside which crystallizes from water as colorless prisms of the monohydrate with an m.p. of 188° ($C_{16}H_{18}O_6 \cdot H_2O$). Anal., calcd.: C, 59.3%; H, 6.2%. Found: C, 59.3%; H, 6.0%.

14. *β-Naphthyl Glucoside* (*Ryan, 1899*)

A solution of acetochloroglucose (70.0 g.) in ethanol (150.0 ml.) is added to 28.0 g. of β-naphthol and KOH (11.0 g.) dissolved in absolute ethanol, the total volume of the well-cooled mixture being about 300.0 ml. After several minutes the solution becomes turbid due to the separation of KCl. The solution is allowed to stand for 3 days at room temperature and is then heated for 45 minutes under reflux, cooled, and filtered. After removal of the ethanol on a water bath, a small amount of water is added. Upon cooling, the solution solidifies. The product which contains some free naphthol is recrystallized from boiling water and finally ethanol to form long needles with an m.p. of 184–186° ($C_{16}H_{18}O_6$). Anal., calcd.: C, 62.74%; H, 5.88%. Found: C, 62.67%; H, 5.81%.

15. *6-Bromo-2-naphthyl-β-D-glucopyranoside* (*Cohen et al., 1952b*)

According to the Helferich technique (Helferich and Schmitz-Hillebrecht, 1933; Montgomery *et al.*, 1942), β-D-glucopyranoside pentaacetate (10.0 g., 0.026 *M*), 6-bromo-2-naphthol (16.0 g., 0.072 *M*), and 50.0 mg. of *p*-toluenesulfonic acid are fused *in vacuo* for 30 minutes at 100°. The melt is taken up in 100.0 ml. of benzene and washed with ice-cold water, 2% NaOH solution, again with water, and dried over anhydrous sulfate (Drierite). The filtered dried solution is evaporated under reduced pressure to give the crude product which is recrystallized from ethanol and has a m.p. of 142–143°.

The technique of Zemplen and Kunz (1923) is employed for deacetylation. Three grams of the tetraacetate are suspended in anhydrous methanol, a piece of sodium 2–3 mm. in diameter added, and the mixture allowed to stand at 4° for 2 days. The product separates as fine white needles with an m.p. of 206–207°. Upon recrystallization from methanol the compound was found to contain 1 mole of methanol which could be removed by drying at 100° over phosphorus pentoxide for several hours ($C_{16}H_{17}O_6Br$). Anal., calcd.: C, 49.88%; H, 4.45%. Found: C, 49.98%; H, 4.44%.

16. *Naphthol AS Glucosides* (*Rath and Feess, 1954*)

Glucosides of naphthol AS compounds may be synthesized by the action of acetobromoglucose on the corresponding naphthol in the presence of quinoline and silver oxide. The tetraacetate obtained is re-esterified to the glucoside and methyl acetate in absolute methanol with the aid of sodium methylate as a catalyst (Zemplen and Müller, 1929; Zemplen *et al.*, 1936). Naphthol AS glucosides prepared according to this procedure crystallize nicely and exhibit only a moderate degree of water solubility.

17. *6-Bromo-2-naphthyl-β-D-galactopyranoside* (*Cohen et al., 1952a*)

This substrate for β-D-galactosidase is prepared according to the techniques originally described by Helferich and Schmitz-Hillebrecht (1933) and Montgomery *et al.* (1942).

β-D-Galactopyranoside pentaacetate is prepared from D-galactose (100.0 g.), acetic anhydride (540.0 ml.), and anhydrous sodium acetate (44.0 g.) by heating and stirring the mixture for 20 minutes at 100°. The mixture is then cooled and poured into 1600.0 ml. of ice water and stirred for 1 hour. The product is extracted with benzene, washed, dried, and evaporated to a syrup from which the pentaacetate crystallizes. It is then recrystallized from 95% ethanol to give an m.p. of 142–143°.

Five grams (0.013 *M*) of the pentaacetate, 8.0 g. (0.036 *M*) of 6-bromo-2-naphthol, and 20.0 mg. of *p*-toluenesulfonic acid are fused *in vacuo* for 30 minutes at 100°. The melt is taken up in benzene and the solution is washed with water, cold 2% NaOH, again with water, and dried over anhydrous calcium sulfate (Drierite). After removal of the solvent under reduced pressure a syrup is obtained which is dissolved in 25.0 ml. of 75% methanol. An amorphous solid separates upon standing under refrigeration. It is separated by decantation and dried in a vacuum desiccator over phosphorus pentoxide. Attempts to recrystallize this solid were unsuccessful.

Deacetylation is performed according to the catalytic method of Zemplen and Kunz (1923). Three grams of the pentaacetate are dissolved in 30.0 ml. of methanol and a piece of sodium 2–3 mm. in diameter is added to the solution which is allowed to stand overnight at 4°. The product is collected and recrystallized from methanol as a white powder which exhibited an m.p. of 205° ($C_{16}H_{17}O_6Br$). Anal., calcd.: C, 49.88%; H, 4.55%. Found: C, 49.60%; H, 4.39%.

## APPENDIX 8B

# HISTOCHEMICAL METHODS

1. *β-Glucuronidase, Chelation Method (Friedenwald and Becker, 1948; Fishman and Baker, 1956; Hayashi and Fishman, unpublished data)*

Thin slices of fresh tissue are fixed in cold formol-calcium (containing 10.0 ml. of formaldehyde plus 1 g. of anhydrous calcium chloride and distilled water to make a total volume of 100.0 ml.). After fixation for 24 hours at 4°, the tissues are blotted carefully on filter paper and are transferred to Holt's cold (4°) hypertonic gum sucrose (Appendix 1, Section 1) for 24 hours. Thin sections (5 $\mu$) are cut on a freezing microtome and incubated in the substrate solution.

*Preparation of substrate solution.* A saturated solution of ferric-8-hydroxyquinoline is prepared by dissolving 50.0 mg. of 8-hydroxyquinoline in 30.0 ml. of 0.1 $N$ acetate buffer (pH 5.0) by warming in a boiling water bath. To this hot solution are added 20.0 ml. of hot ferric sulfate solution and the mixture shaken and placed in a water bath at 37° for 2 hours. It is then centrifuged at 2000 RPM for 15 minutes and decanted into a solution prepared by dissolving 100.0 mg. of 8-hydroxyquinoline glucuronide in 2.0 ml. of $M$ acetate buffer pH 5.0. The resulting solution is centrifuged for 15 minutes and the supernatant decanted through No. 2 Whatman filter paper into a glass-stoppered bottle. One gram of finely powdered gum acacia is added to each 100.0 ml. of freshly prepared substrate. The powdered gum is dropped on the surface of the mixture without stirring and placed in a refrigerator until the next day. Shaking should be avoided. An amount for immediate use is centrifuged and the remainder is stored under refrigeration.

Sections are incubated for periods of 30 minutes to 7 hours depending on the activity of the tissue. Following incubation the surface deposit is removed from the substrate with a piece of filter paper. The section is rinsed in distilled water, mounted on a clean glass slide, and examined microscopically. At this stage a positive reaction is usually visible; but if this is not seen the section may be returned to the incubating solution. The sections are placed in 0.5 $M$ oxalate buffer for 15 minutes at room temperature, washed well in distilled water for 2–3 minutes in order to remove traces of oxalate, and transferred to distilled water. The sections are mounted on albuminized slides, carefully blotted face down on fluffless bilulous paper, and dried on a hot plate for 5–10 minutes at 40°. They are then rinsed in distilled water and covered with 6 drops of 2% aqueous potassium ferrocyanide. After the sections have been in contact with the ferrocyanide for

10 seconds, 6 drops of *N* HCl are added and allowed to remain for 2 minutes. The slide is washed with distilled water, retreated with the ferrocyanide-HCl as just described, and washed in several changes of distilled water in order to remove all traces of excess ferrocyanide-HCl. The sections may be then counterstained with 1% neutral red for ½–2 minutes, dehydrated in ethanol, cleared in xylene, and mounted in a synthetic mounting medium. Sites of glucuronidase are delineated by a precipitate of Prussian blue.

2. *β-Glucuronidase, Post-Coupling Method* (*Seligman et al., 1954*)

This procedure is used in conjunction with unfixed frozen sections mounted on glass slides. The mounted sections are briefly fixed (10 minutes) in cold neutral formalin, washed in cold water for 15 minutes, and incubated at 37° for 4–6 hours in the following solution:

| | |
|---|---|
| 6-Bromo-2-naphthyl-β-D-glucopyruronoside | 30.0 mg. dissolved in 5.0 ml. of methanol |
| Phospho-citrate buffer, 0.15 *M*, pH 4.95 | 20.0 ml. |
| Distilled water | 75.0 ml. |

Following incubation the sections are rinsed in tap water for one minute and immersed in a solution of Diazo Blue B prepared immediately before use by dissolving the salt (1.0 mg./ml.) in cold phosphate buffer, 0.02 *M* pH 7.5. After several minutes the sections are washed in two changes of cold distilled water followed by a rinse in 0.1% acetic acid and then mounted in glycerol-gelatin. As with all other techniques employing 6-bromo-2-naphthol as the basis of the substrate, control slides should be incubated in solutions of this naphthol followed by treatment with a diazonium salt in order to objectively evaluate the degree of non-specific naphthol "staining."

3. *β-D-Galactosidase, Post-Coupling Method* (*Cohen et al., 1952a; Rutenburg et al., 1958*)

Thin tissue specimens (3–5 mm. thick) are fixed in cold neutral formalin overnight, rinsed several times in cold distilled water, and sectioned on a freezing microtome. The sections are transferred directly from the knife to 5–10 ml. of substrate solution in a beaker and incubated at 37° for approximately 4 hours. Alternatively unfixed frozen sections may be employed.

The substrate solution is prepared by dissolving 100.0 mg. of 6-bromo-β-D-galactopyranoside in 15.0 ml. of methanol, and adding 200.0 ml. of hot distilled water (70°). The solution is allowed to cool to room temperature and then 85.0 ml. of phospho-citrate buffer, pH 4.95, and 100.0 ml. of distilled water are added. The solution is reported to be stable for a number of months when kept under refrigeration. Control sections, some of which are heat inactivated, are simultaneously incubated in a solution of 6-bromo-2-naphthol prepared in the same fashion except containing no substrate.

Following incubation the sections are washed in several changes of cold distilled water and transferred to a cold, freshly prepared solution of Diazo Blue B (1.0 mg./ml.) in water made alkaline with sodium bicarbonate. The sections are gently agitated for several minutes during which time a deep bule color is formed. The sections are then mounted on glass slides with glycerol-gelatin.

4. *N-Acetyl-β-Glucosaminidase* (*Pugh and Walker, 1961a, b*)

The technique utilizes a simultaneous coupling azo-dye procedure employing either α-naphthyl or naphthol AS-LC *N*-acetyl-β-glucosaminide in conjunction with Garnet GBC salt. Small blocks of tissue are fixed overnight in cold formol-calcium. Although the authors recommend impregnation with gum-sucrose solution or embedding in gelatin, there does not appear to be any reason why, with careful handling, tissues cannot be sectioned directly following formalin fixation. The blocks are rinsed briefly in distilled water prior to sectioning, in order to remove excess formalin. Frozen sections (5–15 μ thick) are transferred to a substrate solution containing 1.0 ml. of α-naphthyl-*N*-acetyl-β-glucosaminide solution (aqueous, saturated at 0°) (alternatively the AS-LC substrate may be employed), 0.1 ml. of a freshly prepared filtered solution containing 15.0 mg./ml. of Garnet GBC salt, and 0.1 ml. of 0.2 *M* citrate buffer. The final pH of the solution is approximately 5.5. Sections are incubated at room temperature for 10–60 minutes. Controls consist of heat inactivated sections or sections incubated in the aforementioned substrate solution containing *N*-acetyl-glucosaminonolactone (0.5–1.0 μ*M*).

There is little doubt that improved techniques for the demonstration of this enzyme employing paraffin embedded tissues in conjunction with new substrates will be evolved in the near future.

## Commercial Sources of Reagents

1. Dajac Laboratories, The Borden Chemical Company, Philadelphia, Pennsylvania
2. General Aniline and Film Corp., New York, New York
3. Pfister Chemical Works, Ridgefield, New Jersey
4. Sigma Chemical Company, St. Louis, Missouri

## References


Aizawa, K. (1939) *J. Biochem.* **30,** 89–100.

Anlyan, A. J., Gamble, J., and Hoster, H. A. (1950) *Cancer* **3,** 116–123.

Artz, N. E., and Osman, E. M. (1950) "Biochemistry of Glucuronic Acid." Academic Press, New York.

Baker, B. R., Joseph, J. P., Schaub, R. E., and Williams, J. H. (1954) *J. Org. Chem.* **19,** 1786–1792.

Barker, S. A., Bourne, E. J., and Stacey, M. (1951) *Chem. and Ind.* (*London*) **45,** 970.

Bembry, T. H., and Powell, G. (1942) *J. Am. Chem. Soc.* **64,** 2419–2420.

Berenbom, M., and Young, L. (1951) *Biochem. J.* **49,** 165–169.

Billing, B. H., Cole, P. G., and Lathe, G. H. (1957) *Biochem. J.* **65,** 774–784.

Bollenback, G. N., Long, J. W., Benjamin, D. G., and Lindquist, J. A. (1955) *J. Am. Chem. Soc.* **77,** 3310–3315.

Borooah, J., Leaback, D. H. and Walker, P. G. (1961) *Biochem. J.* **78,** 106–110.

Boyland, E., Wallace, D. M., and Williams, D. C. (1954) *Biochem. J.* **56,** XXIX.

Brahm, C. (1899) *Z. physiol. Chem. Hoppe-Seyler's* **28,** 439–451.

Bray, H. G. (1953) *Advances in Carbohydrate Chem.* **8,** 251–275.

Burstone, M. S. (1957) *J. Natl. Cancer Inst.* **18,** 167–173.

Burstone, M. S. (1958) *J. Histochem. and Cytochem.* **6,** 322–339.

Burton, J. F., and Pearse, A. G. E. (1952) *Brit. J. Exptl. Pathol.* **33,** 87–97.

Campbell, J. G. (1949) *Brit. J. Exptl. Pathol.* **30,** 548–554.

Campbell, J. G., and Levvy, G. A. (1950) *Nature* **166,** 783–784.

Cohen, R. B., Tsou, K.-C., Rutenburg, S. H., and Seligman, A. M. (1952a) *J. Biol. Chem.* **195,** 239–249.

Cohen, R. B., Rutenburg, S. H., Tsou, K.-C., Woodbury, M. A., and Seligman, A. M. (1952b) *J. Biol. Chem.* **195,** 607–614.

Cohen, S. L., and Bittner, J. J. (1951) *Cancer Research* **11,** 723–726.

Conchie, J., and Levvy, G. A. (1957) *Brit. J. Cancer* **11,** 487–493.

Conchie, J., Levvy, G. A., and Marsh, C. A. (1957) *Advances in Carbohydrate Chem.* **12,** 157–187.

Conchie, J., Findlay, J., and Levvy, G. A. (1959) *Biochem. J.* **71,** 318–325.

de Duve, C., Pressman, B. C., Gianetto, R., Wattiaux, R., and Appelmans, F. (1955) *Biochem. J.* **60,** 604–617.

di Somma, A. (1940) *J. Biol. Chem.* **133,** 277–284.

Dixon, M., and Webb, E. C. (1958) "Enzymes" Academic Press, New York.

Dodgson, K. S., Garton, G. A., Stubbs, A. L., and Williams, R. T. (1948) *Biochem. J.* **42,** 357–565.

Drouin, R. (1895) *Bull. soc. chim. Paris.* [III] **13,** 5.

Dutton, G. J. (1956) *Biochem. J.* **64,** 693–701.

East, M. E., Madinaveitia, J., and Todd, A. R. (1941) *Biochem. J.* **35,** 872–876.

Fernández-García, R., Amorós, L., Blay, H., Santiago, E., Soltero-Díaz, H., and Colón, A. A. (1950) *El Crisol* **4,** 40–44.

Findlay, J., and Levvy, G. A. (1960) *Biochem. J.* **77,** 170–175.

Findlay, J., Levvy, G. A., and Marsh, C. A. (1958) *Biochem. J.* **69,** 467–476.

Fischer, E. H., and Stein, E. A. (1960) *In* "The Enzymes" (P. D. Boyer, H. Lardy, and K. Myrbäck, eds.), 2nd ed., Vol. 4, pp. 301–312. Academic Press, New York.

Fishman, W. H. (1947) *Science* **105,** 646–647.

Fishman, W. H. (1950) *In* "The Enzymes" (J. B. Sumner and K. Myrbäck, eds.), Vol. I, Part 1, pp. 635–652. Academic Press, New York.

FISHMAN, W. H. (1955) *Advances in Enzymol.* **16,** 361–409.
FISHMAN, W. H. (1956) *Ann. Rev. Biochem.* **25,** 659–684.
FISHMAN, W. (1957) *In* "Methods in Enzymology" (S. P. Colowick and N. O. Kaplan, eds.), Vol. III, pp. 55–57. Academic Press, New York.
FISHMAN, W. H. (1959) *In* "Physiopathology of Cancer" (F. Homburger, ed.), Hoeber, New York, pp. 732-760.
FISHMAN, W. H., AND BAKER, J. R. (1956) *J. Histochem. and Cytochem.* **4,** 570–587.
FISHMAN, W. H., AND GREEN, S. (1957) *J. Biol. Chem.* **225,** 435–452.
FISHMAN, W. H., AND MITCHELL, G. W. (1959) *Ann. N. Y. Acad. Sci.* **83,** 105–121.
FISHMAN, W. H., SPRINGER, B., AND BRUNETTI, R. (1948) *J. Biol. Chem.* **173,** 449–453.
FISHMAN, W. H., BAKER, J. R., AND BORGES, P. R. F. (1959) *Cancer* **12,** 240–245.
FRIEDENWALD, J. S., AND BECKER, B. (1948) *J. Cellular Comp. Physiol.* **31,** 303–309.
GARTON, G. A., ROBINSON, D., AND WILLIAMS, R. T. (1949) *Biochem. J.* **45,** 65–67.
GLASER, E., AND WULWEK, W. (1924) *Biochem. Z.* **145,** 514–534.
GOEBEL, W. F., AND BABERS, F. H. (1935) *J. Biol. Chem.* **111,** 347-353.
GOLDFARB, S., AND BARKA, T. (1960) *J. Histochem. and Cytochem.* **8,** 226-227.
GOMORI, G. (1952) "Microscopic Histochemistry." Univ. Chicago Press, Chicago, Illinois.
HAYNES, L. J., AND NEWTH, F. H. (1955) *Advances in Carbohydrate Chem.* **10,** 207-256.
HEILSKOV, N. S. CHR. (1951) *Acta Physiol. Scand.* **24,** 84–89.
HELFERICH, B., AND ILOFF, A. (1933) *Z. physiol Chem. Hoppe-Seyler's* **221,** 252–258.
HELFERICH, B., AND SCHMITZ-HILLEBRECHT, E. (1933) *Ber.* **66,** 378–383.
HICKINBOTTOM, W. J. (1957) "Reactions of Organic Compounds," 3rd. ed. Longmans, Green, London.
HOFFMANN-OSTENHOF, O. (1953) *Advances in Enzymol.* **14,** 219–260.
KAMIL, I. A., SMITH, J. N., AND WILLIAMS, R. T. (1951) *Biochem. J.* **50,** 235–240.
KARUNAIRATNAM, M. C., AND LEVVY, G. A. (1949) *Biochem. J.* **44,** 599–604.
KERR, L. M. H., CAMPBELL, J. G., AND LEVVY, G. A. (1950) *Biochem. J.* **46,** 278–284.
KOENIGS, W., AND KNORR, E. (1901) *Ber.* **34,** 957–981.
KÜLZ, E. (1890) *Z. Biol.* **27,** 247–258.
KUHN, R., AND TIEDEMANN, H. (1954) *Ber.* **87,** 1141–1147.
KUN, E., TALALAY, P., AND WILLIAMS-ASHMAN, H. G. (1951) *Cancer Research* **11,** 855–863.
LARNER, J. (1960) *In* "The Enzymes" (P. D. Boyer, H. Lardy, and K. Myrbäck, eds.), 2nd ed., Vol. 4, pp. 369–378. Academic Press, New York.
LEABACK, D. H., AND WALKER, P. G. (1957) *J. Chem. Soc.* pp. 4754–4760.
LEABACK, D. H., AND WALKER, P. G. (1961) *Biochem. J.* **78,** 151–156.
LESNIK, M., AND NENCKI, M. (1886) *Ber.* **19,** 1534–1538.
LEVVY, G. A. (1952) *Biochem. J.* **52,** 464–472.
LEVVY, G. A. (1956) *Vitamins and Hormones* **14,** 267–303.
LEVVY, G. A., AND MARSH, C. A. (1952) *Biochem. J.* **52,** 690-694.
LEVVY, G. A., AND MARSH, C. A. (1959) *Advances in Carbohydrate Chem.* **14,** 381–428.
LEVVY, G. A., AND MARSH, C. A. (1960) *In* "The Enzymes" (P. D. Boyer, H. Lardy, and K. Myrbäck, eds.), 2nd ed., Vol. 4, pp. 397-407. Academic Press, New York.
MAIO, J. J., AND RICKENBERG, H. V. (1960) *Biochim. et Biophys. Acta* **37,** 101–106.
MARSH, C. A. (1951) *Nature* **168,** 602–603.
MARSH, C. A. (1952) *J. Chem. Soc.* pp. 1578–1582.
MARSH, C. A., AND LEVVY, G. A. (1956) *Nature* **178,** 589–590.
MARSH, C. A., AND LEVVY, G. A. (1958a) *Biochem. J.* **68,** 610–617.
MARSH, C. A., AND LEVVY, G. A. (1958b) *Biochem. J.* **68,** 617–621.

Masamune, H. (1933) *J. Biochem. (Tokyo)* **18,** 259–276.
Masamune, H. (1934) *J. Biochem. (Tokyo)* **19,** 353–375.
Mead, J. A. R., Smith, J. N., and Williams, R. T. (1954) *Biochem. J.* **58,** XXXV.
Mead, J. A. R., Smith, J. N., and Williams, R. T. (1955) *Biochem. J.* **61,** 569–574.
Michael, A. (1879) *Am. Chem. J.* **1,** 305–312.
Monis, B., Banks, B. M., and Rutenburg, A. M. (1960) *Cancer* **13,** 386–393.
Montgomery, E. M., Richtmyer, N. K., and Hudson, C. S. (1942) *J. Am. Chem. Soc.* **64,** 690–694.
Müller, A. (1929) *Ber.* **62,** 2793–2811.
Neuberg, C., and Neimann, W. (1905) *Z. physiol. Chem. Hoppe-Seyler's* **44,** 114–126.
Neuberger, A., and Pitt Rivers, R. V. (1939) *Biochem. J.* **33,** 1580–1590.
Odell, L. D., and Burt, J. C. (1950) *J. Am. Med. Assoc.* **142,** 226–230.
Oshima, G. (1934) *J. Biochem. (Tokyo)* **20,** 361–370.
Oshima, G. (1936) *J. Biochem. (Tokyo)* **23,** 305–318.
Pearse, A. G. E. (1960) "Histochemistry, Theoretical and Applied," 2nd ed. Little, Brown, Boston, Massachusetts.
Pearson, B., Andrews, M., and Grose, F. (1961) *Proc. Soc. Exptl. Biol. Med.* **108,** 619–623.
Pugh, D., and Walker, P. G. (1961a) *J. Histochem. and Cytochem.* **9,** 105–106.
Pugh, D., and Walker, P. G. (1961b) *J. Histochem. and Cytochem.* **9,** 242–250.
Pugh, D., Leaback, D. H., and Walker, P. G. (1957) *Biochem. J.* **65,** 464–469.
Quick, A. (1924) *J. Biol. Chem.* **61,** 667–677.
Rath, H., and Feess, E. (1954) *Melliand Textilber.* **35,** 267–269.
Rauramo, L. (1959) *Scand. J. Clin. and Lab. Invest.* **11,** 285–289.
Robertson, A., and Waters, R. B. (1930) *J. Chem. Soc.* pp. 2729–2733.
Robertson, A., and Waters, R. B. (1933) *J. Chem. Soc.* p. 30.
Robinson, D., Smith, J. N., and Williams, R. T. (1953a) *Biochem. J.* **53,** 125–129.
Robinson, D., Smith, J. N., and Williams, R. T. (1953b) *Biochem. J.* **55,** 151–155.
Roseman, S. (1957) *J. Biol. Chem.* **226,** 115–124.
Roseman, S., and Dorfman, A. (1951) *J. Biol. Chem.* **191,** 607–620.
Rutenburg, A. M., Rutenburg, S. H., Monis, B., Teague, R., and Seligman, A. M. (1958) *J. Histochem. and Cytochem.* **6,** 122–129.
Rutenburg, A. M., Goldbarg, J. A., Rutenburg, S. H., and Lang, R. T. (1960) *J. Histochem. and Cytochem.* **8,** 268–272.
Rutenburg, S. H., and Seligman, A. M. (1953) *J. Biol. Chem.* **203,** 731–741.
Ryan, H. (1899) *J. Chem. Soc. (Trans.)* **75,** 1054–1057.
Ryan, H. (1901) *Sci. Proc. Roy. Soc. Dublin,* [2] **9,** 508–518.
Ryan, H., and Mills, W. S. (1901) *J. Chem. Soc. (Trans.)* **79,** 704–707.
Salkowski, E., and Neuberg, E. (1906) *Biochem. Z.* **2,** 307–311.
Schapiro, E. (1939) *Biochem. J.* **33,** 385–388.
Seidman, M., and Link, K. P. (1950) *J. Am. Chem. Soc.* **72,** 4324.
Seligman, A. M., Nachlas, M. M., Manheimer, L. H., Friedman, O. M., and Wolf, G. (1949) *Ann. Surg.* **130,** 333–341.
Seligman, A. M., Tsou, K.-C., Rutenburg, S. H., and Cohen, R. B. (1954) *J. Histochem. and Cytochem.* **2,** 209–229.
Sellinger, O. Z., Beaufay, H., Jacques, P., Doyen, A., and de Duve, C. (1960) *Biochem. J.* **74,** 450–454.
Sera, Y. (1914) *Z. physiol Chem. Hoppe-Seyler's* **92,** 261–275.
Sumner, J. B., and Somers, G. F. (1953) "Chemistry and Methods of Enzymes." Academic Press, New York.

TAKAHASHI, R. (1925) *J. Pharm. Soc. Japan* No. 525, pp. 969–986.
TALALAY, P., FISHMAN, W. H., AND HUGGINS, C. (1946) *J. Biol. Chem.* **166**, 757–772.
TAUROG, A., ABRAHAM, S., AND CHAIKOFF, I. L. (1953) *J. Am. Chem. Soc.* **75**, 3473–3477.
TOUSTER, O., AND REYNOLDS, V. H. (1952) *J. Biol. Chem.* **197**, 863–868.
TSOU, K.-C., AND SELIGMAN, A. M. (1952a) *J. Am. Chem. Soc.* **74**, 3066–3069.
TSOU, K.-C., AND SELIGMAN, A. M. (1952b) *J. Am. Chem. Soc.* **74**, 5605–5608.
TSOU, K.-C., AND SELIGMAN, A. M. (1953) *J. Am. Chem. Soc.* **75**, 1042–1044.
VEIBEL, S. (1950) *In* "The Enzymes" (J. B. Sumner and K. Myrbäck, eds.), Vol. I, Part 1, pp. 583–634. Academic Press, New York.
WALKER, P. G., WOOLLEN, M. E., AND PUGH, D. (1960) *J. Clin. Pathol.* **13**, 353–357.
WALLENFELS, K., AND MALHOTRA, O. P. (1960) *In* "The Enzymes" (P. D. Boyer, H. Lardy, and K. Myrbäck, eds.), 2nd ed., Vol. 4, pp. 409–430. Academic Press, New York.
WATANABE, K. (1936) *J. Biochem. (Tokyo)* **24**, 297–326.
WESTPHAL, O., AND SCHMIDT, H. (1951) *Ann.* **575**, 84–90.
WILLIAMS, R. T. (1943) *Biochem. J.* **37**, 329–333.
WILLIAMS, R. T. (1959) "Detoxication Mechanisms," 2nd ed. Wiley, New York.
WOTIZ, H. H., SMAKULA, E., LICHTIN, N. N., AND LEFTIN, J. H. (1959) *J. Am. Chem. Soc.* **81**, 1704–1708.
ZEMPLEN, G., AND KUNZ, A. (1923) *Ber.* **56**, 1705–1710.
ZEMPLEN, G., AND MÜLLER, A. (1929) *Ber.* **62**, 2107–2118.
ZEMPLEN, G., GERECS, A., AND HADÁCSY, I. (1936) *Ber.* **69**, 1827–1829.

CHAPTER 9

# TRANSGLYCOSYLASES

## I. GENERAL CONSIDERATIONS

Transglycosylases are transferases which catalyze the transfer of glycosyl groups according to the following general reactions (Hehre, 1951; Edelman, 1956; Hoffmann-Ostenhof, 1953):

$$\text{R—glycosyl} + \text{R'—OH} \rightleftharpoons \text{R—OH} + \text{R'—glycosyl}$$

These enzymes catalyze a number of reversible phosphorolytic and synthetic reactions involving a number of polysaccharide compounds. The term "phosphorolysis" was originally coined to indicate the fact that compounds were being both enzymatically synthesized and decomposed (Sumner and Somers, 1953, p. 321). Cori *et al.* (1937) and Cori and Cori (1937) found that glycogen in animal tissues undergoes a reversible esterification reaction with phosphate to form glycose-1-phosphate:

$$\text{Glycogen} + \text{phosphate} \rightleftharpoons \text{glucose-1-phosphate}$$

Of interest is the fact that this reaction must be primed by the addition of a small amount of starch, dextrin, or glycogen; this suggests that the enzyme does not merely condense individual glucose phosphate units, but transforms glucose from glucose-1-phosphate to build up a pre-existing polysaccharide chain. The relationship between phosphorylase and glycogen metabolism is discussed in an excellent review by Stetten and Stetten (1960).

In addition to phosphorylase which involves the 1,4-glycosidic linkages, the synthesis of higher carbohydrates is also controlled by other enzymes including glucosan transglycosylase ("Q" enzyme, branching enzyme) which transfers part of a 1,4-glucosan chain from a 4- to a 6-position. Other transglycosylases occurring in animal tissues include phospholactose UDP-transgalactosylase, UDP-transglucuronylase, and adenine transphosphoribosylase (Dixon and Webb, 1958, p. 220; Kornberg, 1957; Hassid, 1960, p. 251).

The general reaction catalyzed by phosphorylases may be indicated as follows:

$$X—Y + P \rightleftharpoons X—P + Y$$

Transfer of the group X may occur in either direction, with X—Y as donor and P as acceptor in one direction, and X—P as donor and Y as acceptor in another direction (Cohn, 1961, p. 179). Two general types of phosphorolytic reactions are known to occur, ester bond formation and anhydride bond formation. With reference to the former, phosphorylases in this group catalyze the reversible phosphorolysis of polysaccharides, disaccharides, and nucleosides.

The previously mentioned reaction discovered by Cori *et al.* (1937) which catalyzed the formation of glucose-1-phosphate is associated with a polysaccharide phosphorylase, glucosan phosphorylase (starch phosphorylase) (P enzyme), which synthesizes glucose-1-phosphate from 1,4-α-D-glucopyranosyl units represented by either glycogen or starch. The equilibrium of the reaction is influenced by the pH. At a more acid pH, a larger proportion of starch and inorganic phosphate is formed, while at pH 7 the percentage of ester phosphate markedly increases. Polysaccharide phosphorylase has been isolated as two interconvertible forms, *a* and *b*, from both skeletal and heart muscle. The interconversion is the result of a specific phosphatase which splits phosphate from phosphoserine groups, and a specific kinase which transfers phosphate from ATP to serine (Brown and Cori, 1961, p. 207). The relationships between the two forms of phosphorylase are shown in Chart 1.

Polysaccharides synthesized by animal phosphorylase resemble the amylose fraction of starch in that the product consists of linear molecules exhibiting 1,4-glycosidic linkages. Since naturally occurring glycogen and starches also contain the 1,6-linkages, the action of *in vivo* phosphorylases must be augmented by another enzyme-catalyzed reaction through which branching is induced. The evidence for such a "branching" enzyme which accounts for the synthesis of 1,6-linkages was first described by Cori and Cori (1943). The so-called "Q" enzyme which is capable of converting linear amylose to branched amylopectin is also a transglucosidase which synthesizes 1,6-linkages but which acts upon amylose as contrasted with the Cori "branching" enzyme (Hassid, 1960, p. 278). Of considerable significance from a histochemical standpoint is the observation that relatively non-branched polysaccharides synthesized by phosphorylase give a very intense blue color with iodine. On the other hand, the effect of the "branching" enzyme results in the formation of a polysaccharide which closely resembles glycogen and which gives a reddish-brown coloration with iodine (Bear and Cori, 1941; Bear, 1942).

| | | |
|---|---|---|
| Phosphorylase *a* (phosphoserine) | —— phosphatase ——→ | Phosphorylase *b* (serine) + phosphate |
| Phosphorylase *b* (serine) | —— $ATP + Mg^{++}$, kinase ——→ | Phosphorylase *a* (phosphoserine) + ADP |
| Phosphorylase *a* (phosphoserine) | —— trypsin ——→ | Phosphorylase *b'* + phosphohexapeptide |

CHART 1. Interrelationship between forms of phosphorylase (Brown and Cori, 1961, p. 208).

Further studies on the relationship between branching and the type of iodine staining reaction have been reported by Swanson (1948).

Disaccharide phosphorylases such as sucrose phosphorylase are found only in bacteria.

Nucleoside phosphorylases are widely distributed in animal tissues including liver, brain, thymus, and in human erythrocytes (Friedkin and Kalckar, 1961, p. 237). Two general groups of these enzymes are known, pyrimidine nucleoside phosphorylases:

$$\text{Uridine} + H_3PO_4 \rightleftharpoons \text{ribose-1-phosphate} + \text{uracil}$$

$$\text{Thymidine} + H_3PO_4 \rightleftharpoons \text{deoxyribose-1-phosphate} + \text{thymine}$$

and purine nucleoside phosphorylases:

$$\text{9-}\beta\text{-D-Ribofuranosylpurine} + H_3PO_4 \rightleftharpoons \alpha\text{-D-ribofuranose-1-phosphate} + \text{purine}$$

Among the purine components may be adenine, guanine, nicotinamide, or azaguanine.

In addition to the formation of ester bonds, phosphorylases also catalyze reactions leading to the formation of anhydrides or pyrophosphate bonds in nucleoside diphosphates (Imsande and Handler, 1961, p. 281). These enzymes have also been termed nucleotidyl transferases, transadenylases, or uridyl transferases. Diphosphopyridine nucleotide pyrophosphorylase has been the most extensively studied enzyme of this group (Schneider and Hogeboom, 1956):

$$\text{DPN} + \text{PPi} \rightleftharpoons \text{nicotinamide mononucleotide} + \text{ATP}$$

Flavin adenine dinucleotide pyrophosphorylase has been found in red cells and the cytoplasmic fraction of rat and mouse liver homogenates. Cytidine diphosphate choline pyrophosphorylase has also been found in mammalian liver. Uridine diphosphate glucose pyrophosphorylase, which occurs in animal tissues (Villar-Palasi and Larner, 1958; Robbins *et al.*, 1958; Leloir *et al.*, 1959), catalyzes the following reaction:

$$\text{UDP-D-glucose} + \text{PPi} \rightleftharpoons \text{UTP} + \alpha\text{-D-glucose-1-phosphate}$$

This reaction is associated with another one involving the conversion of UDPG to glycogen by a transglycosylase:

$$\text{UDPG} + \text{glycogen (primer)} \rightleftharpoons \text{glycogen} + \text{UDP}$$

The reaction sequence may play an important role in glycogen synthesis *in vivo.* It should be mentioned that phosphorylases A and B cannot catalyze the aforementioned reaction. Uridine diphosphate acetylglucosamine adenylic acid and deoxyribonucleic acid pyrophosphorylases have also been described. From a physiological standpoint Ismande and Handler (1961, p. 304) point out that only a few of these enzymes appear to be effective catalysts of pyrophosphorolysis in the living cell. Thus the physiological role of these enzymes appears to be related to synthesis of a nucleotide complex with concurrent liberation of inorganic pyrophosphate, with the latter, of course, being abundant in the mitochondrial, microsomal, and cytoplasmic fractions of the cell.

## II. HISTOCHEMICAL METHODS

In 1947 Yin and Sun reported the first histochemical method for the demonstration of amylophosphorylase activity. They were able to obtain the *in vivo* synthesis of starch in thin slices of soy beans by incubating the sections in a solution containing glucose-1-phosphate. The polysaccharide was then revealed by treatment of the sections with iodine-potassium iodide solution to delineate blue-dyed starch granules. Yin and Sun noted that sodium fluoride, which had been added to inhibit phosphatase activity, had no effect upon starch synthesis. The technique was further explored with reference to plant tissues by Dyar (1950) and Stocking (1952).

The first report on the histochemical distribution of phosphorylase activity in animal tissues was made by Goldberg *et al.* (1952) who applied the method of Yin and Sun in a study of frozen sections of rat liver and kidney, rabbit uterus, and human endometrium and cervix (Appendix 9, Section 1). Goldberg *et al.* (1952) incorporated 0.1 *M* fluoride in their substrate solution in order to inhibit non-specific acid phosphatase which might hydrolyze the substrate. As previously noted, fluoride ions do not inhibit phosphorylase (Cowgill and Cori, 1955). In fact, the presence of fluoride may directly exert a beneficial effect upon the histochemical reaction, since the ion is known to inhibit conversion of phosphorylase type *a* to type *b*. The synthesized polysaccharide was revealed by treating sections with periodic acid-Schiff (pa-S) reagent, Best's carmine, or with iodine vapor over an $I_2$-KI solution. Pre-existing glycogen was removed by incubation of sections with saliva or with a buffered solution containing malt diastase. These authors noted, however, that removal of pre-existing glycogen resulted in considerable inhibition of phosphorylase activity. The addition of adenylic acid as a possible activator, as well as the introduction of ethanol in an attempt to precipitate formed polysaccharide, were not found to produce any improvement in the staining pattern.

Cobb (1953) employed the Yin and Sun technique in conjunction with frozen-dried paraffin embedded undecalcified sections of cartilage and bone. The synthesized glycogen was revealed by means of the pa-S reaction. Takeuchi and Kuriaki (1955), Takeuchi *et al.* (1955), and Takeuchi (1956) have conducted the most detailed studies on the histochemical demonstration of phosphorylase in animal tissues. These authors avoided the detrimental effects associated with enzymatic removal of pre-existing glycogen in tissue sections by demonstrating the formed polysaccharide with Gram's iodine instead of Best's carmine or the pa-S reagent. As previously noted, linear amylose of the type synthesized by phosphorylase produces a distinct blue stain in the presence of iodine, as contrasted with the reddish-brown glycogen stain (Bear and Cori, 1941; Bear, 1942). Also of interest are the biochemical findings which demonstrate that a combination of phosphorylase and "Q" enzyme acting upon glucose-1-phosphate yields a polysaccharide which gives a purple-red color with iodine (Hassid *et al.*, 1951, p. 1023). Taking these factors into consideration Takeuchi and Kuriaki (1955) specifically recommended that the histochemically synthesized polysaccharide be revealed by Gram's iodine solution. Their substrate solution contained muscle adenylic acid and insulin as activators, and also a small amount of glycogen as a primer (Appendix 9, Section 3). With this modified technique amylophosphorylase activity was visualized in skeletal (Fig. 1), cardiac, and smooth muscle; and esophageal epithelium. Nuclei were not stained. A modification of the technique which included the presence of EDTA in the substrate mixture has been reported to increase the sensitivity of the method (Godlewski, 1960a).

The Takeuchi phosphorylase method has been applied in the histochemical demonstration of the enzyme in the heart (Carbonell, 1955), the skin (Braun-Falco, 1956), blood and bone marrow cells (Takeuchi and Kinoshita, 1956), developing rat brain (Shimizu and Okada, 1957), and rat uterus (Herbener and Atkinson, 1959; Godlewski and Skurzak, 1961). Although hereditary absence of phosphorylase (McArdle syndrome) has been described in certain individuals (Schmid and Hammaker, 1961), there have been no reports on histochemical studies of this disease entity. Engel *et al.* (1961) employed the phosphorylase method in a study of "central core disease," a rare muscle cell abnormality, and found virtually no synthesized polysaccharide in the "core" regions. Engel (1961) has also described alterations in phosphorylase activity in the so-called "target fibers" which are associated with skeletal muscle fiber denervation (Fig. 2).

Takeuchi (1958) employed the phosphorylase procedure for the demonstration of the "branching" enzyme amylo-1,4 → 1,6-glucosidase in animal tissues. The enzyme was differentiated from phosphorylase by the previously described differences in color which occur when iodine is reacted with an amylose type polysaccharide as contrasted with an amylopectin type. Thus sites of "branching" enzyme activity are revealed by a violet-red to brown-purple stain, which does not occur in sections which are predigested by α-amylase but is unaffected by β-amylase (Appendix 9, Section 4).

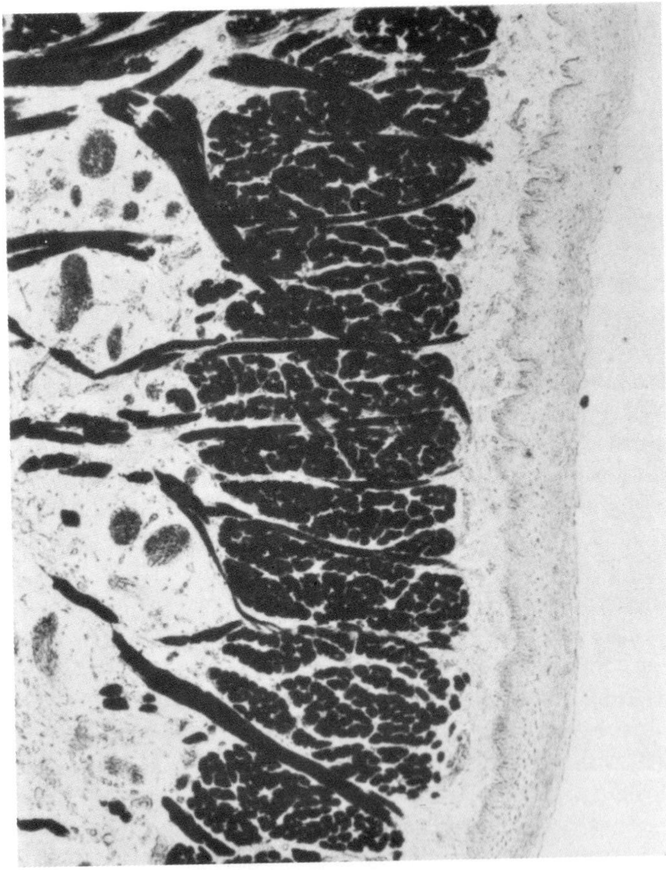

FIG. 1. Amylophosphorylase activity of tongue muscle (× 70).

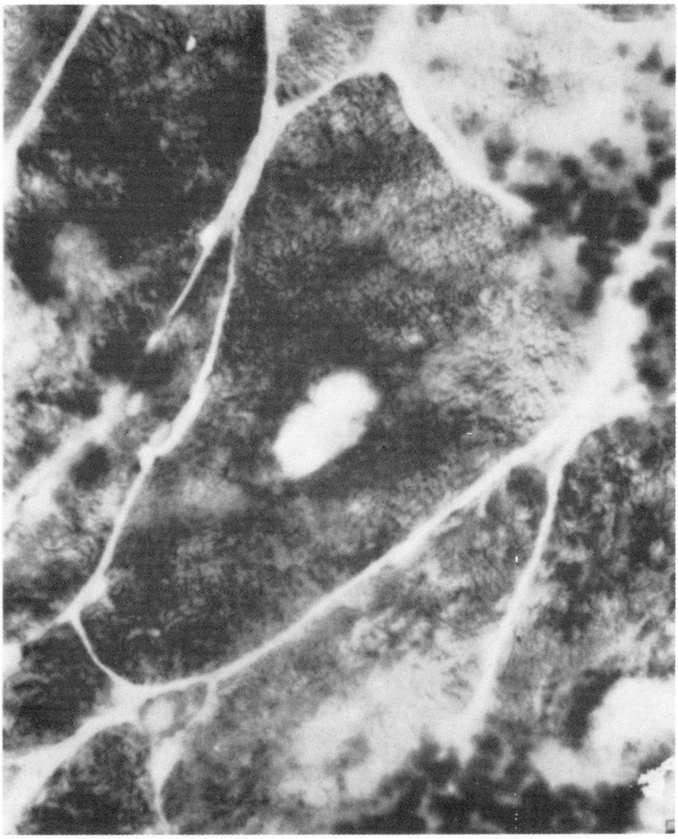

FIG. 2. Amylophosphorylase activity of a "target fiber" associated with denervated human skeletal muscle (× 560) (courtesy Dr. W. King Engel).

The distribution of amylo-1,4 → 1,6-transglucosidase in animal tissues is essentially the same as that of amylophosphorylase with distinct reactions seen in liver, skeletal and cardiac muscle, nervous tissues, and various epithelia.

Takeuchi and Glenner (1960, 1961) reported the histochemical demonstration of UDPG transferase in animal tissues, which had been previously described in the biochemical literature (Villar-Palasi and Larner, 1958; Robbins *et al.*, 1958; Leloir *et al.*, 1959). The technique employed was similar to the original amylophosphorylase method. The histochemically synthesized polysaccharide derived from UDPG as a substrate was considered to exhibit a branched structure because of its red-brown coloration with iodine, its complete digestion by $\alpha$-amylase and incomplete digestion by $\beta$-amylase. Since the red-brown color cannot be differentiated from that of naturally occurring glycogen, a suitable control slide must always be prepared. Of interest was the finding that glucose-6-phosphate intensely stimulated the UDPG-glycogen transferase reaction. Thus on the basis of the histochemical staining reaction, it would appear that only a branched polysaccharide is synthesized from UDPG, although Leloir *et al.* (1959) indicate that the compound produced under biochemical conditions exhibits 1,4-glycosidic linkages, and thus would be expected to stain a blue or blue-violet color. The histochemical distribution of this transferase is similar to that of amylophosphorylase.

Hess and Pearse (1961) applied the aforementioned technique in a study of rat muscle and suggested that there was a difference in distribution as contrasted with amylophosphorylase. These findings obviously require confirmation.

Engel (1961) has found that UDPG transferase activity is completely absent in the previously mentioned "target fibers."

## III. ACTIVATORS AND INHIBITORS

Although muscle phosphorylase A is active in the absence of AMP, addition of the latter compound nevertheless results in an increased activity of about 50%. On the other hand, muscle phosphorylase B is inactive in the absence of AMP. The interconversion of phosphorylase A to B is inhibited by fluoride, while the transformation of B to A is inhibited by EDTA (Krebs *et al.*, 1958, p. 41). Muscle phosphorylase is inhibited by glucose, phlorizin and phloretin, ammonium sulfide, and sodium $\beta$-glycerophosphate. Heating tissue sections to 60° for 30 minutes is known to abolish the histochemical phosphorylase reaction. Takeuchi (1958) notes that 20% ethanol produces a strong inhibition of branching enzyme but has little or no effect upon amylophosphorylase. Mercuric chloride ($10^{-4}$ *M*) produces a relatively complete inhibition of branching enzyme but likewise has little or no effect upon phosphorylase.

Takeuchi and Glenner (1961) found that UDPG-glycogen transferase activity is destroyed by fixation in buffered formalin solution, but is relatively well maintained during fixation in acetone, ethanol, or following air drying. The reaction is stimulated by glucose-6-phosphate and $MgCl_2$ but not by AMP. Among the inhibitors of the reaction are mercuric chloride, *p*-chloromercuriphenylsulfonic acid, cobaltous nitrate, sodium fluoride, potassium iodate, potassium cyanide, sodium borate, glucose, and phlorizin.

# IV. TRANSGLYCOSYLASES IN NEOPLASMS

Amylophosphorylase in tumors and in the livers of tumor-bearing animals has been studied by Godlewski (1960b, 1961) and Godlewski and Penar (1961). Reactions for the demonstration of both phosphorylase and branching enzyme were carried out on smears from three kinds of ascites tumor cells: Ehrlich carcinoma and amytal sarcoma in mice, and Novikoff hepatoma in female rats. The mouse ascites tumor cells usually were completely inactive, although occasionally a single reactive cell was seen. About 15% of the Novikoff hepatoma cells exhibited enzyme activity, particularly of the branching type. No correlation could be made between the distribution of phosphorylase activity and the glycogen content of the tumor cells. Godlewski and Penar (1961) found that in the majority of tumor-bearing animals there was an associated decrease in liver phosphorylase.

## APPENDIX 9

# HISTOCHEMICAL METHODS

1. *Phosphorylase* (*Goldberg et al., 1952*)

Frozen sections are incubated in the following substrate solution:

| | |
|---|---|
| Distilled water | 7.5 ml. |
| Glucose-1-phosphate (K salt) | 80.0 mg. |
| NaF | 63.0 mg. |
| Saturated aqueous sodium acetate solution | 7.5 ml. |

The pH of the solution is adjusted to ca. 6.0 with a few drops of HCl. The saturated sodium acetate solution may inhibit the enzyme, and it is probably advisable to modify the technique by employing 0.1 *M* acetate buffer. Following incubation, sections are treated with the pa-S reagent, Best's carmine, or iodine vapor over an $I_2$-KI solution.

Control slides consisted of heat inactivated sections, and sections inactivated by phlorizin (0.01 *M*). Other slides were incubated with saliva or malt diastase in order to remove pre-existing glycogen.

2. *Phosphorylase* (*Cobb, 1953*)

Cobb employed paraffin embedded undecalcified frozen-dried sections of rat bones.

1. Sections mounted on glass slides are dewaxed and incubated with amylase for 1½ hours at 37° to remove glycogen.
2. The sections are then soaked in distilled water overnight to remove soluble bone salts which might impede the reaction.
3. Tissue phosphoglucomutase is inactivated by heating the sections covered with a few drops of water (in a petri dish containing a wet blotter) to 60° for 15 minutes.
4. The sections are incubated for 6 hours in an acetate buffer, pH 9.0 containing glucose-1-phosphate and barium chloride.
5. Following incubation, sections are placed in absolute ethanol overnight to fix the glycogen, and then the section is stained by the pa-S method.

Since tissue glycogen is removed by amylase, comparison of an amylase-treated section with a similar one incubated according to the aforementioned procedure should permit the localization of phosphorylase activity through the presence of synthesized polysaccharide.

3. *Phosphorylase* (*Takeuchi and Kuriaki, 1955; Takeuchi, 1956*)

Unfixed frozen sections approximately 20 μ thick are used. The substrate solution contains glycogen as a primer, insulin as an activator, and ethanol to augment precipitation of the amylose.

| | |
|---|---|
| Glucose-1-phosphate (Na) | 50.0 mg. |
| Adenosine-5′-phosphate | 10.0 mg. |
| Acetate buffer (0.1 *M*) pH 5.7 | 20.0 ml. |
| Ethanol | 5.0 ml. |
| Glycogen | 2.0 mg. |
| Insulin solution | 1–2 drops (10–20 units) |

The sections are incubated in the substrate solution for 3–4 hours at 37°, and then immersed for several minutes in dilute Gram's iodine-iodide solution. The Gram's stock solution contains $I:KI:H_2O/1:2:300$ and is diluted about 10 times with distilled water. The sections are mounted in iodine-glycerol. (Iodine-glycerol is prepared by adding Gram's stock solution to glycerol in a ratio of 1:9.)

The presence of phosphorylase in tissue sections is revealed by a purplish-blue or deep blue dye. Even if native glycogen is present in tissue sections it is stained a red-brown and not the characteristic blue color of the specific phosphorylase reaction. Thus, the removal of pre-existing glycogen would appear to be unnecessary.

4. *"Branching" Enzyme* (*Takeuchi, 1958*)

Amylo-1,4 → 1,6-transglucosidase may be demonstrated in unfixed frozen sections ca. 15–40 μ thick. The substrate medium contains the following:

| | |
|---|---|
| Potassium glucose-1-phosphate | 50.0 mg. |
| Muscle adenylic acid | 10.0 mg. |
| Glycogen, water-soluble | 2.0 mg. |
| Distilled water | 15.0 ml. |

Ten ml. of acetate buffer (pH 5.7–6.0) and a drop of insulin solution are added to the mixture. For optimal results it is advisable to warm the solution to 37° prior to use.

Control as well as inhibitory solutions are prepared by omitting glucose-1-phosphate and adenylic acid from the aforementioned substrate, or by adding 5.0 ml. of ethanol to the substrate solution.

Gram's iodine solution is employed and is diluted 10–15 times with distilled water prior to use. The amylase (0.5%) is dissolved in distilled water. The α-amylase is buffered at pH 5.5–6.0 while the β-amylase solution is adjusted to pH 4.0–5.7 with acetate buffer.

Frozen sections are incubated for 1–2 hours at 37° employing the aforementioned incubating, control, and inhibitor-containing substrate solutions. The sections are then run quickly through 40% ethanol to distilled water, mounted on clean slides, and then dried in a 37° oven. Following drying, each section is fixed for a few minutes in absolute ethanol and dried at room temperature. Each procedure is carried out on one of three serial sections:

a. One section is immersed for a few minutes in the iodine solution until the newly synthesized polysaccharide stains, and then the section is mounted in iodine-glycerol or glycerol.

b. A second serial section is incubated at 37° for a period of 1–10 hours in α-amylase solution, and washed quickly in distilled water. The section is then immersed in iodine solution for a few minutes and mounted as previously described. Control sections are prepared according to the same procedure, incubating sections in acetate buffer without α-amylase.

c. At the same time, the third serial section is incubated in β-amylase solution and stained and mounted as above.

Amylo-1,4 → 1,6-transglucosidase activity is delineated by a staining reaction which results in the formation of a deep violet or reddish-purple color with iodine according to the procedure (a), disappears with complete digestion by α-amylase in (b), and is unaffected by the β-amylase treatment in procedure (c). On the other hand, the reaction which results in the formation of a deep blue dye with iodine in procedure (a) and disappears with the complete digestion by both α-amylase and β-amylase does not show sites of the transglucosidase enzyme but only enables visualization of amylophosphorylase. The use of an alcohol-containing substrate serves to demonstrate amylophosphorylase alone as a blue dye, since ethanol selectively inhibits the branching enzyme.

5. *Uridine Diphosphate Glucose-Glycogen Transferase* (*Takeuchi and Glenner, 1961*)

Unfixed frozen sections prepared according to the Adamstone-Taylor cold knife method are incubated for 1 hour at room temperature (ca. 25°). In certain cases, sections mounted on glass slides may also be employed, although the use of free-floating sections is preferable. The substrate solution contains the following:

| | |
|---|---|
| Uridine diphosphate glucose (UDPG) | 50.0 mg. |
| Glycogen | 10.0 mg. |
| Versene (EDTA) | 20.0 mg. |
| Glucose-6-phosphate | 10.0 mg. |

These compounds are dissolved in 14.0 ml. of distilled water, and then 10.0 ml. of 0.2 *M* tris buffer (pH 7.4) is added to this mixture, followed by 1.0 ml. of absolute ethanol. The control solution is identical except for the absence of UDPG.

Following incubation the sections are transferred directly to dilute Gram's iodine solution until the iodine reaction appears.

The sections are then mounted on glass slides, covered with iodine-glycerol, and sealed with paraffin. A red-purple stain in tissue sections indicates sites of uridine diphosphate glucose-glycogen transferase activity. These authors note that if their original substrate mixture is used (Takeuchi and Glenner, 1960), a red-brown stain, which cannot be distinguished from naturally occurring glycogen in tissues appears at sites of enzyme activity. However, they note that under most circumstances, with the exception of cartilage cells, naturally occurring glycogen in tissue sections is not demonstrated by the aforementioned technique.

## Commercial Sources of Reagents

1. Dajac Laboratories, The Borden Chemical Company, Philadelphia, Pennsylvania
2. Mann Research Laboratories, New York, New York
3. Nutritional Biochemicals Corp., Cleveland, Ohio
4. Sigma Chemical Company, St. Louis, Missouri

## References

Bear, R. S. (1942) *J. Am. Chem. Soc.* **64,** 1388–1392.

Bear, R. S., and Cori, C. F. (1941) *J. Biol. Chem.* **140,** 111–118.

Braun-Falco, O. (1956) *J. Histochem. and Cytochem.* **4,** 300.

Brown, D. H., and Cori, C. F. (1961) *In* "The Enzymes" (P. D. Boyer, H. Lardy, and K. Myrbäck, eds.), 2nd ed., Vol. 5, pp. 207–228. Academic Press, New York.

Carbonell, L. M. (1955) *J. Histochem. and Cytochem.* **3,** 419.

Cobb, J. D. (1953) *Arch. Pathol.* **55,** 496–502.

Cohn, M. (1961) *In* "The Enzymes" (P. D. Boyer, H. Lardy, and K. Myrbäck, eds.), 2nd ed., Vol. 5, pp. 179–206. Academic Press, New York.

Cori, C. F., Colowick, S. P., and Cori, G. T. (1937) *J. Biol. Chem.* **121,** 465–477.

Cori, G. T., and Cori, C. F. (1937) *Proc. Soc. Exptl. Biol. Med.* **36,** 119–122.

Cori, G. T., and Cori, C. F. (1943) *J. Biol. Chem.* **151,** 57–63.

Cowgill, R. W., and Cori, C. F. (1955) *J. Biol. Chem.* **216,** 133–140.

Dixon, M., and Webb, E. C. (1958) "Enzymes." Academic Press, New York.

Dyar, M. T. (1950) *Am. J. Botany* **37,** 786–792.

Edelman, J. (1956) *Advances in Enzymol.* **17,** 189–232.

Engel, W. K. (1961) *Nature* **191,** 389–390.

Engel, W. K., Foster, J. B., Hughes, B. P., Huxley, H. E., and Mahler, R. (1961) *Brain* **84,** 167–185.

Friedkin, M., and Kalckar, H. (1961) *In* "The Enzymes" (P. D. Boyer, H. Lardy, and K. Myrbäck, eds.), 2nd ed., Vol. 5, pp. 237–255. Academic Press, New York.

Godlewski, H. G. (1960a) *Bull. acad. polon. sci.* **8,** 441–444.

Godlewski, H. G. (1960b) *J. Histochem. and Cytochem.* **8,** 353.

Godlewski, H. G. (1961) *Acta Histochem.* **11,** 58–67.
Godlewski, H. G., and Penar, B. (1961) *Bull. acad. polon. sci.* **9,** 195–199.
Godlewski, H. G., and Skurzak, H. (1961) *Folia Biolog.* (*Warsaw*) **9,** 101–110.
Goldberg, B., Wade, O. R., and Jones, H. W. (1952) *J. Natl. Cancer Inst.* **13,** 543–557.
Hassid, W. Z. (1960) *In* "Metabolic Pathways" (D. M. Greenberg, ed.), Vol. I, pp. 251–300. Academic Press, New York.
Hassid, W. Z., Doudoroff, M., and Barker, H. A. (1951) *In* "The Enzymes" (J. B. Sumner and K. Myrbäck, eds.), Vol. I, Part 2, pp. 1014–1039. Academic Press, New York.
Hehre, E. J. (1951) *Advances in Enzymol.* **11,** 297–337.
Herbener, G., and Atkinson, W. B. (1959) *Proc. Soc. Exptl. Biol. Med.* **100,** 736–737.
Hess, R., and Pearse, A. G. E. (1961) *Proc. Soc. Exptl. Biol. Med.* **107,** 569–571.
Hoffmann-Ostenhof, O. (1953) *Advances in Enzymol.* **14,** 219–260.
Imsande, J., and Handler, P. (1961) *In* "The Enzymes" (P. D. Boyer, H. Lardy, and K. Myrbäck, eds.), 2nd ed., Vol. 5, pp. 281–304. Academic Press, New York.
Kornberg, A. (1957) *Advances in Enzymol.* **18,** 191–240.
Krebs, E. G., Kent, A. B., Graves, D. J., and Fischer, E. H. (1958) *In* "Proceedings of the International Symposium on Enzyme Chemistry, Tokyo and Kyto, 1957" (K. Ichihara, ed.), pp. 41–43. Academic Press, New York.
Leloir, L. F., Olavarría, J. M., Goldemberg, S. H., and Carminatti, H. (1959) *Arch. Biochem. Biophys.* **81,** 508–520.
Robbins, P. W., Traut, R. R., and Lipmann, F. (1958) *Proc. Natl. Acad. Sci. U. S.* **45,** 6–12.
Schmid, R., and Hammaker, L. (1961) *New Engl. J. Med.* **264,** 223–225.
Schneider, W. C., and Hogeboom, G. H. (1956) *Ann. Rev. Biochem.* **25,** 201–219.
Shimizu, N., and Okada, M. (1957) *J. Histochem. and Cytochem.* **5,** 459–471.
Stetten, D., and Stetten, M. (1960) *Physiol. Revs.* **40,** 505–537.
Stocking, C. R. (1952) *Am. J. Botany* **39,** 283–287.
Sumner, J. B., and Somers, G. F. (1953) "Chemistry and Methods of Enzymes," 3rd ed. Academic Press, New York.
Swanson, M. A. (1948) *J. Biol. Chem.* **172,** 805–814.
Takeuchi, T. (1956) *J. Histochem. and Cytochem.* **4,** 84.
Takeuchi, T. (1958) *J. Histochem. and Cytochem.* **6,** 208–216.
Takeuchi, T., and Glenner, G. G. (1960) *J. Histochem. and Cytochem.* **8,** 227–230.
Takeuchi, T., and Glenner, G. G. (1961) *J. Histochem. and Cytochem.* **9,** 304–316.
Takeuchi, T., and Kinoshita, K. (1956) *Blood* **11,** 375–379.
Takeuchi, T., and Kuriaki, H. (1955) *J. Histochem. and Cytochem.* **3,** 153–160.
Takeuchi, T., Higashi, K., and Watanuki, S. (1955) *J. Histochem. and Cytochem.* **3,** 485–491.
Villar-Palasi, C., and Larner, J. (1958) *Biochim. et Biophys. Acta* **30,** 449.
Yin, H. C., and Sun, C. N. (1947) *Science* **105,** 650.

*CHAPTER 10*

# PROTEOLYTIC ENZYMES

## I. INTRODUCTION

Proteolytic enzymes are those which hydrolyze proteins and amino acid compounds, and may be divided into two groups, endopeptidases and exopeptidases. Exopeptidases hydrolyze peptide bonds adjacent to terminal $\alpha$-amino or terminal $\alpha$-carboxyl groups and include enzymes such as aminopeptidase and carboxypeptidase. On the other hand, endopeptidases hydrolyze centrally located peptide bonds as well as terminal peptide bonds (Bergmann, 1942). Among the endopeptidases are pepsin, trypsin, chymotrypsin, and cathepsins. The activity of many of the exopeptidases is dependent upon bound metal ions such as $Mg^{++}$, $Mn^{++}$, $Co^{++}$, and $Fe^{++}$. A characteristic of some of the endopeptidases, including trypsin and chymotrypsin, is their susceptibility to inhibition by organophosphorus compounds such as diisopropyl fluorophosphate (DFP). The biochemical characteristics and classifications of proteolytic enzymes have been the subject of a number of extensive reviews, including those by Sumner and Somers (1953, p. 166), Mehler (1957, p. 13), Smith (1951, p. 793; 1960, p. 1).

The physiological and pathological roles of proteolytic enzymes have stimulated their study from both biochemical and histochemical standpoints. It has been suggested, for example, that cathepsins serve to cleave the proteins of dead cell fragments which subsequently are utilized for protein synthesis in surviving cells (Fruton, 1960, p. 233). The enzyme plasmin (fibrinolysin), which is capable of hydrolyzing the internal peptide bonds of fibrin, may participate in certain metabolic processes such as the dissolution of pathologic fibrin, the destruction of plasma proteins, the resolution of inflammatory processes, and wound healing (Ablondi and Hagan, 1960, p. 175; Cohen and Warren, 1961a, b). Ungar and Damgaard (1954) have studied the influence of thermal injury upon protein breakdown in skin.

Their findings, including those of other workers, revealed an increase in proteolytic activity. The laboratory procedure employed by Ungar and Damgaard (1954) for the determination of proteolytic activity employed fibrinogen as a substrate. The amount hydrolyzed was estimated by the ultraviolet absorption of the non-clottable fraction. Howes *et al.* (1955) have reviewed the literature pertaining to proteolytic enzymes in the healing wound. These authors studied several enzymes in the healing wounds of rabbits and found high peptidase activity in pus and granulation tissue. They indicate that the formation of amino acids by peptidase enzymatic hydrolysis may be related to the synthesis of proteins or may represent a detoxifying mechanism. Rothman (1957) has reviewed some of the proteolytic enzyme systems of the skin and their possible relation to dermatological lesions. The activation of proteolytic enzymes in skin following injury was related to epidermal-dermal separation and vesicle formation.

From a histochemical standpoint the demonstration of proteolytic enzymes is limited to aminopeptidase, although there have been several suggestions relating to the possible identification of chymotrypsin and trypsin-like enzymes. The histochemical aminopeptidase reaction, which has been evolved only during the last six years, depends upon the hydrolysis of a peptide bond adjacent to a terminal $\alpha$-amino group, which results in the liberation of a chromogenic moiety. In addition to the hydrolysis of a peptide or amide bond, many proteolytic enzymes are capable of attacking analogous ester bonds. For example, chymotrypsin can hydrolyze acetyl-L-tyrosine ethyl ester (Doherty and Vaslow, 1952) and naphthol AS chloroacetate (Benditt, 1956). This type of ester hydrolysis may prove to be of value in the design of newer histochemical substrates.

Proteolytic enzymes are also known to catalyze the synthesis of peptide bonds. Chymotrypsin, for example, will catalyze the following reaction (Bergmann and Fruton, 1941; Fruton *et al.*, 1951):

$$\text{Benzoyl-L-tyrosine} + \text{glycinanilide} \rightarrow \text{benzoyl-L-tyrosylglycinanilide} + H_2O$$

The possible demonstration of the synthetic effect of proteolytic enzymes has not as yet received even cursory consideration.

## II. SYNTHETIC APPROACHES AND SUBSTRATE TYPES

There have been several recent and comprehensive reviews dealing with peptide synthesis (Wieland, 1951; Smith, 1957, p. 528; Greenstein and Winitz, 1961, p. 763; Goodman and Stueben, 1959). For purposes of the present discussion, only techniques which have possible application in the preparation of histochemical substrates will be considered. Historical aspects of the development of synthetic peptide chemistry have been dealt with by Fruton (1949), and Greenstein and Winitz (1961, p. 763). One of the earliest methods of peptide synthesis involved the coupling of an $\alpha$-halogen acid halide to an amino acid, followed by amination of the resulting compound (Fischer and Suzuki, 1904). Simple amino acids such as glycine may be synthesized by treating chloroacetic acid with ammonia (Robertson, 1927; Boutwell and Kuick, 1930), and leucine and valine prepared by the amination of $\alpha$-bromoisocaproic and $\alpha$-bromovaleric acids, respectively (Marvel, 1955, pp. 523 and 848). This type of amination procedure has also been used to prepare glycine anilides (Pfeiffer and Saure, 1941); compounds of this type were subsequently employed as histochemical reagents.

In 1932 Bergmann and Zervas reported the development of the "carbobenzoxy method" of peptide synthesis. This highly significant technique, which facilitated the synthesis of a variety of new and complex peptides, involved the condensation of carbobenzoxy amino acid chlorides with other amino acids or amines to form the corresponding carbobenzoxy peptide. The free peptide was subsequently obtained by catalytic hydrogenation. Excellent reviews on the development and application of the carbobenzoxy technique have been given by Fruton (1949), Smith (1951, p. 793), Wieland (1951), and Greenstein and Winitz (1961, p. 763). Carbobenzoxy amino acids have been employed in the synthesis of a number of aminopeptidase substrates via the mixed anhydride technique (Appendix 10A, Section 7).

In recent years, a number of new methods for peptide synthesis, several of which employ carbobenzoxy amino acids, have been reported. Wieland and Sehring (1950) reported the application of mixed carboxylic acid anhydrides in the synthesis of peptides, and found that anhydrides of benzoic or acetic acid with $N$-substituted amino acids react readily with aqueous solutions of a salt of a second amino acid to form a peptide derivative. Vaughan and Osato (1951) found that anhydrides with isovaleric acid were particularly advantageous in the preparation of a number of peptides including anilides. A similar approach was utilized by Boissonnas (1951) who employed ethyl chlorocarbonate in the preparation of the mixed anhydrides. Another technique which has found application in the preparation of histochemical substrates is the carbodiimide procedure (Sheehan and Hess, 1955; Sheehan *et al.*, 1956; Sheehan and Hlavka, 1956). With this mild procedure a carbobenzoxy amino acid may be coupled to an arylamine in aqueous or non-aqueous media (Appendix 10A, Section 8):

$$RCOOH + NH_2R' + C_6H_{11}N{=}C{=}NC_6H_{11} \rightarrow RCONHR' + C_6H_{11}NHCONHC_6H_{11}$$

Peptide synthesis via amino acid active ester hydrobromides (Goodman and Steuben, 1959), and by the trifluoroacetic anhydride method (Weygand and Csendes, 1952; Weygand and Leising, 1954; Weygand and Reiher, 1955), have also been reported. In the latter procedure an $N$-trifluoroacetyl amino acid anhydride, prepared by treating an amino acid with trifluoroacetic anhydride, is reacted with an arylamine to form an amino acid anilide. The *N*-trifluoroacetyl grouping may be removed with aqueous ammonia or with an ion exchange resin. Schallenberg and Calvin (1955) have employed ethyl thioltrifluoroacetate as an acetylating agent in a similar type of peptide synthesis.

Although, as previously noted, carbobenzoxy groups may be removed by catalytic hydrogenation, the method of Ben-Ishai and Berger (1952) is of value in laboratories not equipped with hydrogenation apparatus. According to their procedure, the carbobenzoxy radical is removed by treatment of the appropriate carbobenzoxy compound with HBr-glacial acetic acid (Appendix 10A, Sections 9 and 12).

Relatively few histochemical substrates for the demonstration of proteolytic enzymes have been reported. The Seligman group described several unsuccessful attempts to demonstrate proteolytic enzymes with synthetic substrates. Ravin and Seligman (1951) found that carbonaphthoxy-DL-phenylalanine, which possesses the basic structural requirements of a substrate for carboxypeptidase, and has, in addition, a potential chromogenic moiety, was hydrolyzed by crystalline carboxypeptidase and by carboxypeptidase produced from procarboxypeptidase in pancreas by tryptic activation. Attempts to apply this substrate in the histochemical demonstration of carboxypeptidase were unsuccessful, presumably due to the destruction of cellular detail by the proteolytic enzyme system required to activate the enzyme. The synthesis of carbonaphthoxyphenylalanine as well as that of 1-ketomethyl-1,2-dihydro-2-naphthoyl-DL-phenylalanine and acetylphenylalanine β-naphthyl ester have been reported by Wolf and Seligman (1951a, b) and Seligman and Wolf (1951). Ravin *et al.* (1954) found that the β-naphthyl ester of *N*-benzoyl-DL-phenylalanine was readily hydrolyzed by pancreatic tissue homogenates in which chymotrypsin had been activated by trypsin. Although a suitable colorimetric method for chymotrypsin was developed, the procedure was not applicable for histochemical use. A more successful approach to the demonstration of chymotrypsin has utilized the chloroacetyl ester of 3-hydroxy-2-naphthoic acid anilide (CANAS) first synthesized by Gomori (1953) (Table 1) (see Chapter 6, p. 2). Although this substrate is readily hydrolyzed by non-specific esterases, Benditt (1956, 1958) and Benditt and Arase (1959) have reported that CANAS is also hydrolyzed by crystalline chymotrypsin, and that the substrate is capable of delineating a chymotrypsin-like enzyme in mast cells. Naphthol AS-D chloroacetate (Burstone, 1957) may be similarly applied.

Certain halogenated acid chlorides react with β-naphthylamine to form compounds which may be directly applicable as amidase substrates or serve as intermediates in the synthesis of aminopeptidase substrates. Among the

TABLE 1

POTENTIAL SUBSTRATES FOR PROTEOLYTIC ENZYMES[a]

| Chemical name and formula | Mol. wt. | Reference | Melting point (° C.) |
|---|---|---|---|
| 1. Chloroacetyl-β-naphthylamide, $C_{12}H_{10}NOCl$ | 219.68 | Johnson, 1903; Gomori, 1954a | 117–118 |
| 2. α-Bromopropionyl-β-naphthylamide, $C_{13}H_{12}NOBr$ | 278.16 | Tigerstedt, 1892 | 174 |
| 3. Glycyl-β-naphthylamide, $C_{12}H_{12}N_2O$ | 200.24 | Gomori, 1954b | 114–115 |
| 4. DL-Alanyl-β-naphthylamide, $C_{13}H_{14}N_2O$ | 214.27 | Gomori, 1954b; Burstone and Folk, 1956 | 99–100 |
| 5. L-Prolyl-β-naphthylamide, $C_{15}H_{16}ON_2 \cdot HCl$ | 276.77 | Folk and Burstone, 1956 | 272 |
| 6. Hydroxy-L-prolyl-β-naphthylamide, $C_{15}H_{16}O_2N_2 \cdot HCl$ | 292.77 | Folk and Burstone, 1956 | 228–230 |
| 7. L-Leucyl β-naphthylamide, $C_{16}H_{20}N_2O \cdot HCl$ | 292.82 | Green *et al.*, 1955; Folk and Burstone, 1955 | 240–241<br>248–252 |
| 8. L-Cystine di-β-naphthylamide, $C_{26}H_{26}N_4O_2S_2$ | 490.66 | Tuppy and Nesvadba, 1957 | 167–169 |
| 9. L-Cysteine β-naphthylamide, $C_{13}H_{14}N_2OS$ | 246.34 | Tuppy and Nesvadba, 1957 | 130–132 |
| 10. DL-Methionine β-naphthylamide, $C_{15}H_{18}N_2OS \cdot HCl$ | 310.86 | Tuppy and Nesvadba, 1957 | 172–173 |
| 11. L-Leucyl-4-methoxy-β-naphthylamide, $C_{17}H_{22}N_2O_2$ | 286.38 | Rosenblatt *et al.*, 1958 | 96.5–97.5 |
| 12. DL-Alanyl derivative of 3-aminocarbazole, $C_{15}H_{15}N_3O$ | 253.31 | Burstone and Weisburger, 1961a | 207–208 |
| 13. DL-Alanyl derivative of 3-amino-9-ethylcarbazole, $C_{17}H_{20}N_3ClO$ (hydrochloride) | 317.83 | Burstone and Weisburger, 1961a | 156 |
| 14. Naphthol AS chloroacetate, $C_{19}H_{14}O_3NCl$ | 339.79 | Gomori, 1953 | — |
| 15. Naphthol AS-D chloroacetate, $C_{20}H_{16}O_3NCl$ | 353.81 | Burstone, 1957 | 162 |
| 16. *N*-α-Benzoyl-DL-arginine-β-naphthylamide, $C_{23}H_{25}N_5O_2 \cdot HCl$ | 439.96 | Riedel and Wünsch, 1959 | 193–194 |

[a] Most substrates are for aminopeptidase. (5) and (6) are potential substrates for prolidase, (14) and (15) have been employed in the demonstration of a chymotrypsin-like enzyme, while (16) has been used to demonstrate a trypsin-like enzyme.

latter are chloroacetyl and α-bromopropionyl β-naphthylamides which were originally synthesized by Tigerstedt (1892) and by Johnson (1903), who treated β-naphthylamine with either chloroacetyl chloride or bromopropionyl bromide (Table 1). Chloroacetyl-β-naphthylamine was employed by Gomori (1954a) as a histochemical substrate for the demonstration of an undefined enzyme, presumably an amidase. His studies were influenced by the previous work of Fodor *et al.* (1950), Birnbaum *et al.* (1952), and Rao *et al.* (1952) who synthesized a number of chloroacetyl amino acids and noted that they were hydrolyzed more readily by tissue extracts than their corresponding acetyl and glycyl derivatives. Gomori found that chloroacetyl-β-naphthylamine was not appreciably hydrolyzed by pepsin, chymotrypsin, trypsin, or papain.

Upon treatment of either chloroacetyl-β-naphthylamine or α-bromopropionyl-β-naphthylamide with ammonia, Gomori (1954b) obtained glycine and DL-alanyl-β-naphthylamides (Table 1) (Appendix 10A, Section 5) which are useful substrates for the demonstration of aminopeptidase, since these compounds have a free α-amino group on the terminal glycine or alanine grouping. Mottet (1961) has stated that the enzyme hydrolyzing L-leucyl-β-naphthylamide is an amidase. This statement fails to take into consideration the biochemical studies of Green *et al.* (1955), Folk and Burstone (1955), and Smith and Hill (1960, p. 37) which have characterized the enzymatic hydrolysis of the leucyl as well as similar substrates from a biochemical standpoint.

Green *et al.* (1955) and Folk and Burstone (1955) reported the synthesis of L-leucyl-β-naphthylamide by the carbobenzoxy mixed anhydride method. It should be noted that Green *et al.* (1955) stated that the use of this substrate as a histochemical reagent in conjunction with fresh, unfixed frozen, or formalin fixed sections did not reveal sufficiently dense staining to warrant employing the substrate for histochemical purposes. This is in contrast to the paper by Burstone and Folk (1956) which demonstrated that L-leucyl-β-naphthylamide is a highly useful histochemical substrate for the demonstration of aminopeptidase. Burstone (1956a) prepared DL-alanyl and DL-phenylalanyl β-naphthylamides by the carbobenzoxy mixed anhydride method. The carbobenzoxy grouping was removed with HBr-glacial acetic acid. Subsequently, Folk and Burstone (1956) reported the synthesis of prolyl- and hydroxyprolylnaphthylamides but were unsuccessful in attempts to employ these substrates as histochemical reagents. Tuppy and Nesvadba (1957) reported several new and interesting substrates which were employed in the study of serum aminopeptidase in pregnant women. Among the compounds were L-cystine di-β-naphthylamide, L-cysteine β-naphthylamide and DL-methionine β-naphthylamide. Rosenblatt *et al.* (1958) reported the synthesis of L-leucyl-4-methoxy-β-naphthylamide, a variation of the aforementioned leucyl compound.

More recently, Burstone and Weisburger (1961a) prepared a number of DL-alanyl arylamide compounds (Chart 1; Table 1) (Appendix 10A,

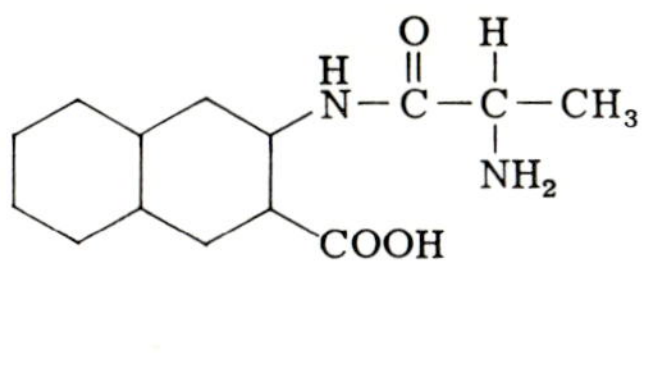

1

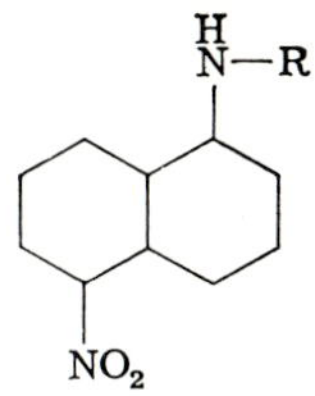

2

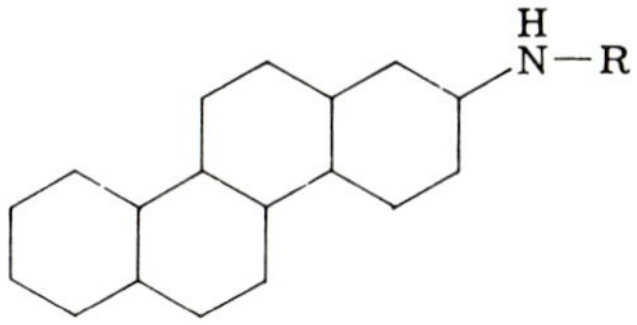

3

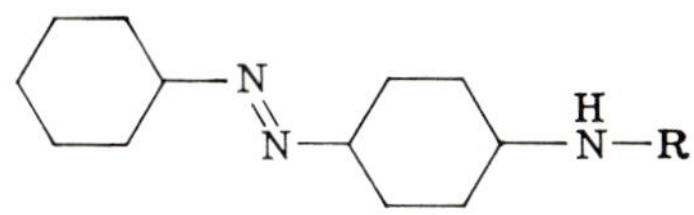

4

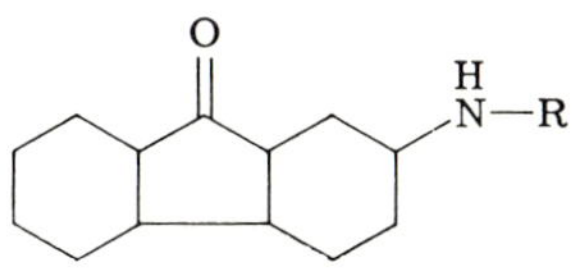

5

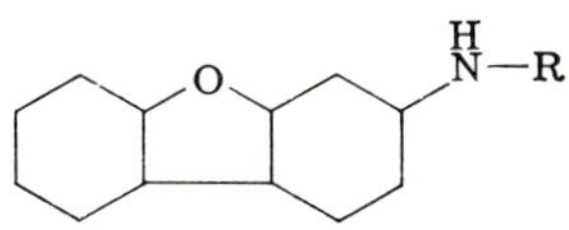

6

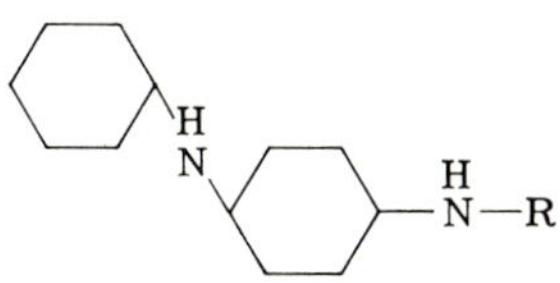

7

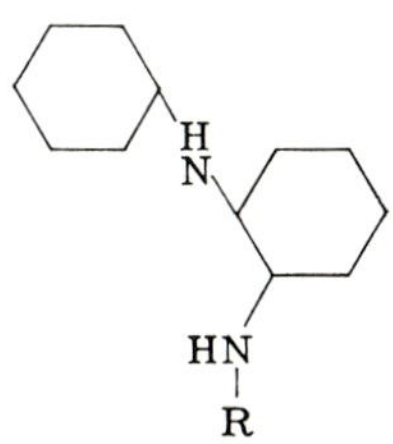

8

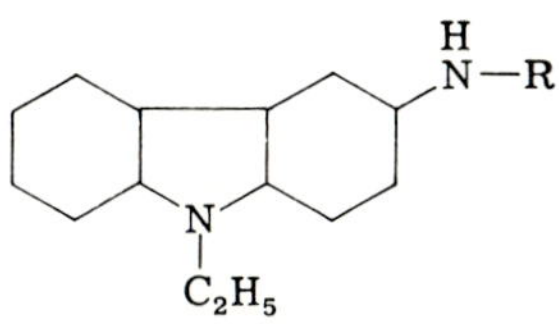

9

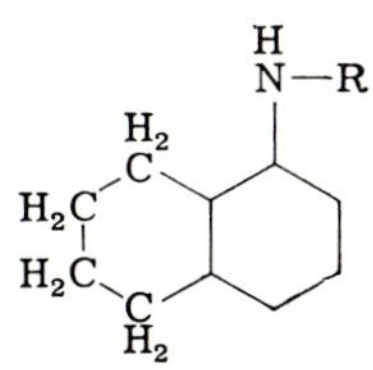

10

CHART 1. Complex DL-alanyl arylamide aminopeptidase substrates (Burstone and Weisburger, 1961a).

Section 6), as aminopeptidase substrates. Among the amines which were used as the basis of these substrates were: 2-amino-3-naphthoic acid (1), 5-nitro-1-naphthylamine (2), 2-aminochrysene (3), *p*-aminoazobenzene (4), 2-aminofluorenone (5), 3-aminodibenzofuran (6), *p*-aminodiphenylamine (7), *o*-aminodiphenylamine (8), 3-amino-9-ethylcarbazole (9), 1-amino-5,6,7,8-tetrahydronaphthalene (10), 3-aminocarbazole, 4-aminodiphenylamine-2-sulfonic acid, and 7-amino-2-fluorenesulfonic acid. Several of these substrates have proven to be of value because they form less soluble azo-dyes as compared with those derived from the previously employed simpler naphthylamines. In addition, azo-dyes derived from the more complex arylamines are less lipid soluble.

In 1959 Riedel and Wünsch synthesized a new substrate for the measurement of trypsin activity, α-benzoyl-L-arginine-β-naphthylamide (BANA). While this compound was readily hydrolyzed by crystalline trypsin, it was completely unaffected upon incubation with human serum. Application of this substrate as a histochemical reagent in the study of mast cells has been reported by Glenner and Cohen (1960).

## III. HISTOCHEMICAL METHODS

The histochemical demonstration of proteolytic enzymes has been primarily limited to the demonstration of aminopeptidase, although there have been a few reports describing the localization of a chymotrypsin and a trypsin-like enzyme.

Gomori (1954a) reported that chloroacetyl-β-naphthylamide, when employed in a simultaneous azo-dye method in conjunction with Garnet GBC salt at neutral pH, was capable of giving histochemical localizations in tissue sections including liver and intestinal mucosa of the guinea pig, intestinal mucosa of the rat, and bronchial mucosa and renal tubular epithelium of the cat. Although homogenates of human tissues were quite active, incubation of paraffin sections failed to reveal any staining. Subsequently Gomori (1954b) employed glycyl and alanyl β-naphthylamides and obtained a reaction in a number of tissues. Details of the histochemical technique, however, were not given. Among the more active tissues were convoluted tubules of the kidney, bile canaliculi of the liver, and duodenal villi. Gomori specifically noted that acetone fixed paraffin embedded tissues were almost completely inactive as contrasted with the relatively intense reactivity of formalin fixed frozen sections.

In 1956 Burstone and Folk compared the DL-alanyl-β-naphthylamide substrate with the newly synthesized L-leucyl-β-naphthylamide in a detailed histochemical study of the specificity and application of these compounds in the demonstration of aminopeptidase (Chart 2). It was found that the enzyme was well preserved in frozen-dried paraffin embedded tis-

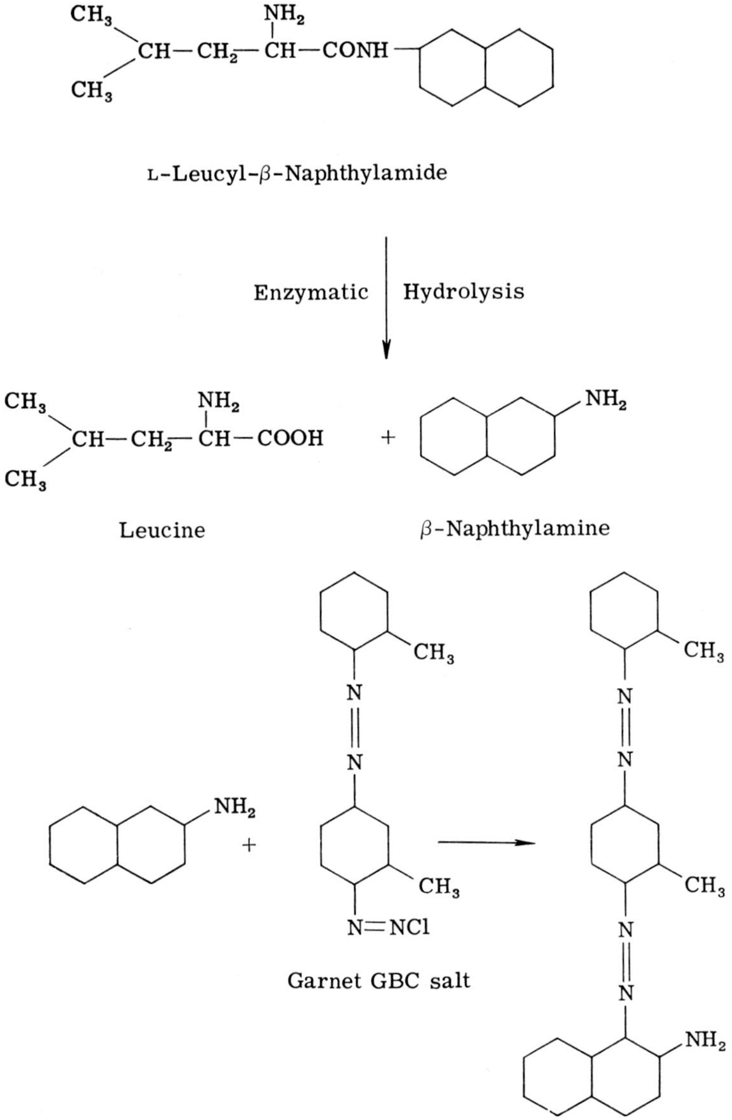


CHART 2. Hydrolysis of histochemical substrate by tissue aminopeptidase to produce an azo-dye (Burstone and Folk, 1956).

sues, and the tendency for the final dye to crystallize was minimized. Among the tissues studied were monkey liver, kidney, pancreas, spleen, and small intestine; human pancreas, spleen, stomach, small intestine, parotid and submaxillary glands and lymph nodes; rat kidney (Fig. 1), liver, pancreas, spleen, stomach, and intestine. The procedure was also employed in

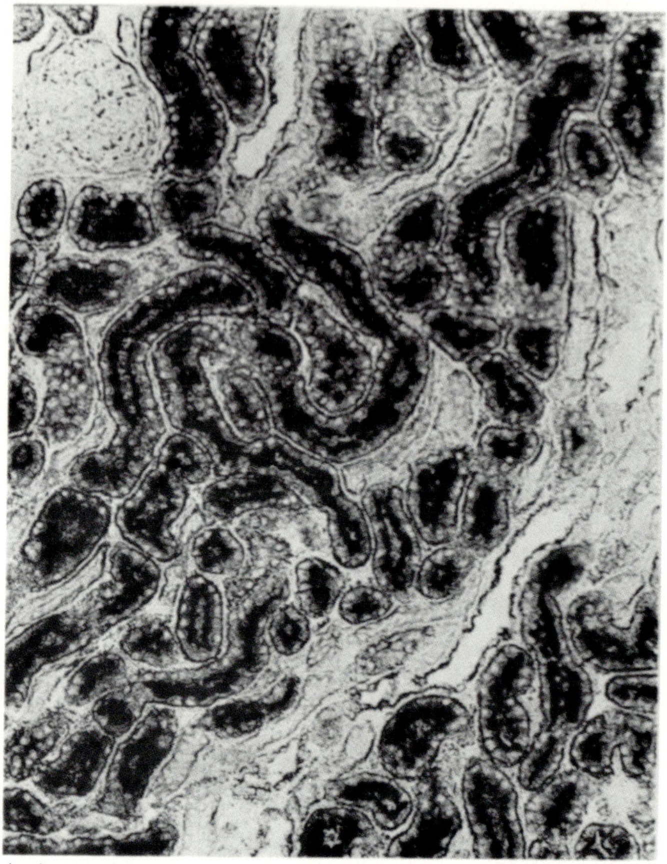

FIG. 1. Aminopeptidase activity of frozen-dried rat kidney, L-leucyl-β-naphthylamide method, Garnet GBC salt (Burstone and Folk, 1956) (× 143).

the demonstration of aminopeptidase in a large series of frozen-dried human neoplasms (Burstone, 1956b). Details of the method are given in Appendix 10B, Section 1. The histochemical reaction was characterized by marked inhibition produced by addition of $Zn^{++}$, $Cu^{++}$, $Pb^{++}$, and $Cd^{++}$, which are known aminopeptidase inhibitors. Cyanide in low concentrations (0.0001–0.0003 $M$) behaved as an activator in tissue sections, while higher concentrations exhibited a distinct inhibitory effect (Fig. 2). At approximately the same time Green *et al.* (1955) reported unsuccessful attempts to use the same leucyl substrate for histochemical purposes, and noted that a simultaneous coupling method was not employed because of the possible reaction of the diazonium salt with the amino group of the substrate.

Following the successful application of the leucyl substrate by Burstone and Folk (1956), the Seligman group, who had originally reported that L-leucyl-β-naphthylamide was unsatisfactory as an aminopeptidase substrate (Green *et al.*, 1955), reported a modification of the technique employing Diazo Blue B instead of Garnet GBC salt in a simultaneous coupling procedure (Nachlas *et al.*, 1957). The photomicrographs published in this paper revealed gross diffusion artefacts; and although the claim was made to have demonstrated "leucine" aminopeptidase, no experimental data was presented to support this assumption. With reference to the type of diazonium salt employed, Nachlas *et al.* (1957) reported that Diazo Blue B (tetrazotized diorthoanisidine) produced more of an inhibitory effect upon the enzyme than did Garnet GBC salt (diazotized *o*-aminoazotoluene). Diazo Blue B is stabilized as a double zinc salt, and $Zn^{++}$ is known to be a powerful inhibitor of aminopeptidase. On the other hand, Garnet GBC salt is stabilized as a diazonium sulfate and is relatively free of diluents as compared with other commercially available diazotates (Chapter 3, Chart 2) (Chapter 3, Table 2). Glenner *et al.* (1959) also confirmed the undesirable inhibitory and slow coupling properties of Diazo Blue B with reference to the histochemical procedure.

More recently, Nachlas *et al.* (1960) reported on the histochemical application of a modified leucyl substrate which contained a methoxy group in the 4-position of the naphthalene nucleus. The enzymatically released 4-methoxy-2-naphthylamine was claimed to couple 40 times as fast as β-naphthylamine. It is difficult to understand how the introduction of a methoxy group can result in such a remarkable increase in coupling rate,

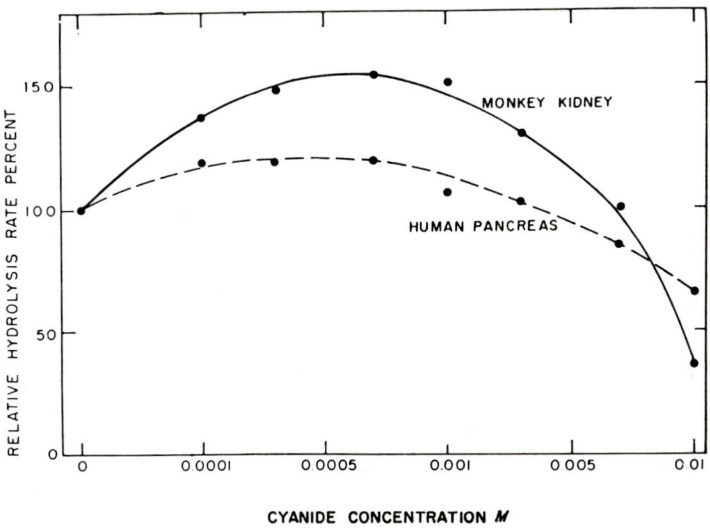


FIG. 2. Effects of varying concentrations of cyanide on aminopeptidase activity.

but in any case some of the photomicrographs shown in the paper exhibit diffusion artefacts and there is little evidence at present to recommend the methoxy substrate over the original leucyl or alanyl naphthylamides.

Burstone and Weisburger (1961a) applied a large number of new DL-alanyl arylamine substrates (Table 1) in an attempt to improve the histochemical procedure. Of the group, the alanyl derivatives of 3-aminocarbazole and 3-amino-9-ethylcarbazole in conjunction with Garnet GBC salt proved to be most satisfactory (Fig. 3). The azo-dyes derived from these substrates exhibited a diminished tendency to crystallize, presumably because of their relatively insoluble nature. Although the remaining substrates were hydrolyzed more slowly, they contain structural features which may prove of value in peptidase substrate design. These structural features include, in addition to enlargement of the molecule, presence of adjacent amino and carboxyl groups as coordinating groups for metal chelation, and the presence of the nitro, azo, and keto grouping. A number of complex diazonium salts were also synthesized and applied. Of these, anthraquinone derivatives and diazonium fluoborates proved to be inhibitory. On the other hand, the diazonium sulfate of 4-phenylazo-1-naphthylamine was found to be a useful coupling component especially with the older leucyl and alanyl substrates. There is little question that the diazonium components of choice, for use in simultaneous coupling reactions, are of the aminoazobenzene type (e.g. Garnet GBC salt).

Burstone and Weisburger (1961b) also described the histochemical demonstration of aminopeptidase by a new oxidation technique. The incubating medium contained the DL-alanyl derivative of 3-amino-9-ethylcarbazole in the presence of potassium ferricyanide (Appendix 10B, Section 2). Upon enzymatic hydrolysis the substrate releases 3-amino-9-ethylcarbazole which is oxidized to form a complex azine type dye which is simultaneously chelated with iron. Alternatively an additional component, 5,6,7,8-tetrahydro-1-naphthylamine, may be added to the aforementioned substrate mixture (Fig. 4). The latter compound alone does not form dyes in the presence of ferricyanide but does form characteristic oxidation dyestuffs in the presence of certain other amines. The oxidation techniques for the demonstration of aminopeptidase are of interest because they represent an additional approach based upon a different principle of dye formation than previously employed for the demonstration of this enzyme.

Burstone and Folk (1956) originally noted that both the leucyl and alanyl substrates gave essentially the same histochemical localizations, although in neoplasms the stromal elements were more intensely stained with the alanyl compound (Burstone, 1956b). Based upon the similarity of staining as well as the effect of potential activators and inhibitors, there was no evidence to support the view that the leucyl substrate was capable of demonstrating the so-called "leucine" aminopeptidase. The lack of

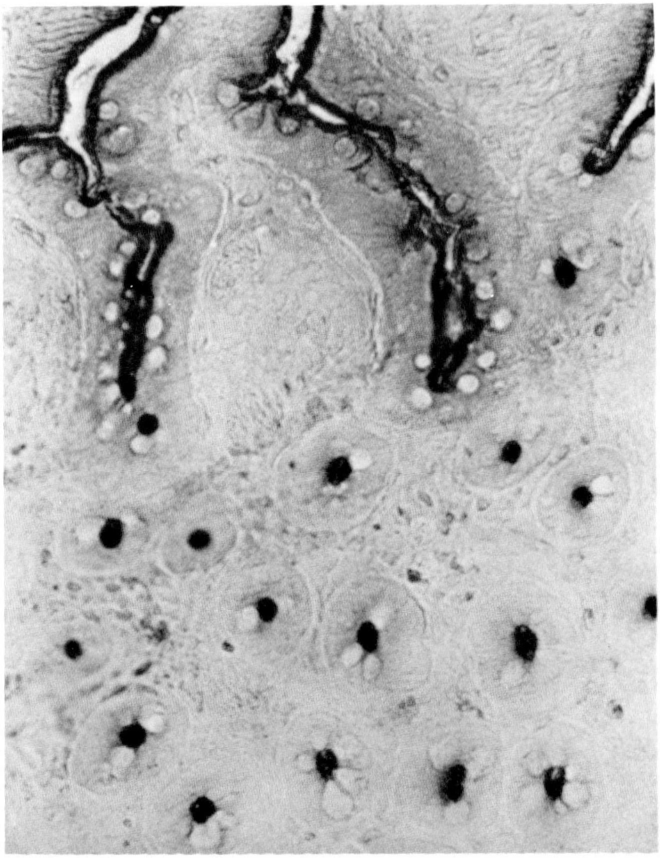

FIG. 3. Aminopeptidase activity of frozen-dried human intestine (ileum). Alanyl derivative of 3-aminocarbazole and Garnet GBC salt (× 285).

specificity of the reaction with reference to the nature of the amino acid grouping was further mentioned by Burstone (1958) who again indicated that no evidence had been presented to support the view that incubation of tissue sections with the leucyl substrate would reveal the presence of "leucine" aminopeptidase. Moreover, biochemical findings have shown that other peptidases are known to hydrolyze leucyl peptides and these have not been identified as "leucine" aminopeptidase (Fleisher, 1954). Patterson *et al.* (1961) have furnished additional biochemical data which clearly indicate that the enzyme which hydrolyzes L-leucyl-β-naphthylamide is not the classical aminopeptidase, and these authors state that *confusion might be avoided if a name other than "leucine" aminopeptidase were used to designate the enzyme hydrolyzing* L-*leucyl-β-naphthylamide.* Recent studies by Waldschmidt-Leitz and Keller (1957) have shown that the so-called "leucine aminopeptidase" is actually a mixture of peptidases, so it would indeed be remarkable if any histochemical substrate were to exhibit preference for such a heterogeneous system. Therefore if "leucine" aminopeptidase does not exist as a specific enzyme, and if no histochemical evidence can be provided to characterize it as a specific entity, the term should be abandoned in favor of the more general term "aminopeptidase" when referring to enzymes which hydrolyze various amino acid $\beta$-naphthylamides.

Despite the relatively recent development of aminopeptidase techniques, they have been applied in the study of this enzyme in a number of normal and pathological tissues including mast cells (Braun-Falco and Salfeld, 1959), skin in normal and pathological states (Braun-Falco, 1961), cir-

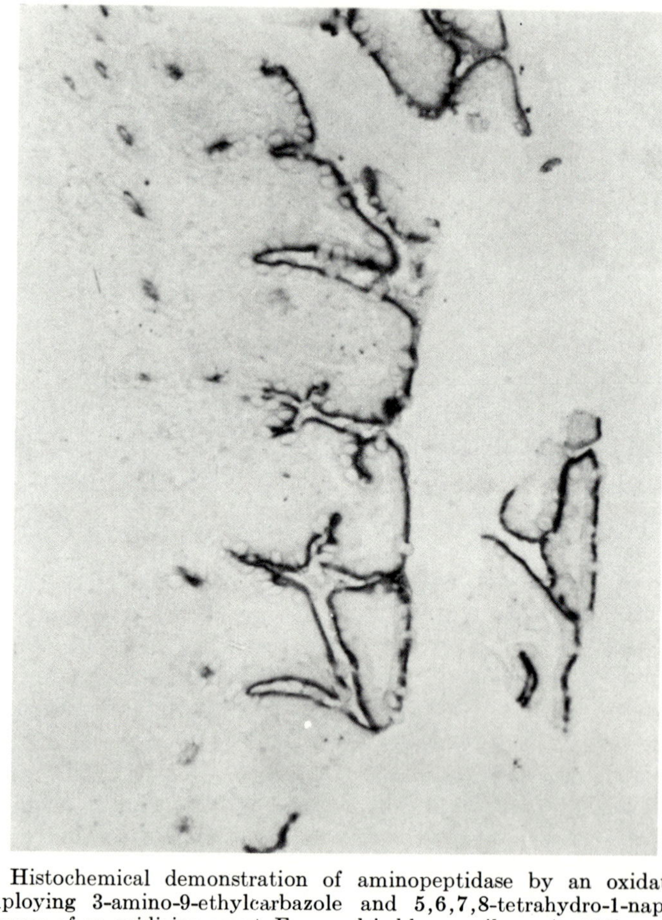

FIG. 4. Histochemical demonstration of aminopeptidase by an oxidation technique employing 3-amino-9-ethylcarbazole and 5,6,7,8-tetrahydro-1-napthylamine in the presence of an oxidizing agent. Frozen-dried human ileum (× 120).

rhotic livers (Rosenholtz, 1960; Rosenholtz and Wattenberg, 1961), pineal gland (Niemi and Ikonen, 1960), rheumatoid nodule (Fig. 5) (Burstone, 1956b), and parathyroid gland (Pearse and Tremblay, 1958). Raekallio (1961) has studied aminopeptidase activity in vital and postmortem skin wounds and notes that it is not surprising that fibroblasts respond to injury by producing high enzymatic activity in the peripheral zone of the wound. Jasmin and Gareau (1961) found aminopeptidase activity in histiocytes associated with damaged muscle fibers.

The histochemical demonstration of a chymotrypsin-like enzyme in mast cells has been reported by Benditt (1956, 1958) and Benditt and Arase (1959) who employed naphthol AS chloroacetate as a substrate.

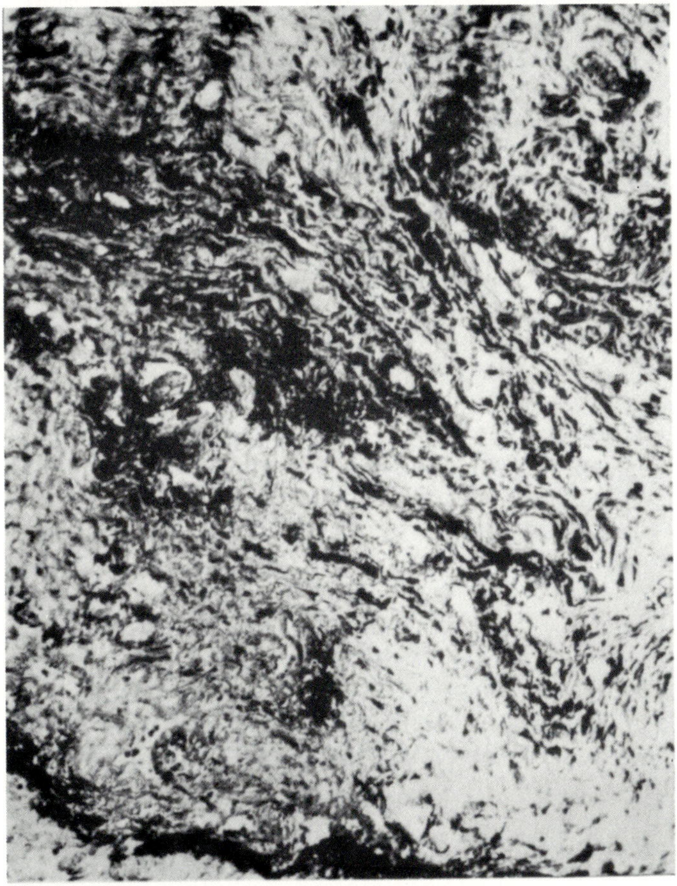

FIG. 5. Aminopeptidase activity of frozen-dried human rheumatic nodule showing high activity in foci of "fibrinoid" necrosis and also in areas of necrosis involving small vascular channels. DL-Alanyl-$\beta$-naphthylamide, Garnet GBC salt (× 143).

Similarly Glenner and Cohen (1960) reported the histochemical demonstration of a trypsin-like enzyme in mast cells by the use of *N*-α-benzoylarginine β-naphthyl ester. Glenner and Folk (1961) have described the histochemical demonstration of γ-glutamyl transpeptidase-like activity in the brush border of kidney tubules and also in the acinar cells of the pancreas. The substrate, *N*-(γ-L-glutamyl)-β-naphthylamide, was incorporated in a substrate solution containing glycylglycine, tris buffer pH 7.2, and Fast Garnet GBC salt. These authors conclude that the enzyme which releases β-naphthylamine from *N*-(γ-L-glutamyl)-β-naphthylamide is not the same as glutaminase I since the system was not inhibited by either *p*-chloromercuribenzoate or L-glutamine nor activated by phosphate ions. These authors also have attempted to characterize α-glutamyl peptidase activity by means of a simultaneous azo-dye technique utilizing *N*-(α-L-glutamyl)-β-naphthylamide and Garnet GBC salt at pH 6.2. Albert *et al.* (1961) have also reported the demonstration of the same transferase by means of the α-naphthylamide analogue.

Karnovsky and Himmelhoch (1961) have described an interesting histochemical method for the demonstration of glutaminase I in unfixed frozen sections. The tissue sections are incubated in a substrate medium containing glutamine, exogenous glutamic dehydrogenase, DPN, phosphate, nitro-BT, veronal buffer pH 7.75, and a number of activators. Among the activators were EDTA and nicotinamide. Application of the method to hamster kidney revealed a very high activity of glutaminase I in the papillary collecting ducts, where ammonia is known to appear in the urine. Strong staining reactions were also visible in other segments of the hamster nephron.

Recently, the histochemical demonstration of proteolytic activity has been extended by the development of a technique which involves the digestion of a gelatin-silver film (Adams and Tuqan, 1961; Adams and Bayliss, 1961). These workers mounted formalin fixed frozen sections on a blackened photographic plate, and found that protease activity was revealed as clear areas where the gelatin film had been digested and the silver granules washed out. Marked activity was observed around the zymogen granules of human, rat, and mouse pancreas and in the region of the Paneth cells. The epithelium of the convoluted tubules of the rat kidney cortex, human venous endothelium, and peripheral nerve myelin also exhibited considerable reactivity. The enzyme activity associated with myelin was found to be thermostable, SH dependent, and resistant to chloroform, methanol, and acetone.

## IV. ACTIVATORS AND INHIBITORS

Cursory mention will be made only of potential activators and inhibitors of aminopeptidase. Observations on the hydrolysis of the leucyl and alanyl substrates as influenced by various potential activators and inhibitors have been reported by Folk and Burstone (1955) and Burstone and Folk (1956). Although the so-called "leucine" aminopeptidase has been shown to be activated by $Mg^{++}$ and $Mn^{++}$ (Smith, 1951, p. 810; Smith and Hill, 1960, p. 45), this activation effect cannot be demonstrated in tissue sections under histochemical conditions. Presumably, the sections already contain the metallic ions necessary for activation, and addition of more ions appears to have no appreciable effect since it has already been shown that the optimal concentration of $Mn^{++}$ and $Mg^{++}$ is critical (Burstone and Folk, 1956).

A number of known aminopeptidase inhibitors completely or almost completely inhibit the reaction when added to the substrate mixture in a concentration of 0.001 *M*. Among these are $Ni^{++}$, $Zn^{++}$, $Cd^{++}$, $Cu^{++}$, $Hg^{++}$, and $Pb^{++}$. The reaction is also inhibted by 0.01 *M* cyanide (Fig. 2), although lower concentrations of this ion (0.0001–0.0003 *M*) behave as an activator. Iodoacetate (0.001 *M*), which is a known prolidase and cathepsin inhibitor, has no effect upon the histochemical reaction when introduced into the substrate solution. Known esterase inhibitors such as sodium fluoride (0.002 *M*) and diisopropyl fluorophosphate (DFP) ($5 \times 10^{-4}$ *M*) are likewise without effect. Because of the sensitivity of the aminopeptidase reaction to metallic ions such as zinc, it is especially important to select the proper diazonium salt for use in the azo-coupling reaction. This limits the types of diazonium salts which may be employed since the large majority of commercially available diazotates are stabilized as double zinc salts. Fortunately, Garnet GBC salt is readily available as a stabilized diazonium sulfate.

## V. AMINOPEPTIDASE IN NEOPLASMS

As soon as the successful application of L-leucyl-β-naphthylamide was reported (Burstone and Folk, 1956), this substrate was applied together with DL-alanyl-β-naphthylamide in a comparative study on the localization of aminopeptidase activity in human neoplasms (Burstone, 1956b). The study included fresh surgical specimens from normal, hyperplastic, inflammatory, and neoplastic tissues which were frozen-dried, infiltrated, and embedded in paraffin. None of the tumors had been irradiated. It was found that carefully processed frozen-dried tissues gave uniform results and good histological localizations, as contrasted with formalin fixed frozen sections which exhibit a marked inhibition of peptidase activity. Furthermore, as previously mentioned, the localizations of the azo-dye are more diffuse in frozen sections, presumably because of the lipid-soluble nature of the dye. One of the findings of the aforementioned study was the intense reactivity of connective tissue stroma, including ground substance, adjacent to invasive epidermoid carcinomas. This did not indicate that fibroblasts in other sites were not reactive, since it was clearly stated that stained fibroblasts were seen in the lamina propria of normal skin and gingiva. Polymorphonuclear leukocytes and macrophages associated with inflammatory infiltrates were also intensely stained. With reference to epidermoid carcinomas, a strong aminopeptidase reaction was also found in the basement membranes and subjacent borders of groups of tumor cells (Fig. 6).

In 1955 Sylvén and Malmgren reported on the distribution of aminopeptidase in tumors using Linderstrøm-Lang micro-techniques, and reported a high peptidase activity in peripheral portions of tumors associated with invasion. Subsequently Sylvén and Malmgren (1957), in a quantitative study of hundreds of experimental tumors, evaluated the relationship

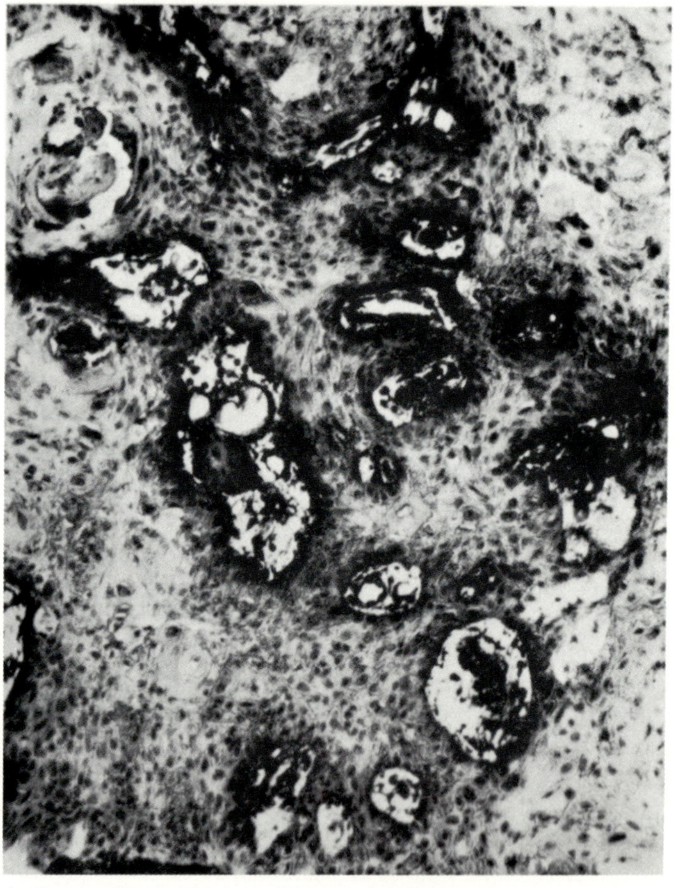

FIG. 6. Aminopeptidase activity of human epidermoid carcinoma. Note activity of periphery of tumor as well as that of degenerating stroma. DL-Alanyl-$\beta$-naphthylamide, Garnet GBC salt (× 143).

of tumor invasiveness and the activity of various proteolytic enzymes. They state, "the present basis of the problem of the destructiveness [of tumors] appears to be the microscopical observations suggesting some kind of extracellular proteolysis *in vivo* of fibrous proteins adjacent to malignant tumors."

With reference to the histochemical findings, it is of interest that the zones of high peptidase activity are closely adjacent to the tumor and not necessarily associated with an inflammatory response. Braun-Falco (1957a) confirmed the aforementioned histochemical findings of Burstone (1956b) and suggested that aminopeptidase activity in different epithelial tumors might be indicative of the grade of invasiveness of these neoplasms. In other studies he found high peptidase activity in certain dermatological lesions characterized by breakdown of stroma and ground substance (e.g. bullous lesions) (Braun-Falco, 1956, 1957b). In a very recent review on aminopeptidase in dermatological diseases including neoplasms, Braun-Falco (1961) specifically relates aminopeptidase to the phenomenon of vesicle formation in pemphigus. In acute dermatitis, Braun-Falco (1961) finds very high activity in epidermal zones associated with spongiosis and incipient vesicle formation as contrasted with the relatively low peptidase activity of normal skin. This author concludes that one obtains the distinct impression that aminopeptidase is liberated in the epidermis and associated tissues only following injury.

Monis *et al.* (1959) studied aminopeptidase in neoplastic and inflammatory tissues with a substrate which had not been used previously, L-leucyl-4-methoxy-β-naphthylamide. These authors reported aminopeptidase activity in connective tissue cells (which had already been described in the literature) and categorically state that aminopeptidase is not involved in the invasion by neoplasms. Several inherent limitations in this paper should be pointed out. First, although their substrate was chemically different from those previously used, comparative studies were not made employing the older substrates. Second, frozen sections were employed, and these as evidenced by the photomicrographs published as well as the general experience of most workers, are not entirely satisfactory for definitive studies of aminopeptidase. It should also be pointed out that the physiological role of aminopeptidase is unknown, so that any comments as to the specific or non-specific role of this enzyme in neoplasms are purely speculative. More recently, Hess (1960) and Mottet (1961) studied aminopeptidase activity in specialized types of animal tumors employing L-leucyl-β-naphthylamide and indicated that the peptidase activity represented a reaction to the invading neoplasm and was not part of the mechanism by which malignant neoplasms invade host tissues. Hess (1960) specifically noted that the highest hydrolytic activity in transplanted rodent tumors is associated with macrophages, mast cells, and fibroblasts. As in the previous study, no experimental evidence was presented to indicate exactly what functional role was played by aminopeptidase associated with tumors studied.

From a historical standpoint, considerable interest has been expressed by some of the German workers in the possible role of proteolytic enzymes in tumor invasion, and pertinent references have been enumerated in the extensive review of Sylvén and Malmgren (1957). In their discussion of the problem, these authors state, "microscopical changes usually appear to begin at a distance of one or a few millimeters from the tumor periphery and become more marked close to the tumor border. The changes include interstitial edema, swelling and dispersal of pre-formed collagen and muscle fiber bundles, and various cytological changes of fibroblasts, ranging from regeneration to cytolysis." These morphological observations would certainly suggest the presence of some type of proteolytic mechanism intimately connected with the invasive process. In a general analysis of the problem of tumor invasion, Berenblum (1958, pp. 454 and 456) states, "even as a mere contributory factor rapidity of growth only affords the scope for invasion by providing new tissues. The mechanism left [for tumor invasion] involves other factors." "The only possible mechanism left [for the invasive ability of malignant tumors] is that the malignant tumor cell elaborates some extracellular lytic products, capable of dissolving or partially destroying the tissue being invaded."

The possible role of enzymatic factors in tumor spread and invasion is suggested by the recent report of Dao and Moore (1961) who found that pretreatment of mice with cortisone results in a preferential growth of metastases in the liver. Reference is also made to a similar phenomenon in cancer patients treated with corticoids. These authors also note that metastases or implants (or both) have been reported to be more frequent in animals following x-ray irradiation, chemotherapeutic agents, formalin, hypochlorite, and other chemicals at sites of injured tissue. Some clinical observations parallel the aforementioned experimental studies. For example, certain patients with cancer of the breast, who had been treated with x-ray, subsequently have developed hundreds of metastases limited to the areas of the superficial skin that had been exposed to the radiation. The aforementioned physical and chemical agents may influence the production of cathepsins and other proteolytic enzymes which, in turn, may affect the ability of the ground substance to retard the invasion by tumor cells.

With reference to the role of aminopeptidase, in view of the fact that the biological function of this enzyme has not been delineated, statements to the effect that the enzyme is completely unrelated to tumor invasion have not been completely established. This may also be said of attempts to associate the enzyme with tumor invasion, although in this instance there is an appreciable amount of evidence which clearly correlates aminopeptidase activity with tissue breakdown. In fact, the very presence of the proteolytic enzymes in leukocytes and macrophages may play an important role in the local spread of tumor.

Aminopeptidase activity has not only been delineated to tumor stroma and peripheral portions of neoplasms. Glenner *et al.* (1959) in a study of over a hundred and twenty surgically removed tumor specimens noted activity in tumor cells from adenocarcinoma of the colon (Fig. 7) and stomach, melanoma, and meningioma (Fig. 8).

Willighagen and Planteydt (1959) studied aminopeptidase activity in lymph node metastases of gastric cell carcinomas with the method of Burstone and Folk (1956). They noted high activity of the metastatic tumor and indicated that it is worthwhile to apply the histochemical procedure in the detection of tumor cells in regional lymph nodes and surgical margins. Wattenberg (1959) similarly observed intense aminopeptidase activity in gastric cell carcinomas metastatic to lymph nodes, and notes that high aminopeptidase activity is found in tumor cells of approximately one-third of gastric carcinomas, but no activity is observed in the epithelium of normal-appearing gastric mucosa from any portion of the stomach.

Rosenholtz and Wattenberg (1961) observed that of the malignant tumors studied, carcinomas of the gall bladder and biliary ducts exhibit the highest percentage of peptidase active cells, while carcinomas of the colon are uniformly negative.

Holmberg (1961) has studied the *in vitro* release of cytoplasmic enzymes from ascites tumor cells as compared with strain L cells. It was found that catheptic activities in the incubation medium increased 2–3 times during 2 hour incubation of ascites tumor cells, while in the case of strain L cells, no measurable change in enzyme activity in the incubation medium was observed. The preferential release of cathepsins from the tumor cells was found to occur under both aerobic and anaerobic conditions, and Holmberg (1961) suggests that the release of enzymes (specifically proteinases) may form the physiological basis of the destructive activity of tumor cells at the invasion frontier.

## APPENDIX 10A

# SYNTHESIS OF PEPTIDES

### *1. Glycyl and Alanyl Acid Chlorides (Levine, 1954)*

This technique has been reported to give high yields of glycyl and alanyl chloride hydrochlorides. Although these compounds had been synthesized in the past by treatment of amino acid solutions with acetyl chloride, the low yields as well as the tendency for side reactions to occur limited the general application of this procedure.

*a. Glycyl Chloride Hydrochloride.* The reaction is conducted in a glass-stoppered vessel. Five grams of glycine are suspended in 200.0 ml. of purified carbon tetrachloride. Fifteen grams of phosphorus pentachloride are added and the tightly stoppered (wired) vessel is shaken vigorously for 10 hours at room temperature. The product is filtered on a long-necked sintered glass filter. During filtration, the top of the sintered glass filter is protected with a stopper containing a Drierite filled tube. The compound is washed three times with carbon tetrachloride, three times with anhydrous petroleum ether, and finally with anhydrous ether, and dried on the filter. The product is then recrystallized from redistilled acetyl chloride (yield, 75–80%) ($C_2H_5ONCl_2$).

*b. Alanyl Chloride Hydrochloride.* This compound is prepared by the same method given for the glycyl derivative (yield, 92%) ($C_3H_7NOCl_2$).

### *2. Chloroacetyl-β-naphthylamide (Johnson, 1903; Gomori, 1954a)*

This compound is prepared by treating β-naphthylamine (ca. 5.0 g.) dissolved in an acetone-pyridine mixture with a slight excess of chloroacetyl chloride. The resulting compound may be recrystallized from 60% methanol-water or ethanol-water. Johnson (1903) reported an m.p. of 117–118° which corresponds closely with the m.p. reported by Gomori (1954a) ($C_{12}H_{10}ONCl$). Anal., calcd.: N, 6.38%. Found: N, 6.42%.

### *3.* DL-*Alanyl-β-naphthylamide (Gomori, 1954b; Burstone and Folk, 1956)*

Five grams of β-naphthylamine (carcinogen) are dissolved in 150.0 ml. of acetone and 15.0 ml. of α-bromopropionyl bromide are added. The mixture is placed in a container of ice water and 25.0 ml. of pyridine added slowly drop by drop until the reaction subsides. The resulting precipitate is then poured into cold water, collected, and recrystallized from methanol-water. The α-bromopropionyl-β-naphthylamide is dissolved in methanol, an equal volume of concentrated $NH_4OH$ added, and the mixture kept in a pressure bottle for 24 hours at 60°. The mixture is then evaporated to a volume of approximately 50.0 ml., and 100.0 ml. of ethanol and 1.0 ml. of *N* acetic acid added. The mixture is brought to a boil, 2.0 g. of charcoal added, and then filtered. Sufficient *N* NaOH is added to the filtrate to make it slightly alkaline to phenolphthalein and the solution is then refrigerated. White platelets of the alanyl compound precipitate, and are collected and recrystallized from 50% ethanol.

### *4.* DL-*Alanyl Derivatives of Complex Arylamines (Burstone and Weisburger, 1961a)*

Alanyl derivatives of 3-aminocarbazole, 3-amino-9-ethylcarbazole, and 3-aminodibenzofuran may be synthesized according to the following technique.

Twenty millimoles (ca. 3.0–4.0 g.) of the aromatic amine is dissolved in a mixture of 15.0–25.0 ml. of pyridine and 10.0–25.0 ml. of dry acetone. The solution is stirred in an ice-water bath (10–15°), and 5.0 ml of freshly distilled α-bromopropionyl bromide added dropwise over a period of 30 minutes. After stirring for an additional 60 minutes, the mixture is poured into ice water; the precipitate is filtered off, washed, and in most cases recrystallizes from ethanol. The bromoamide is then dissolved in methanol or *N*,*N*-dimethylformamide (DMF), an equal volume of concentrated $NH_4OH$ added, and the mixture kept in a pressure bottle at 60° overnight. The solution is then evaporated under a stream of nitrogen, the precipitate filtered off, washed, dissolved as quickly as possible in warm dilute HCl, filtered, and reprecipitated with ammonium hydroxide. Substrates are usually crystallized from ethanol:water (1:1 v/v) or benzene-petroleum ether. The characteristics of complex alanyl derivatives prepared according to this procedure are given in Chart 1 (see Table 1).

### *5. Peptide Synthesis, Mixed Anhydride Method (Vaughan and Osato, 1951)*

The mixed anhydride technique has been used in the preparation of histochemical aminopeptidase substrates (Appendix 10A, Section 7).

A mixed anhydride is formed between carbobenzoxyglycine and isovaleryl chloride in the presence of triethylamine under anhydrous conditions. The anhydrides are then treated with aniline to form the carbobenzoxyglycinanilide. Synthesis of the latter compound is as follows: a solution of 5.23 g. (0.025 *M*) of carbobenzoxyglycine and 2.55 g. (0.025 *M*) of triethylamine in 50.0 ml. of toluene is cooled to 0°, and 3.01 g. (0.025 *M*) of isovaleryl chloride is added. After 2 hours at this temperature, during which time triethylamine hydrochloride separates, 2.34 g. (0.025 *M*) of aniline are added. The reaction mixture is then stored at 8° overnight. Carbobenzoxyglycinanilide crystallizes from the reaction mixture and is filtered off together with triethylamine hydrochloride, washed with water, dilute sodium hydroxide solution, and dilute hydrochloric acid, and dried (yield, 83%).

*6. Peptide Synthesis, Carbodiimide Method (Sheehan and Hess, 1955; Sheehan and Hlavka, 1956; Sheehan et al., 1956)*

This relatively mild reaction for the formation of peptide bonds is insensitive to moisture, and may be carried out in an aqueous medium without racemization. The compounds to be coupled, one containing a free amino group and the other a free carboxyl group, couple directly and rapidly upon treatment with *N,N'*-dicyclohexylcarbodiimide at room temperature. The co-product, *N,N'*-dicyclohexylurea, is poorly soluble in most organic or aqueous solvents and is easily separated by filtration.

The simplicity and efficiency of the technique is illustrated by the synthesis of a tripeptide derivative. A tetrahydrofuran solution of carbobenzoxyglycyl-L-phenylalanine containing a slight excess of crystalline *N,N'*-dicyclohexylcarbodiimide and ethyl glycinate is allowed to stand at room temperature for 4 hours and is then treated with a small amount of acetic acid to decompose the excess reagent. The insoluble urea is removed, the solvent replaced by ethyl acetate, and the solution washed with dilute acid and aqueous potassium bicarbonate. Upon addition of petroleum ether, crystalline carbobenzoxyglycyl-L-phenylalanylglycine ethyl ester is precipitated (yield, 87%). In a similar fashion, a variety of peptide derivatives may be synthesized.

Sheehan and Hlavka (1956) have reported the synthesis of several aliphatic carbodiimides containing tertiary or quaternary amine substituents, and have described their application in peptide bond formation.

*7. L-Leucyl-β-naphthylamide (Green et al., 1955; Folk and Burstone, 1955; Tuppy and Nesvadba, 1957)*

This compound is readily prepared by the mixed anhydride synthesis. Carbobenzoxy-L-leucine (oil) (2.75 g.) is mixed with 1.4 ml. of triethylamine in 4.0 ml. of tetrahydrofuran, and reacted with 1.12 g. of ethyl chlorocarbonate at −5°. After standing for 10 minutes, 1.55 g. of β-naphthylamine dissolved in 3.0 ml. of tetrahydrofuran are added to the aforementioned mixture which then becomes gelatinous. The semi-solid reaction mixture is allowed to stand at room temperature overnight, and then dissolved in 50.0 ml. of ethyl acetate and washed with water, 0.1 *M* HCl, water, and sodium bicarbonate solution. After drying, the carbobenzoxy-L-leucyl-β-naphthylamide is obtained as the hemihydrate by evaporation of the ethyl acetate (m.p. 127–129°) (this form dehydrates after drying *in vacuo* at 100° in the presence of phosphorus pentoxide) ($C_{24}H_{26}O_3N_2$).

The carbobenzoxy group may be removed by catalytic hydrogenation or by treatment of the compound with HBr-glacial acetic acid (38%), the latter procedure offering a certain convenience for laboratories not equipped for hydrogenations. The carbobenzoxy compound (1.23 g.) is mixed with 5.0 ml. of HBr-glacial acetic acid. After standing at room temperature for 1 hour, the hydrobromide of leucyl-β-naphthylamide is precipitated in an oily form by the addition of 100.0 ml. of absolute ether, and is dissolved in the water after the ether is decanted. Dilute acetic acid is added and the warm solution is treated with decolorizing carbon and filtered. Dilute aqueous ammonia is added to the filtrate, which results in the precipitation of the naphthylamide. The latter compound is then washed with water and dried over $P_2O_5$ (m.p. 123–125°) ($C_{16}H_{20}N_2O$). This base is then dissolved in methanol-ether (1:1) and saturated with dry HCl gas to give L-leucyl-β-naphthylamide HCl which crystallizes as slender needles (m.p. 239–241°) ($C_{16}H_{20}N_2O \cdot HCl$).

*8. Prolyl- and Hydroxyprolylnaphthylamides (Folk and Burstone, 1956)*

*a.* L-*Prolyl-β-naphthylamide hydrochloride.* A solution of 6.25 g. (0.025 *M*) of carbobenzoxy-L-proline (oil) and 2.55 g. (0.025 *M*) of triethylamine in a mixture of 25.0 ml. of toluene and 25.0 ml. of chloroform is cooled to −5°, and 3.42 g. (0.025 *M*) of isobutyl chlorocarbonate are added with stirring. After 15 minutes at this temperature a warm (40°) solution containing 3.58 g. (0.025 *M*) of β-naphthylamine in 25.0 ml. of chloroform is added with stirring. Evolution of carbon dioxide is noted. The reaction mixture is kept at 5° overnight, washed with water, dilute sodium bicarbonate, dilute hydrochloric acid, and water. It is then dried with sodium sulfate and evaporated to dryness *in vacuo*. The resulting oil crystallizes under pentane. Recrystallization from ethyl acetate-pentane gives needles of carbobenzoxy-L-prolyl-β-naphthylamide (yield, 77%) (m.p. 131°) ($C_{22}H_{22}O_3N_2$).

A solution containing 3.74 g. (0.01 *M*) of the aforementioned carbobenzoxy compound in 150.0 ml. of methanol containing one equivalent of 5 *N* HCl and 500.0 mg. of palladium black is hydrogenated for 3 hours. The liquid is separated from the catalyst by hot filtration and taken to dryness *in vacuo*. The crystalline product is dissolved in hot methanol and cooled, and then treated with ethanol to induce crystallization. The recrystallized L-prolyl-β-naphthylamide hydrochloride forms shining rectangular crystals (yield, 68%) (m.p. 272°) ($C_{15}H_{16}ON_2 \cdot HCl$). Anal., calcd.: C, 65.1%; H, 6.2%; N, 10.1%. Found: C, 64.8%; H, 6.5%; N, 10.1%.

*b. Hydroxy-L-prolyl-β-naphthylamide hydrochloride.* A solution containing 3.31 g. (0.0125 *M*) of carbobenzoxyhydroxy-L-proline (oil), 3.1 g. (0.015 *M*) of *N*,*N'*-dicyclohexylcarbodiimide, and 1.79 g. (0.0125 *M*) of β-naphthylamine in 25.0 ml. of tetrahydrofuran is allowed to stand at room temperature for 4 hours. A small amount of acetic acid is added to destroy the excess reagent, and the mixture is filtered. The solvent is removed and replaced by ethyl acetate, and the solution is washed with dilute sodium bicarbonate, dilute hydrochloric acid, and water. It is then dried with sodium sulfate and the compound precipitated with pentane. The residue is washed well with hot ether and recrystallized from ethyl acetate-pentane to give carbobenzoxyhydroxy-L-prolyl-β-naphthylamide (yield, 43%) (m.p. 173–174°) ($C_{23}H_{22}O_4N_2$).

A solution containing 1.95 g. (0.005 *M*) of the carbobenzoxy derivative in 250.0 ml. of methanol containing one equivalent of 5 *N* HCl and 250.0 mg. of palladium black is hydrogenated. After 3 hours the catalyst is removed by filtration, and the methanol removed by evaporation *in vacuo*. The resulting crystalline residue is recrystallized from ethanol-ether to give hydroxy-L-prolyl-β-naphthylamide hydrochloride (yield, 65%) (m.p. 228–230°) ($C_{15}H_{16}O_2N_2 \cdot HCl$). Anal., calcd.: C, 61.5%; H, 5.9%. Found: C, 61.1%; H, 6.1%.

*9. L-Cystine Di-β-naphthylamide (Tuppy and Nesvadba, 1957)*

Two synthetic approaches may be employed, the mixed anhydride (A) and (B), and the carbodiimide methods (C).

*Method A.* Ethyl chlorocarbonate (0.63 g.) is added drop by drop, and with shaking, to a cooled (−5°) solution containing 1.6 g. of dicarbobenzoxy-L-cystine dissolved in 3.5 ml. of tetrahydrofuran and 0.9 ml. of triethylamine. The resulting suspension is allowed to stand for 10 minutes at −5° and then 1.05 g. of β-naphthylamine dissolved in 3.0 ml. of tetrahydrofuran are added in small portions. After standing overnight at room temperature the gelatinous mass is suspended in 20.0 ml. of ethyl acetate, the suspension washed with water, 0.1% HCl, again with water, and finally with 1% sodium bicarbonate. The crude product is purified by dissolving it in hot pyridine and then adding methanol drop by drop. The gelatinous dicarbobenzoxycystine di-β-naphthylamide which precipitates is thoroughly washed with ether (m.p. 225–227°) ($C_{42}H_{38}N_6O_2S \cdot H_2O$).

*Method B.* To a solution containing 2.0 g. of dicarbobenzoxy-L-cystine in 5.0 ml. of tetrahydrofuran and 0.8 g. of triethylamine (−5°) is slowly added, with stirring, 0.95 g. of isovaleryl chloride. After standing for 3 hours at 0°, 1.13 g. of β-naphthylamine in 3.0 ml. of tetrahydrofuran are added. The reaction mixture is allowed to stand overnight at room temperature and then treated with ethyl acetate as previously described (m.p. 248–250°).

*Method C.* Dicarbobenzoxy-L-cystine (0.8 g.) and β-naphthylamine (0.53 g.) are dissolved in 3.5 ml. of tetrahydrofuran and treated, drop by drop, with a solution of 0.55 g. dicyclohexylcarbodiimide dissolved in 1.0 ml. of tetrahydrofuran. Heat is developed in the reaction during which dicyclohexylurea and dicarbobenzoxy-L-cystine di-β-naphthylamide are precipitated. The reaction mixture is allowed to stand overnight and after addition of ethyl acetate, the reaction products which crystallize out are filtered with suction. The products are separated by boiling the mixture several times in ethanol. The urea goes into solution, while the naphthylamide is insoluble and is collected, washed with ether, and dried (m.p. 258–260°, dec.) ($C_{42}H_{38}N_4O_6S_2$).

In order to remove the carbobenzoxy radicals from the compounds prepared by methods A, B, or C, 2.33 g. of finely pulverized carbobenzoxy compound are suspended in 15.0 ml. of HBr-glacial acetic acid. The evolution of gas, which begins immediately, ceases after approximately 90 minutes and a clear solution is obtained. Addition of 100.0 ml. of ether results in precipitation of the L-cystine di-β-naphthylamide HBr in crystalline form. The compound is washed with ether and dissolved in 30.0 ml. of 50% aqueous ethanol. When dilute ammonia is added to bring the pH to approximately 4.0, part of the free naphthylamide precipitates; but again goes into solution when heated except for a resinous impurity from which the solution is decanted. Continued addition of ammonia (to pH 8.0) precipitates the dinaphthylamide completely. For further purification the compound is dissolved in dilute acetic acid, treated with charcoal, and precipitated again with ammonia, after which the charcoal is filtered off. When filtered with suction, washed with dilute methanol, and dried at 100° *in vacuo* over $P_2O_5$, the compound forms a white powder (yield, 1.18 g.) (m.p. 167–169°) ($C_{26}H_{26}N_4O_2S_2$). Anal., calcd.: C, 63.64%; H, 5.34%; S, 13.07%. Found: C, 63.05%; H, 5.35%; S, 13.00%.

### *10.* L-*Amino Acid Derivatives of 2-Aminofluorene* (*Greenstein et al., 1955*)

Greenstein *et al.* (1955) synthesized, by the mixed anhydride procedure, a number of amino acid derivatives of the carcinogenic 2-aminofluorene. Among the amino acid compounds described were the glycyl, L-alanyl, norvalyl, norleucyl, valyl, leucyl, isoleucyl, and methionyl derivatives. The simplicity of the procedure recommended it for possible application in the synthesis of histochemical substrates.

One-tenth mole of the appropriate carbobenzoxy acid is dissolved in 200.0 ml. of chloroform containing 14.0 ml. of triethylamine. The solution is chilled to 0–5° in an ice bath and subsequently treated with 0.1 *M* solution of isovaleryl chloride or isobutyl chlorocarbonate. After standing for 45 minutes at 4°, a solution of 0.1 *M* of the aromatic amine in 200.0 ml. of chloroform is added to the resultant mixed anhydride. The mixture is allowed to stand at room temperature overnight and 5–10 volumes of petroleum ether are then added to precipitate the reaction product. The precipitate is filtered with suction, washed copiously with water, and dissolved by boiling in 1–3 liters of ethanol. The solution is filtered hot and the filtrate is then refrigerated at −10° for several hours resulting in the formation of a precipitate. Final recrystallization is accomplished by precipitation of the compound from a 50:50 mixture of DMF-chloroform by addition of petroleum ether.

The carbobenzoxy group is removed by the method of Ben-Ishai and Berger (1952). The carbobenzoxy aminofluorene derivative is dissolved with heating in a 20-fold amount of glacial acetic acid, and the solution is then cooled to 25°. In most cases the compounds crystallize. Dry HBr gas is then introduced into the mixture which is maintained at a temperature not exceeding 40°. In general, most compounds go into solution during the HBr gassing procedure which is carried on for a period of 30–60 minutes. The gas mixtures are allowed to stand for 60 minutes at 25° and then are evaporated to dryness *in vacuo* or blown down to dryness with a stream of dry air. The residues which are in the form of either crystals or oils are washed with ether to remove adherent benzyl bromide. If difficulty is encountered in recrystallization of the HBr salts, the compounds are dissolved in a minimun of methanol and the solution treated with an excess of ammonium hydroxide to convert the compound to the free base. In this form the amino acid aminofluorene derivative may be readily recrystallized.

### *11.* L-*Leucyl-4-methoxy-β-naphthylamide* (*Rosenblatt et al., 1958*)

This substrate for aminopeptidase is prepared by the standard carbodiimide procedure (Appendix 10A, Section 8). A solution containing 2.51 g. of *N*,*N'*-dicyclohexylcarbodiimide in 15.0 ml. of dichloromethane is added to 3.22 g. of carbobenzoxy-L-leucine and 2.10 g. of 4-methoxy-2-naphthylamine (Chapter 2, Table 7) dissolved in 10.0 ml. of the same solvent. An immediate reaction occurs. After 1 hour the mixture is filtered and the precipitate of dicyclohexylurea is washed with ether. The combined filtrate and washings are boiled down to give a pinkish solid residue. The product is triturated with ether to remove the unreacted starting material. The product is dissolved in 35.0 ml. of ethyl acetate, washed with *N* HCl, *N* potassium bicarbonate, and water, filtered, and evaporated nearly to dryness. A precipitate of carbobenzoxyl-L-leucyl-4-methoxy-2-naphthylamide ($C_{25}H_{28}N_2O_4$) is formed by the addition of petroleum ether. The carbobenzoxy group is removed by hydrogenolysis of the aforementioned compound (1.52 g. in 150.0 ml. of methanol) for 3 hours at 2 atmospheres of hydrogen in the presence of 10% palladium-charcoal catalyst (0.1 g.). The mixture is filtered from the catalyst and boiled down to form a mobile oily residue which upon extraction with boiling petroleum ether gives (upon chilling) an oily precipitate which changes to a highly crystalline solid (m.p. 96.5–97.5°) ($C_{17}H_{22}N_2O_2$). Anal., calcd.: C, 71.3%; H, 7.75%; N, 9.8%. Found: C, 71.3%; H, 7.62%; N, 9.9%.

## APPENDIX 10B

# HISTOCHEMICAL METHODS

1. *Aminopeptidase, Azo-Dye Method* (*Gomori, 1954b, Burstone and Folk, 1956*)

Frozen-dried paraffin embedded tissues give the best results with this procedure, although unfixed frozen, acetone fixed frozen, and formalin fixed frozen sections have been successfully employed.

A 1% aqueous stock solution of L-leucyl-$\beta$-naphthylamide is prepared. One ml. of the stock solution is added to 40.0 ml. distilled water and 10.0 ml. of 0.2 *M* tris buffer pH 7.1 (the buffer must be adjusted as carefully as possible). Thirty mg. of diazotized *o*-aminoazotoluene (Garnet GBC salt) are added, the mixture shaken and filtered into a Coplin jar. Alternatively, ca. 5.0 mg. of the leucyl or alanyl substrates may be directly dissolved in distilled water and buffer instead of employing the stock solution. If the base is used instead of the hydrochloride, it is necessary to initially dissolve the substrate in a little, ca. 0.25 ml., methanol or DMF.

Slides are incubated for minutes to several hours until the desired staining reaction is obtained. They are then briefly washed in running tap water, transferred to distilled water, and mounted in glycerol-gelatin. In frozen-dried paraffin embedded specimens the stability of the azo-dye is good, but in frozen sections, the dye exhibits a marked tendency to crystallize rapidly.

2. *Aminopeptidase, Oxidation Method* (*Burstone and Weisburger, 1961b*)

Thus far, only frozen-dried paraffin embedded tissues have been successfully utilized with the oxidation procedure.

The incubating solution is prepared by dissolving 5.0 mg. of substrate (3-amino-9-ethylcarbazole) in 0.5 ml. of ethanol followed by 25.0 ml. of distilled water and 2.0 ml. of 0.05 *M* potassium ferricyanide. Alternatively the substrate may contain 5,6,7,8-tetrahydro-1-naphthylamine which is not oxidized by ferricyanide but which forms characteristic oxidation dyes in the presence of the aminocarbazole compound. In addition to ferricyanide, other oxidizing agents including periodate and persulfate may be employed.

Following incubation, the slides are briefly washed in water and mounted in glycerol-gelatin or a water-soluble synthetic mounting medium.

3. *Aminopeptidase, Azo-Dye Method* (*Burstone and Weisburger, 1961a*)

Approximately 5.0 mg. of the DL-alanyl derivative of 3-aminocarbazole or 3-amino-9-ethylcarbazole are dissolved in 0.5 ml. of acetone, and 35 ml.

of distilled water and 15.0 ml. of 0.2 *M* tris buffer pH 7.1 are added. Approximately 20.0 mg. of Fast Garnet GBC salt are then added and the mixture shaken and filtered into a Coplin jar. Incubations are carried out at room temperature. Alternatively the diazonium sulfate of 4-phenylazo-1-naphthylamine may be substituted for Garnet GBC salt.

4. *Glutaminase I, Tetrazolium Method (Karnovsky and Himmelhoch, 1961)*

Unfixed frozen sections mounted on glass slides may be employed in conjunction with this technique.

The substrate solution contains the following:

| | |
|---|---|
| Glutamine | 0.35 *M* |
| DPN | 0.01 *M* |
| Glutamic dehydrogenase | 45.0 units/ml. |
| Nicotinamide | 0.03 *M* |
| EDTA | 0.001 *M* |
| Sodium phosphate, pH 7.75 | 0.1 *M* |
| Nitro-BT | 0.5 mg./ml. |
| Veronal buffer, pH 7.75 | 0.07 *M* |

The molarity of the aforementioned constituents indicates their final concentration in the substrate solution.

Sections may be incubated for periods of 10-60 minutes, rinsed in distilled water, and mounted in a water-soluble medium or else dehydrated and mounted in an organic solvent-miscible medium.

## Commercial Sources of Reagents

1. Aldrich Chemical Company, Milwaukee, Wisconsin
2. Dajac Laboratories, The Borden Chemical Company, Philadelphia, Pennsylvania
3. Mann Research Laboratories, New York, New York
4. Nutritional Biochemicals Corp., Cleveland, Ohio
5. Sigma Chemical Company, St. Louis, Missouri

## References

Ablondi, F. B., and Hagan, J. J. (1960) *In* "The Enzymes" (P. D. Boyer, H. Lardy, and K. Myrbäch, eds.), Vol. 4. pp. 175–192. Academic Press, New York.

Adams, C. W., and Bayliss, O. B. (1961) *J. Histochem. and Cytochem.* **9,** 473–476.

Adams, C. W., and Tuqan, N. A. (1961) *J. Histochem. and Cytochem.* **9,** 469–472.

Albert, Z., Orlowski, M., and Szewczuls, A. (1961) *Nature* **191,** 767–798.

Benditt, E. P. (1956) *Federation Proc.* **15,** 507.

Benditt, E. P. (1958) *Ann. N. Y. Acad. Sci.* **73,** 204–211.

Benditt, E. P., and Arase, M. (1959) *J. Exptl. Med.* **110,** 451–460.

Ben-Ishai, D., and Berger, A. (1952) *J. Org. Chem.* **17,** 1564–1570.

Berenblum, I. (1958) *In* "General Pathology" (H. Florey, ed.), 2nd ed., pp. 442-459. Lloyd-Lake, London.

BERGMANN, M. (1942) *Advances in Enzymol.* **2,** 49–68.

BERGMANN, M., AND FRUTON, J. S. (1941) *Advances in Enzymol.* **1,** 63–98.

BERGMANN, M., AND ZERVAS, L. (1932) *Ber.* **65,** 1192–1201.

BIRNBAUM, S. M., LEVINTOW, L., KINGSLEY, R. B., AND GREENSTEIN, J. P. (1952) *J. Biol. Chem.* **194,** 455–470.

BOISSONNAS, R. A. (1951) *Helv. Chem. Acta* **34,** 874–879.

BOUTWELL, P. W., AND KUICK, L. F. (1930) *J. Am. Chem. Soc.* **52,** 4166–4167.

BRAUN-FALCO, O. (1956) *Dermatol. Wochschr.* **134,** 1341–1349.

BRAUN-FALCO, O. (1957a) *J. Histochem. and Cytochem.* **5,** 94.

BRAUN-FALCO, O. (1957b) *Dermatol. Wochschr.* **135,** 417–428.

BRAUN-FALCO, O. (1961) *In* "Dermatologie und Venerologie" (H. A. Gottron and W. Schönfeld, eds.), Vol. 1[1], pp. 366–472. Georg Thieme, Stuttgart, Germany.

BRAUN-FALCO, O., AND SALFELD, K. (1959) *Nature* **183,** 51–52.

BURSTONE, M. S. (1956a) *J. Histochem. and Cytochem.* **4,** 421–422.

BURSTONE, M. S. (1956b) *J. Natl. Cancer Inst.* **16,** 1149–1161.

BURSTONE, M. S. (1957) *Arch. Pathol.* **63,** 164–167.

BURSTONE, M. S. (1958) *J. Histochem. and Cytochem.* **6,** 322–339.

BURSTONE, M. S., AND FOLK, J. E. (1956) *J. Histochem. and Cytochem.* **4,** 217–226.

BURSTONE, M. S., AND WEISBURGER, E. K. (1961a) *J. Histochem. and Cytochem.* **9,** 349–355.

BURSTONE, M. S., AND WEISBURGER, E. K. (1961b) *J. Histochem. and Cytochem.* **9,** 712–713.

COHEN, S. I., AND WARREN, R. (1961a) *New Engl. J. Med.* **264,** 79–84.

COHEN, S. I., AND WARREN, R. (1961b) *New Engl. J. Med.* **264,** 128–134.

DAO, T., AND MOORE, G. E. (1961) *Surg. Gynecol. Obstet.* **112,** 191–195.

DOHERTY, D. G., AND VASLOW, F. (1952) *J. Am. Chem. Soc.* **74,** 931–936.

FISCHER, E., AND SUZUKI, U. (1904) *Ber.* **37,** 2842–2848.

FLEISHER, G. A. (1954) *J. Biol. Chem.* **206,** 637–640.

FODOR, P. J., PRICE, V. E., AND GREENSTEIN, J. P. (1950) *J. Biol. Chem.* **182,** 467–470.

FOLK, J. E., AND BURSTONE, M. S. (1955) *Proc. Soc. Exptl. Biol. Med.* **89,** 473–476.

FOLK, J. E., AND BURSTONE, M. S. (1956) *Arch. Biochem. Biophys.* **61,** 257–262.

FRUTON, J. S. (1949) *Advances in Protein Chem.* **5,** 1–82.

FRUTON, J. S. (1960) *In* "The Enzymes" (P. D. Boyer, H. Lardy, and K. Myrbäck, eds.), Vol. 4, pp. 233–241. Academic Press, New York.

FRUTON, J. S., JOHNSTON, R. B., AND FRIED, M. (1951) *J. Biol. Chem.* **190,** 39–53.

GLENNER, G. G., AND COHEN, L. A. (1960) *Nature* **185,** 846–847.

GLENNER, G. G., AND FOLK, J. E. (1961) *Nature* **192,** 338–340.

GLENNER, G. G., BURSTONE, M. S., AND MEYER, D. B. (1959) *J. Natl. Cancer Inst.* **23,** 857–873.

GOMORI, G. (1953) *J. Histochem. and Cytochem.* **1,** 469–470.

GOMORI, G. (1954a) *Proc. Soc. Exptl. Biol. Med.* **85,** 570–572.

GOMORI, G. (1954b) *Proc. Soc. Exptl. Biol. Med.* **87,** 559–561.

GOODMAN, M., AND STUEBEN, K. C. (1959) *J. Am. Chem. Soc.* **81,** 3980–3983.

GREEN, M. N., TSOU, K. C., BRESSLER, R., AND SELIGMAN, A. M. (1955) *Arch. Biochem. Biophys.* **57,** 458–474.

GREENSTEIN, J. P., AND WINITZ, M. (1961) "Chemistry of the Amino Acids," Vol. 2. Wiley, New York.

GREENSTEIN, J. P., WINITZ, M., AND BIRNBAUM, S. M. (1955) *J. Am. Chem. Soc.* **77,** 5721–5725.

HESS, R. (1960) *Cancer Research* **20,** 940–943.

Holmberg, B. (1961) *Cancer Research* **21,** 1386–1393.

Howes, E. L., Armitage, C. M., and Mandl, I. (1955) *Surg. Forum Proc. 40th Congr. Am. Coll. Surgeons* **6,** 54–58.

Jasmin, G., and Gareau, R. (1961) *Brit. J. Exptl. Pathol.* **42,** 592–596.

Johnson, T. B. (1903) *J. Am. Chem. Soc.* **25,** 483–491.

Karnovsky, M. J., and Himmelhoch, S. R. (1961) *Am. J. Physiol.* **201,** 786–790.

Levine, S. (1954) *J. Am. Chem. Soc.* **76,** 1382.

Marvel, C. S. (1955) *In* "Organic Synthesis" (E. C. Horning, ed.), Vol. 3, pp. 523–525 and pp. 848–850. Wiley, New York.

Mehler, A. H. (1957) "Introduction to Enzymology." Academic Press, New York.

Monis, B., Nachlas, M. M., and Seligman, A. M. (1959) *Cancer* **12,** 601–608.

Mottet, N. K. (1961) *Am. J. Pathol.* **39,** 17–26.

Nachlas, M. M., Crawford, D. T., and Seligman, A. M. (1957) *J. Histochem. and Cytochem.* **5,** 264–278.

Nachlas, M. M., Monis, B., Rosenblatt, D., and Seligman, A. M. (1960) *J. Biophys. Biochem. Cytol.* **7,** 261–264.

Niemi, M., and Ikonen, M. (1960) *Nature* **185,** 928.

Patterson, E. K., Keppel, A., and Hsiao, S.-H. (1961) *J. Histochem. and Cytochem.* **9,** 609.

Pearse, A. G. E., and Tremblay, G. (1958) *Nature* **181,** 1532–1533.

Pfeiffer, P., and Saure, S. (1941) *J. prakt. Chem.* [2] **157,** 97–123.

Raekallio, J. (1961) *Ann. Med. Exptl. et Biol. Fenniae (Helsinki)* **39,** (6) *Suppl.*

Rao, K. R., Birnbaum, S. M., Kingsley, R. B., and Greenstein, J. P. (1952) *J. Biol. Chem.* **198,** 507–524.

Ravin, H. A., and Seligman, A. M. (1951) *J. Biol. Chem.* **190,** 391–402.

Ravin, H. A., Bernstein, P., and Seligman, A. M. (1954) *J. Biol. Chem.* **208,** 1–15.

Riedel, A., and Wünsch, E. (1959) *Z. physiol. Chem. Hoppe-Seyler's* **316,** 61–70.

Robertson, G. R. (1927) *J. Am. Chem. Soc.* **49,** 2889–2894.

Rosenblatt, D. H., Nachlas, M. M., and Seligman, A. M. (1958) *J. Am. Chem. Soc.* **80,** 2463–2465.

Rosenholtz, M. (1960) *Gastroenterology* **38,** 794.

Rosenholtz, M., and Wattenberg, L. W. (1961) *Arch. Pathol.* **71,** 63–75.

Rothman, S. (1957) *Arch. Dermatol.* **76,** 277–281.

Schallenberg, E. E., and Calvin, M. (1955) *J. Am. Chem. Soc.* **77,** 2779–2783.

Seligman, A. M., and Wolf, G. (1951) *J. Am. Chem. Soc.* **73,** 2086–2089.

Sheehan, J. C., and Hess, G. P. (1955) *J. Am. Chem. Soc.* **77,** 1067–1068.

Sheehan, J. C., and Hlavka, J. J. (1956) *J. Org. Chem.* **21,** 439–441.

Sheehan, J. C., Goodman, M., and Hess, G. P. (1956) *J. Am. Chem. Soc.* **78,** 1367–1369.

Smith, E. L. (1951) *In* "The Enzymes" (J. B. Sumner and K. Myrbäck, eds.), Vol. I, Part 2, pp. 793–872. Academic Press, New York.

Smith, E. L. (1957) *In* "Methods in Enzymology," (S. P. Colowick and N. O. Kaplan, eds.), Vol. III, pp. 528–540. Academic Press, New York.

Smith, E. L. (1960) *In* "The Enzymes" (P. D. Boyer, H. Lardy, and K. Myrbäck, eds.), Vol. 4, pp. 1–10. Academic Press, New York. 2nd edition.

Smith, E. L., and Hill, R. L. (1960) *In* "The Enzymes" (P. D. Boyer, H. Lardy, and K. Myrbäck, eds.), Vol. 4, pp. 37–62. Academic Press, New York. 2nd edition.

Sumner, J. B., and Somers, G. F. (1953) "Chemistry and Methods of Enzymes," 3rd ed. Academic Press, New York.

Sylvén, B., and Malmgren, H. (1955) *Exptl. Cell Research* **8,** 575–577.

Sylvén, B., and Malmgren, H. (1957) *Acta Radiol. Suppl.* **154,** 1–124.
Tigerstedt, A. (1892) *Ber.* **25,** 2919–2931.
Tuppy, H., and Nesvadba, H. (1957) *Monatsh. Chem.* **88,** 977–988.
Ungar, G., and Damgaard, E. (1954) *Proc. Soc. Exptl. Biol.* **87,** 378–383.
Vaughan, J. R., and Osato, R. L. (1951) *J. Am. Chem. Soc.* **73,** 5553–5555.
Waldschmidt-Leitz, E., and Keller, L. (1957) *Z. physiol. Chem., Hoppe-Seyler's* **309,** 228–238.
Wattenberg, L. W. (1959) *Arch. Pathol.* **67,** 281–286.
Weygand, F., and Csendes, E. (1952) *Angew. Chem.* **64,** 136.
Weygand, F., and Leising, E. (1954) *Ber.* **87,** 248–256.
Weygand, F., and Reiher, M. (1955) *Ber.* **88,** 26–34.
Wieland, T. H. (1951) *Angew. Chem.* **63,** 7–14.
Wieland, T., and Sehring, R. (1950) *Ann.* **569,** 122–129.
Willighagen, R. G. J., and Planteydt, H. T. (1959) *Nature* **183,** 263–264.
Wolf, G., and Seligman, A. M. (1951a) *J. Am. Chem. Soc.* **73,** 2080–2082.
Wolf, G., and Seligman, A. M. (1951b), *J. Am. Chem. Soc.* **73,** 2082–2085.

*Chapter 11*

# OXIDASES

Oxidases or aerobic transelectronases (Hoffmann-Ostenhof, 1953) are enzymes which catalyze the transfer of electrons from a donor substrate to oxygen. Anaerobic transelectronases such as TPN-cytochrome c reductase will not be included in the present discussion. Despite the fact that molecular oxygen generally serves as the ultimate selectron acceptor in biological oxidations, many oxidative enzymes are unable to utilize oxygen directly and therefore electrons or hydrogen atoms must be transferred by means of intermediate compounds. Among the comprehensive biochemical reviews of oxidases and metabolically associated compounds are those of Paul (1951, p. 357; 1960, p. 277), Dawson and Tarpley (1951, p. 454), Zeller (1951, p. 536), Sumner and Somers (1953, p. 213), Mehler (1957, p. 206), Mason (1957), and Dixon and Webb (1958, pp. 198 and 207). Oxidases usually contain either iron or copper. Those containing the former include cytochrome oxidase, peroxidase, verdoperoxidase, and catalase. The copper oxidases found in mammalian tissues include tyrosinase and a kidney polyphenol oxidase (Dawson and Tarpley, 1951, p. 488). Other oxidative enzymes include monoamine oxidase (tyraminase), homogentisate oxidase, which is involved in phenylalanine metabolism, 3-hydroxyanthranilate oxidase, which functions in the intermediary metabolism of tryptophan, lipoxidase, which catalyzes aerobic formation of peroxides from unsaturated fatty acids, and dihydrofumaric acid and tryptophan oxidases, which are involved in the metabolism of dihydroxyfumaric acid and tryptophan, respectively. With the exception of lipoxidase, the aforementioned enzymes are believed to be ferrous proteins or ferroprotoporphyrin proteins.

## I. CYTOCHROME OXIDASE

Cytochrome oxidase is an iron-porphyrin enzyme which catalyzes the reduction of oxygen to water at rates that account for the most of the oxygen utilized by mammalian tissues. Synonyms for cytochrome oxidase are Atmungsferment, cytochrome $a_3$, indophenol oxidase, respiratory enzyme, saüerstoffübertragendes Ferment, and oxygen activating enzyme. In addition to the presence of iron, cytochrome oxidase is also reported to contain copper (Sumner and Somers, 1953, p. 232; Sands and Beinert, 1959; Griffiths and Wharton, 1961). Griffiths and Wharton (1961) do not feel that copper is a functional component of cytochrome oxidase. Recent studies on the porphyrin and hemin prosthetic groups of cytochrome oxidase have been reported by Tuppy and Paléus (1955a, b), Morell *et al.* (1961), and Clezy and Barrett (1961). General biochemical aspects of the enzyme are to be found in the previously noted reviews of Paul (1951, p. 357) and Sumner and Somers (1953, p. 213) as well as the interesting discussions by Keilin and Slater (1953) and Theorell (1956). The enzyme is closely associated with cytochrome c, a hemoprotein which is capable of undergoing a reversible oxidation-reduction in which accepted electrons may be transferred to oxygen.

Keilin and Slater (1953) have reviewed the historical development of concepts with reference to cytochrome oxidase and cytochrome c. According to these authors, MacMunn, during the period of 1884–1886, described tissue pigments from different species which could undergo a reversible oxidation-reduction, and which he felt were related to oxygen consumption. Unfortunately, MacMunn's important discovery was very severely criticized by the Hoppe-Seyler school which resulted in its disappearance from the scientific literature for more than forty years. In 1925 Keilin rediscovered MacMunn's respiratory pigment which he renamed "cytochrome." Subsequent work by Keilin and associates demonstrated the ubiquitous nature of the pigment in both animal and plant world, and also its important role in cellular metabolism (Keilin, 1925, 1930; Keilin and Hartree, 1938, 1939).

There is a direct relationship between the concentration of cytochrome in a tissue and its oxygen requirement. High concentrations are found in muscles capable of rapid contraction such as mammalian cardiac muscle, pectoral muscle of birds, and wing and leg muscle of flying insects. High concentration of cytochrome is also found in the central nervous system. Reduced cytochrome exhibits four distinct absorption bands at 605 mμ, 562 mμ, 550 mμ, and 520 mμ. The first three bands represent portions of distinct pigments (a, b, and c) while the fourth corresponds to fused bands of at least two components. Reduced cytochrome c does not react with oxygen in neutral solution, but readily reacts with oxygen in the presence of cytochrome oxidase. Cytochrome c is reduced by a number of compounds including sodium dithionate, cysteine, ascorbic acid, hydroquinone, and TPNH.

The primary and presumably the sole function of cytochrome oxidase is to oxidize cytochrome c according to the following reaction (Sumner and Somers, 1953, p. 233):

1. Reduced cytochrome c + $H^+$ + ¼ $O_2$ → oxidized cytochrome c + ½ $H_2O$
2. Oxidized cytochrome c + H → reduced cytochrome c + $H^+$

Among the compounds which are oxidized by oxidized cytochrome c are *p*-phenylenediamine and analogues (Slater, 1949; Borei and Björklund, 1953), as well as a variety of complex arylamines and methylene compounds (Person *et al.*, 1962).

The biological function of respiration is to furnish energy for processes including synthesis of biological materials and osmotic or mechanical work. The pathways of oxidation and hydrogen transfer involving the cytochrome system are shown in Chart 1 (Keilin and Slater, 1953). With reference to specific metabolic activities, the cytochrome system has been related to functional activity of the thyroid and adrenal (Paul, 1951, p. 391). Allen *et al.* (1960) have reported a decreased cytochrome oxidase activity in muscle from vitamin E deficient rabbits, and suggest that disruption of the cytochrome system may be part of the biochemical alteration which occurs in nutritional muscular dystrophy.

In 1885 Ehrlich discovered the indophenol reaction by noting that a blue dye was formed upon injection of α-naphthol and *N*,*N*-dimethyl-*p*-phenylenediamine into living tissue. The formation of the dye, indophenol blue, has been referred to as the "nadi" reaction from the first two letters of "naphthol" and "diamine." The specificity of the "nadi" reaction for

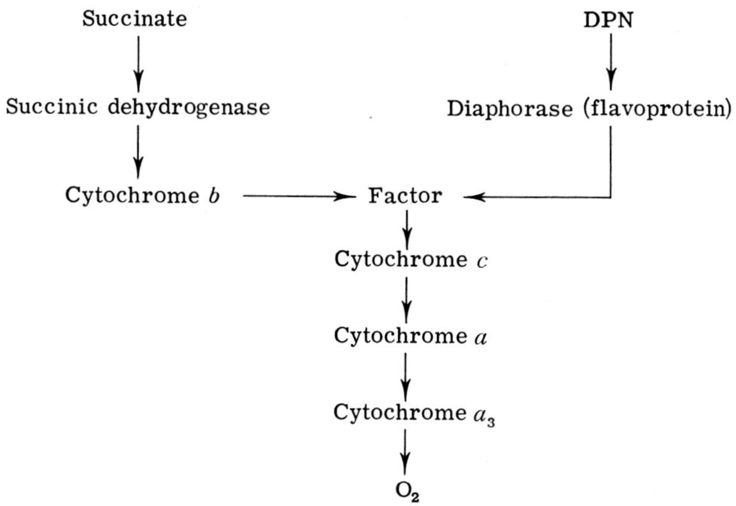


CHART 1. Cytochrome system in biological oxidations (Keilin and Slater, 1953).

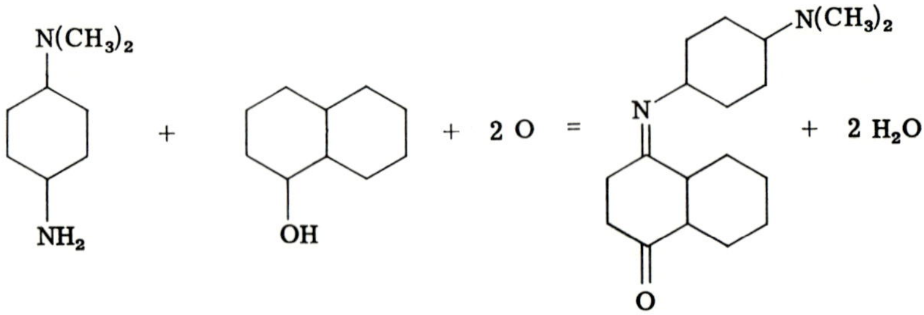


CHART 2. Indophenol oxidase or "nadi" reaction.

cytochrome oxidase was elucidated by Keilin and Hartree (1938). Subsequent studies by Stotz *et al.* (1938) and Straus (1954) also confirmed the specificity of the indophenol cytochrome oxidase reaction from a biochemical standpoint (Chart 2).

In regard to the application of the "nadi" reaction as a tissue stain, it was noted as early as 1895 (Rohmann and Spitzer, 1895) that ground tissues or alcohol-ether dried specimens catalyzed the formation of indophenol blue. In 1907 Winkler found that the "nadi" reagent stained leukocytes in alcohol fixed smears. Schultze (1909a, 1927) also observed an "indophenol oxidase" reaction in leukocytes and other components of fixed smears and tissue sections, and noted the diagnostic value of the method in distinguishing between myeloid and lymphoid cells. Because of the similarity in techniques, the aforementioned application of the reaction in the study of fixed tissues has been referred to as the Winkler-Schultze method. During the same period Fiessinger and Roudowska (1912) reported confirmation of the aforementioned findings of Schultze, and in addition noted that the "nadi" method could be used in conjunction with unfixed frozen sections. Gräff (1922, 1936) confirmed that the distribution of the dye is not the same in fixed as contrasted with unfixed frozen sections treated with the reagent at neutral pH, and concluded that the "nadi" method could demonstrate a labile oxidase (G "nadi" reaction). This reaction did not take place if freshly boiled distilled water was used as a solvent, and it was also found to be readily inhibited by heat, formalin, KCN ($10^{-3}$ *M*) acetone, and other organic solvents. By contrast, the aforementioned reaction of myeloid cells in fixed tissues (M "nadi" reaction) is relatively heat stable and insensitive to prolonged fixation, organic solvents, and alkali. Although the pH optimum is approximately 12.0, the M "nadi" reaction is sensitive to acids. More recently Moog (1943) employed the G "nadi" reaction in a study of chick embryos. Her work is significant in that it represented the first attempt to check the specificity of the method by the use of inhibitors. Sodium azide (0.005 *M*), which was originally reported to be a specific inhibitor of cytochrome oxidase by Keilin (1936), was found by Moog (1943) to completely inhibit the reaction at pH 5.8–7.2. Spiegelman and Moog (1945) also characterized the "nadi" reaction by the use of cyanide, another known inhibitor.

The M "nadi" reaction has been investigated by Person *et al.* (1961) who confirmed one of the original observations of Schultze (1909b) that in addition to the staining of myeloid cells, a reaction could also be localized to acinar cells of the parotid and submaxillary glands. Person *et al.* (1961) treated unfixed frozen sections with 10% formalin for 20 minutes, then carried the sections through four successive 2 minute rinses in distilled water, and finally placed them in the incubating medium. They observed that although the staining reaction of the ducts and tubules was completely abolished by formalin treatment, indophenol blue appeared in the acini of submaxillary and parotid glands but not in sublingual gland acini. Although these authors felt that this demonstrated an M "nadi" reaction, it is possible that the short formalin fixation period employed did not permit the fixative to penetrate the relatively thick frozen sections, and that a more sensitive oxidase was being demonstrated.

Although the work of Keilin and Hartree (1938) showed that the "nadi" reaction could be specific for the demonstration of cytochrome oxidase, its histochemical application has had limited use for a number of reasons. Prior to the aforementioned date many investigators questioned the specificity of the reaction since the formation of indophenol blue occurs spontaneously in the presence of air. In addition, the dye is lipid soluble and readily stains fat containing tissue elements. Interesting reviews of the older material pertaining to the specificity or lack of specificity of the "nadi" reaction are to be found in Gomori (1952, p. 153) and Lison (1960, p. 597). Despite some of the uncertain aspects of the technique, some of the earlier investigators indicated that the reaction would clearly delineate a labile oxidase which in certain cases was assumed to be identical with Atmungsferment (cytochrome oxidase) (Wertheimer, 1926; Gräff, 1936). In addition to the aforementioned studies of Moog (1943) other investigators have employed the "nadi" reaction in a study of several tissues with particular emphasis upon the central nervous system. With reference to the latter, the comprehensive review of Roizin (1955) should be consulted.

The major advance in the histochemical demonstration of cytochrome oxidse has occurred during the last few years, and has been related to the application of new substrates (Nachlas *et al.*, 1958; Burstone, 1959a, 1960a, 1961). Various aspects of the development and applications of the new substrates will be discussed in the forthcoming sections.

## A. Substrate Types and Staining Mechanisms

It has been assumed that the "nadi" reaction is mediated by the oxidation of $\alpha$-naphthol and *N*,*N*-dimethyl-*p*-phenylenediamine by oxidized cytochrome c. Person and Fine (1961) demonstrated that of the two "nadi" reagents only the *N*,*N*-dimethyl-*p*-phenylenediamine was enzymatically oxidized. Thus the α-naphthol assumes only an incidental role in the reaction, one which is related to non-enzymatic oxidative coupling to form indophenol blue. This feature in itself should not militate against the use of α-naphthol as a substrate component, although it would be preferable to employ substrate solutions which did not contain any coupling agent not related to the enzymatically produced oxidation. From a practical standpoint, however, α-naphthol has been shown to strikingly increase the toxic effect of *N*,*N*-dimethyl-*p*-phenylenediamine upon the cytochrome system (Borei and Björklund, 1953). Nachlas *et al.* (1958) described a modification of the "nadi" reaction which employed instead of *N*,*N*-dimethyl-*p*-phenylenediamine (I) its analogue 4-amino-1-*N*,*N*-dimethylnaphthylamine (II) (Table 1) (Appendix 11A, Section 1). The structural similarity

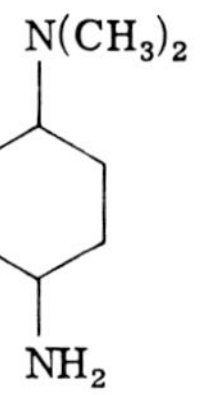


*N*,*N*-dimethyl-
*p*-phenylenediamine
(I)

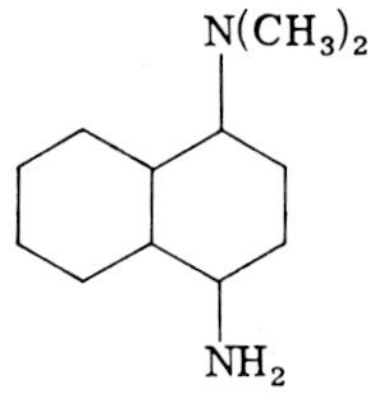


4-amino-1-*N*, *N*-
dimethylnaphthylamine
(II)

of the latter compound (II), which was originally synthesized by Friedländer and Welmans (1888), to the dimethyl-*p*-phenylenediamine of the original "nadi" reagent (Chart 2) is apparent. Although 4-amino-1-*N*,*N*-dimethylnaphthylamine is capable of producing somewhat more insoluble dyes when employed in the "nadi" reaction, this amine suffers from the defect of its previously applied counterpart in that it readily autoxidizes. As previously indicated, the enzymatic oxidation is mediated by way of the diamine and not the α-naphthol (Person and Fine, 1961). The reduction of oxidized cytochrome c by *p*-phenylenediamine and analogues has been studied by Slater (1949) and Borei and Björklund (1953). Keilin and Slater (1953) noted that *p*-phenylenediamine is readily reduced by cytochrome c in oxidase preparations, even in the absence of added cytochrome c. However, the oxidation is augmented by addition of the latter substance. Borei and Björklund (1953) found that more complex amines of the *N*,*N*-alkyl aryl type were increasingly dependent upon added cytochrome c. These authors also noted that the dimethyl-, tetramethyl-, and diethyl-

TABLE 1

NAPHTHOLIC, ARYLAMINE, AND METHYLENE COMPOUNDS EMPLOYED IN THE DEMONSTRATION OF CYTOCHROME OXIDASE

| Chemical name and formula, | Mol. wt | Reference | Melting point (° C) |
|---|---|---|---|
| 1. α-Naphthol, $C_{10}H_7OH$ | 144.17 | — | 95–96 |
| 2. 2,4-Dichloro-1-naphthol, $C_{10}H_5Cl_2OH$ | 213.07 | — | 105–106 |
| 3. 2,4-Dibromo-1-naphthol, $C_{10}H_5Br_2OH$ | 301.99 | — | 108–109 |
| 4. 1-Hydroxy-*N*-octadecyl-2-naphthamide, $C_{29}H_{45}O_2N$ | 439.69 | Burstone, 1959a | — |
| 5. 1-Hydroxy-2-napthanilide, $C_{17}H_{13}O_2N$ | 263.30 | Anschütz *et al.*, 1906; Jadhav and Rao, 1936 | 155–156 |
| 6. *N*,*N*-Dimethyl-*p*-phenylenediamine (*p*-aminodimethylaniline), $C_8H_{12}N_2$ | 136.20 | — | 41, 53 |
| 7. *N*,*N*-Dimethyl-*p*-naphthalenediamine (4-amino-1-*N*,*N*-dimethylnaphthylamine), (ADN) $C_{12}H_{14}N_2$ | 186.26 | Friedländer and Welmans, 1888, Cobenzl, 1915; Hodgson and Crook, 1936; Nachlas *et al.*, 1958 | 22 |
| 8. 8-Amino-1,2,3,4-tetrahydroquinoline, $C_9H_{12}N_2$ | 148.21 | Hazlewood *et al.*, 1938; Elderfield and Claflin, 1952 | Oil |
| 9. *p*-Aminodiphenylamine, $C_{12}H_{12}N_2$ | 184.24 | Meyer, 1920 | 66, 75 |
| 10. *p*-Methoxy-*p*-aminodiphenylamine, $C_{13}H_{14}N_2O$ | 214.27 | — | 102 |
| 11. 3-Aminocarbazole, $C_{12}H_{10}N_2$ | 182.23 | Whitner, 1924 | 254 (Dec.) |
| 12. 3-Amino-9-ethylcarbazole, $C_{14}H_{14}N_2$ | 210.28 | Lindemann, 1924 | — |
| 13. 1-Phenyl-3-methylpyrazolone (3-methyl-1-phenylpyrazolone-5), $C_{10}H_{10}ON_2$ | 174.21 | Tautomeric forms | 127–128 |
| 14. 1-Phenyl-3-(*m*-nitrobenzamido)-5-pyrazolone, $C_{16}H_{12}N_4O_4$ | 324.30 | — | 222 |

substituted *p*-phenylenediamines form oxidation products which are particularly toxic to the oxidase.

Burstone (1959a) also reported a modified "nadi" reaction in which a number of substituted α-naphthol derivatives were employed in place of the usual α-naphthol. The new histochemical reagents (Table 1), which included (see Chart 3 for corresponding numbers) 2,4-dichloro- or dibromo-1-naphthol (1), 2,4-dichloro-5-toluenesulfonylaminonaphthol (2), 1-hydroxy-2-naphthoic acid (3), 1-hydroxy-2-*N*-butylnaphthamide (4), 1-hydroxy-2-naphthanilide (5), and complex alkyl and aryl derivatives of

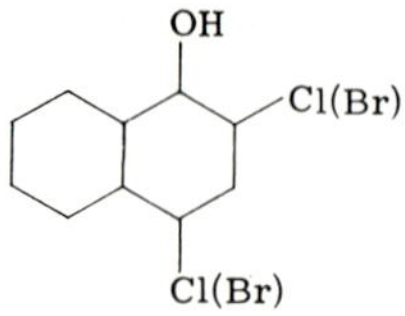


1

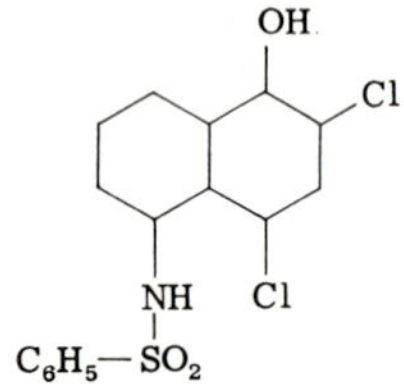


2

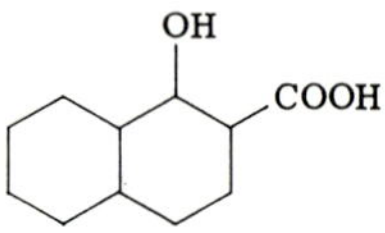


3

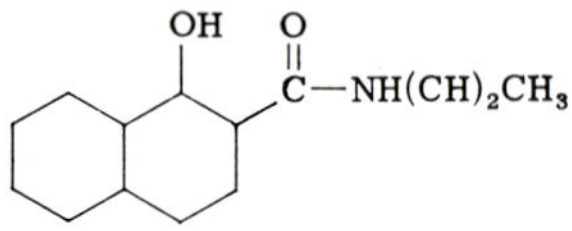


4

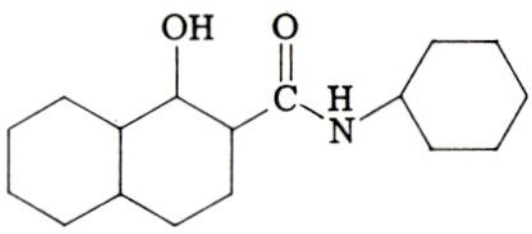


5

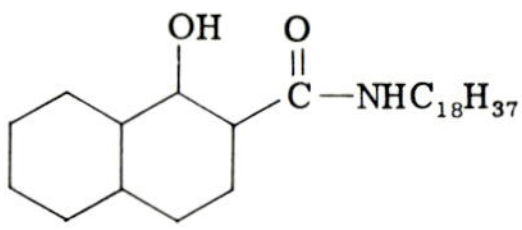


6

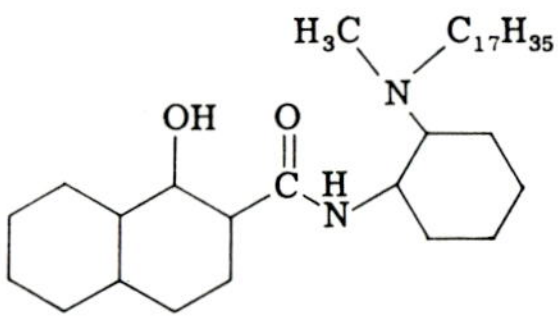


7

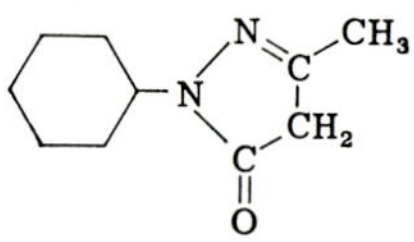


8

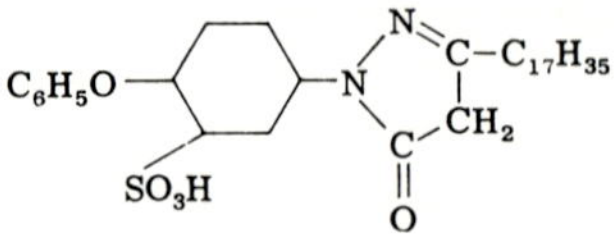


9

CHART 3. Naphtholic and reactive methylene compound cytochrome oxidase substrates.

1-hydroxy-2-naphthoic acid (6 and 7), had the advantage of forming more insoluble and more stable dyes than had been previously reported. The best results were obtained when the aforementioned naphthols were used in conjunction with *p*-aminodiphenylamine (*N*-phenyl-*p*-phenylenediamine) (base). Burstone (1959a, 1960a, 1961) also reported new techniques unrelated to the "nadi" method. These procedures employed either a compound with a reactive methylene group in conjunction with an amine, or else combinations of amines. As with the modified "nadi" procedure, the amines and methylene compounds are used with *p*-aminodiphenylamine [Chart 4 (1); Table 1]. Among the amines employed are aminodiphenylamine analogues and carbazole derivatives (Chart 4; Table 1), while methylene compounds are exemplified by pyrazolones (Chart 3; Table 1). The mechanisms by which amines and naphthols, or amines and reactive methylene compounds (pyrazolones) form oxidation dyes are discussed in the section on *oxidation dyes* in Chapter 4. The specific constitution of oxidation dyes formed from the aforementioned amines has not been elucidated, although it is known that treatment of *p*-aminodiphenylamine with strong oxidizing agents will result in the formation of a tetraquinonoid "perinigraniline" compound (Green and Woodhead, 1910).

Since, by definition, the aforementioned oxidase techniques are not to be classified with the "nadi" reaction, those employing methylene compounds and diamines might be termed "medi," while those employing amines and diamines could be referred to as "adi" procedures. Manometric and spectrophotometric studies have shown that oxidation of *p*-aminodiphenylamine alone, and in conjunction with other amines or methylene compounds, is mediated via cytochrome oxidase (Fig. 1) with the formation of dyes with characteristic spectra (Figs. 2 and 3) (Person *et al.*, 1962).

A novel approach to the demonstration of cytochrome oxidase was reported by Oda *et al.* (1958a) who found that tissue homogenates in the presence of *p*-phenylenediamine, with or without added cytochrome c, reduced cytochrome c which in turn reduced neotetrazolium chloride to produce a formazan dye (Chart 5). With this method colorimetric determinations of cytochrome oxidase activity of kidney, heart, liver, brain, and skeletal muscle were made.

## B. Histochemical Methods

As previously indicated the original "nadi" techniques have not been widely applied because of inherent defects in the procedures including instability of the substrate solution, lipid solubility, crystallization, and fading of the indophenol blue. The modified "nadi" technique of Nachlas *et al.* (1958) offered some improvements, but the amine used (ADN) is readily susceptible to autoxidation. In addition, the final dye

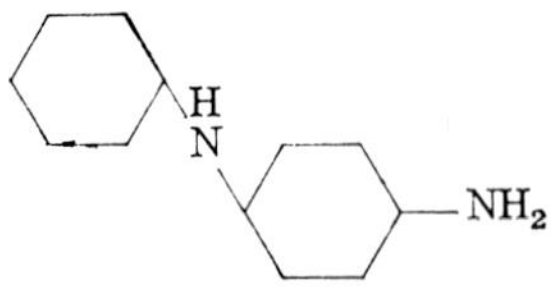

1

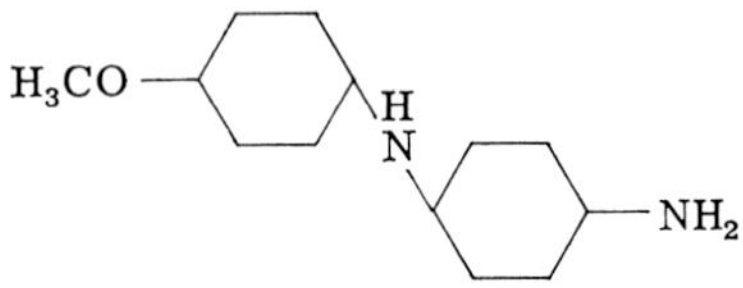

2

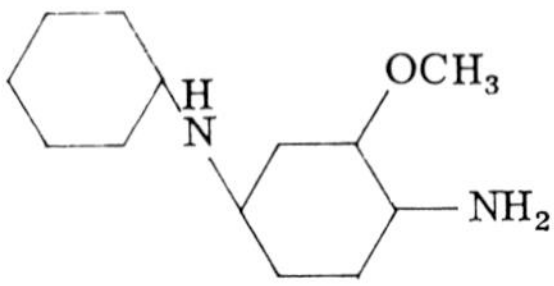

3

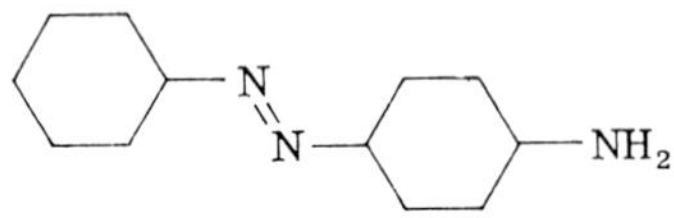

4

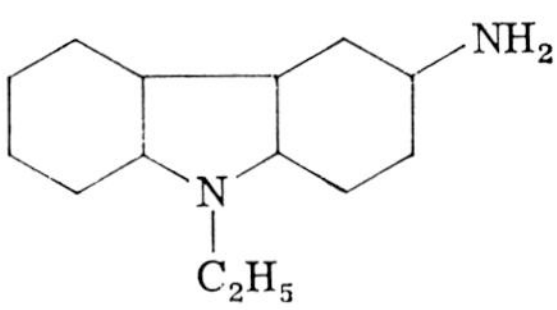

5

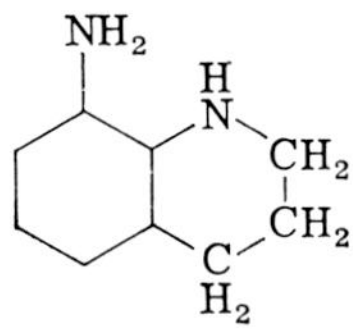

6

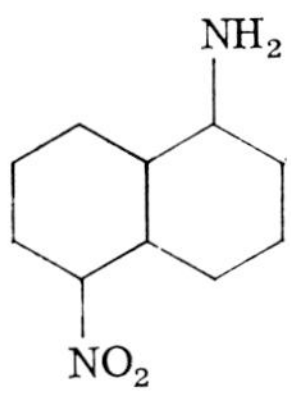

7

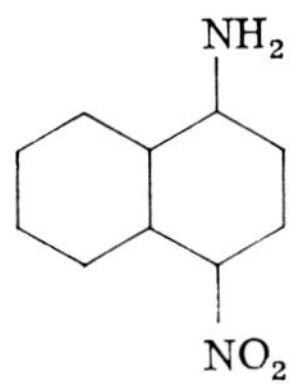

8

CHART 4. Arylamine cytochrome oxidase substrates.

was reported by Nachlas *et al.* (1958) to be lipid soluble, to lack substantivity, and to be susceptible to fading. The methods of Burstone (1959a, 1960a, 1961), which employed either naphthols, reactive methylene compounds, or amines in conjunction with *p*-aminodiphenylamine followed by post-chelation of the final dye with $Co^{++}$, are advantageous in

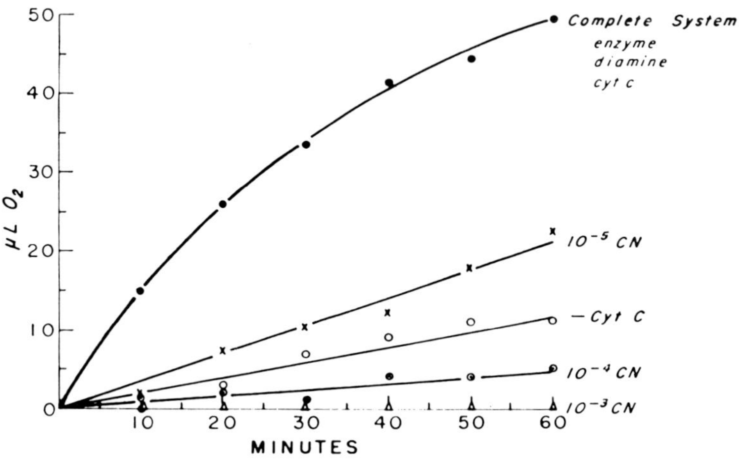


FIG. 1. Oxidation of *p*-aminodiphenylamine by Keilin-Hartree heart muscle preparation (Person *et al.*, 1962).

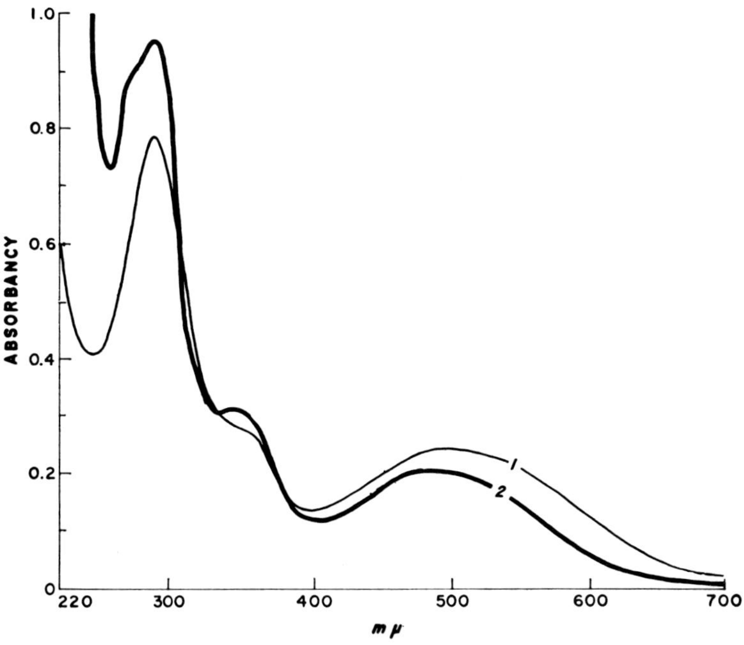


FIG. 2. Absorption spectra of oxidation dyes resulting from action of Keilin-Hartree heart muscle preparation. Curve 1, *p*-aminodiphenylamine; curve 2, *p*-aminodiphenylamine and 3-amino-9-ethylcarbazole (Person *et al.*, 1962).

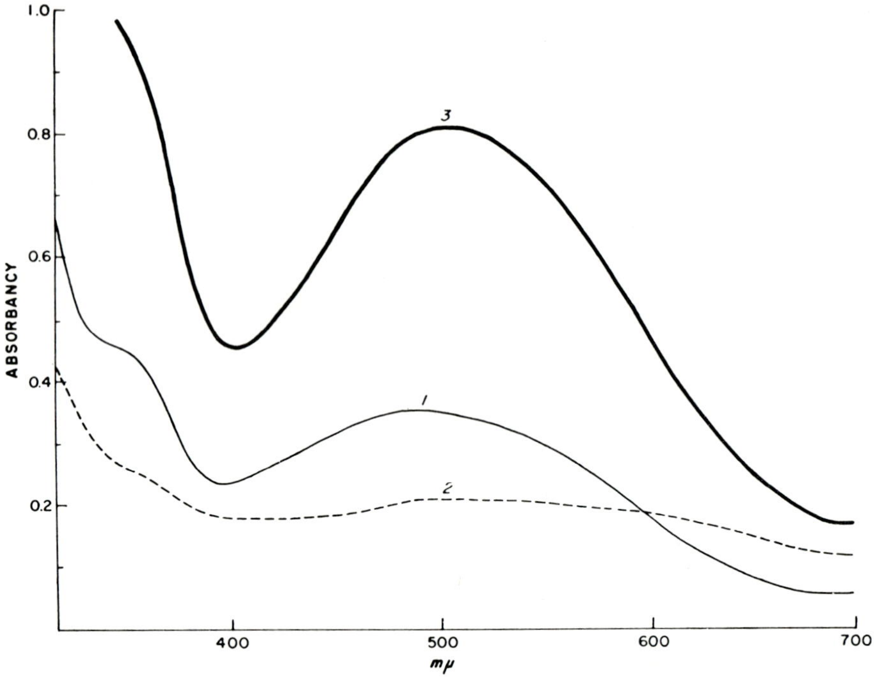


FIG. 3. Comparison of absorption spectra of oxidized *p*-aminodiphenylamine (curve 1), *p*-methoxy-*p*-aminodiphenylamine (curve 2), and a combination of the two amines (curve 3) (Person *et al.*, 1962).

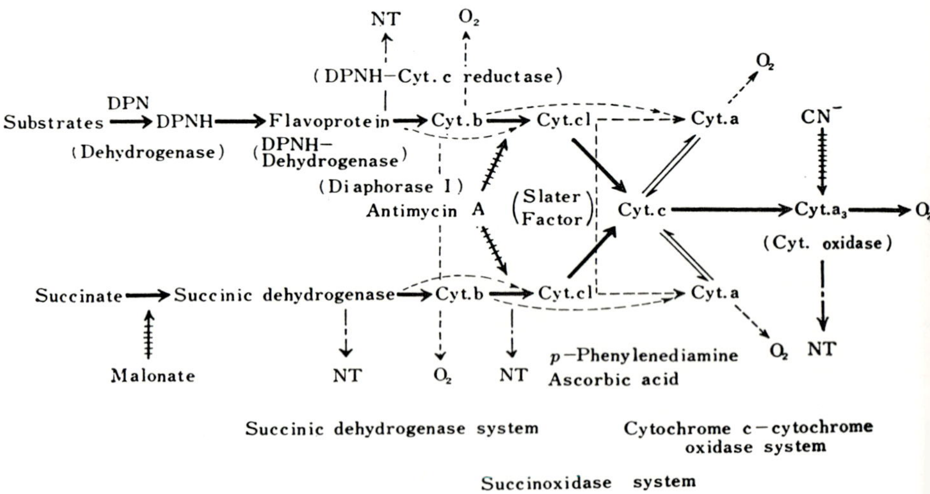


CHART 5. Reduction of neotetrazolium chloride (NT) by the terminal electron transport system (Oda *et al.*, 1958a).

that the localizations are precise and the dyes are permanent (Appendix 11B, Section 3). In addition to cobalt, most of the dyes were found to be capable of complexing with mercury, molybdenum, nickel, cadmium, lead, uranium, and iron, and thus the techniques may have potential application in electron microscopy. In view of the known toxic effect of $\alpha$-naphthol upon the "nadi" reaction, it would appear that the most sensitive substrate mixtures consist of *p*-aminodiphenylamine in combination with another amine or possibly a methylene compound (Charts 3 and 4) (Burstone, 1960a, 1961). The use of *p*-aminodiphenylamine (base) [Chart 4 (1)], in conjunction with *p*-methoxy-*p*-aminodiphenylamine [Chart 4 (2)], 3-amino-9-ethylcarbazole [Chart 4 (5)], or 8-amino-1,2,3,4-tetrahydroquinoline [Chart 4 (6)] has been found to be of particular value in applied histochemistry. In any event, it should be emphasized that with the aforementioned newer techniques, the substrate solutions are remarkably free of autoxidation, and long incubation periods are entirely feasible. Of particular interest is the mitochondrial staining obtained in normal and diseased cardiac muscle (Figs. 4 and 5) (Burstone, 1959b; Burstone and Miller, 1961; Engle, 1961), and the striking pattern seen in insect muscle (Figs. 6 and 7). Other interesting localizations of cytochrome oxidase have been reported in osteoclasts (Fig. 8) (Burstone, 1960b), kidney (Fig. 9), salivary glands (Fig. 10), stomach (Fig. 11), and normal and inflamed gingiva (Burstone, 1960c). Localization of the enzyme has also been described in the nervous system (Figs. 12, 13 and 14) (Burstone, 1960a, 1961). Niemi and Merenmies (1961) have employed the *p*-aminodiphenylamine technique in the study of cytochrome oxidase activity in the retina (Fig. 15). Montagna and Yun (1961) have employed the *p*-aminodiphenylamine cytochrome oxidase method to advantage in the study of the distribution of the enzyme in human skin while Dewey *et al.* (1961) have used the same method in a cytochemical study of L strain fibroblasts.

Oda *et al.* (1958b, c) have reported an interesting new histochemical procedure for the demonstration of cytochrome oxidase which employs *p*-phenylenediamine in conjunction with neotetrazolium chloride. These authors demonstrated that the cytochrome oxidase system is responsible for the neotetrazolium chloride reduction and subsequent formazan dye precipitation. Although the technique has considerable merit, it is limited by the relatively coarse localizations obtained with neotetrazolium chloride. At present, it does not appear that more substantive substrates such as nitro-blue tetrazolium can be employed.

The possibility of employing redox dyes as oxidase substrates warrants study (Chapter 4, p. 129). Crawford and Nachlas (1958) tested a number of leuco dyes but did not find any of them to be satisfactory as histochemical reagents. Pullman and Pullman (1959) have discussed the oxido-reduc-

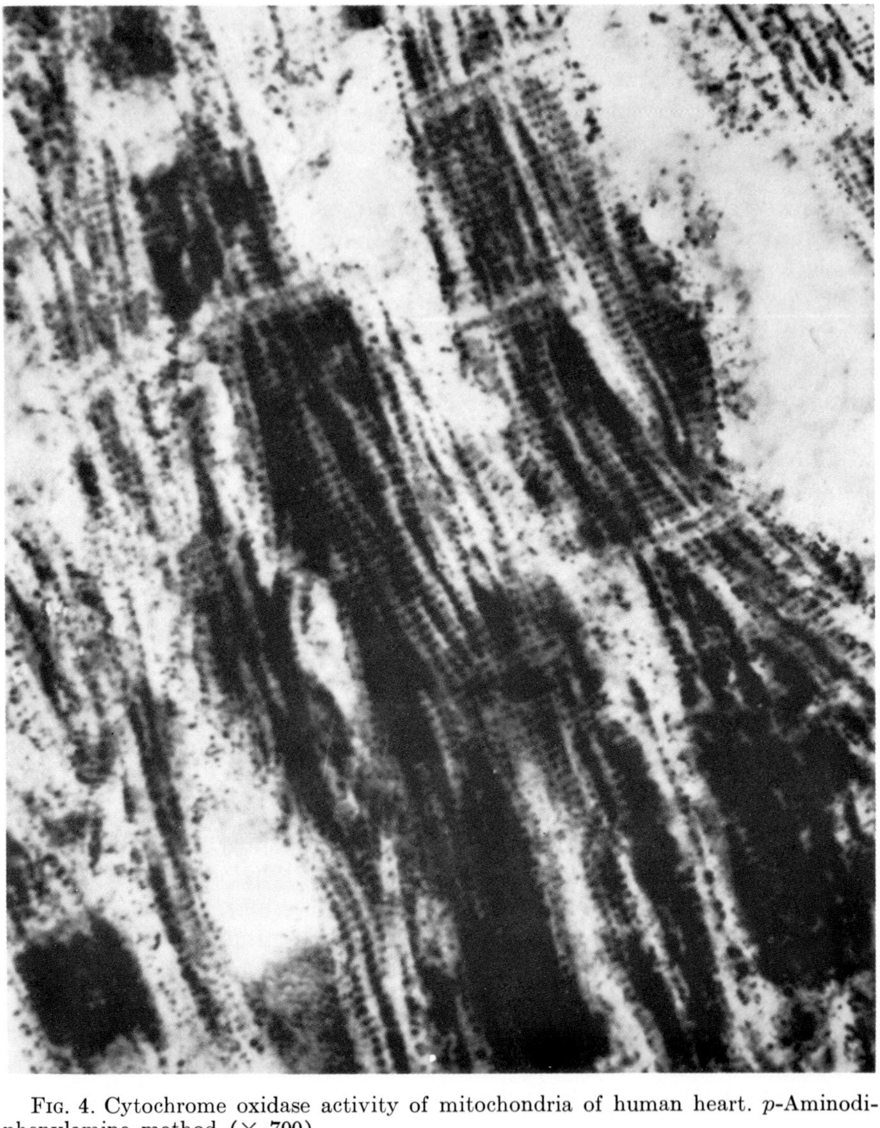

FIG. 4. Cytochrome oxidase activity of mitochondria of human heart. *p*-Aminodiphenylamine method (× 700).

tive properties of several organic dyes with reference to enzyme oxidation-reduction systems.

## C. Activators and Inhibitors

Wainio and Greenlees (1958) reported that the cytochrome oxidase of beef heart mitochondrial fragments which had been inactivated in at-

FIG. 5. Cytochrome oxidase distribution in human myocardial infarction. Note disruption of sarcosomal pattern in damaged fiber. *p*-Aminodiphenylamine method (× 310).

tempts to purify the enzyme could be reactivated by the addition of deoxycholate extract and other phospholipids. The mechanism by which these substances behave as activators is unknown. Person and Fine (1960) have found that protamine sulfate, histone, lysozyme, and ribonuclease are potent inhibitors of mammalian heart muscle cytochrome oxidase. Their inhibitory effect, however, is completely reversed in the presence of a strongly anionic synthetic polyglucose sulfate. Their work constitutes the first report of a reversible inhibition of cytochrome c oxidase by polyionic macromolecules of opposite charge. Keilin (1929) found that the inhibition of the enzyme by carbon monoxide is a light sensitive reversible reaction but attempts to apply this principle to enzyme histochemistry present technical difficulties. Irreversible inhibitors of cytochrome oxidase are cya-

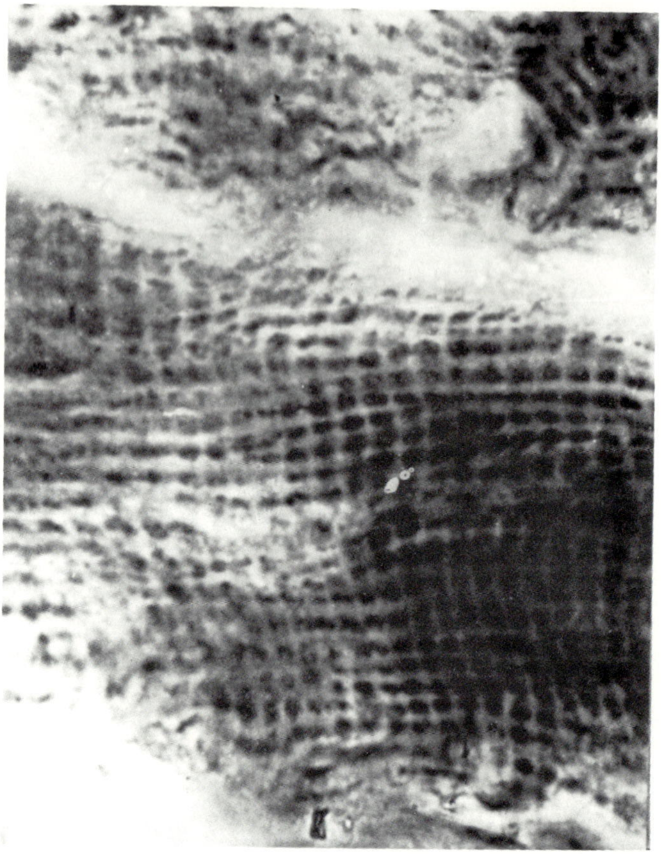

FIG. 6. Localization of cytochrome oxidase in sarcosomes of insect (bumble bee) leg muscle, longitudinal section. *p*-Aminodiphenylamine method (× 1800).

nide, azide, and sulfide in concentration of $10^{-3}$ $M$ or less. With reference to histochemical techniques it is advisable to pre-incubate slides in adequately buffered solutions containing the aforementioned ions rather than to incorporate them in the substrate solution. Cytochrome oxidase in tissue sections is also readily inactivated by heating to approximately 80° or less, and by contact with organic solvents and commonly used fixatives.

## D. Cytochrome Oxidase in Neoplasms

Hoffmann *et al.* (1951) employed the "nadi" reaction in a study of cells from lymph nodes, spleen, bone marrow, and peripheral blood from patients with Hodgkin's disease and leukemia. They noted a low level of enzymatic activity of various cells with the exception of monocytes and

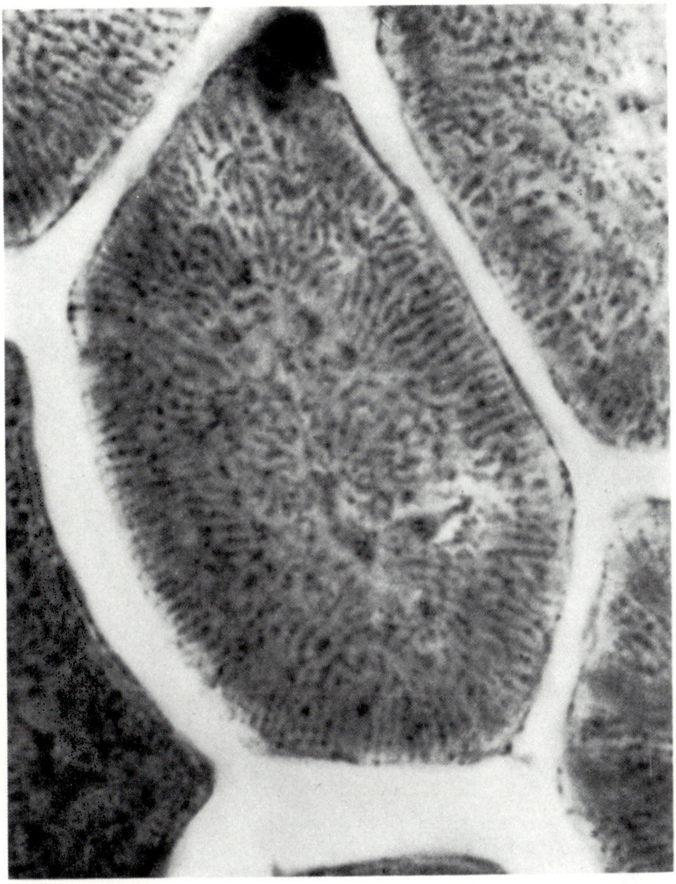

FIG. 7. Specimen similar to that shown in Fig. 6 except shown in cross-section view (× 1800).

macrophages in both normal and abnormal tissues. Wattenberg (1959b) employed the 4-amino-1-*N*,*N*-dimethylnaphthylamine technique in a study of oxidase activity of normal mucosa and proliferative lesions of the large bowel, including focal hyperplasia, benign and malignant adenomatous polyps, and infiltrating carcinoma. A decrease in staining reaction occurred in all proliferate lesions studied, and Wattenberg indicates that this finding appears to reflect an aspect of a metabolic pattern common to both benign and malignant proliferative processes of the large bowel. Burstone (1959a) noted low oxidase activity of a rapidly growing carcinoma of the uterus and a carcinoma of the pharynx. Hannibal *et al.* (1960) studied a large number of human neoplasms with the 4-amino-1-*N*,*N*-di-

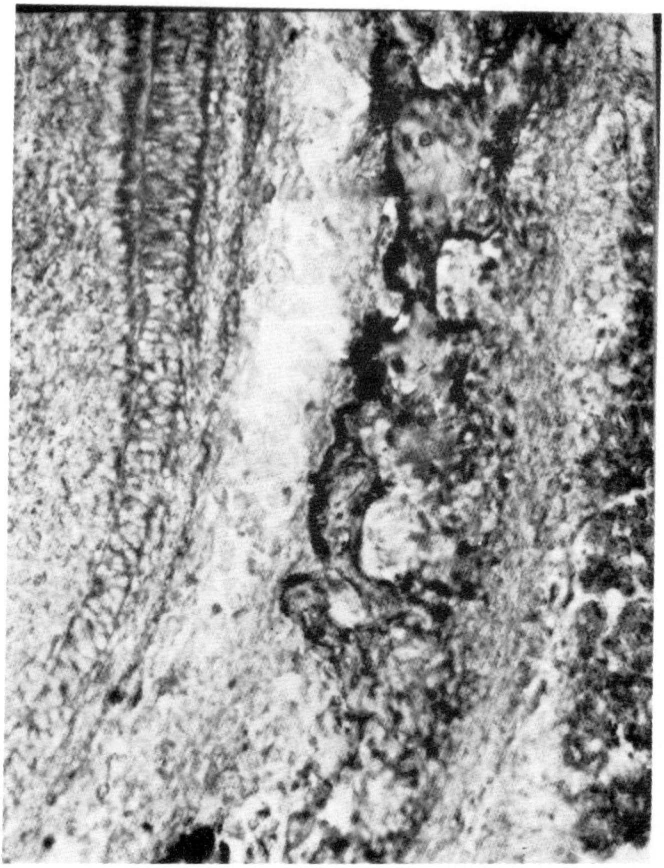

FIG. 8. Cytochrome oxidase activity of osteoclasts in unfixed frozen section through immature hamster head. *p*-Aminodiphenylamine method (× 195).

methylnaphthylamine technique and observed that cytochrome oxidase activity in tumors was usually less than that found in the cells of origin. They also had the impression that undifferentiated carcinomas contained less enzyme than did the well differentiated growths.

The most definitive study dealing with localization of cytochrome oxidase in skin tumors, as well as other dermatological diseases, has been reported by Braun-Falco (1961) who employed the *p*-aminodiphenylamine procedure of Burstone. Braun-Falco noted that acanthotic epithelium exhibited an increased cytochrome oxidase reaction, particularly in the mitotically active basal cell layers and within the rete pegs, while normal hyperkeratotic and parakeratotic layers were unstained. The behavior of cytochrome oxidase in tumors was found to vary. In benign tumors (dermatofibroma and histocytoma) the oxidase reaction corresponded to that of the normal cells from which the tumors derive. Pigmented nevi showed an increase in oxidase activity. In melanosis circumscripta the cells proliferating in the epidermis were considerably more oxidase positive than the adjacent epidermal cells, resulting in a striking contrast. Melanomas were also characterized by an intensive cytochrome oxidase activity which seemed associated with mitochondria. Some variation in the intensity of staining, however, was found in this type of tumor. The activity of basalomas was uniform and roughly corresponded in intensity to that of normal basal cells. In epidermoid carcinomas the mitotically active parenchymal areas in the vicinity of the stroma were characterized by a strongly positive reaction, while toward the center, particularly adjacent to the "pearls," the

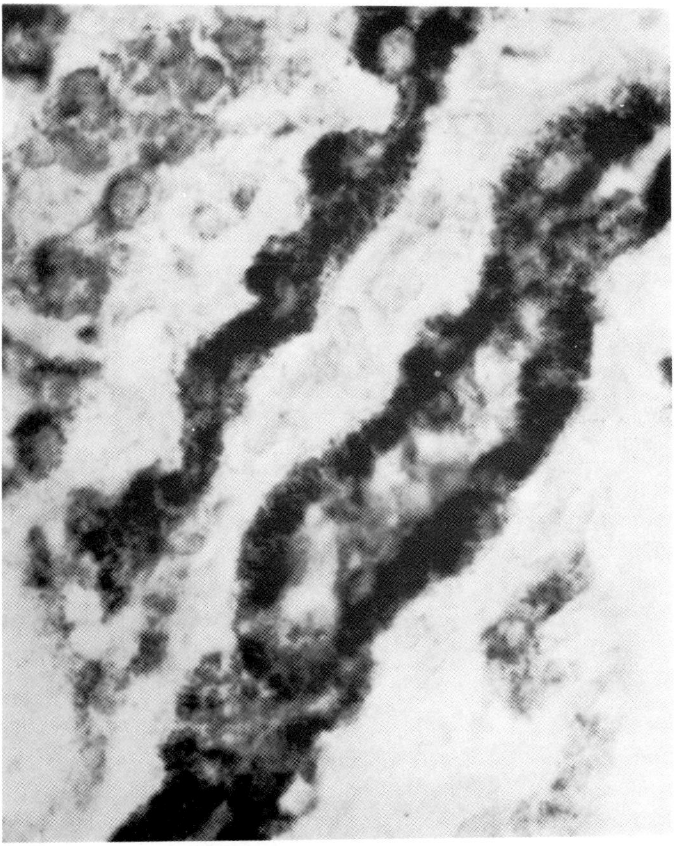

FIG. 9. Cytochrome oxidase activity of rat kidney tubules. *p*-Aminodiphenylamine method (× 300).

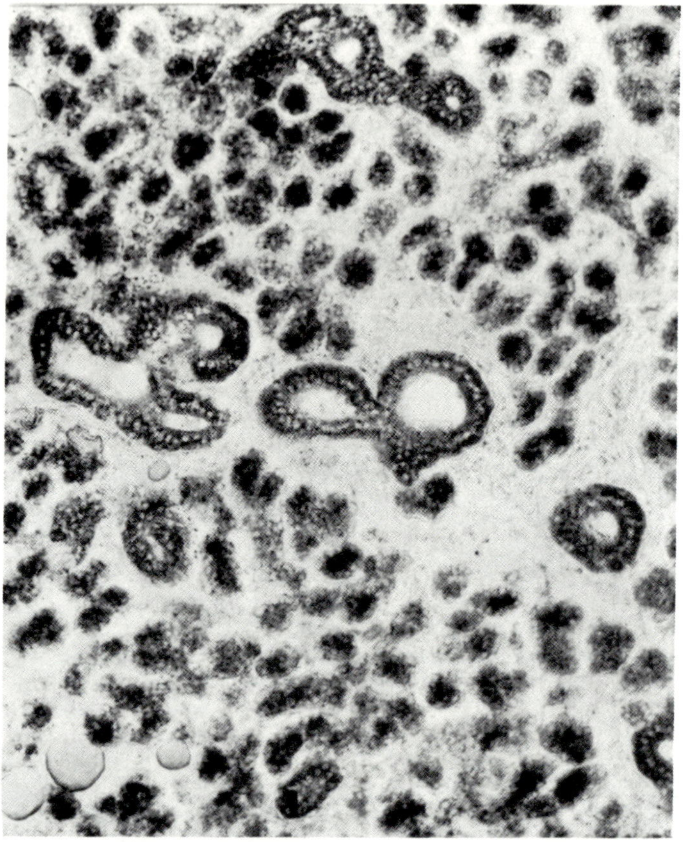

FIG. 10. Human parotid gland. *p*-Aminodiphenylamine method (× 143).

cytoplasmic reaction was considerably weaker. The intense staining in certain areas of keratoacanthomas was noted, and it was indicated that this reaction may offer a suggestion as to the histogenesis of the tumor. Histochemically, the new epithelial proliferations were characterized by a very intensive reaction, which is in complete contrast to the considerably reduced but almost completely absent reaction in the surrounding hyperplastic zones. With increased differentiation of new proliferations, the oxidase reaction fell rapidly while glycogen became concentrated at the same time. The limited degree of growth of these tumors may be related to an early change in anaerobic cell metabolism. In summary, Braun-Falco noted a close correlation between staining patterns of both cytochrome oxidase and succinic dehydrogenase techniques.

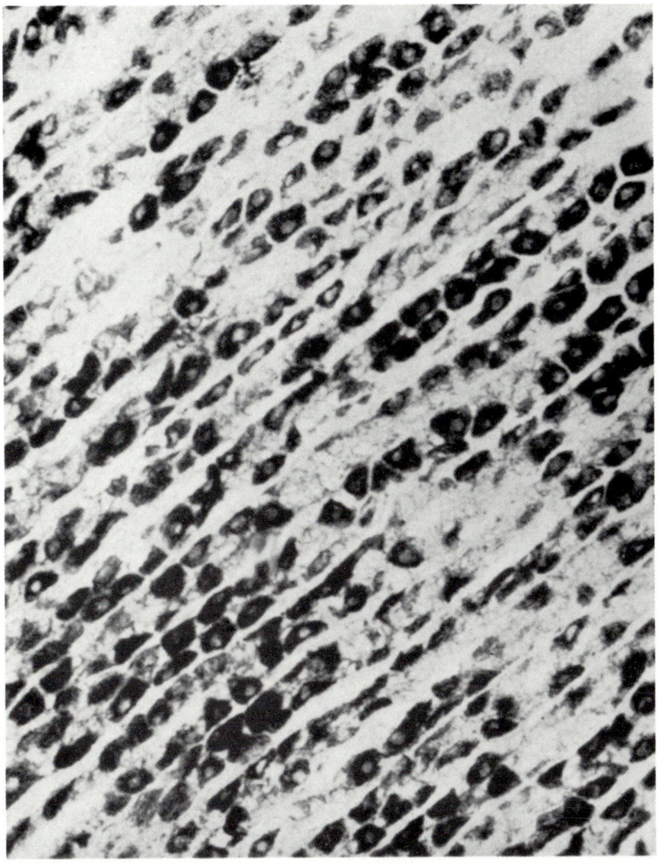

FIG. 11. Cytochrome oxidase activity of parietal cells of rat stomach. *p*-Aminodiphenylamine and 8-amino-1,2,3,4-tetrahydroquinoline (× 285).

## II. PEROXIDASE

Peroxidases are hemoproteins which catalyze the transfer of two electrons from substrates to hydrogen peroxide, forming water and oxidized donors (e.g. dyes). Although peroxidases are widely distributed in plants, their activity is limited to relatively few sites in mammalian tissues including leukocytes and milk. In living cells, peroxidases function by using the hydrogen peroxide which is formed as a result of the reaction of oxygen with a number of autoxidizable substances including flavins and ascorbic acid to oxidize substances which combine with oxygen only with difficulty. Of interest is the observation of Agner (1941) that verdoperoxidase of leukocytes, in the presence of hydrogen peroxide, can detoxify tetanus and

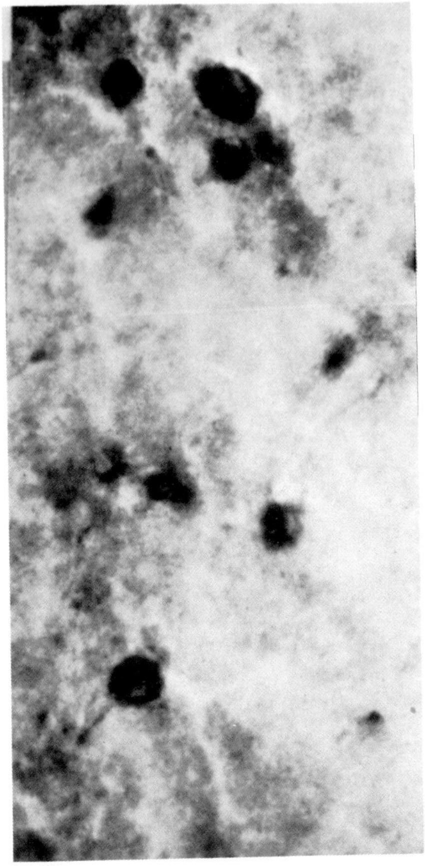

FIG. 12. Cytochrome oxidase activity of nerve cells in nucleus coeruleus of one year old child (× 285) (courtesy Dr. Reinhard Friede).

diphtheria toxins; a finding which may elucidate one of the primary leukocytic functions. Peroxidases do not directly decompose hydrogen peroxide, but "activate" to oxidize a large number of hydroquinones, arylamines, aliphatic compounds including alcohols and aldehydes, and inorganic substances (Theorell, 1951, p. 397; 1956; Sumner and Somers, 1953, p. 220; Mason, 1957; Paul, 1960, p. 277).

The types of compounds which may behave as electron donors for peroxidase are shown in Table 2 (Jayle, 1939; Mason, 1957). Mann and Saunders (1935) have reviewed some of the earlier work which dealt with the peroxidatic oxidation of guaiacol to tetraguaiacol, pyrogallol to purpurogallin, and *o*-phenylenediamine to 2,3-diaminodiphenazine. Mann and Saunders (1935) specifically studied the oxidation of aniline and *p*-

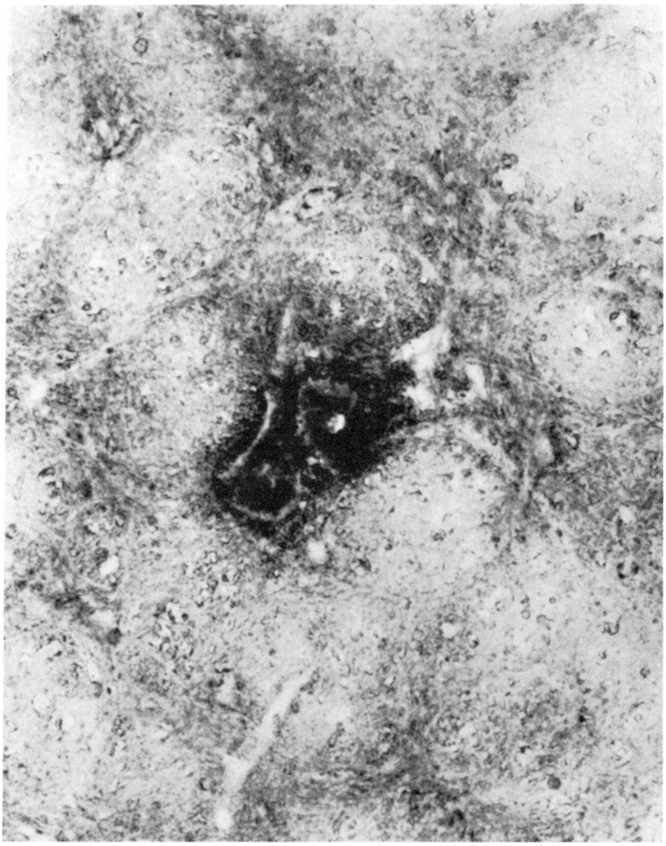

FIG. 13. Cytochrome oxidase activity of large neurons and adjacent fibers from monkey brain (medulla). *p*-Aminodiphenylamine method (× 155).

aminodiphenylamine by peroxidase, and suggested that with both compounds the dyestuff formed exhibited a quinonoid structure similar to that obtained by direct chemical oxidation of the aforementioned amines [Chapter 4 (IX)]. Daniels and Saunders (1951) have studied the effect of peroxidase upon *p*-anisidine and found that during the oxidation the methoxy group is eliminated as methyl alcohol. A similar effect was observed during the oxidation of *p*-chloroaniline, in that the reaction involved a dehalogenation (Daniels and Saunders, 1953). More recently, Booth and Saunders (1956) extended the aforementioned work to include phenolic compounds as substrates. Among the compounds studied and characterized were oxidation products of mesitol and guaiacol. Since two electrons and two protons are ultimately transferred to each molecule of hydrogen peroxide,

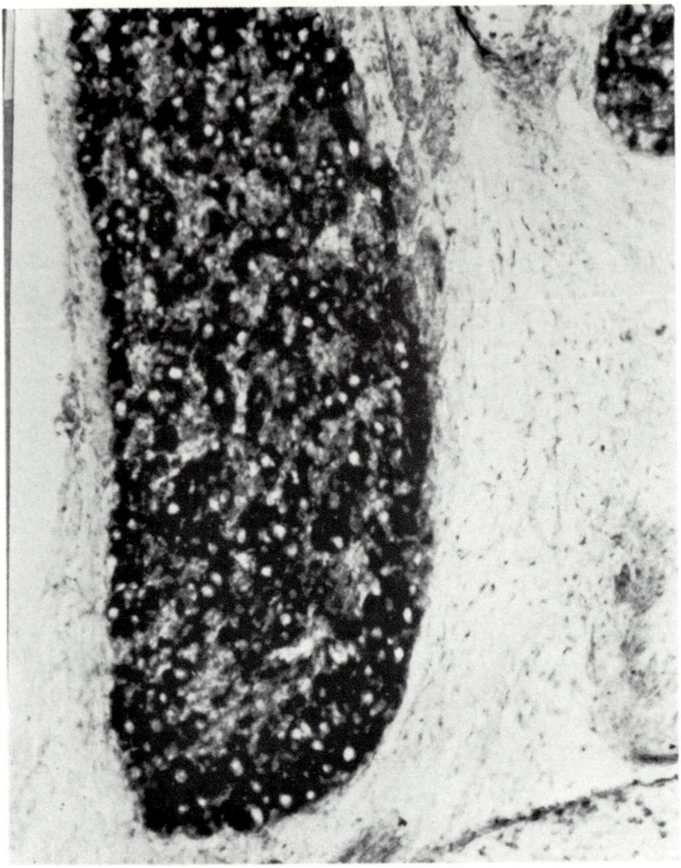

FIG. 14. Cytochrome oxidase activity of ganglion cells from trigeminal ganglion of newborn rat. *p*-Aminodiphenylamine and 8-amino-1,2,3,4-tetrahydroquinoline (× 143).

the peroxidase-peroxide complex undergoes a two-stage reduction with one electron donor. This hypothesis tends to be confirmed by the previously cited work of Saunders and group.

## A. Histochemical Methods

Historical review as well as discussion of more recent methodology with reference to the histochemical demonstration of peroxidases are discussed by Gomori (1952, p. 162) and Lison (1960, p. 589). The histochemical demonstration of mammalian peroxidase has concerned the granular leukocytes which contain verdoperoxidase (Agner, 1941). Agner employed the M "nadi" reaction (see Section I, on cytochrome oxidase) in a study of

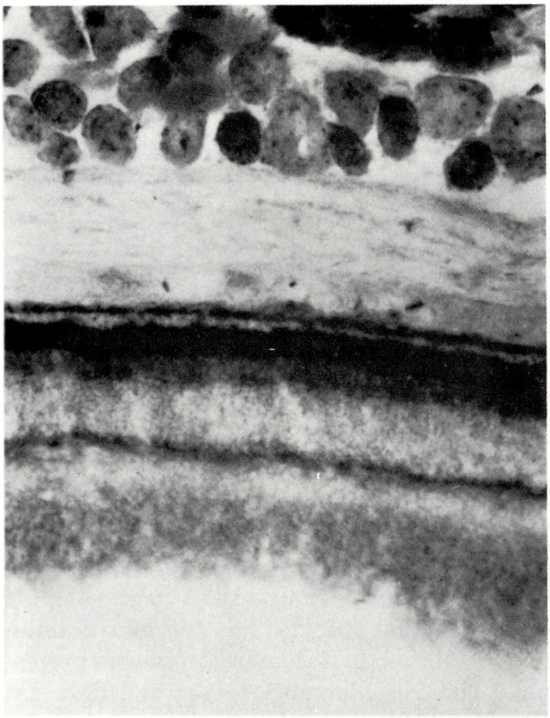

FIG. 15. Cytochrome oxidase activity in retina of rat. The outer rims of the rods and cones are intensely stained. *p*-Aminodiphenylamine method (× 132) (courtesy Drs. M. Niemi and E. Merenmies).

TABLE 2

COMPOUNDS WHICH ARE ELECTRON DONORS FOR PEROXIDATIC OXIDATIONS, AND WHICH ARE CAPABLE OF FORMING OXIDATION DYES[a]

| | |
|---|---|
| I. Aromatic amines: | Aniline, *o*-, *m*-, *p*-toluidine, mesidine, *o*-anisidine, *o*-, *m*-, *p*-chloroaniline, *o*-, *m*-, *p*-phenylenediamine, benzidine, dimethyl-*p*-phenylenediamine, phenyl ethylamine, *p*-bromoaniline, *p*-iodoaniline, tyrosine, tryptophan, indole, indolylacetic, indolylpropionic, indolylbutyric acid |
| II. Phenols: | Catechol, hydroquinone, resorcinol, dihydroxyphenylalamine, *p*-hydroxydiphenyl, guaiacol, vanillan, *p*-cresol, mesitol, durenol, pyrogallol |
| III. Aromatic acids: | Benzoic, salicylic, protocatechuic, gallic, *p*-aminobenzoic acids |
| IV. Miscellaneous substances: | Leucomalachite green, leucophenolphthalein, ferrocytochrome c, TPNH, ascorbic acid, nitrite, tetanus toxin, hypertensin, pepsin, trypsin, chymotrypsin, invertase, uric acid |

[a] Jayle (1939); Mason (1957).

leukocytes from pus and found that the staining could be abolished by addition of catalase. Lillie and Burtner (1953) also confirmed the enzymatic nature of the several reactions given by human neutrophil leukocytes. It is important, however, to note that the M "nadi" reaction of myeloid granules may be related to a non-enzymatic effect (Gomori, 1952, p. 164). Person *et al.* (1961) questioned the specificity of the M "nadi" reaction with reference to peroxidase, and suggested the possibility that other heme-protein enzymes may be demonstrated. Gomori (1952, p. 163) indicates that peroxidase activity is associated with hemoglobin and may be demonstrated in the latter by leuco-dyes (Appendix 11B, Section 5, *c*) or with α-naphthol (Appendix 11B, Section 5, *b*). He points out that while the activity of myeloid granules is rapidly destroyed by heating to 75–80° or by extraction with warm chloroform-methanol, the activity of hemoglobin is entirely resistant to these influences. Gomori (1952, p. 162) has reviewed the older histochemical techniques for the demonstration of peroxidase which employed benzidine, α-naphthol, and various leuco dyes. Lison (1931; 1960, p. 589) suggested that benzidine is oxidized to a blue quinhydrone or black quinone imine dye. It would, however, appear more likely that the benzidine oxidation dye has a more complex linear quinonoid structure similar to that obtained by oxidation of aniline or *p*-aminodiphenylamine. In addition to the aforementioned reagents, other substrates including *o*-phenylenediamine (Mitsui and Kamezawa, 1955) and thioindoxyls (Pearse, 1954) have been recommended. The *p*-aminodiphenylamine cytochrome oxidase techniques of Burstone (1959a, 1960b, 1961) should also lend themselves to the demonstration of peroxidase.

### B. Inhibitors

Peroxidases are completely inhibited by cyanide ($10^{-2}$ *M*), and strongly inhibited by sodium azide ($10^{-3}$ *M*) and hydroxylamine ($10^{-2}$ *M*). Myeloperoxidase is slightly inhibited by fluoride ions ($10^{-2}$ *M*). The enzyme is not significantly affected by 70% ethanol, 10% formalin, or ethanol-formalin mixtures. The optimal concentration of peroxide in the substrate mixture is $10^{-3}$ *M*, with higher concentrations having an inhibitory effect.

## III. MONOAMINE OXIDASE

The biochemistry of monoamine oxidase (tyraminase) (MAO) has been reviewed by Zeller (1951, p. 536), Blaschko (1952), and Sumner and Somers (1953, p. 304). MAO catalyzes the oxidative deamination of certain amines to aldehydes:

$$\text{Amine} + O_2 + H_2O \rightarrow \text{aldehyde} + NH_3 + H_2O_2$$

MAO is an insoluble relatively stable enzyme which is associated with mitochondria, and is widely distributed in mammalian tissues including liver, intestine, kidney, brain, lung, adrenals, and skin. MAO has been related to the detoxication of monoamines of the intestinal mucosa and the possible *in vivo* inactivation of adrenalin. More recently it has been found that 5-hydroxytryptamine (serotonin) is deaminated by MAO to 4-hydroxyindoleacetaldehyde (Udenfriend *et al.*, 1956; Sjoerdsma *et al.*, 1955) and excreted as 5-hydroxyindoleacetic acid.

The enzyme, which has a pH optimum of 7.0–8.0, is known to oxidize both aliphatic and aromatic amines. Biochemical measurement of MAO activity has employed tyramine, adrenaline, and indigo disulfonate as substrates. Blaschko (1952) originally noted that MAO catalyzed the reduction of tetrazolium salts, a finding that was studied in more detail by Weissbach *et al.* (1957). The latter group found that tryptamine and serotonin were the most effective substrates for the reduction of 2-(*p*-iodophenyl)-3-(*p*-nitrophenyl)-5-phenyltetrazolium chloride (INT). The possible reactions by which INT may be reduced through the action of MAO are shown in Chart 6.

## A. Histochemical Methods

Histochemical techniques for MAO are based upon the reduction of tetrazolium salts and also upon the demonstration of the aldehyde produced during the oxidation of amino acids. Oster and Schlossman (1942) devised a procedure which involved treating sections, in which plasma had been made inactive, with tyranide followed by Schiff's reagent. The aldehyde presumably formed by the enzymatic oxidation of tyramine is localized as an indigo-blue pigment. Gomori (1952, p. 161) was critical of the method because he felt that neither the enzyme nor the aldehyde formed was sufficiently insoluble to remain *in situ*. In addition, the oxidation of tyramine by amine oxidase is reported to go beyond the aldehyde stage unless the aldehyde formed is promptly trapped by a substance such as semicarbazide. Gomori also noted that positive reactions are obtained even

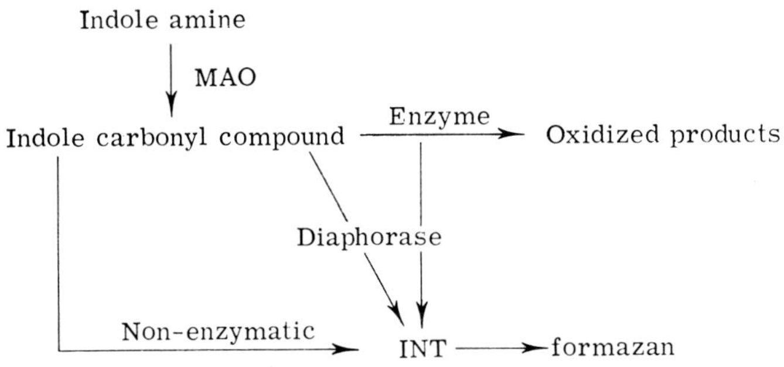


CHART 6. Reactions by which INT may be reduced through the action of MAO (Weissbach *et al.*, 1957).

in the absence of substrate and after treatment of tissue sections with strong enzyme inhibitors such as trichloroacetic acid. Nevertheless, the work of Oster and Schlossman (1942) is significant in that it represents an interesting if not ingenious approach to the demonstration of oxidative enzymes, and one which undoubtedly has had an influence upon subsequent successful modifications. Blaschko and Hellmann (1953) incubated frozen sections in tryptamine or 5-hydroxytryptamine and obtained a brown insoluble pigment which was believed to be a condensation product of the aldehydes. The reaction was abolished by known oxidase inhibitors including pentamidine. Koelle and Valk (1954) found that the aforementioned procedure was insensitive and that there was some question as to its specificity. These authors applied a somewhat different principle in an attempt to visualize the aldehyde formed. Their substrate mixture included tryptamine and 2-hydroxy-3-naphthoic acid hydrazide. The resulting hydrazone was developed into a dye by treating sections with Diazo Blue B (Appendix 11B, Section 6). The reaction was completely inhibited by isonicotinic acid 2-isopropylhydrazide (Marsilid). Koelle and Valk (1954) obtained localizations in a number of sites in cat and rabbit tissues including parietal cells of the stomach, smooth muscle fibers, and hepatic cells. Because of the fact that 2-hydroxy-3-naphthoic acid hydrazide will also react with certain lipids, there exists some danger of artefactual staining. Eder (1957) applied the aforementioned Koelle-Valk method in the study of MAO distribution in a variety of guinea pig tissues.

As previously mentioned, Blaschko (1952) first reported that MAO in the presence of an amine catalyzes the reduction of tetrazolium salts. Francis (1953) employed this principle for the histochemical demonstration of the enzyme in frozen sections which were incubated in a substrate solution containing neotetrazolium and tyramine. A similar approach was reported by Glenner *et al.* (1957) who evolved a tryptamine-tetrazolium procedure in which a number of readily available tetrazolium salts could be utilized (Fig. 16). Among these were blue tetrazolium (BT), 2-(*p*-iodophenyl)-3-(*p*-nitrophenyl)-5-phenyltetrazolium chloride (INT), and nitro-blue tetrazolium (nitro-BT) (Appendix 11B, Section 7). Although Weissbach *et al.* (1957) noted that there are limitations in the use of tetrazolium salts for *in vitro* studies, they nevertheless may be usefully employed in the histochemical localization of MAO (Chart 6).

## B. Inhibitors

Compounds which will inhibit the enzyme have been reported by Friedenwald and Herrmann (1942), Zeller and Barsky (1952), Zeller *et al.* (1955), and Gardner *et al.* (1960, 1961). Among the most effective inhibitors of MAO are Marsilid, phenylhydrazine, *p*-chloromercuribenzoate, alcohols including ethanol, hydroxylamine, semicarbazide, and atabrine.

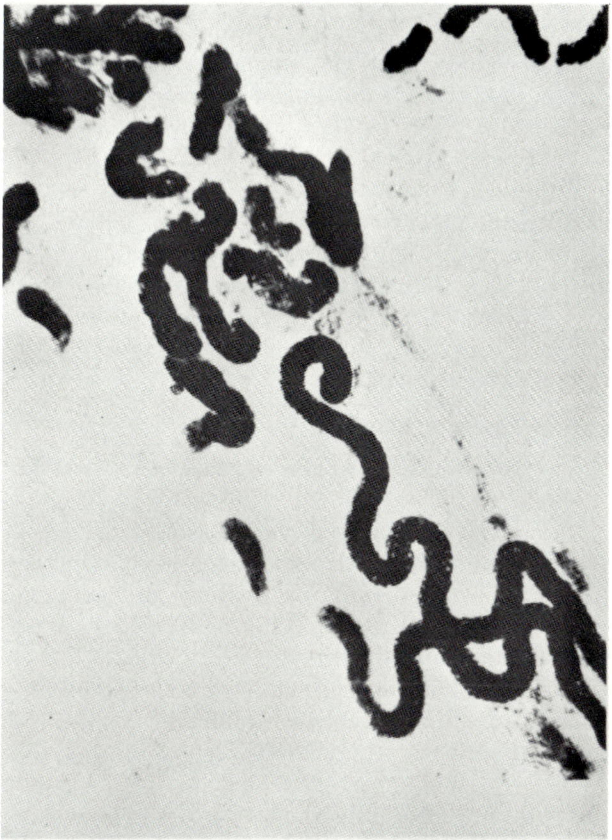

FIG. 16. MAO activity of human eccrine gland. Tetrazolium method (× 150) (courtesy Dr. William Montagna).

With reference to histochemical techniques pre-incubation of sections in 0.01 *M* concentration of the aforementioned compounds will produce complete or almost complete inhibition of the enzyme. Urea (3 *M*) will irreversibly inhibit MAO. More recently, Gardner *et al.* (1960, 1961) reported the synthesis of a number of improved MAO inhibitors which include alkyl and aryl derivatives of picolinoylhydrazine and 5-methyl-3-isoxazolylcarbonylhydrazine. Several reversible MAO inhibitors were also described, but none appear to have been applied as histochemical reagents.

## C. Monoamine Oxidase in Neoplasms

Wattenberg (1959a) reported little or no staining of the colonic mucosa although increased staining was observed in polyps and in some areas associated with malignant changes.

## IV. TYROSINASE AND DOPA-OXIDASE

Tyrosinase is a copper containing phenol oxidase which catalyzes the aerobic oxidation of tyrosine and several phenols (Dawson and Tarpley, 1951, p. 454; Kertész, 1952; Sumner and Somers, 1953, p. 240; Lerner, 1953). When tyrosine is acted upon by tyrosinase, the first product is dihydroxyphenylalanine. The latter is converted to the *o*-quinone, which then condenses to form hallochrome, a red pigment. The red pigment spontaneously is transformed into the black pigment melanin.

In 1917 Bloch found that melanocytes of the basal cell layers of the epidermis were stained by L-3,4-dihydroxyphenylalanine (DOPA). The latter was oxidized to a blackish pigment which appeared to be indistinguishable from natural melanin.

### A. Histochemical Methods

Among the best reviews of the DOPA-oxidase and tyrosinase techniques are those of Gomori (1952, p. 159), Van Duijn (1953), Lillie (1954, p. 229) and Rappaport (1955). Since non-specific phenol oxidase will act upon a number of phenolic and naphtholic compounds to produce dyes, Gomori (1952, p. 160) recommends that control sections be incubated with a substrate other than DOPA (e.g. $\alpha$-naphthol or tyrosine). According to the modification of Laidlaw and Blackberg (1932), unfixed frozen sections or else tissues fixed in formalin for no more than several hours are incubated in a 0.1% solution of dihydroxyphenylalanine at pH 7.3–7.5 (Appendix 11B, Section 8). Fitzpatrick *et al.* (1950) reported that tyrosine could be used in place of DOPA under certain conditions. They incubated 1–2 mm. thick slices of formalin fixed skin which were subsequently embedded in paraffin and sectioned.

Kukita and Fitzpatrick (1955) and Fitzpatrick and Kukita (1956; 1959) reported on a histochemical autoradiographic method for the demonstration of tyrosinase in normal and neoplastic tissues. Tissue sections are incubated with $C^{14}$-labeled radioactive tyrosine which is subsequently incorporated through enzymatic activity into melanin granules, and revealed photographically.

Rappaport (1955) employed frozen-dried paraffin embedded tissues in conjunction with the DOPA-oxidase method, and reported that the staining intensity of melanocytes as well as the number of positive reactions was greater in frozen-dried tissues than those fixed in formalin (Fig. 17). Rappaport noted that fixation in 5% formalin for only 30 minutes decreased the stainability of melanocytes.

### B. Inhibitors

Tyrosinase is inhibited by compounds which form stable complexes with copper, including KCN, diethyldithiocarbamate, hydrogen sulfide,

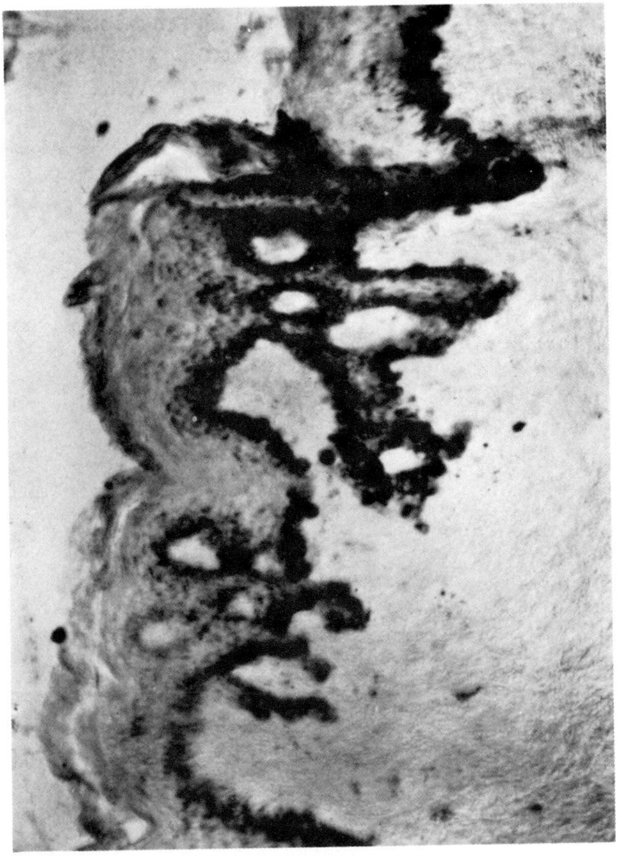

FIG. 17. DOPA oxidase, frozen-dried human skin (× 150) (courtesy Dr. B. Z. Rappaport).

and carbon monoxide. Carbon monoxide inhibition is not light reversible. Other inhibitors include sodium azide, thiourea, thiouracil, cysteine, and 2,3-dimercaptopropanol (BAL). Agents which react with sulfhydryl groups of proteins do not appear to exhibit an inhibitory effect.

## C. DOPA-Oxidase in Neoplasms

Biochemical studies have indicated that animal melanomas contain considerable amounts of the enzyme, with the activity depending upon the type of melanoma (Dawson and Tarpley, 1951, p. 481). Fitzpatrick and Kukita (1956) employed their autoradiographic method using labeled tyrosine and found positive melanocytes in 22 of 24 primary melanomas, in all metastatic melanomas, in 3 of 16 junctional nevi, 2 of 20 compound nevi, and in all blue nevi which were studied. Tyrosinase could not be demonstrated by this technique in dermal nevi and in a number of benign and malignant tumors. Chang *et al.* (1959) employing the aforementioned labeled tyrosinase technique also found a positive reaction in the majority of primary melanomas which were studied, while less than half of the non-pigmented melanomas exhibited a positive reaction. These authors also reported a micro-spectrophotometric method for the semi-quantitative determination of the enzyme and found the highest concentration in viable tumor cells.

APPENDIX 11A

# SYNTHESIS OF AMINE SUBSTRATE

1. *4-Amino-1-N,N-dimethylnaphthylamine (4-Aminodimethyl-1-naphthylamine Hydrochloride) (Friedländer and Welmans, 1888; Gokhlé and Mason, 1931; Hodgson and Crook, 1936; Nachlas et al., 1958)*

This reagent for cytochrome oxidase may be prepared by a nitrosation-reduction technique (Friedländer and Welmans, 1888) or by an azo-coupling-reduction procedure (Hodgson and Crook, 1936).

The starting point is *N,N*-dimethyl-α-naphthylamine which is commercially available or which may be prepared by the method of Gokhlé and Mason (1930).

In the diazotization technique sulfanilic acid (17.3 g.) is diazotized and the suspension, after treatment with urea and sodium acetate, is poured into a solution of 16.0 ml. of dimethyl-α-naphthylamine in 200.0 ml. of glacial acetic acid at 0°. The 4-*p*-sulfobenzeneazodimethyl-α-naphthylamine separates as fine steel-blue needles. The mixture is stirred for 15 minutes, concentrated hydrochloric acid (60.0 ml.) and stannous chloride (54.0 g.) are added, and the mixture is then heated at 70–80° until the red color disappears. The cooled solution is made strongly alkaline with sodium hydroxide and the liberated base is extracted with ether. Dry HCl gas precipitates 4-amino-1-*N,N*-dimethylnaphthylamine dihydrochloride from the extract as a sticky mass which rapidly becomes crystalline (turning brown in air) ($C_{12}H_{14}N_2 \cdot 2HCl$). Anal., calcd.: HCl, 28.2%. Found: HCl, 27.8%.

Gokhlé and Mason (1931) employed the nitrosation procedure of Friedländer and Welmans (1888). Dimethyl-α-naphthylamine (8.5 g.) is dissolved in 15.0 ml. of concentrated HCl with the addition of ice, and a nearly saturated aqueous solution containing 3.5 g. of sodium nitrite is added during a 10 minute period. After 30 minutes, the yellow nitroso compound is filtered off and pressed. The 4-amino-1-*N,N*-dimethylnaphthylamine is readily formed upon reduction of the nitroso compound with zinc dust in acetic acid.

## APPENDIX 11B

# HISTOCHEMICAL METHODS

### 1. *"Nadi" Modification (Moog, 1943; Gomori, 1952, p. 157)*

Frozen sections (preferably thick) or razor-blade slices may be employed.

To about 25.0 ml. of phosphate buffer pH 7.2–7.6 are added 1.0–2.0 ml. of a 1% solution of α-naphthol in 40% ethanol, and an equivalent amount of 1% aqueous *N,N*-dimethyl-*p*-phenylenediamine hydrochloride. The sections are incubated in a shallow dish for 10–60 minutes until dye formation is grossly visible. The sections are then rinsed in water and mounted in glycerol-gelatin. Moog (1943) recommended use of the specific inhibitors, sodium azide (0.005 *M*) and phenylurethan (0.003 *M*), with which sections are treated prior to as well as during incubation.

### 2. *Cytochrome Oxidase, 4-Amino-1-N,N-dimethylnaphthylamine (ADN) Method (Nachlas et al., 1958)*

The following modification of the standard "nadi" mixture was reported by Nachlas *et al.* (1958):

| | |
|---|---|
| 1. Phosphate buffer pH 7.4, 0.1 *M* | 3.0 ml. |
| 2. α-Naphthol (1.0 mg./ml.) | 4.0 ml. |
| 3. 4-Amino-1-*N,N*-dimethylnaphthylamine HCl (or oxalate) (ADN) (2.0 mg./ml.) | 4.0 ml. |
| 4. Cytochrome c (5.0 mg./ml.) | 3.0 ml. |
| 5. Catalase (30.0 μg./ml.) | 1.0 ml. |

The reagents are mixed in a Coplin jar with the exception of ADN, which is added immediately prior to incubation of the sections. Incubation periods range from 10 minutes to 1 hour. In the case of highly active tissues, the cytochrome c may be omitted from the substrate medium. Following incubation the sections are rinsed briefly in saline and mounted in glycerol-gelatin. Nachlas *et al.* (1958) reported that the resulting dye is lipid soluble, susceptible to fading, and crystallizes in the mounted section.

### 3. *Cytochrome Oxidase, p-Aminodiphenylamine Method (Burstone, 1959a, 1960a, 1961)*

The *p*-aminodiphenylamine techniques employ the aforementioned compound (it is important that the base and not the amine hydrochloride be used) in conjunction with either a naphthol, reactive methylene compound, or another amine (Charts 2 and 3). In general, a combination of amines offers the most sensitive substrate for demonstration of the enzyme. Although *p*-aminodiphenylamine may be employed alone, the addition of another amine tends to reduce the particulate size of the final dyestuff and also minimizes crystallization. Among the useful naphthols, methylene compounds, and arylamines are: 1-hydroxy-2-naphthanilide, 2-cyanoacetylcoumarone, 1-phenyl-3-methylpyrazolone, *p*-methoxy-*p*-aminodiphenylamine, 3-amino-9-ethylcarbazole, and 8-amino-1,2,3,4-tetrahydroquinoline. The last compound produces deep blue dyes while the others give reddish-brown shades.

The incubating solutions are prepared as follows:

1. Ten to fifteen mg. of *p*-aminodiphenylamine (base) and 10–15 mg. of another amine, methylene compound, or naphthol are placed in a 50.0 ml. Erlenmeyer flask and dissolved by the addition of 0.5 ml. reagent ethanol. In the case of 8-amino-1,2,3,4-tetrahydroquinoline, which is an oil, one drop is used.
2. Thirty-five ml. of distilled water are added (a cloudy solution results) followed by 15.0 ml. of 0.2 *M* tris buffer pH 7.4.
3. Following addition of buffer the solution is shaken and filtered through folded filter paper into a Coplin jar.

Tissues are incubated for periods of 15–60 minutes. However, in view of the remarkable stability of the aforementioned substrate solutions, longer incubation times may be employed provided that the specificity of the reaction is ascertained by the interposition of known inhibitors such as cyanide.

Following incubation the slides are transferred to a Coplin jar containing 10% cobaltous acetate [$Co(C_2H_3O_2)_2 \cdot 4H_2O$] in 10% formalin (not neutral formalin), plus 5.0 ml. of 0.2 *M* pH 5.2 acetate buffer for 1 hour. A longer chelation-fixation period may be employed. Other metal salts including nickel chloride, ferric ammonium sulfate, cadmium sulfate, uranium nitrate, and lead nitrate may also be used for chelation of the dye. The slides are then washed in running tap water for 5–10 minutes and mounted in glycerol-gelatin or a water miscible mounting medium.

4. *Cytochrome Oxidase, Tetrazolium Method* (*Oda et al., 1958c*)

Unfixed tissue blocks (1–2 mm. thick) and unfixed frozen sections are employed.

The substrate solution is prepared immediately prior to incubation by combining equal volumes (0.2 ml.) of phosphate buffer (0.1 *M*, pH 7.6), *p*-phenylenediamine solution (*p*-PDA) (0.2 *M*), neotetrazolium chloride solution (0.2%) and cytochrome c ($10^{-4}$ *M*). The latter compound may be omitted and substituted by an equal volume of phosphate buffer.

The tissues are incubated at 37° aerobically or anaerobically for 30–60 minutes. The sections are then mounted on slides and post-fixed in 10% formalin. The blocks are similarly fixed and sectioned on a freezing microtome. Oda *et al.* (1958c) mounted sections with glycerol, although it would appear that water-soluble plastic mounting media might be more permanent.

Substrate controls consist of sections incubated in solutions from which *p*-PDA or cytochrome c, or both, have been omitted and replaced by an equal volume of phosphate buffer.

5. *Peroxidase Methods*

Tissues or smears are fixed in acetone, ethanol, or ethanol-formalin (1:10). Gomori (1952, p. 164) states that frozen sections are preferable, although in most cases excellent results are obtained after celloidin or paraffin embedding.

*a. Benzidine method (Washburn, 1928; Gomori, 1952, p. 165)*. A 0.2–0.3% solution of benzidine in 95% ethanol is prepared, and to each 100.0 ml. of this solution is added 0.5 g. sodium nitroprusside dissolved in a few ml. of water. This stock solution will keep for months under refrigeration.

Equal volumes of the stock solution and a 1:50 dilution of 3% hydrogen peroxide are mixed and poured onto the slide and allowed to remain for about 5 minutes. The slide is then washed briefly in water, counterstained, and mounted.

Lillie (1954, pp. 224–227) has reviewed several modifications of this technique which also employ metal salts other than sodium nitroprusside.

*b. M "nadi" method (Lillie, 1954, p. 228)*. Two stock solutions are prepared:

1. α-Naphthol (576 mg.) dissolved in 6.0 ml. of *N* NaOH and diluted with distilled water to 50.0 ml.
2. Dimethyl-*p*-phenylenediamine hydrochloride (690 mg.) dissolved in 50.0 ml. of distilled water.

Equal parts of the two stock solutions are mixed, filtered, and pipetted onto smears or frozen sections. Staining of the myeloid granules develops in about 3 minutes. The sections are then treated with Lugol's solution followed by lithium carbonate. Although no $H_2O_2$ is incorporated in the incubating mixture, the presence of an alkaline α-naphthol probably results in the formation of peroxides.

*c. Lison's zinc-leuco method (Gomori, 1952, p. 164)*. One gram of either acid fuchsin, acid violet, or patent blue is dissolved in 100.0 ml. of a 2% acetic acid solution. Five to ten grams of zinc dust are added and the mixture boiled until the dye is decolorized or bleached to a pale brown shade. An additional 2.0 ml. of concentrated acetic acid are added. The mixture is stored under refrigeration.

For use, 1.0 ml. of 3% $H_2O_2$ is added to 10.0 ml. of the aforementioned filtered stock solution. The mixture is poured onto the slide and allowed to remain for several minutes.

6. *Monoamine Oxidase, Tryptamine-Hydrazone Method* (*Koelle and Valk, 1954*)

Unfixed frozen sections mounted on glass slides are used. Sections are incubated in the following substrate solution:

| | |
|---|---|
| 1. $H_2O$ | 1.35 ml. |
| 2. Tryptamine hydrochloride, 0.1 *M* | 1.0 ml. |
| 3. Hydroxynaphthoic acid hydrazide, 0.005 *M* | 3.0 ml. |
| 4. *N* NaOH | 0.15 ml. |
| 5. Phosphate buffer, 0.2 *M* | 15.0 ml. |
| 6. 40% Sodium acetate | 7.5 ml. |

The reaction may be accelerated by saturating the substrate solution with oxygen prior to use. Following incubation for two hours at 37° the slides are rinsed in distilled water and developed in an alkaline solution of Diazo Blue B for several minutes at room temperature. The slides are then rinsed in distilled water, fixed in 10% formalin, and mounted in glycerol-gelatin. Control slides may be pre-incubated in the presence of a known inhibitor, Marsilid, 0.1 *M* (0.15 ml.).

7. *Monoamine Oxidase, Tetrazolium Method* (*Glenner et al., 1957*)

Unfixed frozen sections mounted on glass slides are used.

Although a number of tetrazolium salts may be employed, the following incubating solution is recommended:

| | |
|---|---|
| 1. Tryptamine hydrochloride | 25.0 mg. |
| 2. Sodium sulfate (may be omitted) | 4.0 mg. |
| 3. Nitro-blue tetrazolium | 4.0 mg. |
| 4. 0.1 *M* Phosphate buffer pH 7.6 | 5.0 mg. |
| 5. Distilled water | 15.0 ml. |

The mixture is warmed to 37° and the sections are incubated at this temperature for 30–45 minutes. They are then washed in running water, fixed overnight in 10% formalin, dehydrated, cleared in graded acetone-xylene mixtures, and mounted in Permount.

8. *DOPA-Oxidase, Modifications of Laidlaw and Blackberg* (*1932*), *Gomori* (*1952, p. 160*), *and Rappaport* (*1955*)

Unfixed frozen sections, tissues fixed in formalin for no more than several hours, or frozen-dried paraffin embedded sections may be employed. Sections are incubated at 20–37° in a substrate solution containing 0.1% dihydroxyphenylalanine (DOPA) buffered with 0.1 *M* of phosphate buffer to pH 7.3–7.5. It is advisable to change the incubating solution once or twice to avoid deposition of melanin precipitate due to spontaneous oxidation of the substrate. Although the reaction is much more rapid at pH 7.7 the danger of non-specific precipitation is also increased. Rappaport (1955) found that the optimum incubation period was approximately 4 hours at room temperature. Following incubation the sections are rinsed in water, counterstained as desired, dehydrated, and mounted in a water-soluble synthetic mounting medium.

## Commercial Sources of Reagents

1. Aldrich Chemical Company, Milwaukee, Wisconsin
2. British Drug House, Poole, England (U.S.A. Distributor, The Ealing Corp., Natick, Massachusetts)
3. Dajac Laboratories, The Borden Chemical Company, Philadelphia, Pennsylvania
4. Eastman Kodak, Rochester, New York
5. Mann Research Laboratories, New York, New York
6. Nutritional Biochemicals Corp., Cleveland, Ohio
7. Sigma Chemical Company, St. Louis, Missouri

## References

Agner, K. (1941) *Acta Physiol. Scand.* **2,** Suppl. 8, 51–62.
Allen, J. R., Sullivan, B. A., and Dobson, H. L. (1960) *Arch. Biochem. Biophys.* **86,** 6–9.
Anschütz, L., Weber, E., and Runkel, K. (1906) *Ann.* **346,** 361–381.
Blaschko, H. (1952) *Pharmacol. Revs.* **4,** 415–458.
Blaschko, H., and Hellmann, K. (1953) *J. Physiol.* **122,** 419–426.
Bloch, B. (1917) *Arch. Dermatol. and Syphilol.* **124,** 129–208.
Booth, H., and Saunders, B. C. (1956) *J. Chem. Soc.* pp. 940–948.
Borei, H. G., and Björklund, U. (1953) *Biochem. J.* **54,** 357–362.
Braun-Falco, O. (1961) *Arch. klin. u. exptl. Dermatol.* **214,** 176–224.
Burstone, M. S. (1959a) *J. Histochem. and Cytochem.* **7,** 112–122.
Burstone, M. S. (1959b) *Nature* **184,** 476–477.
Burstone, M. S. (1960a) *J. Histochem. and Cytochem.* **8,** 63–70.
Burstone, M. S. (1960b) *J. Histochem. and Cytochem.* **8,** 225–226.
Burstone, M. S. (1960c) *Oral Surg. Oral Med. Oral Pathol.* **13,** 1501–1505.
Burstone, M. S. (1961) *J. Histochem. and Cytochem.* **9,** 59–65.
Burstone, M. S., and Miller, F. N. (1961) *Am. J. Clin. Pathol.* **35,** 118–121.
Chang, J. P., Speece, A. J., and Russell, W. O. (1959) *In* "Pigment Cell Biology" (M. Gordon, ed.), pp. 359–387. Academic Press, New York.
Clezy, P. S., and Barrett, J. (1961) *Biochem. J.* **78,** 798–806.
Cobenzl, A. (1915) *Chem. Ztg.* **39,** 859–860.
Crawford, D. T., and Nachlas, M. M. (1958) *J. Histochem. and Cytochem.* **6,** 438–444.
Daniels, D. G. H., and Saunders, B. C. (1951) *J. Chem. Soc.* pp. 2112–2118.
Daniels, D. G. H., and Saunders, B. C. (1953) *J. Chem. Soc.* pp. 822–826.

Dawson, C. R., and Tarpley, W. B. (1951) *In* "The Enzymes" (J. B. Sumner and K. Myrbäck, eds.), Vol. II, Part 1, pp. 454–498. Academic Press, New York.
Dewey, M. M., Kahn, R. H., and Conklin, J. L. (1961) *J. Histochem. and Cytochem.* **9,** 605.
Dixon, M., and Webb, E. C. (1958) "Enzymes." Academic Press, New York.
Eder, M. (1957) *Beitr. pathol. Anat. u. Allgem. Pathol.* **117,** 343–396.
Ehrlich, P. (1885) "Das Sauerstoff Bedürfniss des Organismus." Hirschwald, Berlin.
Elderfield, R. C., and Claflin, E. F. (1952) *J. Am. Chem. Soc.* **74,** 2953–2959.
Engel, W. K. (1961) *Nature* **191,** 389–390.
Fiessinger, N., and Roudowska, L. (1912) *Arch. med. exptl. anat. pathol.* **24,** 585–608.
Fitzpatrick, T. B., and Kukita, A. (1956) *J. Invest. Dermatol.* **26,** 173–183.
Fitzpatrick, T. B., and Kukita, A. (1959) *In* "Pigment Cell Biology" (M. Gordon, ed.), pp. 489–524. Academic Press, New York.
Fitzpatrick, T. B., Becker, S. W., Lerner, A. B., and Montgomery, H. (1950) *Science* **112,** 223–225.
Francis, C. M. (1953) *Nature* **171,** 701–702.
Friedenwald, J. S., and Herrmann, H. (1942) *J. Biol. Chem.* **146,** 411–419.
Friedländer, P., and Welmans, P. (1888) *Ber.* **21,** 3124–3130.
Gardner, T. S., Wenis, E., and Lee, J. (1960) *J. Med. Pharm. Chem.* **2,** 133–145.
Gardner, T. S., Wenis, E., and Lee, J. (1961) *J. Med. Pharm. Chem.* **3,** 241–252.
Glenner, G. G., Burtner, H. J., and Brown, G. W. (1957) *J. Histochem. and Cytochem.* **5,** 591–600.
Gokhlé, B., and Mason, F. A. (1930) *J. Chem. Soc.* pp. 1757–1758.
Gokhlé, B., and Mason, F. A. (1931) *J. Chem. Soc.* pp. 118–126.
Gomori, G. (1952) "Microscopic Histochemistry." Univ. Chicago Press, Chicago, Illinois.
Gräff, S. (1922) *Beitr. pathol. Anat. u. allgem. Pathol.* **70,** 1–19.
Gräff, S. (1936) *In* "Handbuch der biologischen Arbeitsmethoden." (E. Abderhalden, ed.), Vol. IV, Part 1, pp. 93–142. Urban & Schwarzenberg, Berlin.
Green, A. G., and Woodhead, A. E. (1910) *J. Chem. Soc.* **97,** 2388–2403.
Griffiths, D. E., and Wharton, D. C. (1961) *Biochem. Biophys. Research Commun.* **4,** 199–203.
Hannibal, M. J., Nachlas, M. M., and Seligman, A. M. (1960) *Cancer* **13,** 1008–1016.
Hazlewood, S. J., Hughes, G. K., and Lions, F. (1938) *J. Proc. Roy. Soc. N. S. Wales* **71,** 462–474.
Hodgson, H. H., and Crook, J. H. (1936) *J. Chem. Soc.* pp. 1500–1503.
Hoffmann, G. T., Rottino, A., and Stern, K. G. (1951) *Blood* **6,** 1051–1058.
Hoffmann-Ostenhof, O. (1953) *Advances in Enzymol.* **14,** 219–260.
Jadhav, G. V., and Rao, S. N. (1936) *J. Indian Chem. Soc.* **13,** 645–648.
Jayle, M. F. (1939) *Bull. soc. chim. biol.* **21,** 14–47.
Keilin, D. (1925) *Proc. Roy. Soc.* **B98,** 312–339.
Keilin, D. (1929) *Proc. Roy. Soc.* **B104,** 206–251.
Keilin, D. (1930) *Proc. Roy. Soc.* **B106,** 418–444.
Keilin, D. (1936) *Proc. Roy. Soc.* **B121,** 165–173.
Keilin, D., and Hartree, E. F. (1938) *Proc. Roy. Soc.* **B125,** 171–186.
Keilin, D., and Hartree, E. F. (1939) *Proc. Roy. Soc.* **B127,** 167–191.
Keilin, D., and Slater, E. C. (1953) *Brit. Med. Bull.* **9,** 89–96.
Kertész, D. (1952) *Biochim. et Biophys. Acta* **9,** 170–179.

Koelle, G. B., and Valk, A. (1954) *J. Physiol.* **126,** 434–447.
Kukita, A., and Fitzpatrick, T. B. (1955) *Science* **121,** 893–894.
Laidlaw, G. F., and Blackberg, S. N. (1932) *Am. J. Pathol.* **8,** 491–498.
Lerner, A. B. (1953) *Advances in Enzymol.* **14,** 73–128.
Lillie, R. D. (1954) "Histopathologic Technic," (2nd ed.). Blakiston, New York.
Lillie, R. D., and Burtner, H. J. (1953) *J. Histochem. and Cytochem.* **1,** 8–26.
Lindemann, H. (1924) *Ber.* **57,** 555–559.
Lison, L. (1931) *Compt. rend. soc. biol.* **106,** 1266–1268.
Lison, L. (1960) "Histochimie et Cytochimie Animales," 3rd. ed., Vol. 2. Gauthier Villars, Paris.
Mann, P. J. G., and Saunders, B. C. (1935) *Proc. Roy. Soc.* **B119,** 47–60.
Mason, H. S. (1957) *Advances in Enzymol.* **19,** 79–233.
Mehler, A. H. (1957) "Introduction to Enzymology." Academic Press, New York.
Meyer, R. (1920) *Ber.* **53,** 1265–1277.
Mitsui, T., and Kamezawa, S. (1955) *Okajimas Folia Anat. Japon.* **27,** 335–344.
Montagna, W., and Yun, J. S. (1961) *J. Histochem. and Cytochem.* **9,** 694–698.
Moog, F. (1943) *J. Cell Comp. Physiol.* **22,** 223–231.
Morell, D. B., Barrett, J., and Clezy, P. S. (1961) *Biochem. J.* **78,** 793–797.
Nachlas, M. M., Crawford, D. T., Goldstein, T. P., and Seligman, A. M. (1958) *J. Histochem. and Cytochem.* **6,** 445–456.
Niemi, M., and Merenmies, E. (1961) *J. Neurochem.* **6,** 206–209.
Oda, T., Sakai, S., and Okazaki, H. (1958a) *Acta Med. Okayama* **12,** 293–301.
Oda, T., Sakai, S., and Okazaki, H. (1958b) *Acta Med. Okayama* **12,** 205–215.
Oda, T., Okazaki, H., and Seki, S. (1958c) *Acta Med. Okayama* **12,** 302–309.
Oster, K. A., and Schlossman, N. C. (1942) *J. Cellular Comp. Physiol.* **20,** 373–378.
Paul, K.-G. (1951) *In* "The Enzymes" (J. B. Sumner and K. Myrbäck, eds.), Vol. II, Part 1, pp. 357–396. Academic Press, New York.
Paul, K.-G. (1960) *In* "The Enzymes" (P. D. Boyer, H. Lardy, and K. Myrbäck, eds), 2nd ed., Vol. 3, pp. 277–328. Academic Press, New York.
Pearse, A. G. E. (1954) *J. Histochem. and Cytochem.* **2,** 475–476.
Person, P., and Fine, A. (1960) *Science* **132,** 43–44.
Person, P., and Fine, A. (1961) *J. Histochem. and Cytochem.* **9,** 190–196.
Person, P., Schneider, R. M., and Scapa, S. (1961) *J. Histochem. and Cytochem.* **9,** 197–201.
Person, P., Burstone, M. S., and Fine, A. S. (1962) *J. Histochem. and Cytochem.* **10,** 75–78.
Pullman, B., and Pullman, A. (1959) *Biochim. et Biophys. Acta* **35,** 535–537.
Rappaport, B. Z. (1955) *Arch. Pathol.* **60,** 444–450.
Rohmann, F., and Spitzer, W. (1895) *Ber.* **28,** 567–572.
Roizin, L. (1955) *J. Neuropath. Exptl. Neurol.* **14,** 47–69.
Sands, R. H., and Beinert, H. (1959) *Biochim. Biophys. Research Commun.* **1,** 175–178.
Schultze, W. H. (1909a) *Beitr. pathol. Anat. u. allgem. Pathol.* **45,** 127–153.
Schultze, W. H. (1909b) *Verhandl. deut. pathol. Ges.* **13,** 235–241.
Schultze, W. H. (1927) *In* "Enzyklopädie der mikroskopischen Technik" (R. Krause, ed.), Vol. 3, pp. 1777–1786. Urban u. Schwarzenberg, Berlin.
Sjoerdsma, A., Smith, T. E., Stevenson, T. D., and Udenfriend, S. (1955) *Proc. Soc. Exptl. Biol. Med* **89,** 36–38.
Slater, E. C. (1949) *Biochem. J.* **44,** 305–318.
Spiegelman, S., and Moog, F. (1945) *Biol. Bull.* **89,** 122–130.

STOTZ, E., SIDWELL, A. E., JR., AND HOGNESS, T. R. (1938) *J. Biol. Chem.* **124,** 733–744.
STRAUS, W. (1954) *J. Biol. Chem.* **207,** 733–743.
SUMNER, J. B., AND SOMERS, G. F. (1953) "Chemistry and Methods of Enzymes," 3rd. ed. Academic Press, New York.
THEORELL, H. (1951) *In* "The Enzymes" (J. B. Sumner and G. F. Somers, eds.), Vol. II, Part 1, pp. 397–427. Academic Press, New York.
THEORELL, H. (1956) *Science* **124,** 467–472.
TUPPY, H., AND PALÉUS, S. (1955a) *Acta Chem. Scand.* **9,** 353–364.
TUPPY, H., AND PALÉUS, S. (1955b) *Acta Chem. Scand.* **9,** 365–374.
UDENFRIEND, S., TITUS, E., WEISSBACH, H., AND PETERSON, R. E. (1956) *J. Biol. Chem.* **219,** 335–344.
VAN DUIJN, P. (1953) *J. Histochem. and Cytochem.* **1,** 143–150.
WAINIO, W. W., AND GREENLEES, T. (1958) *Science* **128,** 87–88.
WASHBURN, A. H. (1928) *J. Lab. Clin. Med.* **14,** 246–250.
WATTENBERG, L. W. (1959a) *Am. J. Pathol.* **35,** 113–138.
WATTENBERG, L. W. (1959b) *Cancer Research* **19,** 1118–1123.
WEISSBACH, H., REDFIELD, B. G., GLENNER, G. G., AND MITOMA, C. (1957) *J. Histochem. and Cytochem.* **5,** 601–605.
WERTHEIMER, E. (1926) *Fermentforschung* **8,** 497–517.
WHITNER, T. C., JR. (1924) *J. Am. Chem. Soc.* **46,** 2326–2329.
WINKLER, F. (1907) *Folia Haematol.* **4,** 323–328.
ZELLER, E. A. (1951) *In* "The Enzymes" (J. B. Sumner and K. Myrbäck, eds.), Vol. II, Part 1, pp. 536–558. Academic Press, New York.
ZELLER, E. A., AND BARSKY, J. (1952) *Proc. Soc. Exptl. Med.* **81,** 459–461.
ZELLER, E. A., BARSKY, J., FOUTS, J. R., AND LAZANAS, J. C. (1955) *Biochem. J.* **60,** p. v.

CHAPTER 12

# DEHYDROGENASES

## I. INTRODUCTION AND GENERAL CONSIDERATIONS

Biological oxidation-reduction reactions constitute the mechanism by which hydrogen atoms or electrons are transferred from one molecule to another in the living cell. Reactions of this type, which are catalyzed by enzymes, represent the major source of energy in tissues and also are related to synthetic and degradative processes. The living cell contains a series of electron donors and acceptors acting in a chain which, when isolated, appear to be specific with reference to their interaction with other components of the chain or other oxidation-reduction systems. In addition, the cell exhibits a structural orientation of oxidation-reduction systems at the cytological and submicroscopic levels. With reference to oxidation-reduction reactions, the term oxidation indicates loss of an electron, while reduction is characterized by acquisition of an electron. Thus the substance which donates hydrogen or electrons is oxidized while the one which accepts hydrogen is reduced. According to Slater (1958) a number of reactions of the general type $2AH_2 + O_2 \rightarrow 2A + 2H_2O$ (where $AH_2$ is an intermediate metabolite) take place in the cell. The common feature of these reactions is that two hydrogen atoms are removed from the intermediate metabolite and react with molecular oxygen to form water.

Most enzymatically catalyzed oxidation reactions are dehydrogenations or transfers of electrons from the substrate (i.e. the proton donor) to an oxidizing agent, the electron acceptor. Molecular oxygen generally serves as the final electron acceptor in biological oxidations, but certain oxidizing enzymes are unable to use oxygen directly, and thus the hydrogen atoms, or electrons, must be transferred by means of intermediate compounds. Molecules which accept hydrogen in the initial reactions are the respiratory coenzymes flavin adenine nucleotide (FAD), and di- and triphosphopyridine nucleotides (DPN and TPN). Most of the energy derived from the oxidation of a physiological substrate occurs not in the initial reduction of their coenzymes but in the subsequent transfer of the hydrogen to molecular oxygen. In this stepwise process, hydrogen or electrons are transferred in a series of oxidation-reductions from one respiratory enzyme or coenzyme to another, with, as previously mentioned, oxygen acting as the terminal acceptor.

The enzymes which catalyze dehydrogenations but are unable to utilize molecular oxygen as the electron acceptor are designated anaerobic dehydrogenases. These enzymes are involved in the reversible transfer of hydrogen, either to DPN or TPN. On the other hand, enzymes which catalyze oxidation of substrates by molecular oxygen are referred to as aerobic dehydrogenases or oxidases and include the flavoproteins or yellow enzymes. The yellow color is due to the presence of the prosthetic group which may be either flavin mononucleotide (FMN) or flavin adenine binucleotide (FAD). The old yellow enzyme (diaphorase) (Sumner and Somers, 1953, p. 279), which catalyzes the oxidation of reduced TPN, is about 70% as active with FAD as with FMN. The enzyme transfers hydrogen from the substrate to the flavin nucleotide, which is colorless in its reduced state. Hydrogen may subsequently be transferred to oxygen or to another acceptor. Under experimental conditions the reduced flavin may be reoxidized by certain dyes including methylene blue, 2,6-dichlorophenolindophenol, or phenazine methosulfate.

Among the pertinent reviews dealing with respiratory enzymes, the respiratory "chain" and biological oxidation-reductions are those by Racker (1955), Chance and Williams (1956), Mehler (1957, p. 153), Neilands and Stumpf (1958, p. 246), Slater (1958) and White *et al.* (1959, p. 314). As previously indicated, the specific individual components of the respiratory chain are those substances which are actual carriers of hydrogen atoms (or electrons) in the path from substrate to oxygen and thus must alternatively be reduced and oxidized. Dehydrogenases or transhydrogenases (Hoffmann-Ostenhof, 1953) represent the first components of the chain of enzymes which transfer, in successive steps, the hydrogen atoms from substrate to molecular oxygen. The dehydrogenase may either itself act as an intermediate carrier (e.g. succinic dehydrogenase) or it may catalyze the direct transfer from substrate to the first hydrogen acceptor as in the case of the pyridine nucleotide-linked dehydrogenases. As previously discussed in Chapter 11, another series of oxidation-reduction systems include the cytochromes as prosthetic groups. These serve as the final links in the electron transport chain between oxidizable substrates and oxygen. A cytochrome may accept an electron from either a flavoprotein or another cytochrome, and may transfer it to either a cytochrome or oxygen. The transfer of electrons from the cytochrome system to oxygen is more rapid than from flavoproteins to oxygen. Cytochromes may serve as efficient intermediates in the reoxidation of flavoproteins. In addition, water instead of hydrogen peroxide is produced in the reduction of oxygen by cytochromes. With reference to the dehydrogenases, reactions which are catalyzed by pyridine nucleotide and flavin nucleotide enzymes are outlined in Tables 1 and 2 (Neilands and Stumpf, 1958, pp. 249 and 254). A schematic representation of the electron transfer chain (Hartree, 1957), as well as a listing of known oxidation-reduction components of this chain, are shown in Charts 1 and 2.

Specific and selective stages of the electron transfer process are known to be inhibited by certain compounds. The reaction with molecular oxygen catalyzed by the terminal cytochrome a segment of the chain is inhibited by cyanide and sulfide at low concentrations and by azide and carbon monoxide at relatively high concentrations. Antimycin A, an antibiotic substance, inhibits the oxidation of succinate or DPNH by cytochrome c but does not affect the oxidation of reduced cytochrome c by molecular oxygen. The flow of electrons may be completely inhibited by a quantity of antimycin A equivalent to the flavin in a given preparation, thus constituting one of the most precise and efficient inhibitor reactions. Sodium amytal is known to inhibit the oxidation of DPNH at some site beyond the flavoprotein level. On the other hand, the oxidation of succinate is not affected by amytal at levels which inhibit the oxidation of DPNH.

Intimately involved in biological oxidations are cellular organelles known as mitochondria. These structures, which are present in practically all animal and plant cells, range from 0.3–5.0 $\mu$ in length and 0.2–0.5 $\mu$ in diameter. With the advent of the differential centrifugation techniques, the fact that the mitochondrion represents a multienzyme system was soon established. Important features of this work include the findings that both the cytochrome oxidase and succinic oxidase system are mitochondrial. In addition, mitochondria are able to oxidize a large number of substrates including fatty acids and intermediates of the Krebs cycle. These are also able to transform the energy thus released into high energy phosphate bonds by phosphorylating adenine nucleotides. Holter (1952) has reviewed the enzyme systems associated with the aforementioned organelles with particular reference to the pioneering studies in the field. More recent and comprehensive reviews on mitochondrial metabolism and the mitochondrial system of enzymes have been given by Ernster and Lindberg (1958), Schneider (1959), and Green and Fleischer (1960, p. 41).

The mitochondrion may be considered as a complete biochemical unit in that it contains the enzymes, cofactors, and accessory substances required to implement the biochemical reactions which it catalyzes. Carefully prepared preparations of these organelles require the addition only of the appropriate substrates, inorganic phosphate, ADP, and molecular oxygen. The organization of enzymes within the mitochondrion may be related to biochemical reaction rates as well as to the specific function of oxidative phosphorylation.

TABLE 1

REACTIONS CATALYZED BY PYRIDINE NUCLEOTIDE ENZYMES[a]

| Enzyme | Substrate | Product | Coenzyme |
|---|---|---|---|
| Type: $R(R')CHOH + Coenzyme_{Ox} \rightleftharpoons R(R')C{=}O + Coenzyme_{Red}$ | | | |
| Alcohol dehydrogenase | Ethanol<br>Vitamin A | Acetaldehyde<br>Retinine | $DPN^+$ |
| Isocitric dehydrogenase | Isocitrate | Oxalosuccinate or α-Ketoglutarate + $CO_2$ | $TPN^+$ |
| Isocitric dehydrogenase | Isocitrate | α-Ketoglutarate + $CO_2$ | $DPN^+$ |
| α-Glycerophosphate dehydrogenase | α-Glycerophosphate | Dihydroxyacetone phosphate | $DPN^+$ |
| β-Hydroxybutyric dehydrogenase | β-Hydroxybutyrate | Acetoacetate | $DPN^+$ |
| β-Hydroxyacyl coenzyme A dehydrogenase | β-Hydroxyacyl coenzyme A | β-Ketoacyl coenzyme A | $DPN^+$ |
| Lactic dehydrogenase | L(+) Lactate | Pyruvate | $DPN^+$ |
| Malic dehydrogenase | L-Malate | Oxalacetate | $DPN^+$ |
| Malic enzyme | L-Malate | Pyruvate + $CO_2$ | $TPN^+$ |
| Choline dehydrogenase | Choline | Betaine aldehyde | $DPN^+$ |
| Lysine α-keto analogue reductase | Lysine α-keto analogue | Pipecolate | DPNH<br>TPNH |
| Tartrate dehydrogenase | D- or *meso*-Tartrate | Oxaloglycolate<br>Diketosuccinate | $DPN^+$ |
| Galactowaldenase | UDP glucose | UDP glucuronic acid | $DPN^+$ |
| Type: $H_2O + RCHO + Coenzyme_{Ox} \rightleftharpoons RCOOH + Coenzyme_{Red}$ | | | |
| Formic dehydrogenase | Formate | $CO_2$ | $DPN^+$ |
| Glyceraldehyde-3-phosphate dehydrogenase | D-Glyceraldehyde-3-phosphate + acyl acceptor, e.g., $H_3PO_4$ | 1,3-Diphosphoglycerate | $DPN^+$ |
| Glucose dehydrogenase | D-Glucose | D-Gluconate | $DPN^+$ |
| Glucose-6-phosphate dehydrogenase | D-Glucose-6-phosphate | 6-Phospho-D-gluconate | $TPN^+$ |
| Betaine aldehyde dehydrogenase | Betaine aldehyde | Betaine | $DPN^+$ |
| Type: $RCHNH_2COOH + Coenzyme_{Ox} \xrightleftharpoons{H_2O} RCOCOOH + NH_3 + Coenzyme_{Red}$ | | | |
| Glutamic dehydrogenase | Glutamate | α-Ketoglutarate + $NH_3$ | $DPN^+$<br>$TPN^+$ |
| Type: $RSSR + Coenzyme_{Red} \rightleftharpoons 2RSH + Coenzyme_{Ox}$ | | | |
| Glutathione reductase | $Glutathione_{Ox}$ | $Glutathione_{Red}$ | TPNH |
| Cystine reductase | Cystine | Cysteine | DPNH |
| Type: Miscellaneous reactions | | | |
| Ketoglutaric dehydrogenase | Ketoglutarate | Succinyl coenzyme A + $CO_2$ | $DPN^+$ |
| Pyruvic dehydrogenase | Pyruvate | Acetyl coenzyme A + $CO_2$ | $DPN^+$ |
| Quinone reductase | *p*-Benzoquinone | Hydroquinone | DPNH<br>TPNH |
| Nitrate reductase | Nitrate | Nitrite | TPNH |
| Cysteine sulfinate dehydrogenase | L-Cysteine sulfinate | L-Cysteic acid | — |
| Steroid dehydrogenase | HO-steroids | Keto-steroids | $DPN^+$<br>$TPN^+$ |

[a] Neilands and Stumpf (1958).

TABLE 2

REACTIONS CATALYZED BY FLAVIN NUCLEOTIDE ENZYMES[a]

| Enzyme | Substrate | Product | Coenzyme |
|---|---|---|---|
| Type: $RCHNHR'COOH + Flavoprotein_{Ox} \overset{H_2O}{\rightleftharpoons} RCOCOOH + H_2NR' + Flavoprotein_{Red}$ | | | |
| D-Amino acid dehydrogenase | D-Amino acids | Keto acids + $NH_3$ | FAD |
| L-Amino acid dehydrogenase | L-Amino acids | Keto acids + $NH_3$ | FMN |
| Glycine dehydrogenase | Glycine | Glyoxylic acid | FAD |
| D-Aspartate dehydrogenase | D-Aspartate | Oxalacetate | FAD |
| Type: $RCHOHCOOH + Flavoprotein_{Ox} \rightleftharpoons RCOCOOH + Flavoprotein_{Red}$ | | | |
| L-Hydroxy acid dehydrogenase | L-Hydroxy acids | $\alpha$-Keto acids | FMN |
| Glycolic acid dehydrogenase | Glycolate | Glyoxylate | FMN |
| Type: $\frac{DPNH}{TPNH} + H^+ + Flavoprotein_{Ox} \rightleftharpoons \frac{DPN^+}{TPN^+} + Flavoprotein_{Red}$ | | | |
| $DPN^+$-cytochrome c reductase | DPNH | $DPN^+$ | FAD |
| $TPN^+$-cytochrome c reductase | TPNH | $TPN^+$ | FMN<br>FAD |
| Type: 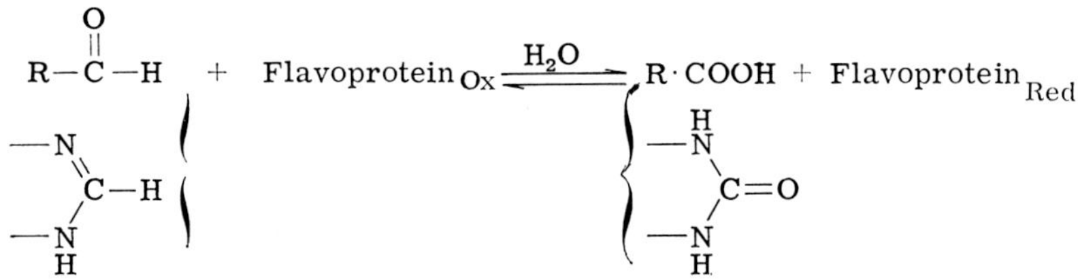 | | | |
| Notatin (glucose dehydrogenase) | D-Glucose | D-Gluconate | FAD |
| Aldehyde dehydrogenase | Aldehydes | Carboxylic acids | FAD |
| Xanthine dehydrogenase | Xanthine<br>Hypoxanthine<br>Aldehydes<br>Xanthopterin | Uric acid<br>Xanthine<br>Carboxylic acids<br>Leucopterin | FAD |
| Quinone oxidase | Aldehydes<br>Quinoline<br>Cinchona alkaloids<br>*N'*-Methyl nicotinamide | Carboxylic acids<br>Carbostyril<br>Carbostyril analogues<br>6-Pyridone of *N'*-methyl nicotinamide | FAD |
| Type: $—CH_2·CH_2— + Flavoprotein_{Ox} \rightleftharpoons —CH{=}CH— + Flavoprotein_{Red}$ | | | |
| Acyl coenzyme A dehydrogenase | Acyl coenzyme A derivatives | $\alpha,\beta$-Unsaturated acyl coenzyme A derivatives | FAD |
| Fumaric reductase | Fumarate | Succinate | FAD |
| Type: Miscellaneous Reactions | | | |
| Hydrogenase | $H_2$ | $2H^+ + 2e$ | FAD |
| Nitrate reductase | Nitrate | Nitrite | FAD |
| Nitrite reductase | Nitrite | Hydroxylamine | FAD |
| Hydroxylamine reductase | Hydroxylamine | Ammonia | — |

[a] Neilands and Stumpf (1958).

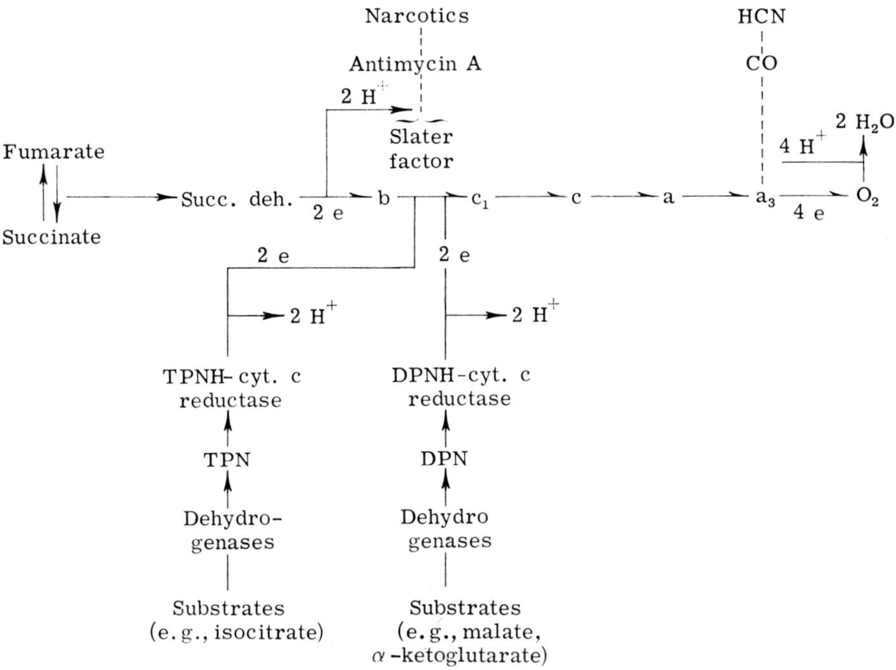


CHART 1. Schematic representation of electron transport chain (Hartree, 1957).

As indicated in the review by Schneider (1959) the term mitochondria is derived from the Greek words for *thread* and *grain*, and has been used for many years by cytologists to designate characteristic intercellular granules resembling bacteria in size and shape. These granules are commonly seen as long filamentous threads and also may occur as rods and spheres. The morphological form of mitochondria in living cells is not stable or constant as revealed by time-lapse photography. From the standpoint of classical cytology, mitochondria have been distinguished from other cellular organelles by their staining reaction. For example, they are not stained vitally with neutral red, but are stained blue by very dilute solutions of Janus Green B. This blue stain becomes decolorized after a period of time and then a red color is seen. This series of staining reactions with Janus Green B is believed to represent an uptake in concentration of the dye followed by its reduction to the colorless leuco compound and subsequently the formation of safranine, a red dye.

Mitochondria, when viewed with a light microscope, are homogeneous in appearance. Recent studies by means of electron microscopy, however, have revealed that mitochondria exhibit a distinct characteristic internal

1 *Lipoic acid* *(5,8-Dithiooctanoic acid)*

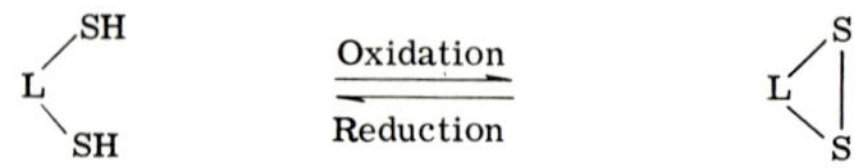


2 *Diphosphopyridine Nucleotide (DPN)*

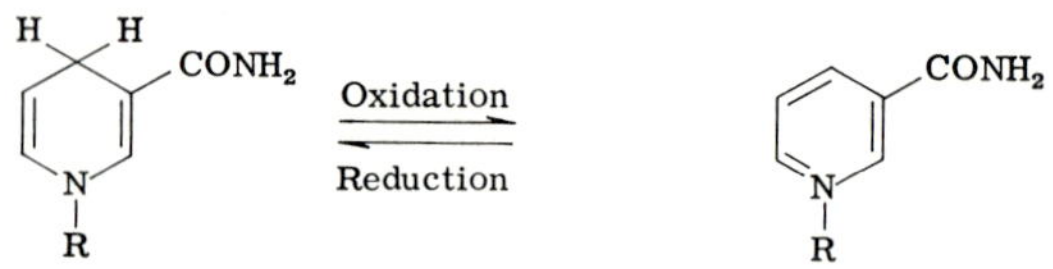


3 *Flavin dinocleotide*

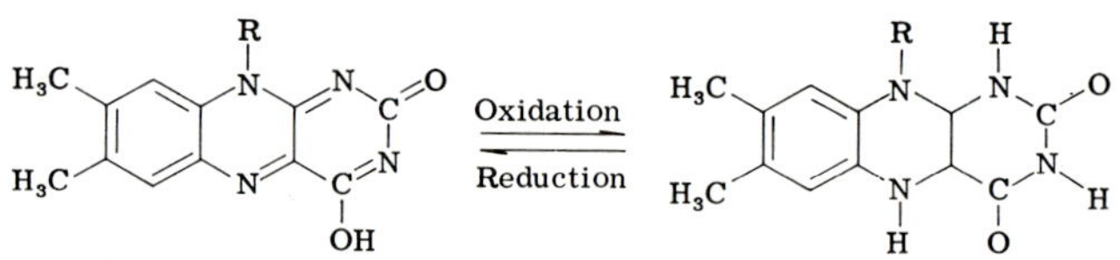


4 *Cytochromes* *(a,b,$c_1$ and c)*

Ferrous heme $\underset{\text{Reduction}}{\overset{\text{Oxidation}}{\rightleftharpoons}}$ Ferric heme

5 *Coenzyme Q*

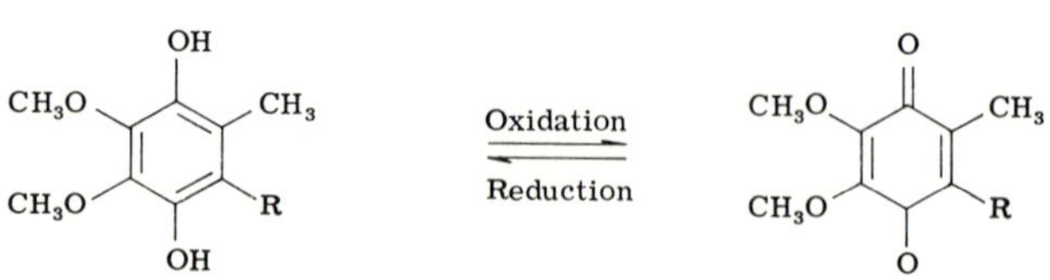


CHART 2. Oxidation-reduction components of electron transport chain.

structure (Palade, 1952, 1953, 1956, p. 185; Sjöstrand, 1953; Sjöstrand and Rhodin, 1953). When studied at the submicroscopic level the periphery of each mitochondrion is bounded by a double membrane ca. 50 Å in thickness separated by a space of ca. 60–80 Å. The outer membrane may be considered as the limiting membrane of the organelle, while the inner membrane is connected with structures which are referred to as *cristae.* The *cristae mitochondriales*, which appear to be invaginations of the innermost membrane, divide the interior of the mitochondrion into compartments or chambers. Certain mitochondria such as those of heart and insect flight muscle contain relatively few cristae. Green and Fleischer (1960, p. 46) indicate that in general it appears that the greater the number of ancillary reactions catalyzed by the mitochondrion, the smaller the number of cristae per organelle.

With newer histochemical procedures, enzymatic activity of mitochondria may readily be seen with the light microscope. Newer advances should make it possible to delineate sites of enzymatic activity to specific mitochondrial structures at the submicroscopic level.

From a histochemical standpoint there are three potential techniques for the demonstration of dehydrogenases, the tetrazole, tellurite, and triazole methods. Of these, only the tetrazolium technique, which involves the production of highly insoluble chromogenic formazan dyes, has achieved widespread application. Oxidation-reduction dyes have also been used in the demonstration of dehydrogenases, but such techniques are primarily of historical interest.

## II. HYDROGEN ACCEPTORS

### A. Tetrazolium Salts

These compounds are characterized by a heterocycle ring structure containing one carbon and four nitrogen atoms, one of which is quaternary. Among the pertinent reviews dealing with the synthesis and biochemistry of tetrazolium salts are those by Ried (1952), Rust (1955), Nineham (1955), Cheronis and Stein (1956), Pádr (1959), and Jámbor (1960). According to these reviews the colored formazans were synthesized independently by von Pechmann and by Bamberger in 1892. In the following year von Pechmann oxidized formazans to close the ring and form tetrazolium salts, of which a large number were prepared during the following years. Tetrazolium salts which are either colorless or pale shades are readily reduced to form intensely colored water-insoluble formazans. This property is in contradistinction to that of many dyes (indophenols, azines, oxazines, and indigoids) which form colorless leuco bases upon reduction.

In the 1940s German investigators first proposed and studied the application of tetrazolium salts as biochemical reagents. Lakon (1942a, b) found that only viable seeds of oats, rye, wheat, and barley were capable of reducing triphenyltetrazolium chloride (TTC) to a red formazan. In 1948 Straus *et al.* first demonstrated the reduction of tetrazolium salts by tumors. Antopol *et al.* (1948, 1950) were also among the first to apply tetrazolium salts in biological research, including the study of neoplasms. Subsequently von Hayek (1950) injected TTC into guinea pigs and found that liver and kidney tissue were intensely stained, and this reaction presumably was related to sites of high enzyme activity. Similarly, certain sites of the central nervous system were found to exhibit a specific staining reaction with tetrazolium salt solutions (Becker and Quadbeck, 1950). A more detailed consideration of the various aspects of the enzymatic reduction of tetrazolium salts will be considered in the sections on *histochemical methods* and *dehydrogenases in neoplasms.*

Black *et al.* (1953) have reported on the application of tetrazolium salts in pathology and have also reviewed the general literature in this field. Kivy-Rosenberg *et al.* (1955) have reported on and reviewed literature relating to the application of tetrazolium salts as indicators of changes in tissue metabolism during functional activity. Their preliminary work has indicated that abnormal changes in vessels, due to such agents as x-radiation or experimentally induced inflammation, may be reflected in the extent of formazan deposits.

Tetrazolium salts may be classified according to three types (Chart 3); monotetrazoles [Chart 3 (1)], N—N ditetrazoles [Chart 3 (2)], and C—C tetrazoles [Chart 3 (3)]. Of these, the first two types have received primary consideration as histochemical reagents. The transformation of a tetrazole to its corresponding formazan occurs through a ring opening as illustrated in Chart 3 (1). Formazans derived from a large number of substituted tetrazolium salts exist as tautomers associated by hydrogen bonding. Under strongly reducing conditions the resultant formazan can be further reduced to yield colorless compounds, although this does not appear to occur in biological systems.

Tetrazolium salts, for the most part, are photosensitive. The action of light in the visible range produces a yellowing of the compound, while short-wave ultraviolet radiation results in a reduction to the formazan in aqueous solution. Upon exposure to visible light certain types of formazans may undergo a reversible color change from red to yellow. If the yellow modification is placed in the dark, the color will gradually return to its original red shade. This phenomenon is believed to be due to a *cis-trans* isomerization. On the other hand, intense radiation with ultraviolet light for long periods produces an irreversible chemical change in tetrazolium salts.

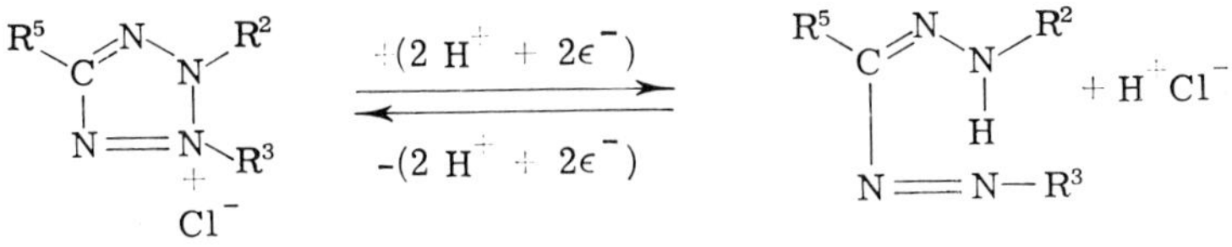

(1)

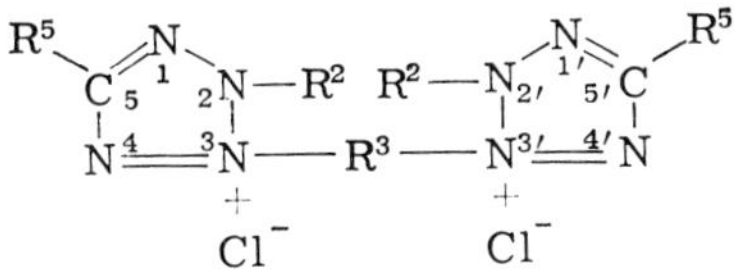

(2)

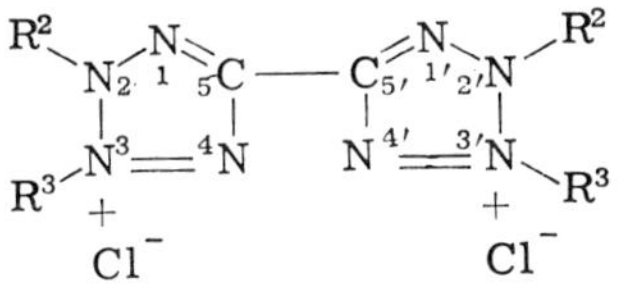

(3)

CHART 3. General types of tetrazolium salts. (1) Monotetrazoles, (2) N—N ditetrazoles; (3) C—C tetrazoles

The first step in the synthesis of tetrazolium salts involves the preparation of the appropriate formazan. The latter are generally prepared by the coupling of an aldehyde hydrazone (aldohydrazone) with a diazonium salt. Other methods for the preparation of formazans are outlined in Chapter 4 (p. 136). In general, substituted arylhydrazones are used. For example, a formazan may be produced by the coupling of the α-naphthalenediazonium chloride with benzaldehyde phenylhydrazone or by the interaction of benzenediazonium chloride with benzaldehyde α-napthylhydrazone (Hunter and Roberts, 1941a, b). The coupling occurs on the active carbon of the aldehyde under controlled conditions and in the presence of acid-binding agents which removed the acid formed as rapidly as it is produced. The reaction is carried out in a solvent which is capable of dissolving each of the components, but is not necessarily a solvent for formazans.

The tetrazolium salt is then produced by oxidation of the corresponding formazan, a process in which the formazan is simultaneously cyclicized and quaternized in the presence of an oxidizing agent at an acid pH. Among the oxidizing agents which have been utilized are alkyl nitrites, sodium nitrite, lead tetraacetate, alkyl hypochlorites, manganese dioxide, hydrogen peroxide, yellow mercuric oxide, and nitric acid (Nineham, 1955). In many of the recently described syntheses of tetrazolium histochemical reagents, isoamyl nitrite in the presence of HCl and a suitable solvent have been employed (Appendix 12A, Sections 1 and 2).

Ashley *et al.* (1953) have outlined specific laboratory procedures for the preparation of tetrazolium salts containing substituted phenyl and heterocyclic rings. Various methods of oxidizing formazans to tetrazoles were evaluated, and it was found that there was no ready way of predicting the suitability of an oxidant other than by trial.

Once the suspension of formazan has been oxidized, the resulting solution contains the tetrazolium salt plus a variety of side products which, in many cases, amount to more than 60% of the expected yield of tetrazolium compound. In other words, the oxidation may result in a reaction mixture which contains unchanged hydrazone, subsituted hydrazine, hydrazidines, and also other high molecular weight condensation products. Thus one of the most difficult aspects of the preparation is to isolate and purify the desired tetrazolium compound. The standard procedure is to concentrate the volume until some separation is obtained and then to add several volumes of ether to precipitate the tetrazolium compound along with other products. Addition of ether often results in oily products that may take weeks or months to crystallize. The crude product is filtered and dissolved in an alcohol and the process of precipitation with ether is repeated.

Although monotetrazolium salts give rise to colored formazans which are of value as indicators, still greater contrast may be achieved from formazans derived from ditetrazoles. For this reason, according to Rust (1955) bis-tetrazolium salts were originally introduced as biochemical reagents by Pannone and Rust in 1947. Nevertheless, there still remain valid reasons for both the application of existing, as well as the synthesis of new, monotetrazolium salts.

Chart 4 illustrates a number of mono- and ditetrazoles which have been employed as histochemical reagents. The first of these, 2,3,5-triphenyltetrazolium chloride, was originally synthesized by von Pechmann and Runge (1894) [Chart 4 (1)]. Its naphthalene-containing analogue is shown in Chart 4 (3).

One of the oldest ditetrazoles, which is presently known as neotetrazolium chloride (NT) [Chart 4 (5)], was synthesized by Wedekind in 1898. More recently a number of new tetrazoles have been described in the chemical literature. Included are iodine-containing compounds (Fox and Atkinson,

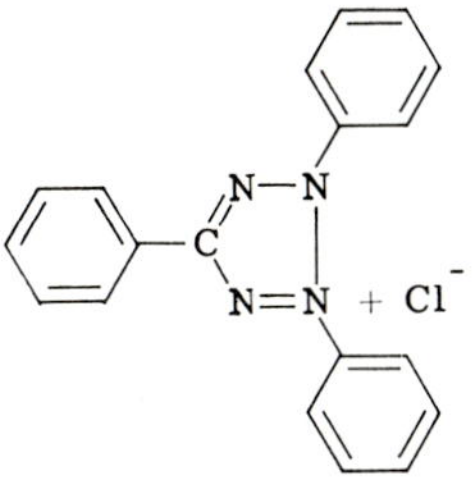

1

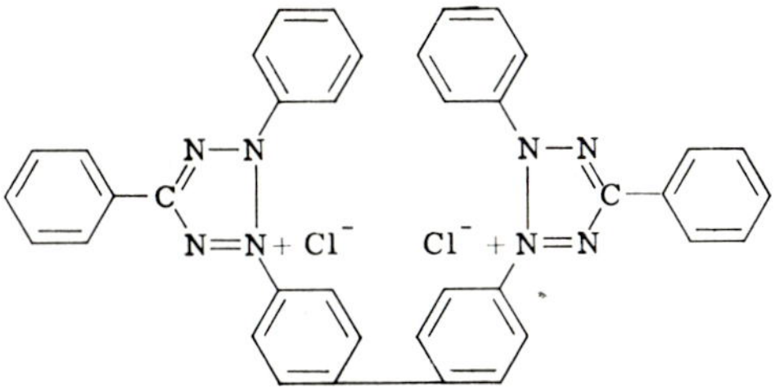

5

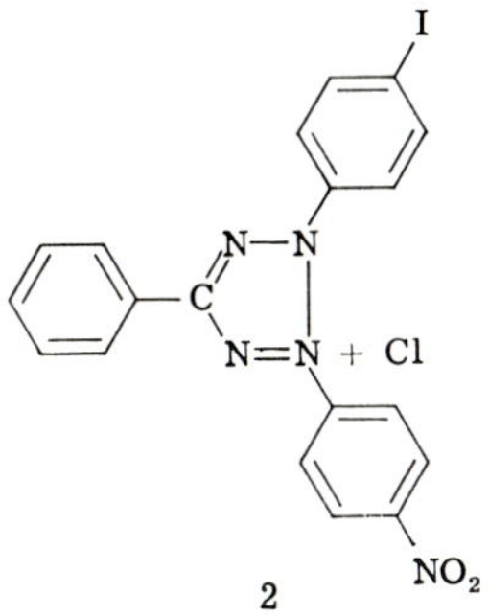

2

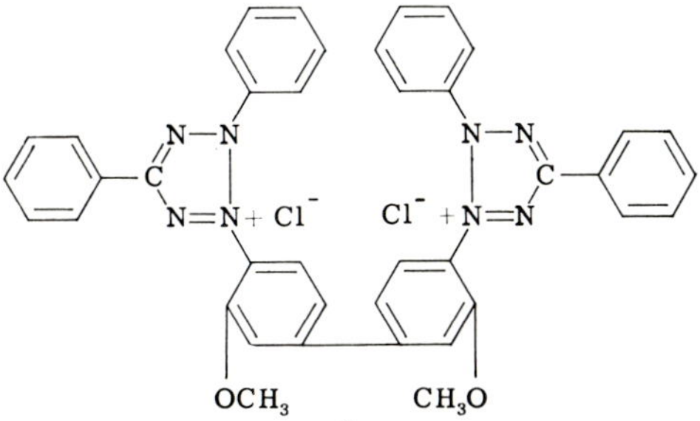

6

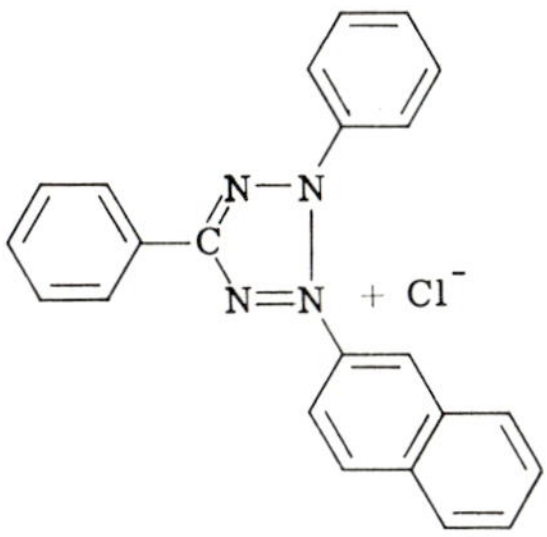

3

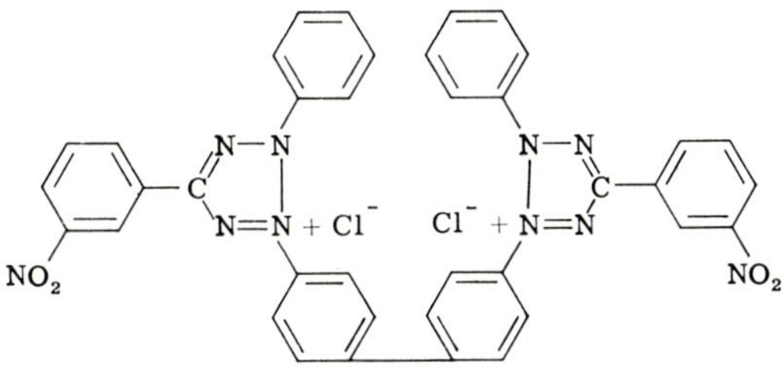

7

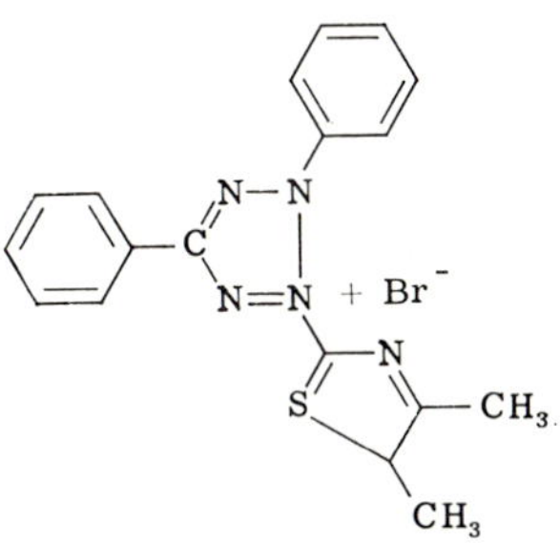

4

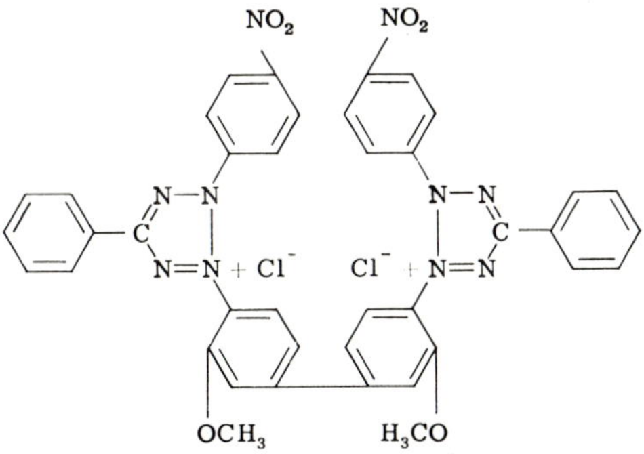

8

CHART 4. Tetrazolium salts commonly used in histochemical applications. (1) 2,3,5-triphenyltetrazolium chloride; (2) iodonitrotetrazolium chloride; (3) tetrazolium violet; (4) MTT; (5) neotetrazolium chloride; (6) blue tetrazolium chloride; (7) nitroneotetrazolium chloride; (8) nitro-BT.

1950; Atkinson *et al.*, 1950; Karmarkar *et al.*, 1958) [Chart 4 (2)]. In 1950, Rutenburg *et al.* described the preparation of a new tetrazolium salt, blue tetrazolium (BT) [Chart 4 (6)], which is analogous to the previously mentioned neotetrazolium, the difference being the presence of two methoxy groups. In 1954 Beyer and Pyl reported the preparation of several formazans and their corresponding tetrazolium salts which contained the thiazole grouping. One of these [Chart 4 (4)] has been applied as a histochemical reagent by Pearse (1960, p. 547) under the designation MTT. Characteristics of histochemically applicable tetrazoles are listed in Table 3, while syntheses are given in Appendix 12A. A significant advance in the synthesis of ditetrazoles was reported by Tsou *et al.* (1956) who prepared nitro-BT [Chart 4 (8)], an excellent substrate which has had wide applications in enzyme histochemistry (Appendix 12A, Section 2).

Another interesting compound, 2,2′,5,5′-tetra-*p*-nitrophenyl-3,3′-(3,3′-dimethoxy-4,4′-5 phenylene-ditetrazolium chloride [TNBT], has been shown to yield a formazan which exhibits a minimal crystalline structure

TABLE 3

TETRAZOLES AND TRIAZOLES

| Chemical name and formula | Mol. wt. | Reference | Melting point (° C.) |
|---|---|---|---|
| 1. 2,3,5-Triphenyltetrazolium chloride (TTC), $C_{19}H_{15}N_4Cl$ | 334.81 | von Pechmann and Runge, 1894; Kuhn and Jerchel, 1941a | 243, 263 |
| 2. 2-(*p*-Iodophenyl)-3-(*p*-nitrophenyl)-5-phenyltetrazolium chloride (INT), $C_{19}H_{13}N_5IO_2Cl_2$ | 541.18 | Fox and Atkinson, 1950 | 229 |
| 3. 3-(4,5-Dimethylthiazolyl-2)-2,5-diphenyltetrazolium bromide (MTT), $C_{18}H_{16}N_5BrS$ | 414.35 | Beyer and Pyl, 1954 | 171 |
| 4. (4,4′-Diphenylene)-bis-2-(3,5-diphenyl)tetrazolium chloride (neotetrazolium chloride) (NT), $C_{38}H_{28}N_8Cl_2 \cdot 2H_2O$ | 703.65 | Wedekind, 1898; Seiler and Schmid, 1954 | 219 |
| 5. 3,3′-Dianisole-bis-4,4′-(3,4-diphenyl)tetrazolium chloride (BT), $C_{40}H_{32}O_2N_8Cl_2$ | 727.67 | Rutenburg *et al.*, 1950; Ried and Gick, 1953 | 245, 247 |
| 6. 2,2′-Di-*p*-nitrophenyl-5,5′-diphenyl-3,3′-(3,3′-dimethoxy-4,4′-biphenylene)ditetrazolium chloride (nitro-BT) $C_{40}H_{30}N_{10}O_6Cl_2 \cdot 3H_2O$ | 871.72 | Tsou *et al.*, 1956 | 156 |
| 7. 2,2′-*p*-Diphenylene-bis-(naphtho-1′,2′,4,5)-triazole chloride $C_{44}H_{30}N_6Cl_2 \cdot 2H_2O$ | 749.72 | Krollpfeiffer *et al.*, 1934; Kuhn and Ludolphy, 1949 | 290 (dec.) |

and lipid solubility when employed under histochemical conditions (Rosa and Tsou, 1961).

The largest series of modified tetrazolium salts, which may have biological application, have been synthesized by Cheronis and associates, but their data have not yet been published.

The most definitive studies dealing with the characterization and purity of formazans and tetrazolium salts have been made by Dr. Nicholas D. Cheronis and associates (unpublished data). They note that one of the most troublesome aspects of the preparation of pure tetrazolium salts involves the purity of the respective formazans. Aside from the fact that a large number of tars (i.e. high molecular weight compounds) are produced by the side reactions of both the diazonium compound and the hydrazone, they also point out that the melting points, including constancy of melting point on successive fractionation, are not a good index of purity. The best method thus far evolved employs chromatographic development with a developer of eight volumes of methanol to one of water. It was also found that formazans which exhibit a constant melting point and also a relatively constant maximum absorption have been shown through paper chromatograms to yield two spots; and on repeated purification to yield only one spot with but a slight alteration in other criteria of purity. Cheronis indicates that the melting points of tetrazolium salts are useless criteria of purity, since they depend upon the extent of solvation either by methanol or by ether (alcoholate and/or etherate). Thus, at the present time the purity of tetrazolium salt can be established by chromatographic development until it gives only one spot upon reduction or by reduction under specified conditions to yield the formazan, which must be identical to the highly purified formazan that is the initial starting compound.

Cheronis and Stein (1956) note that considerable confusion has arisen in the naming of tetrazolium salts. This problem of nomenclature becomes even more confused because biologists and histologists have attempted to eliminate complicated chemical terminology by abbreviating the chemical names and converting them into "trivial" names. Thus, the 2,3,5-triphenyltetrazolium chloride has been designated as TTC and the 2,2′,5,5′-tetraphenyl-3,3′-(*p*-biphenylene) ditetrazolium chloride was first abbreviated to *neotetrazolium* and then to NT. Another ditetrazolium salt which gives a purplish-blue formazan was given the name of *tetrazolium blue* and is frequently abbreviated as BT. This was perhaps reasonable at the time the compound was reported since it was the only tetrazolium compound then known to give a blue formazan; however, a number of new tetrazolium salts also give blue formazans, and the problem arises as to what *blue* names, if any, should be assigned to these. In order to avoid the inevitable confusion in the designation of such substrates, it was suggested that if "trivial" names must be used, some designation of the chemical structure should be included, together with the color of the formazan. Thus, the compound 2,5-diphenyl-3-α-naphthol tetrazolium chloride which produces a violet formazan can be called α-naphthol tetrazolium violet. This serves to differentiate it from over a half dozen other tetrazolium violets such as *o*-diphenyltetrazolium violet.

Formazans generally form metallic chelates because of their coordinated tautomeric structure (Hunter and Roberts, 1941a, b). Hunter and Roberts (1941b) reported copper, nickel, and cobalt complexes of a number of formazans which were obtained by refluxing an acetone solution of the dye with an aqueous ethanolic solution of the appropriate salt. Similar metal complexes of formazans whose *N*-aryl rings contain hydroxyl or carboxyl groupings in the *ortho* position are particularly stable and have been described by Wizinger (1954). An interesting compound of this type is the previously mentioned MTT tetrazole of Beyer and Pyl (1954) [Chart 4 (4)].

### B. Naphthotriazolium Salts

Naphthotriazolium compounds may be derived from the oxidation of arylazo-*N*-phenyl-2-naphthylamines. Thus, upon oxidation, *p*-sulfobenzeneazo-*N*-phenyl-2-naphthylamine [Chart 5 (1)] gives a mixture of the triazolium salt [Chart 5 (2)] and the naphthophenazine [Chart 5 (3)]. Among the reviews dealing with the synthesis and biological application of naphthotriazolium salts are those by Krollpfeiffer *et al.* (1934) and Kuhn and Ludolphy (1949). One of the simplest compounds of this type is the salt of 1,2-diphenyl-(1′,2′,4,5-naphthotriazolium) base [Chart 5 (2)] which was synthesized by chromic acid oxidation of the dye derived from diazotized aniline and *N*-phenyl-2-naphthylamine. As shown in Chart 6, reduction of the colorless triazole results in the formation of the original azo-dye. Characteristics of an aryl-naphtho-1′,2′,4,5-triazolium salt (Chart 7) are given in Table 3. Because of the relatively straightforward synthetic procedures, a thorough investigation of the development and application of triazolium salts as histochemical reagents should be considered. The initial azo-dye is formed by coupling *N*-phenyl-β-naphthylamine with a suitable diazotate. The dye is recrystallized and suspended in glacial acetic acid. Amyl nitrite is added slowly and after the reaction is completed, the cooled solution may be treated with a small amount of concentrated nitric acid. The resulting triazolium nitrate may be precipitated directly and the rest is precipitated by the addition of ether. In general, it is probably best to prepare the more water-soluble triazolium hydrochloride; this may be accomplished by oxidation of the azo-dye with isoamyl nitrite in the presence of HCl gas. An interesting feature of triazolium compounds is their ability to form metal complexes with cobalt,

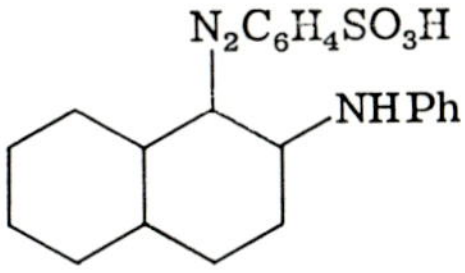


(1)

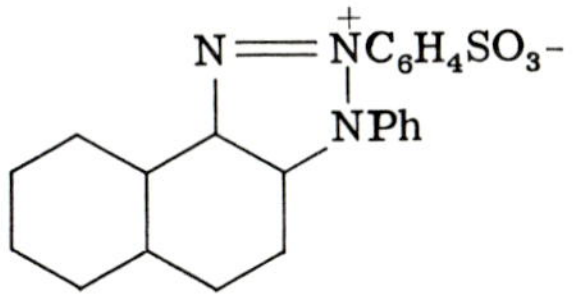


(2)

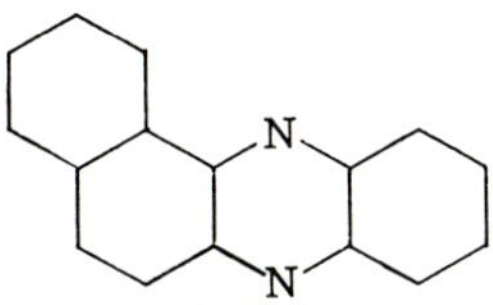


(3)

CHART 5. (1) Oxidation of *p*-sulfobenzeneazo-*N*-phenyl-2-naphthylamine. (2) Triazolium salt. (3) Naphthophenazine (Krollpfeiffer *et al*, 1934.

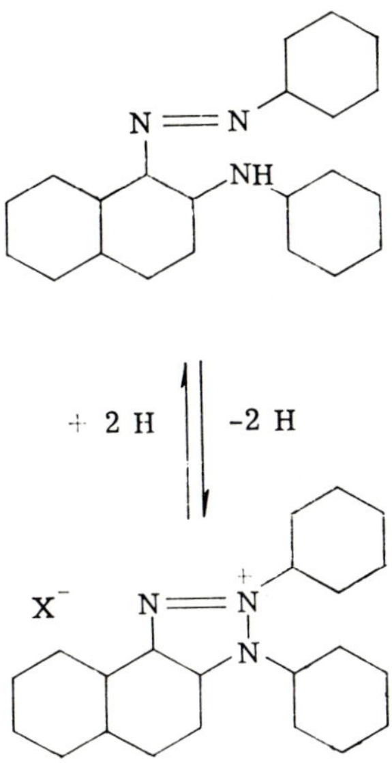


CHART 6. Oxidation-reduction of 1,2-diphenyl-1′, 2′,4,5-naphthotriazolium) salt (Kuhn and Ludolphy, 1949).

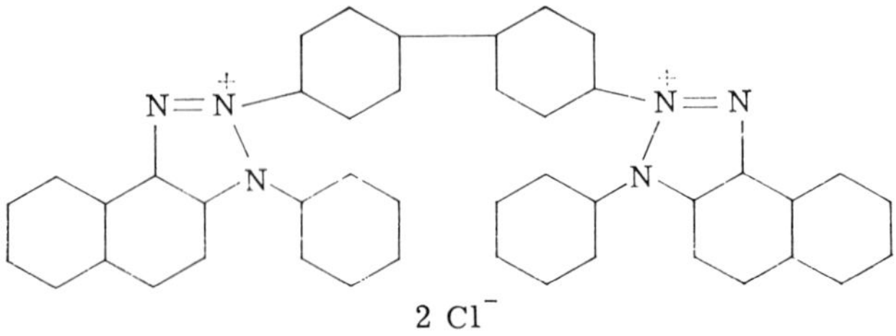


CHART 7. Naphthotriazolium salt derived from benzidine (Kuhn and Ludolphy, 1949).

nickel, mercury, copper, and uranium. In fact, naphthotriazolium salts have been recommended as reagents for cobalt (Kuhn and Ludolphy, 1949).

When treated with alkaline reducing agents such as ammonium sulfide, stannite, alkaline solutions of sodium hydrosulfite, mercaptans, and dextrose, aqueous solutions of triazolium salts regenerate the original azo-dye.

## C. Tellurium Salts

The tellurite method for the demonstration of dehydrogenase systems depends upon the fact that potassium tellurite ($K_2TeO_3$) is reduced by living cells to an insoluble black precipitate of tellurium. The procedure was originally introduced by Klett (1900) as a method for the determination of viability of bacteria. Lakon (1939) was the first to attribute this reaction to the activity of dehydrogenases. More recently, Wachstein (1949) described the application of potassium tellurite for the demonstration of dehydrogenase activity in tissue sections. Gomori (1952, p. 151) notes that although the tellurite method is considerably less sensitive than the tetrazolium procedure, it gives nice sharp localizations.

## D. Oxidation-Reduction Dyes

Certain redox dyes, including methylene blue, have been employed to delineate sites of dehydrogenase activity in tissues (Friedenwald and Steihler, 1938). More recently, the oxidation-reduction properties of organic dyes including methylene blue and its oxygenated analogue have been studied by Pullman and Pullman (1959b).

These authors (Pullman and Pullman, 1959a) have reported that the oxidation-reduction mechanism of the respiratory enzymes, pyridine nucleotide enzymes or flavoproteins, may be related to the energies of the molecular electronic orbitals of the oxidized and reduced forms of the associated enzymes DPN, TPN, FMN, and FAD. In each case a particularly low-lying empty orbital is associated with the oxidized form and a high-lying filled orbital is associated with the reduced form. The oxidized form thus exhibits a great tendency to accept electrons and the reduced form to give them up, the oxidation-reduction being accompanied by the appropriate instantaneous redistribution of these orbitals. Riboflavin-type coenzymes ($FMNH_2$, reduced flavin mononucleotide) possess a very unusual characteristic in that its highest filled molecular orbital was not only a very high-lying one but was an antibonding one. This orbital is generally occupied by electrons only in the excited states of molecules. This factor signifies that the occupation of this orbital in the ground state of the molecule represents a fundamentally unstable arrangement and that $FMNH_2$ will thus have a natural tendency to expel the electrons located at that orbital. This particular finding accounts not only for the outstanding electron-donor properties of this substance but also for its autoxidizability. Non-biological substances such as methylene blue may act as electron carriers in a way similar to that of the coenzymes and may involve a similar mechanism.

## III. HISTOCHEMICAL DEMONSTRATION OF DEHYDROGENASES

### A. Fixation Procedures

In general, unfixed frozen sections mounted on slides are preferable for the demonstration of enzymes of the dehydrogenase system. Shelton and Schneider (1952) found that freezing destroyed the endogenous activity of kidney slices, and concluded that tetrazolium salts could not be used for the demonstration of endogenous dehydrogenase activity in fresh tissue slices, although sites of succinic dehydrogenase can be localized in frozen sections. The freezing and thawing encountered in the preparation of frozen sections, of course, may disrupt the morphological integrity of the specimen, particularly when studied at the cytological level. This type of damage may result in the presence of small formazan crystals which, according to Novikoff (1959, p. 121; 1960c), *are frequently confused with mitochondria by some investigators.* With reference to the staining of mitochondria, these structures may readily be demonstrated in unfixed frozen sections of cardiac muscle (Fig. 1). However, in other tissues the disruptive effects of the freezing and thawing preclude the definitive delineation of mitochondria.

Scarpelli and Frajola (1961) have reported a biochemical study dealing with the effects of formol-calcium fixation upon the activities of various oxidative enzymes. The study was undertaken in an attempt to determine quantitatively the degree of enzyme activity remaining both in purified enzyme systems and in tissues following treatment with buffered formol-calcium (pH 7.4) for 5, 10, and 20 minutes at 4°. The enzyme systems stud-

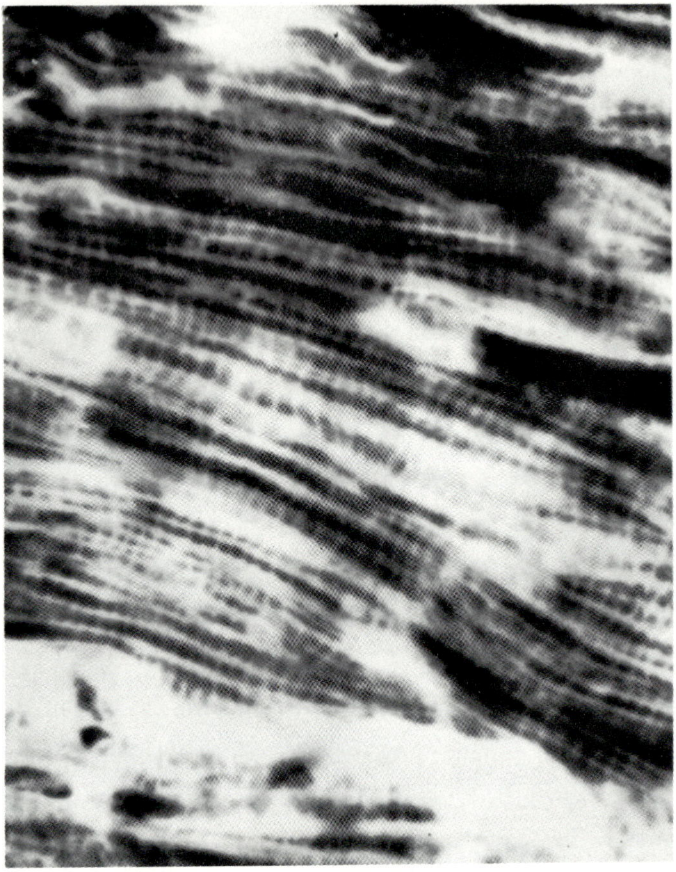

FIG. 1. Demonstration of succinic dehydrogenase activity of mitochondria in unfixed frozen section of rat heart. Nitro-BT (× 1450).

ied were lactic and succinic dehydrogenases. Crystalline lactic dehydrogenase was treated with a fixative for 5 minutes and then dialyzed against phosphate buffer (pH 7.4) until the dialysate was free of formalin. These authors reported that formol-calcium produced marked inhibition of purified lactic dehydrogenase and, to a lesser extent, of lactic rat kidney dehydrogenase. Succinic dehydrogenase activity was decreased fivefold after only 5 minutes of fixation.

### B. Tetrazolium Procedures

The fact that tetrazolium salts have a low redox potential (—0.08 volts) makes them sensitive indicators of reducing enzyme systems in tissues. The ease of reduction of tetrazoles increases with increasing pH, and in some cases color formation occurs spontaneously at pH 9–10. In addition, reduction occurs more readily at elevated temperatures. For the demonstration of reducing agents including reducing sugars, reducing steroids, and SH groups a highly alkaline pH range is employed. By contrast, for the demonstration of enzyme systems a pH around neutrality is utilized. A recent study dealing with the theoretical aspects and quantitation of neotetrazolium reduction is that of Burtner *et al.* (1957). Nachlas *et al.* (1960) report that electronegative groups attached to either the *N*-2 phenyl or *N*-3 phenyl rings are important in determining the readiness of enzymatic transfer of electrons to tetrazolium salts. They also suggest that the degree of influence exerted by nitro, chloro, or cyano groupings varies with the type of dehydrogenase system studied.

In addition to the redox potential of the tetrazole to be employed as a histochemical reagent, another factor is of considerable significance in the applicability of such compounds. The latter relates to the particular size and lipid solubility of the precipitated formazan in the tissue section. Shelton and Schneider (1952) were among the first to evaluate the suitability of tetrazolium salts as histochemical indicators of dehydrogenase activity. Of the four compounds which they tested, BT, TV, TTC and NT, the latter was found to be the most satisfactory salt because of the rapidity with which it is reduced and also the fineness of the formazan precipitate. Nachlas *et al.* (1957) compared the properties of tetrazoles employed in the histochemical demonstration of succinic dehydrogenase. Their findings concerning the characteristics of the formazan precipitate are given in Table 4.

With reference to the validity of tetrazolium reduction as an accurate measure of succinic dehydrogenase activity, Shelton and Rice (1957) compared the reduction of NT and INT by tissue extracts and found that, on a relative basis, the tetrazolium method measures the dehydrogenase system with the same adequacy as the manometric method.

Although the staining of small blocks of tissue by formazan solutions was observed as early as 1941 (Kuhn and Jerschel, 1941a, b), it was not until 1949 that the first use of a tetrazolium salt to determine succinic dehydrogenase activity was reported (Kun and Abood, 1949). Of interest was the observation of Mattson *et al.* (1947) who reported the reduction of 2,3,5-triphenyltetrazolium chloride (TTC) at neutral pH by plant and animal tissues. Kun (1951) indicated that in animal tissues TTC can be reduced by a variety of dehydrogenases. Soluble flavoproteins, like amino

TABLE 4

PROPERTIES OF TETRAZOLES EMPLOYED IN THE HISTOCHEMICAL DEMONSTRATION OF SUCCINIC DEHYDROGENASE[a]

| Tetrazole | Common name | Sensitivity | Advantage of anaerobiosis | Color | Particle size | Lipid solubility |
|---|---|---|---|---|---|---|
| 2,2′,5,5′-Tetraphenyl-3,3′-(3,3′-dimethoxy-4,4′-biphenylene)ditetrazolium chloride | BT | 1+ | + | Red, blue | Fine and coarse | Yes |
| 2,2′-Di-*p*-nitrophenyl-5,5′-diphenyl-3,3′(3,3′-dimethoxy-4,4′-biphenylene) ditetrazolium chloride | Nitro-BT | 4+ | 0 | Blue | Fine | No |
| 2,2′,5,5′-Tetra-*p*-nitrophenyl-3,3′-(3,3′-dimethoxy-4,4′-biphenylene)ditetrazolium chloride | Tetra nitro-BT | 4+ | 0 | Reddish brown | Fine | No |
| 2,2′,5,5′-Tetraphenyl-3,3′-(*p*-biphenylene)ditetrazolium chloride | NT | 2+ | + | Red, purple | Fine and coarse | Yes |
| 2,2′-Di-*p*-nitrophenyl-5,5′-diphenyl-3,3′(*p*-biphenylene)ditetrazolium chloride | Nitro-NT | 4+ | 0 | Red, purple | Fine and coarse | Partially |
| 2,2′,5,5′-Tetra-*p*-nitrophenyl-3,3′-(*p*-biphenylene)ditetrazolium chloride | Tetra nitro-NT | 2+ | 0 | Brown | Fine | No |
| 2-*p*-Iodophenyl-3-*p*-nitrophenyl-5-phenyltetrazolium chloride | INT | 3+ | 0 | Reddish purple | Needle clusters | Yes |

[a] Nachlas *et al.* (1957).

acid dehydrogenases, readily transfer electrons to TTC under anaerobic conditions. Rat kidney extracts, upon addition of amino acids, were found to reduce TTC anaerobically. This tetrazole was also employed by Jensen *et al.* (1951) for the determination of several diphosphopyridine nucleotide (DPN) and triphosphopyridine nucleotide (TPN) linked dehydrogenases in corn embryos. It was not ascertained if the tetrazolium salt was reduced directly by dehydrogenases or indirectly by some other reductase involved in the system. More recently, in the reviews of Brodie and Gots (1951, 1952), it was reported that reduced DPN is capable of reducing TTC only in the presence of a flavoprotein, diaphorase. It was suggested that the tetrazolium salt, like methylene blue, is not directly reduced by dehydrogenases but rather by the flavoprotein intermediates in the enzyme system.

In 1950, Rutenburg *et al.* incubated pieces of freshly excised mouse and rat tissues in a buffered solution of a ditetrazolium salt, blue tetrazolium (BT) [Chart 4 (6)]. Reduction of BT occurred on the surface of the block to a depth of about 0.5 mm. The blocks were then fixed in 10% formalin and frozen sections were prepared for microscopic study. The ability to reduce BT decreased rapidly when tissues were stored at 4° or −5° for several days. It was reported that specific dehydrogenase activity could be demonstrated in extracts of homogenates by addition of specific substrates such as sodium succinate, xanthine, sodium lactate, and sodium malate.

Farber and associates (1956a, b; Farber and Louviere, 1956; Sternberg *et al.*, 1956; Farber and Bueding, 1956) devised several ingenious approaches for the histochemical study of DPN and TPN diaphorases using blue tetrazolium and neotetrazolium. It was found that the staining reaction occurred when frozen sections were incubated in solutions containing tetrazolium salts together with a substrate and cofactor necessary for the activity of any one of several dehydrogenase systems. However, a distinctive staining pattern specific for each dehydrogenase system was not obtained. On the contrary, a single common pattern was obtained with two DPN linked systems and a different pattern with TPN linked systems. Supplementary experiments demonstrated that the actual reduction of the tetrazolium salt in the staining process was not brought about by its direct interaction with dehydrogenase, DPN or TPN, but rather by two diaphorases, one specific for DPN systems and the other for TPN systems. Thus, the staining reactions, even though presumably produced by the activity of single dehydrogenase systems in tissues, are specific only for DPN diaphorase and TPN diaphorase activity. Although the simplified techniques of Farber for the demonstration of DPN and TPN diaphorases are given in Appendix 12B, Section 5, the various dehydrogenase procedures as developed by his group are listed (Appendix 12B, Sections 4 and 5) because of their importance to developmental histochemistry.

Sternberg *et al.* (1956) employed these techniques in the study of the histochemical distribution of DPN diaphorase and TPN diaphorase in the rat kidney. The staining pattern for each enzyme showed a highly reproducible and a distinctly different zonal distribution in the cortex and medulla, and a corresponding regional distribution along the course of the individual nephrons. With these methods, virtually all of the different segments of the nephron could be sharply delineated and identified. Collecting ducts in the papilla, which are unstained with the succinate method, are stained with the DPN and TPN diaphorase procedures. TPN diaphorase gives a good staining reaction of the thin limbs of Henle which are difficult to demonstrate by other methods. Further studies by Farber *et al.* (1956b) elaborated on the specificity and accuracy of their previously described methods for DPN and TPN diaphorases. With reference to cytological localizations, they noted that the formazan deposits derived from either blue tetrazolium or neotetrazolium are satisfactory for study at the histological but not at the cytological level.

Farber and Louviere (1956) described the use of soluble oxidation-reduction dyes as aids in the histochemical localization of oxidative enzymes with tetrazolium salts. The results of this work indicate that several oxidation-reduction indicators, soluble in both the oxidized and reduced forms, can be used in association with blue or neotetrazolium to enhance staining for succinic dehydrogenase systems or for DPN or TPN diaphorase. In practice, the addition of these dyes to the incubation medium results in a considerable increase in the rate of staining with no alteration in the characteristic staining patterns. This type of modification of the tetrazolium procedure interposes a dye as a carrier between the enzyme and the tetrazolium salt and makes it possible to stain for enzymes which do not react directly with the tetrazole. Among the dyes employed were methylene blue, azure C, pyocyanin, and Nile Blue A (Appendix 12B, Section 5).

Farber and Bueding (1956) found that incubation of frozen sections of rat kidney with lipase abolishes the staining reaction for the succinic dehydrogenase system with blue tetrazolium alone, or blue tetrazolium plus methylene blue, pyocyanin, Janus Green B, azure C, or azure B bromide. In contrast, staining with phenazine methosulfate plus blue tetrazolium is unaffected by prior incubation with lipase. Qualitatively similar results were obtained with neotetrazolium. Their results support the conclusion that the succinic dehydrogenase system contains carriers linked to the dehydrogenase. The carriers essential for the reduction of blue or neotetrazolium alone or in combination with several soluble oxidation-reduction dyes are localized in tissue sections by these agents. The dehydrogenase can reduce phenazine methosulfate and this dye coupled to a tetrazolium salt can be used for the histochemical localization of succinic dehydrogenase in tissues.

Oda and Okazaki (1958) have studied the reduction of neotetrazolium chloride by the terminal electron transport system. In order to determine the stages at which the reaction of neotetrazolium chloride reduction involves the terminal electron transport system, an analytical study on the reduction of NT by tissue homogenates was carried out employing various substrates such as sodium succinate, *p*-phenylenediamine, sodium malate, sodium $\beta$-glutamate, and DPN; and inhibitors such as sodium malonate, potassium cyanide, and antimycin A. With sodium succinate as a substrate, the succinate oxidase system was involved in the following steps. The reaction was found to involve about 50% of the succinic dehydrogenase system (succinic dehydrogenase, cytochrome b, cytochrome $c_1$). Of these about 15% involve the step prior to the antimycin A sensitive stage and

35% involve the antimycin A step itself. About 50% of the neotetrazolium reaction was related to cytochrome c oxidase. With the use of the *p*-phenylenediamine substrate the reaction of neotetrazolium reduction was related to activity of cytochrome c oxidase; and when *p*-phenylenediamine was employed with a sufficient amount of cytochrome c, the reaction appeared to be dependent on cytochrome c oxidase activity. With DPN, the neotetrazolium reduction is mainly a reaction conjugating it at the step below antimycin A in the DPNH-cytochrome c reductase system (flavoprotein, cytochrome b, and cytochrome c), probably with the flavoprotein of DPNH-dehydrogenase. The reduction of several tetrazoles, which are employed in histochemistry, by the terminal electron transport system is shown in Chart 8. Oda *et al.* (1959) have also reported on the histochemistry of the succinic dehydrogenase system with reference to the aforementioned analysis. A comparison was made between several commonly used tetrazolium salts, namely NT, nitro-neotetrazolium, and nitro-BT chlorides.

A significant advance in the histochemistry of dehydrogenases occurred with the development of a new tetrazolium substrate 2,2′-di-*p*-nitrophenyl-5,5′-diphenyl-3,3′-(3,3′-dimethoxy-4,4′-biphenylene) ditetrazolium chloride (nitro-BT) by Tsou *et al.* (1956) (Appendix 12A, Section 2). Subsequently the substrate was applied in the demonstration of a number of dehydrogenase systems by Nachlas *et al.* (1957, 1958a, b). In their initial report, Nachlas *et al.* (1957) noted the superiority of nitro-BT over the previously employed tetrazolium salts in that it allowed the cytochemical visualization of sites of enzymatic activity in thin tissue sections after a relatively short incubation period. The same report also outlined characteristics which an ideal tetrazole histochemical substrate should possess:

1. Ready reduction to the formazan and resistance to oxidation by molecular oxygen back to the tetrazole.
2. A strong attraction for hydrogens or electrons liberated from the substrate so that atmospheric oxygen does not act as a significant competitor.
3. Production of a formazan of good pigment quality which appears in the tissue as an amorphous dye rather than as crystalline deposits.
4. Production of a formazan with very little aqueous solubility and insoluble in common organic solvents.

Nitro-BT possesses the qualities enumerated to a large extent, and thus has become one of the valuable new substrates for application in enzyme histochemistry (Appendix 12B, Section 3). Nachlas *et al.* (1957) appear to have been the first to demonstrate succinic dehydrogenase activity in mitochondria of cardiac muscle and possibly in mitochondria of kidney tubular epithelium. In all instances, unfixed frozen sections mounted on

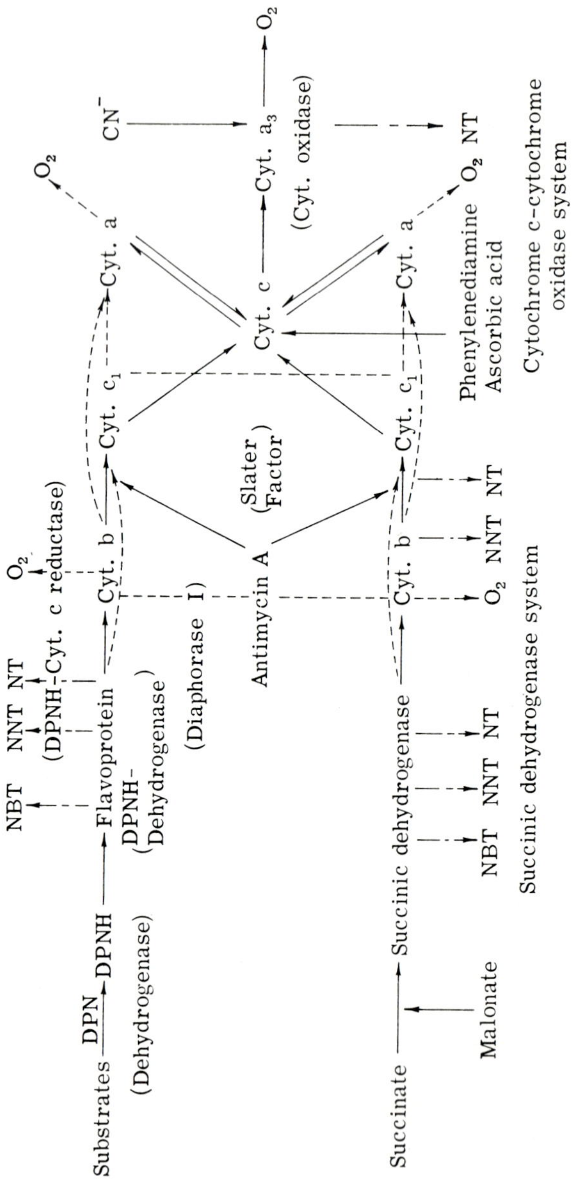


CHART 8. Reduction of several tetrazoles by terminal electron transport system (Oda *et al*, 1959).

glass slides were employed. Subsequently, Nachlas *et al.* (1958a, b) applied nitro-BT to the development of histochemical modifications for the localization of diaphorases. Their techniques, which were based on the approach of the Farber group (see p. 491), are outlined in Appendix 12B, Sections 6 and 7. Among the specific systems which were reported to have been delineated were DPN and TPN diaphorases, β-hydroxybutyric dehydrogenase, isocitric dehydrogenase, and 6-phosphogluconic dehydrogenase.

However, even this new tetrazole possesses certain undesirable qualities in that the formazan exhibits a selective affinity for lipids, especially phospholipids (Wattenberg, 1958; Levy *et al.*, 1959; Sedar and Rosa, 1961). Oda (1961) notes that nitro-BT can be reduced non-enzymatically by DPNH with the mediation of phenazine methosulfate and it is also reduced enzymatically by both intra- and extramitochondrial DPNH dehydrogenases.

Pearse and associates (Pearse, 1958; Pearse and Scarpelli, 1958; Scarpelli and Pearse, 1958a, b; Scarpelli *et al.*, 1958; Hess *et al.*, 1958a, b; Pearse and Scarpelli, 1959; Pearse *et al.*, 1960) have reported on the histochemical demonstration of several enzymes of the dehydrogenase series, including DPN and TPN diaphorase, through the use of the thiazole containing monotetrazolium salt MTT [Chart 4 (4)]. In addition, the possibilities for improved preservation of mitochondrial structure as well as the demonstration of intramitochondrial sites of dehydrogenase activity were also explored. In all of these techniques, frozen cryostat sections mounted on cover slips were allowed to thaw prior to the incubation in substrate medium which contained substances such as sucrose, dextrin, or PVP. Not only were specific intramitochondrial localizations claimed, but it was also stated that the formazan deposits were especially arranged so that areas of enzyme activity are separated by quiescent regions. Novikoff *et al.* (1961) have carefully evaluated some of these claims and concluded that the use of MTT and cobalt is unreliable for the intracellular localizations. On the other hand, nitro-BT was found to be capable of producing good intracellular localizations. Novikoff *et al.* (1961) found that with the MTT-$Co^{++}$ method the formazan is located outside the mitochondria whether the mitochondria are isolated or within cells. Furthermore, the disposition of the formazan was found to be essentially unchanged in the greatly altered mitochondria of the ligated kidney. The great affinity of the MTT-$Co^{++}$ formazan for lipids was also noted. In any case, it would be rather surprising to obtain precise intercellular localizations with formazans derived from monotetrazolium salts of relatively simple structure and poor substantivity.

Cascarano and Zweifach (1959) have published a significant paper dealing with the factors which influence the histochemical demonstration of coenzyme-dependent dehydrogenases and diaphorases. They point out that there is considerable evidence to indicate that the use of different tetrazole substrates alone results in different patterns of distribution of formazan for many of the DPN- and TPN-dependent dehydrogenases, in contrast to the original assumptions advanced by Farber *et al.* (1956a). The data of Cascarano and Zweifach (1959), as well as that of others (Nachlas *et al.*, 1958a, b), bears this out.

Wattenberg (1958), who utilized INT, BT, nitro-BT, and NT, showed that each of these salts reacts directly with the same step in the electron chain. The schematic representation in Chart 9 indicates the transfer of electrons to the tetrazolium salt and shows the final step to be that of the flavoprotein oxidizing DPNH or TPNH. In tissue sections, the dehydrogenases serve primarily to generate the substrate (DPNH or TPNH) required for this reaction. Therefore, Cascarano and Zweifach (1959) indicate that it is highly unlikely that different enzyme patterns are brought out by particular tetrazolium salts because of their reduction by different components on the electron chain. In a reaction which occurs in several stages, the overall rate is determined by the slowest step in the series. When a substrate utilized by a DPN-dependent dehydrogenase is introduced into the tetrazolium incubation medium, at least two variables must be considered. Under one set of conditions, although the cells have comparatively high quantities of flavoprotein, a particular DPN-dependent dehydrogenase may either be absent, or be present only in minimal amounts. In these cells, the production of DPNH becomes the limiting reaction so that the reduction of the tetrazolium salt is a reflection of the quantity of the dehydrogenase and not of flavoprotein. In both kidney and adrenal tissue incubated in malate, β-hydroxybutyrate, or glutamate, there are regions with negligible amounts of reduced tetrazolium, despite the fact that sub-

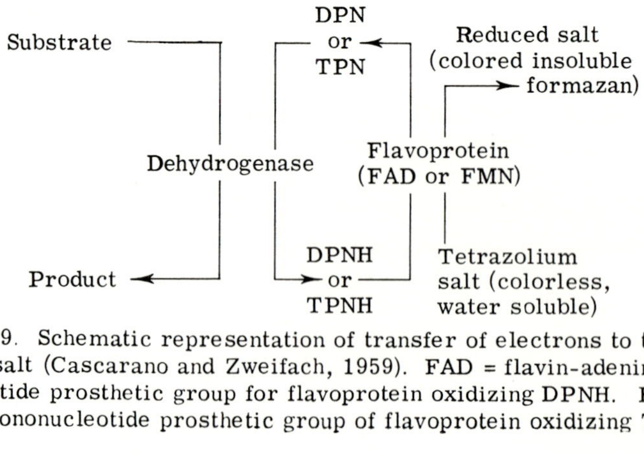


CHART 9. Schematic representation of transfer of electrons to tetrazolium salt (Cascarano and Zweifach, 1959). FAD = flavin-adenine binucleotide prosthetic group for flavoprotein oxidizing DPNH. FMN = flavin mononucleotide prosthetic group of flavoprotein oxidizing TPNH.

stantial formazan is deposited in these same regions when the slices are incubated in DPNH. These considerations lay open to question the assumptions previously made that reactions initiated by dehydrogenases, by necessity, reflect the distribution of flavoprotein. Cascarano and Zweifach (1959) also note that in the development of a procedure for isocitric dehydrogenase, other investigators (Nachlas *et al.* 1958b) indicated the difficulties inherent in allowing aconitase present in the cells of tissues to form the substrate (isocitrate) for the subsequent dehydrogenase reaction. They then apparently disregard the limitations of this factor in formulating a solution for TPN diaphorase activity.

Cascarano and Zweifach (1959) point out that the utilization of tetrazolium salts as a measure of flavoprotein activity will depend upon the use of a system with only one variable. The simplest approach would be to add sufficient DPNH to the incubating medium, a factor which would eliminate the dependency of flavoprotein on substrates generated by other cellular constituents. The procedure recommended by Nachlas *et al.* (1958a), using exogenous dehydrogenase in the incubating medium, should prove equally effective, provided that substrate and exogenous dehydrogenase concentrations are not limiting. The use of DPNH as substrate would make it possible to apply this to the study of the distribution of particular dehydrogenases in areas shown to have flavoprotein activity. Alternate sections, incubated in DPNH and in a substrate of a DPN-dependent enzyme, would then provide information concerning the distribution of the dehydrogenase in these areas.

The various claims for localization of DPN-linked or TPN-linked dehydrogenases do not rest on as firm a footing as the theoretical and practical considerations involved in the localization of other enzyme systems. A prerequisite for the localization of dehydrogenases involves the firm binding of DPN or TPN to the tissue, especially after reduction. It is also necessary to establish that dehydrogenase is not being solubilized and thereby giving a reaction in a cell to which it has diffused, and in which it does not normally occur. Unfortunately, this type of information is lacking for practically every coenzyme-linked dehydrogenase which has been reported to have been demonstrated by histochemical techniques. Although a number of these procedures are listed in Appendix 12B, the most conservative approach would be to view these procedures as tentative until the aforementioned questions are definitely answered.

Wattenberg (1958) has described a histochemical technique for the demonstration of steroid-3$\beta$-ol dehydrogenase which employs DPN in conjunction with either NT or nitro-BT. Subsequently the technique was modified by Allen (1960) who used a substrate solution containing the following: 0.01 *M* $\Delta^5$-androstene-3$\beta$-ol-17-one (dehydroepiandrosterone) dissolved in 1.5 ml. of acetone; DPN, 10.0 mg.; 0.1 *M* phosphate buffer pH 7.5, 8.0 ml.; 0.3 *M* KCN pH 7.5, 5.0 ml.; nitro-BT, 3.5 mg.; and water to make a total volume of 30.0 ml. In general, unfixed frozen sections mounted on glass slides were employed. Allen found that sections incubated in substrate solutions from which DPN had been omitted failed to give any staining reaction. A similar lack of reactivity was observed with substrate solutions from which the dehydroepiandrosterone and acetone were omitted.

Kellogg and Glenner (1960) have described the histochemical localizations of β-estradiol dehydrogenases in the human placenta. Frozen sections were incubated in a substrate solution containing 2.0 μ*M* of steroid substrate, 50.0 μ*M* of oxidized nucleotide, 50.0 mg. of crystalline bovine albumin and 10.0 mg. of nitro-BT. For the histochemical localization of TPN or DPN diaphorase, 50.0 μ*M* of reduced nucleotide was employed to replace the steroid substrate and oxidized nucleotide.

Tetrazolium salts have also been used in the demonstration of D-amino acid dehydrogenase (oxidase) (Dianzani, 1953; Farber *et al.*, 1958). The latter authors reported that phenazine methosulfate is an electron acceptor in the D-amino acid oxidase system and is thus capable of reducing tetrazolium salts (Appendix 12B, Section 8). Choline dehydrogenase (oxidase) may be similarly demonstrated (Appendix 12B, Section 8).

Recently, Wattenberg and Leong (1960) have reported on the effects of coenzyme $Q_{10}$ and menadione on succinic dehydrogenase activity as measured by tetrazolium salt reduction. Both coenzymes $Q_{10}$ and the menadione were shown to enhance the succinic dehydrogenase activity of frozen sections as measured quantitatively and qualitatively by the reduction of INT. Increased formazan deposition as measured by visual examination was also observed in sections of liver in which either of the two quinones was incorporated in a nitro-BT substrate solution. Extraction of tissue sections with acetone decreased the dehydrogenase activity of liver sections by approximately two-thirds. However, addition of coenzyme $Q_{10}$ or menadione restored the activity to essentially that found in the non-extracted sections. TPNH diaphorase and monoamine oxidase showed no significant enhancement in the presence of coenzyme $Q_{10}$, although relatively slight increase was observed in the presence of menadione. This was not the case with α-glycerophosphate dehydrogenase. Wattenberg and Gronvall (1960) observed the effects of coenzyme $Q_{10}$ with reference to the succinate-INT and α-glycerophosphate-INT reductase systems of normal liver, regenerating liver, and hepatoma and noted differences in the effect of the coenzyme.

Ogawa and Hashimoto (1961) have studied the effect of coenzyme $Q_7$ and its substitutes in the electron transfer system of the succinic dehydrogenase system and DPN diaphorases by observing the staining reaction of frozen sections in conjunction with nitro-BT as a substrate. Among the possible substitutes for coenzyme $Q_7$ were *p*-benzoquinone, α-naphthoquinone, phthalimide, *o*-benzoylbenzoic acid, anthraquinone, 9,10-phenanthraquinone, and phloroglucinol. Among these, only phenanthraquinone was found to be of value. However, menadione produced the greatest enhancement of activity for both succinic dehydrogenase and DPN diaphorase.

## C. Applied Histochemistry

Among the earlier applications of tetrazolium staining techniques were the pioneering studies mentioned in the *Introduction* and the section on *Substrates.* Seligman and Rutenburg (1951) observed the distribution of succinic dehydrogenase in a number of rat and mouse tissues by means of the ditetrazolium chloride BT. Rutenburg *et al.* (1953) also studied the comparative distribution of succinic dehydrogenase with the same substrate in tissues from five mammals. More recently, Pearson (1958) has reported improved histochemical localizations of succinic dehydrogenase by use of a nitro-neotetrazolium chloride [Chart 4 (7)]. As previously mentioned, the application of the nitro-BT as originally reported by Nachlas *et al.* (1957) has resulted in definitive localizations of dehydrogenases and diaphorases in a wide variety of tissues, particularly at the cytochemical level.

Friede (1959a, b, c, 1960, 1961) has reported in detail on the localization of succinic dehydrogenase in the central nervous system in the rat and guinea pig. Potanos *et al.* (1959) studied succinic dehydrogenase activity of human neural biopsies and observed cytoplasmic staining in neurons, astrocytes, oliogodendroglia. Becker *et al.* (1960) employed unembedded frozen sections and polyvinyl alcohol embedded sections in the demonstration of DPNH-tetrazolium reductase and TPNH-tetrazolium reductase of rat brain. When these sections were incubated in the presence of nitro-BT, localization of these enzyme systems in mitochondria could be observed. Chason and Pearse (1961) have reported on the localization and comparative activities of five oxidative enzymes of rat brain employing MTT and nitro-BT. In order to demonstrate three DPN-linked enzymes (malate, lactate, and α-glycerophosphate dehydrogenases), they reported that the high DPN-nucleotidase activity of the brain required blocking with high concentrations of nicotinamide, which was incorporated in the substrate solution.

A significant study is that of Rubinstein *et al.* (1962) who described oxidative enzyme activity of glial cells in a local brain injury associated with edema in the white matter in cats. In the area of edema an increase of glutamic dehydrogenase activity was observed in astrocytes within 12 hours after the lesion was produced. Increase of DPN diaphorase, DPN-linked isocitric dehydrogenase and α-glycerophosphate dehydrogenase activity was manifest within 24 hours. TPN diaphorase or TPN-linked dehydrogenase activity was increased after 48 hours. This enzymatic change in astrocytes was preceded by enlargement of these cells and the formation of Holzer fibers which appeared within 6 hours in the edematous white matter. The hypertropic astrocytes of the white matter continue to exhibit an increased tetrazolium reductase activity which was demonstrable for many days following cortical injury.

Kuwabara and Cogan (1960) have described different types of dehydrogenases located in the retina, one represented by the succinic dehy-

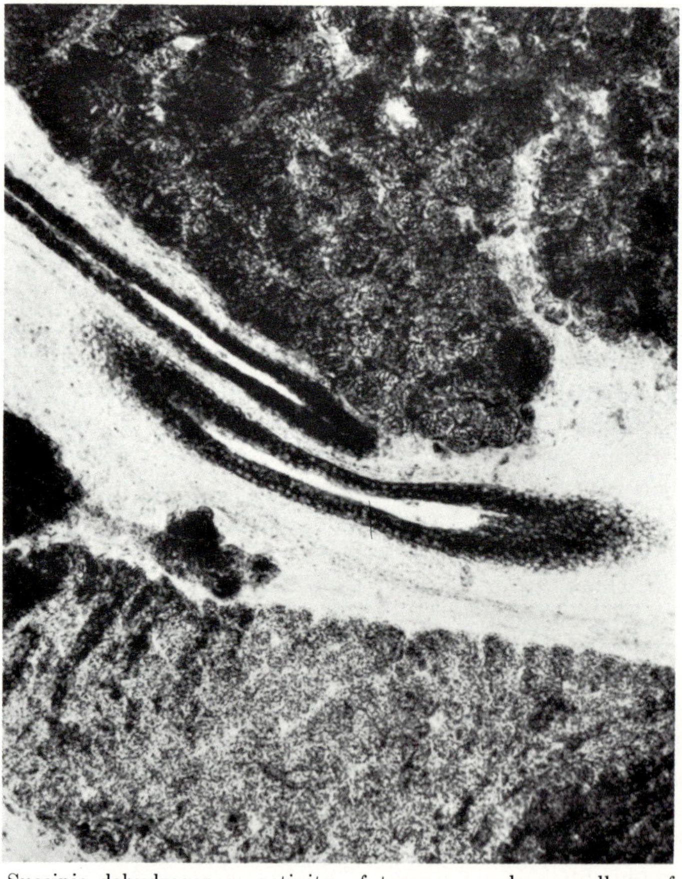

FIG. 2. Succinic dehydrogenase activity of tongue muscle as well as of salivary gland ducts in two day old hamster. Nitro-BT (× 143).

drogenase system and the other by the DPN-linked dehydrogenase. Niemi and Merenmies (1961) have reported on the distribution of pyridine nucleotide diaphorases, succinic dehydrogenase, and five pyridine nucleotide-linked dehydrogenases in the retina of five animal species employing both MTT and nitro-BT as substrates.

Recent reports, including reviews of the literature dealing with the application of tetrazolium methods in the study of dehydrogenases of human leukocytes and platelets, have been made by Balogh and Cohen (1961b) and Marcuse and Cochran (1961). Balogh and Cohen (1961a) have employed the nitro-BT procedure in the demonstration of a variety of oxidative enzymes in the epithelial cells of the human parathyroid.

The distribution of succinic dehydrogenase in human skin employing

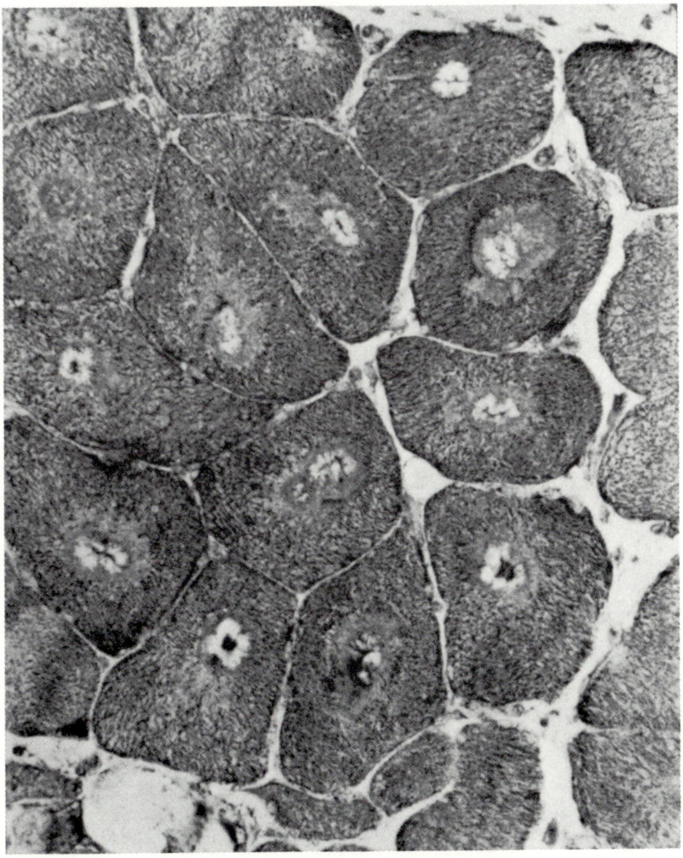

FIG. 3. Lactic dehydrogenase activity of denervated human skeletal muscle showing "target fibers." Nitro-BT (× 225) (Courtesy Dr. W. K. Engel).

NT as a substrate has been reported by Montagna and Formisano (1955) who found that the enzyme is present in the basal layer of the epidermis of the outer sheath and bulb of hair follicles and in sweat glands. The reaction was most intense in the eccrine sweat glands and in their ducts.

Dehydrogenase activity of rat and kitten epiphysial cartilage has been studied by Follis and Melanotte (1956) and Whitehead and Weidmann (1959). However, the substrate employed, TTC, does not favor accurate histological localizations. Burstone (1960a, b) has reported on the localization of succinic dehydrogenase in developing bones, teeth, and cartilage. Distinctive activity was associated with osteoclasts.

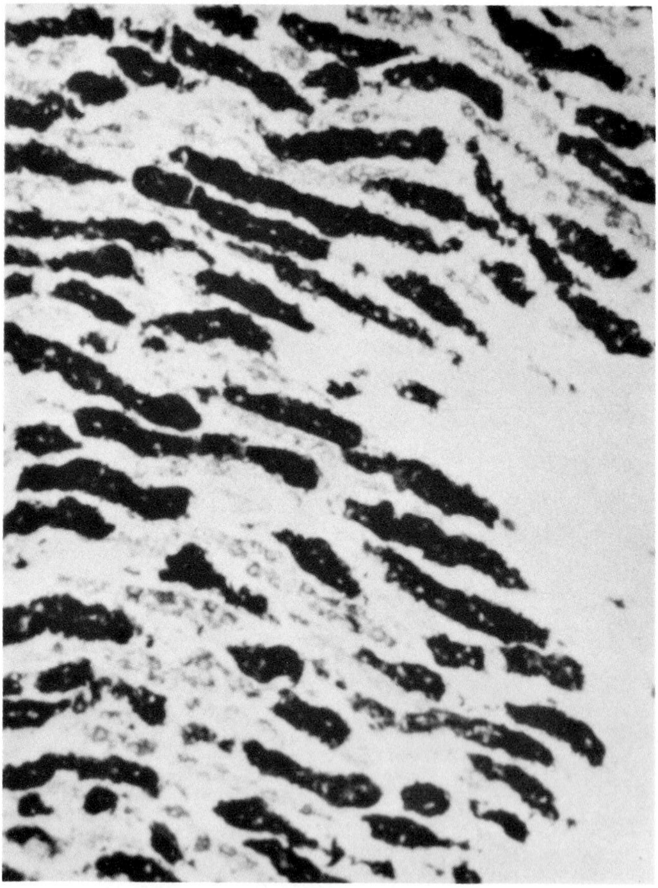

FIG. 4. Succinic dehydrogenase activity of rat kidney tubules. Nitro-BT (× 143).

Allen (1959, 1960) has studied the histochemical localization of DPN-diaphorase activity of the adrenal cortex and also 3$\beta$-steroid dehydrogenase activity of the adrenal cortex of the developing mouse.

Kellogg and Glenner (1960) have reported the histochemical localization of 17$\beta$-estradiol dehydrogenases in human term placental tissues. The staining reaction was most prominent in the cells of the trophoblast and the decidua, and less intense in the stroma.

The dehydrogenase tetrazolium techniques have also been applied in

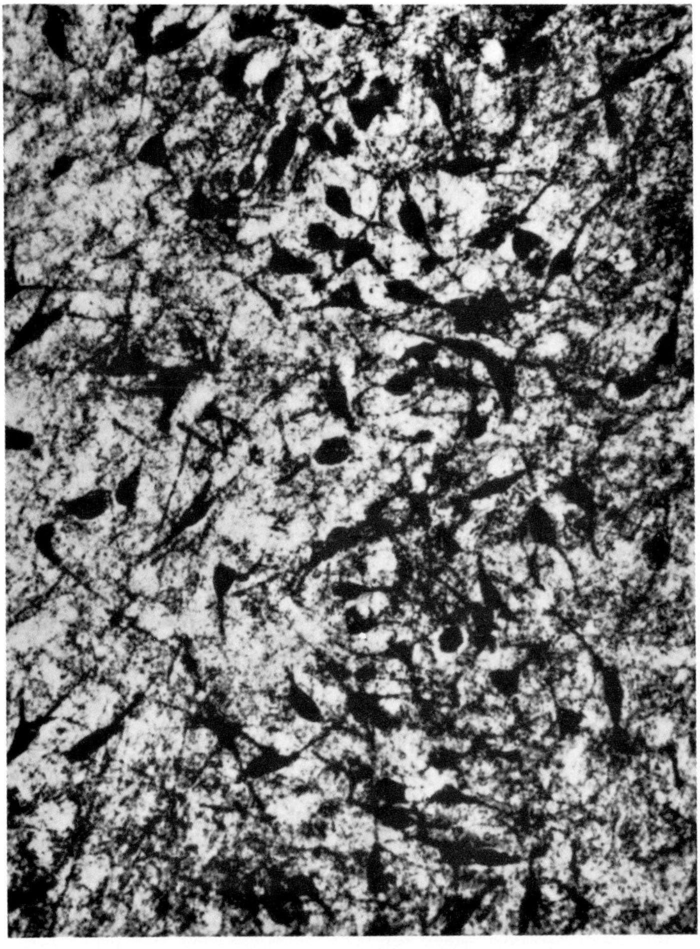

FIG. 5. DPN-diaphorase activity of human brain. One year old child, substantia nigra. NT method of Farber (× 150) (Courtesy Dr. R. Friede).

the study of a number of pathological conditions. Lojda and Felt (1960) have employed nitro-BT in the study of DPN-diaphorase, TPN-diaphorase, lactic dehydrogenase, malic dehydrogenase, glutamic dehydrogenase, and glucose-6-phosphate dehydrogenase in the aorta of rabbits with cholesterol atheromatosis. High activity of the dehydrogenase systems was found in the small plaques.

Hess (1960) has utilized the MTT procedure in the study of pyridine nucleotide-linked dehydrogenases and diaphorases in experimentally produced aminonucleoside nephrosis in the rat. A progressive reduction in the activity of tricarboxylic acid enzymes and $\beta$-hydroxybutyrate dehydrogenase was observed.

Engel (1961) has employed the dehydrogenase techniques in the study of the so-called muscle "target fiber," a newly recognized sign of skeletal

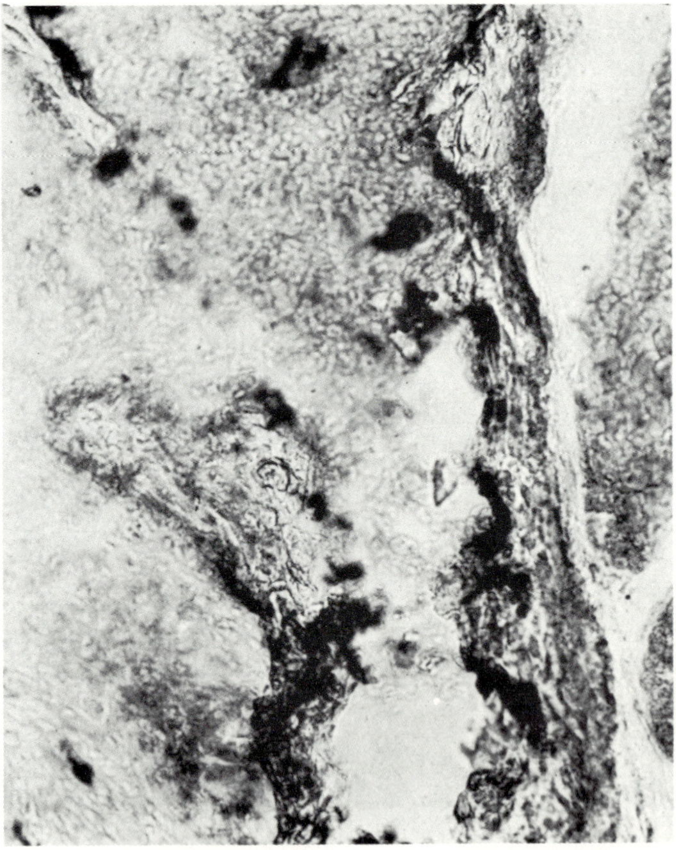

FIG. 6. Succinic dehydrogenase activity of osteoclasts associated with immature hamster bone. Nitro-BT method (× 210).

muscle fiber denervation. In unfixed frozen sections histochemical evidence of succinic dehydrogenase and lactic dehydrogenase activity was found to be completely absent.

The localization of dehydrogenases in a number of unfixed tissues is shown in Figs. 1 through 6.

### D. Triazolium Procedures

As previously indicated, triazole salts have not found application in histochemistry although Kuhn and Ludolphy (1949) observed that tubercle bacilli were capable of reducing the triazole compound derived from benzidine (Chart 7) (Appendix 12A, Section 4).

### E. Tellurite Procedure

As previously indicated, the technique employing sodium tellurite, although capable of yielding sharp localizations, is considerably less sensitive than the standard tetrazolium procedures. Wachstein (1949) found that a 0.1% solution of potassium tellurite in 0.1 *M* phosphate buffer pH 7.4 gave reproducible and consistent results with tissue sections and smears. He noted that in a considerable number of polymorphonuclear leukocytes, reducing activity was revealed by the occurrence of black granules. Kidney tissue exhibited the strongest staining reaction. Wachstein (1949) notes that the distribution of reduced tellurium can be studied in frozen sections prepared from formalin fixed tissues cut at 15–25 $\mu$ which had been incubated for 8–12 hours in the tellurite solution prior to fixation. However, this would appear to be an excessive incubation time so far as the definitive localization of enzyme systems is concerned. As indicated in Chapter 13 (p. 534) the tellurite procedure has been employed in the demonstration of dehydrogenases at the electron microscope level.

## IV. DEHYDROGENASES IN NEOPLASMS

In 1948, Straus *et al.* found that pieces of freshly removed human neoplastic tissue incubated in TTC solution developed a pink color within a matter of minutes, while uninvolved surrounding tissues remained unstained. Among the tumors studied were carcinoma of the breast, pancreatic carcinoma, fibroadenoma, and a parotid mixed tumor. However, Rutenburg *et al.* (1950) found that several animal tumors including sarcoma 37, Bagg lymphosarcoma, and Walker carcinoma were unable to reduce BT *in vivo*. Furthermore, there was no appreciable difference in enzymatic activity of tissues of normal and tumor-bearing mice and rats. Human tumors—including adenocarcinoma of cecum, colon, and rectum; sarcoma of the kidney; hypernephroma of the kidney; carcinoma of the breast; lymphosarcoma; and muscle sarcoma—were also unable to reduce

BT when thin slices of the tumor were incubated in a substrate solution containing the tetrazole.

Goddard and Seligman (1953) studied the distribution of dehydrogenase activity in normal liver and nodules of hepatoma with the BT tetrazolium procedure as well as several standard histological staining methods. The neoplastic tissues were found to exhibit a distinctly lower succinic dehydrogenase activity than normal liver cells.

Novikoff (1960a, b) has reported on the dehydrogenase activity in hepatoma cells, both solid and ascites, with nitro-BT and DPNH or TPNH as substrates. Glutamic dehydrogenase activity was also demonstrable in imprints of solid tumors and in smears of ascites tumors.

Foraker and associates (1954; Foraker, 1956a, b; Foraker and Denham, 1957) have described the dehydrogenase activity of normal and diseased cervices and also activity of the enzyme in carcinoma of the breast.

Wattenberg (1959) observed moderate TPN diaphorase activity of epithelium and crypts of normal colon. However, in carcinomas an intense staining was found in both poorly and well differentiated cells. TPN diaphorase activity was not increased in benign polyps. However, in malignant polyps, areas of intense staining were observed. Wattenberg (1959) observed a decreased α-glycerophosphate dehydrogenase activity of carcinoma of the colon, but the staining reaction was variable. Benign polyps often showed an increased staining reaction associated with atypical glands.

Monis *et al.* (1959) localized succinic dehydrogenase and TPN and DPN diaphorases in a number of neoplasms and noted that succinic dehydrogenase was low in tumors as compared with normal tissues. DPN diaphorase activity was reported to be similar to that seen in normal cells.

Nachlas and Hannibal (1961) reported on the dehydrogenase activity of a number of specimens of human normal colonic mucosa adenomatous polyps, carcinoma *in situ*, and adenocarcinomas. In general, the activities of succinic dehydrogenase, and of DPN and TPN diaphorases were decreased in polyps. However, in the carcinomas, their activities were equal to or greater than that seen in normal epithelial cells. Nachlas and Hannibal (1961) indicate that these studies neither support nor deny the concept that adenomatous polyps are precancerous lesions.

Mossakowski (1962) studied the succinic dehydrogenase activity of astrocytic tumors of varying degrees of anaplasia, and compared the reaction with that in oligodendroglioma and in various elements of normal brain. Dehydrogenase activity, which was present in all glial tumors studied, increased with increasing anaplasia of the tumors. The activity of oligodendroglioma, while high, was less than that of glioblastoma multiforme. The dehydrogenase activity in the cells of piloid astrocytoma appeared similar to that of astrocytoma in the white matter. Succinic dehydrogenase activity in the cells of gemistocytic astrocytoma, malignant glioma, and glioblastoma multiforme exceeded considerably that in normal glia. Reactive glia also showed greater dehydrogenase activity than normal glia.

In the tissue culture studies of Marcuse (1954a, b, 1957) dehydrogenase activity was observed in cultures of most of the malignant tumors and in only some of the benign ones. Wagner *et al.* (1959) and Kaufman and Hill (1959) have employed the tetrazolium method for the demonstration of succinic dehydrogenase in tissue cultures of HeLa cells. Kaufman and Hill (1960) subsequently found an increase in succinic dehydrogenase activity of HeLa cells following inoculation with Newcastle virus, although routine staining procedures failed to reveal any morphological changes in the HeLa cells.

Bitensky *et al.* (1960) studied succinic dehydrogenase activity of liver cells following injection of the carcinogenic 4-dimethylaminostilbene. There was a depression of dehydrogenase activity soon after injection, but this recovered to normal levels in a few hours. These authors suggested that succinic dehydrogenase was directly inhibited, probably by metabolites of the carcinogen. Chang (1960) has reported on a histochemical study of the enzymatic behavior of rat liver during early stages of carcinogenesis induced by 3′-methyl-4-dimethylaminoazobenzene and *N*-2-fluorenyldiacetamide. In general, a decrease in succinic dehydrogenase and DPNH and TPNH diaphorase activity was observed in rats fed the carcinogens for a period of several weeks.

A recent review by Boxer and Devlin (1961) deals with pathways of intracellular hydrogen transport as related to the aerobic glycolysis of malignant tissues. They note the failure of malignant tissues to oxidize DPNH by certain "pathways," and indicate that the absence of hydrogen-carrying systems is a factor in aerobic glycolysis of malignant tissue. The application of dehydrogenases and other histochemical techniques to elucidate the histo- and cytochemical aspects of the various "cycles" or "shuttles" as outlined by Boxer and Devlin (1961) should be of value from both basic research and diagnostic standpoints.

## APPENDIX 12A

# SYNTHESIS OF TETRAZOLES AND TRIAZOLES

### 1. *3,3′-Dianisole-bis-4,4′-(3,4-diphenyl) tetrazolium Chloride* (*BT, Blue Tetrazolium*) (*Rutenburg et al., 1950; Ried and Gick, 1953*)

The synthesis is divided into three parts; the preparation of the phenylhydrazone, coupling of the phenylhydrazone to form a diformazan, and oxidation of the latter to produce the ditetrazolium salt.

Phenylhydrazine (108.0 g.) is dissolved in 95% ethanol (1000.0 ml.), and redistilled benzaldehyde (106.0 g.) is added with stirring. A yellow crystalline product separates. The mixture is cooled by ice, and the crystals of benzal phenylhydrazone (commercially available) are collected and dried.

Benzal phenylhydrazone (157.0 g.) is dissolved in pyridine (1200.0 ml.) and cooled to $-5°$. Tetrazotized *o*-dianisidine (676.0 g.) is added in small portions and with stirring. The temperature, which tends to rise during the coupling reaction, is maintained below 5° by cooling in an ice-salt mixture. After addition of each small portion of diazotate, the reaction is allowed to subside before introducing the next portion. The addition of all of the diazotate should take approximately 2½ hours. The thick, dark blue mixture is then allowed to warm to room temperature and is stirred for 2 hours until gas evolution ceases. Fifty percent ethanol (1000.0 ml.) is added with vigorous stirring. The blue-black diformazan which precipitates is collected on a Buchner funnel, washed once with 50% ethanol (250.0 ml.), dried with suction, and washed thoroughly with hot distilled water. The product is dried, ground to a fine powder, suspended in boiling distilled water, and stirred for 10–15 minutes in order to remove inorganic salts. The blue-black product is filtered with suction and washed ca. 10 times with hot distilled water. The 3,3′-dianisole-bis-4,4′-(3,5-diphenyl) formazan has an m.p. of 204–206°. It was reported that attempts to crystallize the diformazan were unsuccessful. The compound is very soluble in chloroform, carbon tetrachloride, benzene, acetone, and ethyl acetate. Upon dissolving in warm pyridine the formazan may be reprecipitated from water.

The powdered diformazan (66.0 g.) is suspended in 95% ethanol (500.0 ml.) in a 2 liter, 3-necked flask fitted with a mechanical stirrer and a dropping funnel. Amyl or isoamyl nitrite (58.0 g.) is then added. Concentrated HCl (49.0 g.) is slowly added, with mechanical stirring, over a period of 60 minutes at room temperature. Another portion of amyl nitrite (8.0 g.) and of concentrated HCl (6.5 g.) are added dropwise. The diformazan dissolves and its blue color is discharged in the process of oxidation to the ditetrazolium chloride. After being stirred for 24 hours the resulting yellow-brown solution is filtered from a dark-brown precipitate. The residue is washed once with hot ethanol and discarded. The filtrates, concentrated to 500.0 ml., are added to hot water (2000.0 ml.) and the mixture is concentrated to 2000.0 ml. by boiling. During the course of evaporation, a tarry scum is formed on the surface of the solution and is removed and discarded. Charcoal (5.0 g.) is added, and the solution is boiled for 15 minutes and filtered rapidly under suction. The filtrate is concentrated to 1700–1800 ml. and allowed to stand overnight upon which a bright-yellow precipitate of the ditetrazolium chloride (BT) is formed. This product is collected, washed with a little 50% ethanol-ether, and dried. Another yield of the product may be recovered by further concentration of the filtrate. If the aqueous solution of the ditetrazolium chloride is too saturated, the product tends to separate as a gummy tar, which, however, slowly solidifies upon standing. The crude tetrazolium salt is bright-yellow to yellow-brown in color and is very soluble in ethanol, methanol, and chloroform. It is slightly soluble in water and insoluble in ethyl acetate, acetone, and ether. Further purification may be accomplished either by solution in ethanol or chloroform and precipitation with ether, or else by recrystallization from methanol-ethyl acetate. The purified product is pale yellow and decomposes at 242–245°. The insoluble ditetrazolium iodide, nitrate, and thiocyanate may be precipiated from a hot aqueous solution of the chloride by addition of sodium iodide, sodium nitrate, or sodium thiocyanate, respectively ($C_{40}H_{32}O_2N_8Cl_2$).

As an alternative procedure, the diformazan may be oxidized with lead tetraacetate, as recommended by Kuhn and Jerschel (1941a). Lead tetraacetate is added to a chloroform solution of the blue diformazan. The blue color is discharged during the course of an exothermic reaction. However, according to Rutenburg *et al.* (1950) the isolation of the ditetrazolium chloride is difficult, the product is impure, and the yield low. In addition, there is danger that contamination with lead would affect enzymatic activity.

2. *2,2′-Di-p-nitrophenyl-5,5′-diphenyl-3,3′-(3,3′-dimethoxy-4,4′-biphenylene)ditetrazolium Chloride (Nitro-BT) (Tsou et al., 1956)*

A suspension of 12.69 g. (0.04 *M*) of finely powdered *o*-dianisidine HCl is 60.0 ml. of water and 10.0 ml. of concentrated HCl is tetrazotized with a solution of 6.07 g. (0.088 *M*) of sodium nitrite dissolved in 15.0 ml. of water. The resulting mixture is added to a solution of 19.3 g. (0.08 *M*) of benzaldehyde *p*-nitrophenylhydrazone dissolved in 180.0 ml. of tetrahydrofuran, followed by a solution of 8.98 g. (0.08 *M*) of KOH. The reaction mixture is stirred for 30 minutes, 80.0 ml. of water and 180.0 ml. of methanol are added and the mixture filtered to yield a black precipitate with a gold reflex. The precipitate is washed with about 500.0 ml. of methanol until the washing is turned from dark to light red. It is then washed with 600.0 ml. of boiling water and again with methanol, and finally extracted with 340.0 ml. of boiling dioxane. The filtered extract is added to 250.0 ml. of water upon which a black precipitate (monoformazan) separates. The insoluble portion from the dioxane extraction is the diformazan which is dissolved in boiling DMF and precipitated by running the filtered solution into 5 times its volume of acetone. The resulting fine black precipitate of 2,2′-di-*p*-nitrophenyl-5,5′-diphenyl-3,3′-(3,3′-dimethoxy-4,4′-biphenylene) diformazan has an m.p. of 257° (dec.).

The aforementioned compound is oxidized to form the corresponding ditetrazolium salt with isoamyl nitrite. The diformazan (23.74 g., 0.005 *M*) is suspended in a mixture of 15.0 ml. of dioxane and 15.0 ml. of tetrahydrofuran containing 10 drops of isoamyl nitrite are added. The mixture is cooled in an ice bath and saturated with HCl gas. It is then stirred at room temperature for 20 minutes until the purple solution turns greenish and some yellow precipitate appears. Two drops of isoamyl nitrite are added at 10 minute intervals. The use of a magnetic stirrer is recommended since the stirring bar also pulverizes the formazan. The oxidation of the diformazan takes approximately 2 hours at room temperature. The resulting deep-brown solution is treated with carbon black and evaporated to a heavy gum. This is dissolved in 20.0 ml. of hot ethanol, and upon adding 200.0 ml. of anhydrous ether, a light straw color precipitate is obtained. After concentrating the filtrates from the main product, a second crop is obtained. This product (m.p. 156°, dec.) contains 3 molecules of water of crystallization ($C_{40}H_{30}N_{10}O_6Cl_2 \cdot 3H_2O$). The water of crystallization may be replaced by isopropyl alcohol, and subsequently the solvent in the molecule could be removed to form an anhydrous compound. When the hydrate is dried at 100° to a constant weight under vacuum, analysis indicated that one of the two chloride ions was converted to the hydroxide ion ($C_{40}H_{30}N_{10}O_6ClO_8$) (m.p. 134°). Anal., calcd.: C, 60.11%; H, 3.91%; N, 17.52%. Found: C, 60.26%; H, 4.19%; N, 17.82%.

3. *2,2′-(4,4′-Biphenyl)-bis-(α,β-naphtho-3-phenyl-1,2,3-triazolium Nitrate) (Krollpfeiffer et al., 1934)*

The triazolium salt is prepared by oxidation of a dye formed by the coupling of tetrazotized benzidine and *N*-phenyl-β-naphthylamine.

Tetrazotized benzidine is coupled with 2 moles of *N*-phenyl-β-naphthylamine. The dye is recrystallized from benzene and forms greenish-bronze needles which melt at 281–282°.

The dye is suspended in 5 times its weight of glacial acetic acid and the solution heated gently following the addition of a 2-fold excess of amyl nitrite. After brief heating, the mixture is allowed to cool and is treated with a drop or two of concentrated $HNO_3$. Upon standing part of the triazolium nitrate is precipitated directly, while the remainder is precipitated by the addition of ether. If the product forms an oil, the latter will solidify upon trituration with ether or suitable solvents. The triazolium nitrite may be purified by recrystallization from ethanol-ether, water, or glacial acetic acid.

4. *2,2′-p-Biphenylbis(naphtho-1′,2′,4,5-triazolium) Chloride (Kuhn and Ludolphy, 1949)*

This is the chloride of the same compound described in Appendix 12A, Section 3. The starting material, 4,4′-bis(2-anilino-1-naphthalylazo)biphenyl, is prepared by dissolving 10.7 g. of *N*-phenyl-β-naphthylamine in 75.0 ml. of glacial acetic acid at 50° and adding a tetrazotate solution obtained from 4.5 g. of benzidine. The bis-azo compound which precipitates is stirred for approximately 45 minutes, collected by filtration, washed with hot water, and boiled three times in water-glacial acetic acid solution. The compound is recrystallized from 160.0 ml. of xylene and washed with methanol. The dyestuff is poorly soluble in methanol and glacial acetic acid, but is more readily soluble in benzene and pyridine. Upon recrystallization from xylene the compound was obtained as dark red needles with an m.p. of 275–276° ($C_{44}H_{32}N_6$).

For the preparation of the triazolium chloride, 2.0 g. of the bis-azo compound are treated with 1300.0 ml. of chloroform. Part of the dye which remains soluble is dissolved in 2500.0 ml. of benzene. The chloroform solution is treated with 3.0 ml. of isoamyl nitrite and HCl gas, while the benzene solution is treated with 2.0 ml. of isoamyl nitrite and HCl gas. The combined oxidation products, remaining after the two solvents had been removed by distillation, are dissolved in 200.0 ml. of 96% ethanol, boiled in the presence of charcoal, and then precipitated with 500.0 ml. of ether. The triazole is obtained as yellowish flakes which turn dark brown when heated above 290° ($C_{44}H_{30}N_6Cl_2 \cdot 2H_2O$). Anal., calcd.: C, 70.48%; H, 4.57%; N, 11.21%; Cl, 19.46%. Found: C, 69.80%; H, 5.22%; N, 11.48%; Cl, 8.6%.

APPENDIX 12B

# HISTOCHEMICAL METHODS

1. *Succinic Dehydrogenase, Neotetrazolium Modification (Rosa and Velardo, 1954)*

Unfixed frozen sections mounted on glass slides are used.

The incubating mixture contains the following:

| | |
|---|---|
| 0.1 *M* Sodium phosphate buffer containing 0.1% NaCN (pH 8.2) | 30.0 ml. |
| 0.4% Aqueous neotetrazolium chloride | 4.0 ml. |
| 0.5 *M* Sodium succinate | 4.0 ml. |

The phosphate buffer-cyanide solution is prepared by dissolving 1.0 g. of NaCN in 500.0 ml. (0.1 *M*) disodium phosphate solution. It is adjusted to pH 8.2 with 0.1 *M* monosodium phosphate. The total volume is brought up to 1 liter with 0.1 *M* phosphate buffer pH 8.2. The neotetrazolium chloride is brought into solution by initially dissolving it in 0.25 ml. ethanol followed by the addition of water.

Slides are incubated for periods of up to 2 hours at 37°. Following incubation the slides are rinsed briefly in distilled water, fixed from 1 to 24 hours in 10% formalin, and mounted.

2. *Succinic Dehydrogenase, Modified Tetrazolium Method (Burstone, unpublished data)*

Approximately 5–10 mg. of tetrazolium salt are dissolved in 0.5 ml. of ethanol or 0.25 ml. of dimethyl sulfoxide (DMSO). About 20 ml. of distilled water are added slowly and with stirring. The resulting solution is usually milky in appearance. Sodium succinate (500.0 mg.) is dissolved in 20.0 ml. of tris buffer pH 7.4 and added to the substrate solution. Sections mounted on glass slides are incubated at 37°, rinsed in distilled water, and mounted in a suitable mountant.

In the case of tissues exhibiting low activity, phenazine methosulfate (0.5 ml. of a 5.0 mg./ml. solution) is incorporated in the substrate solution.

3. *Succinic Dehydrogenase, Nitro-BT Method (Nachlas et al., 1957)*

Unfixed frozen sections mounted on glass slides or free-floating sections are used.

The substrate solution is prepared from a buffered succinate stock solution which is made by combining equal volumes of phosphate buffer (0.2 *M*) pH 7.6 and sodium succinate (0.2 *M*). Just prior to incubation 10.0 ml. of the buffered substrate reagent are added to 10.0 ml. of an aqueous solution containing 10.0 mg. of nitro-BT. The tetrazolium salt is initially dissolved by the addition of a few drops of ethanol.

A simplified modification of the technique (Burstone, unpublished data) is described as follows. Approximately 5.0 mg. of nitro-BT are placed in a 50 ml. Erlenmeyer flask and dissolved by the addition of 0.25 ml. ethanol or DMF. Twenty ml. of distilled water are added with constant agitation followed by an aqueous solution containing 500.0 mg. of sodium succinate dissolved in 10.0 ml. of distilled water. The solution is brought up to a total volume of 50.0 ml. by the addition of 20.0 ml. of 0.2 *M* tris buffer pH 7.4. The solution is filtered through rapid filtering paper into a Coplin jar.

Slides or sections are incubated for a period of minutes to several hours, rinsed in distilled water, counterstained if desired, and mounted in a water-soluble mountant, or else dehydrated and mounted in ethanolic polyvinyl acetate or Permount.

4. *DPN and TPN Diaphorases* (*Farber et al., 1956a*)

In the original study, frozen sections about 20 μ thick were floated on ice-cold buffer solution and transferred within 3 minutes to 50 ml. glass-stoppered Erlenmeyer flasks containing the cold incubating medium. Different incubating media were employed, each made up to contain substrates for the particular enzyme system under investigation together with appropriate coenzymes and cofactors. The flasks were flushed briefly with $N_2$ (0.5 to 1 minute) and then incubated without shaking in a water bath at 37.5° for periods varying from a few minutes to 2 hours. After incubation, the sections were washed successively in the following solutions: (1) normal saline containing 0.6% acetic acid (two washings), (2) 4% formaldehyde (10% formalin), and (3) distilled water, and were then mounted on slides with glycerin jelly.

For convenience, frozen sections mounted on cover slips or glass slides may be employed and the volume of the substrate solution increased.

*a. DPN diaphorase.*

1. Using malate as substrate:

| | |
|---|---|
| Na L-malate (0.5 *M*) | 0.3 ml. |
| DPN (0.45%) | 0.2 ml. |
| Na L-glutamate (0.5 *M*) | 0.5 ml. |
| NaCN (0.03 *M* in $PO_4$ buffer) | 0.5 ml. |
| Blue tetrazolium (1 mg./ml.) or Neotetrazolium (1 mg./ml.) | 0.7 ml. |
| Phosphate buffer (0.1 *M*, pH 7.4) | 0.7 ml. |
| $H_2O$ to make a total volume of | 3.0 ml. |

2. Using lactate as substrate:

The medium is the same as *a* (1) with K L(+)lactate (0.5 *M*) substituted for malate.

3. Using ethanol and malate as substrate (for optimal demonstration of DPN diaphorase):

| | |
|---|---|
| Alcohol dehydrogenase (1.5 mg. protein/ml.) | 0.1 ml. |
| Ethanol (1.09 *M*) | 0.2 ml. |
| Na L-malate (0.5 *M*) | 0.3 ml. |
| Na L-glutamate (0.5 *M*) | 0.5 ml. |
| DPN (0.45%) | 0.2 ml. |
| Semicarbazide (0.1 *M* in $PO_4$ buffer) | 0.2 ml. |
| Blue tetrazolium (1 mg./ml.) or Neotetrazolium (1 mg./ml.) | 0.7 ml. |
| Phosphate buffer (0.1 *M*, pH 7.4) | 0.8 ml. |

*b. TPN diaphorase.*

1. Using isocitrate or citrate as substrate:

| | |
|---|---|
| Na D-isocitrate (0.1 *M*) or Na citrate (0.5 *M*) | 0.3 ml. |
| TPN (0.2%) | 0.1 ml. |
| $MnCl_2$ (0.005 *M*) | 0.3 ml. |
| L-Cysteine (0.1 *M*) | 0.2 ml. |
| Veronal buffer (0.05 M, pH 7.4) | 1.0 ml. |
| Blue tetrazolium (1 mg./ml.) | 0.7 ml. |
| $H_2O$ to make a total volume of | 3.0 ml. |

2. Using glucose-6-phosphate as substrate:
K glucose-6-phosphate (0.1 *M*) was used in place of isocitrate or citrate and no $MnCl_2$ was added.

3. Using malate as substrate (for malic decarboxylase):
Na L-malate (0.5 *M*) was used in place of isocitrate or citrate.

4. Using citrate and malate as substrates (for optimal demonstration of TPN diaphorase):

| | |
|---|---|
| Mixture of Na citrate and Na malate (each 0.5 *M*) | 0.3 ml. |
| TPN (0.2%) | 0.1 ml. |
| $MnCl_2$ (0.005 *M*) | 0.3 ml. |
| L-Cysteine (0.1 *M*) | 0.2 ml. |
| Veronal buffer (0.05 *M*, pH 7.4) | 0.9 ml. |
| Methocel (15 CPS) (3.0%) | 0.5 ml. |
| Blue tetrazolium (1 mg./ml.) | 0.7 ml. |

The cysteine is freshly prepared (as the hydrochloride) and was neutralized to about pH 7.4 before use.

5. *Succinic Dehydrogenase System, Oxidation-Reduction Dye Modification* (*Farber and Louviere, 1956*)

(1) *Succinic dehydrogenase*

| | |
|---|---|
| Na succinate (0.5 *M*) | 0.3 ml. |
| Potassium $PO_4$ buffer (0.1 *M*, pH 7.4) | 1.0 ml. |

| | |
|---|---|
| $CaCl_2$ (0.004 *M*) | 0.3 ml. |
| $NaHCO_3$ (0.6 *M*, freshly prepared) | 0.15 ml. |
| Blue tetrazolium (1 mg./ml.) | 0.7 ml. |
| Methylene blue (1 mg./ml.) | 0.05 ml. |
| $H_2O$ | 0.5 ml. |

(2) *DPN diaphorase*

| | |
|---|---|
| Na L-malate (0.5 *M*) | 0.3 ml. |
| Ethanol (1.09 *M*) | 0.2 ml. |
| Na L-glutamate (0.5 *M*) | 0.5 ml. |
| Potassium $PO_4$ buffer (0.1 *M*, pH 7.4) | 0.8 ml. |
| Alcohol dehydrogenase (ca. 7.5 mg. protein/ml.) | 0.02 ml. |
| DPN (5 mg./ml.) | 0.2 ml. |
| Semicarbazide (0.1 *M*) | 0.2 ml. |
| Blue tetrazolium (1 mg./ml.) | 0.7 ml. |
| Azure C, azure I, or pyocyanin (1 mg./ml.) | 0.05 ml. |
| $H_2O$ | 0.03 ml. |

(3) *TPN diaphorase*

| | |
|---|---|
| Mixture of Na L-malate (0.5 *M*) and Na citrate (0.5 *M*) | 0.3 ml. |
| Veronal buffer (0.05 *M*, pH 7.4) | 0.9 ml. |
| $MnCl_2$ (0.005 *M*) | 0.3 ml. |
| L-Cysteine (0.1 *M*, freshly prepared) | 0.2 ml. |
| TPN (3 mg./ml.) | 0.1 ml. |
| Blue tetrazolium (1 mg./ml.) | 0.7 ml. |
| Thionine (1 mg./ml.) | 0.05 ml. |
| Nile Blue A (1 mg./ml.) | 0.05 ml. |
| $H_2O$ | 0.4 ml. |

All substrate and cofactor solutions (e.g. semicarbazide, cysteine) are neutralized to pH 7.4. In medium 3, a precipitate of dye, which always formed on adding cysteine, was separated by centrifugation.

Simplified modifications of the TPN and DPN methods are given below:

*DPN diaphorase*

| | |
|---|---|
| 0.2 *M* Phosphate buffer pH 7.4 | 1.0 ml. |
| DPNH (10.0 mg./ml.) | 0.3 ml. |
| Nitro-BT (5.0 mg./ml.) | 0.3 ml. |
| $H_2O$ to make a total volume of | 3.0 ml. |

Incubate sections at 37° for periods up to 40 minutes.

*TPN diaphorase*

| | |
|---|---|
| 0.1 *M* Veronal buffer, pH 7.4 | 0.9 ml. |
| TPNH (10.0 mg./ml.) | 0.3 ml. |
| Nitro-BT (1.0 mg./ml.) | 0.7 ml. |
| $H_2O$ to make a total volume of | 3.0 ml. |

Incubate sections under $N_2$ at 37° for periods up to 40 minutes.

6. *DPN Diaphorase* (*Nachlas et al., 1958a*)

Unfixed frozen sections are incubated in the following substrate solution:

| | |
|---|---|
| Sodium lactate (0.5 *M*) | 0.6 ml. |
| Lactic dehydrogenase (1.5% solution) | 0.2 ml. |
| DPN (5.0 mg./ml.) | 0.3 ml. |
| Nitro-BT (5.0 mg./ml.) | 0.3 ml. |
| Phosphate buffer pH 7.4 (0.2 *M*) | 1.0 ml. |
| Water to make a total volume of | 3.0 ml. |

The sections are incubated at room temperature for 5–30 minutes, rinsed briefly in saline, fixed in formalin for 10 minutes, dehydrated, and mounted in a synthetic medium. Control sections may be immersed in boiling water for 5 minutes prior to incubation.

7. *Triphosphopyridine Nucleotide* (*TPN*) *Diaphorase* (*Nachlas et al., 1958b*)

The substrate mixture contains the following:

| | |
|---|---|
| Sodium DL-isocitrate (0.1 *M*) | 0.6 ml. |
| Sodium L-malate (2.5 *M*) | 0.5 ml. |
| $MnCl_2$ (0.005 *M*) | 0.3 ml. |
| TPN (5.0 mg./ml.) | 0.2 ml. |
| Nitro-BT (5.0 mg./ml.) | 0.3 ml. |
| Veronal buffer (0.05 *M*) pH 7.4 | 1.1 ml. |

At the time of preparation the pH of all stock solutions of substrates is adjusted to 7.4. The stock solutions are stored under refrigeration at ca. 4°, while enzyme preparations are stored at −20°. The incubation of two sections at a time may be carried out on a half-cover slip placed in a test tube. Following incubation the sections may be dehydrated in ethanol, cleared in xylene, and mounted in a synthetic medium.

8. D-*Amino Acid Dehydrogenase* (*Oxidase*) *and Choline Dehydrogenase* (*Oxidase*) (*Farber et al., 1958*)

*a.* D-*Amino acid dehydrogenase.*

| | |
|---|---|
| 0.2 *M* Phosphate buffer 7.4 | 1.0 ml. |
| 2% DL- or D-Methionine | 0.5 ml. |
| Nitro-BT (1.0 mg./ml.) | 0.7 ml. |
| Phenazine methosulfate (5.0 mg./ml.) | 0.05 ml. |
| $H_2O$ to make a total volume of | 3.0 ml. |

Sections are incubated under $N_2$ for 5–20 minutes at 37°.

*b. Choline dehydrogenase.*

The substrate solution is prepared in the same fashion with the exception that 2% choline chloride is substituted for the methionine.

## Commercial Sources of Reagents

1. Dajac Laboratories, The Borden Chemical Company, Philadelphia, Pennsylvania
2. Eastman Kodak Research Laboratories, Rochester, New York
3. Mann Research Laboratories, New York, New York
4. Nutritional Biochemicals Corp., Cleveland, Ohio
5. Sigma Chemical Company, St. Louis, Missouri

## References


Allen, J. M. (1959) *Anat. Record* **134,** 385–395.

Allen, J. M. (1960) *Anat. Record* **137,** 57–62.

Antopol, W., Glaubach, S., and Goldman, L. (1948) *Public Health Repts. (U. S.)* **63,** 1231–1238.

Antopol, W., Glaubach, S., and Goldman, L. (1950) *Trans. N. Y. Acad. Sci.* **12,** 156–160.

Ashley, J. N., Davis, B. M., Nineham, A. W., and Slack, R. (1953) *J. Chem. Soc.* pp. 3881–3888.

Atkinson, E., Melvin, S., and Fox, S. W. (1950) *Science* **111,** 385–387.

Balogh, K., and Cohen, R. B. (1961a) *Lab. Invest.* **10,** 354–360.

Balogh, K., and Cohen, R. B. (1961b) *Blood* **17,** 491–496.

Becker, H., and Quadbeck, G. (1950) *Naturwissenschaften* **37,** 565–567.

Becker, N. H., Goldfischer, S., Shin, W.-Y., and Novikoff, A. B. (1960) *J. Biophys. Biochem. Cytol.* **8,** 649–663.

Beyer, H., and Pyl, T. (1954) *Chem. Ber.* **87,** 1505–1511.

Bitensky, L., Baldwin, R. W., and Chayen, J. (1960) *Brit. J. Cancer* **14,** 690–695.

Black, M. M., Zweifach, B. W., and Speer, F. D. (1953) *Am. J. Clin. Pathol.* **23,** 332–339.

Boxer, G. E., and Devlin, T. M. (1961) *Science* **134,** 1495–1501.

Brodie, A. F., and Gots, J. S. (1951) *Science* **114,** 40–41.

Brodie, A. F., and Gots, J. S. (1952) *Science* **116,** 588–589.

Burstone, M. S. (1960a) *Nature* **185,** 866.

Burstone, M. S. (1960b) *Ann. N. Y. Acad. Sci.* **85,** 431–444.

Burtner, H. J., Bahn, R. C., and Longley, J. B. (1957) *J. Histochem. and Cytochem.* **5,** 127–134.

Cascarano, J., and Zweifach, B. W. (1959) *J. Biophys. Biochem. Cytol.* **5,** 309–318.

Chance, B., and Williams, G. R. (1956) *Advances in Enzymol.* **17,** 65–134.

Chang, J. P. (1960) *In* "Cell Physiology of Neoplasia" (T. G. Hsu, ed.), pp. 471–500. Univ. Texas Press, Austin, Texas.

Chason, J. L., and Pearse, A. G. E. (1961) *J. Neurochem.* **6,** 259–266.

Cheronis, N. D., and Stein, H. (1956) *J. Chem. Educ.* **33,** 120–125.

Dianzani, M. U. (1953) *Nature* **171,** 125–126.

Engel, W. K. (1961) *Nature* **191,** 389–390.

Ernster, L., and Lindberg, O. (1958) *Ann. Rev. Physiol.* **20,** 13–42.

Farber, E., and Bueding, E. (1956) *J. Histochem. and Cytochem.* **4,** 357–362.

Farber, E., and Louviere, C. D. (1956) *J. Histochem. and Cytochem.* **4,** 347–356.

Farber, E., Sternberg, W. H., and Dunlap, C. E. (1956a) *J. Histochem. and Cytochem.* **4,** 254–265.

Farber, E., Sternberg, W. H., and Dunlap, C. E. (1956b) *J. Histochem. and Cytochem.* **4,** 284–294.

FARBER, E., STERNBERG, W. H., AND PEARCE, N. A. M. (1958) *J. Histochem. and Cytochem.* **6,** 389.
FOLLIS, R. H., AND MELANOTTE, P. L. (1956) *Proc. Soc. Exptl. Biol. Med.* **93,** 382–384.
FORAKER, A. G. (1956a) *Ann. N. Y. Acad. Sci.* **63,** 1107–1116.
FORAKER, A. G. (1956b) *Surg. Gynecol. Obstet.* **102,** 1–8.
FORAKER, A. G., AND DENHAM, S. W. (1957) *Am. J. Obstet. Gynecol.* **74,** 13–24.
FORAKER, A. G., DENHAM, S. W., AND CELI, P. A. (1954) *Cancer* **7,** 311–317.
FOX, S. W., AND ATKINSON, E. H. (1950) *J. Am. Chem. Soc.* **72,** 3629–3631.
FRIEDE, R. L. (1959a) *J. Neurochem.* **4,** 101–110.
FRIEDE, R. L. (1959b) *J. Neurochem.* **4,** 111–123.
FRIEDE, R. L. (1959c) *J. Neurochem.* **4,** 290–303.
FRIEDE, R. L. (1960) *J. Neurochem.* **5,** 156–171.
FRIEDE, R. L. (1961) *J. Neurochem.* **6,** 190–199.
FRIEDENWALD, J. S., AND STEIHLER, R. D. (1938) *Arch. ophthalmol. (Paris)* **20,** 761–786.
GODDARD, J. W., AND SELIGMAN, A. M. (1953) *Cancer* **6,** 385–389.
GOMORI, G. (1952) "Microscopic Histochemistry." Univ. Chicago Press, Chicago, Illinois.
GREEN, D. E., AND FLEISCHER, S. (1960) *In* "Metabolic Pathways" (D. M. Greenberg, ed.), Vol. 1, pp. 41-96. Academic Press, New York.
HARTREE, E. F. (1957) *Advances in Enzymol.* **18,** 1–64.
HESS, R. (1960) *Am. J. Pathol.* **37,** 583-597.
HESS, R., SCARPELLI, D. G., AND PEARSE, A. G. E. (1958a) *Nature* **181,** 1531–1532.
HESS, R., SCARPELLI, D. G., AND PEARSE, A. G. E. (1958b) *J. Biophys. Biochem. Cytol.* **4,** 753–760.
HOFFMANN-OSTENHOF, O. (1953) *Advances in Enzymol.* **14,** 219–260.
HOLTER, H. (1952) *Advances in Enzymol.* **13,** 1–20.
HUNTER, L., AND ROBERTS, C. B. (1941a) *J. Chem. Soc.* pp. 820–823.
HUNTER, L., AND ROBERTS, C. B. (1941b) *J. Chem. Soc.* pp. 823–826.
JÁMBOR, B. (1960) "Tetrazoliumsalze in der Biologie." Fisher, Jena, Germany.
JENSEN, C. O., SACKS, W., AND BALDAUSKI, F. A. (1951) *Science* **113,** 65–66.
KARMARKAR, S. S., BARRNETT, R. J., NACHLAS, M. M., AND SELIGMAN, A. M. (1958) *J. Am. Chem. Soc.* **81,** 3771–3775.
KAUFMAN, N., AND HILL, R. W. (1959) *J. Histochem. and Cytochem.* **7,** 144–146.
KAUFMAN, N., AND HILL, R. W. (1960) *Cancer Research* **20,** 335–338.
KELLOGG, D. A., AND GLENNER, G. G. (1960) *Nature* **187,** 763–764.
KIVY-ROSENBERG, E., CASCARANO, J., AND ZWEIFACH, B. W. (1955) *Trans. N. Y. Acad. Sci.* **17,** 402–406.
KLETT, A. (1900) *Z. Hyg. Infektsionskrankh.* **33,** 137–160.
KROLLPFEIFFER, F., MÜLHAUSEN, C., AND WOLF, G. (1934) *Ann. Chem. Liebigs* **508,** 39–51.
KUHN, R., AND JERSCHEL, D. (1941a) *Ber.* **74,** 941–949.
KUHN, R., AND JERSCHEL, D. (1941b) *Ber.* **74,** 949–952.
KUHN, R., AND LUDOLPHY, E. (1949) *Ann. Chem. Liebigs* **564,** 35–43.
KUN, E. (1951) *Proc. Soc. Exptl. Biol. Med.* **78,** 195–197.
KUN, E., AND ABOOD, L. G. (1949) *Science* **109,** 144–146.
KUWABARA, T., AND COGAN, D. G. (1960) *J. Histochem. and Cytochem.* **8,** 214–224.
LAKON, G. (1939) *Ber. deut. botan. Ges.* **57,** 191–203.
LAKON, G. (1942a) *Ber. deut. botan. Ges.* **60,** 299–305.
LAKON, G. (1942b) *Ber. deut. botan. Ges.* **60,** 434–444.
LEVY, H., DEANE, H. W., AND RUBIN, B. L. (1959) *Endocrinology* **65,** 932–943.

Lojda, Z., and Felt, V. (1960) *Experientia* **16,** 514–517.

Marcuse, P. M. (1954a) *Texas Repts. Biol. and Med.* **12,** 126–137.

Marcuse, P. M. (1954b) *Texas Repts. Biol. and Med.* **13,** 874–886.

Marcuse, P. M. (1957) *Lab. Invest.* **6,** 133–141.

Marcuse, P. M., and Cochran, J. (1961) *Blood* **17,** 738–746.

Mattson, A. M., Jensen, C. O., and Dutcher, R. A. (1947) *Science* **106,** 294–295.

Mehler, A. H. (1957) "Introduction to Enzymology." Academic Press, New York.

Monis, B., Nachlas, M. M., and Seligman, A. M. (1959) *Cancer* **12,** 1238–1247.

Mossakowski, M. J. (1962) *J. Neuropathol. Exptl. Neurol.* **21,** 137–146.

Montagna, W., and Formisano, V. (1955) *Anat. Record* **122,** 65–78.

Nachlas, M. M., and Hannibal, M. J. (1961) *Surg. Gynecol. Obstet.* **112,** 534–542.

Nachlas, M. M., Tsou, K.-C., De Souza, E., Cheng, C.-S., and Seligman, A. M. (1957) *J. Histochem. and Cytochem.* **5,** 420–436.

Nachlas, M. M., Walker, D. G., and Seligman, A. M. (1958a) *J. Biophys. Biochem. Cytol.* **4,** 29–38.

Nachlas, M. M., Walker, D. G., and Seligman, A. M. (1958b) *J. Biophys. Biochem. Cytol.* **4,** 467–474.

Nachlas, M. M., Karmarkar, S. S., and Seligman, A. M. (1960) *Proc. Soc. Exptl. Biol. Med.* **104,** 407–409.

Neilands, J. B., and Stumpf, P. K. (1958) "Outlines of Enzyme Chemistry," 2nd ed. Wiley, New York.

Niemi, M., and Merenmies, E. (1961) *J. Neurochem.* **6,** 200–205.

Nineham, A. W. (1955) *Chem. Revs.* **55,** 355–483.

Novikoff, A. B. (1959) *In* "Analytical Cytology" (R. C. Mellors, ed.). McGraw-Hill, New York.

Novikoff, A. B. (1960a) *Acta Unio Intern. contra Cancrum* **16,** 966–972.

Novikoff, A. B. (1960b) *In* "Cell Physiology of Neoplasia" (T. G. Hsu, ed.), pp. 219–268. Univ. Texas Press, Austin, Texas.

Novikoff, A. B. (1960c) *In* "Developing Cell Systems and Their Control" (D. Rudnick, ed.)., pp. 167–203. Ronald Press, New York.

Novikoff, A. B., Shin, W.-Y., and Drucker, J. (1961) *J. Biophys. Biochem. Cytol.* **9,** 47–61.

Oda, T. (1961) *J. Histochem. and Cytochem.* **9,** 605–606.

Oda, T., and Okazaki, H., (1958) *Acta Med. Okayama* **12,** 193–204.

Oda, T., Matsuoka, K., Okazaki, H., and Kawasaki, M. (1959) *Acta Med. Okayama* **13,** 31–44.

Ogawa, K., and Hashimoto, K. (1961) *J. Histochem. and Cytochem.* **9,** 635.

Pádr, Z. (1959) "Tetrazolium Salts; Their Preparation and Use in Biology and Medicine." (Translated by Claudius Mayer.) Stat. zdrav. Naklad, Praha.

Palade, G. E. (1952) *Anat. Record* **114,** 427–451.

Palade, G. E. (1953) *J. Histochem. and Cytochem.* **1,** 188–211.

Palade, G. E. (1956) *In* "Enzymes: Units of Biological Structure and Function" (O. H. Gaebler, ed.), pp. 185–215. Academic Press, New York.

Pearse, A. G. E. (1958) *J. Clin. Pathol.* **11,** 520–534.

Pearse, A. G. E. (1960) "Histochemistry, Theoretical and Applied," 2nd ed. Little, Brown, Boston, Massachusetts.

Pearse, A. G. E., and Scarpelli, D. G. (1958) *Nature* **181,** 702–703.

Pearse, A. G. E., and Scarpelli, D. G. (1959) *Exptl. Cell. Research* **7,** 50–64.

Pearse, A. G. E., Scarpelli, D. G., and Hess, R. (1960) *J. Biophys. Biochem. Cytol.* **7,** 243–249.

Pearson, B. (1958) *J. Histochem. and Cytochem.* **6,** 112–121.

Potanos, J. N., Wolf, A., and Cowen, D. (1959) *J. Neuropathol. Exptl. Neurol.* **18,** 627–635.

Pullman, B., and Pullman, A. (1959a) *Proc. Natl. Acad. Sci.* (*U. S.*) **45,** 136–144.

Pullman, B., and Pullman, A. (1959b) *Biochim. et Biophys. Acta* **35,** 535–537.

Racker, E. (1955) *Physiol. Revs.* **35,** 1–56.

Ried, W. (1952) *Angew. Chem.* **64,** 391–396.

Ried, W., and Gick, H. (1953) *Ann. Chem. Liebigs* **581,** 16–23.

Rosa, C. G., and Tsou, K. -C. (1961) *Nature* **192,** 990–991.

Rosa, C. G., and Velardo, J. T. (1954) *J. Histochem. and Cytochem.* **2,** 110–114.

Rubinstein, L. J., Klatzo, I., and Miquel, J. (1962) *J. Neuropathol. Exptl. Neurol* **21,** 116–136.

Rust, J. B. (1955) *Trans. N. Y. Acad. Sci.* **17,** 379–384.

Rutenburg, A. M., Gofstein, R., and Seligman, A. M. (1950) *Cancer Research* **10,** 113–121.

Rutenburg, A. M., Wolman, M., and Seligman, A. M. (1953) *J. Histochem. and Cytochem.* **1,** 66–81.

Scarpelli, D. G., and Frajola, W. G. (1961) *J. Histochem. and Cytochem.* **9,** 603–604.

Scarpelli, D. G., and Pearse, A. G. E. (1958a) *J. Histochem. and Cytochem.* **6,** 369–376.

Scarpelli, D. G., and Pearse, A. G. E. (1958b) *Anat. Record* **132,** 133–151.

Scarpelli, D. G., Hess, R., and Pearse, A. G. E. (1958) *J. Biophys. Biochem. Cytol.* **4,** 747–752.

Schneider, W. C. (1959) *Advances in Enzymol.* **21,** 1–72.

Sedar, A. W., and Rosa, C. G. (1961) *J. Ultrastruc. Research* **5,** 226–243.

Seiler, H., and Schmid, H. (1954) *Helv. Chim. Acta* **37,** 1–7.

Seligman, A. M., and Rutenburg, A. M. (1951) *Science* **113,** 317–320.

Shelton, E., and Rice, M. E. (1957) *J. Natl. Cancer Inst.* **18,** 117–125.

Shelton, E., and Schneider, W. C. (1952) *Anat. Record* **112,** 61–82.

Sjöstrand, F. S. (1953) *Nature* **171,** 30–31.

Sjöstrand, F. S., and Rhodin, J. (1953) *Exptl. Cell Research* **4,** 426–456.

Slater, E. C. (1958) *Advances in Enzymol.* **20,** 147–199.

Sternberg, W. H., Farber, E., and Dunlap, C. E. (1956) *J. Histochem. and Cytochem.* **4,** 266–283.

Straus, F. H., Cheronis, N. D., and Straus, E. (1948) *Science* **108,** 113–115.

Sumner, J. B., and Somers, G. F. (1953) "Chemistry and Methods of Enzymes," 3rd ed. Academic Press, New York.

Tsou, K.-C., Cheng, C.-S., Nachlas, M. M., and Seligman, A. M. (1956) *J. Am. Chem. Soc.* **78,** 6139–6144.

von Hayek, H. (1950) *Naturwissenschaften* **37,** 262–263.

von Pechmann, H., and Runge, P. (1894) *Ber.* **27,** 2920–2930.

Wachstein, M. (1949) *Proc. Soc. Exptl. Biol. Med.* **72,** 175–178.

Wagner, B. M., Coriell, L. L., McAllister, R. M., and Shapiro, S. H. (1959) *Lab. Invest.* **8,** 939–947.

Wattenberg, L. W. (1958) *J. Histochem. and Cytochem.* **6,** 225–232.

Wattenberg, L. W. (1959) *Am. J. Pathol.* **35,** 113–137.

Wattenberg, L. W., and Gronvall, J. A. (1960) *Proc. Soc. Exptl. Biol. Med.* **104,** 394–396.

WATTENBERG, L. W., AND LEONG, J. L. (1960) *J. Histochem. and Cytochem.* **8,** 296–303.
WEDEKIND, E. (1898) *Ann.* **300,** 239–258.
WHITE, A., HANDLER, P., SMITH, E. L., AND STETTEN, D. (1959) "Principles of Biochemistry." McGraw-Hill, New York.
WHITEHEAD, R. G., AND WEIDMANN, S. M. (1959) *Biochem. J.* **72,** 667–672.
WIZINGER, R. (1954) *Z. Naturforsch.* **9b,** 729–740.

CHAPTER 13

# APPLICATION OF ENZYME HISTOCHEMISTRY TO ELECTRON MICROSCOPY

The successful demonstration of enzyme systems at the submicroscopic level depends, in part, upon proper fixation of the tissue to be studied. Although tissues fixed in cold osmium tetroxide have been shown to retain certain enzyme activities (Sheldon *et al.*, 1955; Brandes *et al.*, 1956; Essner *et al.*, 1958; Mölbert *et al.*, 1960a), the degree of inactivation as well as other alterations of enzyme systems which may occur with this fixative are not known. Pease (1960, p. 51) states that thoughtful electron microscopists have often worried about their virtual dependence upon osmium tetroxide fixation, and have wondered what might be missed by this method or what specific artefacts might be introduced. He further indicates that one might wish to avoid chemical fixatives altogether and preserve tissue by purely physical means such as freezing and drying (Chapter 1, Section IV, E). In view of the relatively early stage of development of the field of enzyme histochemistry as studied by the electron microscope, it would appear that this new approach will not be limited to the application of only one type of fixative.

Barrnett and Palade (1958) have reviewed application of histochemical techniques, including enzyme procedures, to electron microscopy, and note that electron microscopy, of necessity, requires the most stringent conditions for the adequate preservation of the fine structure of cells. At present, there are relatively few fixatives which give satisfactory results. Among the primary ones are osmium tetroxide, potassium dichromate, potassium permanganate, and formalin. Although several of these fixatives may be employed at low temperatures, there is little question that considerable inactivation of enzyme systems results.

One of the most recent and significant advances with reference to the fixation of enzyme systems to be studied at the submicroscopic level has been made by the Barrnett group (Chapter 1, p. 22) who have employed a new series of aldehyde fixatives, some of which facilitate preservation of oxidative enzymes.

The application of physical fixation procedures such as the imaginative freeze-dry techniques of Gersh *et al.* (Chapter 1, Section IV, E) and the freeze substitution methods of Fernández-Morán (Chapter 1, pp. 29 and 56) represent pioneering studies in the development of new methodologies for the preservation of both the enzymatic activity and the integrity of the submicroscopic structure of tissue. A significant factor in the development and execution of submicroscopic histochemistry involves the extreme thinness of the specimens employed, the most useful range being 200–600 Å. This requirement of necessity limits the morphological information, in effect, in a two-dimensional level. Furthermore, it is apparent that the quantity of precipitate or dye in such thin sections is extremely low. At the present time, in order to obtain such thin sections, tissues are usually embedded in a hard plastic such as methacrylate. This type of processing subjects the tissue specimens to a fairly strong solvent action (methacrylate monomer) which readily removes many azo-dyes. Fortunately, however, water-soluble plastic embedding procedures are being developed which may obviate this difficulty (Chapter 1, p. 22). With reference to this approach, Craig *et al.* (1962) have found that a commercially available water-miscible plastic, Epon 812, is a useful embedding medium in the preparation of ultrathin sections for electron microscopy.

The first application of enzyme histochemical methods to electron microscopy was reported by Sheldon *et al.* (1955) who employed the Gomori metal-salt technique for the demonstration of acid phosphatase in mouse intestinal epithelium. In order to avoid complete inactivation of the enzyme by osmium tetroxide, tissues were briefly fixed (8 minutes or less) in 1% buffered isotonic osmium tetroxide, washed in Tyrode's solution, and incubated in the Gomori acid phosphatase substrate at 37° for periods up to 16 hours. The tissues were rinsed in Tyrode's solution, washed for 2 minutes in dilute acetic acid, and then treated with ammonium sulfide. This treatment resulted in an end product of lead sulfide which could be visualized in electron micrographs as a dense homogeneous precipitate. The tissues were routinely dehydrated in graded solutions of ethanol and embedded in methacrylate. The electron micrographs of Sheldon *et al.* (1955) revealed sites of acid phosphatase activity as a dense homogeneous precipitate distributed in the intercylindrical spaces of the brush border projections. In sections through cylindrical projections tangential or parallel to the cell surface, the precipitate was arranged in a honeycomb pattern with the cylindrical projections appearing non-dense. On the other hand, the interspaces exhibited a dense appearance. In the juxtanuclear zone the precipitate appeared as grape-like clusters of round dense bodies, each body being about 500 Å in diameter. Neither the membranes nor the large vacuoles of the Golgi complex appear to be associated with the reaction. The authors noted that fluoride-inhibited controls showed a normal structure without a lead precipitate. They also indicate that their observations do not clarify the precise relation of the reaction sites to the cell surface membranes of the brush border zone; nor is it apparent exactly which of the many components of the juxtanuclear zone are represented by the reactive sites.

This type of approach was also applied in the study of alkaline phosphatase by Brandes *et al.* (1956). Tissues were similarly fixed for 8 minutes or less in 1% buffered isotonic osmium tetroxide, washed in Tyrode's solution, and incubated in the Gomori alkaline phosphatase substrate solution for 30–120 minutes at 37°. The precipitate of calcium phosphate was "stained" by immersion of the block in 0.5% silver nitrate for 10 minutes, after which the tissue was washed several times in distilled water. The specimens were dehydrated in graded solutions of ethanol and subsequently embedded in methacrylate. Electron micrographs of thin sections so prepared exhibited a dense granular precipitate of silver phosphate distributed uniformly in the intercylindrical spaces of the projections of the brush border. In sections through the cylindrical projections parallel or tangential to the cell surface, the granular precipitate was arranged in a honeycomb, the cylindrical projections appearing nondense and the interspaces dense.

The significance of the two studies on alkaline and acid phosphatases, respectively, is twofold. First, it represented a pioneering effort in the field of submicroscopic enzyme histochemistry; and second, it clearly demonstrated that powerful fixatives such as osmium tetroxide, which would normally completely inactivate enzyme systems, could be used to advantage at low temperatures in the preservation of certain enzyme systems.

Mölbert *et al.* (1960a) have published excellent electron micrographs demonstrating the distribution of alkaline phosphatase in mouse kidney tubules. Very small tissue blocks were prefixed with 1% osmium tetroxide at 0° and incubated for 20 minutes at room temperature in the following mixture: 2.0 g. potassium sodium tartrate dissolved in 45.0 ml. of tris buffer pH 7.6 and mixed with 5.0 ml. of 0.04 *M* disodium phenylphosphate and 1.0 ml. each of 2% $MnCl_2$ and $MgCl_2$. Lead nitrate (2.5 ml., 0.1 *M*) was added drop by drop with shaking. The pH of the final mixture was about 7.6–7.7. This substrate solution is stable for approximately 30 minutes. After brief fixation, the tissue blocks are washed several times with the incubation mixture. In order to avoid non-specific precipitation by residual components of the substrate solution, the fixative was changed several times during the post-fixation procedure.

Mölbert *et al.* (1960a) note that the enzyme activity is referable mainly to alkaline phosphatase, but are cautious to indicate that the participation of other phosphatases (including the specific ones) should not be excluded. With their technique, they show excellent preservation of cellular fine structures in which sites of precipitation are revealed in high contrast. A fine precipitate in seen in all outer cellular membranes of the kidney tubule, including basement membrane infoldings, the microvilli of the brush border, and the cell membranes of the endothelium of both peritubular and glomerular capillaries (Figs. 1 and 2). These authors indicate that the sharp localization of the precipitate of lead phosphate suggests a localization of the enzymes within or closely subadjacent to the cell membrane. This finding, of course, has been confirmed by routine light microscopy in the case of carefully fixed tissues processed according to some of the definitive alkaline phosphatase procedures (Chapter 5, p. 160).

Mölbert *et al.* (1960b) described the submicroscopic demonstration of glucose-1-phosphatase and AMPase in cardiac muscle, and histidine decarboxylase and acid phosphatase in the kidney. Particularly striking was their electron micrograph showing the distribution of glucose-1-phosphatase in rat cardiac muscle (Fig. 3).

Holt and Hicks (1961a, b) have carefully evaluated the effect of formalin fixation for electron microscopy with reference to the preservation of the acid phosphatase and esterase in rat liver. They reported that examination of methacrylate-embedded preparations demonstrates that calcium-containing fixatives (e.g. formol-calcium) result in poor preservation of fine structure whereas veronal-treated or phosphate-buffered formalin gives excellent results if the tonicity of the solutions is suitably adjusted with sucrose. Formol-phosphate, to which Versene has been added, caused deterioration of cellular morphology. Phospholipids are retained almost quantitatively in tissues fixed in formol-calcium, and in phosphate, collidine, or triethanolamine buffered formalin. They further note that about 50% of the activity of acid phosphatase and esterase is preserved after 24 hour fixation in the aforementioned solutions at 0–2°, and that the distribution of the enzymes and the phospholipids, as judged by cytochemical staining results, is not altered by any of these formalin solutions. They conclude that consideration of the morphological and biochemical integrity of the fixed tissues suggests that the 4% formaldehyde, buffered at pH 7.2 with 0.067 $M$ phosphate, and containing 7.5% sucrose, is the most suitable of the formalin-containing fixatives for electron microscopical studies. With reference to the Gomori *lead* acid phosphatase tech-

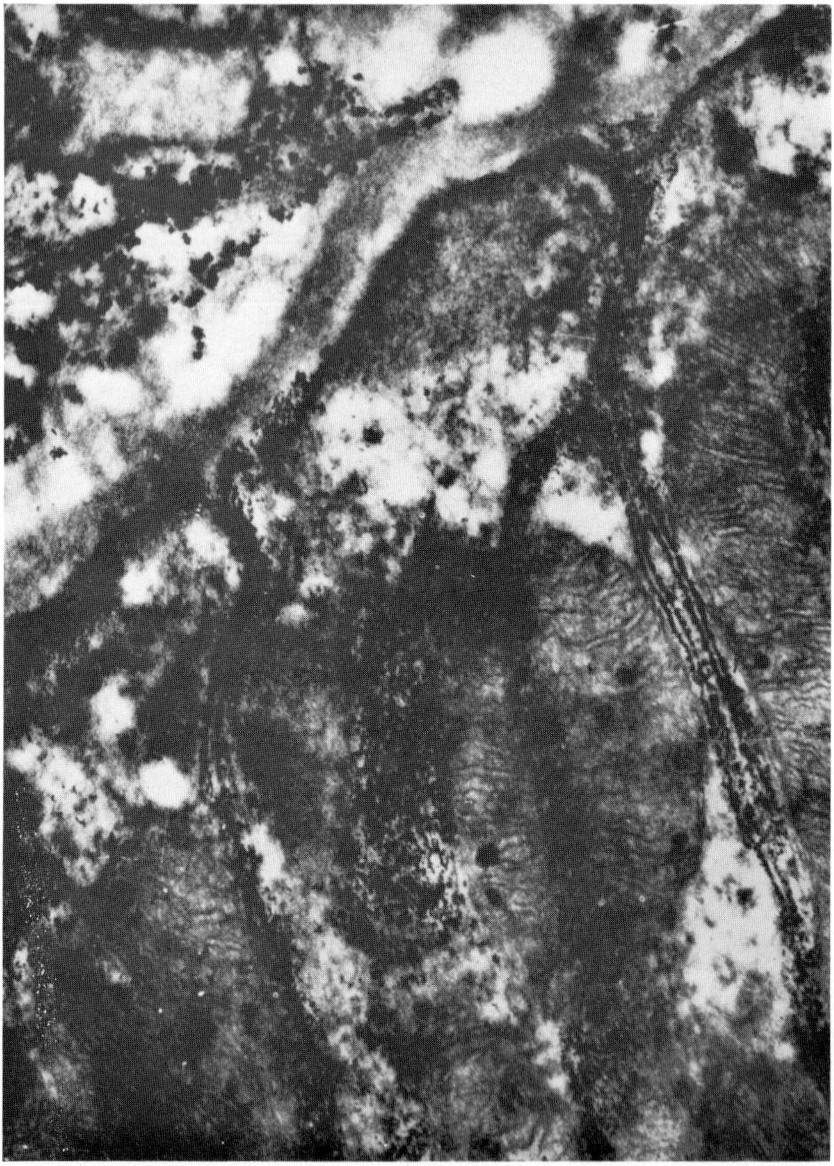

FIG. 1. Electron micrograph showing alkaline phosphate activity of basal zone of epithelial cell of mouse kidney tubule. Phenylphosphate-lead nitrate method. The infoldings of the cell membrane are revealed by a fine crystalline precipitate of lead phosphate. Mitochondria are non-reactive (× 76,000) (courtesy Dr. E. Mölbert).

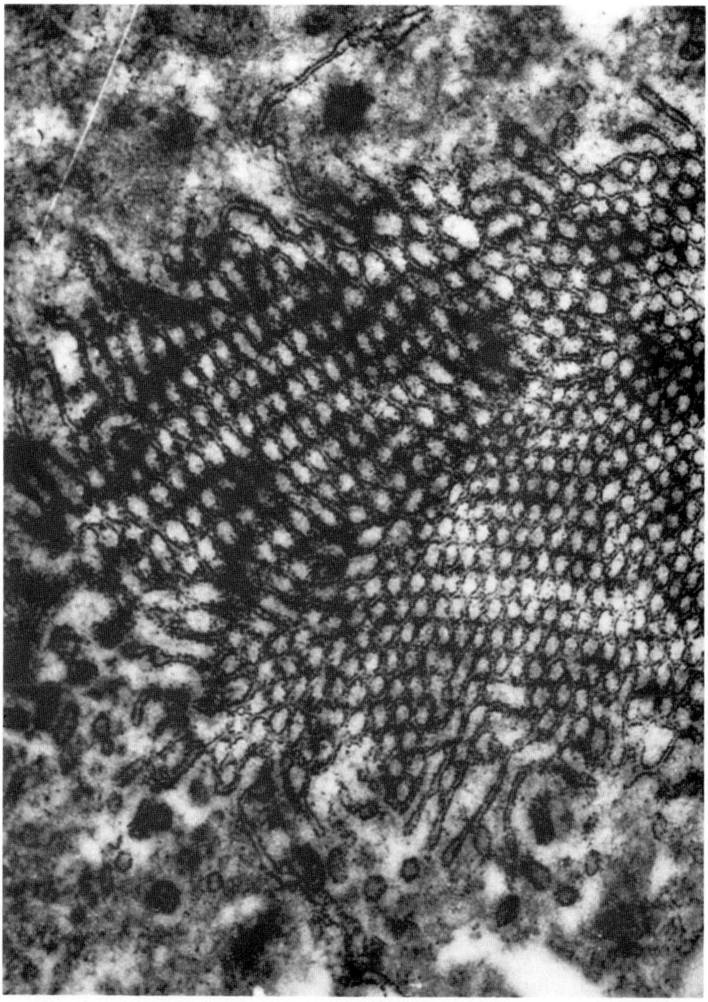

FIG. 2. Electron micrograph of transversely cut brush border of epithelial cell of mouse kidney tubule. Phenylphosphate-lead nitrate method. Phosphatase activity is delineated to the cell membranes of the microvilli (× 56,000) (courtesy Dr. E. Mölbert).

nique, Holt and Hicks (1961b) found that examination by light microscopy of the staining pattern and lead penetration in fixed blocks and slices of various dimensions showed nuclear staining and other artefacts to be present, the latter being produced by the different rates of penetration of the various components of the staining medium into the tissue. A uniform pericanalicular staining pattern could, however, be obtained with sections

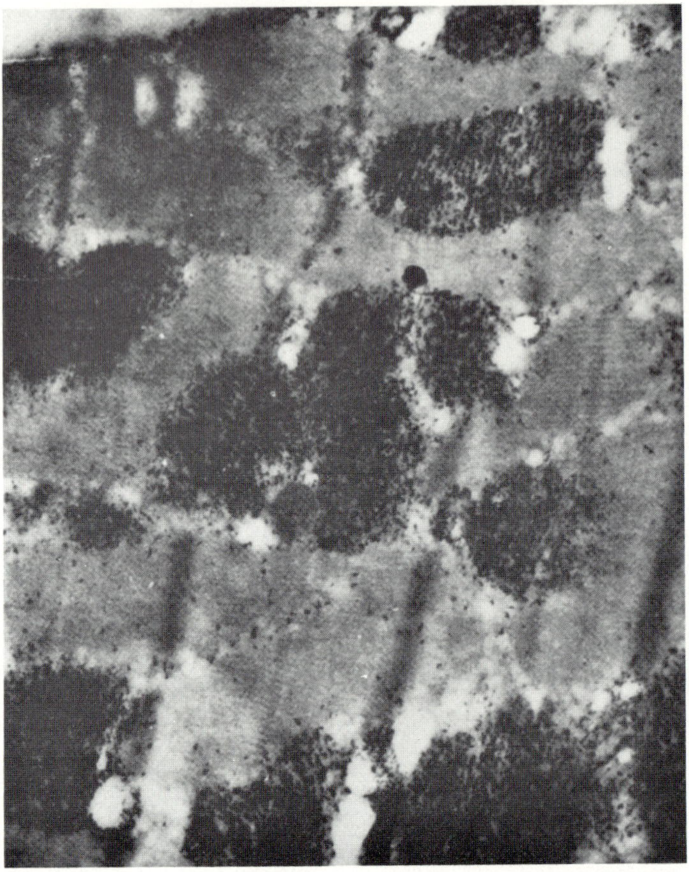

FIG. 3. Electron micrograph showing glucose-1-phosphatase activity of rat cardiac muscle. Note intramitochondrial localization of enzymatic activity (× 43,500) (courtesy Dr. E. Mölbert).

not more than 50 μ thick, into which the substrate solution could rapidly penetrate. The staining pattern produced in the 50 μ slices was the same at pH 5.0 and 6.2, and was not altered by embedding in methacrylate. According to these authors, electron microscopy revealed the fine structures of fixed 50 μ frozen slices to be well preserved, but the sections deteriorated badly when they were incubated in the routine lead-containing substrate medium at pH 5.0 before post-fixing in osmium tetroxide. Following incubation in the substrate medium at pH 6.2, the detailed morphology was substantially preserved. In both cases, the reaction product, lead phosphate, was found in the pericanalicular regions of the cell, but only in the vacuolated dense bodies and never in the microbodies. It was found that not every vacuolated dense body contained lead, and frequently, stained and unstained bodies were sometimes seen adjacent to each other. This heterogeneous distribution of the reaction product within a morphologically homogeneous group of particles is consistent with de Duve's suggestion that there is a heterogeneous distribution of enzymes within the *lysosome* population (de Duve, 1959, p. 128). Holt and Hicks (1961b) thus conclude that the vacuolated dense bodies as visualized by electron microscopy are the morphological counterparts of the *lysosomes* defined biochemically by de Duve.

Novikoff *et al.* have reported on the submicroscopic demonstration of several enzyme systems by application of metal-salt procedures (Novikoff, 1959, p. 69; 1960a, b, c; Essner *et al.*, 1958; Spater *et al.*, 1958).

Essner *et al.* (1958) reported on ATPase and 5-nucleotidase activities of the plasma membranes of liver cells (hepatic) as revealed by electron microscopy. Thin slices of freshly cut rat liver were fixed in cold 1% osmium tetroxide buffer at pH 7.2. The slices were cut into smaller blocks while in the fixative, and after a total fixation period of 7 minutes, they were washed several times in veronal-acetate buffer and incubated in a substrate medium containing ATP and lead ions at pH 7.2. In an analogous procedure, the medium contained 5-adenylic acid instead of ATP. Because the precipitate of lead phosphate is opaque to the electron beam, conversion to lead sulfide was usually omitted. Following incubation, the tissue samples were washed in cold veronal-acetate buffer and stored overnight in cold 1.6% formaldehyde-0.28 *M* veronal-0.28 *M* acetate buffer at pH 7.2. After rapid dehydration in ethanol the specimens were embedded in a 1:5 mixture of methyl-butyl methacrylate and sectioned according to standard techniques.

Essner *et al.* (1958) reported that attempts to incubate unfixed tissue blocks were unsuccessful because of damage to fine structure. The results of their study, which employed formol-calcium as a fixative, indicate that both ATPase and 5-nucleotidase activities are localized in the plasma membrane where it folds to form the microvilli of the bile canaliculus, and that 5-nucleotidase activity is also present in the microvilli at the sinusoidal aspects of the cell.

Spater *et al.* (1958) have studied the cytological localization of ATPase in cell membranes of kidney tubule cells and indicate that the enzymatic activity of the infolded plasma membranes, as seen with ATP as a substrate, is distinctive. However, it should be pointed out that in carefully processed tissues the staining of a variety of basement membranes may be readily seen with a number of azo-dye and related procedures (Chapter 5, p. 162). Whether or not this is due to the presence of ATPase or other specific or non-specific enzymes has not been ascertained.

J. C. H. de Man *et al.* (1960) studied the relationship between acid phosphatase and electron-dense bodies in mouse liver. The localization of acid phosphatase in the parenchymal cells between the nucleus and bile capillaries and in the Kupffer cells agreed well with sites of electron-dense material. A higher acid phosphatase activity was correlated with a larger amount of electron-dense material. Following intraperitoneal injection of dextran, many electron-dense bodies were found between the dextran vacuoles present in parenchymal cells. In the Kupffer cells this association was even more evident, thus indicating that liver acid phosphatase is attached to electron-dense material.

Essner and Novikoff (1961) similarly employed the Gomori acid phosphatase method in a study of livers of rats infused with unconjugated bilirubin, and also in human liver biopsies. Their observations indicate that the pericanalicular bodies of parenchymatous cells, the large spherical bodies of Kupffer cells, the microbodies appearing after bilirubin infusion, and lipofuscin granules all belong to the lysosome group of cellular organelles.

Bullivant (1960) has applied the Gomori acid phosphatase technique in the study of freeze-substituted mouse pancreas. The specimens were rapidly frozen in helium II, substituted in methanol at −75°, and embedded in methacrylate by ultraviolet polymerization in the cold. Sections were floated on a glycerophosphate-lead nitrate solution for periods of several minutes to many hours. It was suggested that the lead deposits which formed at the periphery of zymogen granules might not represent the true site of enzymatic activity.

In 1959, Barrnett and Palade described an enzyme in muscle, presumably a cholinesterase, by means of the thiolacetic acid method (Chapter 6, p. 308). Using the aforementioned substrate in conjunction with lead nitrate, the final product of the histochemical reaction was discretely deposited in the M band of the myofibrils of skeletal and cardiac muscle. In addition, deposits of the reaction product were also found in mitochondria and other cytoplasmic organelles. However, these sites appeared to be enzymatically different from those observed in the M band. The experimental tissues consisted of diaphragmatic muscle and the ventricular portion of rat cardiac muscle. The diaphragm was used either as small blocks, or as intact leaves which, after incubation, were cut into small fragments. Heart muscle was used either as cryostat sections, small blocks, or centrifuged pellets of briefly blendorized muscle (40 seconds in 0.44 *M* sucrose). Within the reactive areas of the specimen, fine, roughly circular, or stellate deposits of highly dense material, approximately 140–200 Å in diameter, were found located on all or nearly all the thick elementary fragments of the myofibrils at the level of the M band (Fig. 4). In specimens which

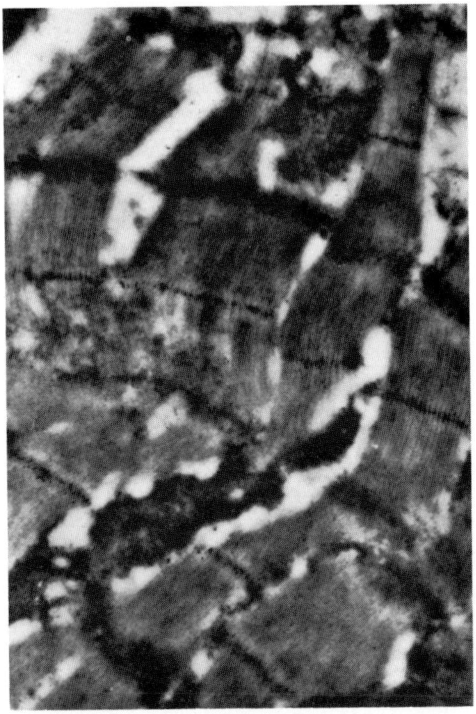

FIG. 4. Electron micrograph showing M band enzyme of rat myocardium as stippled granular pattern. Thiolacetic acid-lead nitrate method (× 50,000) (courtesy Dr. R. J. Barrnett).

were incubated for a short time and, in addition, contained only partly contracted myofibrils, the relationship of the dense deposits to the elemental filaments and cross-striations of the sarcomeres could be better visualized. The deposits were almost entirely confined to the swelling of the thick elemental filaments that occur at the M band level, and the rest of each thick filament through the H and A bands of the sarcomere was essentially free of reaction product, as were the filaments in the I band. Omitting the substrate prevented any reaction as did brief fixation of the muscle in formalin or osmium tetroxide. Furthermore, addition of well known enzymatic inhibitors such as eserin, TEPP, and DFP partly or completely inhibited the histochemical reaction.

Zacks and Blumberg (1961) have also employed the thiolacetic acid method in the study of the histochemical localization of acetylcholinesterase in the fine structure of neuromuscular junctions of mouse and human intercostal muscle. The tissues were initially fixed in formol-sucrose solution followed by osmium tetroxide after the histochemical reaction product had formed. Lead sulfide deposits were found in axonal and sarcoplasmic mitochondria and most prominently in the primary and secondary synaptic clefts of the subneural apparatus.

Davis and Ornstein (1959), in an abstract, indicated that an azo-dye method employing diazotized pararosaniline in conjunction with α-naphthyl acetate or α-naphthyl phosphate was capable of producing localizations of enzymatic activity down to the level of about 100 Å without evidence of diffusion artefacts. Lehrer and Ornstein (1959) reported on the application of the aforementioned technique using α-naphthyl acetate as a substrate in the localization of cholinesterase at the electron microscope level. The tissues were fixed for 1–2 hours in 10% formalin buffered with 1/15 *M* phosphate pH 7.4 at 0–4°. The specimens were then transferred directly to the incubating medium containing *hexazonium pararosaniline* and α-naphthyl acetate, and incubated at or near 0° for a period of 4–6 hours. Following incubation the tissues were washed in cold buffered 10% formalin and embedded in methacrylate. At the electron microscopic level a homogeneous deposit of azo-dye was seen in the synaptic or junctional folds, the synaptic space, and in the space between the telogelia and the axon terminal.

Burstone (1958) has proposed the application in electron microscopy of dyes derived from naphthol AS and analogues. Specific naphthol AS dyes are of large molecular size and are capable of forming metal complexes [Chapter 4, p. 124 (II)]. Among the metal complexes which may prove of value are those containing lead, tungsten, molybdenum, and uranium. The application of salts of these metals as electron microscope stains has been considered by Watson (1958a, b).

Barrnett (1959a), in an abstract, also reported the application of the azo-dye techniques for the demonstration of hydrolases at the submicroscopic level. It was stated that the method employed esters of α-naphthol with certain diazonium salts, the nature of which was not specified.

In a subsequent paper it was stated that diazotates of 1,5-diaminoanthraquinone, Bismarck brown and basic fuchsin could be employed with α-naphthyl phosphate (Barrnett, 1959b).

With reference to the utilization of simultaneous coupling azo-dye procedures for electron microscopy, it should be pointed out that in such techniques the finite coupling time which occurs as the naphthol is released from its corresponding ester may result in diffusion artefacts. In fact, such artefacts may be readily seen at the routine optical level when simpler naphtholic substrates are used (Chapter 4, p. 149). The fact that diffusion artefacts may be less readily evaluated at the submicroscopic level may be the result of several factors. First, the diffused dye, which may be less tightly bound to tissue protein is readily removed by the solvents employed in the methacrylate embedding and thus the section may contain only sharply defined foci of dye deposits. Second, the nature and type of diffusion artefacts at the submicroscopic level have not been definitively evaluated at the present time.

From a theoretical standpoint it would appear that the most desirable type of azo-dye or naphtholic substrate procedure would be a post-coupling or non-coupling method (Chapter 5, p. 196). Thus, a suitable tissue specimen could simply be incubated in the presence of a naphtholic ester, which upon enzymatic hydrolysis would leave an insoluble precipitate *in situ*. Substrates of this type are illustrated in Chapter 5 (Charts 3 and 4). If the released naphthol were sufficiently dense to the electron beam, the specimen could be studied without further treatment. However, such a naphthol may be readily coupled with a variety of diazonium salts, including slow couplers of large molecular size, diazo components containing metals, and those containing coordinating groups for metal chelation. Diazonium salts of this type have been described in a recent paper by Burstone and Weisburger (1961a) (Chaper 3, p. 95).

Included in this general type of approach is the application of a substrate containing the 1-hydroxyanthraquinone residue which is capable of forming intermolecular chelates with a number of metals (Chart 1). In this instance, the metal complexing procedure would follow the incubation of the tissue specimen. Another potential approach involves the formation of "lakes" of 1-nitroso derivatives of 3-hydroxy-2-naphthoic acid arylamides. According to Venkataraman, (1952, p. 404) naphthol AS and its analogues, when treated in ethanolic solution with different molecular proportions of

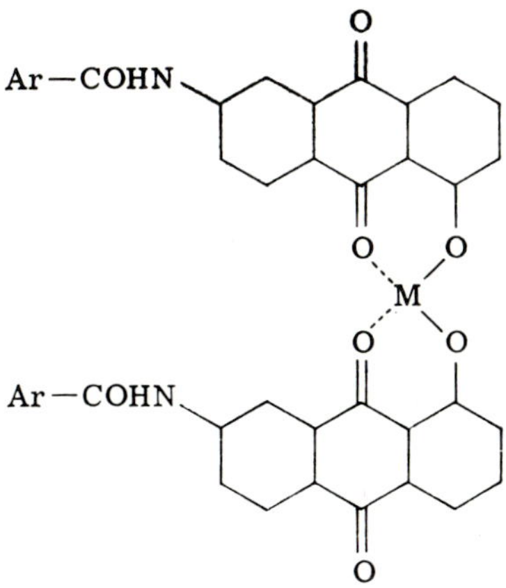


CHART 1. Intermolecular chelation of aromatic compound containing alizarin grouping (M = metal). Ar represents the aryl gouping containing the ester which is hydrolyzed by enzymatic activity.

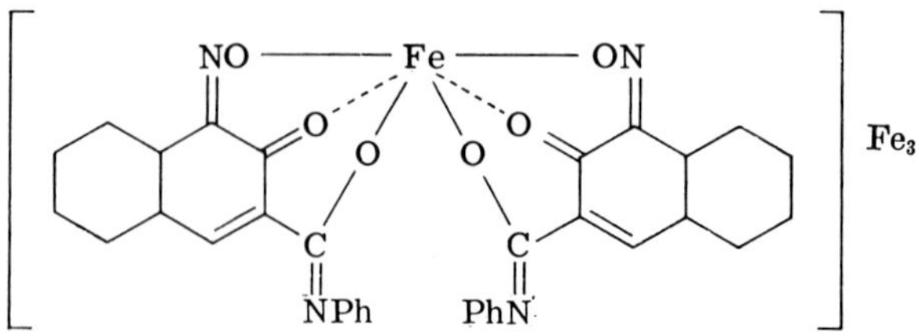


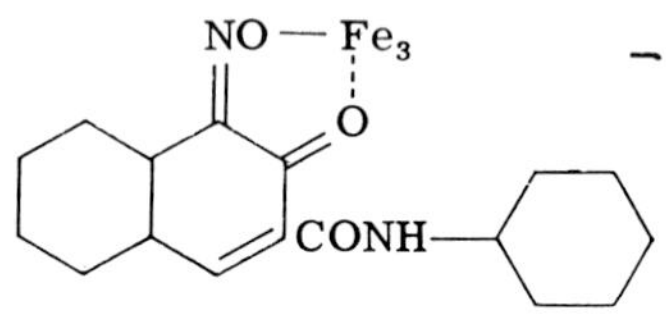


CHART 2. Types of iron complexes formed from naphthol AS compound following enzymatic hydrolysis of esterified hydroxyl groupings.

ferric salt in the presence of nitrous acid, yield two types of iron complexes which are represented in Chart 2. Iron or other metallic complexes of naphthol AS derivatives deposited *in situ* following enzymatic hydrolysis should offer a novel approach to the demonstration of a variety of hydrolytic enzymes at the submicroscopic level. Several new aminopeptidase substrates which contain coordinating groups for metal chelation have been reported by Burstone and Weisburger (1961b) (Chapter 10, p. 400), but as yet these compounds have not been applied to electron microscopy.

Barrnett and Palade (1957) reported on the submicroscopic demonstration of dehydrogenase activity employing potassium tellurite as the hydrogen or electron acceptor. The reagent was used intravitally (intravenously, intraperitoneally, or intraluminally in hollow organs) or supravitally on small blocks of tissue, for the demonstration of endogeneous dehydrogenase activity. Blocks of tissue which had been frozen and thawed, or which had been washed in 0.44 *M* sucrose to prevent endogenous activity, were used to demonstrate the activity of the succinic dehydrogenase system. In the latter case, the incubating medium contained tellurite, succinate, phosphate buffer, sucrose, and activators. The incubation was performed either aerobically (with or without the addition of potassium cyanide) or anaerobically. Reduced tellurite, which is the end product of the reaction, was visualized in thin sections of osmium tetroxide-fixed, methacrylate-embedded tissues as crystals or fine particulate particles of high density. These precipitates were localized on or in close relationship to mitochondrial membranes of rat cardiac muscle.

Sedar and Rosa (1958), in an abstract, reported on the demonstration of succinic dehydrogenase at the electron microscope level by the use of the newer tetrazolium compound, nitro-BT. Small pieces of myocardium were washed in 0.44 *M* sucrose and then incubated at 37° in a solution containing phosphate buffer, sodium succinate, sucrose, and nitro-blue tetrazolium. The tissues were fixed in cold buffered osmium tetroxide and finally embedded in methacrylate. Sedar and Rosa indicate that nitro-BT shows promising possibilities for use as a reagent in the demonstration of enzymatic activity in thin sections of methacrylate-embedded tissues to be studied by means of electron microscopy. Their electron micrographs revealed localization of the enzyme in the form of dense particles limited for the most part to mitochondrial structures.

Further studies by Barrnett (1960) on the demonstration of sites of succinic dehydrogenase activity at the submicroscopic level employed both the tellurite method and the nitro-BT procedure. With reference to the latter, the method was essentially that reported by Sedar and Rosa (1958). Barrnett noted that the formazan derived from nitro-BT was deposited nearly always in mitochondria, but the occurrence of formazan in other sites such as myofibrils and nuclei was more commonly observed than with the tellurite procedure. The precipitates were roughly circular or oval in shape and were sometimes much larger than those obtained in the tellurite experiment (several hundred to over one thousand Å in diameter). These deposits were sometimes so dense as to obscure the underlying mitochondrial fine structure, but at times they were smaller and less opaque and were seen to overlay the cristae. Another tetrazole, iodonitro-BT, was also utilized but deposits were not considered to be more electron-opaque than those of the formazan of nitro-BT despite the presence of the iodine atom. Further evaluation of the submicroscopic demonstration of the dehydrogenases by both the tellurite and tetrazole methods is considered in a more recent report by Barrnett (1959b). More recently, Sedar and Rosa (1961) have carefully evaluated the application of nitro-BT as a reagent in electron microscopy. For the most part, enzymatic activity of the succinic dehydrogenase system was located within the cristae mitochondriales. The formazan particles were approximately 300 Å in diameter, thus imposing a serious limitation upon the accuracy of localization. Sedar and Rosa (1961) also note the selective non-specific affinity of the formazan for lipids (e.g. mitochondrial phospholipids) as an unsatisfactory feature of the nitro-BT procedure.

Nelson (1958) studied the submicroscopic localization of ATPase in frozen-dried preparations of rat epididymal sperm. The enzyme was found to be confined to the nine longitudinal fibers of the outer axial bundle of the flagellum. For fixation, thin films of epididymal fluid on pieces of metal foil were frozen by immersion in liquid propane at −175° and dried *in vacuo* at ca. −35°. The samples were then vacuum-infiltrated with a pH

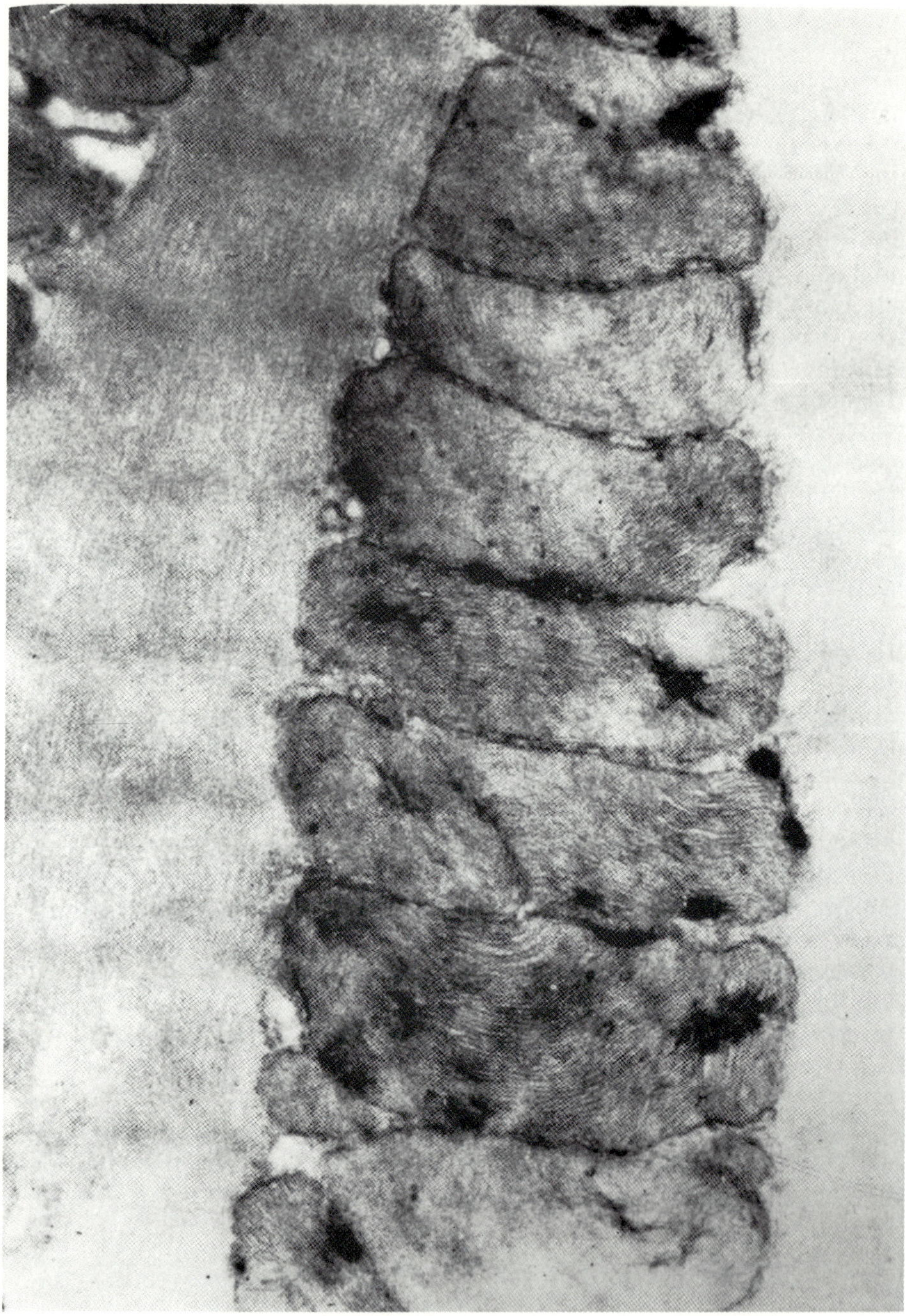

FIG. 5. Electron micrograph of cardiac muscle fixed in hydroxyadipaldehyde solution and incubated in cytochrome oxidase substrate solution (*p*-aminodiphenylamine and 3-amino-9-ethylcarbazole). Specimen post-fixed in osmium tetroxide and embedded in Epon. Dye deposits are limited to mitochondria and, in favorable instances, related to underlying cristae (× 45,000) (courtesy Dr. R. J. Barrnett).

9.0 veronal buffer. Following exposure of the specimens to air, they were incubated in the standard ATP substrate medium, dehydrated in ethanol, embedded in celloidin and then methacrylate.

In a similar fashion, Nelson (1959) studied the succinic dehydrogenase of rat spermatozoa. Three types of dehydrogenase reagents were employed: polymeric nitro-BT tetrazolium, *p*-nitrotetrazolium, and potassium tellurite.

Yaeger (1961) described methods for the localization of succinic dehydrogenase activity in frozen-dried diaphragm muscle with nitro-BT after post-fixation of the dried specimens with anhydrous vapors of iodine, acetaldehyde, and formaldehyde. After appropriate intervals, these vapors were removed from the system and the staining solution was added to the tissue chamber without breaking the vacuum. Electron micrographs of the blue formazan granules, presumably localized in mitochondria, revealed the presence of irregularly distributed groups of microgranules 10–15 m$\mu$ in diameter. This punctate distribution of succinic dehydrogenase activity within mitochondria is not consistent with the view that confines the activity to the surfaces of mitochondrial cristae.

Mitsui (1960) has studied the localization by electron microscopy of the products of the peroxidase reaction in neutrophilic leukocytes employing the standard benzidine method (Chapter 11, p. 454). Following fixation of small pieces of liver tissue for 1 hour in buffered osmium tetroxide, the specimens were treated with a 0.3% benzidine solution and 50% ethanol and then dehydrated and embedded in either methacrylate or epoxy resin. In the electron micrographs, the reaction products of the peroxidase technique were evidenced in the form of dense materials localized in specific granules in the cytoplasm of the neutrophil leukocytes. Neither mitochondria nor other particles exhibited an increased density.

Recently Barrnett (1962) has employed the *p*-aminodiphenylamine cytochrome oxidase technique of Burstone (Chapter 11, p. 441) for the demonstration of this enzyme at the electron microscope level in cardiac muscle. Intramitochondrial localizations of high contrast were clearly delineated (Fig. 5). This oxidase technique is of interest because the oxidation dyes produced may be complexed with a number of heavy metals.

## References


Barrnett, R. J. (1959a) *J. Histochem and Cytochem.* **7,** 298.

Barrnett, R. J. (1959b) *Exptl. Cell Research Suppl.* **7,** 65–89.

Barrnett, R. J. (1960) *In* "Fourth International Conference on Electron Microscopy" (W. Bargmann, G. Möllenstedt, H. Niehrs, D. Peters, E. Ruska, and C. Wolpers, eds.), Vol. II, pp. 91–100. Springer, Berlin.

Barrnett, R. J. (1962) *J. Histochem. and Cytochem.* (in press).

Barrnett, R. J., and Palade, G. E. (1957) *J. Biophys. Biochem. Cytol.* **3,** 577–588.

Barrnett, R. J., and Palade, G. E. (1958) *J. Histochem. and Cytochem.* **6,** 1–12.
Barrnett, R. J., and Palade, G. E. (1959) *J. Biophys. Biochem. Cytol.* **6,** 163–170.
Brandes, D., Zetterqvist, H., and Sheldon, H. (1956) *Nature* **177,** 382–383.
Bullivant, S. (1960) *J. Biophys. Biochem. Cytol.* **8,** 639–647.
Burstone, M. S. (1958) *J. Histochem. and Cytochem.* **6,** 322–339.
Burstone, M. S., and Weisburger, E. K. (1961a) *J. Histochem. and Cytochem.* **9,** 301–303.
Burstone, M. S., and Weisburger, E. K. (1961b) *J. Histochem. and Cytochem.* **9,** 349–355.
Craig, E. L., Frajola, W. G., and Greider, M. H. (1962) *J. Cell Biol.* **12,** 190–194.
Davis, B. J., and Ornstein, L. (1959) *J. Histochem. and Cytochem.* **7,** 297–298.
de Duve, C. (1959) *In* "Subcellular Particles" (T. Hayashi, ed.)., pp. 128–159. Ronald Press, New York.
de Man, J. C. H., Daems, W. Th., Willighagen, R. G. J., and van Rijssel, Th. G. (1960) *J. Ultrastruct. Research* **4,** 43–57.
Essner, E., and Novikoff, A. B. (1961) *J. Biophys. Biochem. Cytol.* **9,** 773–784.
Essner, E., Novikoff, A. B., and Masek, B. (1958) *J. Biophys. Biochem. Cytol.* **4,** 711–716.
Holt, S. J., and Hicks, R. M. (1961a) *J. Biophys. Biochem. Cytol.* **11,** 31–45.
Holt, S. J., and Hicks, R. M. (1961b) *J. Biophys. Biochem. Cytol.* **11,** 47–66.
Lehrer, G. M., and Ornstein, L. (1959) *J. Biophys. Biochem. Cytol.* **6,** 399–406.
Mitsui, T. (1960) *J. Biophys. Biochem. Cytol.* **7,** 251–260.
Mölbert, E., Duspiva, F., and von Deimling, O. H. (1960a) *J. Biophys. Biochem. Cytol.* **7,** 387–390.
Mölbert, E., von Deimling, O., and Duspiva, F. (1960b) *Proc. European Regional Conf. Electron Microscopy* (F. S. Sjöstrand and J. Rhodin, eds.), Vol. 2, pp. 635–640. Academic Press, New York.
Nelson, L. (1958) *Biochim. et Biophys. Acta* **27,** 634–641.
Nelson, L. (1959) *Exptl. Cell Research* **16,** 403–410.
Novikoff, A. B. (1959) *In* "Analytical Cytology" (R. C. Mellors, ed.), pp. 69–186. 2nd ed. McGraw-Hill, New York.
Novikoff, A. B. (1960a) *Acta Unio Intern. contra Cancrum* **16,** 966–972.
Novikoff, A. B. (1960b) *In* "Cell Physiology of Neoplasia" (T. G. Hsu, ed.), pp. 219–268. Univ. Texas Press, Austin, Texas.
Novikoff, A. B. (1960c) *In* "Developing Cell Systems and Their Control" (D. Rudnick, ed.). Ronald Press, New York. pp. 167–203.
Pease, D. C. (1960) "Histological Techniques for Electron Microscopy." Academic Press, New York.
Sedar, A. W., and Rosa, C. G. (1958) *Anat. Record* **130,** 371.
Sedar, A. W., and Rosa, C. G. (1961) *J. Ultrastruct. Research* **5,** 226–243.
Sheldon, H., Zetterqvist, H., and Brandes, D. (1955) *Exptl. Cell Research* **9,** 592–596.
Spater, H. W., Novikoff, A. B., and Masek, B. (1958) *J. Biophys. Biochem. Cytol.* **4,** 765–770.
Venkataraman, K. (1952) "The Chemistry of Synthetic Dyes," Vol. 1. Academic Press, New York.
Watson, M. L. (1958a) *J. Biophys. Biochem. Cytol.* **4,** 475–478.
Watson, M. L. (1958b) *J. Biophys. Biochem. Cytol.* **4,** 727–729.
Yaeger, J. A. (1961) *Exptl. Cell Research* **22,** 493–502.
Zacks, S. I., and Blumberg, J. M. (1961) *J. Histochem. and Cytochem.* **9,** 317–324.

*CHAPTER 14*

# ELECTROPHORETIC PROCEDURES

## I. INTRODUCTION AND GENERAL CONSIDERATIONS

Although the so-called "gel" electrophoresis methods have but recently been employed in enzyme histochemistry, the development and biochemical application of electrophoretic techniques predates their histochemical use by more than a decade.

Electrophoresis may be defined as the phenomenon of migration of colloidal particles in an electrical field. The particles of many aqueous lyophilic sols such as proteins are known to migrate in an electrical field and the direction of movement may be influenced by the pH of the medium. The movement of colloidal particles in an electrical field has been studied microscopically as well as macroscopically. In the latter case, instead of observing several single particles, the motion of the boundary between the sol and the pure dispersion medium is observed under the influence of an applied current. As indicated in the review by Kunkel (1954, p. 141), the term "zone electrophoresis" is a name given by Tiselius to the separation of materials in an electrical field in the presence of some type of supporting medium in addition to the electrolyte solution used to carry the current. Zone electrophoresis should be distinguished from classical electrophoretic techniques utilizing free solutions where the separation of boundaries observed by optical methods is obtained despite only partial separation of the materials. Although electrophoresis was originally limited to the analysis of colloids and submicroscopic particles, in recent years it has been applied to amino acids as well as to more complex proteins including enzymes.

Electrophoretic separation of components of a solution into discrete zones is difficult to achieve due to convection currents. Because of this, the possibility of stabilizing the zones by the application of various solid supporting media has been suggested. Among the supporting media employed have been various gels, glass beads and powders, sands, resins, powdered cellulose, and filter paper. Kunkel (1954, p. 142) notes that much of the earlier work on transference of organic ions was carried out in gels, particularly gelatin, with observation of color boundaries. Included in this work were studies on the migration of diphtheria toxin and antitoxin. In 1939, Coolidge described an electrophoretic column packed with ground glass wool which was utilized in the separation of serum into fractions. This study also reported on a number of principles of column zone electrophoresis which are utilized at the present time. Another advance was the electrophoretic separation of amino acids by a silica gel supporting medium (Consden *et al.*, 1946). A modification of this procedure employed agar instead of silica gel (Gordon *et al.*, 1950).

An advance in the zone electrophoresis procedure occurred with the introduction of filter paper as the supporting medium. Review of the literature with reference to this approach prior to 1954 is indicated by Kunkel (1954, p. 144). The filter paper moistened with a buffer is draped over one or more glass rods with the ends of the paper dipping into electrode vessels (Durrum, 1950; Flynn and de Mayo, 1951). The entire structure is placed under a plastic cover to maintain a moist atmosphere. With this type of separation, the paper is almost in a vertical position, and the test material is often placed at the area adjacent to the glass rod. One difficulty with these techniques lies in the variation of buffer concentration in different portions of the paper due to evaporation and gravity effects. Another type of filter paper electrophoresis procedure employs the filter paper in a stretched, taut, horizontal position with the ends dipping into the electrode vessels (Grassmann *et al.*, 1951). In still another method, the filter paper is placed between glass plates under pressure with the protruding ends of the filter paper dipping into the electrode vessels (Kunkel and Tiselius, 1951). This modification is advantageous in that the evaporation problem is minimized. Slater and Kunkel (1953) indicate that one of the chief disadvantages of the filter paper electrophoresis is a difficulty in reproducibility of the pattern from one experiment to another. Inherent in this are variations in the filter paper itself, in the preparation of the paper for the separation, in the buffer concentration, in the electrical system, in the temperature, and in the final staining.

With reference to the techniques employing various supporting media, Kunkel and Slater (1952) used various materials including ground hard glass, washed sands, Hyflo Super-Cel, anionic and cationic resins, polystyrene, polyvinyl chloride particles (Geon), various types of ground filter paper, agar gels, and potato starch. They noted some success in the separation of serum with a number of these materials including the agar gels, polyvinyl chloride particles, powdered filter paper, and potato starch. According to Kunkel (1954, p. 155) the greatest success in the isolation of serum proteins was achieved with a potato starch supporting medium with which a homogeneous smooth-surfaced solid block could be readily formed containing approximately 50% electrolyte solution by volume (Appendix 14, Section 1). Kunkel (1954, p. 159) also notes that certain purified cellulose fibers may be used to advantage in the separation of peptides, particularly those of the basic type with a large number of aromatic amino acids. In connection with the aforementioned plastics and gels, of interest is the recent application of polyacrylamide gels in the electrophoretic separation of enzyme systems by Barka (1961a) (Appendix 14, Section 4). It should be noted that recent developments in the plastics industry have resulted in the commercial availability of a wide variety of water-soluble plastic materials which may form gels under specified conditions, thus suggesting a wide range of materials which may have application in electrophoretic procedures.

Smithies (1955a, b, 1959) evolved a technique which combined the advantages of the low adsorption characteristics of the starch-grain method with the convenience of protein detection by the staining characteristics of the filter paper method. Serious adsorption effects which may occur with the starch-grain method are minimized. The new procedure utilized a thin starch gel layer supported by a plastic tray, with filter paper bridges to convey the solution to the gel (Appendix 14, Section 2). Following electrophoretic separation, the proteins were detected by staining with Amido-Black 10B. Smithies (1955b) indicates that the greater resolving power of the starch gel method as compared with previous electrophoretic procedures is probably a consequence of the use of a supporting medium, the pore size of which approaches the molecular dimensions of some of the proteins involved. It is found that proteins with closely similar mobilities in free-solution electrophoresis and filter paper electrophoresis may have widely different mobilities in the starch gels, and that two proteins that migrate in a given order in free solution may migrate in the reverse order in the gels. This suggests that proteins of comparatively small size are slowed relatively much less in the gels than proteins of larger size which have dissymmetry; even though in free solution they have the same mobility. Thus, the starch gel electrophoresis method appears to combine resolution by "true" (free solution) mobilities, with resolution according to molecular dimensions, at least over some of the ranges of molecular size present in the serum proteins.

More recently, Poulik and Smithies (1958) have both compared and combined the starch gel and filter paper electrophoretic methods in what they termed a "two-dimensional electrophoresis." Their results suggest that a suitable combination of the two methods in a single procedure would lead to a more complete resolution of the serum-protein components than is possible by any other single method. They specifically described a two-dimensional electrophoretic system, first on filter paper and secondly at right angles in starch gel, by which more than 20 serum components could be demonstrated.

Goldberg (1958) has reported a modified starch gel method which employs a discontinuous buffer system of tris-EDTA-borate/barbital in order to achieve a greater resolving power of the system.

Although potato starch, in order to be used in the gel method, may be hydrolyzed according to several laboratory procedures, a suitable hydrolyzed starch is commercially available (Appendix 14, Section 3).

Several papers have described the application of a dilute agar solution as the anticonvection material for application in obtaining zone electrophoretograms, and these techniques have been reported to be simpler than those employing concentrated gels (Ressler and Jacobson, 1955; Ressler and Zak, 1956; Zak and Sun, 1958; Ressler *et al.*, 1958; Ressler and Moy, 1959; Zak and Jarkowski, 1958). The serum is applied to filter paper which is placed in a contact agar plate, following which the proteins are stained in the usual way. The aforementioned authors stress the simplicity of the technique and the feasibility of its application in clinical laboratories, although the starch gels appear to give higher resolving power. Ressler (1960) has also described a two-dimensional electrophoresis of protein antigens with an antibody containing buffer. This involved a first migration in a starch gel, followed by second migration in a stabilized buffer which contained a dissolved anti-serum. By conducting the second migration at right angles to the first, the overlap of components in complex solutions was avoided.

Scherr (1961) has described the application of oxoid cellulose acetate strips for electrophoresis, and indicates that the unusual sensitivity and clarity of separation provide a new tool where electrophoretic techniques are applicable.

Raymond and Weintraub (1959) have reported an acrylamide gel which is employed as a supporting medium for zone electrophoresis (Appendix 14, Section 4). The commercial product which they utilized, Cyanogum 41, is a mixture of two organic monomers; acrylamide and $N,N'$-methylene-bisacrylamide. Gels may be readily formed in the presence of various buffers Raymond and Weintraub (1959) specifically reported that a 3–5% Cyanogum solution in acid or alkaline buffers (0.3–0.01 $N$) produces a satisfactory gel for use in electrophoresis. The gel is flexible and elastic, stable, and completely insoluble in water. In concentrations of 5% and more it is strong enough to support its own weight but is somewhat brittle unless a plasticizer (e.g. 5% glycerol) is included in the formulation. The rate of migration of serum albumin in 0.1 $N$ tris buffer pH 9 gelled with 5% Cyanogum is approximately 1.2 cm./hour at a field strength of 10.0

volts/cm. A serum pattern can be developed in this medium in 8 hours, fixed in dilute acid, and stained according to standard procedures. A convenient way of preserving the electrophoretic patterns in the acrylamide gel involves the complete drying of the gel. During this process the gel shrinks uniformly producing a thin, flexible, transparent celluloid-like film which preserves the original pattern relationships. The film may be rehydrated to its original dimensions by soaking in water.

A significant advance in electrophoretic techniques has been made by Ornstein and associates (unpublished data) who devised the Disc Electrophoresis. This procedure takes advantage of the adjustability of the pore size of a synthetic gel and automatically produces starting zones of the order of 10 $\mu$ thickness from the initial volumes with thickness of the order of centimeters. Thus high resolution is achieved in relatively brief runs. With disc electrophoresis over 20 serum proteins may be routinely separated from a sample of whole human serum as small as 1 $\mu$l. in a 20-minute run. Details dealing with some of the laboratory aspects of disc electrophoresis may be obtained from the Canalco Company (see Commercial Sources of Reagents, p. 560).

Subsequent to the report of Smithies (1955b), application of his procedure has been utilized in the electrophoretic analysis of serum proteins in both animals and man in normal as well as pathological conditions (Beaton *et al.*, 1961; Engle *et al.*, 1961; Yoon, 1961).

With reference to the biochemical study of enzymes by means of the starch gel procedure, Keiding (1959) has found three different electrophoretic fractions of serum alkaline phosphatase. In the study of sera from normal adults and children, as well as patients with liver, bone, and neoplastic disease, one fraction was found to relate to the involvement of osseous tissue whereas the other two fractions were related to liver disease. Rosenberg (1959) has also reported on variations of alkaline phosphatase activity in the protein fractions of sera from patients with bone and liver diseases. Estborn and Swedin (1959) separated eleven fractions in human semen acid phosphatase by means of the starch gel method. Estborn (1961) has subsequently reported on the localization of prostatic serum acid phosphatase by the starch gel method in conjunction with the Amido-Black B stain.

The most extensive studies dealing with the electrophoretic separation and classification of mammalian and non-mammalian blood plasma esterases employing cellulose columns have been made by Augustinsson (1958, 1959a, b, c), who reported on a comparative study of esterases present in the plasma of twenty-seven vertebrates. On the basis of these studies, it was concluded that each species has its own esterase pattern, and within each of the three groups of esterases, arylesterases (A-esterases), aliesterases (B-esterases), and cholinesterases (C-esterases), there are variations in the properties of the individual enzymes. Arylesterase is the predominant esterase in mammalian plasma and is absent from the plasma of birds, reptiles, amphibians, and fish. Aliesterase is the main esterase of lower vertebrate plasma and is not present in human, monkey, dog, and ruminant plasma. The electrophoretic studies also indicated that a propionyl aliesterase is considered to be the most primitive esterase present in vertebrate plasma, and all other esterases, including the various types of aryl cholinesterases, are suggested to be phylogenetically evolved from such an enzyme.

Plagemann *et al.* (1960a, b, 1961) have described electrophoretically distinct forms of mammalian lactic dehydrogenase in rabbit, human, and dog tissues. Five electrophoretically distinct forms (*isozymes*) of lactic dehydrogenase were isolated from both rabbit and human tissues. Each tissue was found to have characteristic distribution of these isozymes, and quantitative isozyme patterns for nine rabbit tissues and plasma as well as five human tissues and plasma were described. Analogous rabbit and human tissues were found to have similar but not identical isozyme patterns.

Comprehensive reviews dealing with multiple molecular forms of enzymes as delineated by electrophoretic techniques are to be found in the recently published monograph edited by Wróblewski (1961).

## II. ELECTROPHORETIC PROCEDURES AS RELATED TO ENZYME HISTOCHEMISTRY

The recent application of the starch gel electrophoretic techniques in conjunction with histochemical substrates has stimulated a new area of histochemical endeavor which may elucidate not only aspects of substrate specificity, but also the variety of enzyme systems as demonstrated by the aforementioned substrates.

For the detection of enzymes following gel electrophoresis the blocks are incubated employing the standard histochemical techniques, usually azodye procedures. Master (1960) has recommended the application of indophenol esters which form an indophenol dye upon hydrolysis. Specifically, acetyl, butyryl, and caproyl esters of 2,6-dichlorophenylindophenol were employed as substrate solutions at pH 8.6. In areas where the compound is hydrolyzed the intensely colored blue-purple 2,6-dichlorophenylindophenol is formed. An ingenious approach to the visualization of phosphatases separated by starch gel electrophoresis has been reported by Sandler and Bourne (1961) who employed alizarin red S in the incubation mixture. The latter, of course, is known to form a "lake" with calcium ions as produced in the metal-salt technique.

Hunter and Markert (1957) first reported on the histochemical demonstration of enzymes separated by zone electrophoresis in starch gels and proposed the term *zymogram* to refer to the strips in which the location of enzymes is demonstrated by histochemical methods. They homogenized tissue samples and adsorbed the supernatant liquid onto a strip of filter paper which was inserted into the starch block. Satisfactory separation of proteins was achieved in 5 hours at pH 8.6 at room temperature using a 6-volt drop per cm. of travel through the starch block. Migration occurred at about half this rate at 2°. The starch blocks were sliced into two or more thin slabs, permitting several techniques to be applied to the same sample. Azo-dye methods for the demonstration of esterases utilizing $\alpha$-naphthol and naphthol AS acetate were used to visualize the separation of esterase-active proteins from mouse liver and other organs. A surprisingly large and reproducible spectrum of enzymes was obtained. Tyrosinase activity was also studied by reacting the starch slabs with 3,4-dihydroxyphenylalanine to produce pink bands, which later transformed into black melanin. This initial report of Hunter and Markert (1957) stimulated the further application of the zymogram procedure in the elucidation of the complex nature of enzymes as visualized by histochemical methods.

Markert and Hunter (1959) analyzed by means of the zymogram technique the esterases contained in 32 mouse tissues and organs (Appendix 14, Section 2, *b*) (Fig. 1). They were able to resolve more than ten different esterases and partly characterize them by their substrate and their inhibitor specificities. As suggested by previous histochemical findings, they indicate that esterases comprise a family of distinct enzymes with overlapping specificities, and that the total activity in any tissue is the sum of the activities of the particular esterases in that tissue. Markert and Hunter (1959) note the small effect of age, sex, genotype, and nutrition upon the esterase pattern. However, they find that striking changes occur during the differentiation of organs.

They also point out that only about 50% of the total esterase complement of mouse liver moves through the starch column to be resolved into bands as visualized by the zymogram. Thus there exists the possibility that the portion of the enzyme remaining at the origin in the electrophoretic column may contain esterases which could interfere with the interpretation of results inferred from examination of a final pattern.

Estborn (1959) visualized acid and alkaline phosphatases following starch gel electrophoresis of seminal plasma, serum, and bile by means of the $\alpha$-naphthyl phosphate azo-dye method. Following completion of the electrophoresis, a strip of filter paper moistened with buffer (pH 5.0 or 9.8, for acid and alkaline phosphatases, respectively) is placed in contact with the gel for 30 seconds. The filter paper is then transferred to a spe-

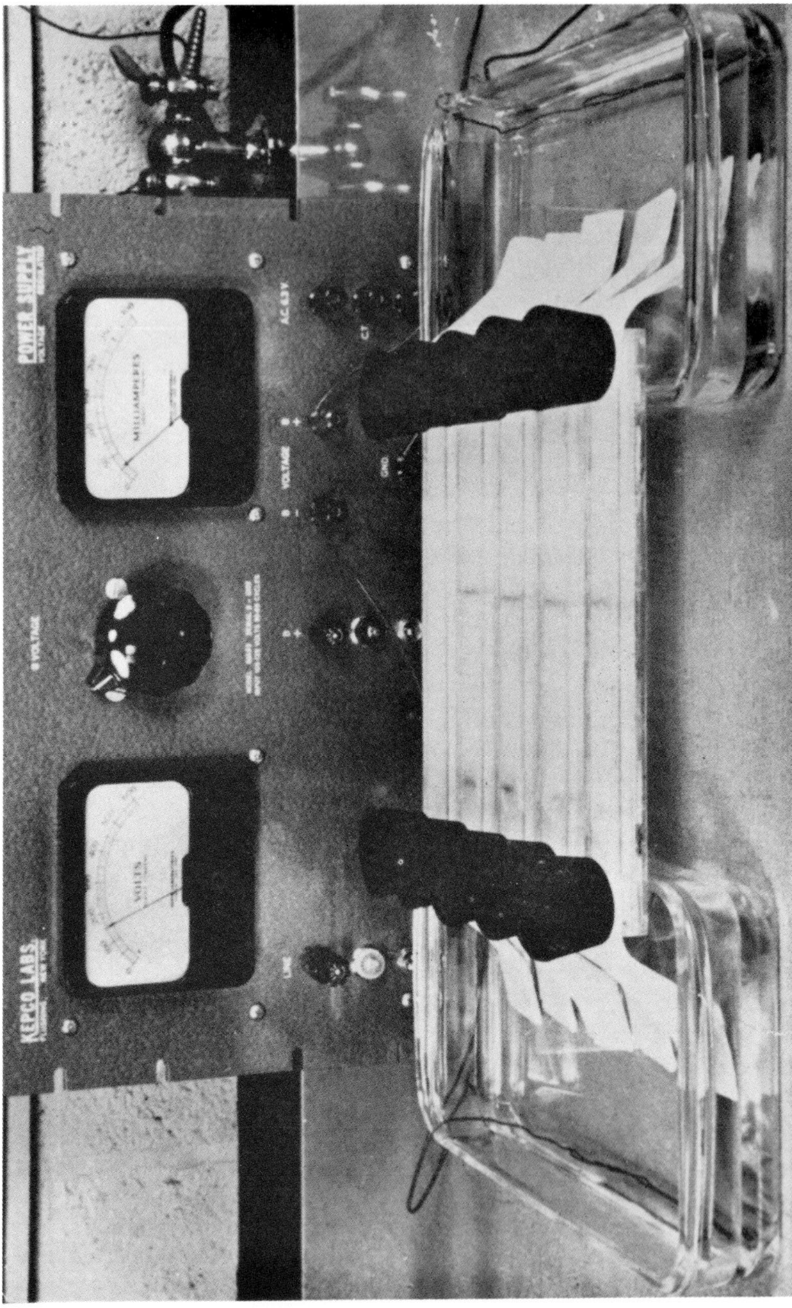


FIG. 1. The figure illustrates the organization of materials and technique for carrying out starch gel electrophoresis. Five starch trays are shown with a buffer tray at each end and power supply behind (courtesy Dr. C. L. Markert).

cially prepared moistened test paper containing α-naphthyl phosphate and the diazonium salt. In a short period of time, a purple band appears on the strip corresponding to the zone in the gel which contains phosphatase.

Hunter and Burstone (1960) compared the substrate specificities of 30 esterase and phosphatase substrates using mouse liver as a primary source of esterases (Figs. 2A, B, C, and D). In addition, kidney, epididymis, prostate, and skin of the mouse were characterized with reference to certain specific substrates. The results of this study are shown in Chart 1 which shows 23 zymograms illustrating the location of esterases following 4½ hours of electrophoresis. Although there are pronounced differences in the reactivity of individual members of the esterase family and of enzymes among the many substrates tested, these differences result primarily from the rate of reaction, and not from absolute substrate specificity which would or which would not permit a given reaction to occur. Nevertheless,

FIG. 2A. Modified starch gel electrophoresis set-up (courtesy Dr. Robert Hunter). Power supply, voltmeter, platinum wire for electrodes, buffer trays, and trays for starch gel.

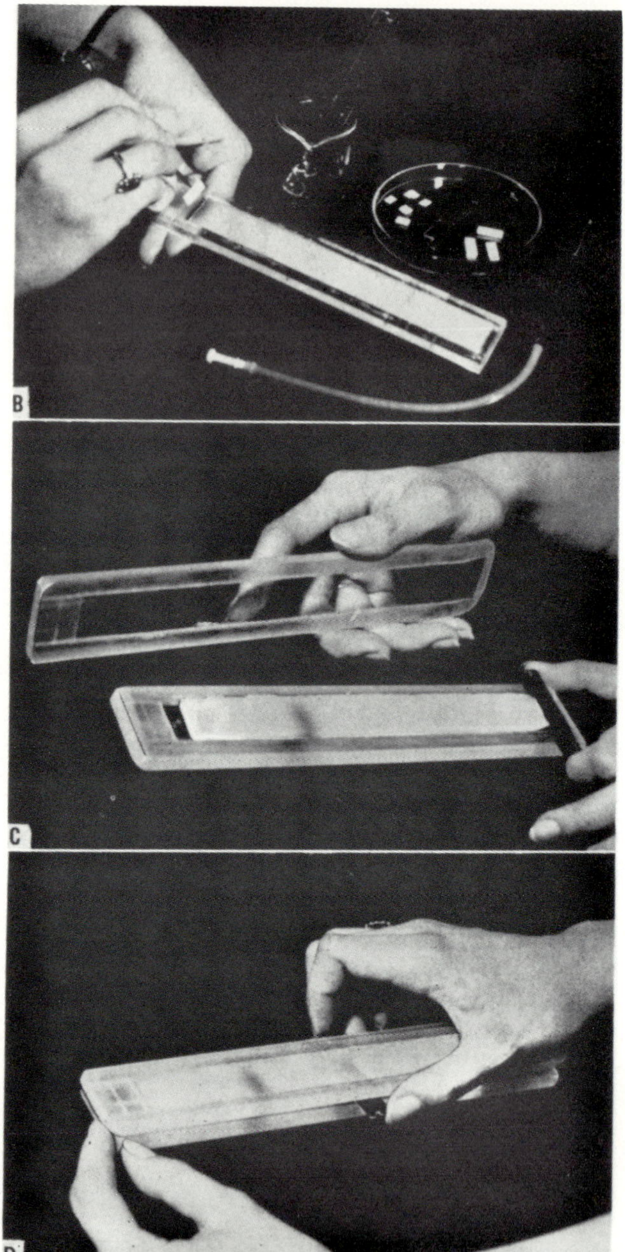


FIGS. 2B–D. Modified starch gel electrophoresis set-up (courtesy Dr. Robert Hunter). B, Insertion of sample into column. C and D, Slicing of gel following electrophoresis.

there are specific and distinct cases in which the specificity of a particular compound may be delineated by the location of the bands.

With reference to Chart 1, the phosphatases demonstrated in zymograms 9, 14, 15, 21, and 22 fall essentially into two bands, one of which migrates rapidly and the other slowly. On the basis of other work involving the demonstration of these enzymes on columns at different hydrogen-ion concentrations, it has been shown that the rapidly moving phosphatase is an acid phosphatase while the band near the origin is an alkaline phosphatase. Another phosphatase migrating between these two bands can be seen in rat duodenum. This is in addition to the more strongly reacting phosphatase band near the origin.

Lawrence *et al.* (1960) used the zymogram technique for a species comparison of enzymes including aminopeptidase, β-glucuronidase, cholinesterase, non-specific esterase, alkaline and acid phosphatases, cytochrome oxidase, succinic dehydrogenase, lactic dehydrogenase, β-hydroxybutyric dehydrogenase, and glutamic dehydrogenase. All of the aforementioned procedures utilize substrates which are generally employed as histochemical reagents. These authors noted that highest esterase activity, as judged by width of bands and staining intensity, was found in the mouse. Cholinesterase was observed in fractions with low mobility in all species except the dog. Cytochrome oxidase exhibited the greatest divergence with respect to number of bands varying from one in the guinea pig to five in the rabbit. β-Glucuronidase activity was found only in one band in all species. Although there was some variability in the intensity of various zones following the enzyme stains among various members of the same species, the overall pattern for that particular species was constant.

Markert and Møller (1959) described multiple forms of enzymes, separated electrophoretically by the starch gel techniques, which they termed *isozymes.* Three dehydrogenases, lactic, malic, and isocitric, of tissue homogenates were resolved into their component dehydrogenase isozymes on the starch gel and visualized by ultraviolet fluorescence. Since there is a loss of fluorescence as DPNH is oxidized to DPN, the dehydrogenases show up as dark bands on a fluorescent background. Markert and Møller (1959) also resolved esterase, peroxidase, and phosphatase into separate molecular types by electrophoresis and indicated that these enzymes probably represent a family of enzymes, the members of which have characteristic overlapping substrate specificities. Genetic, embryological, physiological, and serological aspects relating to the specific pattern of isozymes were also considered by the aforementioned authors.

In a similar vein, Hunter and Strachan (1961) studied the multiple esterases present in mouse blood in order to determine whether similar isozymatic groups could be identified. Among the substrates employed were

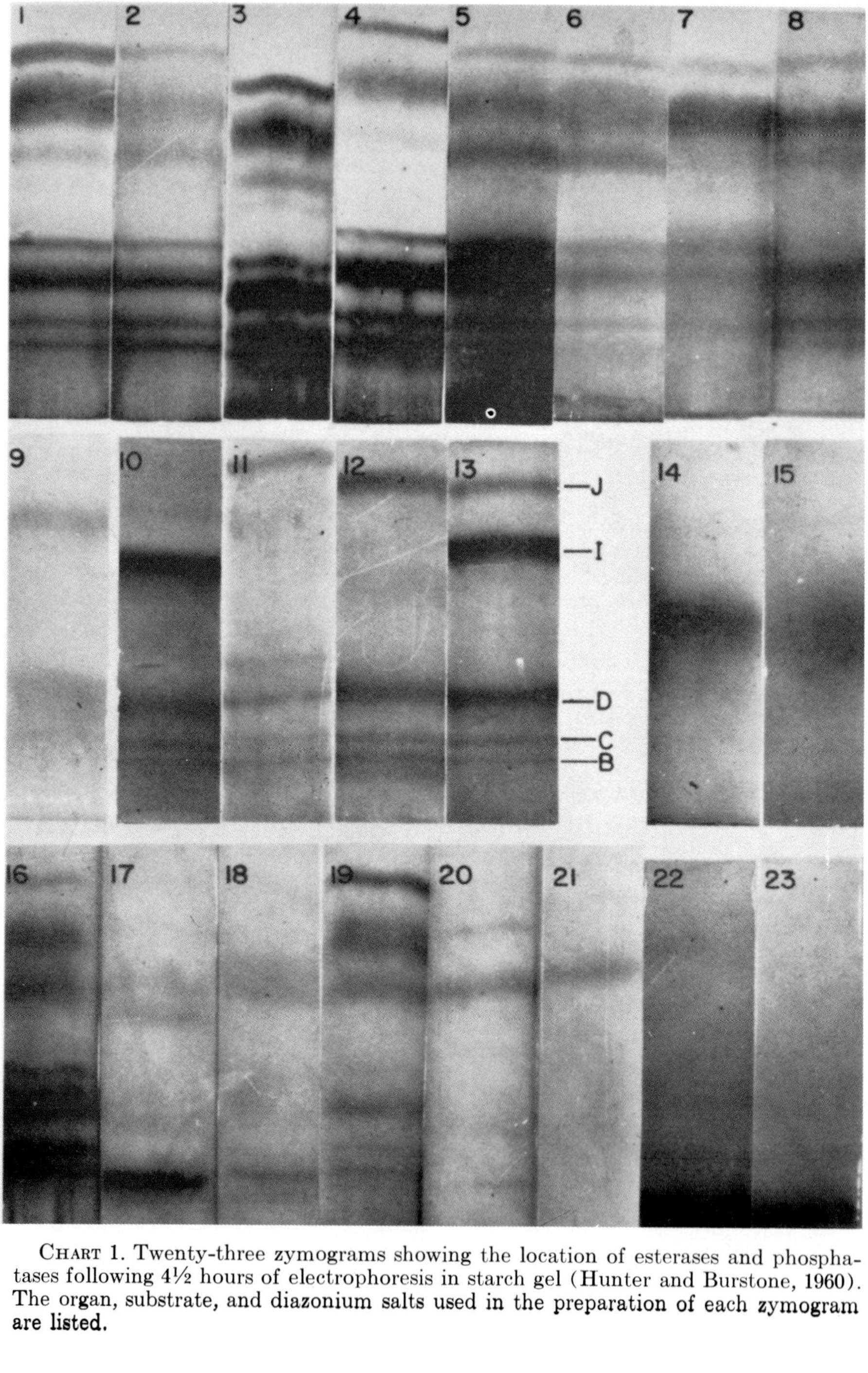


CHART 1. Twenty-three zymograms showing the location of esterases and phosphatases following 4½ hours of electrophoresis in starch gel (Hunter and Burstone, 1960). The organ, substrate, and diazonium salts used in the preparation of each zymogram are listed.

KEY TO CHART 1

| Organ | Substrate | Diazonium salt |
|---|---|---|
| (1) Mouse liver | α-Naphthyl butyrate | Fast Blue BB |
| (2) Mouse liver | α-Naphthyl butyrate | Fast Garnet GBC |
| (3) Mouse liver | α-Naphthyl butyrate | Fast Blue RR |
| (4) Mouse liver | α-Naphthyl isovalerate | Fast Blue RR |
| (5) Mouse liver | α-Naphthyl acetate | Fast Blue RR |
| (6) Mouse liver | β-Naphthyl acetate | Fast Blue RR |
| (7) Mouse liver | Indoxyl acetate | — |
| (8) Mouse liver | Naphthol AS-MX-acetate | Fast Blue RR |
| (9) Mouse prostate | Naphthol AS-MX-phosphate | 5-Chloro-*o*-toluidine |
| (10) Mouse prostate | Naphthol AS-D-acetate | Fast Blue RR |
| (11) Mouse prostate | α-Naphthyl cinnamate | Fast Blue RR |
| (12) Mouse kidney | α-Naphthyl acetate | Fast Blue RR |
| (13) Mouse kidney | Naphthol AS-MX-acetate | Fast Blue RR |
| (14) Mouse kidney | Sodium α-naphthyl acid phosphate | 5-Chloro-*o*-toluidine |
| (15) Mouse kidney | Naphthol AS-MX-phosphate | 5-Chloro-*o*-toluidine |
| (16) Mouse liver | α-Naphthyl propionate | Fast Blue RR |
| (17) Mouse epididymis | α-Naphthyl propionate | Fast Blue RR |
| (18) Mouse skin | α-Naphthyl propionate | Fast Blue RR |
| (19) Mouse liver | 4-Benzoyl 1-naphthyl acetate | Fast Blue RR |
| (20) Mouse epididymis | 4-Benzoyl-1-naphthyl acetate | Fast Blue RR |
| (21) Mouse skin | 4-Benzoyl-1-naphthyl acetate | Fast Blue RR |
| (22) Mouse liver | Sodium α-naphthyl acid phosphate | 5-Chloro-*o*-toluidine |
| (23) Mouse liver | Naphthol AS-MX phosphate | 5-Chloro-*o*-toluidine |

α-naphthyl acetate, α-naphthyl propionate, α-naphthyl butyrate, β-naphthyl laurate, naphthol AS acetate, and indoxyl acetate. Nine esterases in plasma and two in red blood cells were demonstrated. Their results indicate that the multiple esterases present in mouse blood are not readily classifiable into subgroups exhibiting common properties.

Dubbs *et al.* (1960) have reported on the subfractionation of human serum enzymes by the starch gel method in conjunction with established azo-dye techniques. Among the enzymes studied were cholinesterase, aromatic esterase, and acid phosphatase. Of interest is the separation of acid phosphatase into three zones when naphthol AS phosphate was employed as a substrate. The authors indicate that their work was stimulated by the probability that the blood level of each new subfractionation found, by reflecting the healthy or diseased metabolism of its particular source tissue in the body, can offer a more penetrating and selective means of diagnosis.

Boyer (1961), employing the β-naphthyl phosphate azo-dye method in conjunction with the starch gel procedure, has reported at least 16 bands of non-specific alkaline phosphatase in human serum. A considerable degree of heterogeneity was noted in the electrophoretic patterns from normal subjects, a finding which was confirmed by cellulose chromatography. Since the β-naphthyl phosphate technique is known to give rise to serious diffusion artefacts (at the microscopic level), application of this azo-dye method in the staining of gels should be carefully evaluated.

Barka (1961a, b) described the electrophoretic separation of soluble acid phosphatases of rat liver on polyacrylamide gels, and described three bands exhibiting acid phosphatase activity (Appendix 14, Section 4). Among the substrates employed were α-naphthyl, 6-benzoyl-2-naphthol, and naphthol AS-BI phosphates. Barka notes that the soluble acid phosphatases of the liver represent only 70% of the total enzyme activity, the remaining 30% probably corresponding to another acid phosphatase more firmly bound to cellular organelles. Barka (1961b) subsequently described the separation of soluble acid phosphatases of rat liver by anion exchange chromatography and found four acid phosphatase fractions which are suggested to represent distinct molecular types of the enzyme.

More specific applications of the starch gel technique involve the study of esterases of the adrenal medulla of the rat (Allen *et al.*, 1958) and esterases and phosphatases of the epididymis of normal, castrated, and hormone-replaced castrated mice (Allen and Hunter, 1960). The previously described work of Sandler and Bourne (1961) employed a variety of tissue extracts and tissue fluids which, following electrophoretic separation, were incubated in the presence of a number of substrates including sodium β-glycerophosphate and ATP in the presence of calcium ions. The precipitated calcium phosphate was revealed by "lake" formation with sodium alizarin sulfonate (alizarin red S). This type of approach should also facilitate the identification of substrate-specific phosphatases, particularly those which hydrolyze known phosphorylated metabolic intermediates.

Dr. John M. Allen has also employed a metal-salt method in conjunction with gel electrophoresis (unpublished data). After electrophoresis and incubation in Gomori type substrates for the localization of alkaline phosphatase, gels, either starch or acrylamide, are washed briefly in distilled water and are then placed in a tris-maleate buffer (Gomori) at pH 7.0, 0.08 *M* containing 0.003 *M* lead nitrate. After remaining in this mixture for 30 minutes at room temperature the gels are washed repeatedly in distilled water for 1 hour. The sites of phosphatase activity are then visualized by placing the gels in 3% ammonium sulfide for 1 minute. It is important to wash out the excess sulfide in several changes of distilled water. When used with acrylamide gels the method appears capable of visualizing about 3 millimicromoles of phosphate.

Barron *et al.* (1961) have reported on the starch gel electrophoresis of

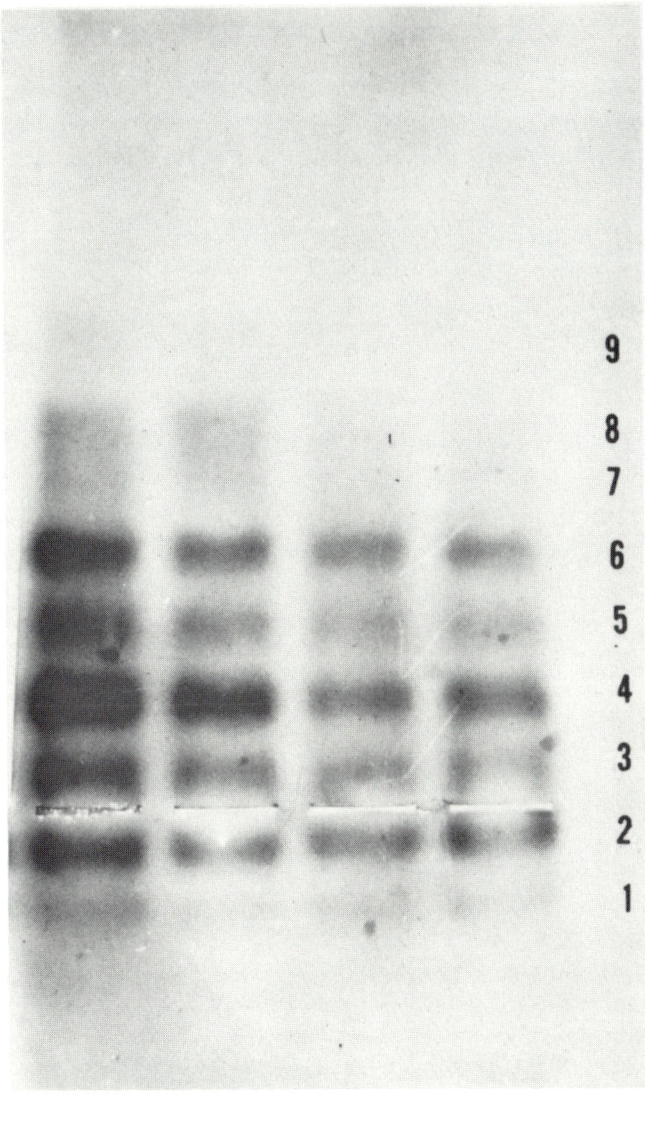


CHART 2. Zymograms obtained with α-naphthyl acetate at pH 7.0 in conjunction with Fast Blue B salt. The pair of zymograms on the left represent normal white matter, while that on the right is of multiple sclerosis plaque tissue (courtesy Dr. Kevin D. Barron).

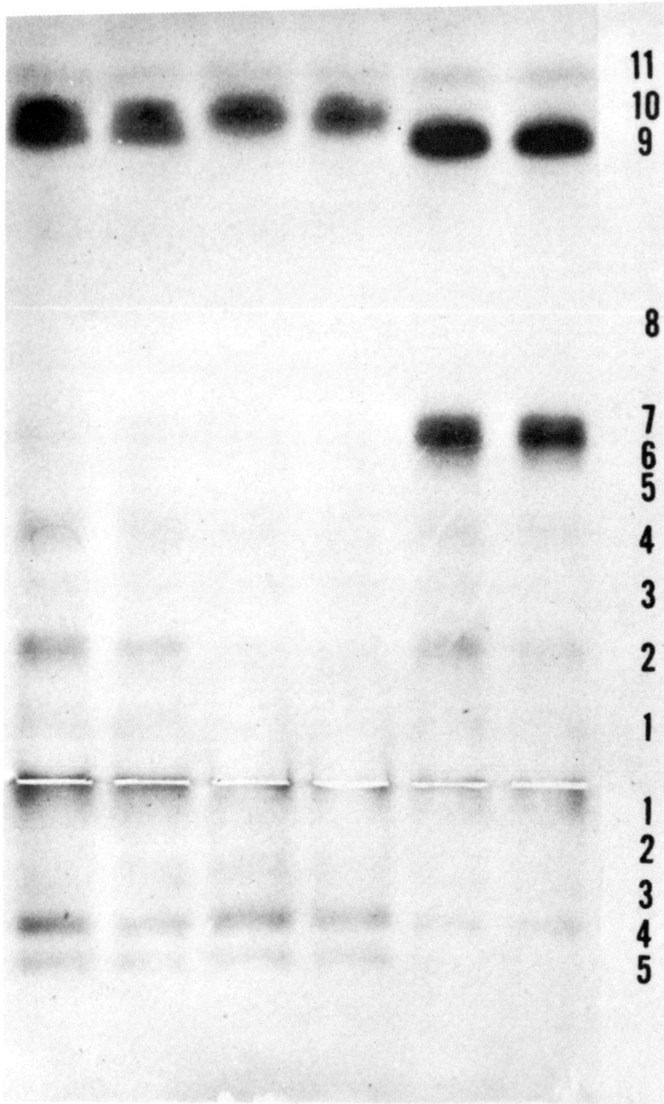


CHART 3. Zymograms obtained with α-naphthyl acetate at pH 7.0. The pair of zymograms on the far right are obtained from normal human temporal white matter. The middle and left hand pairs are from multiple sclerosis tissue. The band at 8 is not visible in the photograph (courtesy Dr. Kevin D. Barron).

brain esterases. Nine esterases which hydrolyze α-naphthyl acetate and naphthol AS acetate were separated from rat brain homogenates. One of these was identified as acetylcholinesterase. Another enzyme, presumably a pseudocholinesterase, which hydrolyzed butyryl thiocholine iodide was also demonstrated. With reference to human brain, homogenates of white matter yielded nine esterases which were active against naphtholic substrates. Acetylcholinesterase was also demonstrable in human material. Chart 2 illustrates zymograms from human normal white matter and multiple sclerosis plaque tissue, respectively. Chart 3 shows zymograms from normal human temporal white matter and also from multiple sclerosis tissue.

# LABORATORY METHODS

## APPENDIX 14

### 1. *Zone Electrophoresis, Starch Block Starch-Grain Method* (*Kunkel, 1954, p. 141; Rosenberg, 1959*)

The following procedure has been employed in obtaining electrophoretic patterns of alkaline phosphatase activity in sera from patients with disorders associated with increased serum phosphatase. The starch blocks may be prepared from slurries of commercially available potato starch. Although the degree of resolution of this method is not satisfactory for histochemical purposes, the procedure is of value in the quantitation of electrophoretically separated enzymes.

If routinely available brands of starch (e.g. Matheson, Coleman and Bell) are employed, the slurries are repeatedly washed by decantation in pH 8.6 barbital buffer. The blocks are prepared in a rectangular Lucite trough ($40 \times 10 \times 2$ cm.) lined with wax paper; the trough also supports the blocks during electrophoresis. Usually three parallel blocks, each containing 70.0 g. of starch and each $35 \times 3.3 \times 1$ cm., are prepared in the same trough, the sides of their wax paper envelopes separating one from the other. In this manner, electrophoresis of three different serum samples may be performed simultaneously. For preparative work a large single block (10.0 cm. wide, and containing 210.0 g. starch) may be used. Buffer saturated plastic sponges and filter paper strips provide electrical continuity between the blocks and buffer vessels. The latter are connected by buffer-filled U-tubes to the vessels containing platinum electrodes. Either barbital or tris buffers pH 8.6, 0.1 $N$ may be used. Test samples, diluted with an equal volume of buffer, are delivered from the needle of the syringe to a transverse trough (ca. 1.0 mm. wide) cut in the starch block midway between the center and the cathode end. The volume of material applied is 0.7 ml. in the case of the small blocks, and 3.5–5.0 ml. for the larger blocks. The solution in the buffer vessel is used repeatedly for a number of experiments before being discarded, the polarity of the electrodes being reversed after each run. The starch block and associated vessels are refrigerated at 4° throughout the period of electrophoresis. A potential difference of 2.5–3.0 volts/cm. is maintained between the ends of the block and the current is 1.4 ma./cm. width.

At the end of the electrophoretic period (36–48 hours), under which conditions the albumin migration distance is ca. 15–20 cm., the blocks are sectioned transversely into serial segments 5.0 mm. wide from a point 1–2 cm. on the cathode side of the origin to a point just beyond the albumin zone. The latter is generally visible because of its yellowish color. A section to be used as a blank is taken from the region close to the cathode end of the block, well outside the range of protein migration. Each starch segment is suspended in 5.0 ml. of distilled water, the mixture shaken gently, and then filtered through Munktell 00 paper. Aliquots of the filtrate may be analyzed for alkaline phosphatase by the standard nitrophenylphosphate procedure. The filtrate (0.5–2.0 ml.) is added to an equal volume of buffered substrate solution and incubated at 38° for 30–120 minutes, depending upon the activity of the sample. Aliquot volumes and incubation times should be the same for all sections of a particular block. At the end of the incubation period, NaOH solution (0.05–0.10 *N*) is added to each tube to make a total volume of 7.0 ml. The optical density is measured at 410 m$\mu$ before and after acidification with 2–3 drops of concentrated HCl. From the calibration curve relating *p*-nitrophenol concentration to optical density, phosphatase activity is expressed as m$\mu$ nitrophenol liberated per hour per ml. of filtrate. The buffer-substrate solution consisted of a mixture of equal volumes of 0.3% aqueous *p*-nitrophenylphosphate ($Na_2$) and pH 10.3 ammediol buffer.

2. *Electrophoresis, Starch Gel Method* (*Smithies, 1955b; 1959, p. 65; Markert and Hunter, 1959; Hunter and Burstone, 1960*)

*a. Equipment* (*Figs. 2A through 2D*). Power supply, voltmeter, platinum wire for electrodes, buffer trays, and trays for starch gel. (Power supply PS-4 and voltmeter V-7 from the Heath Company, Benton Harbor, Michigan, are satisfactory; Fig. 2A.) Starch tray dimensions are 2.0 × 0.6 × 22 cm. Dimensions of buffer trays will vary depending upon the number of gel trays that are to be accommodated at the same time. It is important that the sides of these trays are low (2.5 cm.) in order that they can be used full of buffer with a short filter paper wick to the gel columns.

*b. Preparation of sample.* Protein containing fluids such as serum or saliva may be used without pretreatment. Tissue extracts are made by homogenizing a weighed sample of tissue with a volume of distilled water which for convenience is usually equal to a multiple of the weight. Tissue weighing 250 mg., for example, may be homogenized in 0.75 ml. of water. It is often desirable to freeze and thaw the sample six times preceding the extraction in order to more fully break up subcellular particles. Following homogenization the sample is centrifuged at 10,000 g. for 10 minutes and the supernatant fluid is used as the sample. When quantities are limited, the centrifugation can be carried out in a capillary tube 1 mm. in diameter and the supernatant fluid obtained by breaking the tube at the appropriate level. Even smaller samples may be run without centrifugation but a smearing will be seen between the bands. This smearing arises due to the continuing dissolving of the proteins from cellular particles during the electrophoretic run.

3. *Separation of Proteins with Starch Gel Electrophoresis*

1. Add approximately 11.6 g. of hydrolyzed potato starch to 100 ml. of .03 *M* of sodium borate buffer adjusted to pH 8.6. (Hydrolyzed potato starch is available from Connaught Laboratories packaged in 2000 g. units. Directions indicating exact starch and buffer concentrations are given on the label and should be followed.)
2. The starch buffer mixture is heated in an Erlenmeyer vacuum flask over a Bunsen burner with constant agitation. Four hundred milliliters of solution in a 2 liter flask, for example, would be heated for 5 minutes with agitation. During this period the slurry will become viscous for a brief period, but with continued heating it becomes more fluid. No attempt should be made to bring the solution to a boil.
3. After removing the flask from the flame, a rubber stopper is placed in the flask and the side arm is attached to a vacuum line for about 30 seconds. During this period air bubbles will be removed. The solution will continue to boil as long as the vacuum is applied so no attempt should be made to wait out the boiling process.
4. The mixture is then poured into the gel trays. The trays should be slightly overfilled so that when the 1/6 inch plastic top is placed over the tray a small amount of gel will be expressed and no air bubbles trapped under the top.
5. The columns are ready for use after setting for 90 minutes or overnight at refrigerator temperatures, depending upon which is more convenient. The gels undergo continuing changes after preparation and should not be used after 24 hours.
6. A .01 ml. sample is then pipetted onto a small piece of filter paper measuring 5 × 13 mm. which is placed in the column as illustrated in Fig. 2B, 10 cm. from the cathodal end of the column. Care should be taken that there are no air bubbles between the paper and the gel. When comparisons are to be made within a column, filter paper pieces measuring 5 × 7 mm. are used and placed about 1 mm. apart in the column.
7. The entire column is then wrapped with a non-porous, transparent film (Saran Wrap, Dow Chemical Corporation) except for an area 3.0 cm. long at each end of the column.
8. A filter paper wick made up of six layers of filter paper measuring approximately 2.5 × 8.0 cm. is wetted in the concentrated buffer solution in the trays and placed at each end of each column in such a way as to form an electrical connection to the buffer. This is illustrated in Fig. 2A.

9. The current is turned on and the voltage is adjusted such that there is a 6-volt drop per centimeter along the column. This is accomplished by measuring the voltage drop over a known distance and adjusting the output from the power supply until the desired difference in potential is achieved. This is being done in Fig. 2A.
10. The time required for adequate separation depends upon the material being examined and the voltage being applied. Initially, samples are run for 4 hours at 6 volts per centimeter at room temperature. At 4° samples run at approximately half the speed they do at room temperature.
11. The gel is removed from the trays, using a wet spatula specially ground to fit the trays, and sliced, as shown in Figs. 2C and 2D.
12. Each half is then stained for protein or incubated in an appropriate substrate solution.

The proteins present in the starch gel column are undergoing two changes which adversely affect the observable enzyme bands and their intensity as they are pretreated with inhibitor solutions prior to the demonstration of their location with histochemical enzyme methods. First, as they stand undeveloped, they are undergoing diffusion, and the "sharpness" of the bands becomes reduced with time. Second, the solution containing the inhibitor may leak out some of the protein in the columns and thus, without regard to the specific inhibitor involved, there would be an apparent reduction in the activity of the enzyme. To minimize these effects, the sample should be run as promptly as possible and preincubated only 10 minutes before the elements of the histochemical reaction are added. Control columns should always be run.

4. *Electrophoretic Separation of Acid Phosphatases, Polyacrylamide Gel Method* (*Raymond and Weintraub, 1959; Barka, 1961a*)

The polyacrylamide gel is prepared by mixing equal amounts of four refrigerated stock solutions in the following order:

1. Acrylamide (30%) and *N,N'*-methylenebisacrylamide (organic monomers) (0.8%) in distilled water. This solution is prepared at about 3 week intervals and stored at about 4°.
2. β-Dimethylaminopropionitrile (DMAPN) (accelerator) (1.6%) in distilled water prepared daily.
3. Potassium ferricyanide (0.03%).
4. Ammonium persulfate (0.48%) diluted freshly from a 2.4% stock solution. The stock solution is prepared at weekly intervals.

The acrylamide and *N,N'*-methylenebisacrylamide may be used without further purification while the solution of monomers is filtered after preparation.

The above-mentioned mixture forms a satisfactory gel within 30–45 minutes at 37°. The gel, which is strong enough to support its own weight, is used in the form of vertical columns. Glass cylinders, 50.0 mm. in length and 6.0 mm. inner diameter, are inserted vertically into a plasticine bed, deep enough to provide a water tight seal. The cylinders are filled to within 10.0 mm. of the top of the liquid catalyzed monomer solution. Water is then carefully overlayed to produce a flat surface of the gel, and the cylinders are kept at 37° for 30–45 minutes. After the polymerization is complete, the overlayer of water is poured off and replaced by buffer.

For electrophoresis the glass cylinders containing the gel are connected to the upper buffer compartment by Tygon tubing. After both compartments are filled with buffer (400.0 ml., 0.05 *N* glycine buffer, pH 8.6) and air bubbles, occasionally present in the tube above the gel, are removed, the upper compartment is lowered to make contact with the lower one. Samples (15–100 $\mu$l.) are applied by layering on the upper surface of the gel beneath the buffer and the circuit is completed through a power supply. Since acid phosphatases may be relatively heat sensitive in solution, a voltage gradient of 20 volts/cm. is employed, and the electrophoresis performed in a cold room (4°). Under these conditions, four columns conduct approximately 1.2 ma. each at 100 volts. The running time for enzymes is reported to be approximately ca. 90 minutes.

Following separation, the columns are stained for periods of 2–24 hours without fixation and at low temperature (ca. 4°). Minimal diffusion of proteins in the gel is observed at this temperature. However, enzyme activity may be relatively low at ice box temperature. The standard azo-dye technique employing $\alpha$-naphthyl phosphate or naphthol AS phosphates, as well as the metal-salt technique, may be employed in conjunction with the acrylamide gels.

## Commercial Sources of Reagents

1. Hydrolyzed potato starch. Connaught Medical Research Laboratories, University of Toronto, Toronto, Canada
2. Saran Wrap. Dow Chemical Co., Midland, Michigan
3. Cyanogum 41. American Cyanamid Co., New York, New York
4. Acrylamide and *N,N'*-methylenebisacrylamide. The Borden Chemical Co., Philadelphia, Pennsylvania
5. Oxoid cellulose acetate strips. Consolidated Laboratories, Inc., Chicago Heights, Illinois
6. Canalco, Bethesda, Maryland

## References

Allen, J. M., and Hunter, R. L. (1960) *J. Histochem. and Cytochem.* **8,** 50–57.
Allen, J. M., Eränkö, O., and Hunter, R. L. (1958) *Am. J. Anat.* **102,** 93–116.
Augustinsson, K-B. (1958) *Nature* **181,** 1786–1789.
Augustinsson, K-B. (1959a) *Acta. Chem. Scand.* **13,** 571–592.

Augustinsson, K-B. (1959b) *Acta. Chem. Scand.* **13,** 1081–1096.
Augustinsson, K-B. (1959c) *Acta. Chem. Scand.* **13,** 1097–1105.
Barka, T. (1961a) *J. Histochem. and Cytochem.* **9,** 542–547.
Barka, T. (1961b) *J. Histochem. and Cytochem.* **9,** 564–571.
Barron, K. D., Bernsohn, J. I., and Hess, A. (1961) *J. Histochem. and Cytochem.* **9,** 656–660.
Beaton, G. H., Selby, A. E., and Wright, A. M. (1961) *J. Biol. Chem.* **236,** 2001–2004.
Boyer, S. H. (1961) *Science* **134,** 1002–1004.
Consden, R., Gordon, A. H., and Martin, A. J. P. (1946) *Biochem. J.* **40,** 33–41.
Coolidge, T. B. (1939) *J. Biol. Chem.* **127,** 551–553.
Dubbs, C. A., Vivonia, C., and Hilburn, J. M. (1960) *Science* **131,** 1529–1531.
Durrum, E. L. (1950) *J. Am. Chem. Soc.* **72,** 2943–2948.
Engle, R. L., Jr., Woods, K. R., Castillo, G. B., and Pert, J. H. (1961) *J. Lab. Clin. Med.* **58,** 1–22.
Estborn, B. (1959) *Nature* **184,** 1636–1637.
Estborn, B. (1961) *Clin. Chim. Acta* **6,** 22–24.
Estborn, B., and Swedin, B. (1959) *Scand. J. Clin. & Lab. Invest.* **11,** 235–237.
Flynn, F. V., and de Mayo, P. (1951) *Lancet* **261,** 235–239.
Goldberg, C. A. J. (1958) *Clin. Chem.* **4,** 484–495.
Gordon, A. H., Keil, B., Sebesta, K., Knessl, O., and Sorm, F. (1950) *Collection Czechoslov. Chem. Commun.* **15,** 1–16.
Grassmann, W., Hannig, K., and Knedel, M. (1951) *Deut. med. Wochschr.* **76,** 333–336.
Hunter, R. L., and Burstone, M. S. (1960) *J. Histochem. and Cytochem.* **8,** 58–62.
Hunter, R. L., and Markert, C. L. (1957) *Science* **125,** 1294–1295.
Hunter, R. L., and Strachan, D. S. (1961) *Ann. N. Y. Acad. Sci.* **94,** 861–867.
Keiding, N. R. (1959) *Scand. J. Clin. & Lab. Invest.* **11,** 106–112.
Kunkel, H. G. (1954) *In* "Methods of Biochemical Analysis" (D. Glick, ed.), Vol. 1, pp. 141–170. Interscience, New York.
Kunkel, H. G., and Slater, R. J. (1952) *Proc. Soc. Exptl. Biol. Med.* **80,** 42–44.
Kunkel, H. G., and Tiselius, A. (1951) *J. Gen. Physiol.* **35,** 89–118.
Lawrence, S. H., Melnick, P. J., and Weimer, H. E. (1960) *Proc. Soc. Exptl. Biol. Med.* **105,** 572–575.
Markert, C. L., and Hunter, R. L. (1959) *J. Histochem. and Cytochem.* **7,** 42–49.
Markert, C. L., and Møller, F. (1959) *Proc. Natl. Acad. Sci. U. S.* **45,** 753–763.
Master, R. W. P. (1960) *Biochim. et Biophys. Acta* **39,** 159–160.
Plagemann, P. G. W., Gregory, K. F., and Wróblewski, F. (1960a) *J. Biol. Chem.* **235,** 2282–2287.
Plagemann, P. G. W., Gregory, K. F., and Wróblewski, F. (1960b) *J. Biol. Chem.* **235,** 2288–2293.
Plagemann, P. G. W., Gregory, K. F., and Wróblewski, F. (1961) *Biochem. Z.* **334,** 37–48.
Poulik, M. D., and Smithies, O. (1958) *Biochem. J.* **68,** 636–643.
Raymond, S., and Weintraub, L. (1959) *Science* **130,** 711.
Ressler, N. (1960) *Clin. Chim. Acta* **5,** 795–800.
Ressler, N., and Jacobson, S. D. (1955) *Science* **122,** 1088–1089.
Ressler, N., and Moy, T. (1959) *Clin. Chim. Acta* **4,** 901–904.
Ressler, N., and Zak, B. (1956) *Clin. Chim. Acta* **1,** 392–400.
Ressler, N., Nelson, N. A., and Oster, H. D. (1958) *J. Lab. Clin. Med.* **51,** 623–629.
Rosenberg, I. N. (1959) *J. Clin. Invest.* **38,** 630–644.

Sandler, M., and Bourne, G. H. (1961) *Exptl. Cell Research* **24,** 174–177.
Scherr, G. H. (1961) *Trans. N. Y. Acad. Sci.* **23** [11], 519–530.
Slater, R. J., and Kunkel, H. G. (1953) *J. Lab. Clin. Med.* **41,** 619–631.
Smithies, O. (1955a) *Nature* **175,** 307–308.
Smithies, O. (1955b) *Biochem. J.* **61,** 629–641.
Smithies, O. (1959) *Advances in Protein Chem.* **14,** 65–113.
Wróblewski, F. (1961) *Ann. N. Y. Acad. Sci.* **94,** 655–1030.
Yoon, C. H. (1961) *Science* **134,** 1009–1010.
Zak, B., and Jarkowski, T. L. (1958) *Clin. Chim. Acta* **3,** 494–496.
Zak, B., and Sun, K. M. (1958) *Am. J. Clin. Pathol.* **29,** 69–79.

CHAPTER 15

# THE RELATIONSHIP BETWEEN CARCINOSTATIC AGENTS AND HISTOCHEMICAL SUBSTRATES

The development and application of carcinostatic or cancer chemotherapeutic agents which are hydrolyzed *in vivo* by enzymes to liberate compounds toxic to the tumor are intimately related to the general problem of cancer chemotherapy. Among the pertinent reviews on chemotherapy including cancer chemotherapy are those by Work and Work (1948), Ross (1953), Greenstein (1954, p. 285), St. Whitelock (1958), Danielli (1959, p. 167), Mandel (1959), Woolley (1959), Albert (1960), Quastel (1960, p. 167), Zubrod (1960), and Welch (1961).

Carcinostatic agents may be classified according to three groups: compounds which are directly toxic to the tumor; antimetabolites, which resemble in chemical structure some naturally occurring compounds and which specifically antagonize the biological action of these (Woolley, 1959); and compounds which are hydrolyzed *in vivo* by enzymatic action to release substances toxic to tumors.

Friedman and Boger (1956), in a review of pertinent literature, point out that one of the promising approaches to cancer chemotherapy is based upon the toxigenic substrates: compounds from which highly cytotoxic agents would be liberated by the action of enzymes. In this way, a lethal dose of a cytotoxic agent might be delivered to the cells of a tumor, whereas a relatively smaller non-lethal dose would reach the cells of normal tissue in which less of the enzyme was present. Of interest are the findings of Ross (1958, p. 669) who reported a positive correlation between the biological activity of alkylating agents and their hydrolysis rates.

With reference to the so-called enzyme-activated drugs, initial studies employed esterified nitrogen mustard derivatives. The latter are agents related to the vesicant war gas, di-2-chloroethyl sulfide. Among the first carcinostatic agents of this type to be employed were bis-2-chloroethylamine

(HN2), tris-2-chloroethylamine (HN3), $\beta$-naphthyl-di-2-chloroethylamine (Ross, 1953). The *p*-amino derivative of *N*,*N*-di-2-chloroethylamine is known to be an effective inhibitor of transplanted Walker rat carcinoma (Ross *et al.*, 1955). Philips (1950) and Biesele *et al.* (1950) suggest that cytotoxic action is dependent upon the presence in the compound, or its active intermediate, of an unstable three-membered heterocycle ring system. On the other hand, Loveless and Ross (1950) consider that the ability to react with nucleophilic centers through a carbonium ion mechanism is a more general feature of these reagents.

Impetus to the study of enzyme-activated carcinostatic compounds was given by the original observation of Gomori, who in 1948 first reported that a large number of human neoplasms exhibited an increased phosphamidase activity (Chapter 5, p. 231). The fact that the N—P bond of certain compounds is readily hydrolyzed by certain neoplasms stimulated the development of carcinostatic agents containing a phosphoamide linkage, which upon hydrolysis *in vivo* might be expected to release a toxic compound. Subsequently, drugs containing ester and amide linkages, which were presumably capable of being hydrolyzed by other enzyme systems, were developed. The most extensive reviews dealing with these types of agents which are activated both by constitutive and adaptive enzymes are those by Danielli (1954, p. 263; 1959, p. 167; 1961, p. 1). Reports dealing with specific aspects of this approach have been made by Friedman and Seligman (1954a), Ross *et al.* (1955), Friedman and Boger (1956), Arnold and Bourseaux (1958), and Brock (1958).

The first large series of N—P containing carcinostatic agents were the ethylene phosphoramides synthesized by Parker *et al.* (1952a, b, c) and further described by Crossley *et al.* (1953). Among these compounds are *N*,*N'*,*N''*-triethylenephosphoramide (1), *N*,*N*-diethyl-*N'*,*N''*-diethylenephosphoramide (2), *N*-pentamethylene-*N'*,*N''*-diethylenephosphoramide (3), *N*-(3-oxapentamethylene)-*N'*,*N''-diethylenephosphoramide* (4) which are illustrated in Chart 1. These compounds are prepared by reacting a trihalophosphoric acid such as phosphorous oxychloride with a molecular equivalent *C*-substituted ethylenimine. The intermediate *N*-(substituted ethylene) amidodihalophosphoric acid is then reacted with two molecular equivalents of an ethylenimine to produce the appropriate triamide. Although the aforementioned phosphoramides have been employed in animal as well as clinical studies, it is not known if the N—P bond is susceptible to *in vivo* enzymatic hydrolysis.

Friedman and Seligman (1954a, b) reported the synthesis of *N*-phosphorylated derivatives of nitrogen mustards as chemotherapeutic agents. Since the chemical activity and presumably the biological potency of nitrogen mustards is known to depend upon the basicity of the nitrogen atom,

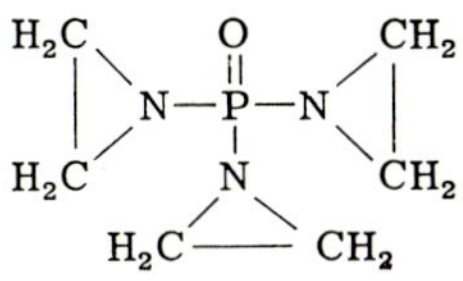

1

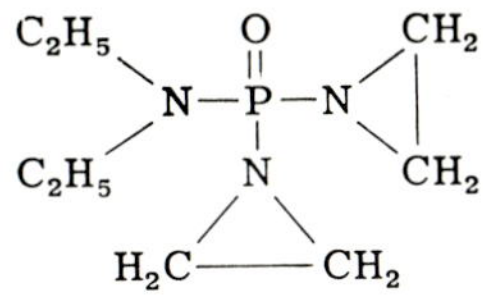

2

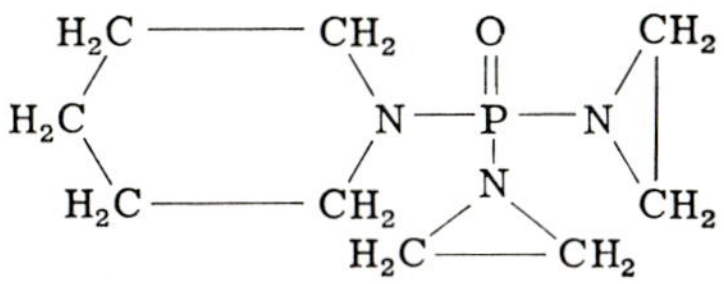

3

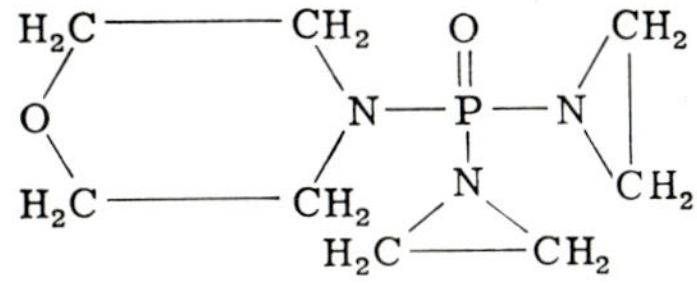

4

CHART 1. Ethylene phosphoramides

*N*-phosphorylation might be expected to give products significantly less toxic than the parent mustard. Loss of the basicity of the nitrogen atom in the phosphoamide link according to the inductive effect of the $PO_3$ or $PON_2$ group and possible resonance would inhibit intermolecular cyclization to the ethylene immonium ion necessary for high chemical activity. Among the compounds synthesized were *N,N*-di-(2-chloroethyl) phosphorodiamidic acid (I) and phenyl-*N,N*-di-(2-chloroethyl) phosphorodiamidate (II)

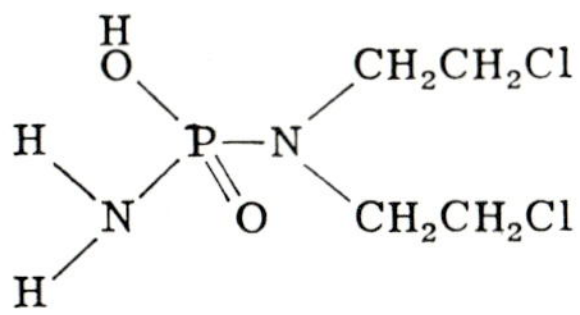

*N*, *N*-di-(2-chloroethyl) phosphorodiamidic acid

(I)

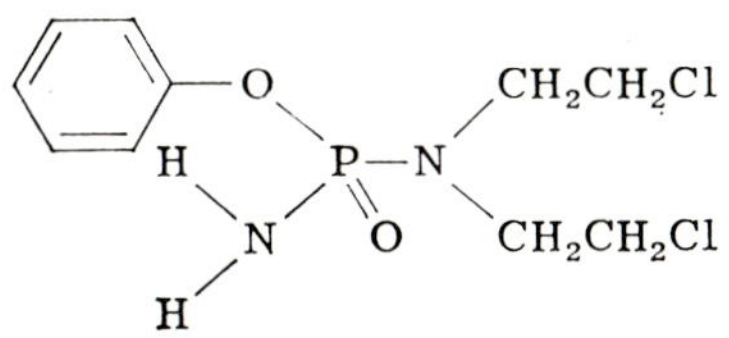

Phenyl-*N*, *N*-Di-(2-chloroethyl)-phosphorodiamidate

(II)

(Friedman and Seligman, 1954a). Although the former compound was unstable at pH 5, evidence of enzymatic hydrolysis was obtained in tissue homogenates. On the other hand, the phenyl-substituted derivative was not hydrolyzed (Foley *et al.*, 1961). Friedman and Seligman (1954b) prepared *N,N*-di-(2-chloroethyl) phosphorotriamidate and found that this compound exhibited only a fraction of the toxicity of the parent bis ($\beta$-chloroethyl) amine in mice. As in the case of the other phosphoamide mustards, they indicate that the parent mustard should be released in tumor cells if the enzyme phosphamidase were present. Rutenburg *et al.* (1954) tested the toxicity of *N*,*N*-di-(2-chloroethyl) phosphorotriamidate in mice. Based on the same theoretical approach Friedman *et al.* (1954) prepared *N*-phosphorylated derivatives of diethanolamine including dibenzyl di-(2-hydroxyethyl) phosphoramidate and the corresponding dichlorophosphoramidate. Ross *et al.* (1955) have reported the synthesis of phosphoramidates from *N*,*N*-di-(2-chloroethyl)-*p*-phenylenediamine. Friedman and Boger (1956) have also described other *N*-phosphorylated nitrogen mustards which are claimed to be biologically inactive per se, but which liberate secondary amine mustards capable of intramolecular cyclization to form potent chemotherapeutic agents.

More recently, Arnold and Bourseaux (1958), Arnold *et al.* (1958a, b), and Brock (1958) have reported on the synthesis and experimental study of a new chemotherapeutic cyclic nitrogen mustard phosphoamide ester, Cytoxan {2*H*-1,3,2-oxazaphosphorine, 2-[bis (2-chloroethyl) amino] tetrahydro-2-oxide} which is similar to some of the aforementioned bis-$\beta$-(chloroethyl) amines. They suggest that Cytoxan represents a "transport" form of the drug which is presumably transformed in the "active" structure *in vivo*. The marked biological inactivity of the "transport" form is the result of combination of the nitrogen with the electronegative phosphoryl group which reduces the reactivity of the $\beta$-chloroethyl groupings. They postulate that the "active" form is produced under special conditions prevailing in tissues during which enzymatic cleavage of the ring occurs (Chart 2). It is also possible that an enzymatic cleavage of the N—P bond may also occur (Chart 2).

The suggestion that the previously discussed compounds represent enzyme-activated anti-tumor agents which are activated by phosphamidase or a similar enzyme represents a stimulating hypothesis in light of the high phosphamidase activity of human neoplasms (Chapter 5, p. 231). Theoretically, a lethal dose of the cytotoxic agent would be delivered to the tumor cells whereas a relatively smaller non-lethal dose would reach cells of normal tissue in which less phosphamidase was present. Unfortunately, there are several defects to this type of interpretation. First, the enzymatic hydrolysis of phosphoamides is not solely limited to the action of phosphamidase (Chapter 5, p. 229). Phosphoamides are readily hydrolyzed by non-specific alkaline and acid phosphatases over a wide pH range, and since these non-specific phosphatases are ubiquitously distributed throughout the body one would tend to question the selective hydrolysis of phosphorylated nitrogen mustards by tumor cells. An interesting approach to

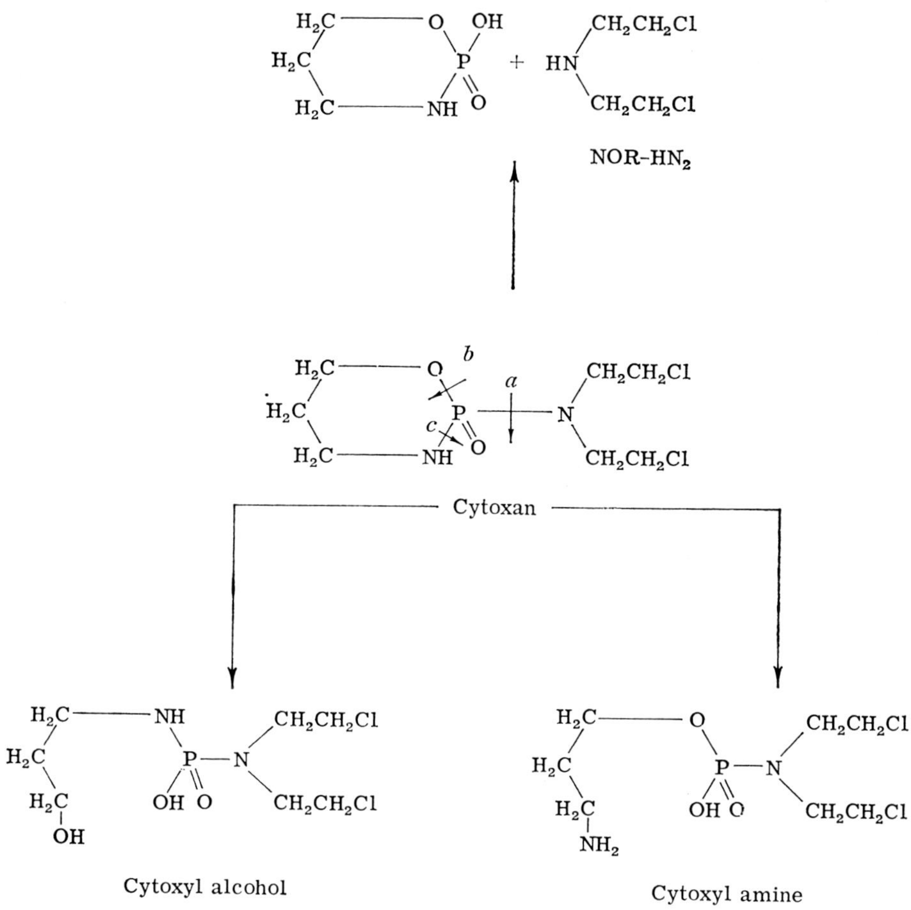


**CHART 2.** Possible mechanisms of activation of Cytoxan (Arnold and Bourseaux, 1958 Foley *et al.*, 1961).

the testing of the enzyme-activated anti-tumor agent hypothesis could employ the various compounds as specific histochemical reagents in conjunction with sections of normal and neoplastic tissue in order to delineate histological sites, if any, at which hydrolysis of the compound occurred (Foley *et al.*, 1961).

In 1948 Friedman and Seligman reported the synthesis of derivatives of 10-methyl-1,2-benzanthracene related to nitrogen and sulfur β-chloroethyl vesicants. The carcinogenic hydrocarbon was selected as the nucleus because of its special carcinogenic influence on cells, possibly indicating a special reaction with cellular constituents, and because of its known ability to inhibit the growth of tumors in rats and mice. The nitrogen and sulfur mustard groupings were attached to the 10-methyl group through the readily available 10-chloromethyl-1,2-benzanthracene. In addition to direct nitrogen and sulfur linkages, amide, urea, and urethan linkages were used. These authors note that although there is no experimental data to suggest that the amide, urea, and urethan linkages in such compounds are cleaved *in vivo*, this type of hydrolysis would release a potent mustard derivative.

Seligman *et al.* (1952) have also reported on the effects of hemisulfur mustard (2-chloro-2-hydroxyethyl sulfide) derivatives. Among these were an acetate and a phosphate derivative which presumably could be hydrolyzed by enzymatic action to produce the more potent hemisulfur compound.

The most extensive reviews on anti-tumor drugs which contain ester or amide linkages and which are capable of being activated by enzymes are those by Danielli (1954, p. 263; 1959, p. 167; 1961, p. 1). In 1954, Danielli described the enzymatic activation of drugs derived from the *N*-acylation of *N*,*N*-di-(2-chloroethyl)-*p*-phenylenediamine (III). It was found that the *N*-acetyl derivative exhibited greatly reduced toxicity as contrasted with the free amino compound, but yet produced a distinct carcinostatic effect, presumably due to the *in vivo* enzymatic hydrolysis of the acetyl grouping. The *N*-benzoyl compound was less toxic but exhibited a negligible carcinostatic effect. According to Danielli, the tumor (Walker rat carcinoma 256) was found to contain an enzyme which would liberate the parent substance from its acetyl derivative but not from its benzoyl derivative. The enzyme responsible for the hydrolysis of the *N*-acetyl compound was presumed to be a peptidase or an amidase. Hebborn and Danielli (1958) and Danielli (1959, 1961) reported on the application of a number of compounds prepared by acylation of the aforementioned *N*,*N*-di-(2-chloroethyl)-*p*-phenylenediamine and also *N*,*N*-di-(2-chloroethyl)-*p*-aminophenol (IV). A variety of derivatives were obtained by substituting the terminal amino and phenolic groups.

Details of the synthesis of esters and amides as well as phosphates of chloroethylamines including *N*,*N*-di-(2-chloroethyl)-*p*-phenylenediamine and *N*,*N*-di-(2-chloroethyl)aminophenol have been reported by Davis and Ross (1950) and Ross *et al.* (1955).

An investigation of the enzymes which would catalyze the hydrolysis of the substituent groups from the parent mustards was conducted on a quantitative basis using model substrates derived from aniline and β-naphthol, the amount of aniline liberated from a substrate in the presence of a tissue homogenate being estimated colorimetrically after diazotization and coupling with a sulfonated naphthol. The types of potential anti-tumor agents

TABLE 1

ACTIVATION OF *N*-ACYLATED MUSTARD BY PEPTIDASES IN WALKER RAT TUMOR 256[a]

| | Presence of enzyme in tumor | I[b] | PR[c] | CR[d] |
|---|---|---|---|---|
| $R \cdot NH_2$ | | + | − | − |
| $R \cdot NH \cdot CO \cdot CH_3$ | ++ | + | + | 20 |
| $R \cdot NH \cdot CO \cdot Ph$ | − | − | − | − |
| $R \cdot NH \cdot CO \cdot CCl_3$ | ++ | + | + | − |
| $R \cdot N \cdot CO \cdot CF_3$ | + | + | + | − |

[a] Danielli (1959).
[b] I = inhibition.
[c] PR = partial regression.
[d] CR = complete regression.

derived from the acylation of the nitrogen mustard shown in (III), and their effect upon the Walker rat tumor 256 is shown in Table 1. In the case of the acetylated and benzoylated hydroxyl containing nitrogen mustards derived from (IV), the effect of esterase activity in the tumor is presumably to hydrolyze the ester bond and activate the drug as shown in Table 2.

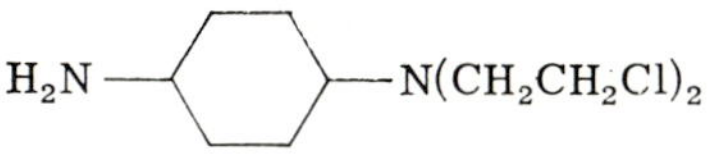

*N*, *N*-Di-(2-chloroethyl)-
-*p*-phenylenediamine

(III)

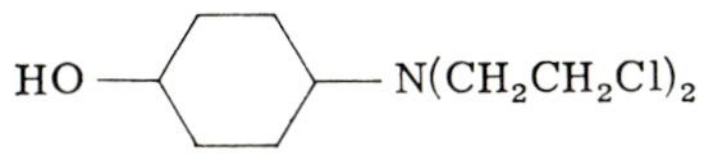

*N*, *N*-Di-(2-chloroethyl)-
*p*-aminophenol

(IV)

Ross and Warwick (1955) described the experimental application of arylazo nitrogen mustards which could be reduced by the xanthine oxidase enzyme system to the more cytotoxic hydrazo compounds. It is also possible that various dehydrogenases may play a significant role in the "activation" of arylazo nitrogen mustards.

TABLE 2

ACTIVATION OF ESTERIFIED HYDROXYL-CONTAINING MUSTARD BY ESTERASES OF WALKER RAT TUMOR 256[a]

| | Presence of enzyme in tumor | I[b] | PR[c] | CR[d] |
|---|---|---|---|---|
| $R \cdot OH$ | | + | + | − |
| $R \cdot O \cdot CO \cdot CH_3$ | ++ | + | + | 55 |
| $R \cdot O \cdot CO \cdot Ph$ | + | + | + | 35 |

[a] Danielli (1959).
[b] I = inhibition.
[c] PR = partial regression.
[d] CR = complete regression.

Kramer *et al.* (1959) employed a somewhat different approach in the synthesis and application of cytotoxic esters capable of detoxification by esterase. To be effective against carcinomas, such esters would need to be more toxic than any of their hydrolytic products. Thus they would be potentially effective in tumors which exhibited deficient esterase activity. Many tumors, however, including epidermoid carcinomas and carcinomas of the lung, exhibit appreciable esterase activity. These authors prepared a series of alkyl and aryl-*N*-(2-chloroethyl)-*N*-nitrosocarbamates and postulated that the urethan linkage would be hydrolyzed in normal vital organs to yield less cytotoxic hydrolysis products. Although the approach of Kramer *et al.* (1957) is a unique one, they indicate that the class of compounds which they employ is not recommended for further investigation in patients and concluded that the drugs did not suit the original aim of providing a toxic ester which could be detoxified by enzymes in normal tissue.

A similar principle has been employed in the development of hemisulfur mustards which could be phosphorylated in an attempt to produce relatively non-toxic derivatives (Rutenburg and Seligman, 1956).

The previously mentioned drugs presumably depend upon activation by fixed or constitutive enzymes. Danielli (1961, p. 4), noting previous findings that a urethan initially inhibits the growth of tumors (Haddow and Sexton, 1946, Greenstein, 1954, p. 285) but is soon followed by resistance to the drug, suggested that the resistance was due to the induction of an enzyme capable of splitting the urethan. Presumably, the same enzyme, once induced, could be used for drug activation. Thus by employing a specific urethan derivative Danielli (1959, p. 167; 1961, p. 6) found that this type of compound would behave as a potentiator for a urethan mustard.

It is apparent that the development and application of anti-tumor agents which are activated by enzyme systems is still in its earliest stages of development, and is by no means limited to nitrogen mustard type of compounds. Certain inherent difficulties to this general approach should be clearly recognized. For example, until a chemotherapeutic drug is found which is capable of being "activated" by a specific enzyme limited to a specific type of neoplasm, there is no particular reason to assume that enzyme-activated cancer chemotherapeutic agents will not be readily activated in tissues known to possess high enzymatic activities such as liver and kidney. Of course, these drugs may be readily applied in the direct perfusion of tumor tissue in an attempt to minimize the aforementioned systemic effects.

Several new approaches may also be suggested (unpublished data). The first involves the application of radioactive naphthyl phosphate containing $P^{32}$. Upon perfusion of a tumor rich in either acid or alkaline phosphatase (e.g. prostatic carcinoma and osteogenic sarcoma), the substrate would be hydrolyzed at a rapid rate, thus resulting in a continuous release of the radioactive material in the neoplastic cells. Another approach is based upon the use of specialized complex naphthols which have been employed in the post-coupling and fluorescence procedures (Chapter 5, pp. 196 and 197). These naphthols are unique in that although their esters are relatively soluble, the enzymatically released naphthol is precipitated *in situ*. Chart 3 illustrates the general type of carcinostatic agent which

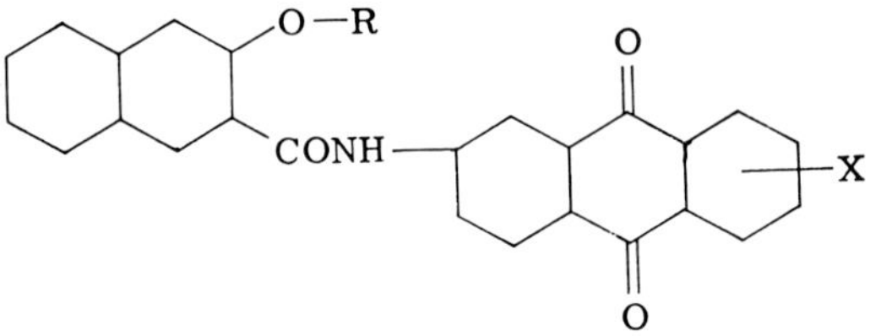


1

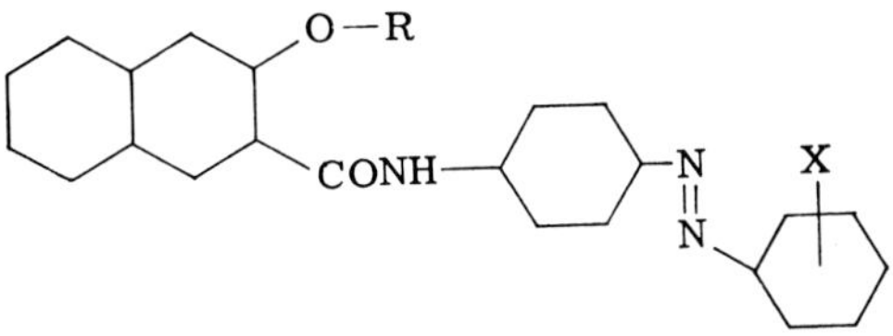


2

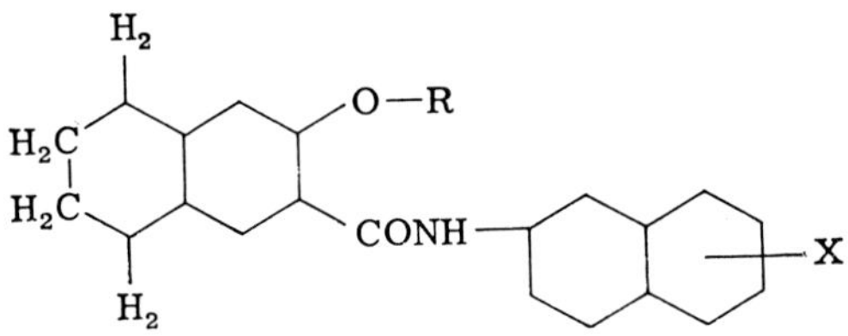


3

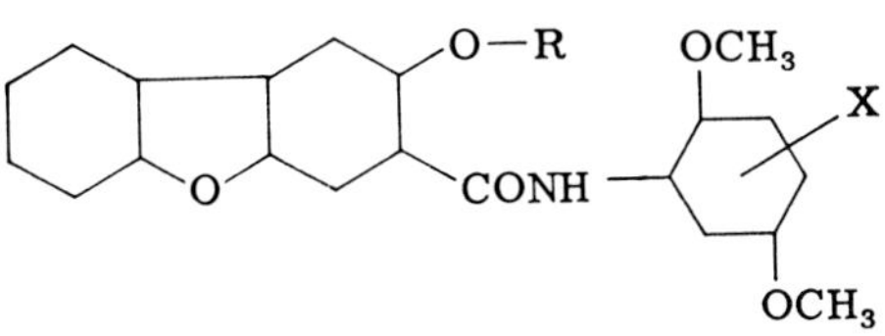


4

**CHART 3. Potential carcinostatic compounds based on complex naphtholic esters which release highly insoluble products upon enzymatic hydrolysis. (R) = ester grouping. (X) indicates the presence of a nitrogen mustard residue or else a grouping containing a radioactive material.**

might be developed from such naphthols. "X" indicates the presence of a nitrogen mustard residue or else a grouping containing a radioactive material. "R" indicates an ester group (e.g. acetate or phosphate). This type of drug is of interest, presumably because uptake of the soluble ester by the tumor cell would result in the *in situ* release of insoluble naphthol containing the carcinostatic grouping; where it would remain as a non-diffusible compound for prolonged therapeutic effect.

Although the aforementioned work of Danielli (1961, p. 6) has indicated that a nitrogen mustard may be activated by the cleavage of an amide bond, potentiation of this effect could be achieved by the synthesis of

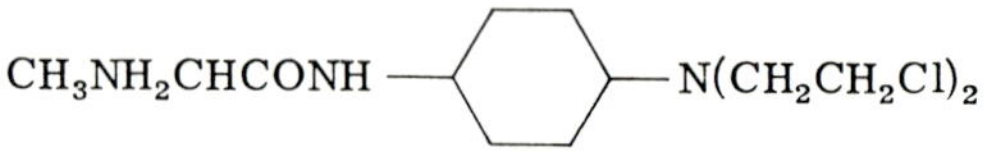


Alanyl derivative of *N*, *N*-di-(2-chloroethyl)-*p*-phenylenediamine.

(V)

a nitrogen mustard containing a specific amino acid residue such as the alanyl derivative of *N*,*N*-di-(2-chloroethyl)-*p*-phenylenediamine (V).

## REFERENCES

ALBERT, A. (1960) "Selective Toxicity," 2nd ed. Methuen, London.

ARNOLD, H., AND BOURSEAUX, F. (1958) *Angew. Chem.* **70,** 539–544.

ARNOLD, H., BOURSEAUX, F., AND BROCK, N. (1958a) *Nature* **181,** 931.

ARNOLD, H., BOURSEAUX, F., AND BROCK, N. (1958b) *Naturwissenschaften* **45,** 64–66.

BIESELE, J. J., PHILIPS, F. S., THIERSCH, J. B., BURCHENAL, J., BUCKLEY, S. M., AND STOCK, C. C. (1950) *Nature* **166,** 1112–1113.

BROCK, N. (1958) *Arzneimittel-Forsch.* **8,** 1–9.

CROSSLEY, M. L., ALLISON, J. B., PARKER, R. P., KUH, E., AND SEEGER, D. R. (1953) *Proc. Soc. Exptl Biol. Med.* **83,** 438–447.

DANIELLI, J. F. (1954) The designing of selective drugs. *In* "Ciba Foundation Symposium on Leukemia Research" (G. E. W. Wolstenholme and M. P. Cameron, eds.), pp. 263–274. Churchill, London.

DANIELLI, J. F. (1959) *In* "The Centennial Lectures Commemorating the One Hundredth Anniversary of E. R. Squibb and Sons," pp. 167–180. Putnam, New York.

DANIELLI, J. F. (1961) *In* "Biological Approaches to Cancer Chemotherapy" (R. J. C. Harris, ed.), pp. 1–9. Academic Press, New York.

DAVIS, W., AND ROSS, W. C. J. (1950) *J. Chem. Soc.* pp. 3056–3062.

FOLEY, G. E., FRIEDMAN, O. M., AND DROLET, B. P. (1961) *Cancer Research* **21,** 57–63.

FRIEDMAN, O. M., AND BOGER, E. (1956) *J. Am. Chem. Soc.* **78,** 4659–4662.

FRIEDMAN, O. M., AND SELIGMAN, A. M. (1948) *J. Am. Chem. Soc.* **70,** 3082–3086.

FRIEDMAN, O. M., AND SELIGMAN, A. M. (1954a) *J. Am. Chem. Soc.* **76,** 655–658.

FRIEDMAN, O. M., AND SELIGMAN, A. M. (1954b) *J. Am. Chem. Soc.* **76,** 658–661.

FRIEDMAN, O. M., KLASS, D. L., AND SELIGMAN, A. M. (1954) *J. Am. Chem. Soc.* **76,** 916–917.

Gomori, G. (1948) *Proc. Soc. Exptl. Biol. Med.* **69,** 407–409.

Greenstein, J. P. (1954) "Biochemistry of Cancer," 2nd ed. Academic Press, New York.

Haddow, A., and Sexton, W. A. (1946) *Nature* **157,** 500–503.

Hebborn, P., and Danielli, J. F. (1958) *Biochem. Pharmacol.* **1,** 19–24.

Kramer, S. P., Miller, J. I., Gaby, S. D., Reeves, A. M., Vail, S. J., Witten, B., Rutenburg, A. M., and Seligman, A. M. (1957) *Cancer* **10,** 971–975.

Kramer, S. P., Seligman, A. M., Gaby, S. D., Solomon, R. D., Miller, J. I., Williamson, C., and Witten, B. (1959) *Cancer* **12,** 446–462.

Loveless, A., and Ross, W. C. J. (1950) *Nature* **166,** 1113–1114.

Mandel, H. G. (1959) *Pharmacol. Revs.* **11,** 743–838.

Parker, R. P., Seeger, D. R., and Kuh, E. (1952a) U. S. Patent 2,606,900.

Parker, R. P., Seeger, D. R., and Kuh, E. (1952b) U. S. Patent 2,606,901.

Parker, R. P., Seeger, D. R., and Kuh, E. (1952c) U. S. Patent 2,606,902.

Philips, F. S. (1950) *Pharmacol. Revs.* **99,** 281–323.

Quastel, J. H. (1960) *In* "Enzymes in Health and Disease" (D. M. Greenberg and H. A. Harper, eds.), pp. 167–190. Charles C Thomas, Springfield, Illinois.

Ross, W. C. J. (1953) *Advances in Cancer Research* **1,** 397–449.

Ross, W. C. J. (1958) *Ann. N. Y. Acad. Sci.* **68,** 669–681.

Ross, W. C. J., and Warwick, G. P. (1955) *Nature* **176,** 298–299.

Ross, W. C. J., Warwick, G. P., and Roberts, J. J. (1955) *J. Chem. Soc.* pp. 3110–3116.

Rutenburg, A. M., and Seligman, A. M. (1956) *New Engl. J. Med.* **255,** 361–368.

Rutenburg, A. M., Persky, L., Friedman, O. M., and Seligman, A. M. (1954) *J. Pharmacol. Exptl. Therap.* **111,** 483–488.

St. Whitelock, O. V., ed. (1958) "Comparative Clinical and Biological Effects of Alkylating Agents." Ann. N. Y. Acad. Sci., Vol. 68, pp. 657–1266.

Seligman, A. M., Rutenburg, A. M., Persky, L., and Friedman, O. M. (1952) *Cancer* **5,** 354–363.

Welch, A. D. (1961) *Cancer Research* **21,** 1475–1490.

Woolley, W. (1959) *Science* **129,** 615–621.

Work, T. S., and Work, E. (1948) "The Basis of Chemotherapy." Oliver & Boyd, Edinburgh.

Zubrod, C. G. (1960) *A.M.A. Arch. Internal Med.* **106,** 663–678.

Chapter 16

# BUFFERS

The following tables indicate the pH ranges of the various buffers which may be employed in conjunction with histochemical reagents. The data are primarily derived from the work of Gomori (1946; 1955, p. 138). Unless otherwise indicated, bibliographic data on the buffers is to be found in the latter paper of Gomori (1955, p. 138). The accuracy of the following data is ca. 0.05 pH at 23°, and in most cases the values will not deviate more than ca. 0.12 pH at 37°.

## Acetate Buffer

*Stock Solutions*

A: 0.2 *M* solution of acetic acid (11.55 ml. in 1000 ml.).
B: 0.2 *M* solution of sodium acetate (16.4 g. of $C_2H_3O_2Na$ or 27.2 g. of $C_2H_3O_2Na \cdot 3H_2O$ in 1000 ml.).
$x$ ml. of A + $y$ ml. of B, diluted to a total of 100 ml.

| pH | $x$ | $y$ |
|---|---|---|
| 3.6 | 46.3 | 3.7 |
| 3.8 | 44.0 | 6.0 |
| 4.0 | 41.0 | 9.0 |
| 4.2 | 36.8 | 13.2 |
| 4.4 | 30.5 | 19.5 |
| 4.6 | 25.5 | 24.5 |
| 4.8 | 20.0 | 30.0 |
| 5.0 | 14.8 | 35.2 |
| 5.2 | 10.5 | 39.5 |
| 5.4 | 8.8 | 41.2 |
| 5.6 | 4.8 | 45.2 |

## 2-Amino-2-methyl-1,2-propanediol (Ammediol) Buffer

*Stock Solutions*

A: 0.2 *M* solution of 2-amino-2-methyl-1,3-propanediol (21.03 g. in 1000 ml.).
B: 0.2 *M* HCl.
50 ml. of A + $x$ ml. of B, diluted to a total of 200 ml.

| pH | $x$ | pH | $x$ |
|---|---|---|---|
| 10.0 | 2.0 | 8.8 | 22.0 |
| 9.8 | 3.7 | 8.6 | 29.5 |
| 9.6 | 5.7 | 8.4 | 34.0 |
| 9.4 | 8.5 | 8.2 | 37.7 |
| 9.2 | 12.5 | 8.0 | 41.0 |
| 9.0 | 16.7 | 7.8 | 43.5 |

## Carbonate–Bicarbonate Buffer

*Stock Solutions*

A: 0.2 *M* solution of anhydrous sodium carbonate (21.2 g. in 1000 ml.).
B: 0.2 *M* solution of sodium bicarbonate (16.8 g. in 1000 ml.).
$x$ ml. of A + $y$ ml. of B, diluted to a total of 200.

| pH | $x$ | $y$ |
|---|---|---|
| 9.2 | 4.0 | 46.0 |
| 9.3 | 7.5 | 42.5 |
| 9.4 | 9.5 | 40.5 |
| 9.5 | 13.0 | 37.0 |
| 9.6 | 16.0 | 34.0 |
| 9.7 | 19.5 | 30.5 |
| 9.8 | 22.0 | 28.0 |
| 9.9 | 25.0 | 25.0 |
| 10.0 | 27.5 | 22.5 |
| 10.1 | 30.0 | 20.0 |
| 10.2 | 33.0 | 17.0 |
| 10.3 | 35.5 | 14.5 |
| 10.4 | 38.5 | 11.5 |
| 10.5 | 40.5 | 9.5 |
| 10.6 | 42.5 | 7.5 |
| 10.7 | 45.0 | 5.0 |

**Tris(hydroxymethyl)aminomethane (Tris) Buffer (Burstone, unpublished data)**

Prepare a solution of tris(hydroxymethyl)aminomethane (24.2 g. in 500 ml. $H_2O$). To obtain pH values shown in table add indicated amount of *N* HCl and bring the total volume up to 1000 ml. with distilled $H_2O$.

| pH | ml. *N* HCl | pH | ml. *N* HCl |
|---|---|---|---|
| 7.1 | 189 | 8.3 | 70 |
| 7.2 | 183 | 8.5 | 50 |
| 7.4 | 170 | 8.7 | 16.5 |
| 7.8 | 150 | 9.2 | 5.75 |
| 8.1 | 90 | | |

**Tris(hydroxymethyl)aminomethane-maleate (Tris-maleate) Buffer**

*Stock Solutions*

A: 0.2 *M* solution of tris acid maleate (24.2 g. of tris(hydroxymethyl)-aminomethane + 23.2 g. of maleic acid or 19.6 g. of maleic anhydride in 1000 ml.).

B: 0.2 *M* NaOH.

50 ml. of A + $x$ ml. of B, diluted to a total of 200 ml.

| pH | $x$ | pH | $x$ |
|---|---|---|---|
| 5.2 | 7.0 | 7.0 | 48.0 |
| 5.4 | 10.8 | 7.2 | 51.0 |
| 5.6 | 15.5 | 7.4 | 54.0 |
| 5.8 | 20.5 | 7.6 | 58.0 |
| 6.0 | 26.0 | 7.8 | 63.5 |
| 6.2 | 31.5 | 8.0 | 69.0 |
| 6.4 | 37.0 | 8.2 | 75.0 |
| 6.6 | 42.5 | 8.4 | 81.0 |
| 6.8 | 45.0 | 8.6 | 86.5 |

## Maleate Buffer

*Stock Solutions*

A: 0.2 *M* solution of acid sodium maleate (8 g. of NaOH + 23.2 g. of maleic acid or 19.6 g. of maleic anhydride in 1000 ml.).

B: 0.2 *M* NaOH.

50 ml. of A + $x$ ml. of B, diluted to a total of 200 ml.

| pH | $x$ | pH | $x$ |
|---|---|---|---|
| 5.2 | 7.2 | 6.2 | 33.0 |
| 5.4 | 10.5 | 6.4 | 38.0 |
| 5.6 | 15.3 | 6.6 | 41.6 |
| 5.8 | 20.8 | 6.8 | 44.4 |
| 6.0 | 26.9 | | |

## Barbital Buffer

*Stock Solutions*

A: 0.2 *M* solution of sodium barbital (veronal) (41.2 g. in 1000 ml.).

B: 0.2 *M* HCl.

50 ml. of A + $x$ ml. of B, diluted to a total of 200 ml.

| pH | $x$ | pH | $x$ |
|---|---|---|---|
| 9.2 | 1.5 | 7.8 | 22.5 |
| 9.0 | 2.5 | 7.6 | 27.5 |
| 8.8 | 4.0 | 7.4 | 32.5 |
| 8.6 | 6.0 | 7.2 | 39.0 |
| 8.4 | 9.0 | 7.0 | 43.0 |
| 8.2 | 12.7 | 6.8 | 45.0 |
| 8.0 | 17.5 | | |

Solutions more concentrated than 0.05 *M* may crystallize on standing, especially in the cold.

## Boric Acid–Borax Buffer

*Stock Solutions*

A: 0.2 $M$ solution of boric acid (12.4 g. in 1000 ml.).
B: 0.05 $M$ solution of borax (19.05 g. in 1000 ml.; 0.02 $M$ in terms of sodium borate).
50 ml. of A + $x$ ml. of B, diluted to a total of 200 ml.

| pH | $x$ | pH | $x$ |
|---|---|---|---|
| 7.6 | 2.0 | 8.7 | 22.5 |
| 7.8 | 3.1 | 8.8 | 30.0 |
| 8.0 | 4.9 | 8.9 | 42.5 |
| 8.2 | 7.3 | 9.0 | 59.0 |
| 8.4 | 11.5 | 9.1 | 83.0 |
| 8.6 | 17.5 | 9.2 | 115.0 |

## Borax–NaOH Buffer

*Stock Solutions*

A: 0.05 $M$ solution of borax (19.05 g. in 1000 ml.; 0.02 $M$ in terms of sodium borate).
B: 0.2 $M$ NaOH.
50 ml. of A + $x$ ml. of B, diluted to a total of 200 ml.

| pH | $x$ | pH | $x$ |
|---|---|---|---|
| 9.28 | 0.0 | 9.7 | 29.0 |
| 9.35 | 7.0 | 9.8 | 34.0 |
| 9.4 | 11.0 | 9.9 | 38.6 |
| 9.5 | 17.6 | 10.0 | 43.0 |
| 9.6 | 23.0 | 10.1 | 46.0 |

## Citrate Buffer

*Stock Solutions*

A: 0.1 *M* solution of citric acid (21.01 g. in 1000 ml.).
B: 0.1 *M* solution of sodium citrate (29.41 g. $C_6H_5O_7Na_3 \cdot 2H_2O$ in 1000 ml.; the use of the salt with 5½ $H_2O$ is not recommended).
$x$ ml. of A + $y$ ml. of B, diluted to a total of 100 ml.

| pH | $x$ | $y$ |
|---|---|---|
| 3.0 | 46.5 | 3.5 |
| 3.2 | 43.7 | 6.3 |
| 3.4 | 40.0 | 10.0 |
| 3.6 | 37.0 | 13.0 |
| 3.8 | 35.0 | 15.0 |
| 4.0 | 33.0 | 17.0 |
| 4.2 | 31.5 | 18.5 |
| 4.4 | 28.0 | 22.0 |
| 4.6 | 25.5 | 24.5 |
| 4.8 | 23.0 | 27.0 |
| 5.0 | 20.5 | 29.5 |
| 5.2 | 18.0 | 32.0 |
| 5.4 | 16.0 | 34.0 |
| 5.6 | 13.7 | 36.3 |
| 5.8 | 11.8 | 38.2 |
| 6.0 | 9.5 | 41.5 |
| 6.2 | 7.2 | 42.8 |

## Citrate–Phosphate Buffer

*Stock Solutions*

A: 0.1 *M* solution of citric acid (19.21 g. in 1000 ml.).

B: 0.2 *M* solution of dibasic sodium phosphate (53.65 g. of $Na_2HPO_4 \cdot 7H_2O$ or 71.7 g. of $Na_2HPO_4 \cdot 12H_2O$ in 1000 ml.).

$x$ ml. of A + $y$ ml. of B, diluted to a total of 100 ml.

| pH | $x$ | $y$ |
|---|---|---|
| 2.6 | 44.6 | 5.4 |
| 2.8 | 42.2 | 7.8 |
| 3.0 | 39.8 | 10.2 |
| 3.2 | 37.7 | 12.3 |
| 3.4 | 35.9 | 14.1 |
| 3.6 | 33.9 | 16.1 |
| 3.8 | 32.3 | 17.7 |
| 4.0 | 30.7 | 19.3 |
| 4.2 | 29.4 | 20.6 |
| 4.4 | 27.8 | 22.2 |
| 4.6 | 26.7 | 23.3 |
| 4.8 | 25.2 | 24.8 |
| 5.0 | 24.3 | 25.7 |
| 5.2 | 23.3 | 26.7 |
| 5.4 | 22.2 | 27.8 |
| 5.6 | 21.0 | 29.0 |
| 5.8 | 19.7 | 30.3 |
| 6.0 | 17.9 | 32.1 |
| 6.2 | 16.9 | 33.1 |
| 6.4 | 15.4 | 34.6 |
| 6.6 | 13.6 | 36.4 |
| 6.8 | 9.1 | 40.9 |
| 7.0 | 6.5 | 43.6 |

### Phosphate Buffer

*Stock Solutions*

A: 0.2 *M* solution of monobasic sodium phosphate (27.8 g. in 1000 ml.).
B: 0.2 *M* solution of dibasic sodium phosphate (53.65 g. of $Na_2HPO_4 \cdot 7H_2O$ or 71.7 g. of $Na_2HPO_4 \cdot 12H_2O$ in 1000 ml.).
$x$ ml. of A + $y$ ml. of B, diluted to a total of 200 ml.

| pH | $x$ | $y$ | pH | $x$ | $y$ |
|---|---|---|---|---|---|
| 5.7 | 93.5 | 6.5 | 6.9 | 45.0 | 55.0 |
| 5.8 | 92.0 | 8.0 | 7.0 | 39.0 | 61.0 |
| 5.9 | 90.0 | 10.0 | 7.1 | 33.0 | 67.0 |
| 6.0 | 87.7 | 12.3 | 7.2 | 28.0 | 72.0 |
| 6.1 | 85.0 | 15.0 | 7.3 | 23.0 | 77.0 |
| 6.2 | 81.5 | 18.5 | 7.4 | 19.0 | 81.0 |
| 6.3 | 77.5 | 22.5 | 7.5 | 16.0 | 84.0 |
| 6.4 | 73.5 | 26.5 | 7.6 | 13.0 | 87.0 |
| 6.5 | 68.5 | 31.5 | 7.7 | 10.5 | 90.5 |
| 6.6 | 62.5 | 37.5 | 7.8 | 8.5 | 91.5 |
| 6.7 | 56.5 | 43.5 | 7.9 | 7.0 | 93.0 |
| 6.8 | 51.0 | 49.0 | 8.0 | 5.3 | 94.7 |

### Succinate Buffer

*Stock Solutions*

A: 0.2 *M* solution of succinic acid (23.6 g. in 1000 ml.).
B: 0.2 *M* NaOH.
25 ml. of A + $x$ ml. of B, diluted to a total of 100 ml.

| pH | $x$ | pH | $x$ |
|---|---|---|---|
| 3.8 | 7.5 | 5.0 | 26.7 |
| 4.0 | 10.0 | 5.2 | 30.3 |
| 4.2 | 13.3 | 5.4 | 34.2 |
| 4.4 | 16.7 | 5.6 | 37.5 |
| 4.6 | 20.0 | 5.8 | 40.7 |
| 4.8 | 23.5 | 6.0 | 43.5 |

**Phthalate–Sodium Hydroxide Buffer**

*Stock Solutions*

A: 0.2 *M* solution of potassium acid phthalate (40.84 g. in 1000 ml.).
B: 0.2 *M* NaOH.
50 ml. of A + $x$ ml. of B, diluted to a total of 200 ml.

| pH | $x$ | pH | $x$ |
|---|---|---|---|
| 4.2 | 3.7 | 5.2 | 30.0 |
| 4.4 | 7.5 | 5.4 | 35.5 |
| 4.6 | 12.2 | 5.6 | 39.8 |
| 4.8 | 17.7 | 5.8 | 43.0 |
| 5.0 | 23.9 | 6.0 | 45.5 |

**Phthalate–Hydrochloric Acid Buffer**

*Stock Solutions*

A: 0.2 *M* solution of potassium acid phthalate (40.84 g. in 1000 ml.).
B: 0.2 *M* HCl.
50 ml. of A + $x$ ml. of B, diluted to a total of 200 ml.

| pH | $x$ | pH | $x$ |
|---|---|---|---|
| 2.2 | 46.7 | 3.2 | 14.7 |
| 2.4 | 39.6 | 3.4 | 9.9 |
| 2.6 | 33.0 | 3.6 | 6.0 |
| 2.8 | 26.4 | 3.8 | 2.63 |
| 3.0 | 20.3 | | |

**Hydrochloric Acid–Potassium Chloride Buffer**

*Stock Solutions*

A: 0.2 *M* solution of KCl (14.91 g. in 1000 ml.).
B: 0.2 *M* HCl.
50 ml. of A + $x$ ml. of B, diluted to a total of 200 ml.

| pH | $x$ | pH | $x$ |
|---|---|---|---|
| 1.0 | 97.0 | 1.7 | 20.6 |
| 1.1 | 78.0 | 1.8 | 16.6 |
| 1.2 | 64.5 | 1.9 | 13.2 |
| 1.3 | 51.0 | 2.0 | 10.6 |
| 1.4 | 41.5 | 2.1 | 8.4 |
| 1.5 | 33.3 | 2.2 | 6.7 |
| 1.6 | 26.3 | | |

## Glycine–NaOH Buffer

*Stock Solutions*

A: 0.2 *M* solution of glycine (15.01 g. in 1000 ml.).
B: 0.2 *M* NaOH.
50 ml. of A + $x$ ml. of B, diluted to a total of 200 ml.

| pH | $x$ | pH | $x$ |
|---|---|---|---|
| 8.6 | 4.0 | 9.6 | 22.4 |
| 8.8 | 6.0 | 9.8 | 27.2 |
| 9.0 | 8.8 | 10.0 | 32.0 |
| 9.2 | 12.0 | 10.4 | 38.6 |
| 9.4 | 16.8 | 10.6 | 45.5 |

## Glycine–HCl Buffer

*Stock Solutions*

A: 0.2 *M* solution of glycine (15.01 g. in 1000 ml.).
B: 0.2 *M* HCl.
50 ml. of A + $x$ ml. of B, diluted to a total of 200 ml.

| pH | $x$ | pH | $x$ |
|---|---|---|---|
| 3.6 | 5.0 | 2.8 | 16.8 |
| 3.4 | 6.4 | 2.6 | 24.2 |
| 3.2 | 8.2 | 2.4 | 32.4 |
| 3.0 | 11.4 | 2.2 | 44.0 |

## Cacodylate Buffer

*Stock Solutions*

A: 0.2 *M* solution of sodium cacodylate (42.8 g. of $Na(CH_3)_2AsO_2 \cdot 3H_2O$ in 1000 ml.).
B: 0.2 *M* HCl.
50 ml. of A + $x$ ml. of B, diluted to a total of 200 ml.

| pH | $x$ | pH | $x$ |
|---|---|---|---|
| 7.4 | 2.7 | 6.0 | 29.6 |
| 7.2 | 4.2 | 5.8 | 34.8 |
| 7.0 | 6.3 | 5.6 | 39.2 |
| 6.8 | 9.3 | 5.4 | 43.0 |
| 6.6 | 13.3 | 5.2 | 45.0 |
| 6.4 | 18.3 | 5.0 | 47.0 |
| 6.2 | 23.8 | | |

## Collidine Buffer

The pH values of 0.05 *M* buffer mixtures, obtained by mixing 25 cc. of a 0.2 *M* solution of the base (collidine, 2.64 cc. in 100 cc.) with varying volumes of 0.1 *N* HCl and diluting the mixtures to the final volume of 100 cc. are given in the table.

| 0.1 *N* HCl (cc.) | Collidine | |
|---|---|---|
| | 23° C | 37° C |
| 5.0 | 8.35 | 8.28 |
| 7.5 | 8.18 | 8.10 |
| 10.0 | 8.00 | 7.94 |
| 12.5 | 7.88 | 7.80 |
| 15.0 | 7.77 | 7.70 |
| 17.5 | 7.67 | 7.60 |
| 20.0 | 7.57 | 7.50 |
| 22.5 | 7.50 | 7.40 |
| 25.0 | 7.40 | 7.32 |
| 27.5 | 7.30 | 7.23 |
| 30.0 | 7.22 | 7.14 |
| 32.5 | 7.13 | 7.05 |
| 35.0 | 7.03 | 6.95 |
| 37.5 | 6.92 | 6.84 |
| 40.0 | 6.80 | 6.72 |
| 42.5 | 6.62 | 6.54 |
| 45.0 | 6.45 | 6.37 |

### References

Gomori, G. (1946) *Proc. Soc. Exptl. Biol. Med.* **62,** 33–34.

Gomori, G. (1955) *In* "Methods in Enzymology," (S. P. Colowick and N. O. Kaplan, eds.), Vol. I, pp. 138–146. Academic Press, New York.

# AUTHOR INDEX

Numbers in italic indicate the pages on which the complete references are listed

## A



## B

## C

## D

E



F

## G

## H

**L**

## M

**Y**

## Z

# SUBJECT INDEX

Italicized page numbers refer to specific procedures given in Appendixes

## N